TechOne: Automotive Electricity and Electronics

TechOne: Automotive Electricity and Electronics

Al Santini

Retired, College of DuPage
Glen Ellyn, Illinois

Jack Erjavec, Series Editor

Professor Emeritus,
Columbus State Community College
Columbus, Ohio

THOMSON
DELMAR LEARNING™

Australia Canada Mexico Singapore Spain United Kingdom United States

TechOne: Automotive Electricity and Electronics

Al Santini

Vice President, Technology and Trades SBU:
Alar Elken

Editorial Director:
Sandy Clark

Acquisitions Editor:
David Boelio

Developmental Editor:
Matthew Thouin

Marketing Director:
Cyndi Eichelman

Channel Manager:
Fair Huntoon

Marketing Coordinator:
Sarena Douglass

Production Director:
Mary Ellen Black

Production Editor:
Barbara L. Diaz

Art/Design Coordinator:
Cheri Plasse

Technology Project Manager:
David Porush

Technology Project Specialist:
Kevin Smith

Editorial Assistant:
Jill Carnahan

Printed in the United States of America
4 5 XX 06

For more information contact
Delmar Learning
Executive Woods
5 Maxwell Drive, PO Box 8007,
Clifton Park, NY 12065-8007
Or find us on the World Wide Web at
http://www.delmarlearning.com

Library of Congress Cataloging-in-Publication Data:

Santini, Al.
TechOne : automotive electricity and electronics / Al Santini.
p. cm.
Includes index.
ISBN 1-4018-1394-1 (pbk.:alk. paper)—ISBN 1-4018-1395-X (instructor's manual : pbk./CD-ROM)
1. Automobiles—Electric equipment.
I. Title.
TL272.S34624 2003
629.25'4—dc21 2003053744

NOTICE TO THE READER

Publisher does not warrant or guarantee any of the products described herein or perform any independent analysis in connection with any of the product information contained herein. Publisher does not assume, and expressly disclaims, any obligation to obtain and include information other than that provided to it by the manufacturer.

The reader is expressly warned to consider and adopt all safety precautions that might be indicated by the activities herein and to avoid all potential hazards. By following the instructions contained herein, the reader willingly assumes all risks in connection with such instructions.

The publisher makes no representation or warranties of any kind, including but not limited to, the warranties of fitness for particular purpose or merchantability, nor are any such representations implied with respect to the material set forth herein, and the publisher takes no responsibility with respect to such material. The publisher shall not be liable for any special, consequential, or exemplary damages resulting, in whole or part, from the readers' use of, or reliance upon, this material.

Contents

Preface

THE SERIES

Welcome to Delmar Learning's *TechOne*, a state-of-the-art series designed to respond to today's automotive instructor and student needs. *TechOne* offers current, concise information on ASE and other specific subject areas, combining classroom theory, diagnosis, and repair into one easy-to-use volume.

You'll notice several differences from a traditional textbook. First, a large number of short chapters divide complex material into chunks. Instructors can give tight, detailed reading assignments that students will find easier to digest. These shorter chapters can be taught in almost any order, allowing instructors to pick and choose the material that best reflects the depth, direction, and pace of their individual classes.

TechOne also features an art-intensive approach to suit today's visual learners: images drive the chapters. From drawings to photos, you will find more art to better understand the systems, parts, and procedures under discussion. Look also for helpful graphics in features like You Should Know and Interesting Fact that draw attention to key points.

Just as importantly, each *TechOne* starts off with a section on safety and communication, which stresses safe work practices, tool competence, and familiarity with workplace "soft skills" such as customer communication and the roles necessary to succeed as an automotive technician. From there, learners are ready to tackle the technical material in successive sections, ultimately leading them to the real test—an ASE practice exam in the Appendix.

THE SUPPLEMENTS

TechOne comes with an **Instructor's Manual** that includes answers to all chapter-end review questions and a complete correlation of the text to NATEF standards. A **CD-ROM**, included with each **Instructor's Manual**, consists of **PowerPoint Slides** for classroom presentations and a **Computerized Testbank** with hundreds of questions to aid in creating tests and quizzes. Chapter-end review questions from the text have also been redesigned into adaptable **Electronic Worksheets**, so instructors can modify questions if desired to create in-class assignments or homework.

Flexibility is the key to *TechOne*. For those who would like to purchase jobsheets, Delmar Learning's NATEF Standards jobsheets are a good match. Topics cover the eight ASE subject areas and include:

- Automatic Transmissions and Transaxles
- Automotive Brakes
- Automotive Electrical and Electronic Systems
- Automotive Engine Repair
- Automotive Engine Performance
- Automotive Heating and Air Conditioning
- Automotive Suspension and Steering
- Manual Drive Trains and Axles

Visit **http://www.autoed.com** for a complete catalog.

OTHER TITLES IN THIS SERIES

TechOne is Delmar Learning's latest automotive series. We are excited to announce these future titles:

- Advanced Engine Performance
- Air Conditioning
- Automatic Transmissions
- Automotive Computer Systems
- Brakes
- Engine Performance
- Fuels and Emissions
- Manual Transmissions
- Steering and Suspension

Check with your sales representative for availability.

A NOTE TO THE STUDENT

There are now more computers on a car than aboard the first spacecraft, and even gifted backyard mechanics long ago turned their cars over to automotive professionals for diagnosis and repair. That's a statement about the nation's need for the knowledge and skills you'll develop as you continue your studies. Whether you eventually choose a career as a certified or licensed technician, a service writer or manager, an automotive engineer, or even if you decide to open your own shop, hard work will give you the opportunity to become one of the 840,000 automotive professionals providing and maintaining safe and efficient automobiles on our roads. As a member of a technically proficient, cutting-edge workforce, you'll fill a need, and even better, you'll have a career to feel proud of.

Best of luck in your studies,

The Editors of Delmar Learning

About the Author

Al Santini has been involved in automotive education for 35 years. He is semiretired and is currently teaching technician seminars. Before retirement, he taught automotive electricity and electronics for 22 years for the College of DuPage in Glen Ellyn, Illinois. He has also taught at an area vocational center and a local high school.

Al has been married to his wife, Carol, for 34 years. They have two adult children, Amy and Keith. Amy is married to Jeff and they have a daughter, Alexandra, who is one year old. Keith teaches automotive technology at Addison Trail High School, in Addison, Illinois.

Al has been the author of *Automotive Electricity and Electronics* since the first edition and is also the author of the *L1 Prep Guide* for Mitchell Manuals. He is an active member of the Society of Automotive Testers (SAT), the Illinois College Automotive Instructor's Association (ICAIA), the Council of Advanced Automotive Trainers (CAAT), and the North American Council of Automotive Teachers (NACAT).

When he is not teaching, Al enjoys sailing, camping, and bike riding.

Acknowledgments

TechOne: Automotive Electricity and Electronics has been written to fill a void in post-secondary textbooks. Many books lack the required breadth or depth needed to cover this complex subject, and often textbooks reproduce only factory circuits, which dates the material and the book. For this reason, I have generally avoided specific specialized circuits, instead concentrating on component testing from a generic standpoint. Because we frequently find the same component on a variety of vehicles, the testing procedure can be universal—plus this approach greatly enhances the student's ability to comprehend the material.

But the real strength of this text is its ability to teach, in a logical sequence, transferrable skills, from the fundamentals to more advanced procedures. The knowledge that will be needed to repair tomorrow's vehicles is rooted in the basics, and this text begins with an education in those essential skills. Circuit design, various applicable laws, wiring principles, and a hands-on practical approach in the use of test equipment have all been incorporated. That means that the automotive electricity student will develop a strong foundation to build upon in the remainder of the text, which covers batteries, starting, charging, ignition (both distributed and distributorless), and accessories. On-vehicle testing is also stressed throughout the book, especially the use of digital multimeters, scanners, and digital storage oscilloscopes.

Finally, an emphasis has been placed on readability in this text. Most students will find it easy to comprehend, and the relaxed style of writing should improve the enjoyment of a basic electricity class. If the students' first exposure to our industry is a positive experience, I believe those students will become the technicians of the future that our industry needs.

Many individuals have helped with this text. To mention all would fill pages. I have retired from full-time teaching at the College of DuPage in Glen Ellyn, Illinois, and am thankful that it was the kind of community college not only where students could learn, but where instructors were encouraged to create positive student-oriented activities, such as this text. Vetronix and specifically Mike Gustafson have been very supportive of my efforts. Not only do they manufacture some of the best automotive test equipment available, but they are willing to help instructors learn how to improve their teaching. I also need to mention all the great people I have come into contact with at Delmar Learning. San Rao, my acquisitions editor, was a steady force that made the series a reality, and Alison Weintraub, my developmental editor, was always willing and able to answer questions and keep the project on track.

The following individuals reviewed every word of this text. Their input has been extremely helpful:

Si Acuna
Texas State Technical College
Sweetwater, Texas

James Armitage
Waubonsee Community College
Sugar Grove, Illinois

Gerard DiCola
School of Cooperative Technical Education
New York, New York

David Dosser
Lamar State College
Port Arthur, Texas

George Generke
College of DuPage
Glen Ellyn, Illinois

Phil Krolick
Linn-Benton Community College
Albany, Oregon

John Thornton
Pro-Tec Auto Repair, Inc.
Naperville, Illinois

Last, I want to thank my wonderful family. It grew a bit this year. Our first granddaughter, Alexandra, was born to our daughter, Amy, and her husband, Jeff. Our son, Keith, started teaching automotive service technology at our local high school. My wife, Carol, helped a great deal with this text, organizing all of the illustrations. Without her, you would not be reading this. I thank her and love her more and more every day.

Features of the Text

TechOne includes a variety of learning aids designed to encourage student comprehension of complex automotive concepts, diagnostics, and repair. Look for these helpful features:

Section Openers provide students with a **Section Table of Contents** and **Objectives** to focus learners on the section's goals.

Interesting Facts spark student attention with industry trivia or history. Interesting facts appear on the section openers and are then scattered throughout the chapters to maintain reader interest.

Section 3

Vehicle Circuits

SECTION OBJECTIVES

At the conclusion of this section, you should be able to:

- Understand the components that are part of a working circuit.
- Have a working knowledge of Ohm's Law.
- Analyze the voltage, current, and resistance of a series circuit.
- Analyze the voltage, current, and resistance of a parallel circuit.
- Analyze the voltage, current, and resistance of a series-parallel circuit.
- Have a working knowledge of voltage division in a series circuit.
- Have a working knowledge of current division in a parallel circuit.
- Use a 12-volt test light to diagnose an open circuit.
- Use a DMM to diagnose an open circuit.
- Understand the characteristics of a short circuit.
- Understand the characteristics of an open circuit.
- Calculate the wattage developed within a circuit.
- Use Ohm's Law to calculate the relationship between voltage, current, and resistance in circuits.

The vehicle of today has a mix of series, parallel, and series-parallel circuits. Knowledge of the fundamentals of circuits and a working knowledge of Ohm's Law are the foundations for success as a technician.

Working As an Electricity/ Electronics Technician

Introduction

Whether you can be a technician who works only on the electrical system of the vehicle will depend on the size of your service center. Most technicians work on most systems but have a few specialties. For example, brake work or engine work might usually be assigned to the same technicians in a large shop because they find the most success with it.

For smaller service centers, electrical work is expected of everyone. It is considered to be part of the basic knowledge of the technician in some service centers. As the electrical system has become increasingly more complicated, the idea that everyone should be capable of electrical work is changing. What does the future hold? For the answer to this, look to the past. Thirty years ago, many systems on the vehicle were mechanical only. Electricity or electronics was not involved in the control of engines, transmissions, braking, or steering. Some technicians worked daily without finding it necessary to understand electricity.

If you look at the modern vehicle, electricity or electronics is in charge of most engine and transmission functions. In addition, nowadays active suspension and antilock braking systems rely on electronics to function. Many accessories are now power assisted. The use of electronics will increase because it is reliable, safe, and relatively inexpensive.

This chapter looks at some information that you need to know to function as an electricity/electronics technician. Much of the information will be safety oriented, and all of it is important.

YOUR TOOLBOX

The toolbox of a general technician contains a tremendous assortment of tools. As the vehicle changes, the tools required to repair it also change. Talk to technicians who have been repairing vehicles for twenty years and ask them if they have any tools that they do not use daily. It is likely that they will have an entire drawer filled with outdated tools that have not been used since a specific vehicle model disappeared off the streets.

If the technicians you are talking to generally repair electrical systems, their toolboxes will include the usual assortment of wrenches, pliers, sockets, and screwdrivers. They will, however, include some specialized test equipment. The first tool that you might notice is a **digital multimeter (DMM)**, capable of measuring voltage, current flow, and resistance **(Figure 1)**. Thirty years ago the meter would have been analog with a swinging-needle display. DMMs took over the test market in the early 1980s as the manufacturers began placing computer-type modules on vehicles. The old analog multimeter was not compatible with the vehicle and so was placed in the bottom drawer of the toolbox. We will find many applications for the DMM to be used within this text.

The second tool that you might notice is the **digital storage oscilloscope (DSO) (Figure 2)**. The DSO is the tool of choice for many diagnostic and repair technicians because it allows us to monitor a device over a period of time. The DSO can be set up to look at voltage changes or current changes with the addition of a current probe. Again, as a comparison, the relatively new DSO replaced the larger ignition scope in many service centers. The advantages to the battery-powered or vehicle-powered DSO

An **Introduction** orients readers at the beginning of each new chapter. **Technical Terms** are bolded in the text upon first reference and are defined.

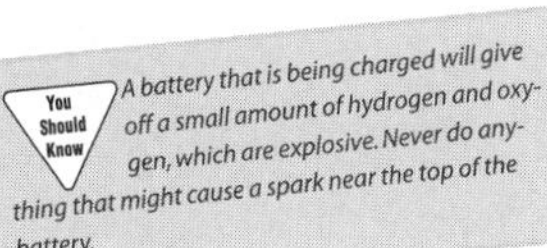

A battery that is being charged will give off a small amount of hydrogen and oxygen, which are explosive. Never do anything that might cause a spark near the top of the battery.

Prevention is the key to battery safety. Never do anything that might cause a spark near a battery. This includes the improper use of jumper cables or charging cables. If a connection is made near the battery and causes a spark, the battery might blow up. Do not let it happen to you. In Chapter 31, Chapter 32, and Chapter 33, we look at more specific examples of what not to do.

Summary

- The image that you project as a professional technician includes not only your abilities but your work clothes.
- Make sure that your clothes and hair cannot get caught in moving components.
- Safety shoes protect your toes from being crushed by heavy objects.
- ANSI 87 safety glasses should be worn at all times.
- You should have available a full face shield and safety goggles for certain automotive jobs.
- You should have available a respirator and wear it when fumes or chemicals are being used.
- Service centers or school labs should have multiple class ABC fire extinguishers available that have been periodically checked.
- MSDS should be prominently displayed and referred to before exposure to a new chemical.
- Chemical sensitivity can be prevented by the use of gloves and other related safety equipment.
- A battery can explode if a spark occurs near its top.
- Know where the eyewash fountains are located before beginning work.

Review Questions

1. Technician A states that a spark near the top of a charging battery might cause it to explode. Technician B states that batteries contain acid, which could damage your eyes. Who is correct?
   A. Technician A only
   B. Technician B only
   C. Both Technician A and Technician B
   D. Neither Technician A nor Technician B
2. A class B fire is one that involves
   A. burning wood
   B. electrical short circuits
   C. flammable liquids
   D. none of the above
3. Chemical sensitivity can be caused by
   A. repeated exposure to chemicals without taking precautions
   B. ignorance of the MSDS information
   C. the absorption of chemicals into the bloodstream through the hands
   D. all of the above
4. A class A fire extinguisher should be used in a service center environment.
   A. True
   B. False
5. It is OK to run a vehicle for an extended period of time in a closed shop.
   A. True
   B. False
6. An exploding battery showers the area with acid and pieces of plastic.
   A. True
   B. False
7. How is chemical sensitivity best prevented?
8. What types of fire extinguishers are in use in your school lab, and when were they last checked?
9. Why is the type and condition of your work clothes important?
10. What type of safety equipment is available to you?

You Should Know informs the reader whenever special safety cautions, warnings, or other important points deserve emphasis.

A **Summary** concludes each chapter in short, bulleted sentences. **Review Questions** are structured in a variety of formats, including ASE style, challenging students to prove they've mastered the material.

An **ASE Practice Exam** is found in the **Appendix** of every *TechOne* book, followed by a **Bilingual Glossary**, which offers Spanish translations of technical terms alongside their English counterparts.

Appendix

ASE PRACTICE EXAM FOR ELECTRICAL SYSTEMS

1. Technician A states that more work is accomplished with parallel circuitry. Technician B states that series resistances add up. Who is correct?
   A. Technician A only
   B. Technician B only
   C. Both Technician A and Technician B
   D. Neither Technician A nor Technician B
2. Adding additional resistance to a series circuit will
   A. have no effect on the current flow
   B. reduce the current flow
   C. increase the wattage
   D. increase the current flow
3. A taillight bulb is glowing dimly. A voltmeter across the bulb reads 7.4 volts. Technician A states that there must be a voltage drop somewhere else in the circuit. Technician B states that resistance in parallel must be present. Who is correct?
   A. Technician A only
   B. Technician B only
   C. Both Technician A and Technician B
   D. Neither Technician A nor Technician B
4. Technician A states that control current for a relay will be small in comparison to the contact current. Technician B states that the relay coil will have high resistance and the contacts will have no resistance. Who is correct?
   A. Technician A only
   B. Technician B only
   C. Both Technician A and Technician B
   D. Neither Technician A nor Technician B
5. Technician A says that a 12.6-volt OCV reading indicates that a battery needs recharging. Technician B says that a 1.265 hydrometer reading indicates that a battery needs recharging. Who is correct?
   A. Technician A only
   B. Technician B only
   C. Both Technician A and Technician B
   D. Neither Technician A nor Technician B
6. A 450-CCA battery is being load tested. The correct load will be
   A. 1,350 amps
   B. 450 amps
   C. 225 amps
   D. 990 amps
7. Technician A says that during a load test the battery voltage must not fall below 9.6 volts. Technician B says that the load applied should be three times the A/H rating. Who is correct?
   A. Technician A only
   B. Technician B only
   C. Both Technician A and Technician B
   D. Neither Technician A nor Technician B
8. Technician A states that slowing down the starter armature will result in increased amperage draw. Technician B states that speeding up the armature will result in decreased amperage draw. Who is correct?
   A. Technician A only
   B. Technician B only
   C. Both Technician A and Technician B
   D. Neither Technician A nor Technician B

Bilingual Glossary

AC Alternating current.
AC *Corriente alterna.*

Advance Delivering the spark earlier to the cylinder.
Avance *Producir la chispa tempranamente hacia el cilindro.*

Advanced timing A spark that is delivered before the piston reaches TDC.
Avance de la regulación de tiempo *Chispa que se produce cuando el pistón alcanza el punto muerto superior.*

Alternator An AC (alternating current) generator that produces electrical current and forces it into the battery to recharge it.
Alternador *Generador de corriente alterna (AC) que produce corriente eléctrica y la fuerza en la batería para recargarla.*

Ammeter The instrument used to measure electrical current flow in a circuit.
Amperímetro *Instrumento que se usa para medir el flujo de corriente eléctrica en un circuito.*

Ampere The unit for measuring electrical current; usually called an amp.
Amperio *Unidad para medir la corriente eléctrica. Generalmente se conoce como amp.*

Armature The assembly that becomes the rotating magnetic field inside a motor.
Armadura *El ensamble que se convierte en el campo magnético giratorio dentro de un motor.*

ASD Automatic shutdown relay.
ASD *Relé de desconexión automática.*

ATDC After top dead center.
ATDC *Después del punto muerto superior.*

Auto-ranging A DMM that will automatically range to give the most accurate reading.
Automedición *DMM que mide de forma automática para proporcionar la lectura más precisa.*

Base The setting for ignition timing when the engine is at normal engine speed, load, and temperature.
Base *El ajuste para sincronización del encendido en el cual el motor está a velocidad, carga y temperatura normales.*

Base timing The timing that usually has the lowest advance that the vehicle requires. Rotating the distributor usually sets it.
Sincronización base *Sincronización que generalmente tiene el avance más bajo que requiere el vehículo. Hacer girar el distribuidor generalmente lo ajusta.*

BCM Body control module. The module that is used to control various body functions.
BCM *Módulo de control de la carrocería. Módulo que se usa para controlar funciones diversas de la carrocería.*

BTDC Before top dead center.
BTDC *Antes del punto muerto superior.*

Bullet connector A single wire connector.
Conector de bala *Conector de un solo alambre.*

Burn time How long the arc across the spark plug gap lasts; normally measured in milliseconds.
Tiempo de quemado *El tiempo de duración del arco al cruzar de lado a lado la brecha de la bujía, medido normalmente en milisegundos.*

Case ground A ground that is part of the metal component.
Tierra a chasis *Tierra que es parte del componente de metal.*

Cell The individual part of a battery.
Celda *Pieza individual de una batería.*

CEMF Counterelectromotive force; counter voltage.
CEMF *Fuerza contraelectromotriz; contravoltaje.*

Chip A group of electronic devices; an integrated circuit.
Chip *Grupo de dispositivos electrónicos; circuito integrado.*

A comprehensive **Index** helps instructors and students pinpoint information in the text.

Index

Section 1

Safety and Communication

SECTION OBJECTIVES

At the conclusion of this section, you should be able to:

- Discuss how to ensure your eye safety in a service environment.
- Understand chemical sensitivity and be capable of protecting yourself against it.
- Understand the different types of fires.
- Know which fire extinguisher best fights a specific type of fire.
- Know the benefit of and understand the use of material safety data sheets (MSDS).
- Prevent the explosion of batteries.
- Understand the use of the eyewash fountain.
- Recognize the special tools necessary to diagnose and repair electrical systems.
- Understand the dangers of the supplemental restraint system (SRS), or air bag system.
- Prevent static electricity from damaging components.
- Discuss repair needs with a customer.

The most important job of the technician is to protect himself against injury and chemical sensitivity.

Chapter 1 Safe Working Practices

Introduction

Professional technicians realize that their own safety is extremely important. Over the years, the environment of the shop has changed from the dark, dingy garage to the modern, well-lit service bays of today. We even find some service centers that are air-conditioned. It is common to see computers at some workstations. Times have changed. Within this chapter, we look at some personal safety issues, which become more important each year. To present to you what you will need to do to remain safe for years is the goal of this chapter. It is not difficult information but it is extremely important. Make sure you understand it and its implication for your safety as a technician.

WORK CLOTHES

Many shops will supply work uniforms for you to wear, but keep in mind a few important points about how you dress. The first point has nothing to do with personal safety, but everything to do with image. Wearing an old, ripped T-shirt full of grease and stains does nothing for the image of our trade. A clean, professional-looking uniform or coveralls will help dispel the image of the old-style mechanic and replace it with the image of a technician. If you are a properly trained technician who looks professional, your ability to gain the customer's confidence will be improved. Customers look at you and base their opinion of your ability to diagnose and repair vehicles and your professionalism on your appearance **(Figure 1)**.

Figure 1. A clean work uniform projects a positive image of the technician.

The second point concerns your safety and the safety of the vehicle that you are working on. Loose-fitting shirts or long sleeves can get caught in moving components and draw you into a serious situation. Tuck in anything that might get caught. If you wear a jacket, keep it zipped up so that it stays tight to your body. In addition, remember that you will be driving the vehicle and possibly working in its interior. Nothing drives customers away from your service center as fast as leaving behind grease or grime that came off your clothes or hands.

If you wear a belt, wear one that does not have a buckle that might scratch the side of the vehicle. Your boss will not be pleased if he has to paint the side of a vehicle because your belt scratched the paint.

The safety of your fingers is vitally important. Do not wear rings; they could get caught in moving components. From an electrical standpoint, the ring is conductive and could short a circuit and damage the vehicle or your finger. If, for instance, starting current were to run through your ring, the ring would develop sufficient heat to seriously burn your finger. Rings, watches, pendants, and other jewelry should remain in your locker. If your hair is long, tie it back or cut it so that it does not get caught in moving components. Getting your hair caught in a drive belt that is spinning can cause serious injury to your scalp. Make sure that these easily preventable accidents do not happen to you.

Depending on the type of work you do, your feet may need to be protected. Good-quality insulated boots with steel toes are a must for any heavy work such as transmission or engine removal.

EYE PROTECTION

The use of safety glasses has become common practice in most service centers. Every technician should own a comfortable pair and wear them almost continuously. Most tool vendors sell comfortable safety glasses **(Figure 2)**. If you normally wear glasses, make sure your doctor realizes that you are a technician so that your glasses may be made

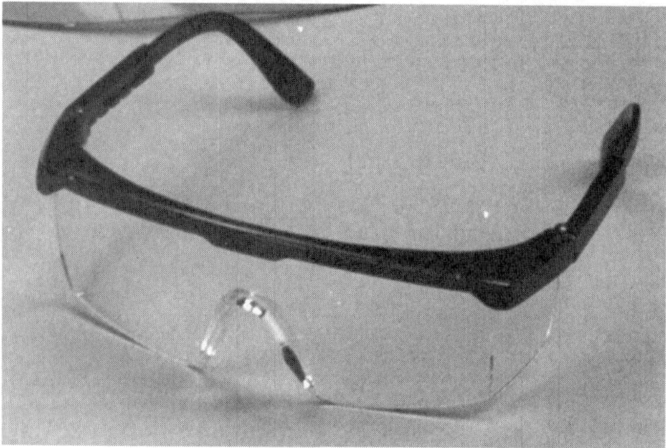

Figure 2. Safety glasses protect your eyes only if they are worn.

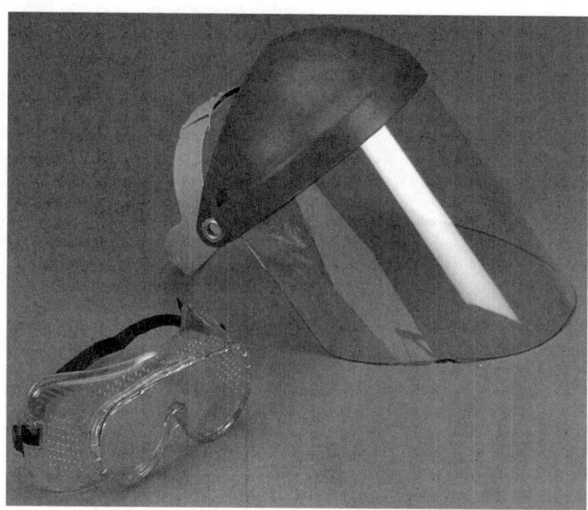

Figure 3. Safety goggles or a full face shield should be used if chemicals could get sprayed or splashed up to the face.

to safety standards. Safety glasses and frames will have the ANSI 87 label on them. Good safety glasses will also have side shields to protect you from objects that would otherwise injure you from the side.

In addition to safety glasses, your toolbox should contain a full face shield and wraparound safety goggles, or the service center should have them available for you to use **(Figure 3)**. Wear these when working around chemicals or solvents.

THE AIR WE BREATHE AND NOISES WE HEAR

The environment of even the most modern service center can contain harmful dust, fumes, and vapors, including exhaust from running engines. Your service center's vehicle exhaust system must be capable of removing the harmful by-products of combustion from the air and exhausting them to the outside. Running the exhaust hose under the shop door is not an acceptable practice. The vehicle exhaust could come back into the building. Repeated exposure to combustion gases will have an effect on your health over time. Make sure the exhaust system in the building can remove the gases. You need to breathe fresh, clean air if you expect to stay healthy.

Your employer should make available a respirator for you to use when you are using solvents, especially those that are sprayed, or you should have one of your own **(Figure 4)**. Do not hesitate to use it.

A working automotive shop is a noisy place, especially if more than one technician is at work. Air tools, drills, grinders, and even engine noise can produce hearing loss if you are exposed to them repeatedly. Make sure that you have ear protection available and that you wear it **(Figure 5)**.

Figure 4. The use of certain chemical agents requires that you use a respirator.

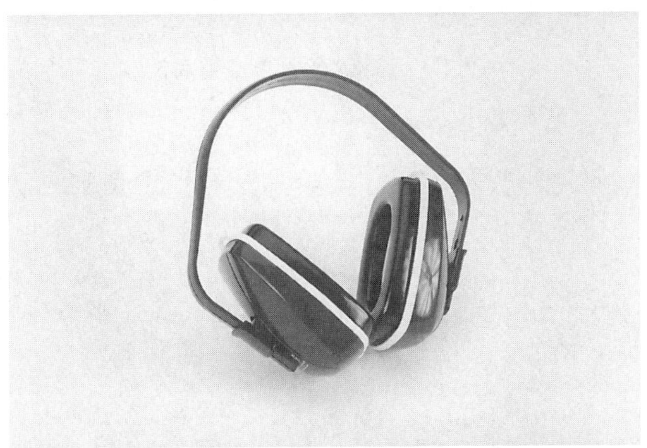

Figure 5. It is important to protect your hearing.

FIRE PROTECTION

The potential for fire in an automotive service center is always present. With gasoline, solvents, oil, and grease present, all it takes is sufficient heat or a spark and a fire can start.

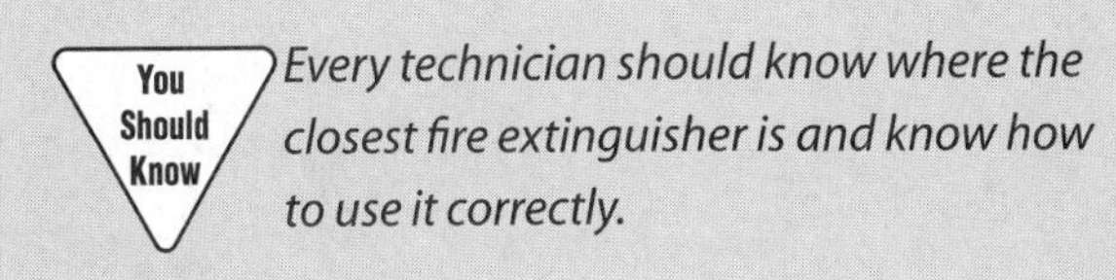

You Should Know *Every technician should know where the closest fire extinguisher is and know how to use it correctly.*

Fire extinguishers have labels on them indicating the type of fire that they are designed to fight. Most fire extinguishers will be effective on more than one type of fire. The most frequently found are the class ABC types **(Figure 6)**.

- Class A fires involve wood, plastic, paper, or other materials that are considered combustible. Many combustible materials are in vehicles and in the service center.

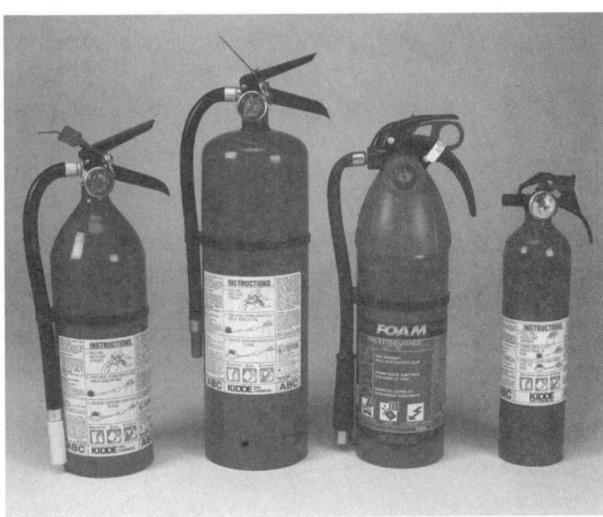

Figure 6. You should know the location of the nearest fire extinguishers.

- Class B fires involve flammable liquids such as gasoline, oil, solvents, and cleaning solutions. These are the most common types of fires and have the potential for the most damage.
- Class C fires involve electricity and generally start because of a short circuit or the overheating of an electrical component.

In the environment that you will be working in, the potential for any of these three types of fires is present. This is why a class ABC fire extinguisher is the most common type found. Make sure that it is adequate in size and current in date. There is nothing worse than running to an extinguisher only to find that it is empty, too small, or nonfunctional. Fire extinguishers need to be checked periodically and refilled if necessary. Fire extinguisher companies will help an employer with supplying adequate fire protection.

MATERIAL SAFETY DATA SHEETS

Material safety data sheets, or **MSDS**, are available for all of the chemicals found in the workplace. Your parts house or chemical supplier should have one available for each separate chemical used in your shop. Frequently MSDS are found in a rack in full view. Make sure that you read them for the chemicals that are routinely used in your work environment. Do not forget that chemicals can float from one bay to another, so be aware of what the technician next to you is using **(Figure 7)**.

Our industry is experiencing an increase in occurrence of chemical sensitivity. Chemical sensitivity can be as simple as a rash on your hand or as deadly as respiratory arrest. As the industry uses increasingly more chemicals, you should be vigilant of your sensitivity to them. Before you

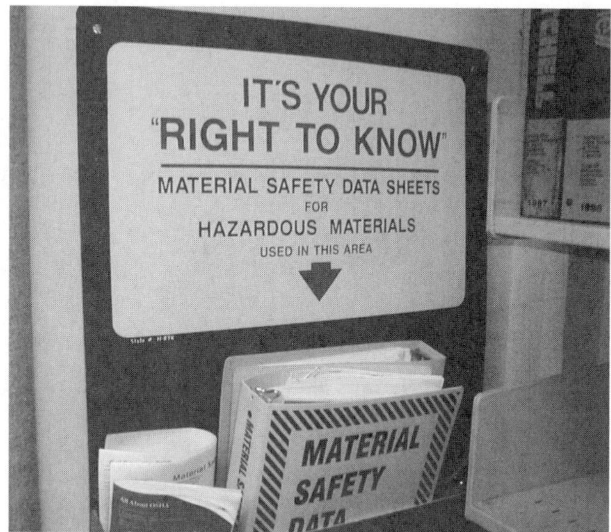

Figure 7. MSDS information is extremely important to the technician.

use something, read the MSDS about the product and take protective measures. Every technician should have disposable latex gloves available. Latex gloves will not only help to keep your hands clean but will prevent the absorption of chemicals into your bloodstream.

Vehicle waste products are also considered toxic. Used crankcase oil, gasoline, and antifreeze must be correctly handled and properly disposed of. Parts cleaning solvents are also considered toxic, requiring that you wear protective gloves and possibly a respirator.

> **You Should Know** *Chemical sensitivity is becoming a large problem within the automotive industry. Always protect yourself from repeated exposure to chemicals and fumes. You should have all of the required personal safety equipment and use it.*

BATTERY SAFETY

Batteries need to be specifically addressed because of the danger that they pose. When a battery is being charged, either by a separate charger or by the vehicle charging system, hydrogen and oxygen are produced. Hydrogen is explosive, especially in the presence of oxygen. A spark near the top of a battery could ignite the hydrogen and blow up the battery.

The inside of the battery is filled with a solution containing sulfuric acid. If it explodes, a battery will shower the area with acid and pieces of plastic from the case. **Figure 8** shows what happens if a spark is made near the top of a charging battery. If you happen to get the acid in your eyes, it can cause permanent damage unless you can get to an eyewash station immediately and flush your eyes with lots of clean water. Your service center or school lab should have emergency eyewash stations placed within reach of the bays **(Figure 9)**. Make sure that you know where they are located and that they function. They should be tested monthly.

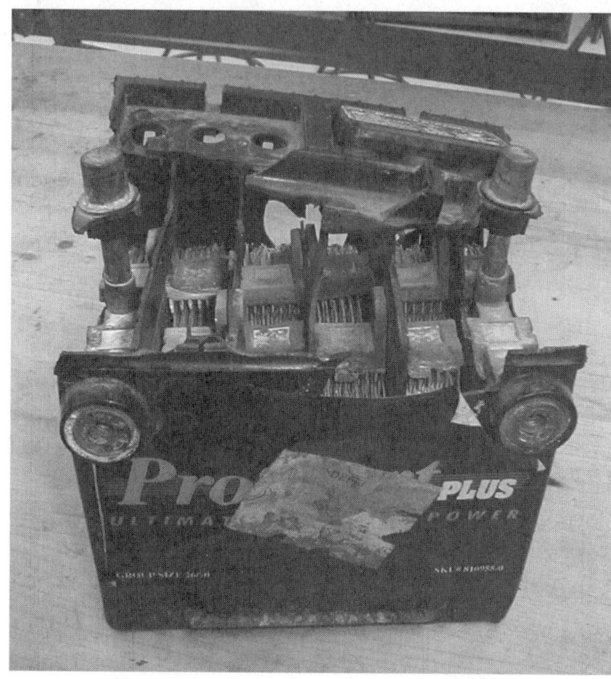

Figure 8. A spark caused this battery to explode.

Figure 9. You should know where the nearest eyewash station is.

You Should Know *A battery that is being charged will give off a small amount of hydrogen and oxygen, which are explosive. Never do anything that might cause a spark near the top of the battery.*

Prevention is the key to battery safety. Never do anything that might cause a spark near a battery. This includes the improper use of jumper cables or charging cables. If a connection is made near the battery and causes a spark, the battery might blow up. Do not let it happen to you. In Chapter 31, Chapter 32, and Chapter 33, we look at more specific examples of what not to do.

Summary

- The image that you project as a professional technician includes not only your abilities but your work clothes.
- Make sure that your clothes and hair cannot get caught in moving components.
- Safety shoes protect your toes from being crushed by heavy objects.
- ANSI 87 safety glasses should be worn at all times.
- You should have available a full face shield and safety goggles for certain automotive jobs.
- You should have available a respirator and wear it when fumes or chemicals are being used.
- Service centers or school labs should have multiple class ABC fire extinguishers available that have been periodically checked.
- MSDS should be prominently displayed and referred to before exposure to a new chemical.
- Chemical sensitivity can be prevented by the use of gloves and other related safety equipment.
- A battery can explode if a spark occurs near its top.
- Know where the eyewash fountains are located before beginning work.

Review Questions

1. Technician A states that a spark near the top of a charging battery might cause it to explode. Technician B states that batteries contain acid, which could damage your eyes. Who is correct?
   A. Technician A only
   B. Technician B only
   C. Both Technician A and Technician B
   D. Neither Technician A nor Technician B
2. A class B fire is one that involves
   A. burning wood
   B. electrical short circuits
   C. flammable liquids
   D. none of the above
3. Chemical sensitivity can be caused by
   A. repeated exposure to chemicals without taking precautions
   B. ignorance of the MSDS information
   C. the absorption of chemicals into the bloodstream through the hands
   D. all of the above
4. A class A fire extinguisher should be used in a service center environment.
   A. True
   B. False
5. It is OK to run a vehicle for an extended period of time in a closed shop.
   A. True
   B. False
6. An exploding battery showers the area with acid and pieces of plastic.
   A. True
   B. False
7. How is chemical sensitivity best prevented?
8. What types of fire extinguishers are in use in your school lab, and when were they last checked?
9. Why is the type and condition of your work clothes important?
10. What type of safety equipment is available to you?

Chapter 2 Working As an Electricity/Electronics Technician

Introduction

Whether you can be a technician who works only on the electrical system of the vehicle will depend on the size of your service center. Most technicians work on most systems but have a few specialties. For example, brake work or engine work might usually be assigned to the same technicians in a large shop because they find the most success with it.

For smaller service centers, electrical work is expected of everyone. It is considered to be part of the basic knowledge of the technician in some service centers. As the electrical system has become increasingly more complicated, the idea that everyone should be capable of electrical work is changing. What does the future hold? For the answer to this, look to the past. Thirty years ago, many systems on the vehicle were mechanical only. Electricity or electronics was not involved in the control of engines, transmissions, braking, or steering. Some technicians worked daily without finding it necessary to understand electricity.

If you look at the modern vehicle, electricity or electronics is in charge of most engine and transmission functions. In addition, nowadays active suspension and antilock braking systems rely on electronics to function. Many accessories are now power assisted. The use of electronics will increase because it is reliable, safe, and relatively inexpensive.

This chapter looks at some information that you need to know to function as an electricity/electronics technician. Much of the information will be safety oriented, and all of it is important.

YOUR TOOLBOX

The toolbox of a general technician contains a tremendous assortment of tools. As the vehicle changes, the tools required to repair it also change. Talk to technicians who have been repairing vehicles for twenty years and ask them if they have any tools that they do not use daily. It is likely that they will have an entire drawer filled with outdated tools that have not been used since a specific vehicle model disappeared off the streets.

If the technicians you are talking to generally repair electrical systems, their toolboxes will include the usual assortment of wrenches, pliers, sockets, and screwdrivers. They will, however, include some specialized test equipment. The first tool that you might notice is a **digital multimeter (DMM)**, capable of measuring voltage, current flow, and resistance **(Figure 1)**. Thirty years ago the meter would have been analog with a swinging-needle display. DMMs took over the test market in the early 1980s as the manufacturers began placing computer-type modules on vehicles. The old analog multimeter was not compatible with the vehicle and so was placed in the bottom drawer of the toolbox. We will find many applications for the DMM to be used within this text.

The second tool that you might notice is the **digital storage oscilloscope (DSO) (Figure 2)**. The DSO is the tool of choice for many diagnostic and repair technicians because it allows us to monitor a device over a period of time. The DSO can be set up to look at voltage changes or current changes with the addition of a current probe. Again, as a comparison, the relatively new DSO replaced the larger ignition scope in many service centers. The advantages to the battery-powered or vehicle-powered DSO

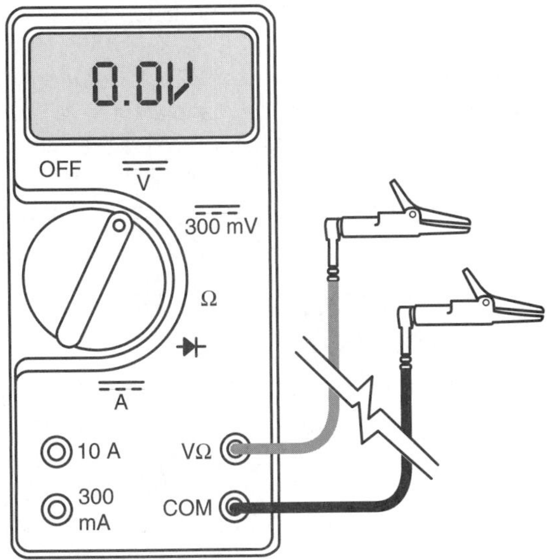

Figure 1. The DMM is the basic electrical tester.

over the massive analyzer that needs to be plugged into the wall are many. Many of the newer DSOs are also software driven, which makes updating as easy as plugging the unit into a computer and loading the latest changes found on a CD. As a technician, you will find that replacing an outdated piece of equipment with the one you need is expensive and a waste of resources. Having the DSO be software updatable is a definite plus. Each year the newest information can be loaded rather than purchasing an entire new unit. After we introduce DSOs to you, we will use them in about two-thirds of the chapters. It is the tool of choice for many of the diagnostic and repair challenges of today. You will need to have a good background in its application to be successful with it.

The DSO has become the tool of choice for many drivability, emission, or electrical problems. You need to understand how it works and how to interpret the patterns that it displays.

The third piece of equipment that you will find is the **scanner (Figure 3)**. A scanner allows the technician to view data from the vehicle's **powertrain control module (PCM)**. The scanner also allows the technician to control a system or component and monitor the results. The technician who diagnoses and repairs the PCM systems is required to have a scanner. Many technicians have more than one, especially if their scanners are dedicated to specific manufacturers. Like the DSO, most scanners are software driven so that vehicle

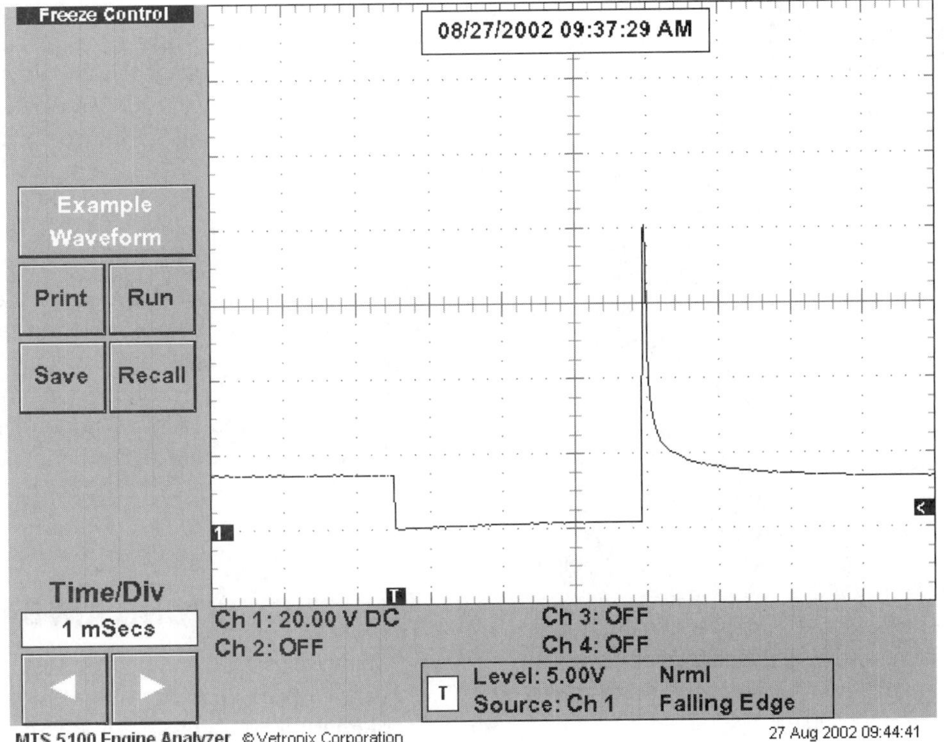

Figure 2. The DSO can show voltage changes over time.

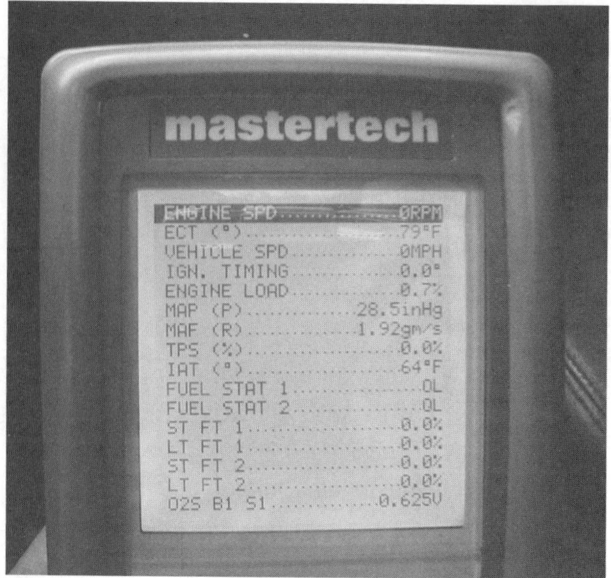

Figure 3. The scanner allows the technician to monitor PCM functions.

data for each model year can be loaded into the scanner. The scanner is an invaluable piece of test equipment for electricity or electronics technicians, because it gives them the opportunity to look inside the PCM and its circuits. We will find application for the scanner in this text.

ACCESS TO WIRING DIAGRAMS AND REPAIR INFORMATION

The vehicle of today is virtually impossible to diagnose and repair if you do not have information. On the modern vehicle, it is difficult to follow a single wire as it winds its way around different sections. The wiring diagram specific to the vehicle will give the details necessary to figure out just where the wire goes, if it changes color, and where the connectors are located. Component location and the wiring diagram go hand in hand and are usually included together. In addition, system diagnostic aides are necessary to simplify the diagnosis and repair. Specification values to compare against the actual are also necessary. One last item designed to simplify the technician's procedure is the **technical service bulletin (TSB)**. An example of a TSB is shown in **Figure 4**. It is through the TSB that technicians find out if more vehicles than just the one they are working on have the same problem. The amount of time that a TSB can save a technician is tremendous. There is no sense in reinventing the wheel when the manufacturer, realizing that all of a certain category of vehicles has the potential for a similar problem, makes that information available.

The ability to diagnose and repair a modern vehicle is dependent on the technician having the correct information. Where you get the information is not as important as having it available. Aftermarket information systems, such as Mitchell On-Demand and All Data, supply information on-line through a subscription or on CDs. In addition, if you are working in a dealership, the manufacturer will supply the required information. Another source is a technical help line. For a monthly fee, you can phone a person who will supply the needed information for diagnosis and repair.

COMMUNICATING WITH THE CUSTOMER

Depending on the size of the service center that you are working in, you may or may not talk directly with the customer. A service advisor or manager might be in between the consumer and you, the technician. In either case, getting the correct information from the customer cannot be overemphasized. The customer is likely the person who was driving the vehicle when the problem showed up. The conversation that someone has with the customer can be extremely useful and save hours of fruitless work. A repair ticket that states "drivability problem" will require the technician to figure out the driving conditions that were present when the problem occurred. It is possible that the technician will not be able to duplicate the conditions and not observe the problem. The vehicle is returned to the consumer with the note of "no problem found." Think about how frustrating this could be, especially if the consumer experiences the problem the very next day. The conversation with the customer should have revealed important information for the technician. When did the problem occur? What specifically did the vehicle do or not do? What was the outside temperature? Was the vehicle warm or cold? What driving conditions produced the problem? Can you duplicate the problem with the technician in the vehicle?

Think about how much more information the technician has in the second example. It is likely that less time will be necessary to fix the vehicle because the technician has a starting point. Once you get out in the field, try to develop your communication skills and especially your ability to listen to the customer. The customer's information, if you have heard it, will save you countless hours of frustration and no doubt result in a better repair. Better repairs bring customers back to the service center the next time they need service.

WORKING AROUND AIR BAGS

Technicians need to recognize that most vehicles they work on have air bags. The air bag system is part of the **supplemental restraint system (SRS)**. If the air bag were to deploy when you were working on the vehicle, you could be seriously injured. In addition, the air bag or bags are expensive to replace. Your employer and the customer

ARTICLE BEGINNING

TECHNICAL SERVICE INFORMATION

FALSE DTC P0121 (REPROGRAM PCM)

Model(s):	1997 Buick Century, Skylark 1997 Chevrolet Lumina Monte Carlo Malibu, Venture 1997 Oldsmobile Achieva, Cutlass, Cutlass Supreme, Silhouette 1997 Pontiac Grand Am, Grand Prix, Trans Sport with 3100/3400 V6 Engine (VlNs M, E - RPOs L82, LA1)
Section:	6E - Engine Fuel & Emission
Bulletin No.:	77-65-14A
Date:	May, 1998

NOTE: This bulletin is being revised to add additional models and calibration numbers. Please discard Corporate Bulletin 77-65-14 (Section 6E - Engine Fuel & Emission).

CONDITION

Some owners may experience a MIL (Malfunction Indicator Lamp) light illuminated on the vehicle's instrument panel. Additionally, the engine's normal controlled idle speed may be slightly elevated when the MIL is illuminated.

CAUSE

The current DTC (Diagnostic Trouble Code) P0121 is too sensitive. The rational check that the diagnostic calibration performs has been changed. Part of those changes involve eliminating the defaulted higher idle.

CORRECTION

Check the calibration identification number utilizing a scan tool device. Re-flash with the updated calibration if the current calibration is not one listed in this bulletin. If the vehicle already has the most recent calibration, then refer to the appropriate service repair manual to diagnose and repair for DTC P0121. Test drive the vehicle after repair to ensure that the condition has been corrected. The new calibrations are available from the GM Service Technology Group starting with CD number 6 for 1998.

IMPORTANT: Do not attempt to order the calibrations from GMSPO. The calibrations are programmed into the vehicles PCM via a Techline Tool device.

Figure 4. TSBs allow the technician to see if other vehicles are experiencing the same difficulty and to outline a repair procedure.

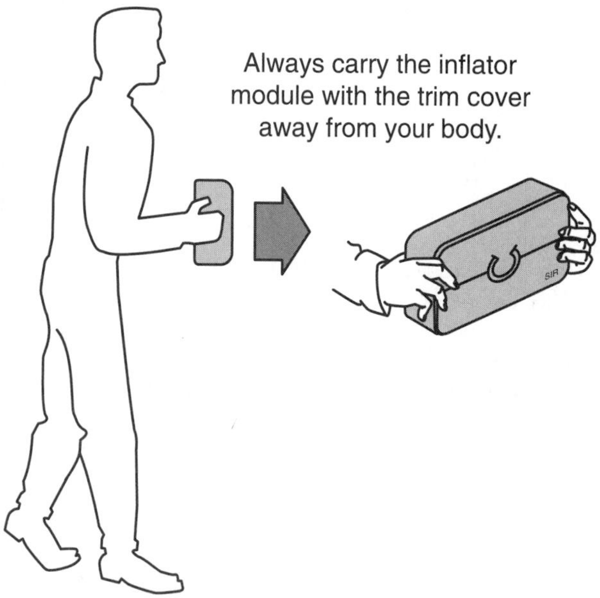

Always carry the inflator module with the trim cover away from your body.

Figure 5. Proper handling of the air bag is vitally important.

> **You Should Know** *The air bag system is extremely dangerous to the unsuspecting technician. Always disarm the system before working in the area of the air bag.*

will not be pleased if they have to replace a deployed air bag **(Figure 5)**.

The air bag can deploy even with the battery disconnected. Most systems can still blow the bag after ten minutes or more. If the vehicle requires service in the area of the air bag, follow the manufacturer's recommendations on disconnecting the system. Frequently a specific connector needs to be opened. This shuts the system down, preventing the accidental deployment of the bag. Do not forget that the vehicle might have multiple air bags or curtains. Before you begin a diagnosis or repair, make sure that the system has been disabled. The air bag warning light should be lit. This ensures that the system is not functional **(Figure 6)**. It is now safe to work on the vehicle.

Figure 6. The SRS cannot function if the air bag light is on.

STATIC ELECTRICITY

Any discussion of protecting electronic components during service should include a discussion of static electricity. You are familiar with static charges in one form or another. As you slide your feet across a carpeted floor and then touch something, you might feel and see a slight static discharge. The action of sliding your feet across the carpet placed a slight electrical charge on you. A change in the number of electrons on you puts you at a different charge level than the objects around you. When you touch them, there is a discharge between you and the object. Although this discharge generally does nothing to you other than to perhaps surprise you, it can do potentially great damage to electronic circuitry. Static electricity has become a topic of many articles written with the automotive technician in mind. As more microprocessors are added to the vehicle, more concern is expressed by the manufacturers for the environment that we typically work in. Our shops are generally not as static free as an electronics repair shop, and yet we work on sophisticated integrated circuitry.

Technicians of today must realize that the static electricity that they have ignored for so long will have to be discharged safely before they begin working on an electronic

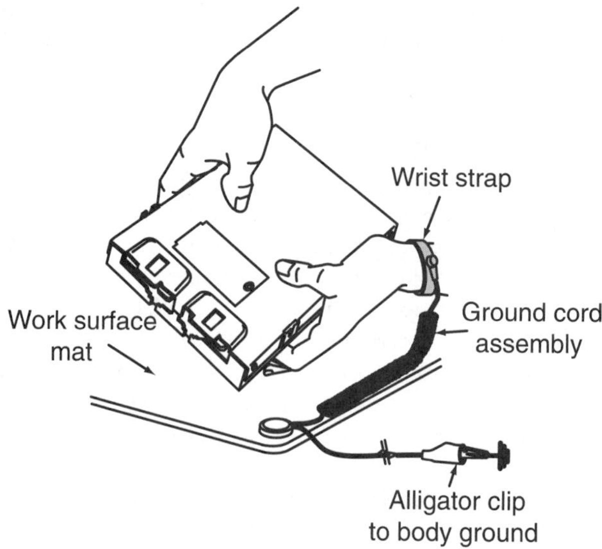

Figure 7. Connecting the technician, the component, and the vehicle together to control static.

component or processor: dashboards, trip computers, memory circuits, processors for fuel and ignition control, antilock braking, body computers—the list grows each year. To effectively work on these circuits, some precautions are necessary. Generally these can be summarized by the statement that you must be at the same electrical potential as the

Interesting Fact

The simple act of sliding across the front seat of the vehicle could generate sufficient static electricity to damage an electronic module.

component you are working on and the vehicle you are working in.

We are seeing increased use of grounding straps, which connect the technician, the component, and the ground system of the vehicle together **(Figure 7)**. The theory behind this is to place all things that will touch at the same electrical potential so that a discharge will not take place. Even if you do not have all the special grounding equipment, run jumper ground wires between the components and the vehicle, and ground yourself to the vehicle before you begin working. In addition, follow any precautions printed on the packaging materials of the replacement components. Frequently, special packaging protects the component from static discharge. Leaving the components in their original packaging until they are ready to be installed is the best insurance.

Summary

- The electricity technician's toolbox will include a scanner, DSO, and DMM.
- Repairing a vehicle is next to impossible without the correct wiring diagram.
- A TSB is designed to help the technician by indicating if other vehicles in the same class have experienced similar problems.
- Asking the correct questions of the customer and listening to the answers help guide the technician through the diagnostic process.
- The air bag system should be disabled if a technician is working near the bag.
- Static electricity can destroy components before their installation.
- The best method to prevent static electricity is to connect yourself, the vehicle, and the component together.

Review Questions

1. Technician A states that the DSO is the tool for viewing the signal from certain components. Technician B states that the DMM is the tool for checking out certain components. Who is correct?
   A. Technician A only
   B. Technician B only
   C. Both Technician A and Technician B
   D. Neither Technician A nor Technician B
2. A scanner allows the technician to look at
   A. resistance
   B. current flow
   C. PCM information
   D. battery functions
3. Before working on or near the SRS system, the technician should
   A. determine if any TSBs are applicable
   B. look up the wiring diagram and repair information available
   C. disconnect the SRS
   D. all of the above
4. Static electricity can
   A. cause a component to not function correctly
   B. cause a component to burn out before its installation
   C. not do damage to automotive components
   D. short circuit an entire system
5. The SRS could blow the air bag even with the battery disconnected.
   A. True
   B. False
6. If the air bag warning light is on, the SRS is disabled.
   A. True
   B. False
7. What precautions should be observed when working in the area of the air bag?
8. What precautions should be observed to prevent static electricity damage?
9. A customer complains of a hesitation during acceleration. What questions should be asked to help in the technician's diagnosis?

Circuit Fundamentals and Basic Test Equipment

SECTION OBJECTIVES

At the conclusion of this section, you should be able to:

- Understand how basic circuits are wired.
- Discuss the differences between circuit types.
- Measure current with an ammeter.
- Measure voltage with a voltmeter.
- Measure resistance with an ohmmeter.
- Recognize the basics of Ohm's Law and its application to circuits.
- Recognize and discuss the specifics of a series circuit.
- Recognize and discuss the specifics of a parallel circuit.
- Know what makes a material a conductor.
- Know what makes a material an insulator.
- Know the basic components of a circuit.

Your window into the circuit will be the voltmeter, ammeter, and ohmmeter. A working knowledge of how to use these tools is fundamental to your success.

Chapter 3 Voltage, Current, and Resistance

Introduction

The ability to diagnose many modern systems found on the vehicles of today lies in the technician's ability to understand what electricity has done or is doing in a particular circuit. Frequently, diagnosis is simple because the technician can mentally figure out what is happening before the tools even leave the toolbox. To understand how electricity behaves within a circuit and to be able to analyze the circuit on the vehicle, one must have a good background of the fundamentals. This chapter contains the building blocks of the rest of the text. Learn these principles now, and the remainder of the material will be easier. Keep reading the information until it becomes second nature to you. The section objectives, summaries, and review questions will help test your ability to comprehend the material.

Within this chapter, we attempt to build your vocabulary and your comprehension of the necessary ideas and components that you will use when diagnosing or repairing a circuit. We discuss the sections of a circuit and build some simple, yet practical, examples of working circuits. New words are in bold text.

ELECTRICITY

Electricity is difficult to define unless split into its elements. To be able to do this, take a brief look at the atom. All matter is composed of atoms. Everything, including you, contains atoms. They are the building blocks of everything you see. Refer to **Figure 1** and notice that the center of the atom, called the nucleus, contains both positive-charged (+) particles and negative-charged (–) particles surrounded by moving electrons in orbits, like the planets around the sun. These orbiting electrons are what we will be concerned with as we look at electricity. If we could examine a length of copper wire very closely, we would notice that the outer electron orbit of the copper atom has only one electron in it. Notice, as shown in **Figure 2**, a total of 29 electrons in four separate orbits around the nucleus. The first three orbits do not concern us. The outermost orbit, or **valence orbit**, is where the action is. With a little "push" we will be able to move this electron out of its orbit. The fewer the electrons in the valence orbit, the easier they move or travel between atoms. These electrons are sometimes referred to as being "free." They normally move between atoms that are very close together. This characteristic makes copper (or another atom with one or two electrons in the valence orbit) a good electrical wire. These free electrons can be bumped or pushed into adjacent or nearby atoms easily. When this bumping occurs, the second atom, which now has two electrons in its outermost orbit, will then give up one electron to a third atom, and so on. The idea of pushing a series of electrons along a wire from one atom to the

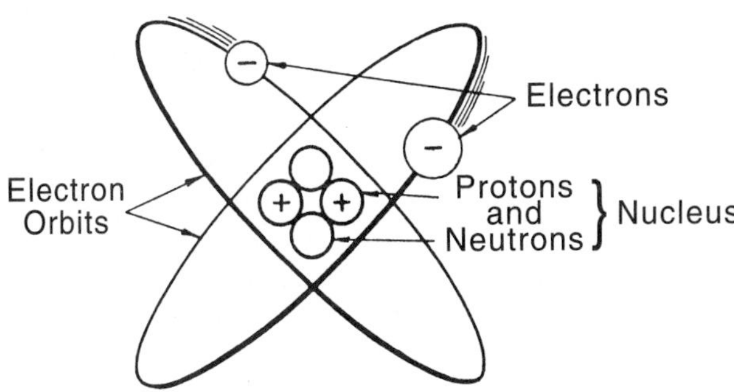

Figure 1. Electrons revolve around the nucleus.

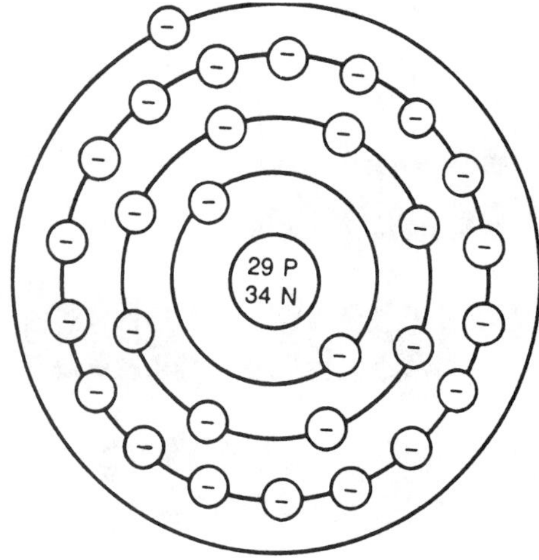

Figure 2. A copper atom has one electron in the valence orbit.

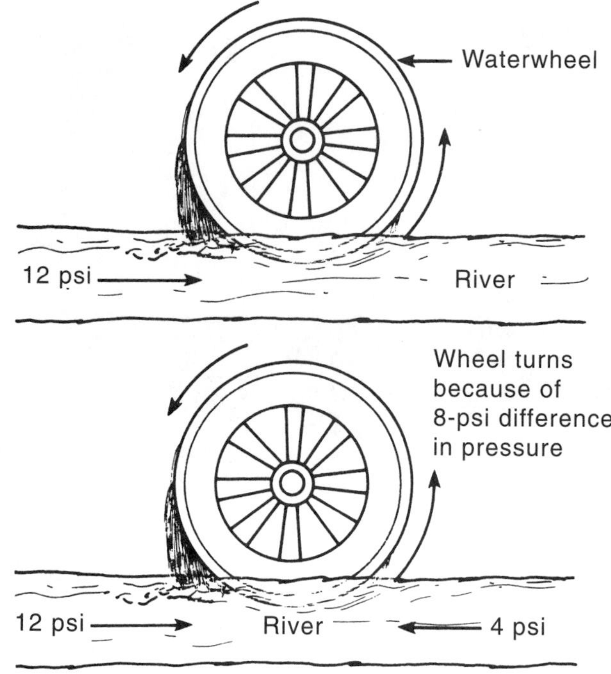

Figure 3. A difference in water pressure causes the waterwheel to turn.

next leads us to a good simple definition of electricity: moving electrons. The more electrons that move, the more electricity. Copper or any other atom with one or two electrons in its valence orbit is called a **conductor**. Simply defined, a conductor is a material in which electrons move easily. We typically use copper, aluminum, or steel for conductors in automobiles.

The opposite of a conductor is an **insulator**. Insulators are composed of materials whose atoms contain nearly full valence orbits. Full valence orbits tend to stabilize the atom and eliminate the movement of outer electrons between atoms. Insulators are used in electrical systems to ensure that electricity arrives where we want it to go without being sidetracked. Paper, plastic, rubber, and cloth are frequently used as insulators in automobiles.

VOLTAGE

Up to this point, we have been talking about moving electrons and have defined them as electricity. Voltage is the reason why the electrons were moving in the first place. Voltage is electrical pressure.

Electrons move when we apply a pressure against them. This pressure, voltage, can be measured. Defining voltage as pressure allows us to compare it to the pressure that forces water out of a faucet when you open it. Water pressure measured in pounds per square inch, or psi, is the push behind the water. Without some push behind the water, opening the faucet would not accomplish anything. The water would remain in the pipes and not flow. Electricity, defined as moving electrons, would react in the same manner. Without pressure, or voltage, behind the electrons, they would not move and would remain in the wire, or conductor.

Voltage should be thought of as a pressure difference, which will cause electron movement. Look at **Figure 3**. In the top example, water flows because there is a pressure of 12 pounds per square inch from the left. But notice in the bottom example that water will also flow where a pressure of 12 pounds per square inch is pushing against 4 pounds per square inch and still pushes or moves the water to the right. The pressure difference will move the water from a higher to a lower pressure. Electricity is much the same. If a pressure of 12 volts is applied against a pressure of 4 volts, the electrons will move from the higher 12 volts toward the lower 4 volts because of the 8-volt difference.

We should also realize that voltage can be positive or negative. This is called **polarity** and is simply the type of charge or pressure on the electrons. Let us try some examples. Refer to **Figure 4**. Notice that the pressure difference determines the direction of electron movement. If there is a pressure differential, electrons will move. If, however, we apply equal pressure of the same polarity (+12 against +12), the end result is the same as if we applied no pressure (0 volts): no electron movement. Later you will see that the major source of voltage or pressure is the battery or the charging system.

Keep in mind that the amount of pressure behind water will determine just how much or how fast the water flows. We can increase the amount of water flowing by increasing the pressure. Electricity is the same. Increasing the pressure behind the electrons will cause more of them to flow. Reducing the pressure will cause less of them to flow. Remember that it is the pressure difference, not just the pressure, that causes the movement.

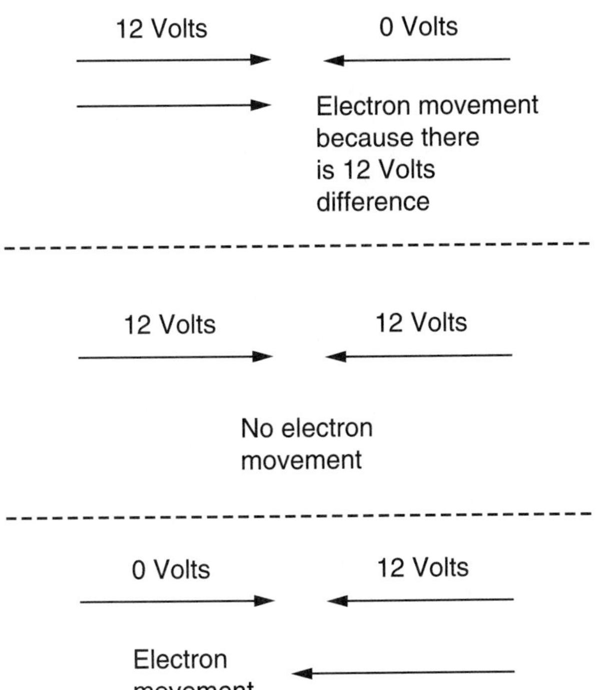

Figure 4. Current moves because of voltage differences.

CURRENT

Up to this point, we have been referring to moving electrons without trying to count or define them. This movement of electrons is called a **current**. Much the same as a current in a river, it flows through the wires. We frequently want to know the amount or quantity of electrons flowing, because it is this current that will do the work for us. It would be impractical to count electrons because the number needed to do a reasonable amount of work is extremely large. A small automotive light bulb, for example, would require six billion billion electrons per second to light—not a very easy number to work with. It would be similar to being paid each week in pennies. Scientists have helped us by defining the six billion billion electrons as an **ampere**, or simply an amp. Remember that the quantity of electrons moving is what will do the work we want done and that this mass of electrons is called current and is measured in amperage, or amps.

Returning to the analogy of water coming out of the faucet because of the pressure behind it **(Figure 5)**, what is the term for the electrical current equal to the flow of water? If you said electrons or amperage, you were correct. Remember that another term for this electron mass is current. You will sometimes see "amps of current" printed. This is just another way of telling you how many electrons are actually working for you. Another way to remember current is to think of the current in a river. It is the moving water. The electrical current then becomes the moving electrons. Why

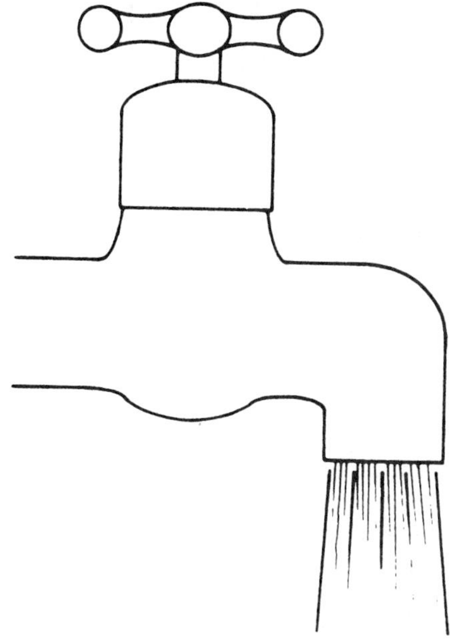

Figure 5. Water and current react the same way.

are they moving? The pressure or voltage behind them is pushing them from one atom to the next atom.

It then becomes simple to realize that it is this quantity of pressurized electrons that will do work for us. Let us define work. Electricity working will give us light, heat, or magnetic field **(Figure 6)**. It is important to realize that any time work is being done we will be producing one or more of these results. Most of the time, this work is desired or engineered into the circuit, but sometimes it is not. A bad connection, for example, might get warm as electrons flow through it. The warmth is an indication of work being done even though we did not design the bad connection into the vehicle.

You might be thinking about all of the devices on the car that do work for us. All of them have some current or amperage flowing through them and are doing work to produce light, heat, a magnetic field, or a combination of two or more of these. Certain devices, however, also give us motion. Starter motors, heater blower motors, windshield wipers, and so on, all give us motion, but they do so by setting up magnetic fields as we will see in future chapters.

Work cannot be done by just voltage or just amperage. It takes both. If we need a waterwheel turned, it will take both water and pressure to get the work accomplished. Pressure only or water only will not turn the wheel. Electricity is the same. Voltage without current or current without voltage accomplishes nothing. When you leave your headlights on overnight and run the battery down, you are removing the pressure or voltage. When you try to start the car the next morning, you have electrons available because the whole vehicle is made up of them, but without voltage or pressure behind the electrons there will not be any

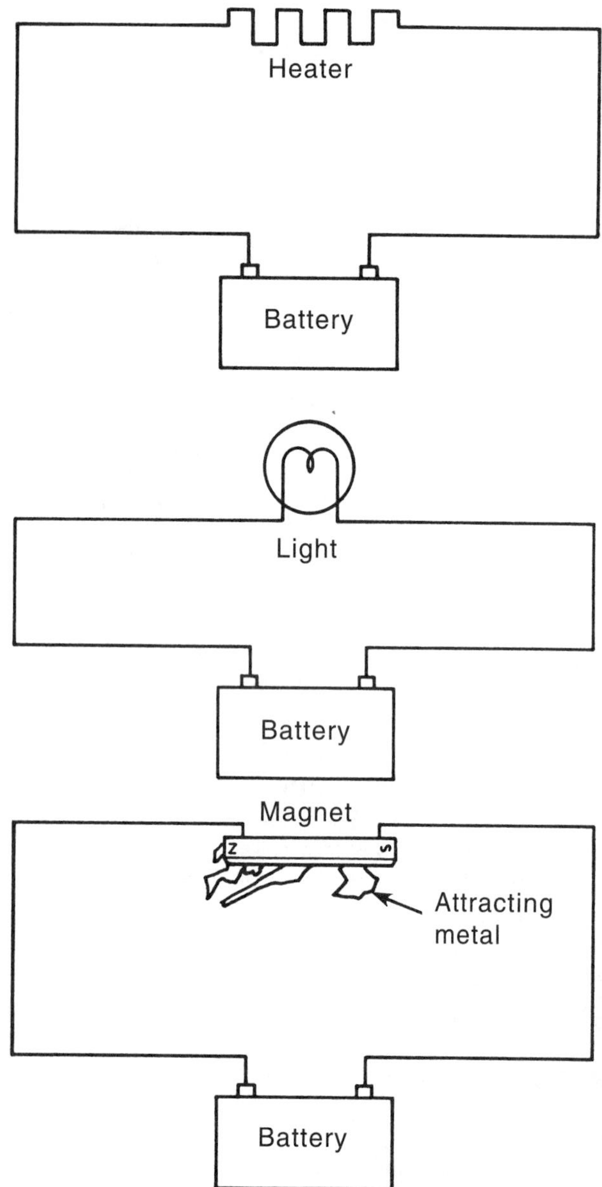

Figure 6. A load will develop heat, light, or a magnetic field.

current flow and nothing happens when you turn the key. No pressure with lots of electrons will not do any work.

WATTAGE

Another way of looking at work done, besides amperage, is to consider **wattage**. Wattage is the electrical means of monitoring how much work is being done. Consider the light bulb in your home. It is rated in watts. A 60-watt bulb delivers one-half the light of a 120-watt bulb. Wattage is the result of multiplying the voltage applied by the current flowing. Because it takes both pressure and current to do work, it makes sense that an indication of the amount of work accomplished would use both in its calculations. Your 60-watt bulb at home actually draws ½ ampere of current at 120 volts, for example:

120 V × 0.5 A = 60 watts **(Figure 7)**

We usually do not use the wattage formula when we repair vehicles, but knowing that work accomplished can be measured in wattage is useful, especially when dealing in some of the experiments we will try in Chapter 3 and Chapter 4. We also use wattage when trying to understand starting systems. If a certain amount of work must be accomplished, such as cranking over an engine, the amperage used will be based on the voltage applied. Mathematically, it figures out like this: if 2000 watts of work will be needed to crank over an engine, 200 amps will be needed if the voltage can be kept up to 10 volts:

$$200 \times 10 = 2000$$

However, if the voltage drops down to 5 volts, the amperage will have to go up to 400 amps to crank the engine over at the same speed. Our discussion of conductors later in this chapter shows why it will be harder to deliver 400 amps than 200 amps.

If you look at vehicles today, you will see mostly 12-volt systems, and yet years ago many systems were 6 volt. Getting the same amount of light, heat, and magnetic field out of a 6-volt system requires twice the amperage. Delivering and generating twice the amperage was not only expensive but difficult to control. Another example of wattage at work can be seen when looking at large diesel engines. Most crank over at 24 volts rather than 12 volts because the starting system will need to draw half the current at 24 volts than it will at 12 volts to accomplish the same amount of work. Do not forget:

voltage × amperage = wattage

RESISTANCE

The last item necessary (voltage and amperage being the other two) is **resistance**. Resistance, measured in **ohms**, is the force that opposes the flow of electrons. This back pressure to electrons, along with the voltage, will dictate the number of electrons able to flow through a circuit. When current is moving through electrical components, such as bulbs and motors, the amount of current flowing will be dependent on two other conditions: the voltage, or pressure on the electrons, and the resistance that the circuit puts up against the flow of electrons. In this way, resistance is the opposition to the flow of current. If we decrease the resistance, it will be easier for electrons to flow. If we decrease the resistance, more electrons will flow through the circuit and more work will be accomplished. Let us go back to our water example. Pressure against the water moved a current and turned the waterwheel. That waterwheel will take some effort to turn. It will offer some resistance to the water trying to push it. In essence, this waterwheel is equal to the electrical resistance that a light bulb offers to the flow of current. Resistance is measured in

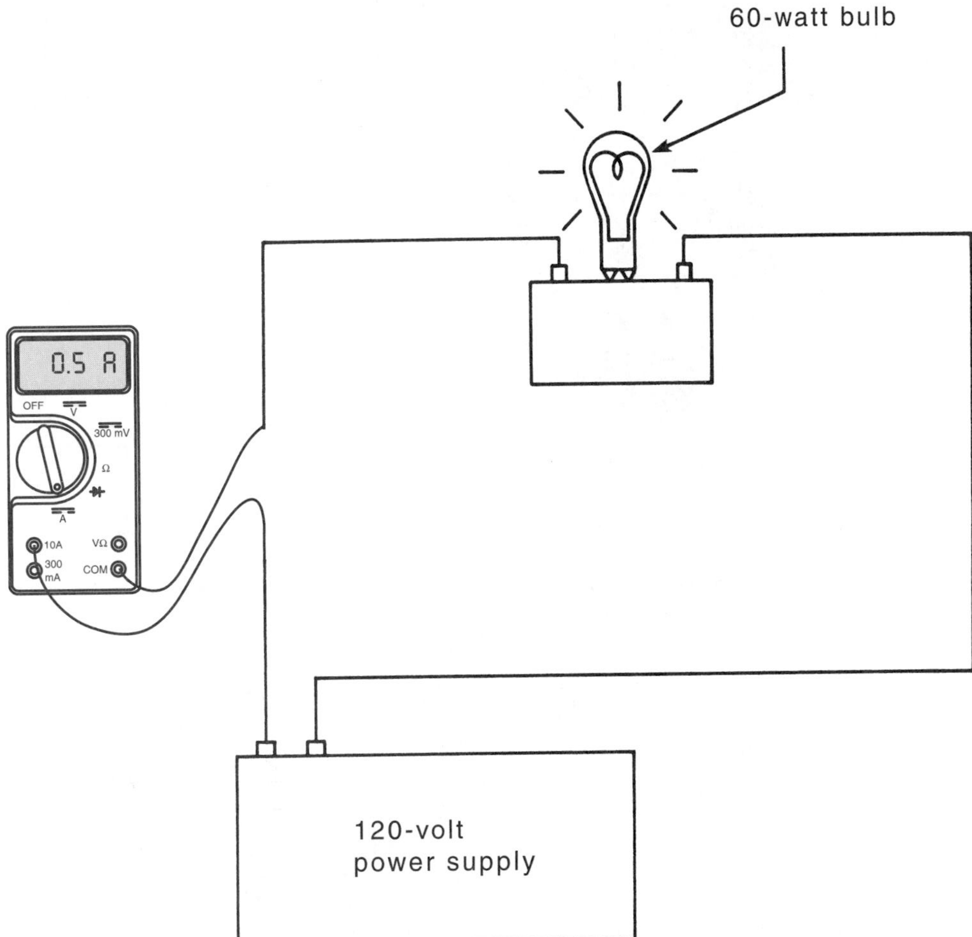

Figure 7. The voltage multiplied by the current equals the wattage (V × A = W).

ohms and has the Greek symbol of the omega, Ω. As you look at resistance and measure it, always keep in mind that it is the opposition, or force, that will make it harder for current to flow.

Consider now the interrelationship of voltage, amperage, and resistance. This relationship is called **Ohm's Law**, and, simply stated, it tells us what the third, or unknown, value will be in a circuit if we know the remaining two. If you know the voltage and the resistance, you can determine the current flowing. Or, if you know the current and the voltage, you will be able to find the resistance. This relationship must be clear in your mind because diagnosis of most circuits involves looking for changes that occur in voltage, amperage, and resistance. To be practical, we seldom sit alongside the automobile and use Ohm's Law mathematically to repair a problem. We do, however, use the principles of Ohm's Law in virtually every electrical repair. Let us look closely at this interrelationship.

If we understand that resistance is opposition, increasing or decreasing this opposition will have an inverse, or opposite, effect on current flow (assuming the voltage remains the same). In other words, if we increase the resistance in a circuit whose voltage remains the same, we will see that more opposition will allow less current to flow. It will now be harder for current to flow through the circuit and therefore less will be able to get through the resistance. The opposite is also true. If we decrease the resistance within a circuit, the current will find itself in an easier path and more will flow. This inverse relationship assumes that the voltage remains the same. Remember our discussion of work. We discussed the amount of work done by the amount of current. The more current flowing, the more work accomplished. With this in mind, answer the following question: which component has the greatest resistance, a taillight bulb or a starter motor? If you said taillight bulb, you were correct. Let us think through this example together. Which component is doing the least amount of work? The taillight is. This then means that it must be drawing the least current, and Ohm's Law says that if the current is down, the resistance must be up. Let us try the same example, but look at it from the starter side. The starter is turning over the engine. This represents a lot of work accomplished. Lots of work equals lots of current. Go back to Ohm's Law. Lots of current can flow only if the resistance

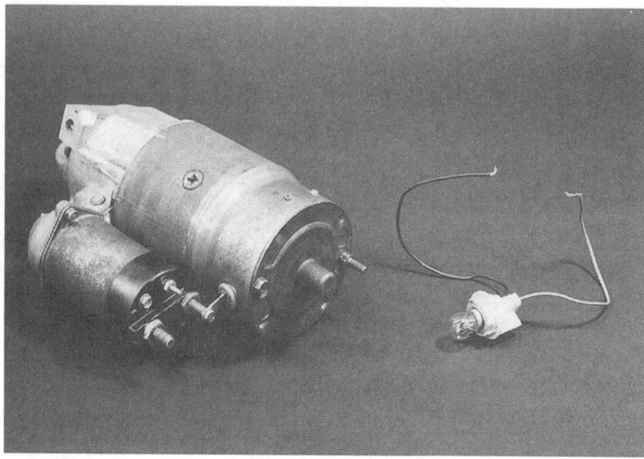

Figure 8. The size of the load and its resistance are not equal.

is very low. The starter and a taillight bulb have been used as examples to emphasize that physical size has nothing to do with electrical resistance **(Figure 8)**.

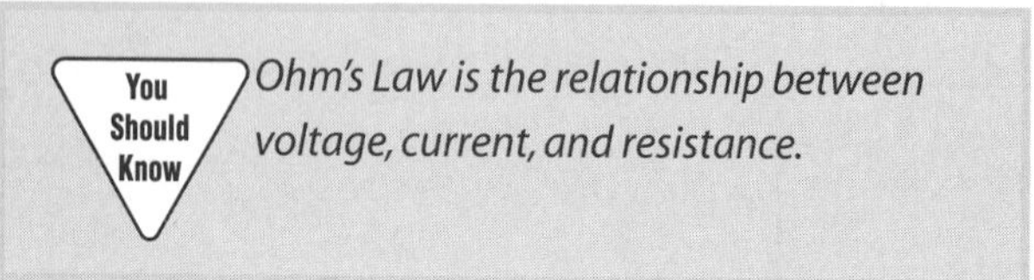

You Should Know
Ohm's Law is the relationship between voltage, current, and resistance.

To be specific, Ohm's Law is best understood by looking at **Figure 9**. Putting your finger over the unknown value will leave uncovered the formula that you use to determine what the unknown is. If the voltage and the amperage are known, the resistance can be easily figured. Blocking over the resistance, the unknown, leaves behind voltage over amperage. For example, if a 12-volt circuit has 2 amps flowing, a simple fraction with 12 on top and 2 on the bottom is the result of placing the known voltage and amperage into the formula. Dividing 12 by 2 shows that the circuit must have 6 ohms of resistance. How many watts of work are being accomplished? Remember watts? Wattage equals voltage multiplied by amperage. That is:

12 volts × 2 amps = 24 watts of work

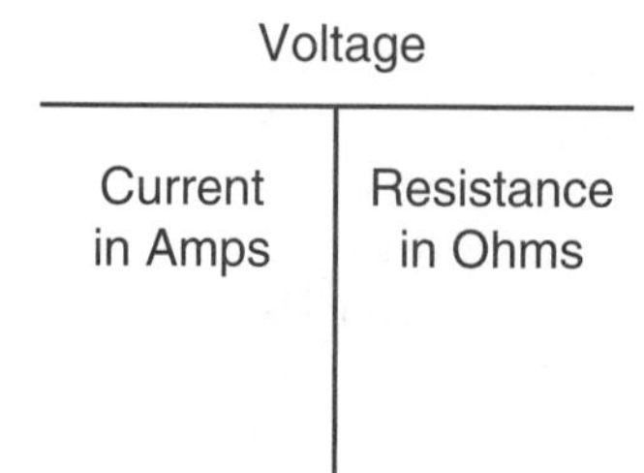

Figure 9. Ohm's Law is the interrelationship of voltage, current, and resistance.

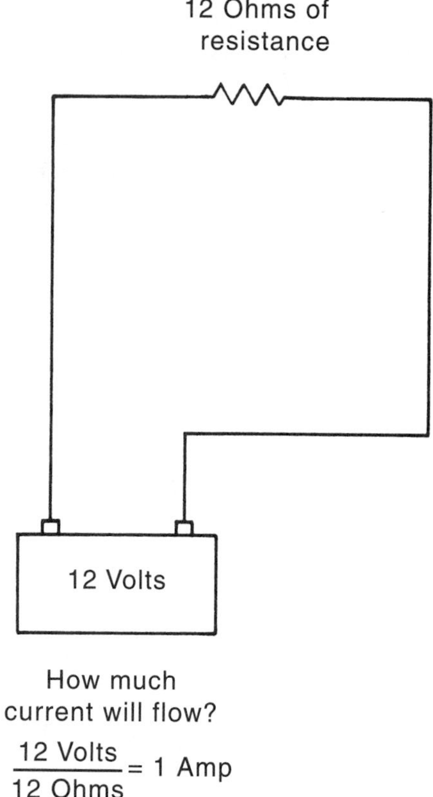

Figure 10. Ohm's Law at work.

Let us try another example. If 12 volts is applied to 12 ohms of resistance, how much current should be flowing? Cover the unknown value of amperage. This leaves voltage over resistance, or 12 over 12. Dividing 12 volts by 12 ohms will equal 1 amp flowing **(Figure 10)**. Most of the actual examples on the modern vehicle are with approximately 12 volts applied because most components are powered by the vehicle's 12-volt battery, but circuits with more or less voltage can just as easily be figured using Ohm's Law. We use Ohm's Law again when we discuss types of circuits toward the end of this chapter.

INSULATORS AND CONDUCTORS

We will now begin to put the various principles that we have discussed into workable circuits. We should realize that any discussion of amps, ohms, watts, and so on would not make much sense unless we apply it to actual circuits on the vehicle. To be able to analyze what happens with Ohm's Law also requires a working knowledge of circuits. What is a circuit? Simply stated, a circuit is a path electricity travels from a source to a component that will do work and back to the source. A circuit does work. Voltage pushing current through resistance results in light, heat, or a magnetic field, which is defined as work. Let us consider a circuit and analyze its components or parts. At this point, let us use

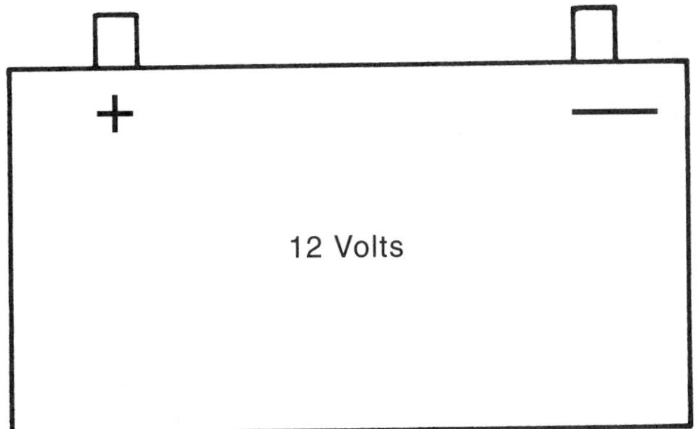

Figure 11. The battery is a source of power.

a battery as a source of power **(Figure 11)**. We will study batteries in detail, discussing their construction and service, in Chapter 30, Chapter 31, and Chapter 32. For now, just remember a battery as a source of electrical pressure when charged. If the work to be accomplished will be to light a bulb, the next thing we need are conductors, or wires. Remember our discussion of atoms and valence electrons. A conductor was defined as a substance that has one or two electrons in its valence, or outermost, orbit. This substance accepts and gives up electrons easily, which is another way of saying it conducts electricity.

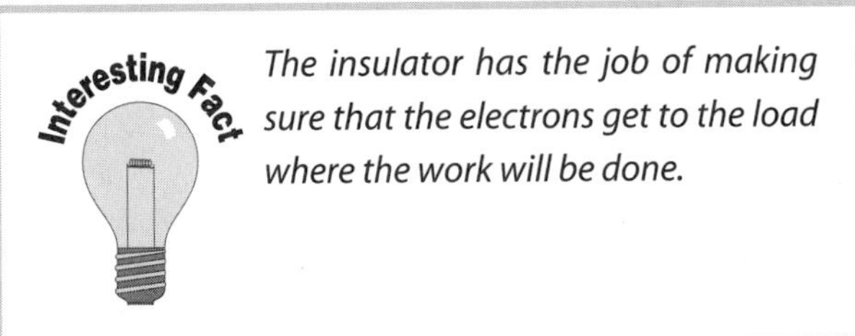

The insulator has the job of making sure that the electrons get to the load where the work will be done.

The opposite (electrically) of a conductor is an insulator. An insulator is defined as a substance whose atoms do not easily accept and give up electrons. It is important to note that the difference between conductors and insulators is sometimes a fine line. Consider the human body. If you put your hands across a positive (+) and a negative (–) battery terminal, the high resistance of your body prevents the electrons from flowing through you. Yet, when you touch the end of a spark plug wire, you can feel those electrons flowing right through you. The difference is the pressure. We are insulators at low pressure; yet, if the pressure, or voltage, is raised high enough, we become conductors. The 12.6 volts of the battery is low enough so that the resistance of your body could not be overcome. However, raise this voltage up to 30,000 volts, and your high resistance will not prevent the flow of current. The ignition system currently in use today has in excess of 30,000 volts available and can easily pass small amounts of current through your body. Another example can be seen during any lightning storm. The

Figure 12. Very high voltage at work.

extremely high voltage of lightning passes current right through the center of normal insulating air **(Figure 12)**. Given a high enough voltage, electrons can be pushed through almost anything. This dictates that the insulation in use must match the voltage applied in any circuit. Low voltage requires less insulation than higher voltage.

If you were to cut a spark plug wire in half, you would observe that the greatest percentage of the wire is insulation. Sometimes as high as 90 percent of the wire is insulation. Cutting a taillight wire in half would reveal the opposite **(Figure 13)**. The greatest percentage of this wire would be the conductor, or center, which is the part that will be carrying the current. The size of the conductor will be determined by the amount of current flowing and the length of the circuit **(Table 1)**. The larger the number of the wire, the thinner it is. Notice that a larger wire (smaller number) is needed to carry the same amount of current longer distances. The greater length of the wire adds more resistance. The additional resistance must be compensated for with a larger conductor. Greater current or additional length will require larger conductors. Do not forget this when you repair a circuit. Always use an adequate-size wire for the current and the length of the circuit, with adequate insulation for the voltage. Remember, it is the job of the insulation to

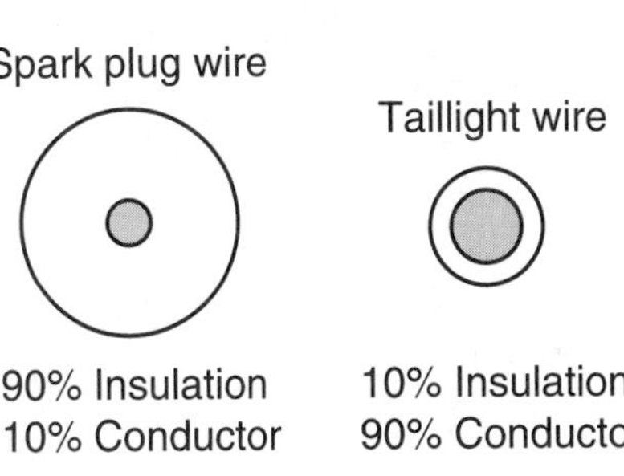

Figure 13. The size of the conductor determines current capacity, whereas the amount of insulation determines voltage capacity.

Total Approximate Circuit Amperes	Wire Gauge (for Length in Feet)											
12 V	3'	5'	7'	10'	15'	20'	25'	30'	40'	50'	75'	100'
1.0	18	18	18	18	18	18	18	18	18	18	18	18
1.5	18	18	18	18	18	18	18	18	18	18	18	18
2	18	18	18	18	18	18	18	18	18	18	16	16
3	18	18	18	18	18	18	18	18	18	18	14	14
4	18	18	18	18	18	18	18	18	16	16	12	12
5	18	18	18	18	18	18	18	18	16	14	12	12
6	18	18	18	18	18	18	16	16	16	14	12	10
7	18	18	18	18	18	18	16	16	14	14	10	10
8	18	18	18	18	18	16	16	16	14	12	10	10
10	18	18	18	18	16	16	16	14	12	12	10	10
11	18	18	18	18	16	16	14	14	12	12	10	8
12	18	18	18	18	16	16	14	14	12	12	10	8
15	18	18	18	18	14	14	12	12	12	10	8	8
18	18	18	16	16	14	14	12	12	10	10	8	8
20	18	18	16	16	14	12	10	10	10	10	8	6
22	18	18	16	16	12	12	10	10	10	8	6	6
24	18	18	16	16	12	12	10	10	10	8	6	6
30	18	16	16	14	10	10	10	10	10	6	4	4
40	18	16	14	12	10	10	8	8	6	6	4	2
50	16	14	12	12	10	10	8	8	6	6	2	2
100	12	12	10	10	6	6	4	4	4	2	1	1/0
150	10	10	8	8	4	4	2	2	2	1	2/0	2/0
200	10	8	8	6	4	4	2	2	1	1/0	4/0	4/0

Table 1. The size of the conductor is determined by the amount of current and wire length.

ensure that the current gets to where it is supposed to go. Without insulation, current would take the path of least resistance back to the battery, bypassing, or shorting, around the bulb or motor. This is called a **short circuit**. Because all of the metal used to make our vehicles is conductive, without insulation we would have one large short circuit.

LOADS

Following this current, which we now are assured will go where we want it to go, we eventually come to the component that will do the work the circuit was designed for: the load. The **load** in a circuit is the component whose resistance will produce light, heat, or a magnetic field when current is pushed through it **(Figure 14)**. A load does work.

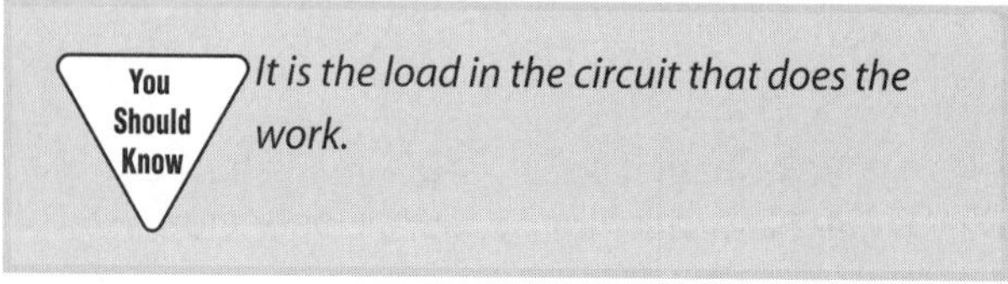

It is the load in the circuit that does the work.

As mentioned previously, voltage pushing current through resistance forces the resistance to do work. It is the resistance of the circuit that controls how much current will flow and how much work will be done. We know this from Ohm's Law. It is this resistance that we are concerned with. All circuits require a load if they are to do productive work. However, we should realize that sometimes unproductive work is done. The corroded connection ahead of the headlight is an example of electrical resistance. Not only will it have an effect on the brightness of the light, causing it to be dimmer, but, because it is resistance, when voltage pushes current through it, it will do

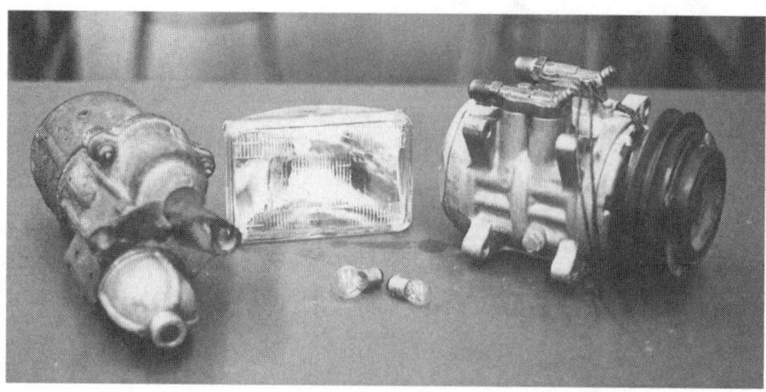

Figure 14. Anything that has resistance and creates a voltage drop is considered a load.

some work—probably producing a slight amount of heat. This unwanted heat was produced because of the unwanted resistance of this load. Unwanted resistance is frequently referred to as a **false load**. The diagnostic process job on vehicles usually involves finding and eliminating false loads. These loads do produce work but usually are not engineered into the vehicle. Instead they tend to appear after a few years and can cause some strange and exasperating results.

The load designed into the circuit will control the current because of its resistance. Decreasing the resistance will increase the current flow and the amount of work. Increasing the resistance will decrease the amount of current flow and the amount of work.

Summary

- Electricity is the movement of electrons from the valence orbit.
- Electrical pressure is measured in volts.
- The quantity of electrons moving is measured in amps.
- The amount of back pressure, or resistance, is measured in ohms.
- The amount of work accomplished by a circuit is measured in watts.
- Insulated conductors carry the current to the load, where it will do work.
- Ohm's Law is the interrelationship of voltage, current, and resistance.
- Current moves because there is a difference in pressure.

Review Questions

1. Voltage
   A. forces current through the circuit.
   B. resists current flow.
   C. is measured in amps.
   D. does the work in the circuit.
2. Current is
   A. the measurement of resistance.
   B. the measurement of electrical pressure.
   C. electricity flowing through the circuit.
   D. equal to the wattage of the circuit.
3. Resistance is
   A. the measurement of resistance.
   B. the measurement of electrical pressure.
   C. electricity flowing through the circuit.
   D. equal to the wattage of the circuit.
4. Current is measured in
   A. ohms.
   B. volts.
   C. amps.
   D. watts.
5. Resistance is measured in
   A. ohms.
   B. volts.
   C. amps.
   D. watts.
6. Electrical pressure is measured in
   A. ohms.
   B. volts.
   C. amps.
   D. watts.
7. The amount of work that is accomplished in a circuit is measured in
   A. ohms.
   B. volts.
   C. amps.
   D. watts.
8. A complete circuit is a path that includes a source, insulated conductors, and a load.
   A. True
   B. False
9. An insulator allows the easy transfer of electrons.
   A. True
   B. False
10. A conductor resists the flow of electrons.
    A. True
    B. False

Chapter 4 Circuits

Introduction

No textbook would be complete without an examination of the circuits that actually accomplish the work. This chapter introduces you to the components that make up circuits and the types of circuits that are common to the vehicle. We devote an entire chapter to how voltage and current behave in each type of circuit (see Chapter 5, Chapter 6, and Chapter 7). Do not forget the principles and terms that we introduced to you in Chapter 3. We are continuing to build the foundation that you will need to be a successful technician. Make sure the material makes sense to you, because it is fundamental to your diagnosis and repair of electrical systems.

CIRCUIT COMPONENTS

Before we look at some sample circuits, we have to look at two necessary components that control and protect most automotive circuits. Without them, we would lose the ability to decide when our circuit does work and run the risk of an electrical fire in case of an overload. Our circuit can handle only a predetermined amount of current. Anything in excess is considered an overload.

Let us examine the **circuit controls** first. Most circuits on the modern vehicle are not energized all of the time. A break, or "open," is designed into the wiring and can be closed whenever we need current to flow through the load. A switch is the best and simplest example of a control **(Figure 1)**. We open the switch when we want the circuit to be off and close the switch when we want the circuit on. Keep in mind that a control is not a load and, therefore, should not have any resistance. If a switch does develop some resistance, it will become a false load and give off some heat. It might also have a detrimental effect on the rest of the circuit. If you ever feel a switch and it is warm to the touch, it should be replaced. The switch contacts have probably corroded inside from the arcing that occurs each time the switch opens and closes. This corrosion is resistance and is giving off heat as current flows through it. The exception to this is the headlight switch. It will normally develop some heat as it dims the dashboard lights. The dimmer function is accomplished by adding some resistance to the dashboard lights. This resistance causes a small amount of heat to be produced. The remainder of the switches used on vehicles should not be warm to the touch.

Relays are also a type of circuit control that allows us to control high current with lower current **(Figure 2)**. We discuss relays in more detail in Chapter 12. Most modern

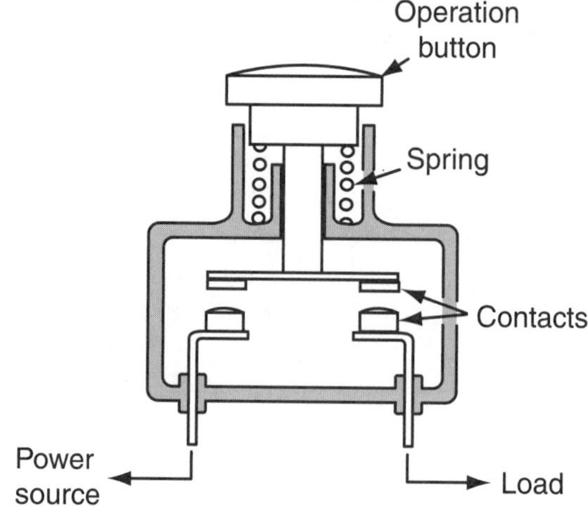

Figure 1. The switch is the simplest form of a control.

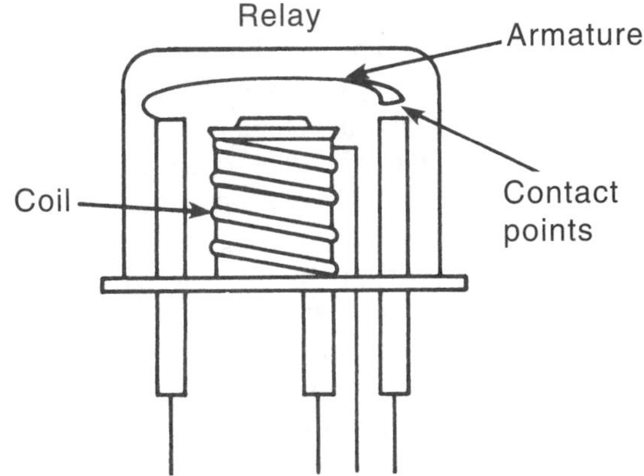

Figure 2. Relays control high current with low current.

Figure 4. Fusible links located near the battery.

vehicles use relays in their circuitry. The modern diagnostician must be capable of understanding how and why relays are used. For now, just remember that they are a type of control and should be treated as such.

If for some reason the total resistance drops too low within a circuit (such as when a bare hot wire touches the metal frame), the current flow will increase. All circuits have a limit of how much current can flow through the controls and conductors. If we exceed this limit, we can turn the conductors into loads, producing heat and eventually causing a fire. For this reason, fuses, fusible links, and circuit breakers protect circuits **(Figure 3)**. Let us examine each, but first the **fuse**.

Probably the most common circuit protector is the in-line fuse. Made of a fusible or meltable material, it is designed to melt from a system overload of current and open the circuit before the conductors feeding the load are melted. The size of the fuse is indicated in amperage and is usually printed on the fuse body and sometimes on the holder or fuse panel. Never, under any circumstances, put a larger fuse into a circuit than the one that the manufacturer designed to protect it. If you install a larger fuse, the circuit might overload and be damaged or destroyed before the fuse melts. Remember that a fuse is a circuit protector. It can do its job only if it is placed in a circuit that will have less current normally flowing than the rating of the fuse.

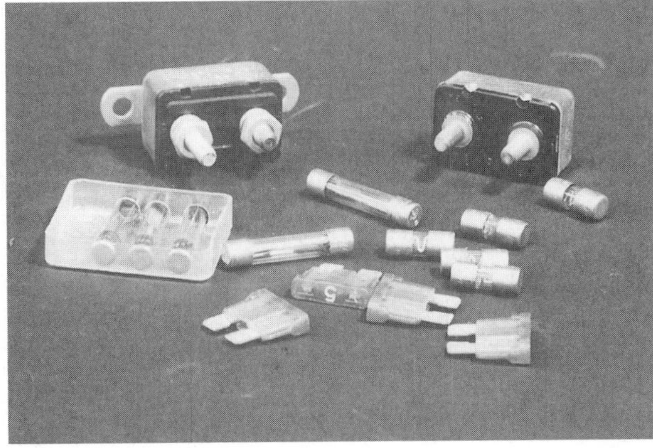

Figure 3. Circuit protection devices.

The second type of circuit protector is the **fusible link (Figure 4)**. It is made of meltable material like the fuse, with heat-resistant insulation. You will usually find the fusible link at a main connection at or around the battery. It protects the main power lines before they are split into smaller circuits and run through the vehicle's fuse box. Some manufacturers use fusible links because they are completely protected from the weather and road splash, which might contain salt or other chemicals. Their amperage capacity is determined by their size. Fusible links are usually four numbers higher than the circuit they protect. Four numbers higher is another way of saying two wire sizes smaller. The smaller the wire, the larger its number. In other words, a no. 14 circuit would be protected by a no. 18 fuse link, which is four sizes smaller. Using a no. 14 fuse link in a no. 14 circuit could result in the whole circuit burning up if a short occurred. Remember that the object of circuit protection is just what the name implies: to protect. Do not overfuse.

When manufacturers are designing a circuit that might experience an overload on a routine basis, they frequently use a **circuit breaker** as the form of protection **(Figure 5)**. Similar in principle to a fuse, a circuit breaker is designed to open the circuit in case of a system overload. It, however, has the advantage that it can be reset either manually or automatically once the overload is gone. A good example of the use of a circuit breaker is the power-window circuit. In climates in which ice is a problem, the window could jam or freeze shut. This could cause a current overload as the window motor fights the frozen window. The breaker should overload and open the circuit before damage to the motor occurs. Once the overload is gone, the breaker closes and is ready for the next try. We look at other examples of circuit breakers in Chapter 14.

Now that we have identified the different components necessary for a practical circuit, let us look at the types of

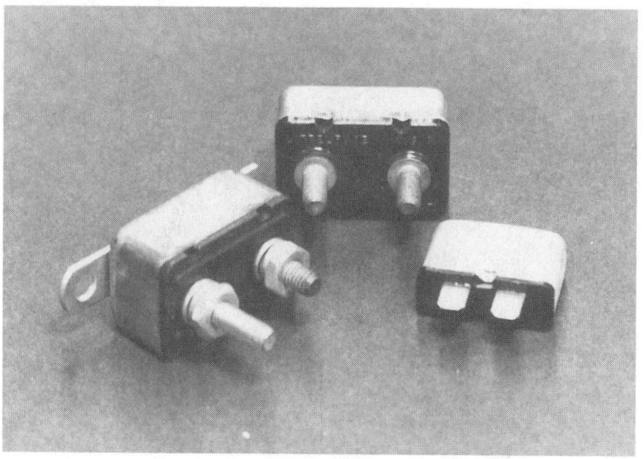

Figure 5. Circuit breakers are used most often in motor circuits.

circuits available. Within the automotive field, we find the use of series, parallel, and series-parallel circuits. Let us look at series circuits first.

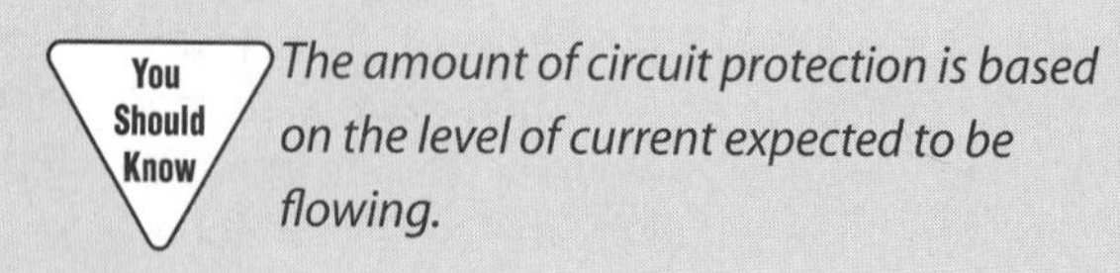

The amount of circuit protection is based on the level of current expected to be flowing.

SERIES CIRCUITS

If we take our insulated conductors, a switch (to control), a fuse (to protect), and a battery (as a source of voltage), along with two light bulbs as loads, and wire them as shown in **Figure 6**, we have a simple series circuit. How does it work? The name implies the function: series. Each component is dependent on all the others if the circuit is to do any work. Bulb 1 is the source of B+ for bulb 2, and bulb 2 is the source of B– for bulb 1. The switch, conductors, and fuse are all dependent on one another. If bulb 2 burns out, the whole circuit will go dead. With only one path for current, an interruption or break anywhere along the path will turn off the whole circuit just as if the switch were opened. This dependence on all other parts is one reason why series wiring is not used too often in automobiles. For instance, we do not want all the lights to go out if the right front headlight burns out. Another characteristic of this type of circuit is the addition of resistances. Because the resistances are all in a row, they have a cumulative effect on the total resistance **(Figure 7)**.

Switch
Fuse
Load 1
Load 2
B+ B-
12.6 V

Figure 6. The components of a simple series circuit.

If we wire a circuit with three 2-ohm resistances and compare its action to that of a circuit with only one 2-ohm resistance, we can see the results in terms of work accomplished. Refer to **Figure 8** for our Ohm's Law formula and wattage formula. How much current will circuit A draw? How many watts of work will be accomplished? Apply the

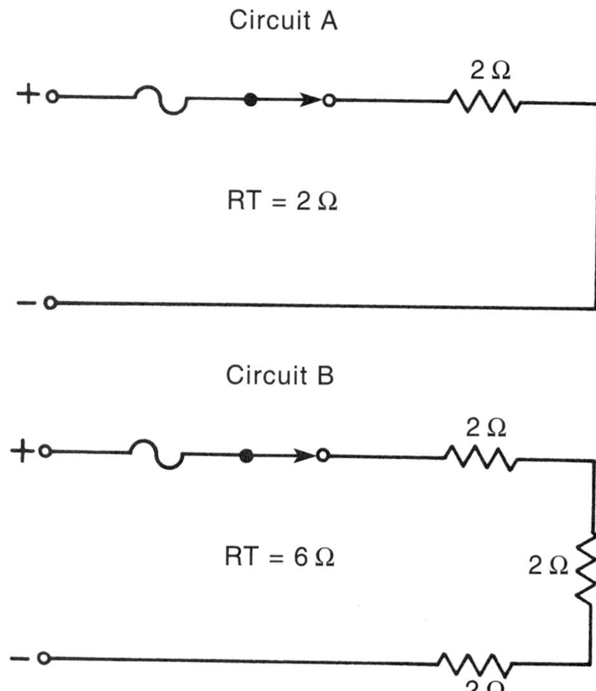

Figure 7. Series resistance will add up. RT = resistance total.

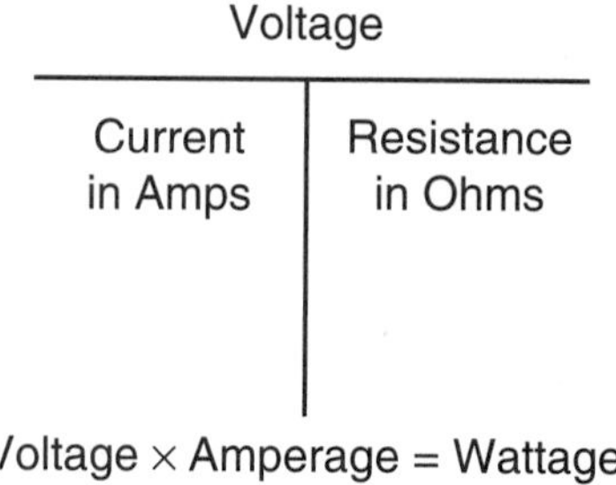

Figure 8. Ohm's Law.

same questions for circuit B. Let us look at them together. First, in circuit A, **Figure 9**, cover up the amperage (the unknown value). This leaves voltage over resistance, or 12/2:

12 volts ÷ 2 ohms = 6 amps of current flowing

Therefore,

6 amps × 12 volts = 72 watts of light, heat, and magnetic field produced

Now let us look at circuit B with three bulbs of 2 ohms each **(Figure 10)**. The three resistances add up (2 + 2 + 2). The battery is pushing current through 6 ohms of resistance now. Return to Ohm's Law to find out the current:

12 volts ÷ 6 ohms = 2 amps of current flowing,

then

12 volts × 2 amps = 24 watts of work being accomplished.

There is less total work because the resistances are wired in series. Series resistance will drop down the current within a circuit. If that resistance is unwanted, as in a corroded connection, the actual load will not receive as much current and will do less work (false loads again).

Circuit A

$$\frac{12\text{ Volts}}{2\text{ Ohms}} = 6\text{ Amps}$$

12 V × 6 A = 72 Watts

Figure 9. 6 amps at 12 volts will do 72 watts of work.

Circuit B

$$\frac{12\text{ Volts}}{6\text{ Ohms}} = 2\text{ Amps}$$

12 Volts × 2 Amps = 24 Watts

Figure 10. 2 amps at 12 volts will do 24 watts of work.

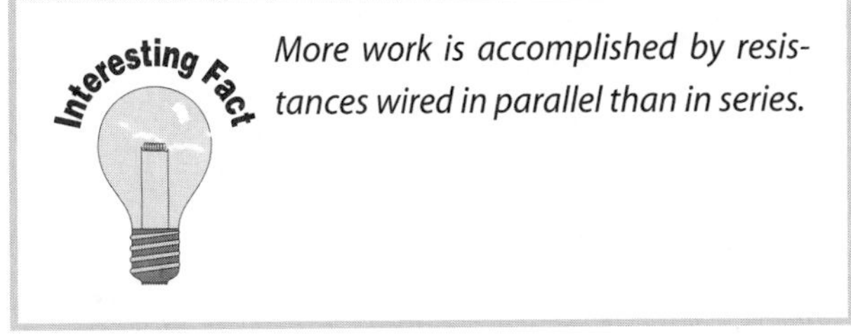

Interesting Fact
More work is accomplished by resistances wired in parallel than in series.

The last trait of a series circuit is the voltage division or dropping that occurs. Battery voltage will have to divide, or drop, down as it pushes current through each resistance. An example of this is seen in **Figure 11**. Notice that the voltage drops down as it pushes current through each bulb. You will see this for yourself when you wire up circuits in Chapter 8 and use the meters that we study in Chapters 5–7. We have demonstrated mathematically what will happen in series circuits. The opportunity to prove Ohm's Law in this chapter will help your overall understanding of circuits.

PARALLEL CIRCUITS

If we realize why most automotive circuits are not series wired, it becomes fairly obvious that certain characteristics are necessary or desirable for the next type of circuit we look at, the **parallel circuit**. The parallel circuit is so named because the drawing of one usually looks like a railroad track with its parallel rails **(Figure 12)**. Notice in **Figure 13** that some of the same components are used. We still have our voltage source, fuse, switch, and insulated conductors. The loads, however, are not wired one after another.

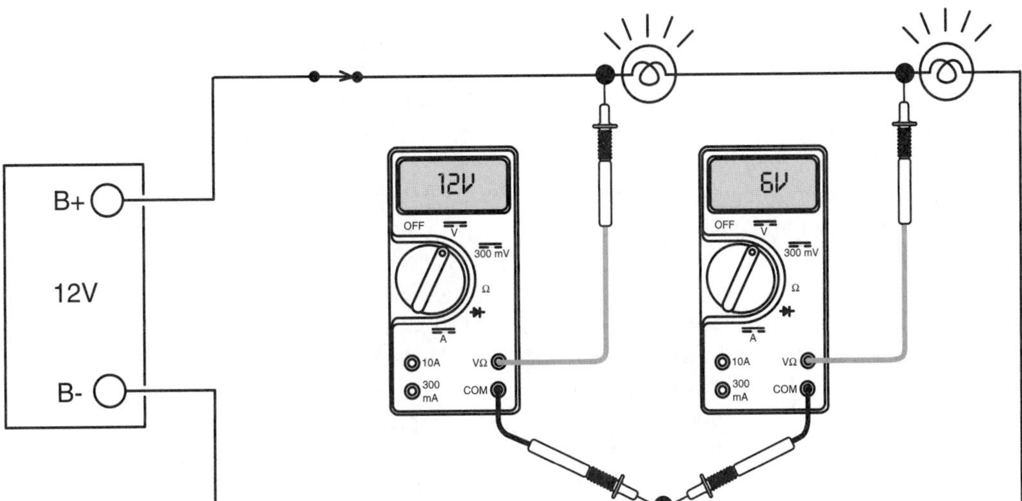

Figure 11. A voltage drop occurs when current flows through resistance.

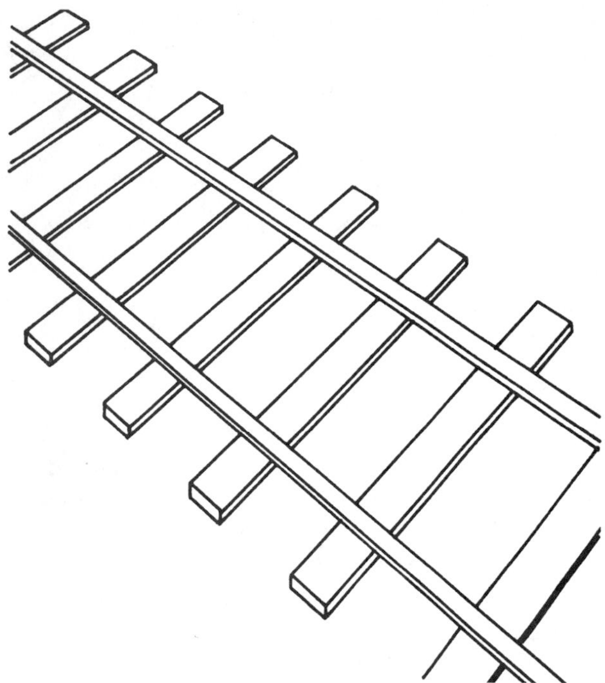

Figure 12. A parallel circuit resembles a set of railroad tracks.

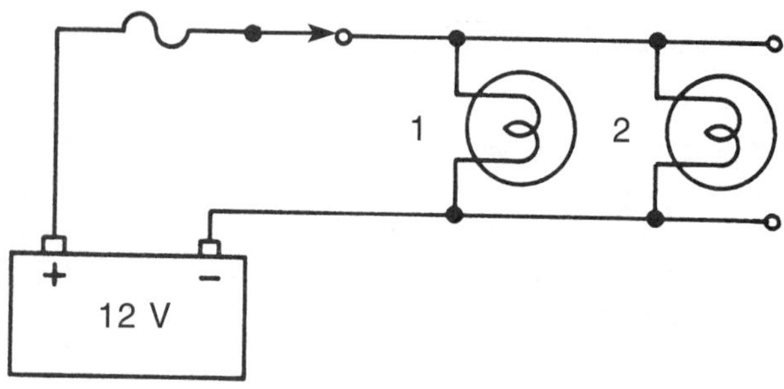

Figure 13. Two bulbs wired in parallel.

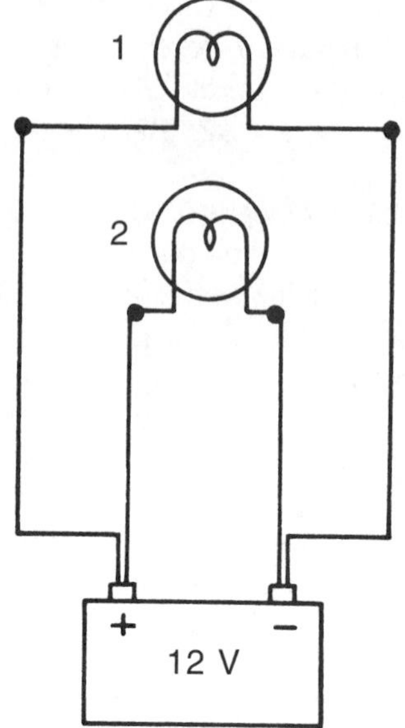

Figure 14. Each load wired in parallel has its own source of power.

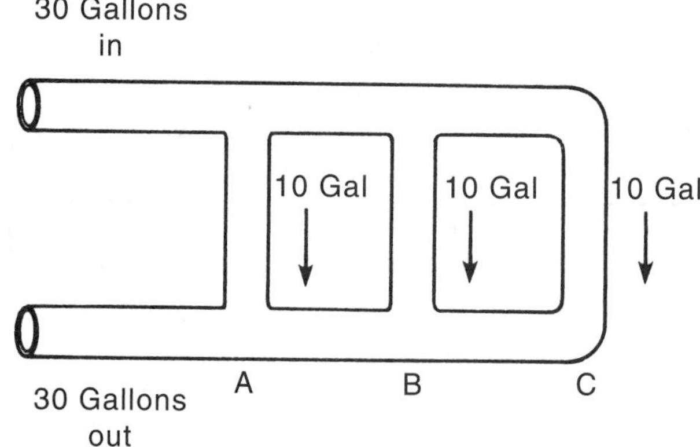

Figure 15. Current flow and water flow react the same.

Instead, they each have their own source of B+ and B–. This will eliminate the dependence all loads have on one another that is characteristic of a series circuit. Independence is gained. Bulb 2 can now burn out without having an effect on bulb 1. Most of the vehicle is wired this way. If the right front headlight burns out, it will not turn off the rest of the lighting system. Independence is achieved because each load has its own path from and back to the source. It is the same as if the wiring were as that shown in **Figure 14**. Parallel loads each receive full B+ voltage and therefore draw full current and produce full work. Current flow through a parallel circuit will divide down each leg. The water example pictured in **Figure 15** shows this best. Remember, current is like water. It is a quantity. If each leg of the water system has 10 gallons of water flowing in it, with three legs, the total water flow will be 30 gallons. The top main water pipe and the bottom main water pipe will have the 30 gallons flowing, with each leg receiving its 10 gallons. How many gallons are flowing at point B? If you said 20 gallons, you are on the right track. How about at point C? You are right if you said 10 gallons.

If we substitute loads in **Figure 16** for each of the water legs and use Ohm's Law and our wattage formula, we will be able to see the main differences between series and parallel circuits. Each leg has 12 volts pushing current through 2 ohms of resistance. Cover up amperage because it is the unknown:

12 volts ÷ 2 ohms = 6 amps; 12 volts × 6 amps = 72 watts

Does this sound familiar? This is the same as our series example, but notice the difference. We have three separate legs to this circuit, and if each leg draws 6 amps the total current flow in the circuit will be 18 (6 + 6 + 6 = 18):

12 volts × 18 amps = 216 watts of light, heat, or magnetic field produced

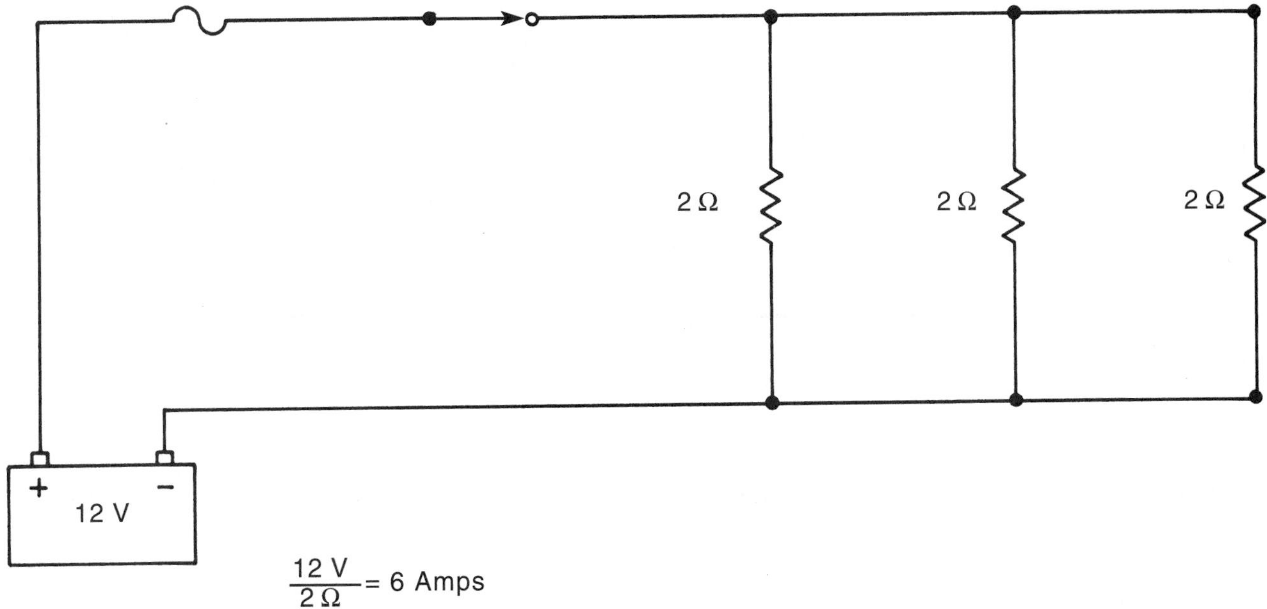

$$\frac{12\ \text{V}}{2\ \Omega} = 6\ \text{Amps}$$

$$12\ \text{V} \times 6\ \text{A} = 72\ \text{W per}\ 2\ \Omega\ \text{Leg}$$

Figure 16. Three loads wired in parallel.

This current flow is quite a bit more than the same loads wired in series. This also means that the main B+ and B− conductors must be capable of handling at least 18 amps, and a fuse rating of probably 25–30 amps will be necessary.

This example is with three equal resistances. What do you think will happen if the three legs have the resistances shown in **Figure 17**? You should realize first of all that each leg will draw different levels of current based on the resistance of the load in each leg. You should also realize that the total will be the total of each leg added together. Let us figure this one together, one leg at a time. The first load of 3 ohms will draw 4 amps of current:

12 volts ÷ 3 ohms = 4 amps

The second load of 4 ohms will draw 3 amps:

12 volts ÷ 4 ohms = 3 amps

The fourth load of 6 ohms will draw 2 amps of current:

12 volts ÷ 6 ohms = 2 amps

If we add up the individual amperages, we will know the total:

4 + 3 + 2 = 9 amps total

The important concept to understand here is that the current through any parallel leg is dependent on the resistance of that leg, and the current will always be the total of all legs added together. We discuss another way of figuring current flow, or resistance, within a parallel circuit in Chapter 9.

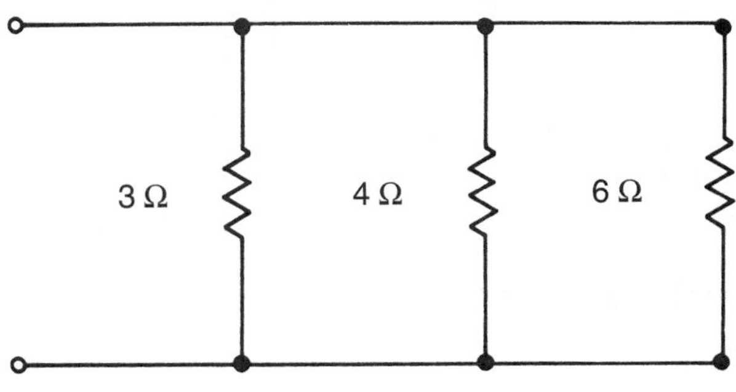

$$\frac{12\ \text{V}}{3\ \Omega} = 4\ \text{Amps}$$

$$\frac{12\ \text{V}}{4\ \Omega} = 3\ \text{Amps} \qquad 4 + 3 + 2 = 9\ \text{Amps}$$

$$\frac{12\ \text{V}}{6\ \Omega} = 2\ \text{Amps}$$

Figure 17. Parallel circuit with different resistances.

SERIES-PARALLEL CIRCUITS

The final type of circuit we examine is the **series-parallel circuit**. You can tell from its name that it shares the characteristics of both series and parallel circuits. To be realistic, we must recognize that virtually all forms of parallel circuits have a series section. **Figure 18** shows the circuit we have been working with. The loads are in parallel with a single control and a single fuse. The switch and fuse are series wired because there is only one path through them.

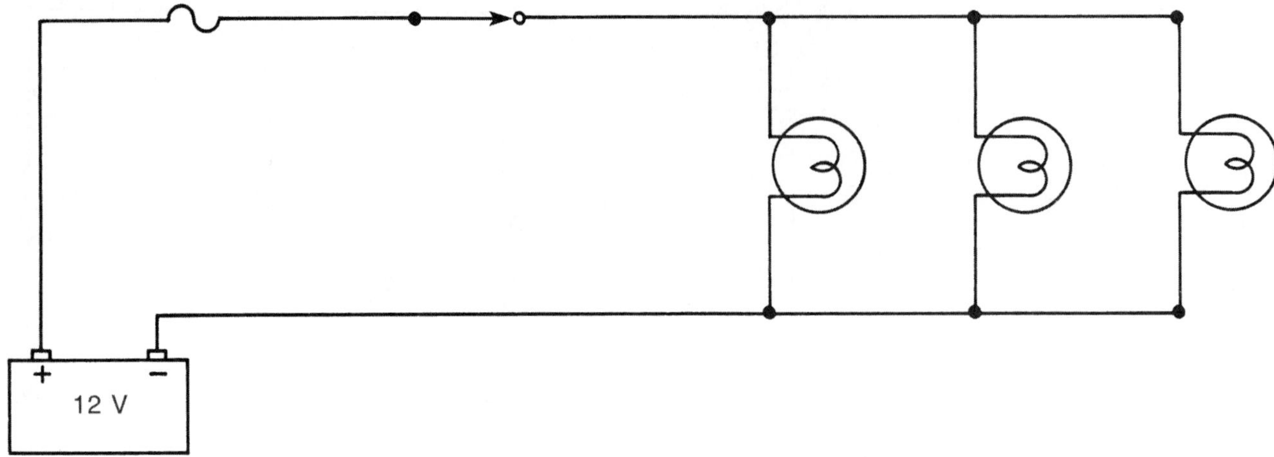

Figure 18. Parallel loads with series control (the switch) and protection (the fuse).

There is no alternative path. Thus, if the switch or fuse were open, the circuit would be dead. The parallel sections are dependent on the series-wired switch and fuse for their current path.

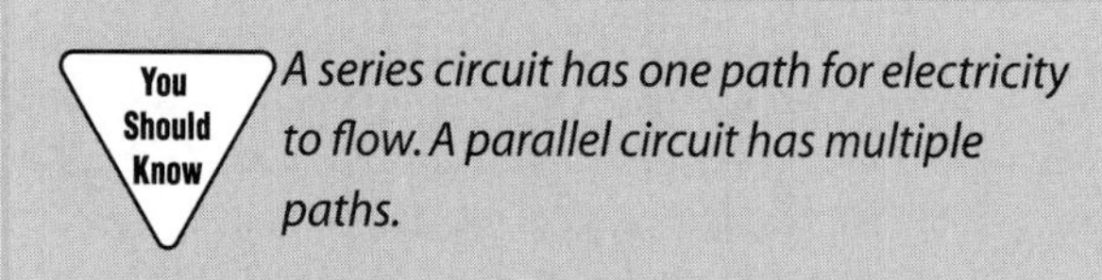

A series circuit has one path for electricity to flow. A parallel circuit has multiple paths.

Another style of series-parallel is shown in **Figure 19**. Notice that an additional load has been placed in series with the three parallel loads. Each parallel load is independent from the others, but dependent on the series load. A burned out parallel bulb would have no other effect, but a burned out series bulb would shut down the whole circuit.

Your dashboard dimmer circuit is the best example of this type of wiring. A single variable resistance (to be discussed in Chapter 9) is wired in series with the parallel-wired dashboard lights. Turning the dimmer increases the resistance in series. What effect do you think this has? Let us think about it. If we increase series resistance, voltage drops or divides and current decreases. This will reduce the available voltage to the parallel section, and the lights will go dimmer. Reducing the resistance will brighten up the dash. The series load now has become a control because of its effect on voltage and amperage. Remember that Ohm's Law has shown that as resistance goes up, current goes down, and as resistance goes down, current goes up. What about the bulbs in the dash? Why are they all dim? They are all in parallel and have the same voltage across them.

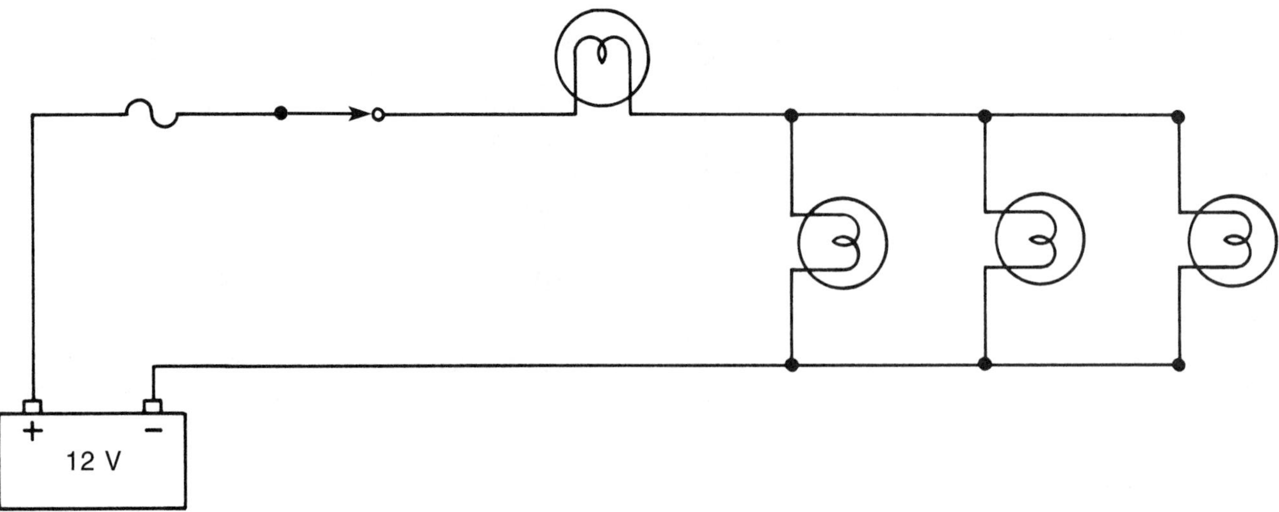

Figure 19. One bulb in series with three parallel bulbs.

The material presented within this chapter is the foundation of the rest of this book and quite possibly the foundation of your electrical career. Do not hesitate to reread it. Make sure you understand the principles presented before you go on. It is easy to get lost from this point on if your background is incomplete. You should be able to easily answer the questions presented after the chapter summary. In addition, you will find sample problems dealing with Ohm's Law for you to work through in subsequent chapters.

Summary

- The typical circuit is protected against overload by a fuse, a fusible link, or a circuit breaker.
- A series circuit has loads wired in a single path.
- Adding loads in series reduces the amount of work (watts) accomplished within the circuit.
- A parallel circuit has loads wired in their own paths.
- Adding loads in a parallel circuit increases the wattage developed.
- The vehicle is composed mostly of parallel circuits.
- A series-parallel circuit has some of the characteristics of a series circuit and some of a parallel circuit.
- Ohm's Law defines the interrelationship among voltage, current, and resistance.

Review Questions

1. Which of the following is not required for a complete circuit?
   A. a source
   B. conductors
   C. a load
   D. a relay
2. Adding resistance in series to a circuit will cause
   A. more current to flow
   B. less current to flow
   C. increased wattage
   D. the total resistance to go down
3. A short circuit could cause a
   A. fuse to blow
   B. discharged battery
   C. circuit breaker to open
   D. all of the above
4. Which of the following is classified as a circuit protector?
   A. a fuse
   B. a circuit breaker
   C. a fusible link
   D. all of the above
5. An example of a load would be
   A. a fuse
   B. a circuit breaker
   C. a battery
   D. a motor
6. If a circuit is said to be open, current will be flowing through it.
   A. True
   B. False
7. Series resistances always add up.
   A. True
   B. False
8. In a parallel circuit, there is only one path for current to flow.
   A. True
   B. False
9. In a series circuit, there are multiple paths for current to flow.
   A. True
   B. False
10. What components are necessary for a circuit?
11. What is the major advantage to a parallel circuit? Why?

Chapter 5 Voltmeters

Introduction

A frequently heard comment from new automotive students is "give me something that I can see and hold and I can repair it. The trouble with electricity is that you cannot see it." Does this sound familiar? This statement emphasizes a common misconception—that you cannot see electricity. The meter must become your eyes into the circuit and let you see what is going on. Knowing what you are looking at and how to interpret the information that the meter is supplying is the goal of Chapters 5–7. The use of the voltmeter, ammeter, and ohmmeter is fundamentally important to your ability to diagnose and repair automotive electrical systems. We start with the voltmeter in this chapter because it is probably the most used of the three. We assume that you have a DMM available. A DMM such as that shown in **Figure 1** is the most common meter. It has voltmeter, ammeter, and ohmmeter capability, is either auto ranging or manually ranged, and has fused inputs for current measurements.

Another type of multimeter, which we will not cover, is the analog meter. Analog meters were popular until the computer arrived. Their use has diminished to the point where most repair facilities do not even have one and most tool salespeople do not even sell one.

PURPOSE OF THE VOLTMETER

Probably the most used and useful meter in the shop is the voltmeter. Its purpose is to read voltage, which as you know is the pressure behind the flow of electrons. The simplest use of the voltmeter is to measure system pressure, or voltage, which is usually 5–16 volts. Let us look at the common voltmeter and analyze the front **(Figure 2)**. The large rotary dial in the middle allows us to select the unit of measurement. This is an example of a multimeter, so it will be capable of measuring DC voltage (the first position clockwise from off), a low amount of DC voltage (300-mV position), ohms (position with the omega symbol), continuity (which we will discuss in Chapter 7), and DC amperage (the last position).

This multimeter and most that are currently in use are **auto-ranging** meters. The range of the meter is its scale or maximum voltage that it can read. The meter will automatically range itself up to the correct level. This makes its use

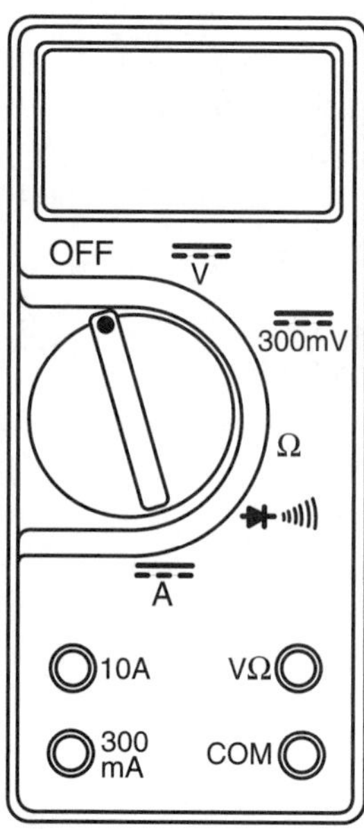

Figure 1. A DMM.

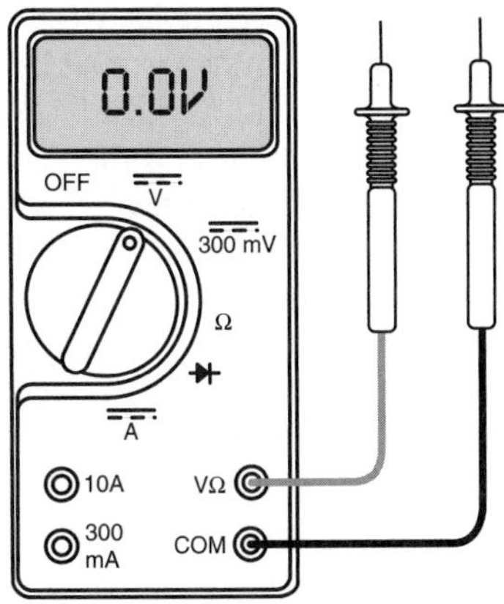

Figure 2. A DMM set up for voltage.

simple and accurate. The button in the middle of the rotary dial is used to lock out the auto-ranging feature, but has the disadvantage that the meter will display "OL" (out of limits) if the voltage is higher than the scale can accept. For example, with the auto range locked out, the lowest range, usually approximately 10 volts, is lower than battery voltage. If we put the meter across a fully charged battery, as diagrammed in **Figure 3**, the meter will display an OL, because the 12.6 volts is above the maximum of 9.99 volts. One of the advantages to a digital multimeter is that the OL shows up if you exceed the maximum voltage setting, but no damage to the meter occurs. One of the major disadvantages to analog meters was that they would be damaged if the range of the meter were exceeded, which is not a problem with digital meters. If the OL shows up and you are on a manually ranged position, a touch of the button raises up the scale (usually to approximately 99.9 volts) so that the battery reading of 12.6 volts is easily displayed.

Allowing the meter to auto range is usually the preferred method. If you connect the meter to the battery, it will start on the lowest value (9.99 volts), automatically range up to the next range (99.9 volts), and display the 12.6 volts of the battery. If you are watching the digital display as

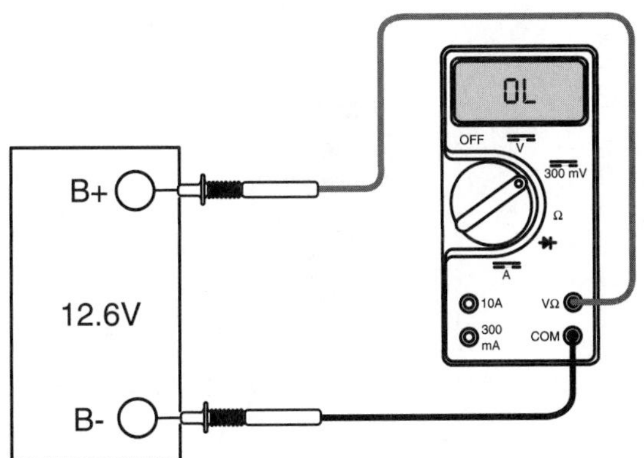

Figure 3. A DMM with the range too low.

you connect it, you will see the OL flash on the screen as the auto ranging takes place.

Many meters also have a millivolt range. A millivolt is a thousandth of a volt and is abbreviated as mV. Our meter has a 300-millivolt setting. This is the lowest setting and is not used very frequently on vehicles. Normally, when you need to look at voltage this low, an oscilloscope is used. The millivolt setting on many voltmeters is the setting that is used when a current probe is connected to measure the amperage of the circuit. We look at current probes in later chapters.

Do not forget that a millivolt is 1/1000 of a volt and is written as 1 mV or 0.001 V. We will look at some conversions later in this chapter.

CONNECTING THE VOLTMETER

The two leads coming out of the front of the meter are color coded and labeled red for positive and black for negative. This is the most common color coding for meters found in the automotive industry and involves the polarity of the meter. Polarity can be described simply as the direction the electrons are flowing or, in the case of voltage, the direction of the pressure **(Figure 4)**.

If a meter is placed into a circuit with the polarity of the meter leads not matching the polarity of the circuit, the meter will display a negative symbol.

Our meter is capable of measuring this pressure in either direction and will display a negative symbol ahead of the reading if the leads are connected backwards. All circuits have a negative and a positive charge. If the red lead is

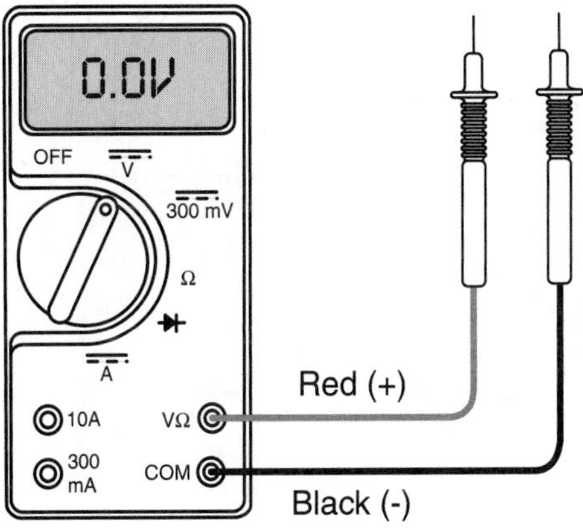

Figure 4. The color and position of the leads indicate polarity.

placed in the most positive part of the circuit and the black lead is placed in the most negative, the polarity of the circuit and the polarity of the meter match. With digital meters, you need not be concerned with doing damage if you reverse the polarity. The meter will just display the negative symbol.

MEASURING VOLTAGE DROP

Sometimes we wish to measure the pressure that has been used in a circuit. This is called voltage drop, and it is in its measurement that we might get confused. The definition of voltage drop is the voltage that is used to push current through resistance. The key to understanding voltage drop is knowing that it occurs only when current is being pushed through resistance. **Figure 5** shows a simple series circuit with three loads.

To measure the voltage or pressure available from the battery, we attach the positive meter lead to point A and the negative lead to point F in the circuit. Point A is the most positive part of the circuit, whereas point F is the most negative. Measuring the voltage used or dropped across each load is a little more complicated. A point in the circuit can be either positive or negative, depending on what we are measuring. If we want to know the voltage drop across load 1, we would connect the positive meter lead to point B because it is the most positive of the two points and the negative lead to point C because it is the most negative. The meter will now read the difference in pressure between these two points. This is the voltage drop across load 1. However, to measure the voltage drop across load 2, the polarity of point C changes from negative to positive. The negative connection point now becomes point D. The reason for this is that point C was the most negative point for load 1 at the same time that it was the most positive point for load 2. Keep this in mind and figure out how to measure the voltage drop across load 3. If you mentally connected your meter's positive lead to point D and your negative lead to point E, you are on the right track. **Figure 6** shows the complete picture with three voltmeters connected to measure voltage drops of the three-load circuit.

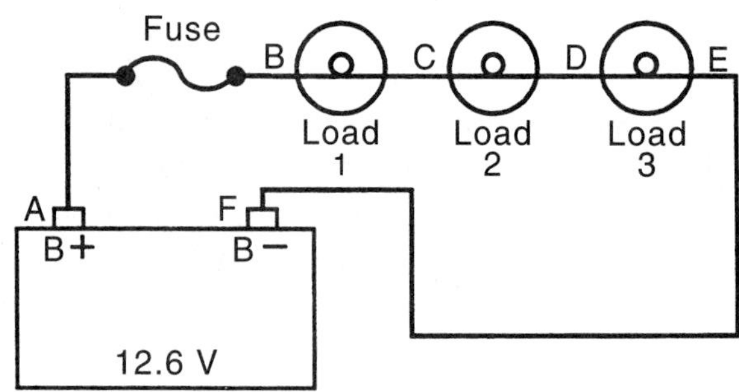

Figure 5. A voltage drop will occur across each load.

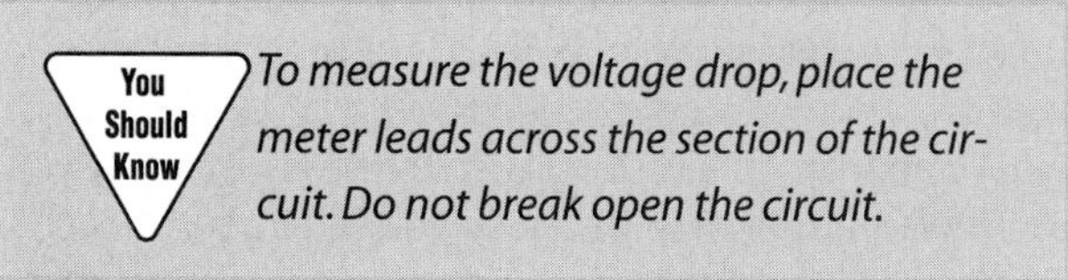

To measure the voltage drop, place the meter leads across the section of the circuit. Do not break open the circuit.

INTERPRETING VOLTAGE READINGS

Voltage readings are straightforward until we get into the millivolt range. Remember that a millivolt is a thousandth of a volt and is written as either 0.001V or 1mV. Milli is abbreviated as a small "m." Do not confuse it with a capital "M," which means million. Your meter might display the small m after the digits, and you need to recognize what it actually means. Let us try an example of a meter reading of 500mV. What is the meter actually telling you? You have 500

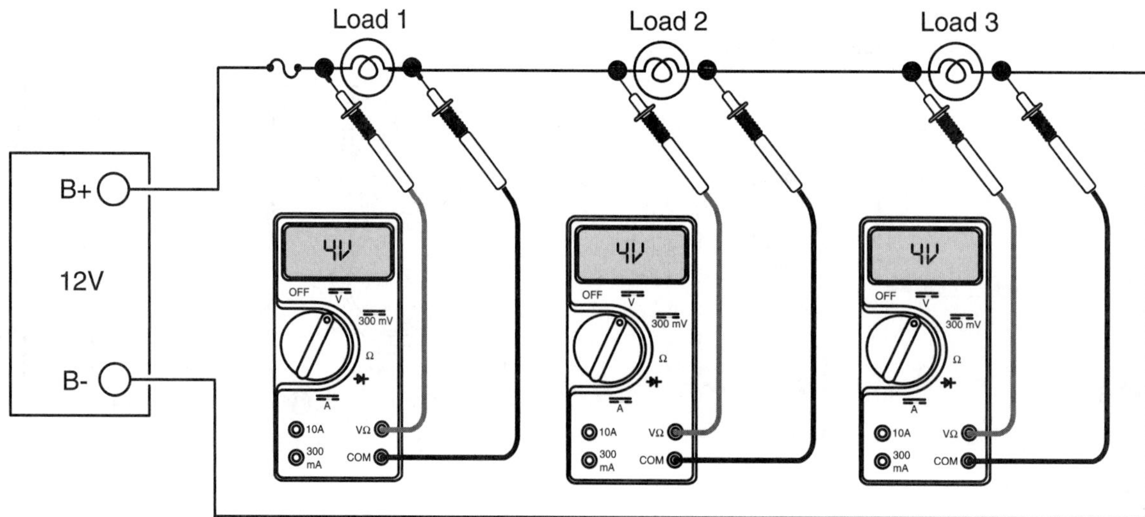

Figure 6. The available voltage will divide based on the resistances of the load.

thousandths of a volt or one-half volt. It is usually written as 0.500V or 500mV. Both forms mean the same thing. How would you convert 32mV to volts? The answer is 32mV becomes 0.032V. How would you convert 1000 mV to volts? This is 1 volt and would be written as 1.0V.

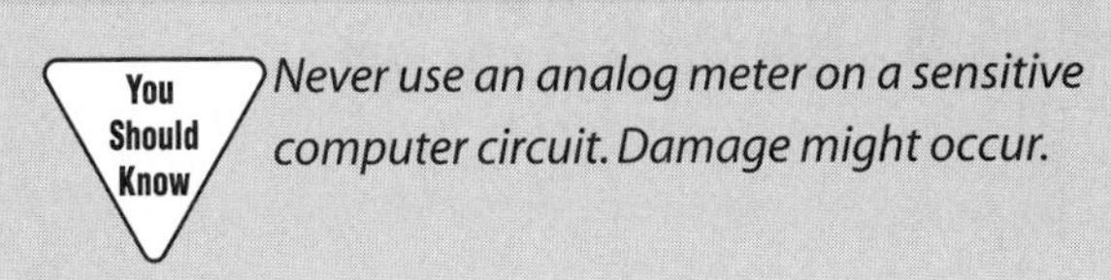

Never use an analog meter on a sensitive computer circuit. Damage might occur.

The voltmeter has a high input resistance. This is frequently referred to as the input impedance of the meter. The input impedance of a meter is a measurement of the input resistance of the meter. Most modern computer systems require an input impedance of at least 10,000,000 ohms. Input impedance is what makes a voltmeter and allows it to sample the pressure without having current flowing through it. A tire pressure gauge acts like a voltmeter in that it measures the pressure without having to have all of the air flowing through it. The input impedance limits the current flow through a voltmeter.

Summary

- Voltmeters measure differences in electrical pressure.
- Voltmeters are polarity sensitive and usually display a negative symbol if the leads are crossed.
- A voltage drop occurs if current is pushed through resistance.
- A millivolt is a thousandth of a volt.
- 50mV is .050V.
- Auto-ranging voltmeters are common.
- OL is displayed when the voltage applied to the meter exceeds the maximum voltage that the meter can display.
- Modern vehicles require the use of a 10,000,000-ohm input impedance voltmeter.

Review Questions

1. A voltmeter is always wired in ________ with the load.
   A. parallel
   B. series
   C. either series or parallel
   D. none of the above
2. Most digital meters display an OL if the voltage input exceeds the scale. What does the OL signify?
   A. The meter has been damaged.
   B. The meter cannot read the voltage because it is out of limits.
   C. The technician is out of luck.
   D. The meter leads are reversed.
3. Technician A states that a digital meter is preferred to an analog meter when dealing with sensitive computer circuits. Technician B states that an analog meter might load down a computer circuit. Who is correct?
   A. Technician A only
   B. Technician B only
   C. Both Technician A and Technician B
   D. Neither Technician A nor Technician B
4. A reading of 62mV will convert to
   A. 620 volts
   B. 62 volts
   C. 0.62 volts
   D. 0.062 volts
5. Voltmeters are usually polarity sensitive.
   A. True
   B. False
6. Voltmeters are normally connected across the load (in parallel).
   A. True
   B. False
7. A voltage drop reading is a measurement of the amount of voltage lost across the load.
   A. True
   B. False
8. A negative symbol on the display indicates that the polarity has been reversed.
   A. True
   B. False

Ammeters

Introduction

There frequently are times when the voltmeter by itself does not give the technician the complete picture. After all, the circuit has current, voltage, and resistance. It is the combination of the three that makes the circuit do the work it was designed for. With this in mind, the next meter that we look at is the ammeter. We examine the digital low-current DMM and the use of the **current probe**. We discuss in Chapter 30, Chapter 31, and Chapter 32 the starting and charging testers that can measure hundreds of amps.

TYPES OF AMMETERS

When we eliminate the battery starting and charging diagnosis types of ammeters, we are left with two varieties, the **direct reading** and the current probe. Let us look at the direct-reading variety first.

A direct-reading ammeter is the type that will have the entire amount of current flowing through the meter. This means that the meter must be capable of handling and measuring the circuit's current. If you recall from Chapter 5, we discussed input impedance and defined it as a high resistance that prevents the meter from drawing much current. As soon as we change the meter over to a current measurement, we need to remove the impedance or the meter will not allow the current to flow through it. Our meter in **Figure 1** will also have to have the meter leads changed over to the left side for current measurement. We have two choices, 300 milliamps or 10 amps. There is limited application for the 300-milliamp range. Most of our current measurement readings will need to have the red (positive) lead in the 10-amp position. Note that if the circuit has in excess of 10 amps flowing and we try to put it through the 10-amp position, the internal fuse would blow and shut the ammeter section of the DMM down.

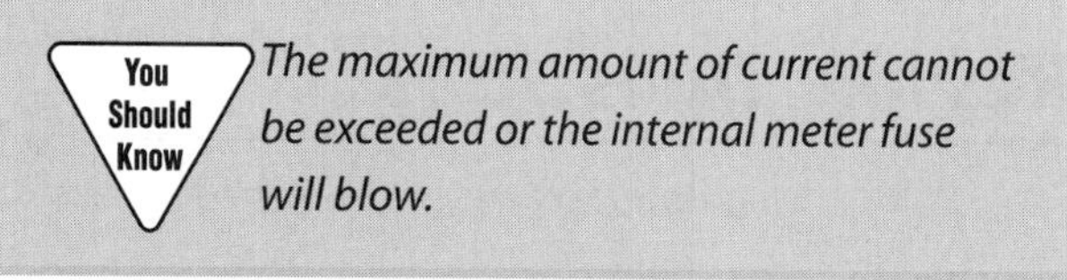

The maximum amount of current cannot be exceeded or the internal meter fuse will blow.

Two important points here for you to think about are having a rough idea of the amount of current flowing before connecting the meter, and, when you purchase your own meter, making sure it has fused current measurement

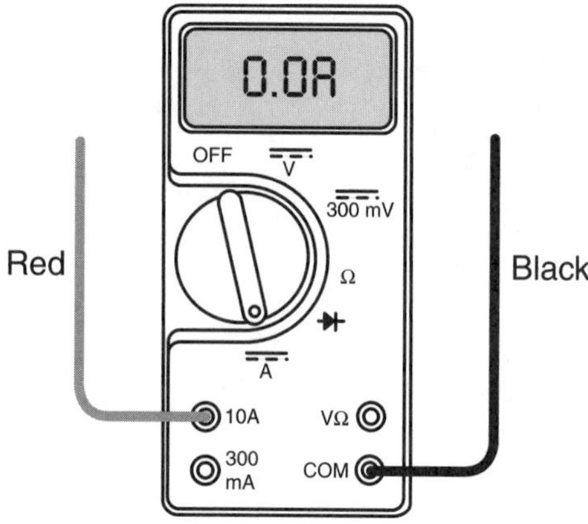

Figure 1. To measure current, the meter's positive lead must be moved to an amperage position.

capability. An ammeter is designed to measure total current flow and will not limit the amount of current. It is more like a counter that will count the electrons as they flow through the meter movement. For this reason, ammeters are designed with virtually no input resistance. No resistance is another way of saying a dead short circuit. If an ammeter is placed across a source of power, such as a battery, the meter will act like a dead short and possibly burn out the fuse because of the tremendous current, which will flow through 0 ohms of resistance. Never place an ammeter across a circuit like we did with the voltmeter. If an ammeter sees a positive and a negative, it will draw maximum current from the source because it does not have any input resistance (impedance).

The correct method for connecting the ammeter into a circuit is to break open the circuit and wire the meter in series with the loads, as shown in **Figure 2**. This ensures that the current flow will be determined by the resistance of the loads rather than by the resistance of the meter. Common methods for the use of this type of meter involve removal of a battery cable and series wiring it between the battery post and cable or replacement of a fuse with the meter and a fused jumper. Remember that at no time should the meter be placed across the load or into a circuit that has no resistance **(Figure 3)**. An ammeter is always connected in series with the circuit's resistance controlling current. As you wire the meter into the circuit, polarity will be indicated on the meter face. If the negative is placed on the positive side and the positive placed on the negative side, the meter will show the negative, or reverse, polarity symbol on the meter face, as **Figure 4** shows. The meter can handle the current in either direction and no DMM damage will result.

Remember the scales, or ranges, that our voltmeter had? Most ammeters do not have multiple ranges because they measure a small band of current. On the 10-amp range, our meter will be capable of measuring from 0.0 amps to 10.0 amps. The measurement is accurate to one-tenth of an amp. If greater accuracy is required for very small measurements, the meter lead can be changed to the 300-milliamp position. You must be sure that the circuit has less than 300 milliamps or the fuse will blow. Three hundred milliamps is not much current. It is less than one-third of an amp. Normally, technicians will start on the highest amperage setting and work down once they are sure that the next lower range will be high enough to handle the current flow.

Once the meter is connected to the circuit, the circuit load should be live because current is now flowing through the meter. The meter will not change the circuit characteristics at all. The circuit load should draw its current as it did before installing the ammeter.

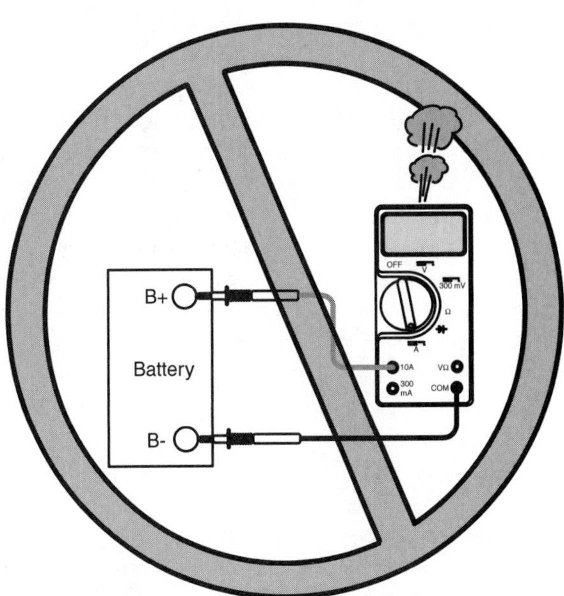

Figure 3. Connecting the DMM set for amperage across a power source will blow the fuse or damage the meter.

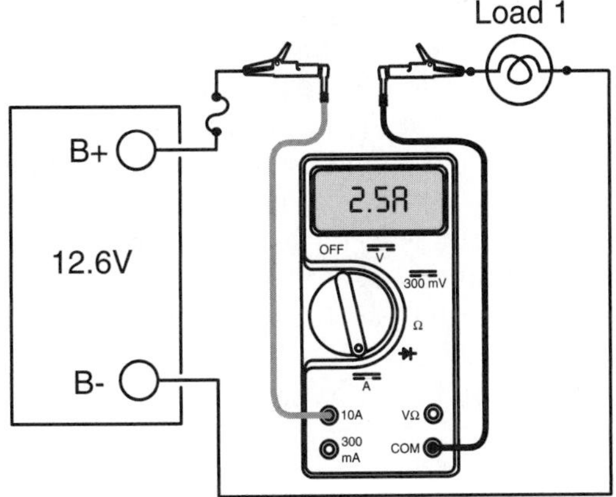

Figure 2. To measure current, the DMM is wired in series with the load.

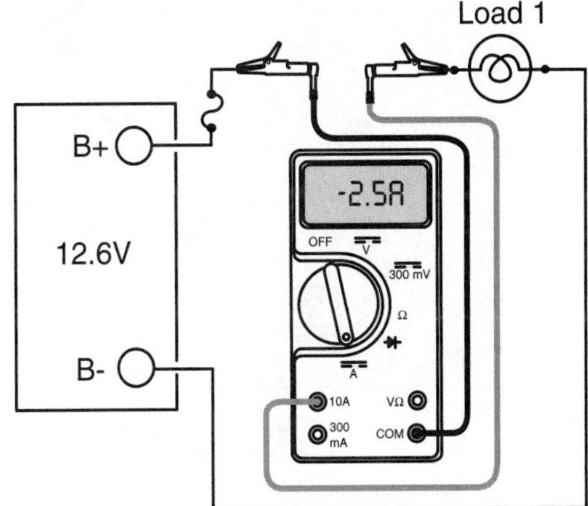

Figure 4. Ammeters are wired in series.

CURRENT PROBES

A variation of the direct-reading ammeter is the inductive pickup design, usually called the current probe. Most applications allow the current probe to be connected into the DMM, usually in the lowest voltage position. If this sounds complicated and does not seem to make sense, read on. A current probe is placed around a wire, as shown in **Figure 5**. The probe will measure the amount of magnetic field around the wire because whenever current flows magnetism is present. The measurement of the field is converted to a small voltage signal, which is read by the meter. In our application, we would use the 300-millivolt setting and plug the current probe into the positive and negative voltmeter connections. Reading the meter face is easy. Our probe generates a 100 millivolt signal for every amp that is flowing. So, if the meter face were to read 250mV, we would have 2.5 amps flowing. You can see that this will greatly increase the adaptability of the meter.

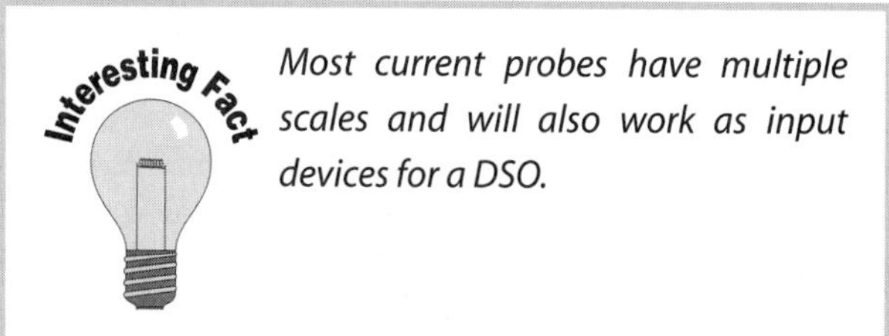

Most current probes have multiple scales and will also work as input devices for a DSO.

We now have the capacity to read up to 300 amps with the current probe and the DMM. The meter face and range setting will also indicate the polarity of the circuit. Do not forget that this style of meter senses the amount of current flowing through a wire by measuring the magnetic field around the conductor. Any time current flows through a conductor, a magnetic field will be present. By sensing the magnetic field around the wire, the meter is able to determine indirectly the amount of current flowing **(Figure 6)**.

Figure 5. A current probe clamped around a wire will measure the current flow.

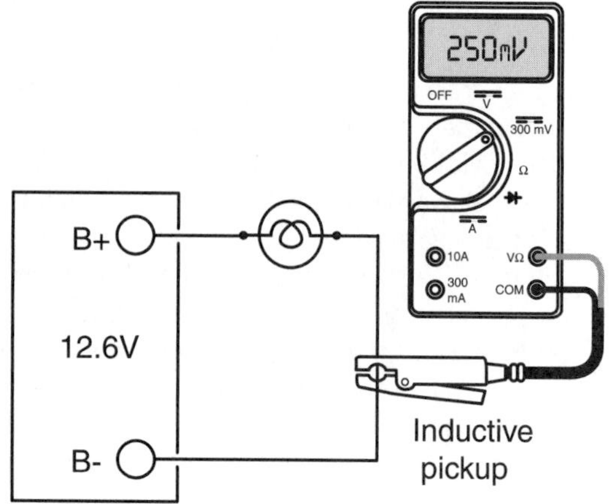

Figure 6. A current probe is connected to the voltage input of the DMM.

The large circle clamp on the end of the cable of the current probe is a very sensitive electronic component and cannot be handled roughly. Dropping it or hitting it against a bench will usually impair the accuracy of the meter. An advantage to the inductive style is its simplicity of connection. There is no need to worry about connecting it in series, because current does not flow through it. Many of these current probes are polarity sensitive and may be marked with a + symbol to help you when making a connection. An additional advantage of the current probe is its ability to take the measurements without having to disconnect the vehicle's battery. Most vehicles today have multiple memory circuits that are powered off the battery. Disconnecting the battery to obtain an amp reading would allow the processor to lose its stored memory. Radio presets, multiple driver's seat and mirror positions, and stored diagnostic codes, in addition to basic vehicle operations information, could be lost. Customers do not like to have to reprogram their radio preset memories, so if at all possible do not disconnect the battery. Use a current probe and the DMM where possible. The typical DMM with a current probe will measure amperage in both directions, which is especially helpful when doing starting and charging testing. The DMM can measure and tell the technician whether current is going into or coming out of the battery. Its use will be explained in the starting and charging section of this text. Keep in mind that this style of probe is still sensitive to an overrange condition. If you are using a 100-amp probe and crank the vehicle over, the approximate 200 amps flowing will have an impact on the sensitive circuitry in the probe.

Most technicians have multiple current probes so that they can match the probe to the current flow, the high-current probe for starting and charging applications and the low-current probe for all other circuits.

There is no difference in reading the DMM on voltage or current. A millivolt is a thousandth of a volt and a milliamp is a thousandth of an amp. The "m" abbreviation means the same thing and is read the same in either application.

Summary

- A direct-reading ammeter will require that the circuit be broken into and the meter wired in series with the load.
- The circuit's resistance, not the meter, will control the amount of current flow.
- Most DMMs have two current scales.
- The DMM will blow a fuse if the circuit current exceeds the meter's capacity.
- All of the current that the circuit is drawing must flow through the direct-reading ammeter.
- A current probe is placed around the wire and will indicate how much current is flowing through the wire.
- Current probes are usually in either the low range (up to 100 amps) or the high range (up to 500 amps).
- The signal from a current probe is a voltage signal fed into the voltage connections of a DMM.

Review Questions

1. An ammeter is connected directly to the positive and negative of a battery. This will result in
   A. accurate reading
   B. an indication of the state of charge of the battery
   C. discharging the battery
   D. a burned-out fuse or meter
2. An ammeter is always wired in ________ with the load.
   A. parallel
   B. series
   C. either series or parallel
   D. none of the above
3. Technician A states that the ammeter is wired in series with the load. Technician B states that the amount of current flowing in the circuit has no impact on the meter. Who is correct?
   A. Technician A only
   B. Technician B only
   C. Both Technician A and Technician B
   D. Neither Technician A nor Technician B
4. The input impedance of a DMM in the amperage position is
   A. 10,000,000 ohms
   B. 0 ohms
   C. variable and based on the design of the DMM
   D. the same as for the voltmeter positions
5. Ammeters should be connected into circuits that have resistive loads already in them.
   A. True
   B. False
6. A current probe will generate a voltage signal based on current flow.
   A. True
   B. False
7. A current probe is placed into a circuit that has been broken open.
   A. True
   B. False
8. A direct-reading ammeter will usually be capable of reading a small current flow.
   A. True
   B. False
9. Is there an advantage to using a current probe over a direct-reading ammeter?
10. Why does a direct-reading ammeter need to be wired in series with the load of the circuit?

Chapter 7 Ohmmeters

Introduction

No discussion of measuring electricity would be complete without looking at the use of an ohmmeter or a DMM on the ohms scale. The major difference between the use of a voltmeter or ammeter and an ohmmeter is in how they are used. The voltmeter or ammeter is usually used on a live circuit, one that has current flowing through it. An ohmmeter is never used on a live circuit. If we want to measure the resistance of a load with an ohmmeter, it must be isolated and tested **(Figure 1)**. This difference is the most important point. We emphasize how to read and interpret the ohmmeter section of the DMM in this chapter. By the end of this chapter, the major meters will have been covered. This should give you sufficient information to begin your study of circuits.

With an ohmmeter, the component or section of a circuit being tested must be isolated or disconnected from the power source. Ohmmeters have their own source of power, usually from an internal battery. If the component or circuit being tested has power from the vehicle's battery, one of two things might happen. There is a chance that the vehicle battery will overpower the internal battery and damage the DMM, or the two power sources could conflict and make the reading inaccurate. Neither is good. A short discussion about how an ohmmeter works might help you understand and interpret the reading. Ohmmeters have their own battery inside the meter and will push current through the resistance or circuit that is being measured. You will see later that the relationship between current flow and resistance is defined in Ohm's Law. When you connect an ohmmeter to a resistance, the meter will push and measure the current flow. It will then use the relationship between current and resistance to calculate the resistance. The calculation of ohms will show on the screen, with the Ω symbol on the screen. The Ω symbol stands for ohms.

The ability to understand the reading also involves knowledge of the two abbreviations that are used in the measurement of ohms. If the resistance exceeds 1000 ohms, the meter will automatically (or manually, if you wish) switch ranges and go to the "k" scale. K ohms is 1000 ohms. The k stands for kilo, or 1000. So if the meter reads 123k ohms, we are looking at a converted value of 123,000 ohms. If the resistance exceeds 1000k ohms, the meter will change the range and the abbreviation to an "M," which stands for million. Do not get the small m and the capital M confused. The small m stands for milli, or thousand, and the capital M stands for million. We do not use the m in

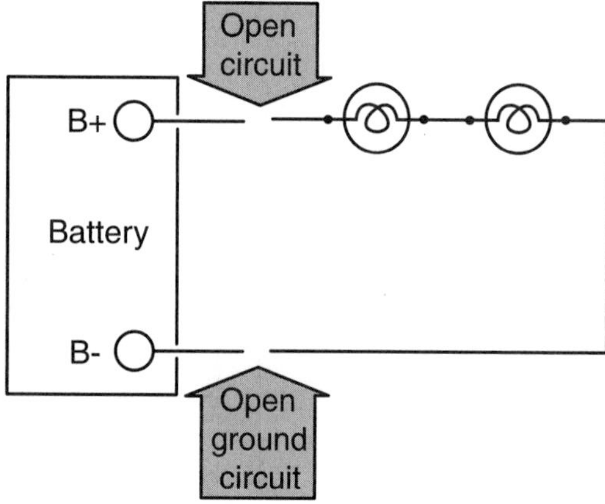

Figure 1. The circuit must be open or isolated from power and ground for resistance checking.

resistance measurements because it is too small, and we do not use the M in voltage because 1,000,000 volts would be lethal.

READING AND INTERPRETING OHMMETER READINGS

DMMs in use today are usually auto ranging. This greatly simplifies the front switch arrangement of the meter but can involve some inaccurate readings if technicians do not understand what they are looking at. With a turn of the main knob to the ohms position, the meter is set for auto ranging from a tenth of an ohm up to 999,000,000 ohms. Before taking any readings, it is important to test the ability of the meter to accurately measure no resistance and display it as zero ohms. Hold the two leads together tightly, with the DMM set on ohms as shown in **Figure 2**. The display should ideally show zero, but a few tenths is acceptable. If the meter is zeroed, then it can be used for measurements. **Figure 3** shows a DMM set on ohms with a resistor between the leads. The display shows 45.8M ohms. Remember that anything to the left of the decimal point is in millions of ohms, so this means 45,800,000 ohms.

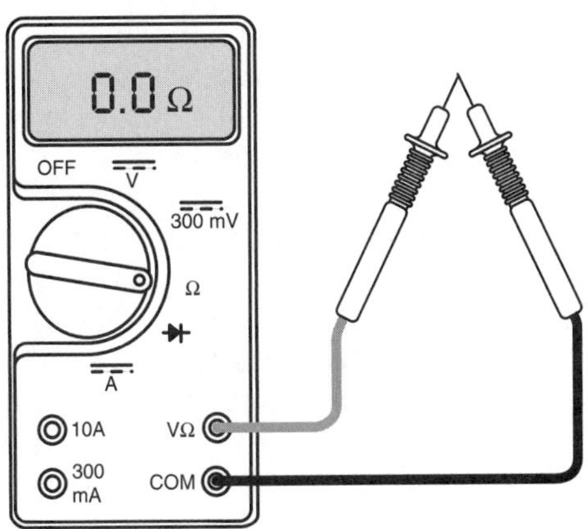

Figure 2. Checking the zero on an ohmmeter.

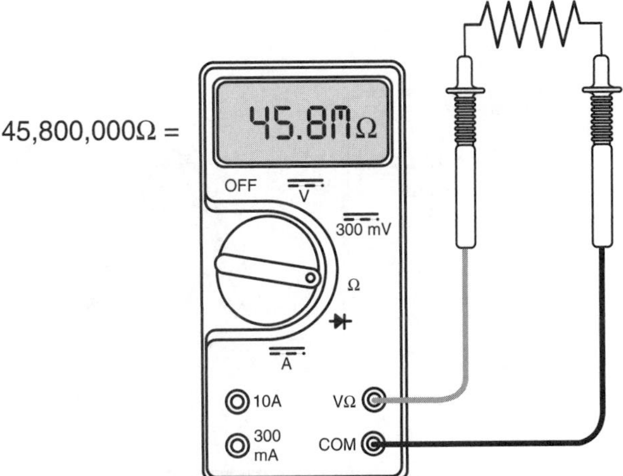

Figure 3. A reading of 45.8M ohms is 45,800,000 ohms.

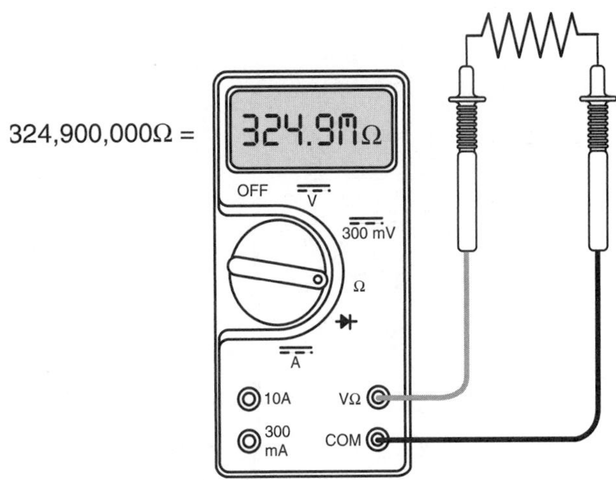

Figure 4. A reading of 324.9M ohms is 324,900,000 ohms.

Figure 4 displays a reading of 324.9M ohms. How many ohms is that? If you answered 324,900,000 ohms, you are correct.

Do not forget that any digits before the decimal point are millions, and anything after the decimal point is less than a million. The M scale is usually not encountered often in typical automotive applications. We do, however, find the use of the k abbreviation frequently in automotive applications. Do not forget that k, or kilo, is 1000, as we have seen before.

Figure 5 shows our DMM set on ohms with a resistor between the leads and a display of 24.65k ohms. Remember that anything to the left of the decimal point is in thousands. When we convert our reading to whole numbers, we get 24,650 ohms.

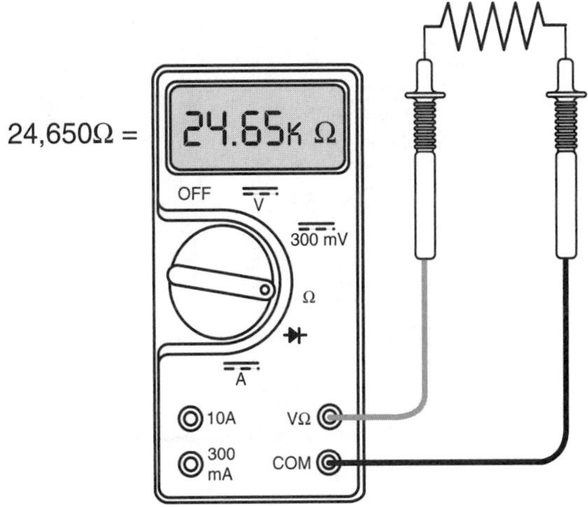

Figure 5. A reading of 24.65k ohms is 24,650 ohms.

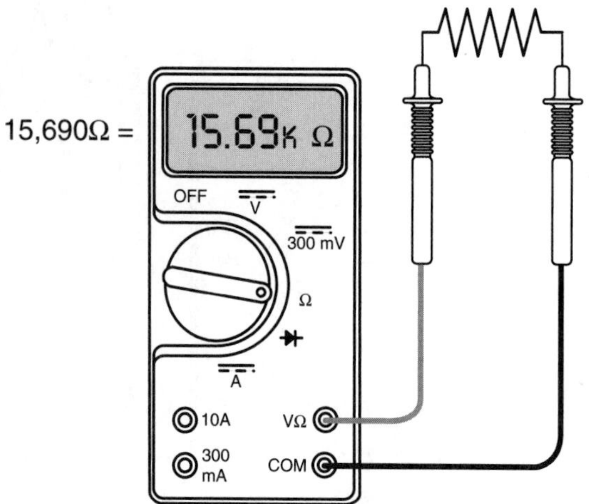

Figure 6. A reading of 15.69k ohms is 15,690 ohms.

Let us try one more example. **Figure 6** shows 15.69k ohms on the display. Have you figured out that this converts to 15,690 ohms? You should be able to make these conversions because that is something that you will have to do frequently as you diagnose and repair a vehicle.

When you are on auto range, the meter will automatically range to give the most accurate reading. As an example, a reading of 456 ohms could be written as 0.456k ohms, but this would be cumbersome and could allow for a misdiagnosis. The auto range will show it as in **Figure 7** as 456 ohms. As long as you keep the meter in the auto-range position, you need not worry about getting readings that need interpretation. Do not be fearful about using a DMM. It is easy to use, accurate, and almost impossible to burn out from over-ranging or reverse polarity. Its use is important on computer circuits because its high input resistance allows an accurate reading without loading down the circuit. If various DMMs are available for your use, do not hesitate to use them all in an effort to become proficient in their use. The more experience you get with DMMs, the easier they will be to read and the more accurate will be the information obtained.

One other concept is important, that of **infinity**. By definition, infinity is an open circuit. It is a circuit that has more resistance than the meter can measure. If the DMM is set on auto range, that means that the circuit has more than 999,000,000 ohms or, in other words, is an open circuit. You may sometimes see the symbol ∞, referring to resistance. The symbol is still used and is a holdover from the analog meter days. It is the infinity symbol.

CONTINUITY TESTING

We have not used the word **continuity** yet, but no discussion of ohmmeters would be complete without it. Continuity means that there is a circuit of very low resistance. A wire has continuity because it can be a path. Most DMMs have a continuity position. With the DMM in this position, as shown in **Figure 8**, when the meter sees continuity it will sound a tone. When you hear the tone, you know that the wire or component that you are testing has continuity; it has a path for current to flow. This is not a specific resistance test; it is a path test. Whatever the meter display shows is not important. Having a tone on or off is what matters. A good example of the use of continuity is in testing the two ends of a wire that is buried within a harness. If there is a break in the wire where we cannot see it and we touch the two ends of the wire, no tone will sound, indicating that the wire is open. If the tone is on, the wire has continuity and current can flow through it.

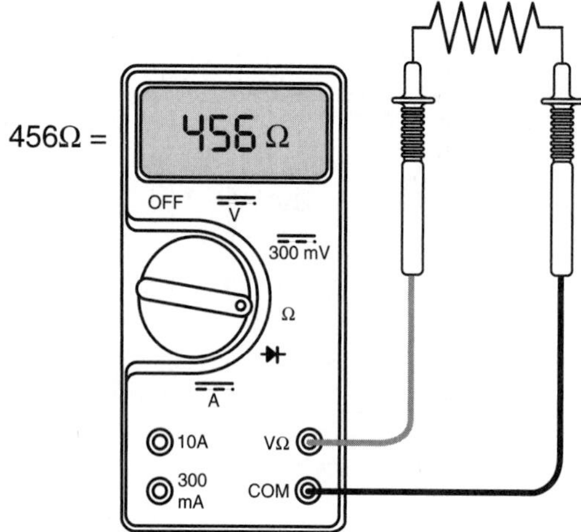

Figure 7. Without a k or an M, the meter reads resistance directly.

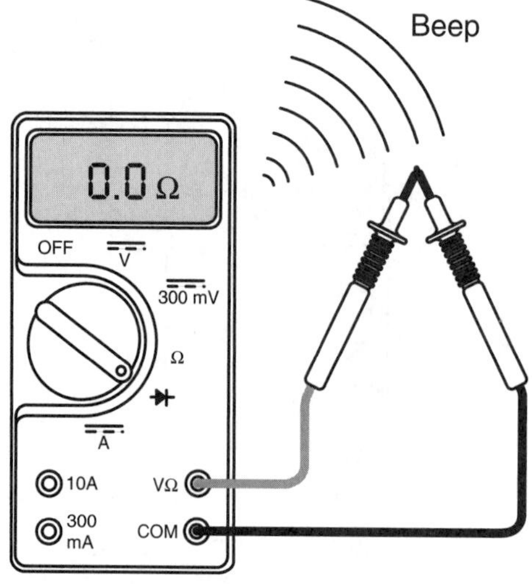

Figure 8. The DMM beeps when continuity is present.

Summary

- To use an ohmmeter, the circuit must be open.
- Ohmmeters should be tested for zero before using them on a circuit.
- An auto-range DMM will range itself to the best setting and display a value.
- k ohms is 1000 ohms.
- M ohms is 1,000,000 ohms.
- 550k ohms is 550,000 ohms.
- 4.3M ohms is 4,300,000 ohms.
- Testing for continuity is testing for an electrical path.

Review Questions

1. An ohmmeter is connected to a conductor and reads infinity. This indicates
   A. nothing
   B. the wire is open
   C. the wire is shorted
   D. none of the above
2. A digital reading of 2.370k ohms is
   A. 2.370 ohms
   B. 2,370,000 ohms
   C. 2370 ohms
   D. 23,700 ohms
3. An ohmmeter is placed on a load and reads OL. The meter is in the auto-range position. Technician A states that this means infinity. Technician B states that no current will be able to flow through the load. Who is correct?
   A. Technician A only
   B. Technician B only
   C. Both Technician A and Technician B
   D. Neither Technician A nor Technician B
4. 767.4 M ohms is
   A. 767.4 ohms
   B. 767.4k ohms
   C. 7674 ohms
   D. 767,400,000 ohms
5. Ohmmeters are wired in series with the power and load.
   A. True
   B. False
6. 0.42 k ohms is 420 ohms.
   A. True
   B. False
7. A wire that has low resistance will turn the continuity tone on.
   A. True
   B. False
8. Why is an ohmmeter used in an isolated circuit?
9. What does the symbol ∞ mean?

Vehicle Circuits

SECTION OBJECTIVES

At the conclusion of this section, you should be able to:

- Understand the components that are part of a working circuit.
- Have a working knowledge of Ohm's Law.
- Analyze the voltage, current, and resistance of a series circuit.
- Analyze the voltage, current, and resistance of a parallel circuit.
- Analyze the voltage, current, and resistance of a series-parallel circuit.
- Have a working knowledge of voltage division in a series circuit.
- Have a working knowledge of current division in a parallel circuit.
- Use a 12-volt test light to diagnose an open circuit.
- Use a DMM to diagnose an open circuit.
- Understand the characteristics of a short circuit.
- Understand the characteristics of an open circuit.
- Calculate the wattage developed within a circuit.
- Use Ohm's Law to calculate the relationship between voltage, current, and resistance in circuits.

The vehicle of today has a mix of series, parallel, and series-parallel circuits. Knowledge of the fundamentals of circuits and a working knowledge of Ohm's Law are the foundations for success as a technician.

Chapter 8 Circuits That Do Work

Introduction

Now that we have taken a look at series, parallel, and series-parallel circuits and have seen how meters are used to look into the circuit, it is time to review the concepts as we look at work being done. This chapter is somewhat of a review because we will not introduce any new material. Instead, we will look at a circuit's components and redefine their purpose and function. As you look at this review of material, make sure that all of it makes sense. If it does not, use it as a guide to direct you to the chapter that you need to spend more time understanding. Chapters 9–11 deeply analyze the interrelationship among voltage, current, and resistance.

CIRCUIT COMPONENTS

As we have seen, for a circuit to be functional, it needs various components. The first is the source. The source of electricity is usually the battery or the charging system if the engine is running. This source will be 12.6 volts with the vehicle off or approximately 14 volts if the charging system is functional. The charging system has the job of running the vehicle electrically and recharging the battery. Without the charging system, the battery would discharge, and in a short period of time the vehicle would stop running.

The second component is the load. It is the load that does the work of the circuit, and it is the resistance of the load that determines the current flow. The higher the resistance, the lower the current will be and the less work the circuit will do.

The third component is the circuit protection. Whether the circuit uses a fuse, fusible link, or a circuit breaker does not matter. What does matter is that the level of circuit protection is based on the amount of current that will be flowing under normal conditions. Typically the circuit protector is rated in amperage and is usually about a third higher than the normal current flowing. A 20-amp fuse is used in a 15-amp circuit and so on. Fusible links are not in use as much as they were in the past. They are rapidly being replaced by the maxi-fuse shown in **Figure 1**. Circuit breakers continue to be used in circuits in which motors are in use. Power windows, seats, and sunroofs, for example, still usually use the circuit breaker.

The next component is the control. Few automotive circuits are on all of the time. Something is used to control their operation. A switch is the most common type of control, with two positions—on and off. When it is in the off position, a switch causes a controlled open circuit with no current flow-

Figure 1. Maxi-fuses have replaced fusible links on some applications.

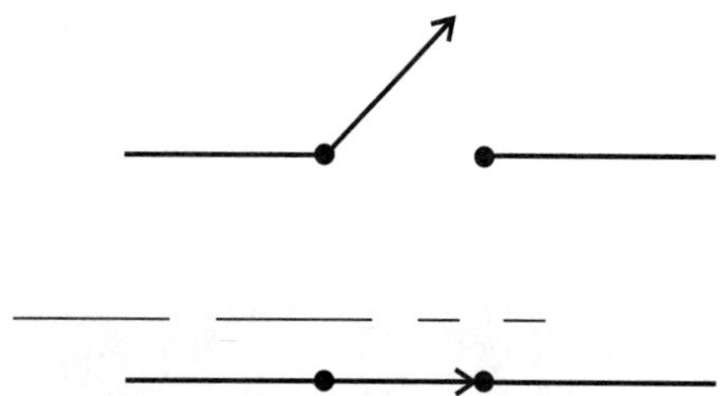

Figure 2. A switch opens and closes to control the flow of current.

ing, as shown in **Figure 2**. Closing a switch causes current flow through the circuit, and the production of work begins. Relays or electronic modules can also act as switches in circuits. Relays are generally found in high-current circuits such as the starting system, whereas modules are found in low-current circuits such as the ignition system.

The last components are the insulated conductors. Conductors have the job of delivering the current to the load, where the work will be accomplished. Their insulation makes sure that the current arrives at the load. Insulators are the opposite of conductors. Insulators resist electron movement, whereas conductors will make electron movement easy. The size of the conductor is dependent on the amount of current that the load draws, whereas the thickness of the insulation is dependent on the voltage applied to the circuit.

Our complete simple circuit is pictured in **Figure 3**. Based on the fact that 12 volts are applied to the 2-ohm load, how much current will flow and how much work will be done? This is where Ohm's Law is applied. The voltage, 12 volts, is divided by the resistance of the load, 2 ohms **(Figure 4)**.

12 volts ÷ 2 ohms = 6 amps of current

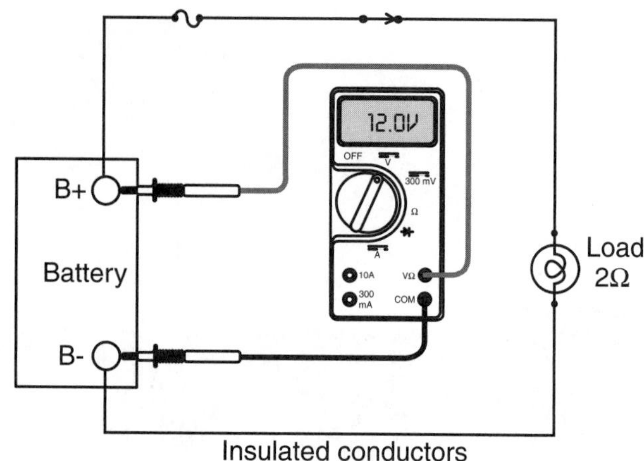

Figure 3. Six amps of current will flow if 12 volts is applied across a 2-ohm load.

If amps is the amount of current, what unit of measure indicates the amount of work? Wattage is our indication and measurement of the amount of work that a load will accomplish. This work will be in light, heat, magnetic field, or a combination of all three. The wattage is figured by multiplying the voltage across the load by the current flowing through the load. There is 12 volts across the load and 6 amps of current flowing, so

12 volts × 6 amps = 72 watts

This circuit is producing 72 watts of light, heat, and magnetic field.

Most circuits are not as simple as our example. We will use three 2-ohm resistors and wire them in series, parallel, and series-parallel and analyze the voltage, current, and total resistance that each circuit offers.

Figure 5 shows the three 2-ohm loads wired in series. The meters show that there is 2 amps of current

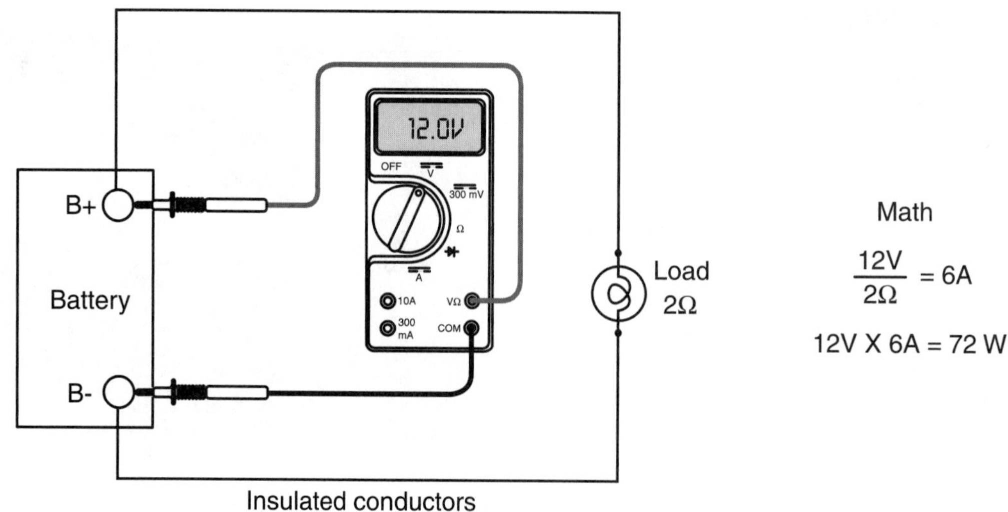

Figure 4. A 2-ohm load will generate 72 watts of work if 12 volts is applied across it.

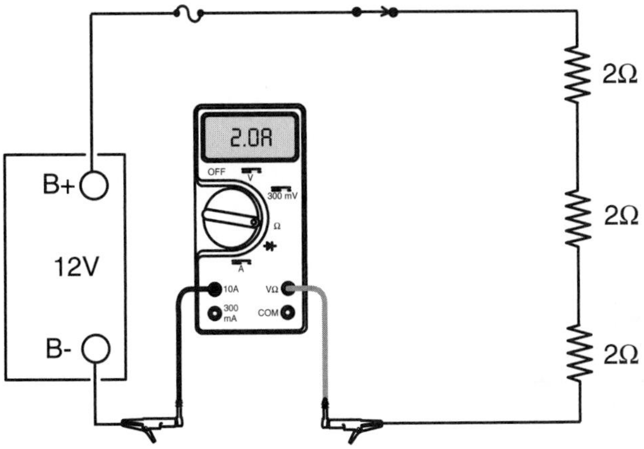

Figure 5. Two amps of current flow through 6 ohms of resistance.

flowing and that there is a 4-volts drop across each load. The total resistance of this circuit must be 6 ohms because resistances in series always add up. Each load is therefore producing 8 watts:

2 amps × 4 volts = 8 watts

and the total wattage produced by the circuit is 24 watts:

8 watts + 8 watts + 8 watts = 24 watts total

> **You Should Know** *A series circuit has only one path for current to flow. The resistances of the loads add up, while the voltage drops across each load.*

Figure 6 shows the same three 2-ohm loads wired in parallel. The meters show that the voltage did not divide and the current flow increased. Each load is drawing the full 6 amps because 12 volts is applied to each leg of the parallel circuit, for an 18-amp total. Each leg is producing 72 watts of work, so the total work of the circuit is 216 watts:

72 watts + 72 watts + 72 watts = 216 watts

Parallel circuitry is used more than series in automotive applications. The total resistance of this circuit is $\frac{2}{3}$ of an ohm, figured by 12 volts divided by 18 amps

12 volts ÷ 18 amps = $\frac{2}{3}$ ohm total

Notice that $\frac{2}{3}$ ohm is lower than the lowest (2 ohms) load in each leg of the parallel circuit.

The last circuit that we look at is that in **Figure 7**, which shows one 2-ohm resistor in series with two 2-ohm resistors in parallel. This is a series-parallel circuit, which will have some of the characteristics of both styles. Notice that the ammeter reads 4 amps for the entire circuit. How did we get that? It is easy. The parallel circuit with two 2-ohm resistors has the equivalent resistance of 1 ohm. Add the 1 ohm to the 2 ohms and you get a total resistance of 3 ohms. Put 12 volts across 3 ohms and you will have 4 amps of current flowing:

12 volts ÷ 3 ohms = 4 amps

The voltmeter readings are the result of current flowing through resistance. In the series part of the circuit, 4 amps flows through 2 ohms:

> **You Should Know** *Parallel circuits allow full voltage to each leg and therefore produce more work than the same loads wired in series. The total resistance of a parallel circuit is always lower than the lowest individual resistance.*

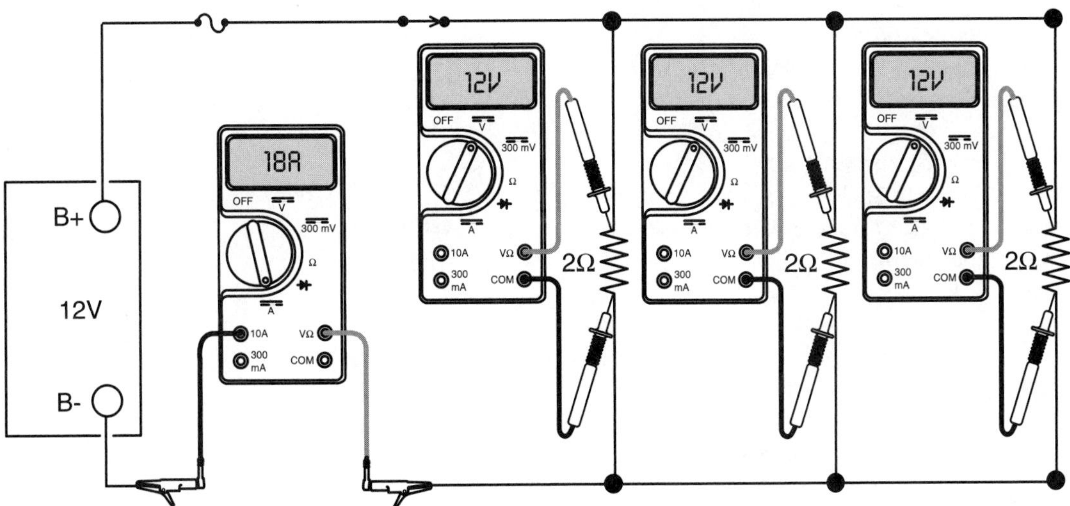

Figure 6. Each leg of a parallel circuit will draw 6 amps.

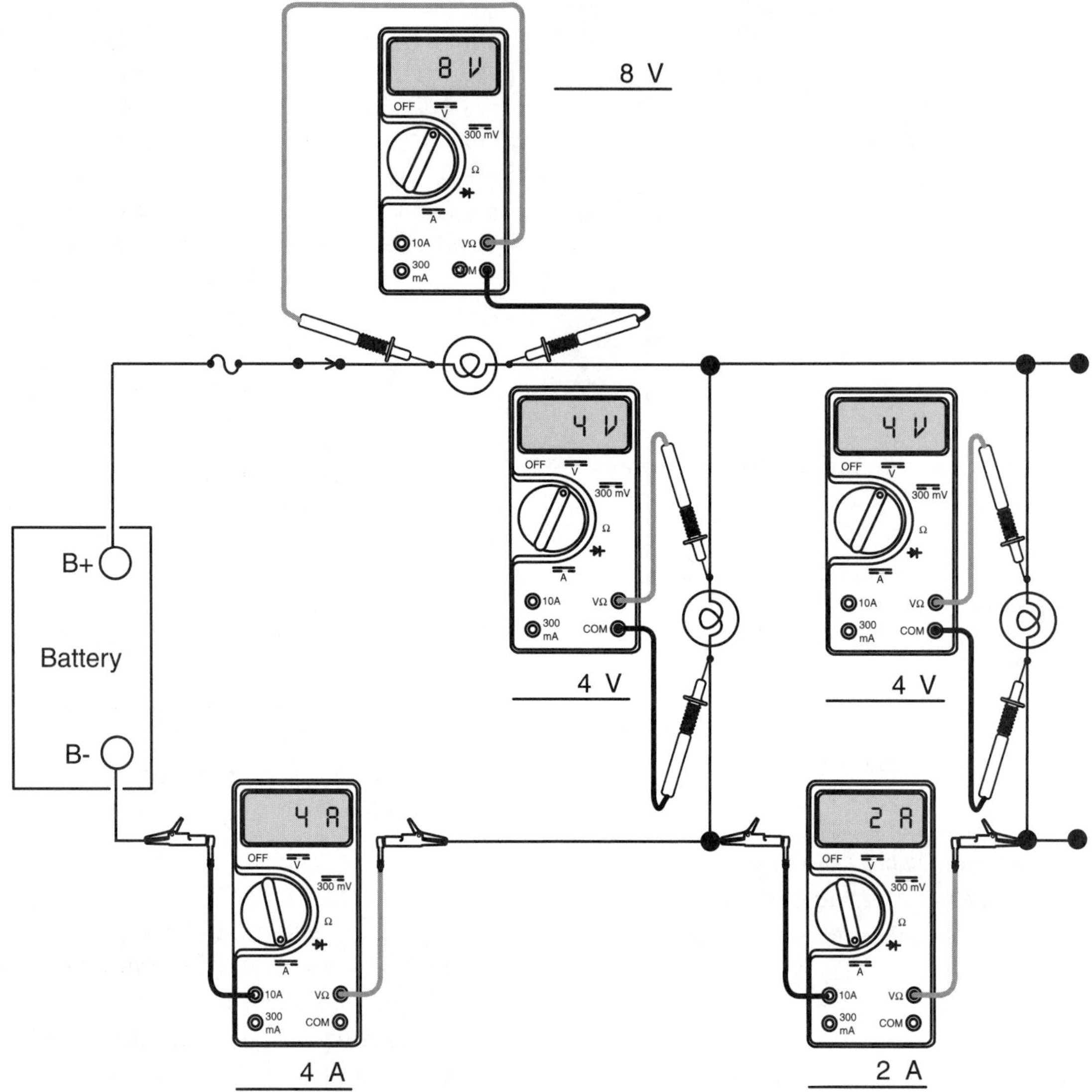

Figure 7. A series-parallel circuit will have some series characteristics and some parallel.

4 amps × 2 ohms = 8 volts

The current splits through the parallel legs so 2 amps flows through each 2 ohm load, producing a voltage drop of 4 volts:

2 amps × 2 ohms = 4 volts

This is the hardest circuit to analyze. Chapter 11 covers it again, in case you are having trouble with the voltmeter and ammeter readings.

> **You Should Know** *A series-parallel circuit has some of the characteristics of both a series and a parallel circuit. The voltage divides in the series section, and the current divides in the parallel section.*

Summary

- A series circuit has one path for current to flow.
- A parallel circuit has multiple paths for current to flow.
- The voltage will divide in a series circuit.
- The current will divide in a parallel circuit.
- More work is accomplished in a parallel circuit.
- The total resistance of a parallel circuit is always lower than the lowest individual resistance.
- Individual loads wired in series will have a voltage drop across each one. The sum of all of the voltage drops will equal the applied voltage.

Review Questions

1. Three 2-ohm resistors are wired in series. The total resistance is
   A. 2 ohms
   B. 4 ohms
   C. 6 ohms
   D. 8 ohms
2. Technician A states that more work is accomplished with parallel circuitry. Technician B states that series resistances add up. Who is correct?
   A. Technician A only
   B. Technician B only
   C. Both Technician A and Technician B
   D. Neither Technician A nor Technician B
3. The total wattage of two 2-ohm resistors wired in parallel is
   A. 72 watts
   B. 4 watts
   C. 24 watts
   D. 144 watts
4. If a circuit is said to be open, current will be flowing through it.
   A. True
   B. False
5. A fusible link is a type of circuit protection.
   A. True
   B. False
6. In a parallel circuit, there is more than one path for current to follow.
   A. True
   B. False
7. The voltage will divide among the resistances (loads) in a series circuit.
   A. True
   B. False
8. What are the characteristics of a series circuit?
9. What are the characteristics of a parallel circuit?
10. What are the characteristics of a series-parallel circuit?

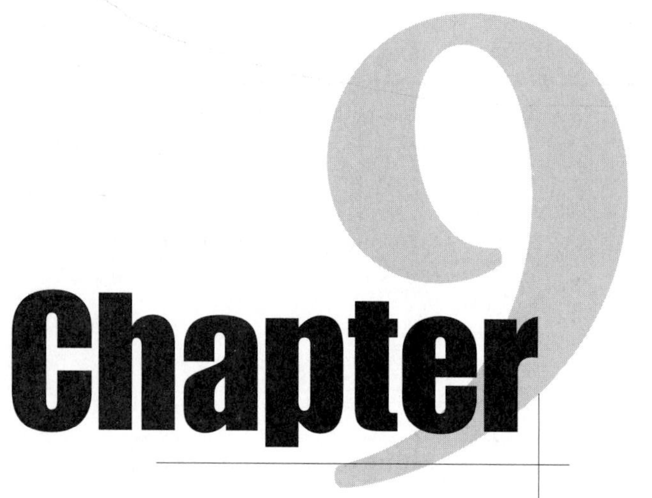

Chapter 9 Analyzing Series Circuits

Introduction

Although series circuits are everywhere, this statement is not exactly correct when applied to the vehicle. The vehicle is mostly a parallel circuit; however, the individual loads, when thought about by themselves, might be considered a series circuit. As a technician, you will find a lot of resistance wired in series that is unwanted. This would be a more correct statement than to say that series circuits are everywhere. Remember our example of the corroded battery cable preventing the cranking of the engine? Unwanted series resistance reduced the current flow and prevented the engine from cranking. In this chapter, you will wire up simple series circuits; measure the voltage, current, and resistance; and do some simple calculations. The math will not be difficult. You need to be comfortable with and have a working knowledge of Ohm's Law in a series circuit. Get a voltmeter, ammeter, ohmmeter, three automotive light bulbs or resistors, switch, fuse, and insulated conductors (preferably with clips on the ends). It is time to see electricity in action.

WIRING UP A SERIES CIRCUIT

Wire up a simple series circuit as shown in **Figure 1** and turn it on. The bulb should light or the resistor should get warm. With the light on, use your voltmeter and record the pressure difference found in the blanks provided. Do not forget that the voltmeter is polarity sensitive. Get in the habit of wiring the positive lead into the most positive part of the circuit and the negative lead into the most negative part of the circuit. As you record your voltmeter readings, do not worry about what the meter is indicating; we will analyze what and why shortly. Once all the blanks are filled in, look at the voltmeter 1 reading. What is it indicating? If you said the available pressure, you are correct. This meter tells us that with current flowing in the circuit (the bulb is on), we have a certain pressure. This is the applied voltage reading. Notice that this number is the same as that in voltmeter 3. If it is not, rewire your circuit. The same reading indicates two different things. First, it tells us the voltage drop across the resistance of the load. Second, because it is the same as the applied voltage, it tells us that the load is the only resistance in the circuit. If, for instance, it were 2 volts lower than the applied voltage, we would know that somewhere in the circuit there must be additional resistance. This resistance was not designed into the circuit by you and is therefore unwanted. Remember that unwanted resistance is referred to as a false load. We can prove that there is no other resistance in the circuit by looking at voltmeters 2 and 4. They should both read 0 (zero). Why? Think about what causes differences in pressure: resistance. No difference in pressure equals no resistance. Our voltmeters are indicating that there is no difference in pressure, and, therefore, as long as current is flowing, there must not be any resistance.

Open the switch and break the circuit open at the source (either positive or negative). Insert an ammeter, as in **Figure 2**, in series so that all current flowing must go through the meter. Do not forget that the ammeter is polarity sensitive. The positive lead must be in the most positive part of the circuit, and the negative lead must be in the most negative part of the circuit. Remember also that an ammeter must be used only in a circuit that has some resistance in it. The resistance of the circuit, not the ammeter, must control the current flow. Our circuit has the resistance of the bulb to control current. In addition, if the ammeter

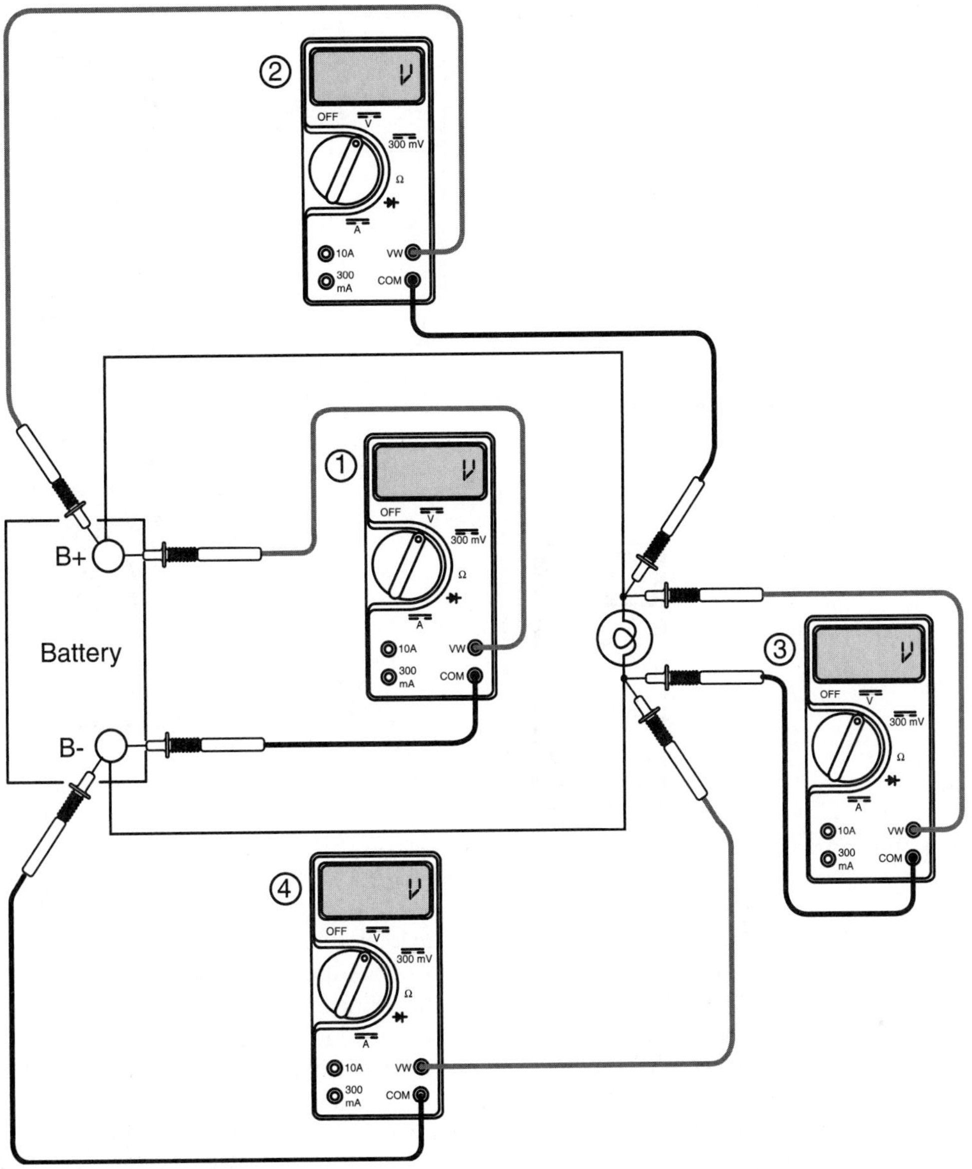

Figure 1. Voltmeters read the difference in pressure.

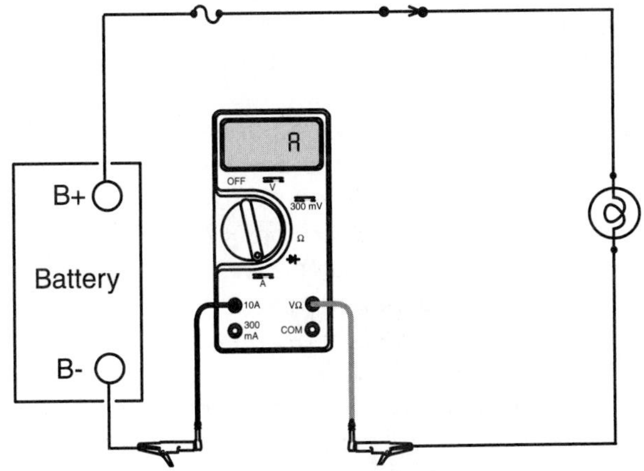

Figure 2. All current must flow through the ammeter.

has multiple scales, use the highest one and work your way down to the lowest one that will measure the circuit current. Record the ammeter reading after you close the switch and reactivate the bulb. We will use this reading as a reference later in this chapter.

VOLTAGE DIVISION IN A SERIES CIRCUIT

Follow **Figure 3** and wire up a simple circuit with two bulbs. Use two bulbs of the same size. The number printed on the two bulbs should match; 1156 is a good size to use because it has one filament. Record the voltmeter readings and the ammeter readings. Let us analyze the meter readings. What does the first one tell you (the first one being the

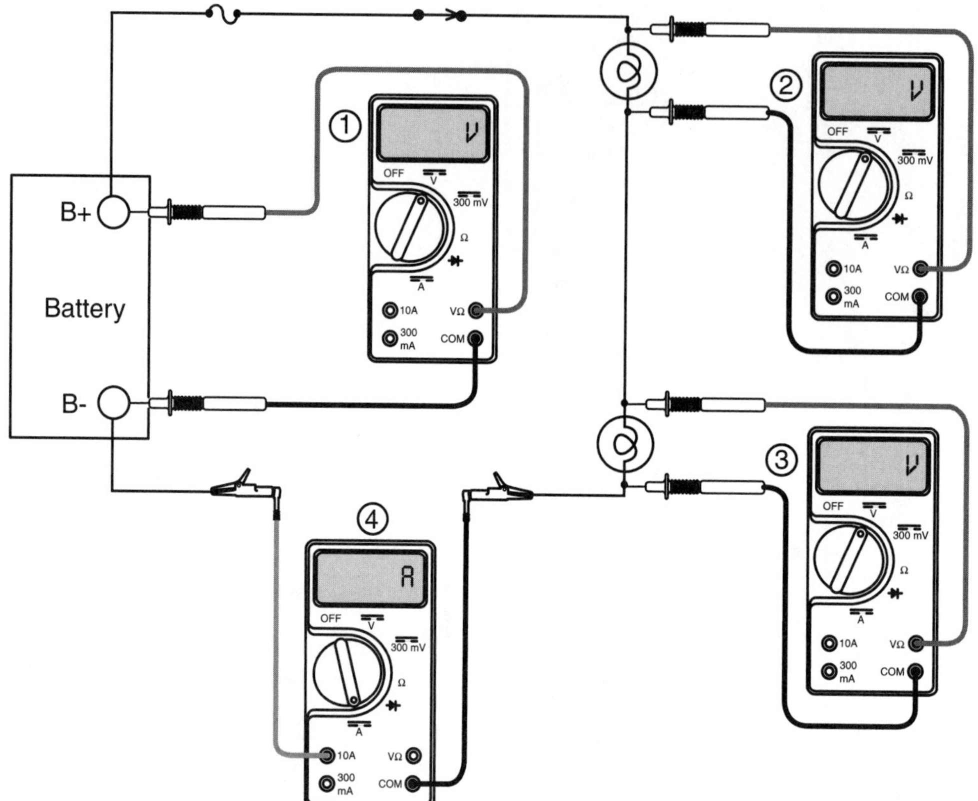

Figure 3. Voltage division will be equal if loads have equal resistance.

applied voltage that pushes the current through the circuit)? This number will be a reference. What does voltmeter 2 show? The voltage drop across the first bulb will probably be a voltmeter reading close to half of the applied voltage, and the same is true with voltmeter 3. What actually happened here was that the pressure dropped or divided equally as it pushed current through the resistances of the bulbs. The resistances wired in series are about the same number of ohms each, so the voltage divided equally. How is it that the ammeter reading is approximately half the reading of Figure 2? You should have realized that the current went down because the resistance went up. Another way of saying this is to say that the current had to be pushed through twice the resistance and, as a result, only half of the current was allowed through. The resistance in the circuit is what controls the current. Increasing the resistance decreases the current.

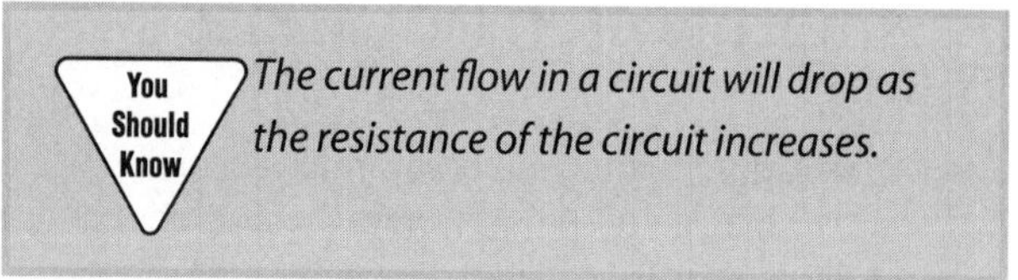

The current flow in a circuit will drop as the resistance of the circuit increases.

Our ammeter reading dropped as the resistance increased. We will look at current flow more closely in Chapter 10 and Chapter 11. If you are using resistors, you can measure the increased resistance of the circuit. Does this sound familiar? It should, because this is Ohm's Law. You just proved it to yourself. We will use the math of Ohm's Law with some additional problems later in this chapter.

Add another bulb (the same size) in series and look at the current flow. Did it drop down like you thought it would? What is the meter telling you? The ammeter reading is probably about one-third the value of the reading in Figure 2. Why? Because the resistance was tripled. Each time we raise the resistance, we drop the current. Did you notice that the bulbs are getting dimmer each time you add another one? The current times the voltage will give you the wattage of work being accomplished. Dropping the voltage down by half and the current down by half reduced the wattage down to one fourth. You can prove this to yourself with the following simple formula:

Example 1 volts × amps = ? watts

If you do not have a live example, we will supply one. A 6-ohm light bulb will draw 2 amps.

12 volts × 2 amps = 24 watts

Example 2 volts × amps = ? watts

Our example continues: two 6-ohm lights have a 12-ohm total resistance and will draw 1 amp of current. Each bulb will have a 6-volt drop across it, so:

6 volts × 1 amp = 6 watts

Six watts is one-fourth of the 24 watts that the single light bulb gave off.

This formula has shown you why the bulbs are getting progressively dimmer. Each time another load is wired in series, the voltage and the amperage drops lower. Multiplying this reduced voltage times the reduced amperage gives us lower wattage levels. Remember that wattage is an indicator of the amount of work actually accomplished in the circuit.

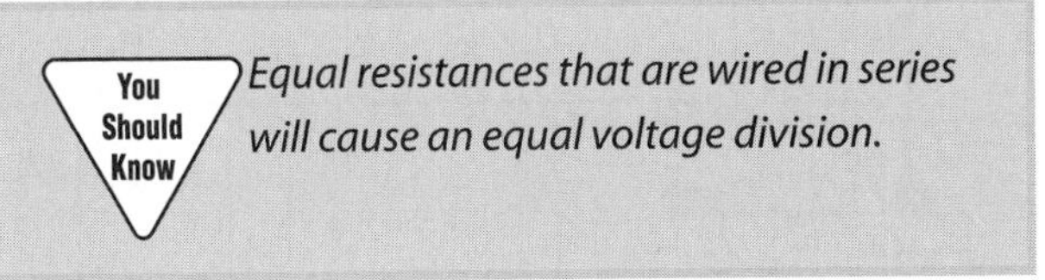

Equal resistances that are wired in series will cause an equal voltage division.

Let us summarize. The voltage dropped or divided in the series circuit and is based on the combined resistances of the loads. Equal resistances give us equal voltage division or drops. Each additional equal resistance wired in series causes the voltage to redivide or drop evenly throughout the rest of the circuit.

Wire up the circuit from **Figure 4**, making sure that you use different-sized bulbs (different numbers). Determine the voltmeter readings below, and analyze the differences.

Voltmeter 1 = ? volts
Voltmeter 2 = ? volts
Voltmeter 3 = ? volts

What happened here? The voltage dropped or divided, but this time it did not do so equally, as in the other examples. What was the difference? Have you realized that the unequal resistances are the reason for the unequal voltage division? Unequal resistance equals unequal voltage division. The total voltage drop still equals the applied voltage, but the individual drops are unequal. Ohm's Law is still at work here. Let us look at the ammeter reading that we were able to obtain in the examples of series circuits wired. We have seen how the current was reduced by the addition of series resistance. Again, this is Ohm's Law in action. Increasing the resistance reduces the current. When we doubled the resistance, we found that the current was halved. Let us use a circuit with specific resistances and prove this to ourselves mathematically.

CURRENT FLOW IN A SERIES CIRCUIT

Figure 5 shows a series circuit with one single 2-ohm resistance and an applied voltage of 12 volts. The current in this circuit will be 6 amps. Ohm's Law states that:

$$\frac{12 \text{ volts}}{2 \text{ ohms}} = 6 \text{ amps}$$

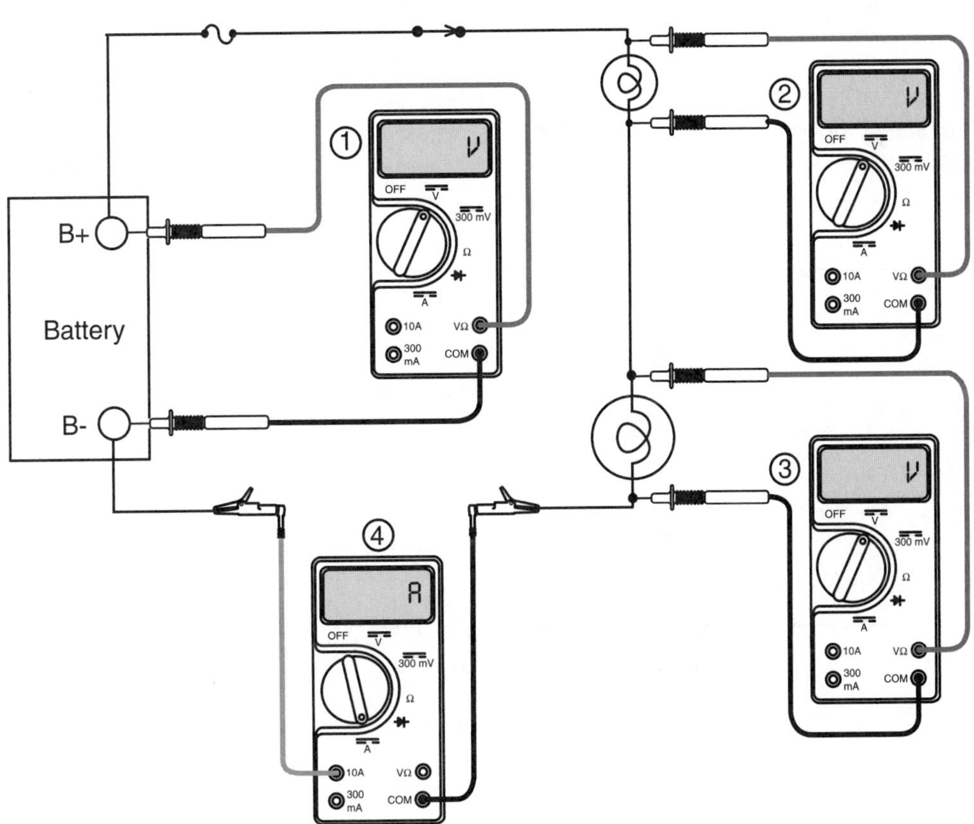

Figure 4. The voltage will divide unequally if the loads have unequal resistance.

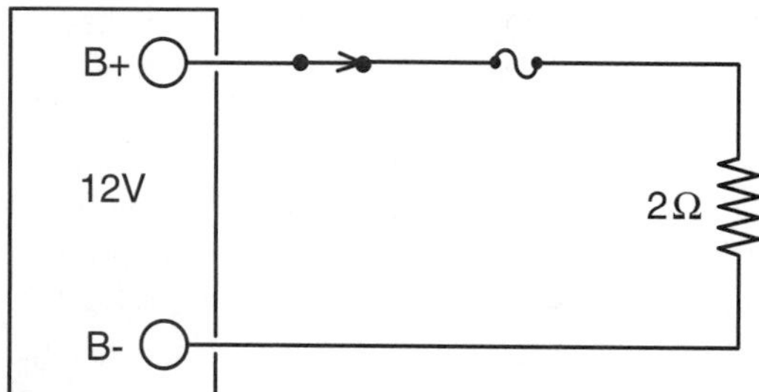

Figure 5. A single 2-ohm load with 12 volts applied.

If, on the other hand, we were presented with the same circuit and did not know the total resistance, we could use a different form of Ohm's Law.

$$\frac{12\text{ volts}}{6\text{ amps}} = 2\text{ ohms}$$

And yet another way is:

$$6\text{ amps} \times 2\text{ ohms} = 12\text{ volts}$$

Notice that Ohm's Law gives us the ability to find the unknown if we have the two other values. Now let us add more resistance to this series circuit. **Figure 6** shows the addition of another 2-ohm resistor, giving us a total of 4 ohms wired in series. Total current flow will now be:

$$\frac{12\text{ volts}}{4\text{ ohms}} = 3\text{ amps}$$

Or again, if we did not know the ohms but knew the current:

$$\frac{12\text{ volts}}{3\text{ amps}} = 4\text{ ohms}$$

Again, the relationship between the known values and the unknown is Ohm's Law.

How much current will flow through a circuit that has five resistors wired in series—a 1-ohm resistor, a 2-ohm resistor, a 3-ohm resistor, a 4-ohm resistor, and another 2-ohm resistor—as seen in **Figure 7**? The approach to this is to first add up the resistances:

$$1 + 2 + 3 + 4 + 2 = 12\text{ ohms total}$$

Now figure the current:

$$\frac{12\text{ volts}}{12\text{ ohms}} = 1\text{ amp of current}$$

The total current flow in this circuit will be 1 amp.

Question: What will the individual voltage drops be across each resistor? This question raises a new point in our discussion of circuits and Ohm's Law. It is easily answered, though, because we know two of three variables. Up to this point, we have looked at only the total

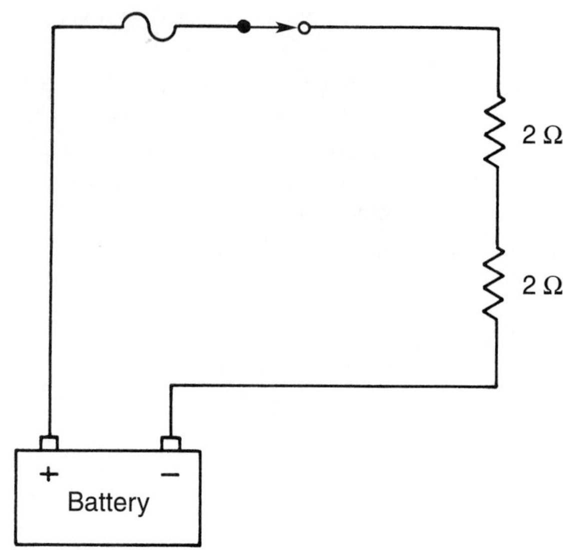

Figure 6. Two 2-ohm loads wired in series.

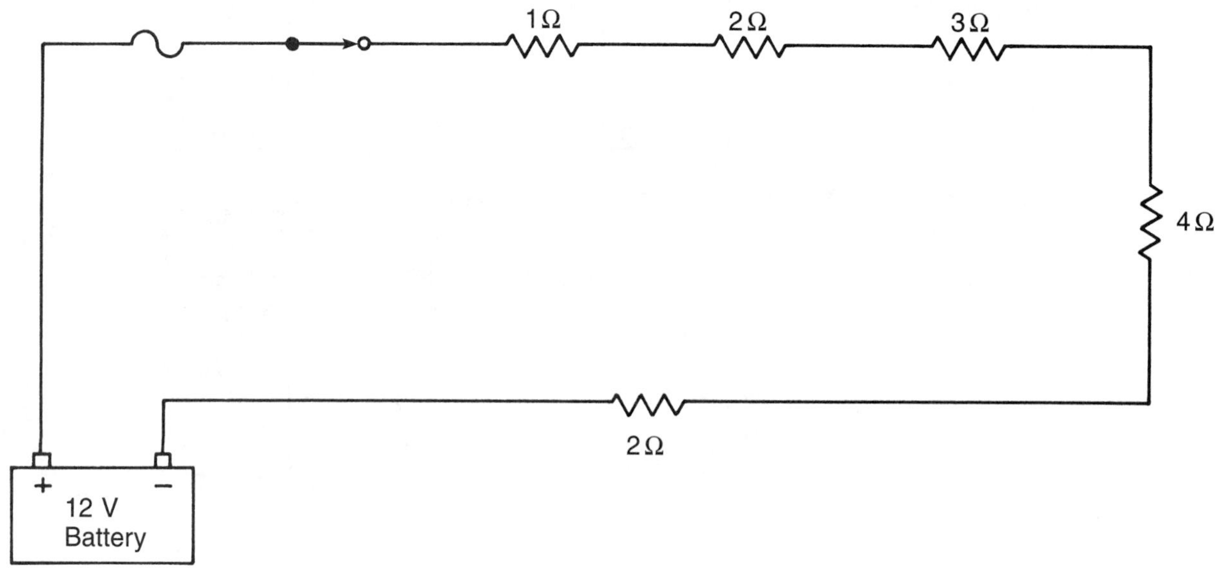

Figure 7. Five unequal loads will cause unequal voltage division.

resistance, voltage, and current. To look at individual drops is just as easy. We know the current flowing through each resistor is the same because this is a series circuit. Current flow through any and all parts of a series circuit is the same because there is only one path for current. If 1 amp is flowing into the first load, 1 amp is flowing out of the last load. We also know the resistance of each load. By applying Ohm's Law to each resistor, we are able to find the voltage drop across each resistor. If our numbers are correct, they will add up to the total applied voltage: 12 volts, in this example. Start with the 1-ohm resistor:

1 ohm × 1 amp = 1 volt

This 1 volt is the voltage drop measured across a 1-ohm resistor with 1 amp of current flowing. Now consider the rest of the resistors:

2 ohms × 1 amp = 2 volts
3 ohms × 1 amp = 3 volts
4 ohms × 1 amp = 4 volts

As a final check, add up the individual voltage drops. Do not forget that there are two 2-ohm resistors.

1 volt + 2 volts + 2 volts + 3 volts + 4 volts = 12 volts

The applied total voltage and the sum of the individual voltage drops was 12 volts, so our circuit follows Ohm's Law perfectly. You will find additional series circuit problems presented at the end of this chapter. You may wish to go to them now and test your knowledge of voltage, current, and resistance in a series circuit before we look at parallel circuits in Chapter 10.

The sum of the individual voltage drops will always equal the applied voltage.

Summary

- Resistance causes voltage drops.
- Increasing series resistance will cause a drop in current.
- The voltage will divide based on the resistance in a series circuit.
- As the resistance increases, the total current decreases.
- Additional series resistance will drop the wattage developed by the load.
- Unwanted resistance is called a false load.
- Equal resistances wired in series will cause equal voltage division.
- Unequal resistances wired in series will cause unequal voltage division.

Review Questions

1. Adding additional resistance to a series circuit will
   A. have no effect on the current flow
   B. reduce the current flow
   C. increase the wattage
   D. increase the current flow
2. Technician A states that the resistance in a series circuit is added together to find the total. Technician B states that the total resistance in a parallel circuit is always lower than the lowest single resistor. Who is correct?
   A. Technician A only
   B. Technician B only
   C. Both Technician A and Technician B
   D. Neither TechnicianA nor Technician B
3. A taillight bulb is glowing dimly. A voltmeter across the bulb reads 7.4 volts. Technician A states that there must be a voltage drop somewhere else in the circuit. Technician B states that resistance in parallel must be present. Who is correct?
   A. Technician A only
   B. Technician B only
   C. Both Technician A and Technician B
   D. Neither Technician A nor Technician B
4. A 2-ohm resistor is wired in series with two 7-ohm resistors in parallel. The total resistance is
   A. 3.5 ohms
   B. unknown because not enough information is given
   C. 5.5 ohms
   D. 6.9 ohms
5. To find the total current flow in any circuit
   A. multiply the amperage times the voltage
   B. divide the voltage by the amperage
   C. multiply the voltage times the resistance
   D. divide the voltage by the resistance

6. Three resistors, 2-, 3-, and 4-ohm, are wired in series. Their combined resistance is
   A. 9.00
   B. 0.93
   C. 1.08
   D. 2.16
7. How many amps will flow through the three resistances of review question 6 if you apply 12 volts to them?
   A. 11.1
   B. 6.4
   C. 3.3
   D. 12.90
8. Voltmeters are always wired in parallel.
   A. True
   B. False
9. Ammeters are always wired in parallel.
   A. True
   B. False
10. The resistance in a series circuit adds up to the total.
    A. True
    B. False
11. Adding resistance in series will increase the current flow.
    A. True
    B. False
12. Why does the wattage decrease when additional resistance is added in series?
13. A vehicle does not crank over. A technician removes the battery cable and cleans it, and the vehicle starts. Why?

Chapter 10 Analyzing Parallel Circuits

Introduction

Now that we have seen what current, voltage, and resistance do in series circuits, let us examine the same voltage, current, and resistance in parallel circuits. You will see that some things remain the same. For instance, the amount of current flow will still be based on the total resistance. However, figuring the total resistance is not the same as that for a series circuit. The ability to diagnose and repair circuits on the modern vehicle is based on your working knowledge of circuits and the interrelationship among voltage, current, and resistance. Wire up the simple two-bulb parallel circuit shown in **Figure 1**. Use same-sized bulbs. Take voltmeter readings and ammeter readings, and fill in the readings directly on the blank meter faces. We will discuss the results after you have them filled in.

CURRENT FLOW IN A PARALLEL CIRCUIT

Let us look at the ammeter reading first. Did you notice that it was about two times the ammeter reading you measured in the first series circuit in Chapter 9? Why is this? Simply, each bulb has the full B+ and B– applied across it. Do you remember the series circuit? With the maximum voltage applied across the load, the maximum current was pushed through the load.

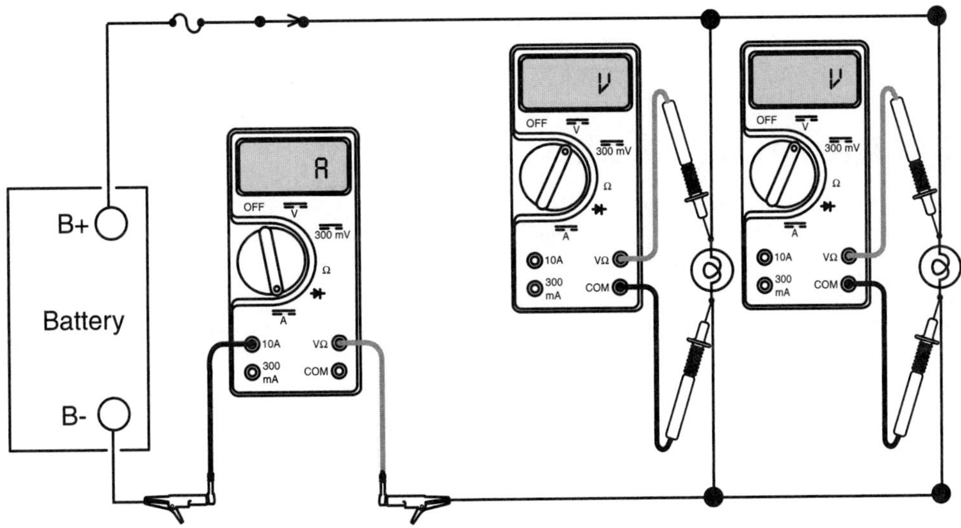

Figure 1. The DMMs set up to measure voltage and current flow.

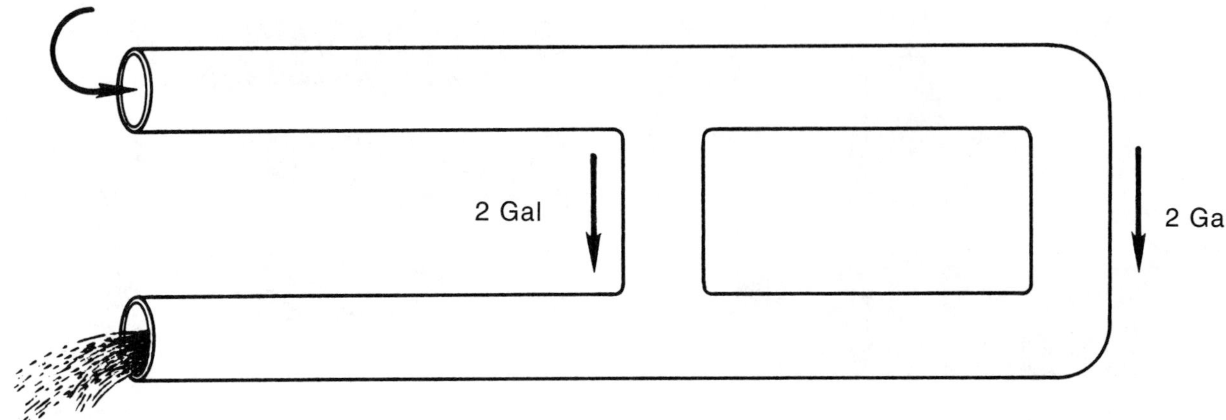

Figure 2. Water and current behave in the same manner.

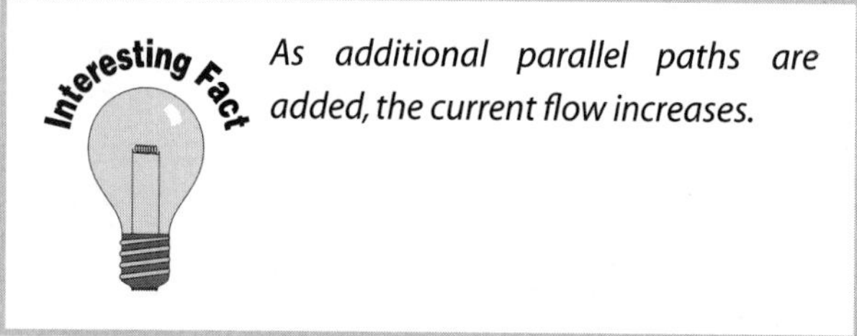

As additional parallel paths are added, the current flow increases.

This means that each leg of the parallel circuit has full current flowing through it and that the total current flow is the addition of all legs together. This relates to the water system shown in **Figure 2**. The top pipe has 4 gallons flowing because each pipe flowing down has 2 gallons flowing: 2 gallons in each leg equals 4 gallons total. Current responds in much the same way. The current in the main leg will have the total of the two parallel legs. If each leg has full B+ and B– applied, it will draw full current, or the same current when it was alone across the power source. Add the current together because two loads are each drawing the current they want.

RESISTANCE IN A PARALLEL CIRCUIT

Let us think about the current in terms of the resistance. If the current went up or doubled, the resistance must have gone down by half. Does this make sense to you? Let us do some resistance problems to prove that the resistance did go down. Do you remember the formula?

$$\frac{\text{voltage}}{\text{amperage}} = \text{resistance}$$

Let us say that we wire one bulb across a 12-volt power source and the ammeter reads 3 amps of current. Using Ohm's Law, let us figure the resistance:

$$\frac{12\text{ volts}}{3\text{ amps}} = 4\text{ ohms}$$

To continue the example, if we were to wire up an additional bulb of the same resistance, it would draw 3 amps of current also. The total current flow for the circuit is now 6 amps (3 amps for each leg). Figure the total resistance for the circuit. Use the total amperage and the applied voltage:

$$\frac{12\text{ volts}}{6\text{ amps}} = 2\text{ ohms total}$$

The resistance is one half. This is the reason that the current doubled.

Now let us use a formula to add up the individual resistances. The total should equal our previous total. If you need help, ask your instructor. We will be using a form of Ohm's Law that states how resistances are totaled when they are wired in parallel. Keep in mind that you cannot just add them together as you did in a series circuit. Here is the formula:

$$\text{Total resistance } (R_t) = \frac{1}{1/R_1 + 1/R_2 + 1/R_3 + \ldots}$$

For our example with two resistances of 4 ohms each:

$$R_t = \frac{1}{1/4 + 1/4}$$

The next thing we must do is add the $1/4$ to the $1/4$:

$$1/4 + 1/4 = 2/4, \text{ or } 1/2$$

This now gives us:

$$R_t = \frac{1}{1/2}$$

Working again with the denominator $1/2$, we can either divide it directly into the numerator and get 2 ohms total:

$$\frac{1}{1/2} = 2$$

or we can convert the $1/2$ to the decimal 0.5:

$$R_t = \frac{1}{0.5}$$

We then divide the 1 by 0.5 to reduce the fraction to its lowest form and find that the total resistance is 2 ohms.

Notice that this is the same total that we found when we figured the total resistance by dividing the applied voltage by the total current flow in amps.

Figuring total resistance by looking at individual resistances and using the previous formula will show that, as additional resistances are added in parallel, the total will always go down to a level lower than the lowest resistance. Add an additional bulb to our two-bulb circuit. Measure the amperage and the applied voltage. Insert the figures into the formula below to find the total resistance.

$$\frac{\text{volts}}{\text{amps}} = \text{ohms of total resistance}$$

Did you notice that the total was less than the individual resistance we figured before? Notice that in parallel circuits the current divides and the voltage remains the same throughout the circuit. This is the opposite of what we found in series circuits. In series circuits, the voltage divided, and the current was the same throughout the circuit.

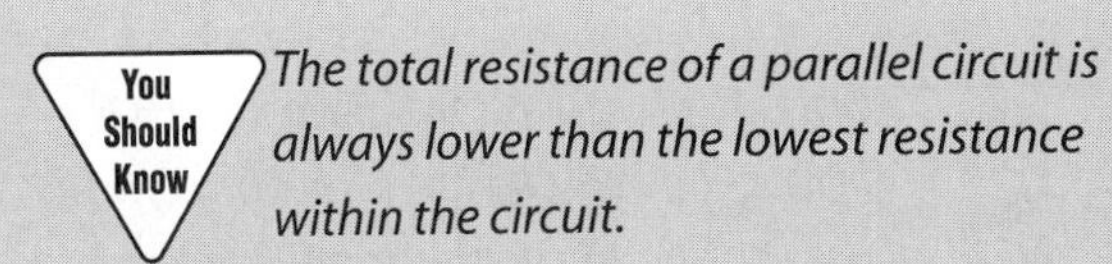

The total resistance of a parallel circuit is always lower than the lowest resistance within the circuit.

Let us try an example with different resistances wired in parallel and determine the current flow. **Figure 3** shows four resistances of 2, 4, 6, and 8 ohms wired in parallel with 12 volts applied. Follow along as we work through the math for total resistances in parallel:

$$(R_t) = \frac{1}{1/2 + 1/4 + 1/6 + 1/8}$$

To be able to add the fractions in the denominator, we will have to find the common denominator for all. This is the number that all fractions can be changed into. For our example, 24 is the lowest common denominator. It is the lowest number that all of our fractions will evenly divide into. This now give us:

$$(R_t) = \frac{1}{12/24 + 6/24 + 4/24 + 3/24}$$

Or

$$R_t = \frac{1}{25/24}$$

By dividing the 25 by 24 we get 1.04 as the denominator. Our fraction now looks like this:

$$R_t = \frac{1}{1.04}$$

By dividing 1 by 1.04, we see that the total resistance of a circuit with 2, 4, 6, and 8 ohms of resistance wired in parallel is 0.96 ohms.

Now that we know the total resistance, we can figure the total current flow:

$$\frac{12 \text{ volts}}{0.96 \text{ ohms}} = 12.5 \text{ amps}$$

Our circuit has 12.5 amps of current flowing through it. We can take this example one step further and look at the current flow through each of the legs of the circuit. If we do it correctly, the individual legs should equal 12.5 amps when added together.

As each leg of the circuit has 12 volts applied, all we have to do is divide the voltage applied by the individual resistances to figure current flow.

$$\frac{12 \text{ volts}}{2 \text{ ohms}} = 6 \text{ amps}$$

$$\frac{12 \text{ volts}}{4 \text{ ohms}} = 3 \text{ amps}$$

$$\frac{12 \text{ volts}}{6 \text{ ohms}} = 2 \text{ amps}$$

$$\frac{12 \text{ volts}}{8 \text{ ohms}} = 1.5 \text{ amps}$$

$$6 + 3 + 2 + 1.5 = 12.5 \text{ amps total}$$

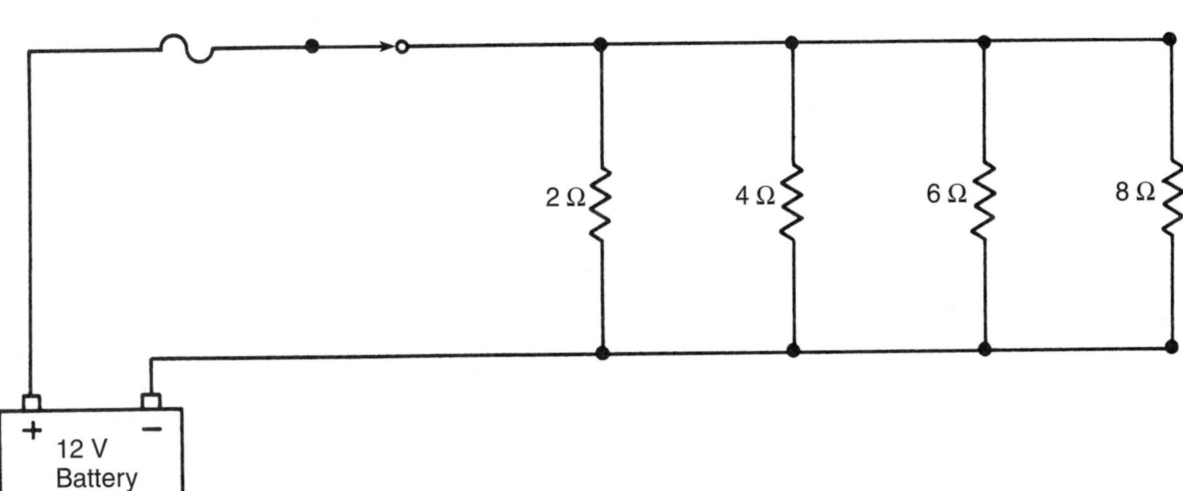

Figure 3. The total resistance of loads in parallel will always be lower than the lowest individual resistance.

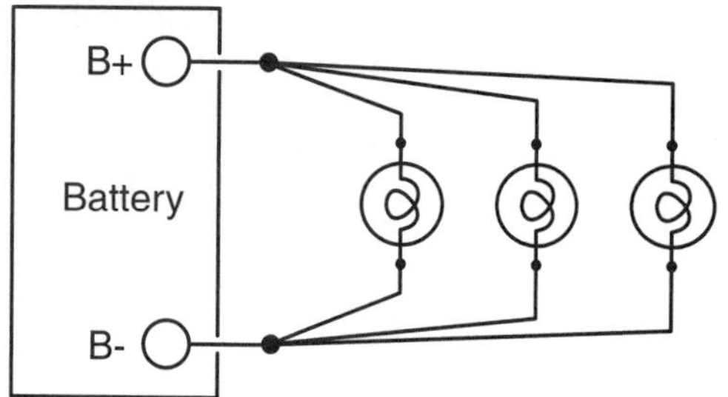

Figure 4. All parallel loads have full power applied.

VOLTAGE IN A PARALLEL CIRCUIT

If you go back to any of the parallel circuits that we have wired up, one thing should have been obvious. Each load in parallel had full voltage applied to it. The explanation of this is simple and involves the principles diagrammed in **Figure 4**. Notice that all loads have their own path back to the battery.

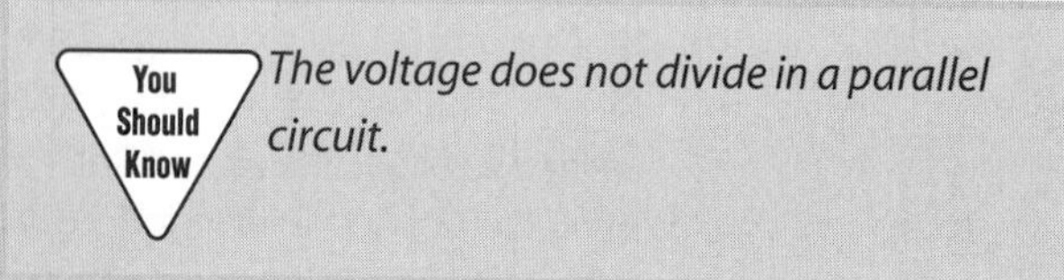

The voltage does not divide in a parallel circuit.

As long as the main conductors can handle the total current flow that all loads demand, a parallel circuit will apply full voltage to all legs. To apply this example to the vehicle is simple. If you turn on the headlights and open the trunk, the taillight will have the same voltage as the battery. This is because the taillight is wired in parallel with the rest of the loads. The conductors that carry current to the taillight are large enough that there is no voltage drop. It is as if the taillights were wired directly with their own conductor to the battery. Conductors should not offer resistance to the flow of current. They should bring the current to the loads where it will do the work. If you suspect there is resistance in the wiring or connections to a load, it is an easy task to measure the voltage across the load and compare it to the battery voltage. Those voltages should be the same or extremely close. Any voltage reduction from battery voltage signifies that there must be resistance present. Your job as a technician will frequently be to find the resistance and correct it. Correcting is another way of saying reduce the resistance so that full current can flow to the load. The load, not the corroded connection that is adding series resistance to the circuit, must determine current flow.

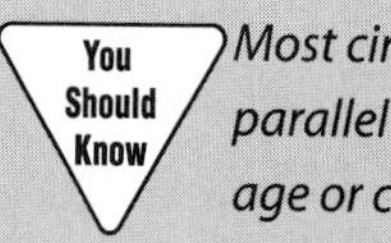

Most circuits on the vehicle are wired in parallel and should have full battery voltage or charging voltage applied to them.

Additional problems for you to try are at the end of this chapter. You may wish to try them now, while the math of parallel circuits is still fresh in your mind. Feel free to use a calculator, because sometimes the numbers can be cumbersome to work with.

Summary

- Current divides in a parallel circuit.
- Voltage remains the same in all legs of a parallel circuit.
- Adding additional resistances in parallel will reduce the total circuit resistance.
- The total resistance in a parallel circuit is always lower than the lowest resistance.
- To calculate the total current flow, add the individual parallel circuit's current.
- To calculate the total resistance, the parallel resistance formula must be used.
- In the vehicle's electrical system, most circuits are parallel.
- The voltage applied to most circuit loads will be battery voltage or charging voltage if the vehicle is running.

Review Questions

1. Three 10-ohm resistors are wired in parallel. What is the total resistance?
   A. 30 ohms
   B. 10 ohms
   C. 3.3 ohms
   D. 33.3 ohms
2. How much current will flow through the circuit described in review question 1 if 12 volts is applied to it?
   A. 3.6 amps
   B. 39.6 amps
   C. 0.275 amps
   D. none of the above

3. A taillight bulb is glowing dimly. A voltmeter across the bulb reads 7.4 volts. Technician A states that there must be a voltage drop somewhere else in the circuit. Technician B states that resistance in parallel must be present. Who is correct?
   A. Technician A only
   B. Technician B only
   C. Both Technician A and Technician B
   D. Neither Technician A nor Technician B
4. To find the total current flow in any circuit,
   A. multiply the amperage times the voltage
   B. divide the voltage by the amperage
   C. multiply the voltage times the resistance
   D. divide the voltage by the resistance
5. Resistors of 2, 3, and 4 ohms are wired in parallel. Their combined resistance is
   A. 9.00
   B. 0.93
   C. 1.08
   D. 2.16
6. How many amps will flow through the three resistances described in review question, if you apply 12 volts to them?
   A. 11.1
   B. 6.4
   C. 3.1
   D. 12.90
7. Voltmeters are always wired in parallel.
   A. True
   B. False
8. Ammeters are always wired in parallel.
   A. True
   B. False
9. The total amperage flowing in a parallel circuit is equal to each of the individual legs.
   A. True
   B. False
10. Why is the total resistance in a parallel circuit always lower than the lowest resistance?
11. Why will current increase when additional resistance is added in parallel?

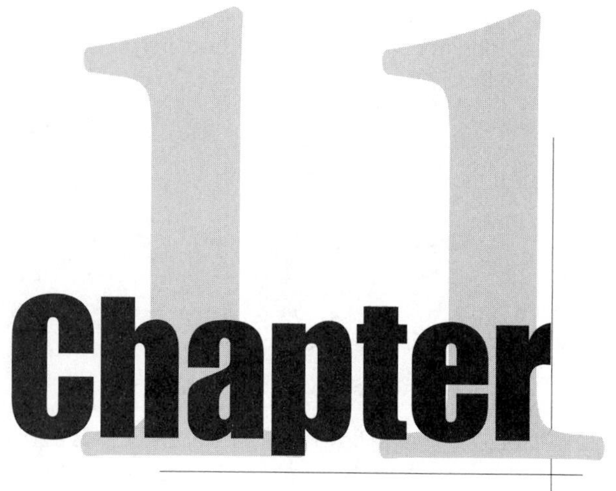

Series-Parallel Circuits

Introduction

We have looked at series circuits and parallel circuits. You can see that parallel circuits are the most common because of the independence that they offer. There are few examples of series circuits in vehicles. The application of series-parallel circuits is even less. However, no discussion of electrical principles would be complete without looking at all three types. We examine how the circuit functions, how the voltage and current behave, and how the resistance combines. We also look at the dashboard dimmer circuit, which is the most common series-parallel circuit found on most vehicles.

SERIES-PARALLEL CIRCUITS

If you look at most circuits, you will find that they have a series-parallel characteristic because they contain some sections of the B+ and B– limited to one or two main feed wires or cables before branching into parallel sections. The battery cables are the series section, and the individual loads are the parallel section. We will wire up a simple series-parallel circuit to see how voltage and current behave.

Wire up the circuit in **Figure 1**. Take your voltmeter and ammeter readings and record them. Once you have completed this, move bulb 1 into the B– series section of the circuit and retake your voltage and ammeter readings. Do not forget to record them, because we are going to refer back to them. Use the same-sized bulbs.

When you have the numbers filled in and are ready to analyze them, let us look at the parallel section first, which is fairly straightforward. The current divided, and the voltage to each leg remained the same. Notice, though, that the voltage across the loads is not the applied circuit voltage. This is because, as far as the applied voltage was concerned, it viewed the parallel two loads as one single resistance. Look at ammeter 2 and one of the parallel voltmeters. Use Ohm's Law to figure out the total resistance of the parallel circuit.

$$\frac{\text{volts}}{\text{amps}} = \text{ohms (total parallel)}$$

If the total resistance of the parallel section looks like one single resistance of the value you have just figured, the voltage will divide between the first load (the series bulb) and the second series load (which is actually two loads in parallel). Look at **Figure 2** for a schematic representation of this.

Going back to the original question, why did the parallel section not have the fully applied circuit voltage across its loads? The voltage had to divide between the load in series and the two loads in parallel because the parallel loads were in series with the first load. Be sure to reread that sentence a couple of times until it makes sense.

Notice in the second series-parallel circuit that the numbers remained the same. In other words, moving the series load around had no effect. Resistance in series will cause the same voltage division if it is in the B+ section or the B– section. Remember that the battery sees two series resistances to divide its voltage between. It does not respond to their positions. The voltage division takes place because of the resistance of each load as a percentage of the total resistance. If a resistance is 25 percent of the total, it will drop the applied voltage by 25 percent. As long as the total does not change, altering the position will have no effect on the results. Prove this to yourself by wiring up

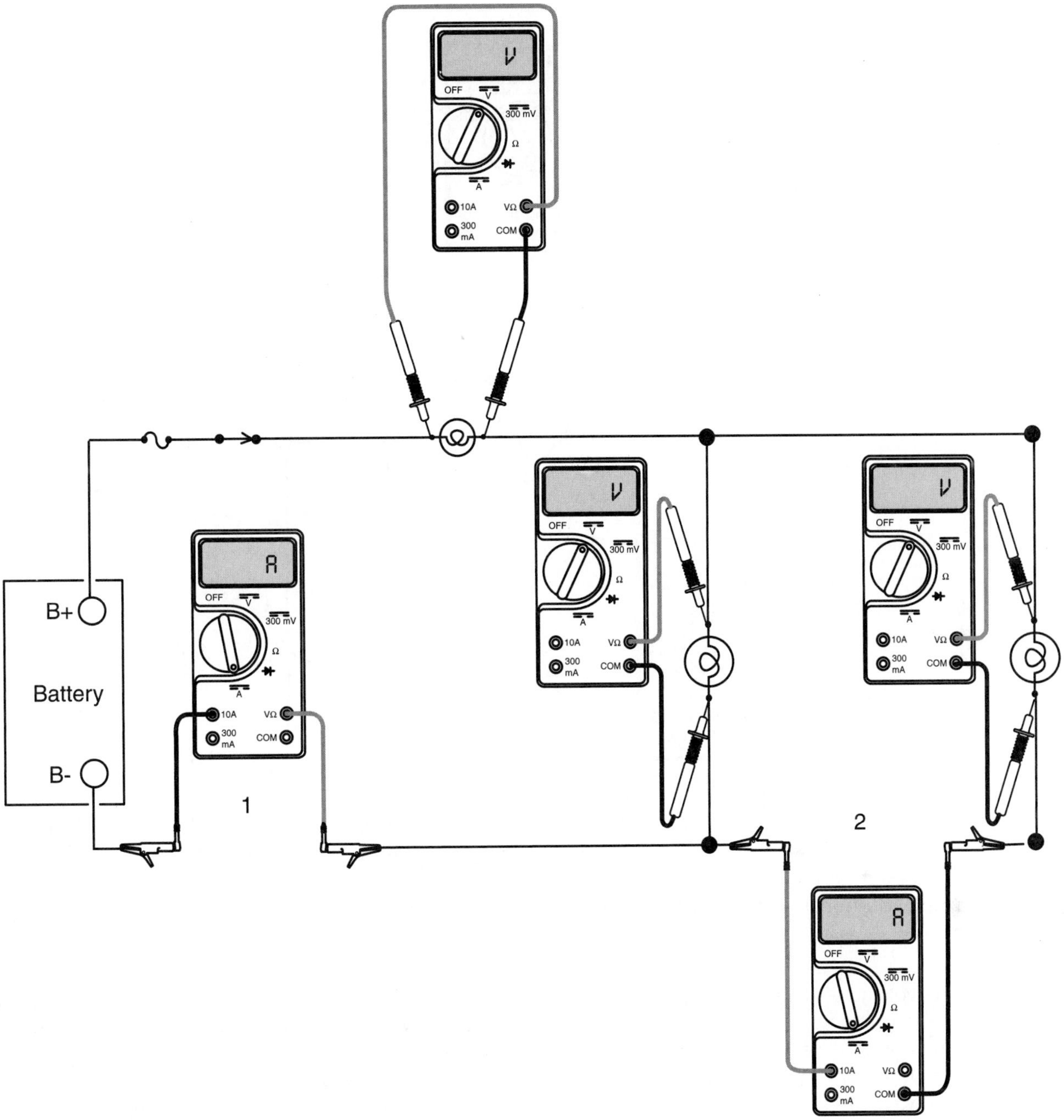

Figure 1. Voltmeters and ammeters installed to indicate circuit conditions.

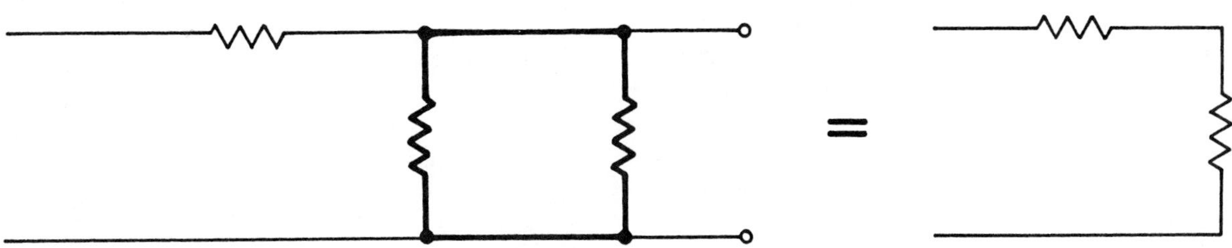

Figure 2. Two resistors in parallel are "seen" as one resistor in series.

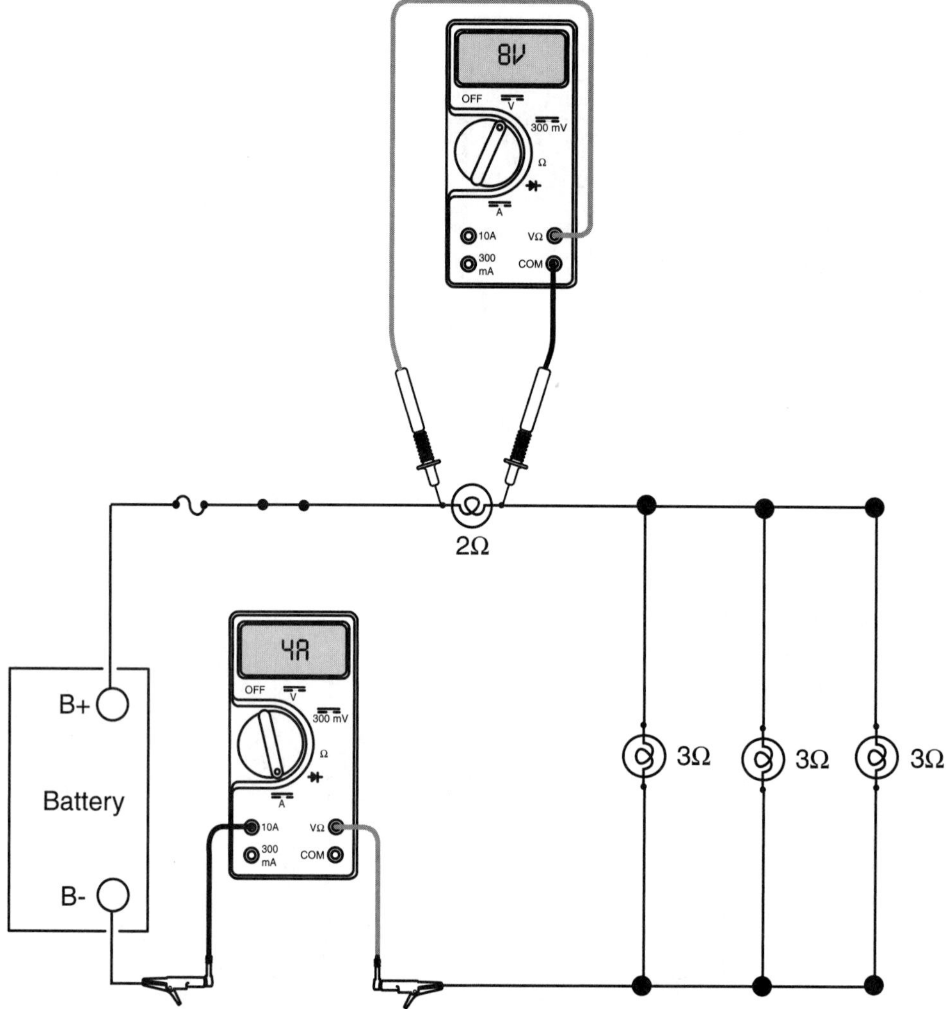

Figure 3. Adding additional parallel resistance will increase the series voltage drop.

Figure 3. Take the voltage readings and figure out what happened. In case you are not wiring up these circuits, we have put some numbers in. The voltage drop of the first load is now greater. What happened? By adding another resistance in parallel, we reduced the total parallel resistance and the total circuit resistance. At the same time, we did not change the resistance of the series section (bulb 1), so it became a greater percentage of the total resistance and experienced a greater voltage drop across it.

The important thing to remember is that resistance, when placed in series, will drop down all applied voltage to the parallel loads and reduce the current that the loads draw. If you experience a voltage drop before a parallel load, examine the wiring diagram and look for a similar problem on the other loads wired in parallel. If all loads have the same reduced voltage, the excessive resistance or false load must be in the series part of the circuit. This little, seemingly simple hint, can save you hours of diagnosis time. Many times we can diagnose the vehicle almost completely from the wiring diagram by being alert to how the circuit is wired and what the voltage reading is across the parallel loads.

Let us try another example with resistance values and see how Ohm's Law would be applied. The circuit shown in **Figure 4** has a 6-ohm load in series with 2- and 4-ohm loads in parallel; 12 volts is applied. What is the total resistance of the circuit? Let us figure the parallel section first:

$$R_t = \frac{1}{1/2 + 1/4}$$

The lowest common denominator is 4:

$$R_t = \frac{1}{2/4 + 1/4} = \frac{1}{3/4}$$

4 is divided into 3 and equals 0.75.

$$R_t = \frac{1}{0.75} = 1.33 \text{ ohms total (parallel section)}$$

Notice that the 1.33 ohms is lower than the lowest resistance (2 ohms).

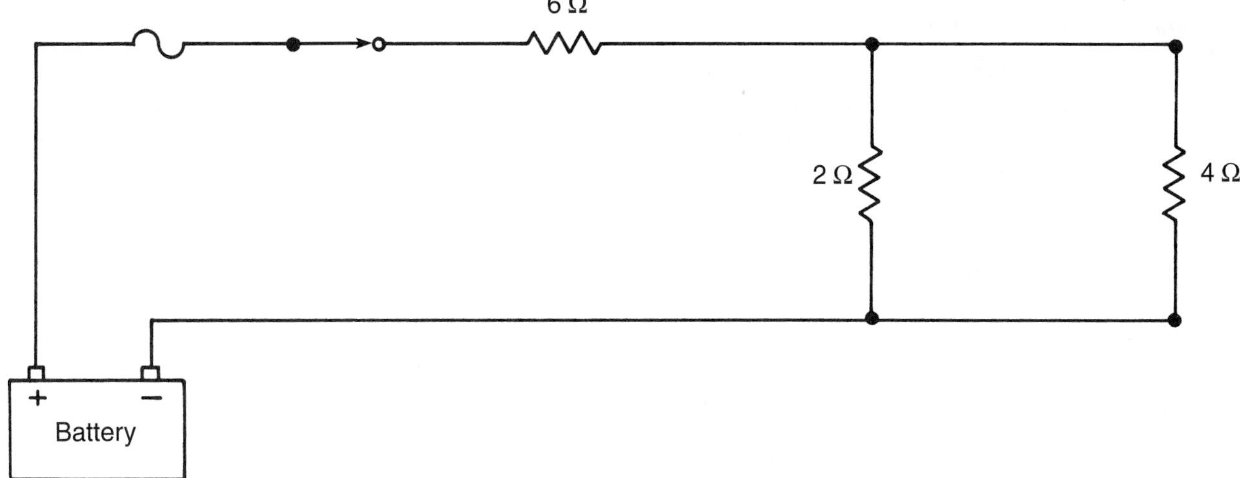

Figure 4. What is the total resistance of the series-parallel circuit?

The total resistance of the circuit is the sum of 1.33 ohms (parallel) plus 6 ohms (series), or 7.33 ohms total in the circuit. Total amperage can now be figured:

$$\frac{12 \text{ volts}}{7.33 \text{ ohms}} = 1.65 \text{ amps}$$

The total resistance of 7.33 ohms allows 1.65 amps of current to flow. Do not forget that this current will flow through the series section of the circuit but will split or divide as it goes through the parallel part of the circuit. With the information we now have, we can figure the rest of the unknowns. What is the voltage applied to the parallel circuit legs? Figuring the voltage to the parallel circuit will involve subtracting the voltage drop of the series from the total applied voltage. We will figure the drop again by using Ohm's Law, our previous 1.65 amps, and the resistance of 6 ohms:

$$6 \text{ ohms} \times 1.65 \text{ amps} = 9.8 \text{ volts}$$

Next, we subtract the series voltage drop from the applied and we have our answer:

$$12 \text{ volts} - 9.8 \text{ volts} = 2.2 \text{ volts}$$

The 2.2 volts is being applied to the parallel sections of the circuit because the series section dropped the voltage down by 9.8 volts.

How many amps of current are flowing through each leg of the parallel section? This question will rely on previous information. We now know that the parallel legs have 2.2 volts applied, and we also know the individual resistances. Ohm's Law again will tell us the individual amperages as follows:

$$\frac{2.2 \text{ volts}}{2 \text{ ohms}} = 1.1 \text{ amps}$$

$$\frac{2.2 \text{ volts}}{4 \text{ ohms}} = 0.55 \text{ amps}$$

Based on those calculations, 1.1 amps plus 0.55 amps equals 1.65 amps total in the parallel section of the circuit. Again, our circuit is complete with all numbers matching.

Additional series-parallel voltage, resistance, and current questions are at the end of this chapter. By working through them you will become more confident in Ohm's Law and will have a greater understanding of what is happening within a circuit.

DASHBOARD DIMMER CIRCUIT

Most vehicles have a good example of a series-parallel circuit in the instrument cluster or dashboard lights. A dimmer allows the driver to make the lights bright or dim or any gradation in between. How does this work? This works by applying Ohm's Law to a series-parallel circuit. **Figure 5** illustrates this. Notice that we have six lights wired in parallel to one resistance in series.

Do you notice anything different with the series resistance? It has an arrow through it. The arrow signifies that this resistance is not fixed but is variable. This is the dimmer control that will be adjusted to change the brightness of the bulbs. In our example with six lights of approximately 24 ohms each, the dimmer will have a range from zero ohms to an open circuit. When the dimmer control is in its open position, the lights are off; when the control is in its zero-ohms

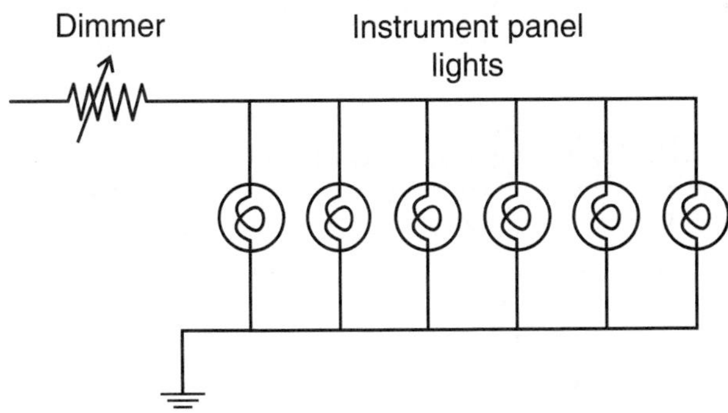

Figure 5. The dimmer adds resistance in series to dim the parallel lights.

The dashboard dimmer control will offer resistance, which means it will generate some heat and magnetic field. The heat is usually sufficient for the control to be warm.

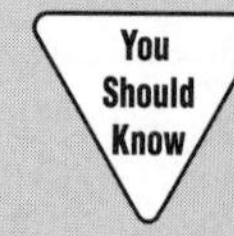

The series section of a series-parallel circuit causes a voltage drop and reduces the current flow to the parallel section of the circuit.

position, the lights are on full brightness. Let us dial up 4 ohms on the dimmer and see what happens. First figure out the total resistance of the parallel circuit. We use a slightly different method here where we figure the total current flow through each leg of the parallel circuit if 12 volts is applied. Then we multiply this total by six, which is the number of bulbs. Once the total current is figured, dividing it into the voltage gives us the total resistance. Each resistance draws 0.5 amps

$$12 \text{ volts}/24 \text{ ohms} = 0.5 \text{ amp}$$
$$0.5 \text{ amp} \times 6 \text{ bulbs} = 3 \text{ amps}$$
$$12 \text{ volts}/3 \text{ amps} = 4 \text{ ohms } R_t$$

The resistance of the parallel section is 4 ohms.

Our parallel circuit has a total resistance of 4 ohms, which follows what we have previously said: it is lower than the lowest resistance of 24 ohms. If the dimmer is set for 4 ohms, then the total resistance of the circuit is 8 ohms. The next thing we need to figure is the amount of current flow. If 12 volts are applied to the circuit, then:

$$\frac{12 \text{ volts}}{8 \text{ ohms}} = 1.5 \text{ amps}$$

How bright are the parallel bulbs with 1.5 amps flowing? Each bulb will get only ⅙ of the total:

$$\frac{1.5 \text{ amps}}{6} = 0.25 \text{ amps per bulb}$$

Our parallel bulbs are receiving only 0.25 amps per bulb. How bright are the bulbs? The answer to this question requires the voltage to the parallel section. We figure it by knowing the voltage applied. Going back to the series section, we know the current flow (1.5 amps) and we know the resistance (4 ohms). Multiplying the amperage by the resistance gives us 6 volts:

$$1.5 \text{ amps} \times 4 \text{ ohms} = 6 \text{ volts}$$

If 6 volts will be dropped through the series circuit, then that will leave 6 volts applied to the entire parallel section. The brightness of the bulbs will be its wattage, which you should remember is the volts multiplied by the amps:

$$0.25 \text{ amps} \times 6 \text{ volts} = 1.5 \text{ watts}$$

This is not very bright. However, we have six bulbs, which means that the total wattage of the dash will be 9 watts:

$$1.5 \text{ watts} \times 6 \text{ bulbs} = 9 \text{ watts}$$

This means that all of the bulbs combined together will generate only 9 watts. This dash should be quite dim. Contrast this to the total with the dimmer offering no resistance. Each bulb will have the full 12 volts across it, so:

$$\frac{12 \text{ volts}}{24 \text{ ohms}} = 0.5 \text{ amp} \times 12 \text{ volts} = 6 \text{ watts each}$$

The total brightness of the dash would now be 36 watts

$$6 \text{ watts} \times 6 \text{ bulbs} = 36 \text{ watts}$$

That is quite a difference, isn't it?

Summary

- A series-parallel circuit has some of the characteristics of a series circuit and some of a parallel circuit.
- The voltage will divide in the series section.
- The current will divide in the parallel section.
- The current will be the same in the series section.
- The voltage will be the same in the parallel section.
- The circuit total resistance is the series resistance plus the total of the parallel resistances.
- A dimmer circuit works by creating a series voltage division and reducing the current flow to the parallel section of the circuit.

Review Questions

1. A 2-ohm resistor is wired in series with two 7-ohm resistors in parallel. The total resistance is
   A. 3.5 ohms
   B. 5.5 ohms
   C. 6.9 ohms
   D. unknown because not enough information is given
2. A taillight bulb is glowing dimly. A voltmeter across the bulb reads 7.4 volts. Technician A states that there must be a voltage drop somewhere else in the circuit. Technician B states that resistance in parallel must be present. Who is correct?
   A. Technician A only
   B. Technician B only
   C. Both Technician A and Technician B
   D. Neither Technician A nor Technician B

3. Technician A states that the resistance in a series circuit is added together to find the total. Technician B states that the total resistance in a parallel circuit is always lower than the lowest single resistor. Who is correct?
   A. Technician A only
   B. Technician B only
   C. Both Technician A and Technician B
   D. Neither Technician A nor Technician B
4. Current flow in the parallel section of a series-parallel circuit will
   A. be the same in each leg
   B. divide, based on the resistance of each leg
   C. will be based on applying full voltage to the parallel section
   D. be the same as the voltage
5. Technician A states that the voltage will divide in the parallel section of a series-parallel circuit. Technician B states that the current will divide in the series section of a series-parallel circuit. Who is correct?
   A. Technician A only
   B. Technician B only
   C. Both Technician A and Technician B
   D. Neither Technician A nor Technician B
6. Increasing the resistance of the dimmer will increase the brightness of the dash.
   A. True
   B. False
7. The dimmer is wired in parallel to the series bulbs in the typical dash circuit.
   A. True
   B. False
8. If one bulb burns out in the dash circuit, the entire circuit is dead.
   A. True
   B. False

Chapter 12 Control Circuits

Introduction

Most of the circuits that we have looked at have had a control. The object of the control is simple: to allow the circuit to be turned on or off as needed. The switch is by far the simplest of controls. In this chapter, we look at the switch again and then look into the use of relays as control devices. We will end with a discussion of some solid-state timers or devices that will electronically turn a circuit on or off.

THE SWITCH AS A CONTROL DEVICE

The simplest of all controls is the switch. At the touch of the switch, the circuit begins to function. The ignition switch is a good example. Turn it to the start position and the starting circuit cranks and starts the engine. Leave it in the run position and the engine continues to run. When you reach your destination, the ignition switch is turned to the off position and the engine stops running. Actually the ignition switch is not just a single switch, as that in **Figure 1**, but a series of switches to control the starting, ignition, fuel, and computer circuits. From a resistance standpoint, consider that the switch as a control has only two positions: no resistance and infinity. In the on position, the resistance is zero; in the off position, the resistance is infinity. Remember that infinity is another way of saying an open circuit, and that no current flows through an open circuit. **Figure 2** illustrates typical ohmmeter readings with the switch open and closed. OL on the display is the indication of infinity, especially in the auto-range position.

It is possible to place the control in just about any position within the circuit. **Figure 3** has three switches—one in series, two in parallel. Of the two in parallel, notice that one controls B+ and the other one is in the ground circuit. Many of the circuits on today's vehicles are ground controlled. It does not matter where the control is, as long as it can turn off

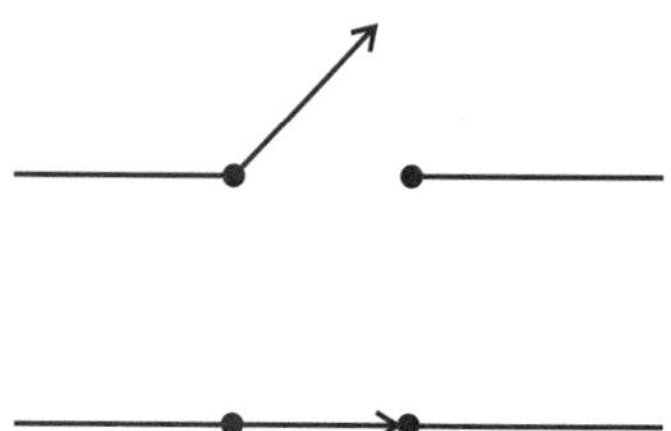

Figure 1. A switch can be either open (off) or closed (on).

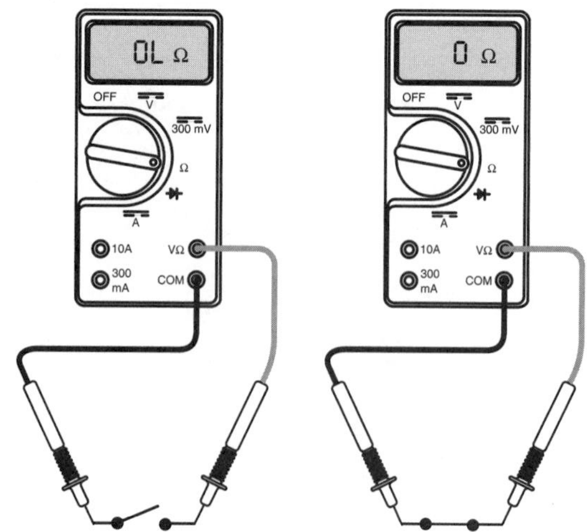

Figure 2. An open switch has an infinity reading. A closed switch has a zero-ohm reading.

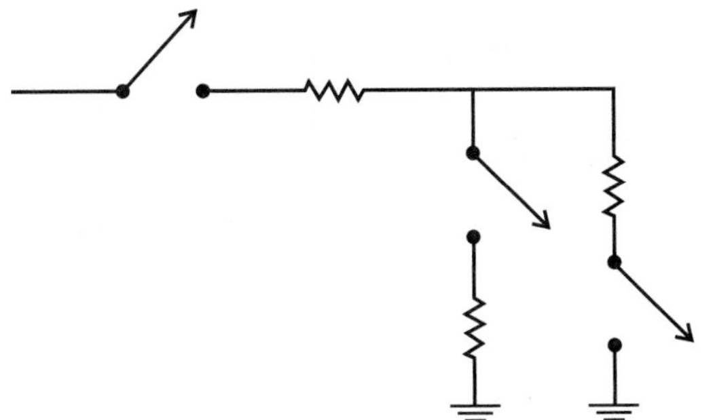

Figure 3. The control can be placed in the power or ground circuit.

one side of the circuit. The circuit will not have any current flow if either the power or the ground is eliminated.

RELAYS AS CONTROL DEVICES

Frequently, on modern vehicles, the current draw of the accessories is beyond the capacity of the ignition switch and sometimes even the normal wiring. Under these circumstances, the manufacturer will use a relay. Simply defined, a relay is a device that uses low current to control high current. Look at **Figure 4**.

Figure 4 is a simple diagram of a common electromechanical relay. The relay coil has very high resistance, which means it will draw very low current. This low current will be used to produce a magnetic field. This field will close two electromechanical contacts. The contacts are designed to carry heavy current. When we apply B+ and B– to the relay coil, the contacts will close and direct heavy B+ to the main load that we want to control.

The simplest application of relays in vehicles is the horn relay circuit. It is quite common to find it wired as in **Figure 5**. Notice that B+ is applied directly to the coil. In the automobile, if you follow the B– side of the relay coil, you will notice that it comes through the fire wall, which separates the engine compartment from the passenger compartment, and goes up the steering column to the horn button. One side of the horn button is grounded, and pushing it in will apply a ground to the B– side of the horn relay coil. With both B+ and B–, current flows through the coil and develops a magnetic field. The field will close the contacts that direct B+ to the horn. Because the other side of the horn is grounded, it will sound. Used in this manner, the horn relay becomes a control of the heavy current necessary to blow the horn. The control circuit can be wired with very thin wire because it will not have much current flowing through it. Horn relays might have as much as 50 ohms of resistance across their coils that will convert to approximately 1/4 amp flowing through the control circuit. By contrast, horns might have less than 1 ohm and might have two wired in parallel that will convert to 24 or more amps flowing through the relay coil contacts.

Another circuit that can use a relay is the blower motor circuit. Many blower motors use a series of voltage-dropping resistors for the lower speeds and a relay for maximum speed. This is frequently done on air-conditioned vehicles where the blower motor has to move large quanti-

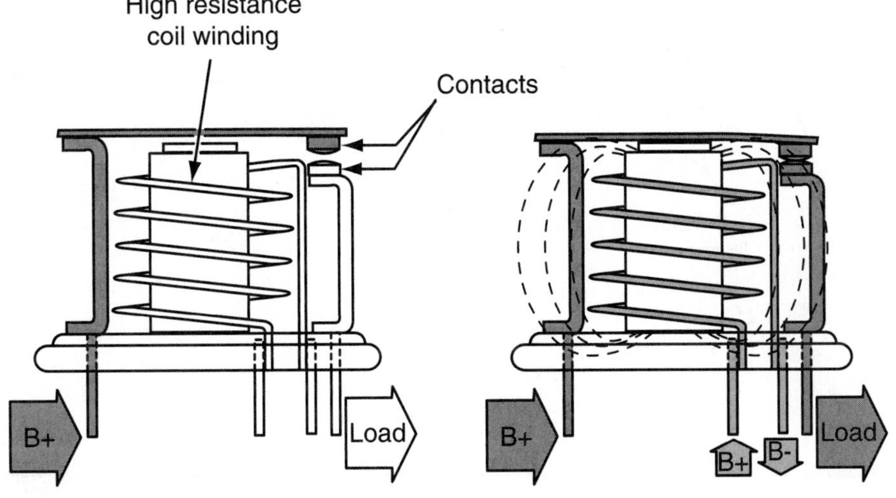

Figure 4. A relay uses low current to control high current.

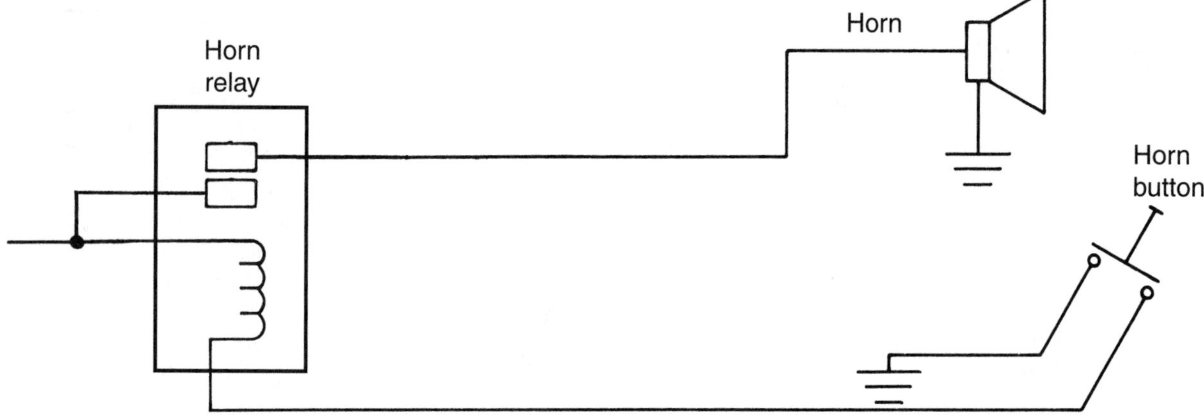

Figure 5. The horn relay contacts carry the high current.

ties of air. Start with the resistor shown in **Figure 6**. It is normal for automotive manufacturers to use a heating coil wired in series to control the lower speeds of the blower motor. Remember that series resistance will drop the voltage and reduce the current to the motor. Lower current will cause the motor to go slower. We use the switch to control how much of the resistance is wired in series. Look at the diagram in **Figure 7**. Notice that we have now used the resistor and the switch in series to control the current. Moving the switch from off to low brings current to the motor through conductor A and all three sections of the resistor. This means that the three sections of the resistor will cause the maximum voltage drop before the motor. Moving the switch up to medium brings current through conductor B and only two of the three resistors. With less resistance and more current flowing, the motor runs faster. Moving the switch to high brings current through conductor C and uses only one resistor in series. You have three speeds by using three different series resistances before the motor.

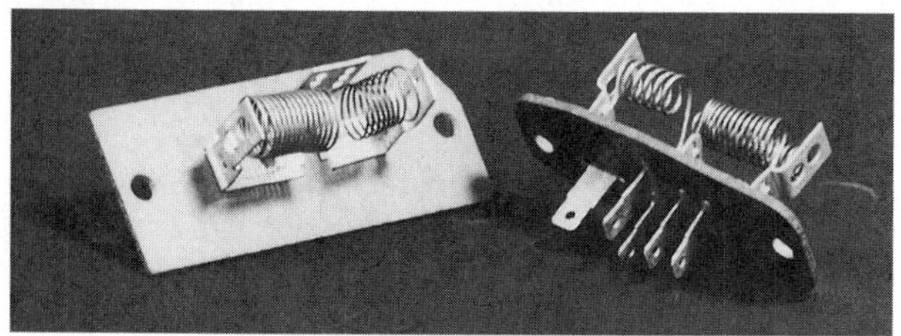

Figure 6. Common blower motor resistors.

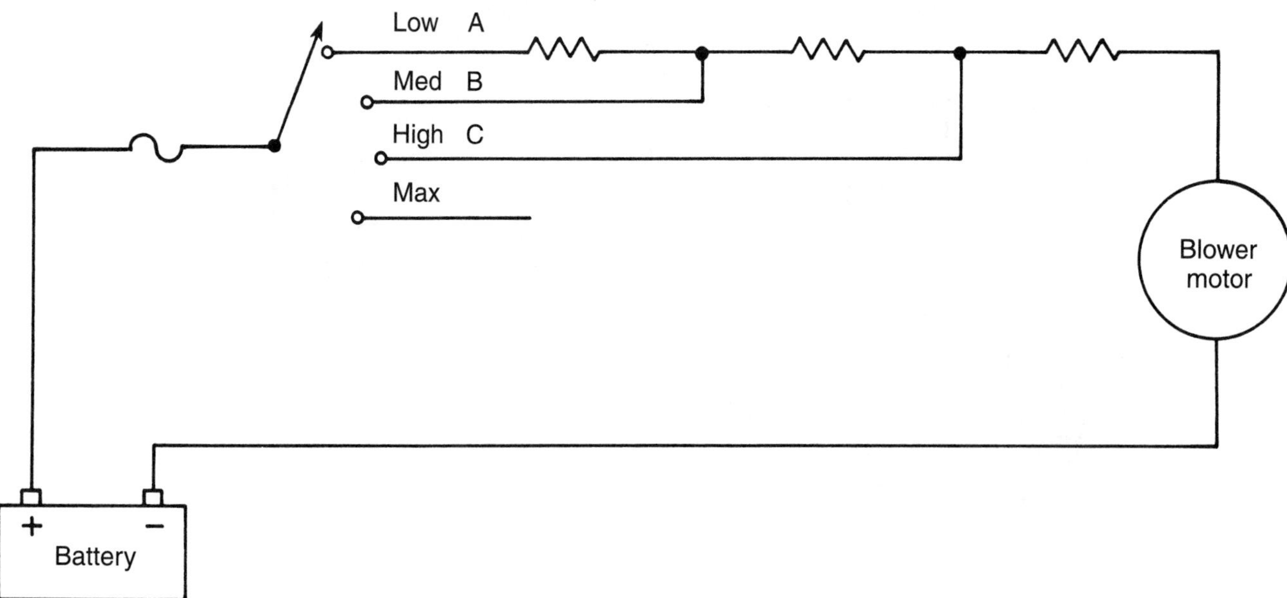

Figure 7. The three resistors vary the speed of the motor.

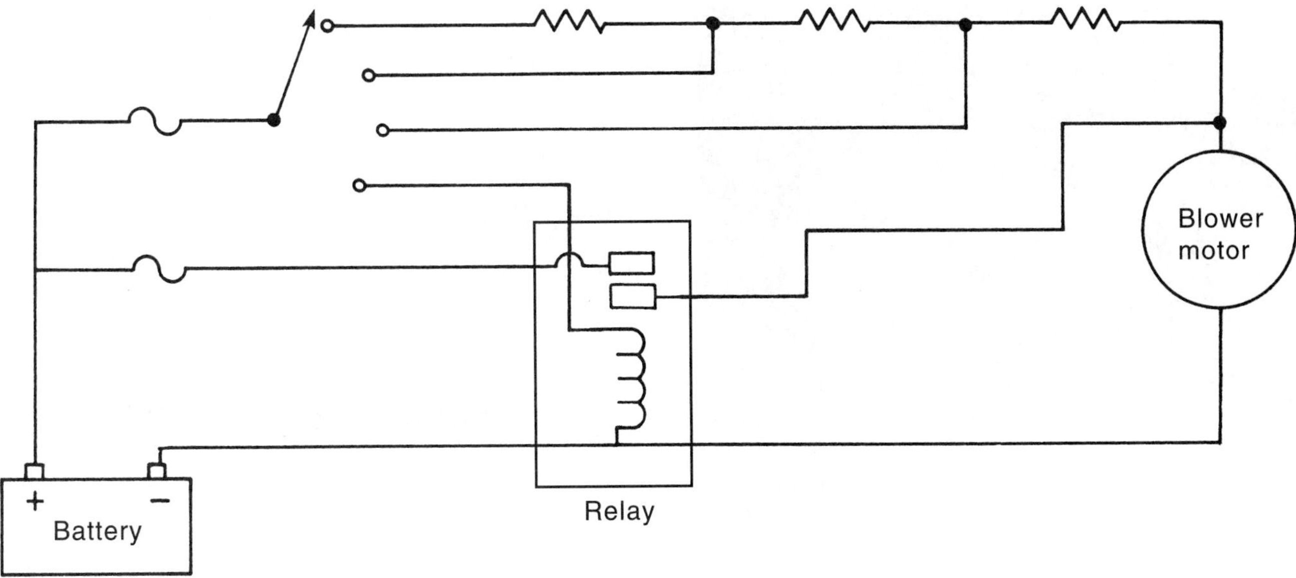

Figure 8. A high-speed relay wired into the blower circuit.

Now for the relay look at **Figure 8**. The conductor from the maximum position of the switch runs to the relay coil. The other end of the coil goes to B–. When the switch is placed in the maximum position, current flows through the coil, produces a magnetic field, and closes the contacts. Let us add a fused B+ conductor to one contact and the motor to the other one and you can see that, when the contacts close, the motor runs at full speed **(Figure 9)**. No series resistance equals maximum current flow.

The current through the switch is now lowest when the switch is in the maximum position. The relay coil has very high resistance, which reduces the current through the switch. Maximum current flow for the motor will go through the relay contacts when they are closed. What this effectively does is eliminate the need for the heavy and expensive contacts and conductors in the dashboard, where the switch is located. The relay is usually placed under the hood in a direct line between the battery and the blower motor in a relay center, as shown in **Figure 10**.

Relays are in frequent use on today's vehicles. With all the heavy current devices being used, there is a move to eliminate some of the current flowing through the switches

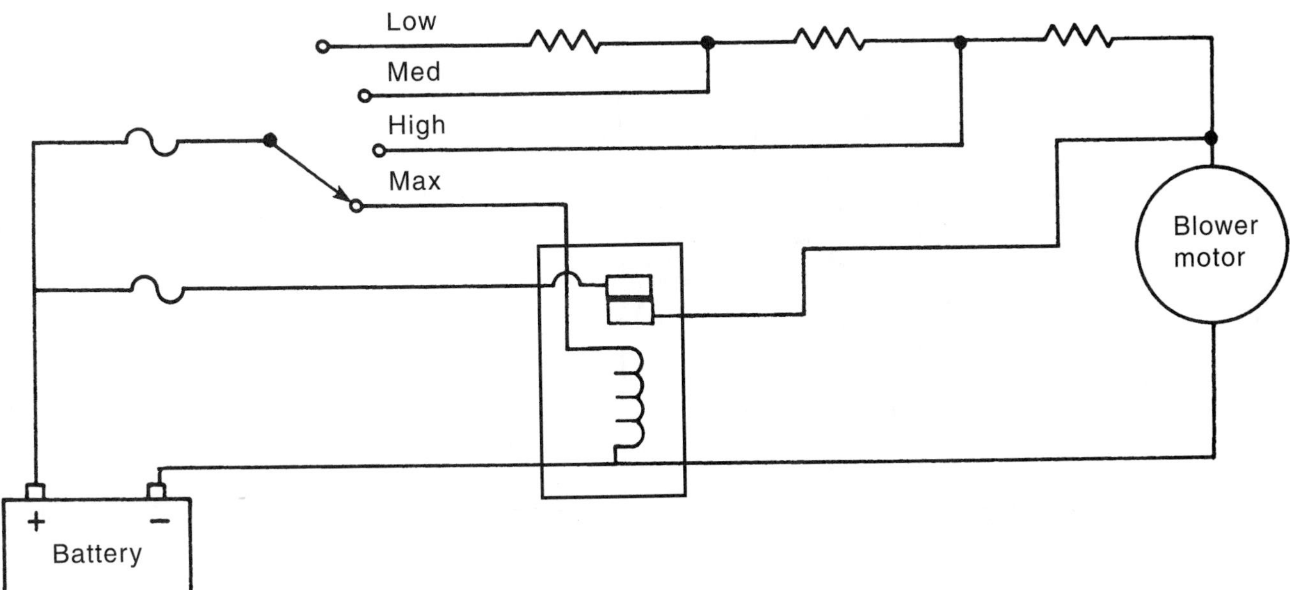

Figure 9. With the contacts closed, the motor has full B+ applied and runs at full speed.

Figure 10. Relays are frequently found in a central location.

and dashboard. Look at a wiring diagram for a modern vehicle and you might be surprised at the number of relays in use.

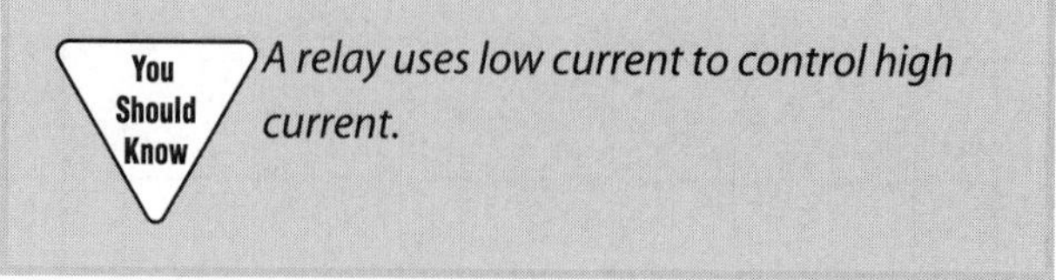

A relay uses low current to control high current.

ELECTRONIC RELAY AS A CONTROL DEVICE

The modern vehicle has the need to control electrical devices for a period of time. The easiest method for the manufacturer is to add electronic control to a standard relay. The rear window defogger found on vehicles today is a good example. Look at **Figure 11**. Notice that the relay coil has a timer attached. This is basically a

Rear defogger grid

Timer

Ignition switch

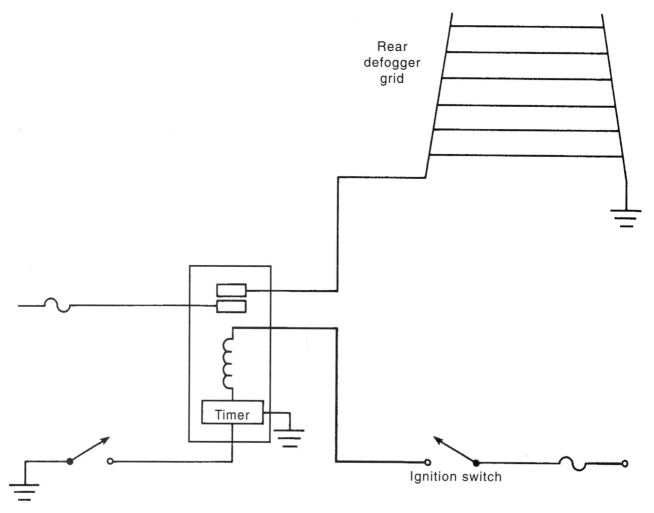

Figure 11. The addition of a timer to the relay coil circuit leaves the relay on for a predetermined time.

countdown electronic timer, similar to a digital watch, that will leave the rear defogger on for about 10 minutes before shutting it down. The relay also uses a momentary contact switch. A momentary switch is just what the name implies. The switch will not stay in the on position but, rather, will spring back to a neutral position. If you follow the lead coming out of the contacts, you will see that the relay coil is powered off the closed side of the contacts.

The driver closes the contacts manually. This also applies B– to the relay coil. The relay coil's magnetic field now holds it closed until the timer removes the B–. Any interruption of either B+ or the timer running out will turn off the heating element at the rear window. Turning off the ignition switch, for example, will remove B+ from the relay coil and allow the contacts to open. The vehicle driver will have to restart the heating by closing the switch momentarily after restarting the car. This rather complicated setup is used because of the large current draw that the rear defogger has. Usually 10 minutes is all that is necessary to melt the snow or defog the inside window. It would be a senseless drag on the charging system to keep the defogger on when it is not necessary. Vehicle mileage is the main consideration here. If the defogger is on longer than is necessary, it will represent a drag on the charging system that will reduce the overall vehicle mileage. Amperage that is being produced by the alternator makes it harder to turn. This equates to a mechanical load. Some smaller engines will have their idle reduced by as much as 100 rpm when the rear defogger turns on—quite a mechanical drag.

Many solid-state control relays are manually turned on.

Summary

- The switch is the simplest and most common type of circuit control.
- When the switch is closed, it offers zero ohms of resistance to the current.
- When the switch is open, it offers infinity, causing no current to flow.
- A switch can be in any part of the circuit.
- A relay is a type of switch that uses low current to control high current.
- Control current supplies a magnetic field that will close the relay's contacts.
- A timer can be added to the relay to give it solid-state control.

Review Questions

1. Technician A states that control current for a relay will be small in comparison to the contact current. Technician B states that the relay coil will have high resistance and the contacts will have no resistance. Who is correct?
   A. Technician A only
   B. Technician B only
   C. Both Technician A and Technician B
   D. Neither Technician A nor Technician B
2. When a switch is closed
   A. it will offer no resistance to the flow of current
   B. another series path will open
   C. it will offer maximum (infinity) resistance to the flow of current
   D. no current flows
3. A relay is a device that will
   A. control low current with high current
   B. have low resistance across its relay coil
   C. be placed in circuits that have low current devices
   D. control high current with low current
4. A rear window defogger is running. Technician A states that the momentary contact switch is closed. Technician B states that the relay coil has power and ground and has closed the contacts. Who is correct?
   A. Technician A only
   B. Technician B only
   C. Both Technician A and Technician B
   D. Neither Technician A nor Technician B

5. Turning the ignition switch off can turn off a timed relay coil.
   A. True
   B. False
6. A relay can be used in most high-current circuits.
   A. True
   B. False
7. Switches are always located on the positive side of the circuit.
   A. True
   B. False
8. Why would a manufacturer use a relay?
9. Are there any disadvantages to the use of a relay?

Chapter 13 Diagnosing Open Circuits

Introduction

The open circuit is a common problem for technicians: the headlight that will not light, the air-conditioning compressor that will not cool, the blower motor that will not turn, or the ignition module that will not fire the spark plugs. Open circuits could cause these problems. If you approach the diagnosis with a systematic procedure, you will be successful. Within this chapter, we examine one acceptable approach, noting that there are many. We start with a discussion of the relative merits of using a test light or a digital voltmeter in various applications and then continue with the actual diagnosis.

USING A 12-VOLT TEST LIGHT

In previous chapters, we have repeatedly mentioned that circuits on the vehicle have limited current flow through them and that the digital voltmeter is the preferred diagnostic tool. When we deal with open circuits, we have to recognize that some conditions prevent the accurate diagnosis of a circuit using a voltmeter. **Figure 1** illustrates this principle. Notice that the test light does not light, because the addition of 200 ohms of resistance at the connection to the 12 ohms of test light resistance allows only 0.056 amps of current to flow. Does the test light indicate that there is current flow? No; the voltage drop across the test light is only 0.67 or slightly higher that half a volt:

12 ohms × 0.056 amps

In this example, the test light shows the problem of excessive resistance wired in series, because it does not light. Two hundred ohms of resistance is just about an open circuit for the vehicle because of the limited amount of current that can flow through it. If we try to diagnose the same difficulty with a digital voltmeter as in **Figure 2** the situation changes drastically. The digital voltmeter is at least 10,000,000 input impedance, which we have discussed previously. When this

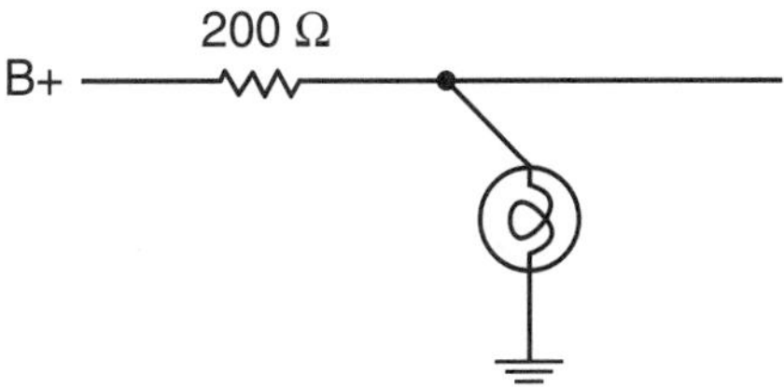

Figure 1. The resistance in series reduces the current flow, so the light is off.

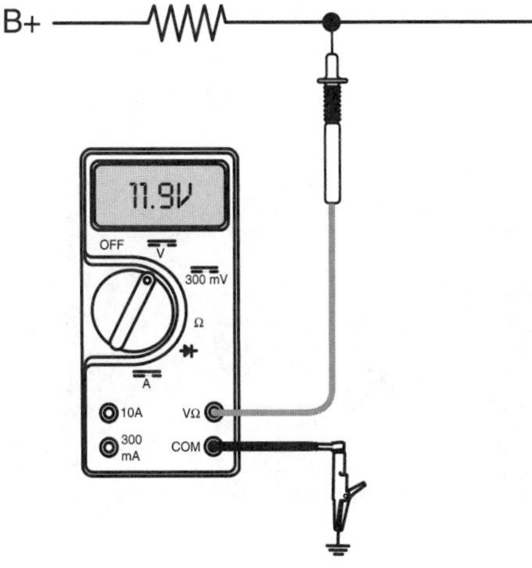

Figure 2. With little current flow, there is little voltage drop.

is added to the 200 ohms of unwanted resistance, the total circuit resistance draws only 0.000012 amps. This is not much current, but notice what happens when we figure the voltage drop across the voltmeter:

0.000012 amps × 10,000,000 ohms = 11.9 volts

The digital meter will read 11.9 volts if you wire it into a circuit with 200 ohms of unwanted resistance. Will this reading tell you that you have a problem? It probably will not. Many circuits that do not have electronic modules and draw substantial current are on the vehicle. These are the circuits to use a test light on. If the circuit is electronic, you should use the digital voltmeter. We will use both in our examples.

OPEN-CIRCUIT DIAGNOSIS USING A 12-VOLT TEST LIGHT

Many different methods exist for diagnosing open circuits. The one presented here works well if you follow it completely. Do not skip or jump around any of the items. Doing the same procedure for every open circuit will ensure that your own repair will not be missing an item or two. Here is the procedure:

1. Verify the defect.
2. Jumper power to the load.
3. Walk through the circuit, using a wiring diagram.
4. Isolate testing for power at connections.
5. Repair the open circuit.
6. Retest.

The procedure is simple if you follow it in the sequence of the numbers.

1. *Verify the defect.* Do not depend on the customer's word for what the problem is, and do not accept anybody else's diagnosis. Insist on doing your own work. Try to duplicate the condition that the customer is complaining about. Let us use an air-conditioning clutch as an example. Think about the circuit and identify the work that is supposed to be done. In this case, it is "engage the compressor clutch" **(Figure 3)**. When current flows through the resistance of the clutch's coil, the strong magnetic field that is generated will engage the crank-

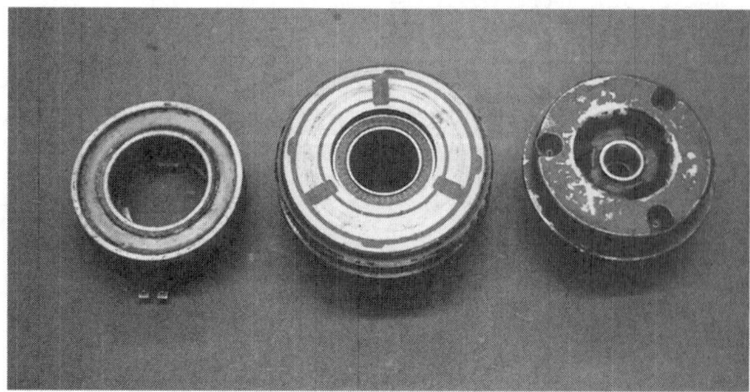

Figure 3. A magnetic AC clutch.

Figure 4. Bring power to the load to isolate it from the circuit.

shaft to the outer pulley. Let us verify this defect. Open the hood, start the engine, and set the controls to maximum cooling. Is the clutch engaged? If not, you have verified the defect. Try all the air-conditioning positions that should engage the clutch. If the clutch will not function, continue to step 2.

2. *Jumper power to the load.* Attach a jumper wire up to the positive battery connection and the other end to the compressor clutch **(Figure 4)**. Break open the circuit and jumper as close to the load as possible. If the load does work when you have it jumped, you have proven two very important things to yourself. First, you have eliminated the load from your list of possible problems, and second, you know that the circuit ahead of the load must be open or contain excessive resistance. Keep in mind that if the load has a ground wire instead of relying on the vehicle frame, you will have to jumper it also. A break in the ground return circuit will appear to be a bad load if you jumper only the B+ and note that the load does not work. Items such as blower motors, headlights, taillights, or anything else mounted on an insulator will have an independent ground wire. Jumper the load on both sides.

3. *Walk through the circuit, using a wiring diagram.* We do not mean walk through the vehicle wiring by hand. This is virtually impossible on modern vehicles. Instead, use the wiring diagram and look at the circuit or circuits affected. The diagram will do two things for you. First, it will identify how the circuit is wired and with what color wires, and second, it will tell you where the connectors and components are located on the vehicle. You might have to use an independent component locator along with the wiring diagram. Refer to **Figure 5**, which is a diagram of an air-conditioning compressor clutch circuit. Notice that B+ for the clutch comes through four series controls. Can you identify them? The ignition switch, the air-conditioning mode control, the cycling switch, and the ambient temperature sensor—these four controls or

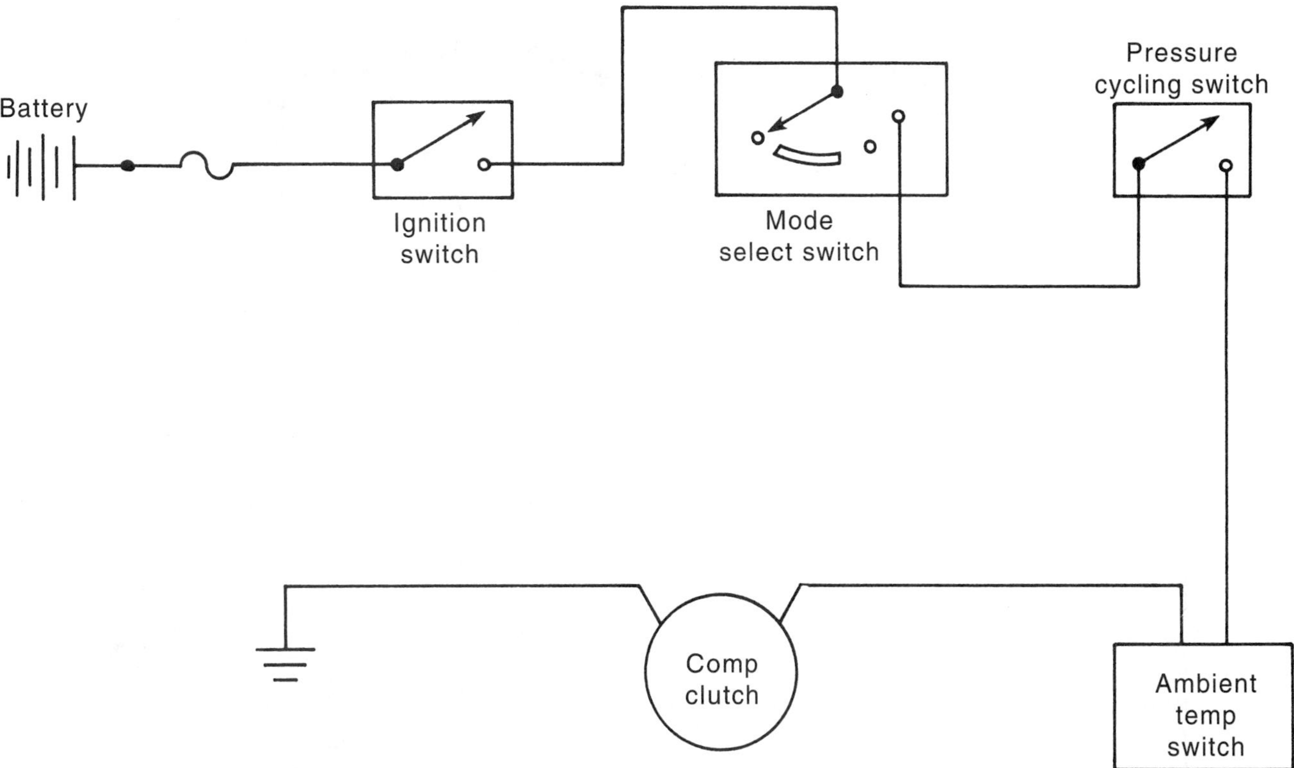

Figure 5. The simplified wiring diagram shows four series controls.

switches must all be closed for the clutch to engage. Without the wiring diagram, you would not know that these controls were wired in series. Now refer to the component locator to determine where the controls are located. **Figure 6** shows you what a component locator for this circuit would look like.

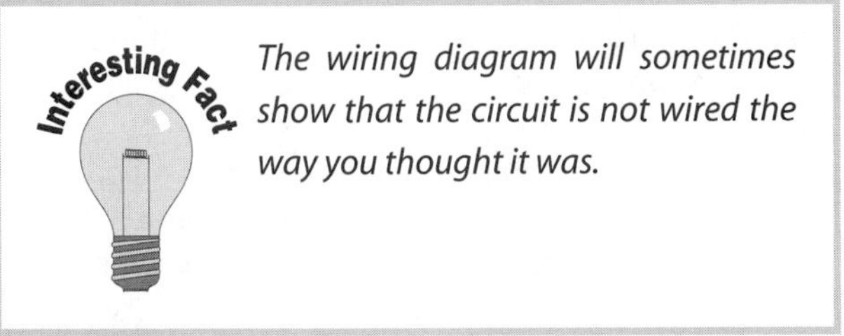

The wiring diagram will sometimes show that the circuit is not wired the way you thought it was.

4. *Isolate testing for power at connections.* This is the heart of the procedure, starting at the simplest end of the circuit (simplest or easiest to get to). There is no need to tear the dashboard apart, unless the underhood testing points to the mode switch. Locate the connector closest to the load, the same one you jumped. Test for B+ with a 12-volt test light like **Figure 7** shows. An open circuit will not pass any current. With the test light grounded and the other end in the feed line to the load, an open circuit will be easy to spot. The test light will be off. No current available equals no light. If no light is present, we move up the B+ line to the next component or connector as identified by the wiring diagram. As we move toward the source, we are looking for B+ and a test light that lights. Once we find a connector that is hot, we know that the open circuit is between the dead connector and the hot one **(Figure 8)**.

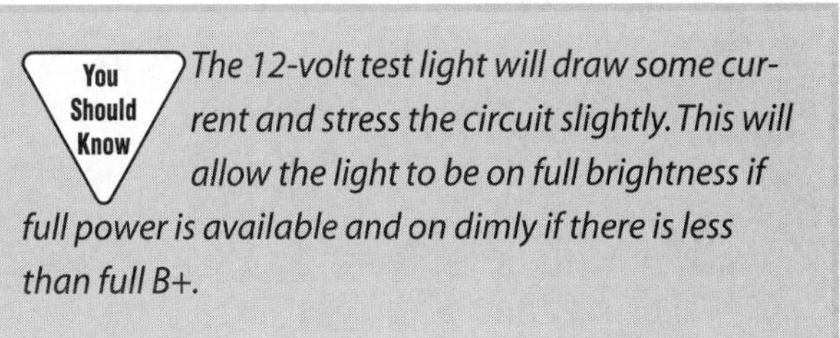

The 12-volt test light will draw some current and stress the circuit slightly. This will allow the light to be on full brightness if full power is available and on dimly if there is less than full B+.

5. *Repair the open by either replacing an open component or repairing a conductor.* This becomes the simplest part of the procedure. Identifying the open circuit is sometimes time consuming because of the length of the circuit. Repairing the open circuit is usually accomplished in a fraction of the diagnosis time. An open wire is repaired with a solderless connector or sometimes soldered and insulated

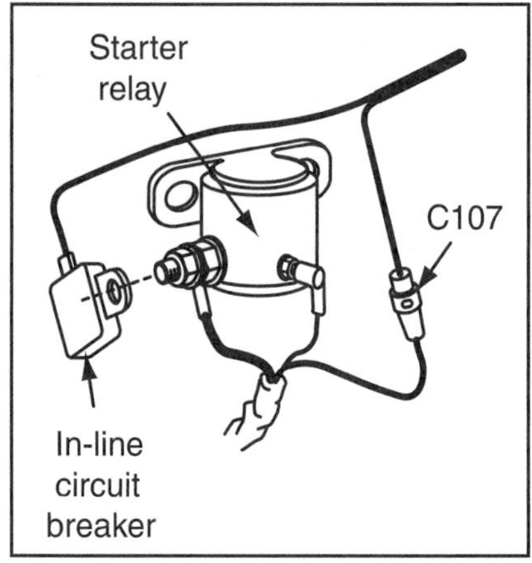

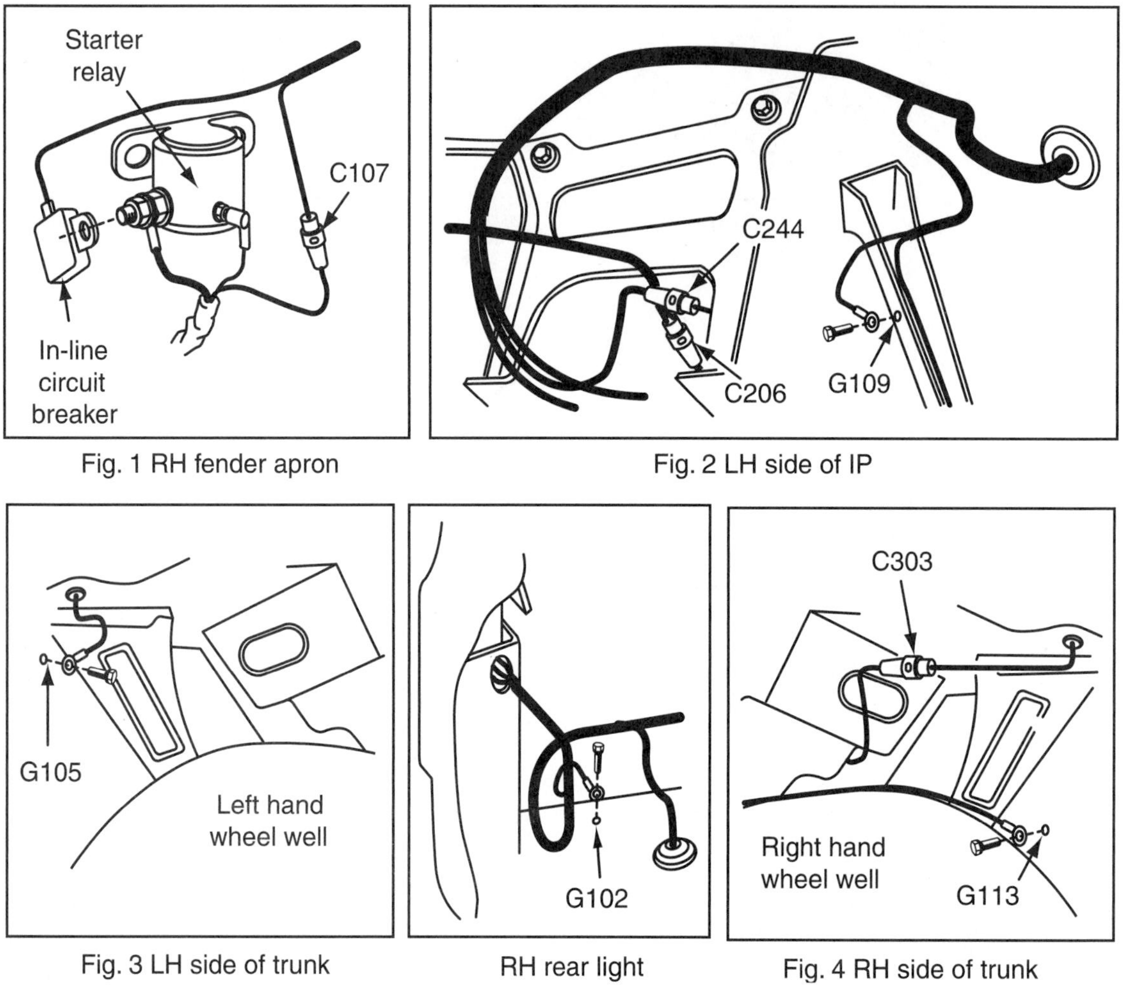

Fig. 1 RH fender apron

C244
C206
G109

Fig. 2 LH side of IP

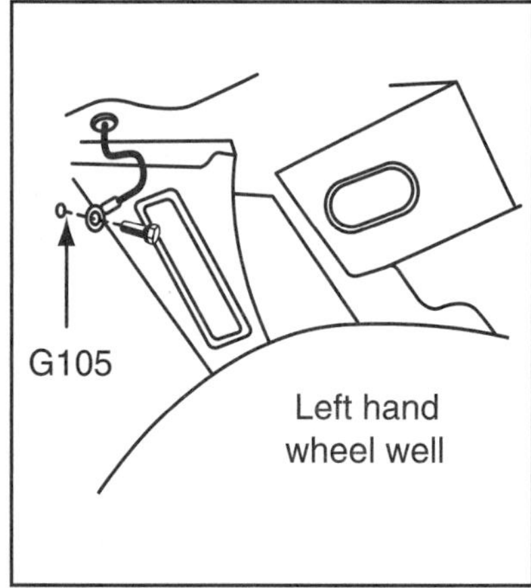

Fig. 3 LH side of trunk

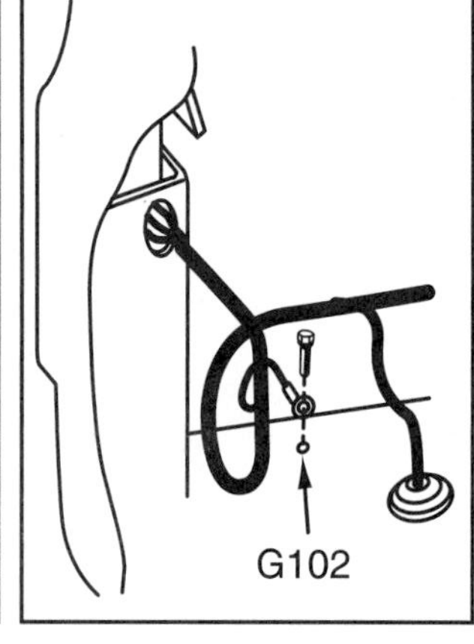

RH rear light

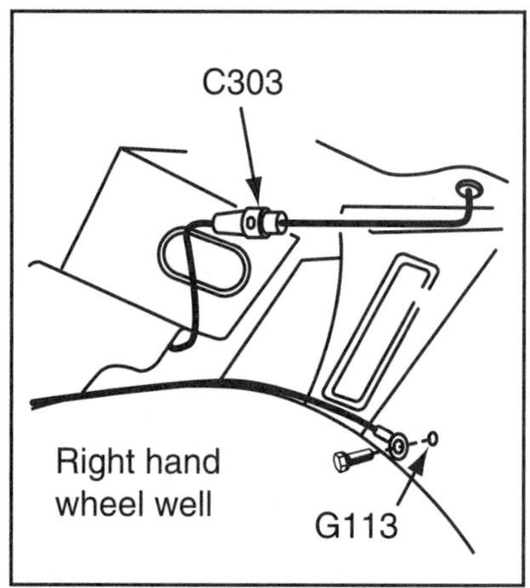

Fig. 4 RH side of trunk

Figure 6. AC circuit component locator. RH, right hand; LH, left hand; IP, instrument panel.

Figure 7. Testing for power at convenient points in the circuit.

with tape or heat-shrink tubing. Repairing the defect so that it will not reappear is important. Take your time and do it right. Remember, if the circuit went open because of a weakness in the harness or excessive movement of the wire, it can happen again. Use wire ties or clamps and eliminate harness flexing so that your repair will last longer than the original.

6. *Retest.* This is the easiest item to accomplish. Turn on the circuit. Make sure it works to specification. Also, measure the applied voltage to the load to make sure that no additional resistance is present **(Figure 9)**. This is extremely important. Remember that series resistance will decrease both the voltage applied and the current used. This, in itself, can cause additional problems with some loads. Our example of a compressor clutch is one of those loads that reduced B+ will burn out. Not enough current will cause it to slip, especially at high speeds. This will greatly reduce its life. If the applied voltage is the same as the battery voltage, your job is finished.

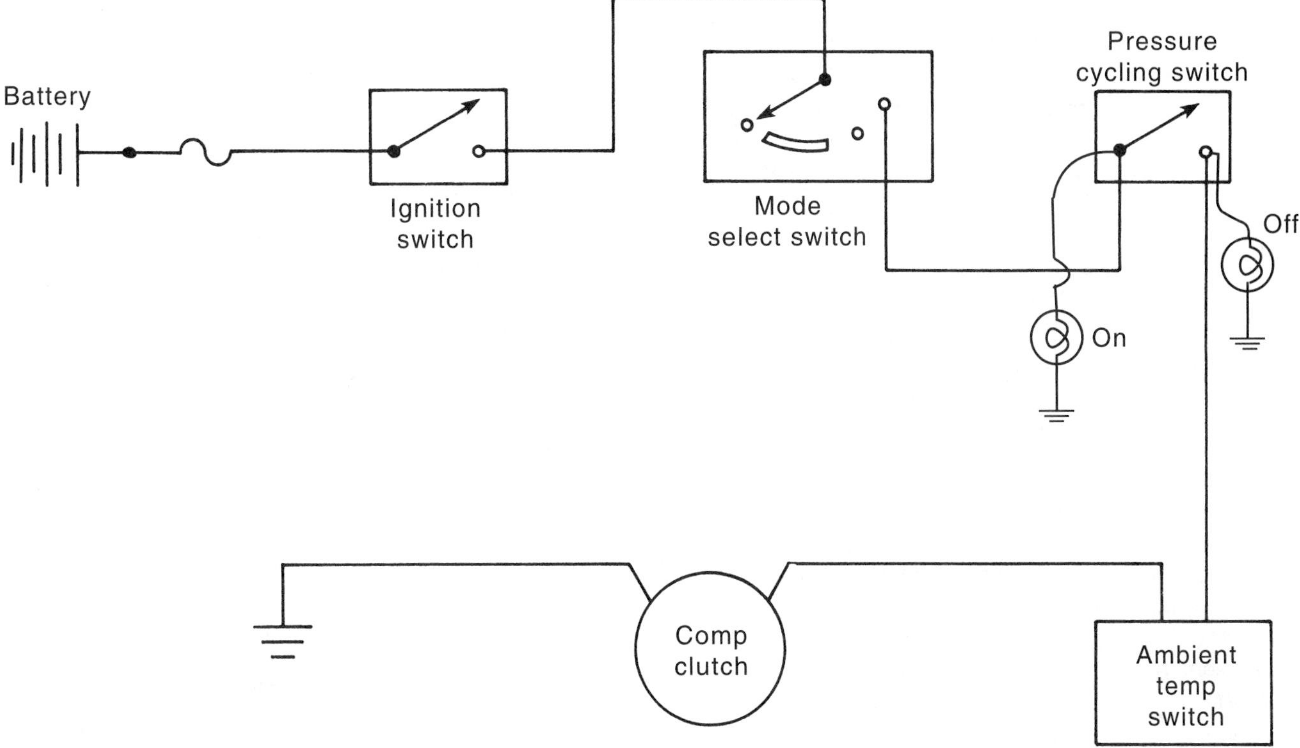

Figure 8. The test light shows an open pressure cycling switch.

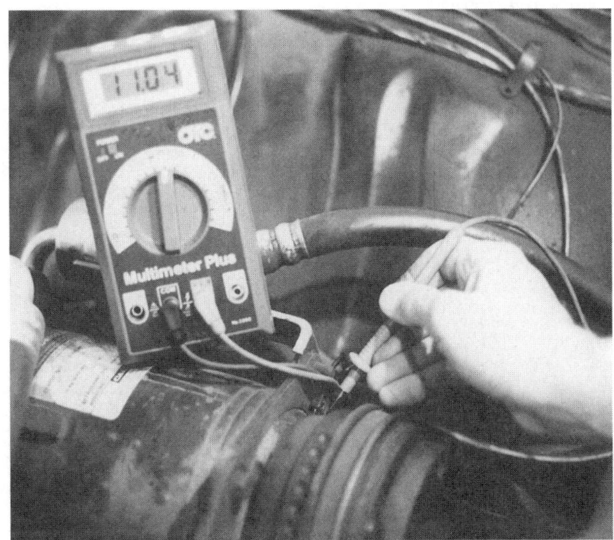

Figure 9. After repair, make sure the load has good applied voltage.

DIAGNOSING OPEN CIRCUITS WITH A VOLTMETER

If the circuit includes electronic controls, modules, or the vehicle's PCM, using a 12-volt test light might cause some damage. Generally, the load will draw sufficient current but the control might use very small levels of current. Getting your 12-volt test light across the control could cause it to draw more current than the control circuit can handle. Now you have two problems. The original one and the one that you just created. Some steps of the procedure will remain the same, and some will change from the previous 12-volt test light series:

1. Verify the defect.
2. Test the voltage drop across the load.
3. Walk through the circuit, using a wiring diagram.
4. Isolate testing for voltage drops at various points in the circuit.
5. Repair the open circuit.
6. Retest.

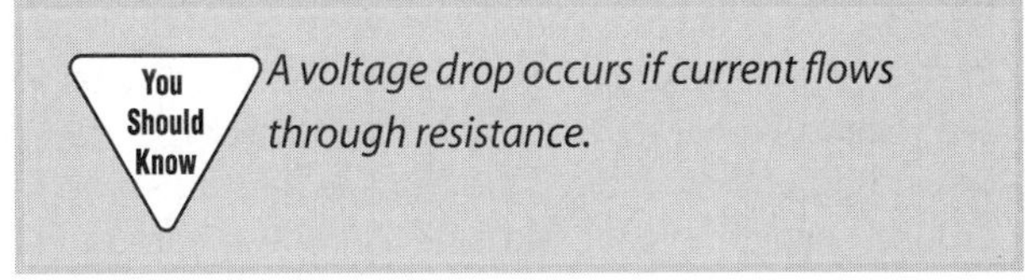

A voltage drop occurs if current flows through resistance.

The only steps we will look at are steps 2 and 4 because they are different from the 12-volt test light procedure. We will use an evaporative solenoid with a module as a control device. Let us analyze the system because

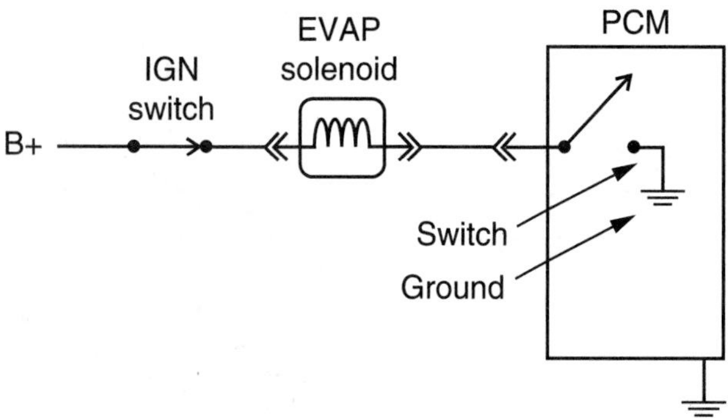

Figure 10. With the ignition switch on, the PCM will apply a ground to open the evaporative (evap) solenoid.

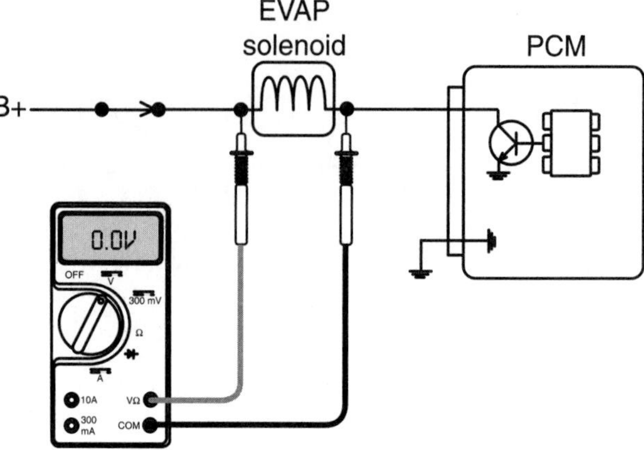

Figure 11. The voltmeter reading indicates that no current is flowing.

there appears to be no current flowing through the evaporative solenoid. Follow along with **Figure 10**. The ignition switch powers up the B+ side of the solenoid, and when the module turns on the ground side, current flows, and the solenoid opens. Remember the principle that a voltage drop occurs when current flows through resistance. No voltage drop means the circuit has no current flowing through it.

2. *Test the voltage drop across the load.* Power up the circuit by turning on the ignition switch and measure the voltage drop across the solenoid. In our example in **Figure 11**, the voltmeter reads zero. There does not appear to be any current flowing.

4. *Isolate testing for voltage drops at various points in the circuit.* Disconnect the solenoid connector and check for voltage at the positive side of the circuit. If there were full battery voltage available, as indicated in the diagram in **Figure 12**, we would use the ohmmeter from our DMM on the solenoid and compare its resistance to the specified resistance. If it has excessive resistance, it needs to be replaced; ours appears to be all right, so we would move to the ground side of the circuit. Reconnect the evaporative solenoid and turn the ignition switch on. Measure the voltage drop across the wire between the solenoid and the module, as shown in **Figure 13**. Our meter is now reading 12.6 volts. What does this indicate? We know that no current is flowing through the solenoid because it did not have a voltage drop, so the voltmeter must *not* be indicating a true voltage drop. Remember that the voltmeter will see voltage if the leads are connected into a circuit that has B+ and ground available. If the module has turned on the ground circuit and B+ is available all the way through the evaporative solenoid, the voltmeter will show 12.6 volts. There must be an open circuit located between the module and the negative of the solenoid winding.

Using a combination of the 12-volt test light and the DMM is probably the best course of action as long as you have a wiring diagram and are sure that you understand

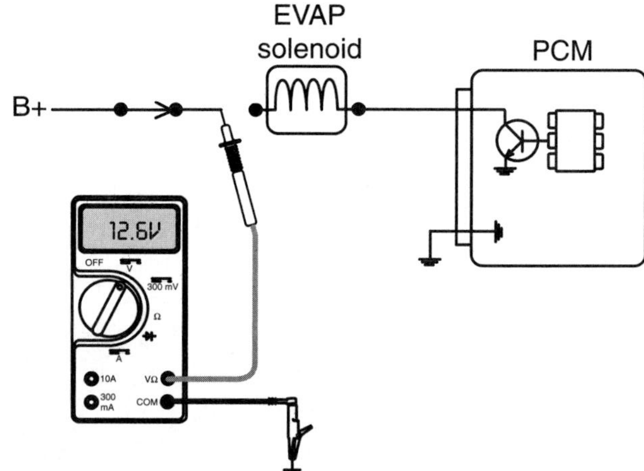

Figure 12. With the solenoid disconnected, measure applied voltage.

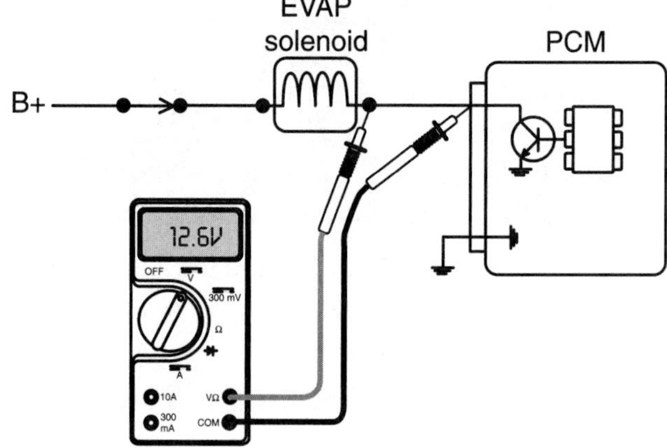

Figure 13. The voltmeter sees the PCM ground and battery voltage indicating a broken wire or poor connection.

current flow. If the circuit is disconnected or isolated from the electronics, feel free to use the light. Do not use a 12-volt light if the circuit is live and contains electronics.

The last thing to make note of here is that you should look for the primary cause of the open. If in your diagnosis you find an open wire, try to figure out why the wire went open. Was there too much current flowing? Was the wire flexing too much? Was it pinched in a hinge? Answers to these and other questions will help to determine just how you will repair the defect. Do not leave the cause of the open circuit still in the vehicle if you can help it. Also, keep in mind that many open circuits are short circuits that have burned out the circuit protector. Finding and eliminating short circuits is our next topic.

Summary

- A 12-volt test light can be used to isolate and find an open on circuits that do not have any electronic modules.
- Power and a ground should be available to the common loads on the vehicle.
- If the test light is on in one section of the circuit and off in the next section, the open is between the two points.
- Using a voltmeter on an open circuit may give you the wrong information.
- The wiring diagram should be used to show how the circuit is wired.
- A voltage drop occurs when current flows through resistance.
- No voltage drop indicates either no current or no resistance is present.
- There should be maximum voltage drop across the load.

Review Questions

1. Open circuit diagnosis involves jumpering the load. This tells the technician
   A. where the open circuit is
   B. why the open occurred
   C. whether the open is before the load or in the load itself
   D. whether the open is intermittent
2. A 12-volt test light is touched into a connector. The light glows. Following the conductor forward (toward the load), the next connector is probed, producing no light. This indicates
   A. nothing
   B. an open circuit between the two points
   C. the load is open
   D. an open before the first connector
3. A voltmeter placed across a load reads 12.6 volts. Technician A states that this indicates that current must be flowing. Technician B states that this indicates that the load has some resistance. Who is correct?
   A. Technician A only
   B. Technician B only
   C. Both Technician A and Technician B
   D. Neither Technician A nor Technician B
4. Bringing power over to the load causes the load to work. Technician A states that this indicates that the open is on the ground side of the circuit. Technician B states that this indicates that the open is in the load. Who is correct?
   A. Technician A only
   B. Technician B only
   C. Both Technician A and Technician B
   D. Neither Technician A nor Technician B
5. Resistance will cause a voltage drop if current is flowing.
   A. True
   B. False
6. A digital voltmeter will draw excessive levels of current.
   A. True
   B. False
7. A 12-volt test light will not draw current.
   A. True
   B. False
8. A DMM should be used in circuits that contain electronics.
   A. True
   B. False
9. Explain the circuit conditions that allow you to use a 12-volt test light for diagnosis.
10. Under what circuit conditions should a digital voltmeter be used for diagnosis?

Chapter 14 Diagnosing Short Circuits

Introduction

In Chapter 13, we took a look at two different methods of diagnosing an open circuit, using a 12-volt test light or a voltmeter. The circuit components determined which method we should use. Short circuits are no different, although their result can be more expensive. When we short out a load, the control circuit for the load can damage electronics. The procedure shown in this chapter utilizes your knowledge of resistance. We also suggest that you build a simple short detector for high-current circuits.

SHORT CIRCUITS AND RESISTANCE

Before we discuss the procedure for finding and eliminating short circuits, let us look at them in terms of resistance. Most short circuits will burn open a circuit protector. A fuse, fusible link, or a circuit breaker should open up the circuit and prevent more extensive damage. Why does the circuit protector open up? It is hoped that by now you realize that a momentary overload of current burns open the element inside. Think about resistance and its effect on current. Remember, the less the resistance, the more the current. We are looking for a reduction in resistance that has caused the circuit protector to open. Two sides of the circuit must have come in contact with one another or the total resistance of the circuit must have dropped. This can happen easily. The entire frame and most of the body on a majority of vehicles is metal and at ground potential. If a B+ conductor touches this ground potential, we have a short circuit around the designed resistance of the load, a circuit protector opens up, and a customer has a dead component or accessory on the vehicle. Always think of short circuits as a reduction in resistance that causes an increase in the amount of current.

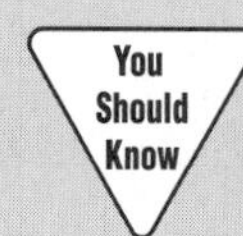

A short will increase the amount of current to the point that it will open the circuit protector.

DIAGNOSING A SHORT CIRCUIT WITH A SHORT DETECTOR

Let us use a procedure that is similar to our open circuit diagnosis:

1. Verify the short and identify the type.
   a. Dead
   b. Intermittent
   c. Cross
2. Walk through the circuit.
3. Control the current with a false load.
4. Separate connectors to pinpoint the short.
5. Repair.
6. Retest.

You can see that our diagnostic procedure is much like open circuit diagnosis, but with a few changes. Let us look at them one at a time.

1. *Verify the short and identify the type.* Plug in a fuse and try to make it blow. Turn on all items on the circuit protector one at a time. Try to pinpoint the circuit that overloads

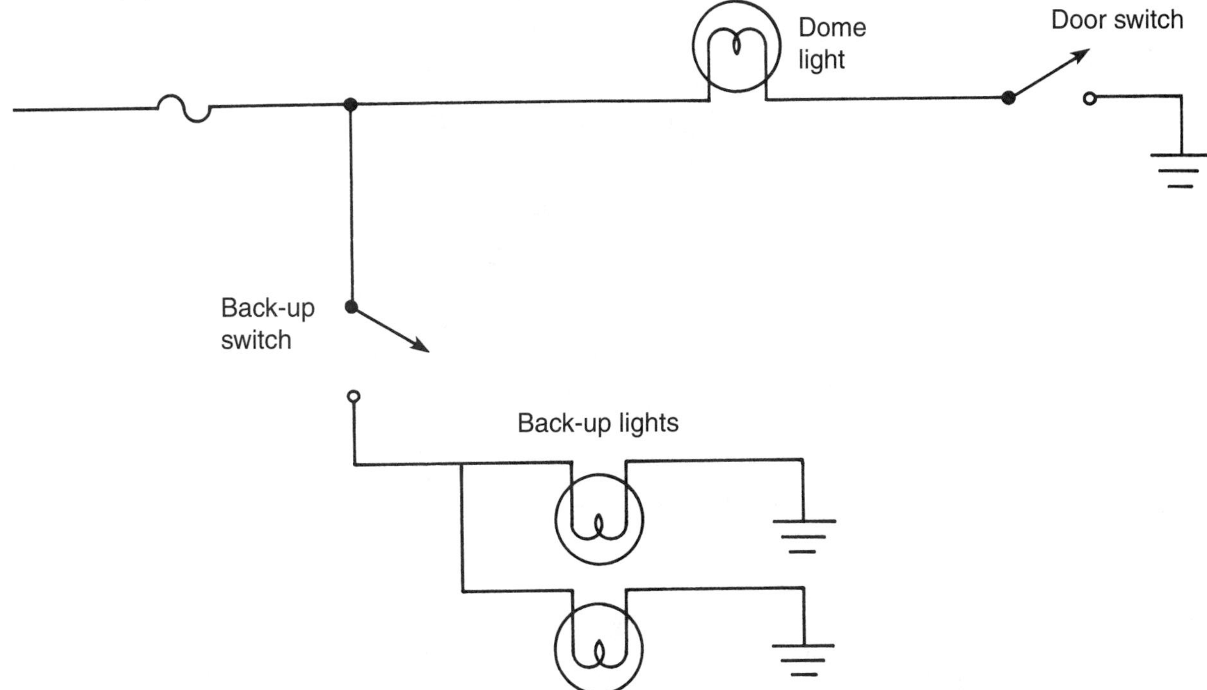

Figure 1. The diagram shows three loads in two separate circuits.

the protector and opens it. **Figure 1** shows two separate circuits, the dome light and the back-up lights both powered off a single fuse. By turning on the dome light first and then the back-up lights individually, we can pinpoint which circuit has the short. If you can make the fuse blow, you are dealing with a dead or constant short to ground. If you cannot make the fuse blow and the vehicle came in the door with a blown fuse, you are dealing with an intermittent short that is not present right now. Cross-shorts, or shorts between two circuits, are easy to identify. The short causes two circuits to be powered off each other.

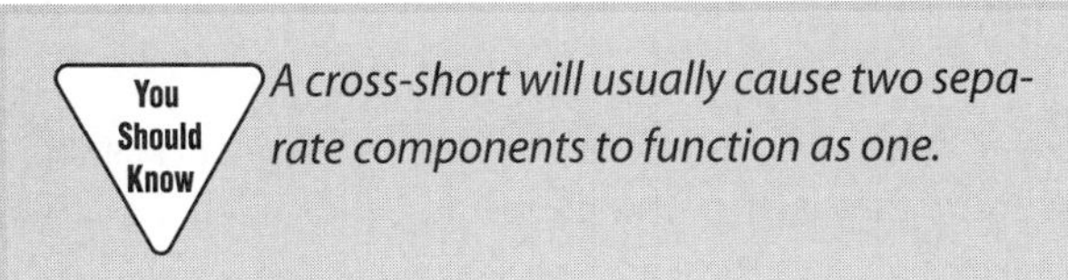

A cross-short will usually cause two separate components to function as one.

2. *Walk through the circuit using the wiring diagram.* With the component locator, identify where the connectors are. Notice what components are on the circuit protector **(Figure 2)**. Many times the manufacturer will wire different circuits together and make this information available on the fuse diagram. This can make the diagnosis either easier or more difficult, depending on how many circuits are involved. Intermittent shorts on multiple load circuits can take a long time to find.

3. *Control the current with a false load.* Here is the most important concept. Whenever a circuit protector opens, it is because of an increase in current. This increase in current is because of the decrease in resistance. What we want is to be able to have time to diagnose our short while it is occurring. This is difficult to do because the short causes too much current to flow. A helpful device is the short detector or loud test light. This device has a headlight and a key warning buzzer wired to a flasher, as diagrammed in **Figure 3**. If it is connected to B+ and B–, it will buzz and light. More important than this, it will control the current if it is wired in place of the burned-out fuse. The light and buzzer are forms of resistance that, when wired in series, will buzz and light because of the B+ available at the fuse box

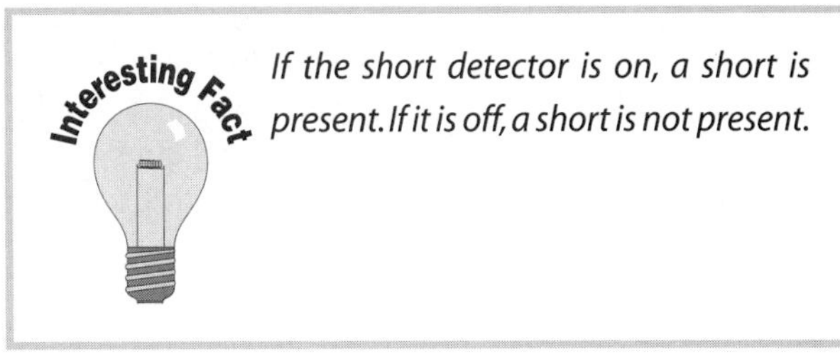

If the short detector is on, a short is present. If it is off, a short is not present.

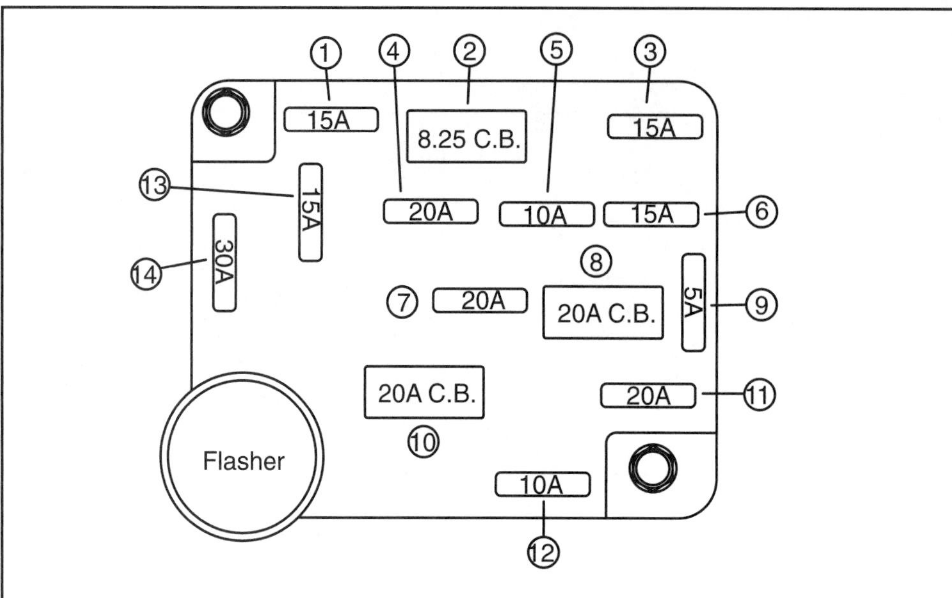

1. Stop lights, emergency warning system, speed control module, cornering light relays, and trailer tow relays
2. Windshield wiper/washer
3. Tail, park, license, coach, cluster illumination, side marker lights, and trailer tow relay
4. Trunk lid release, cornering lights, speed control, chime, heated backlite, and control A/C clutch
5. Electric heated mirror
6. Courtesy lights, clock feed, trunk light, miles-to-empty, ignition key warning chime, garage door opener, autolamp module, keyless entry, illuminated entry system, visor mirror light, and electric mirrors
7. Radio, power antenna, and CB radio
8. Power seats, power door locks, keyless entry system, and door cigar lighters
9. Instrument panel lights and illuminated outside mirror
10. Power windows, sun roof relay, and power window relay
11. Horns and front and instrument panel cigar lighters
12. Warning lights, seat belt chimes, throttle solenoid positioner, autolamps system, and low fuel module
13. Turn signal lights, back-up lights, trailer tow relays, keyless entry module, illuminated entry module, and cornering lights
14. Automatic temperature control blower motor

Figure 2. The fuse panel indicates the circuits that are on a particular fuse.

and the short to ground. This is diagrammed for you in **Figure 4**.

4. *Separate connectors to pinpoint the short.* This is easy. The excessive current has been controlled, and we have a visual and audible indicator of the short. Now we separate the connectors in succession, moving away from the fuse. As you open the connector, if the loud test light stops, you have eliminated the short. In other words, it is beyond the connector you have opened. If the test light still is making noise, the short is ahead of the open connector. In this manner, we can pinpoint the short usually to a small section of the wiring harness and then physically examine the section, looking for the short to ground. **Figure 5** illustrates a short that is after the connector, and a detector that is still on.

5. *Repair.* Repair the short in much the same manner you did in open-circuit diagnosis. Make sure you eliminate the cause of the short **(Figure 6)**. Many times the causes of shorts are found at sharp corners or hinges. Make sure you tape up the wires to hold them away from the cause of the short or you will need to make a repeat repair.

6. *Retest.* Make sure the short is gone. Replace the loud test light with the normal fuse, turn on the circuit, and try to make it blow. If you have done your job, it will not.

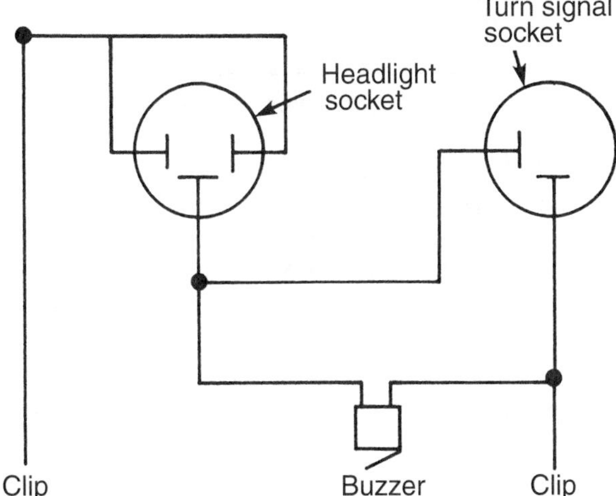

Figure 3. A short detector controls the current that the short caused.

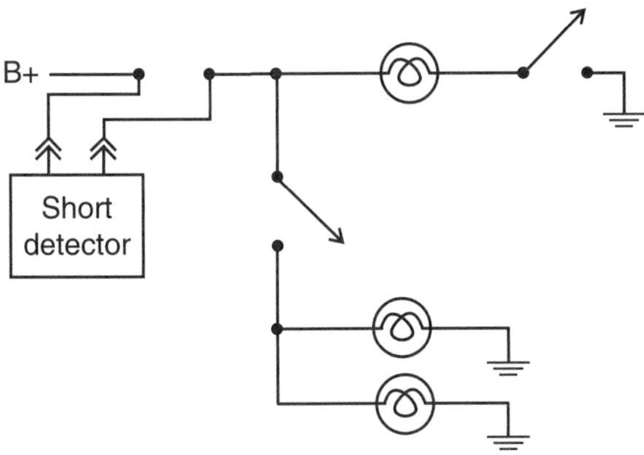

Figure 4. The short detector in place of the fuse.

This procedure really works and can save you tremendous amounts of diagnosis time. Just remember that the goal when dealing with dead shorts will be to turn the tester off. When you have turned the tester off, you have found the short. If the short is intermittent, you will hook up the short detector and try to move the wires until you turn on the tester. Turning on the tester means you have created the short. Now eliminate it in the same manner you did with the dead short.

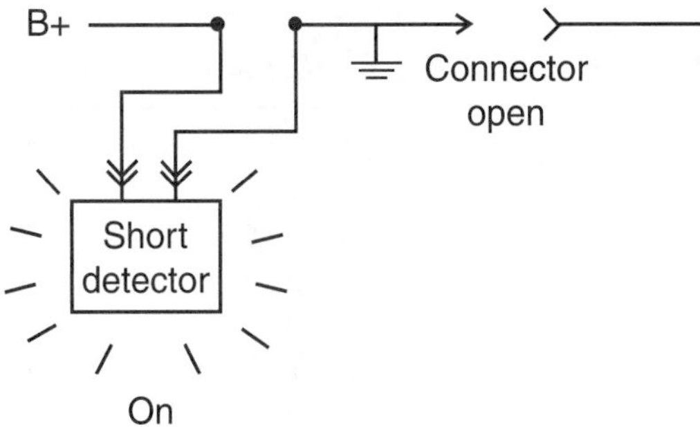

Figure 5. The short is still present with the connector open.

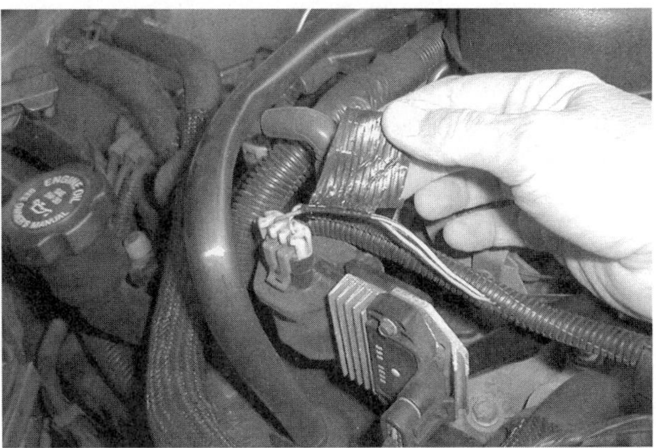

Figure 6. Tape up any repair to prevent future problems.

DIAGNOSING SHORT CIRCUITS WITH ELECTRONIC CONTROL

The only difference in how we treat a short circuit with electronic control is in the load that we will use. If the fuse powering the circuit is less than 5 amps, use a key warning buzzer instead of the short detector. It is possible that the short detector could overload the electronics of the circuit. The key warning buzzer draws so little current that overloading the circuit is not a problem.

Summary

- A short detector can be used to control the flow of current ahead of the short.
- If electronics are involved in the circuit, use a small buzzer to control the current.
- The detector is placed in the fuse box where the burned-out fuse was.
- If the detector is on, the short is present.
- If the detector is off, the short is not present.
- If opening a connector turns the detector off, the short is ahead of the connector.
- If opening the connector does not turn off the detector, the short is after the connector.

Review Questions

1. The object of using a false load in short circuit diagnosis is to
   A. control the current by adding resistance in series
   B. reduce the total resistance
   C. increase the current flow
   D. decrease the current by adding resistance in parallel
2. A short detector is placed in a circuit and is on when a switch is closed and off when the switch is open. Technician A states that this indicates that the short is after the switch. Technician B states that the switch is apparently turning on the short. Who is correct?
   A. Technician A only
   B. Technician B only
   C. Both Technician A and Technician B
   D. Neither Technician A nor Technician B
3. A short detector is placed into a short circuit to
   A. lower the resistance of the circuit
   B. increase the amount of current
   C. increase the resistance of the circuit
   D. turn on when the short has been eliminated
4. A dead short to ground will
   A. turn the detector off
   B. turn the detector on
   C. burn out the load
   D. none of the above
5. Increasing the resistance of the circuit will decrease the current.
   A. True
   B. False
6. A short detector is placed where the load was.
   A. True
   B. False
7. A detector that is off indicates that the short is not present right now.
   A. True
   B. False
8. Why is a short detector a useful diagnostic tool?
9. Explain how the detector indicates a short.

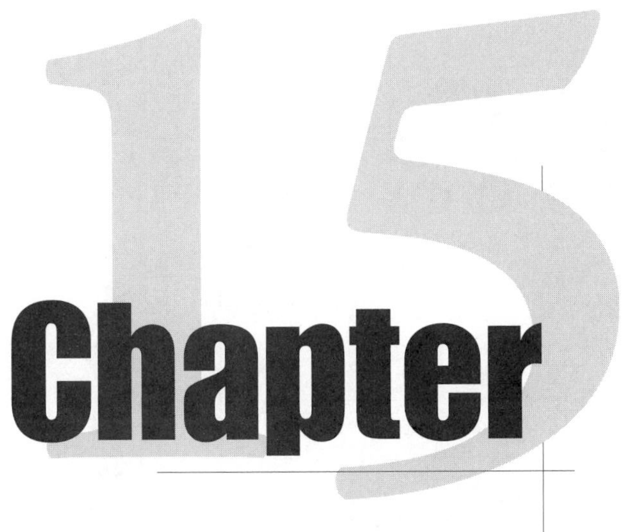

Servicing Open and Short Circuits

Introduction

Once the diagnosis is complete, the repair begins. This short chapter is intended to show you some acceptable methods of repairing circuits. We will cover the replacement of a wire and two forms of repairing a wire.

REPLACING A COMPONENT

Replacing a nonfunctional component is probably the easiest repair to accomplish. The relay is not working; replace it. The switch is open all of the time; replace it. These solutions are easy; however, two considerations must be kept in mind. The first is that the replacement component be a good one, and the second is identifying the cause of the component burning out.

Let us use a high fan relay for an example **(Figure 1)**. We have bypassed the old one and it is not functioning. Before installing the new one, measure the resistance of the coil **(Figure 2)**. Does it match specification? If it does, power up the coil. Does the relay click (only for mechanical relays) and does the resistance of the closed contacts go to zero **(Figure 3)**? If it checks out as all right, install it and measure the current that will be flowing through it. We look at the use of a current probe in Chapter 19. The level of current can then be checked against the specification. Remember that the amount of current flowing through the relay contacts is the basis for which relay the manufacturer chooses to use. If our blower motor is supposed to draw 10 amps on high and it is drawing 14 amps, the life of the relay will be reduced.

Figure 1. A relay out of the relay/fuse center.

By testing the new relay before installation and then checking current flow through the new unit, we are assured that the relay will function normally. We will not see this vehicle next week with a nonfunctional system.

REPLACING A WIRE

If your diagnosis has shown that a wire is broken or has excessive resistance, it may be necessary to completely replace it. Sometimes a wire gets pinched in a door or hinge and is damaged beyond repair. Under these circumstances, replacement of the wire is acceptable practice. It is not acceptable in most cases to replace a complete harness because one or two wires are open. If the wire being replaced is inside a plastic conduit, the job is easy **(Figure 4)**. Open the conduit and pull the wires out so that they can be spread out. Be gentle. We do not want to replace one wire and find that four more were damaged. Locate the bad section or the entire wire. Determine what type of connector is on each end. Premake the wire, with the correct connections, with wire that is the

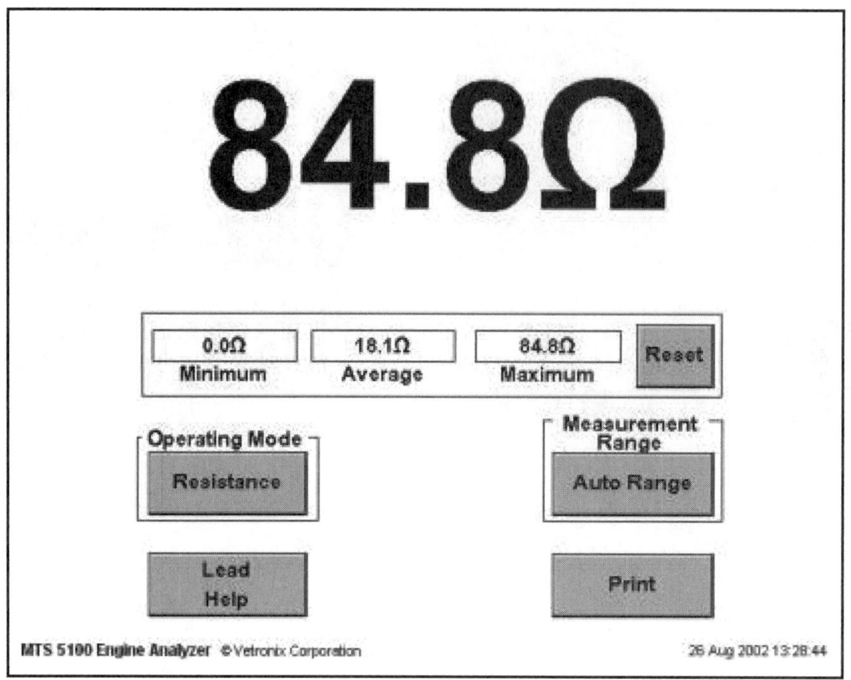

Figure 2. Measure the coil's resistance and compare it to the specification.

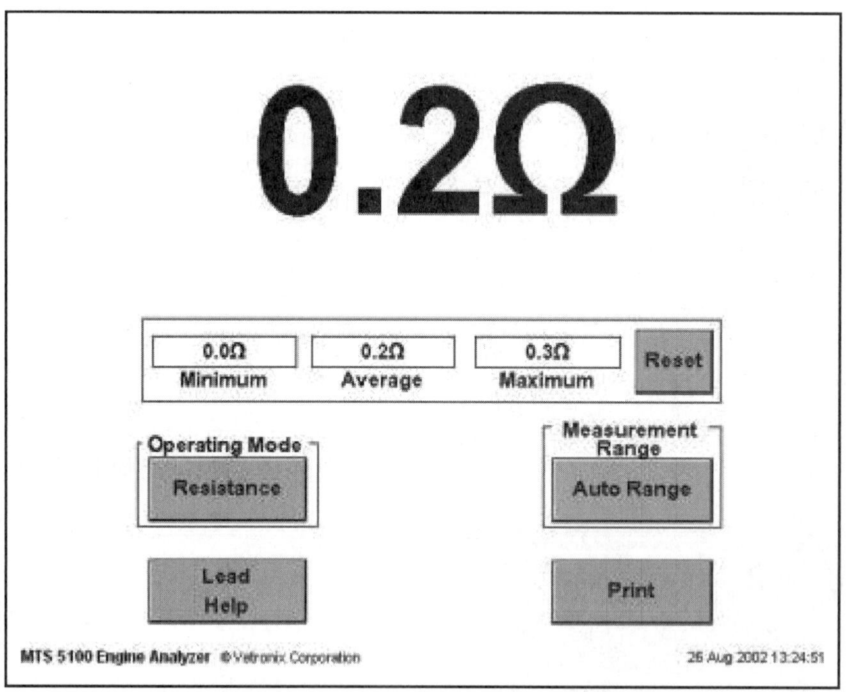

Figure 3. With the relay coil energized, measure the resistance of the closed contacts.

same number as the broken wire and push all of the wires back into the conduit. Make sure the job looks like the original. *Do not* run the wire alongside the outside of the conduit or go point to point where the customer may see it or it may get damaged.

REPAIR CONSIDERATIONS

When you decide that a wire needs to be repaired, take into consideration a few items before deciding what method you will use. The first consideration is, where will

Figure 4. Wires can be pulled out of a conduit for repair.

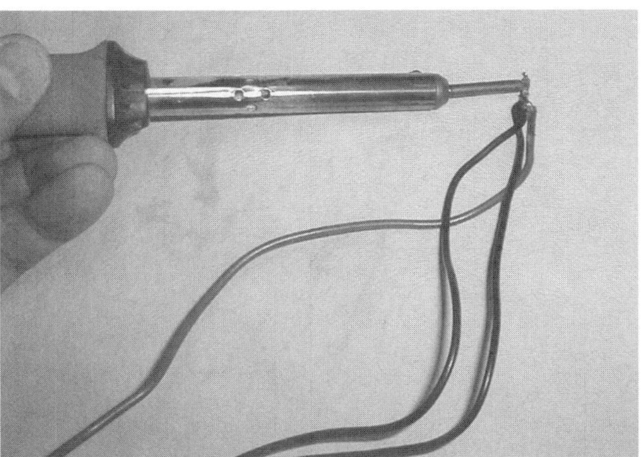

Figure 5. Wire repair might involve soldering a broken wire.

Figure 6. A variety of crimp (solderless) terminals is available.

Figure 7. Use the correct pliers to crimp the terminal.

the repair be? Will it be exposed to road salt or the underhood environment? If yes is the answer, then the repair should be soldered. Soldering makes an effective repair if it is done correctly. It is hoped that you will get the opportunity to practice some soldering in the course you are currently in. Expose approximately 1 inch of wire by removing the insulation. Twist the clean strands of wire together and heat them with the soldering iron **(Figure 5)**. Once they are hot, the solder will flow up the wire. If you use enough heat, the solder will flow and not just sit on the surface of the wire. Soldering a wire will make it stiff, so if the wire has some movement, it might break at the end of the solder joint.

Another possible repair method is the use of **crimp terminals**. They come in a variety of shapes, sizes, and applications **(Figure 6)**. They are not intended to be used in any place that is exposed to the environment. Some technicians consider them temporary repairs. If the crimp connector is the correct size and the wire is stripped back about 1/4 of an inch, the crimp will hold **(Figure 7)**. A crimp connector has the added benefit that it insulates the repair. The standard crimp connectors are not waterproof. They are available with shrink tubing, however, which greatly decreases the likelihood that the connection will fail.

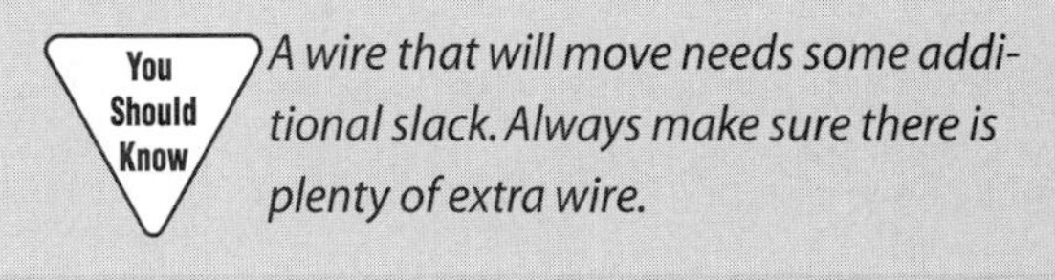

A wire that will move needs some additional slack. Always make sure there is plenty of extra wire.

Summary

- Always test the new component to make sure that it is good before its installation.
- Try to determine what caused the component to fail.
- If a new wire is installed, run it inside the conduit.
- Soldering a broken wire is an acceptable repair.
- Soldering a wire stiffens it where it might break if flexed.
- Crimp connectors are acceptable if they are used correctly and in the correct environment.

Review Questions

1. A broken wire is being replaced. Technician A states that the wire should be run inside the conduit, if possible. Technician B states that the wire could be soldered or crimped into place, depending on the application. Who is correct?
   A. Technician A only
   B. Technician B only
   C. Both Technician A and Technician B
   D. Neither Technician A nor Technician B
2. A relay is being installed. Before installation, the technician should check
   A. the resistance of the relay coil
   B. whether the contacts will close
   C. why the original relay failed
   D. all of the above
3. After a component has been replaced, the technician should
   A. measure the current flow through the device
   B. make sure the installation looks like the original
   C. thoroughly test the system
   D. all of the above

Section 4

Digital Storage Oscilloscope (DSO) Use

SECTION OBJECTIVES

At the conclusion of this section, you should be able to:

- Set or change the volts per division.
- Set or change the time per division.
- Obtain a DSO pattern on the screen and save it.
- Use the signal-finder feature.
- Change from AC to DC and back again.
- Read the voltage and time on the DSO.
- Connect the DSO to a sensor to obtain a pattern.
- Understand why and how to use trigger and slope.
- Obtain and interpret an AC pattern.
- Obtain and interpret a DC pattern.
- Obtain and interpret a variable frequency pattern.
- Obtain and interpret a duty cycle pattern.
- Obtain and interpret a pulse width pattern.
- Understand the various uses of the current probe and how to calibrate it.
- Obtain and interpret a current waveform.
- Know when multiple traces might help with a diagnosis.
- Obtain multiple traces.
- Set the trigger and slope with multiple traces.

Some DSOs are capable of having two input and two output traces on the same screen, which greatly increases the technician's diagnostic ability.

Chapter 16 Digital Storage Oscilloscopes

Introduction

Every few years the automotive repair industry sees a "new" or "improved" tool come from the back of the tool salesperson's truck. Generally, the new tools are successful in making the technician's job easier or more accurate. However, few tools have had the impact on the technician of the DSO. As the vehicle systems have become increasingly more complicated and sophisticated, diagnosis has become more technical. When faced especially with intermittent problems, more technicians are turning to the DSO as their tool of choice, so the DSO is becoming more of a basic tool for technicians every year. Even though computer control is not within the scope of this text, within this chapter we look at how to set up a typical DSO and use it to analyze various computer inputs and outputs. The DSO is as important and innovative now as the DMM was a few years ago. Every good technician has a DMM in his toolbox, and the future will no doubt see a DSO alongside it.

VOLTAGE SETTINGS

DSOs display a changing voltage over time. We have seen voltage changes using our DMM. One of the major differences between a DMM and a DSO is the ability to change around the time and voltage settings so that we can view just about any voltage over just about any period of time. The scope screen is usually comprised of a series of divisions: 10 across and eight high, as **Figure 1** shows. The scope will display voltage vertically up and down. If the voltage rises, so does the waveform. If the voltage goes down, so does the waveform. Time is measured and displayed horizontally from left to right. Notice that the scope screen has been divided in 80 squares, or divisions. Each square will be a certain amount of time and a certain amount of voltage, which you will set from the front panel controls. Let us set the scope to some common voltages and times as an example. If we want to measure DC charging voltage that we expect to be approximately 14 volts, we would touch the voltage arrow buttons until 5 V DC appears just below the left corner of the screen, as **Figure 2** shows. This means that each vertical division would be equal to a reading of 5 volts. If zero volts is exactly in the center of the screen, where the number 1 is, then we would have four divisions above zero to register positive voltages and four divisions below zero to register negative voltages. The total available voltage would be 20 positive to 20 negative, or 40 volts total. However, generally DSOs allow us to move the zero point up or down to suit the situation. Think about the charging system. Would we ever need to be able to read 20 volts negative? That is not likely. So let us move the zero down to one division from the bottom, as shown in **Figure 3**. Now what do we see? If the scope is still set on 5 V/Div, we will be able to see 35 volts positive and 5 volts negative. Again, using our example of charging system voltage, we have more volts available than we need. If we touch the down arrow under Ch 1 Setup, we will reduce the scale. Our next lowest choice is 2 V/Div. Will this be a workable setting? With seven divisions above the zero and one below, we will have a range of 14 volts positive and 2 volts negative.

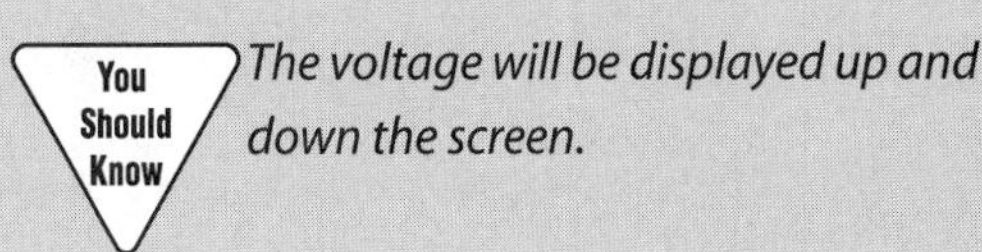

The voltage will be displayed up and down the screen.

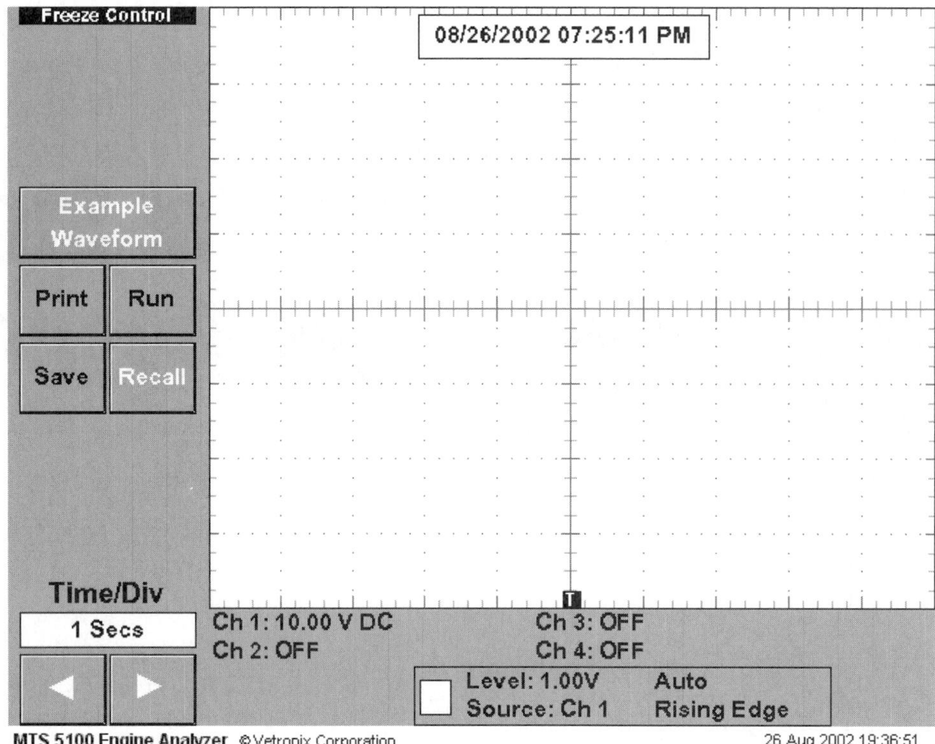

Figure 1. The typical DSO has eight divisions of voltage up and down and 10 divisions of time across.

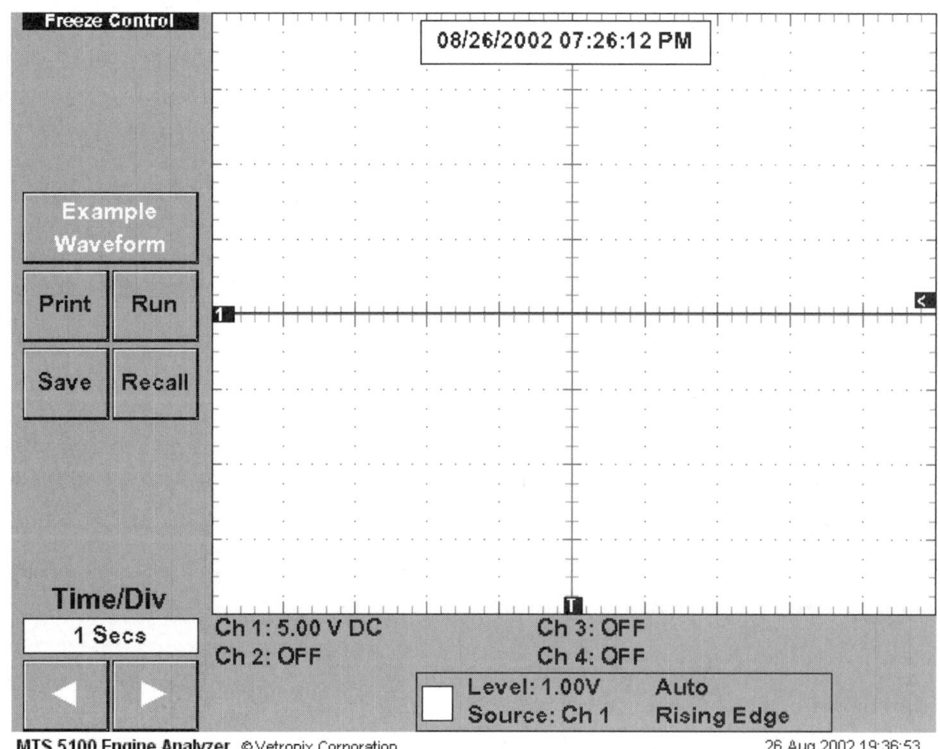

Figure 2. The DSO set for 5.0 V DC per division.

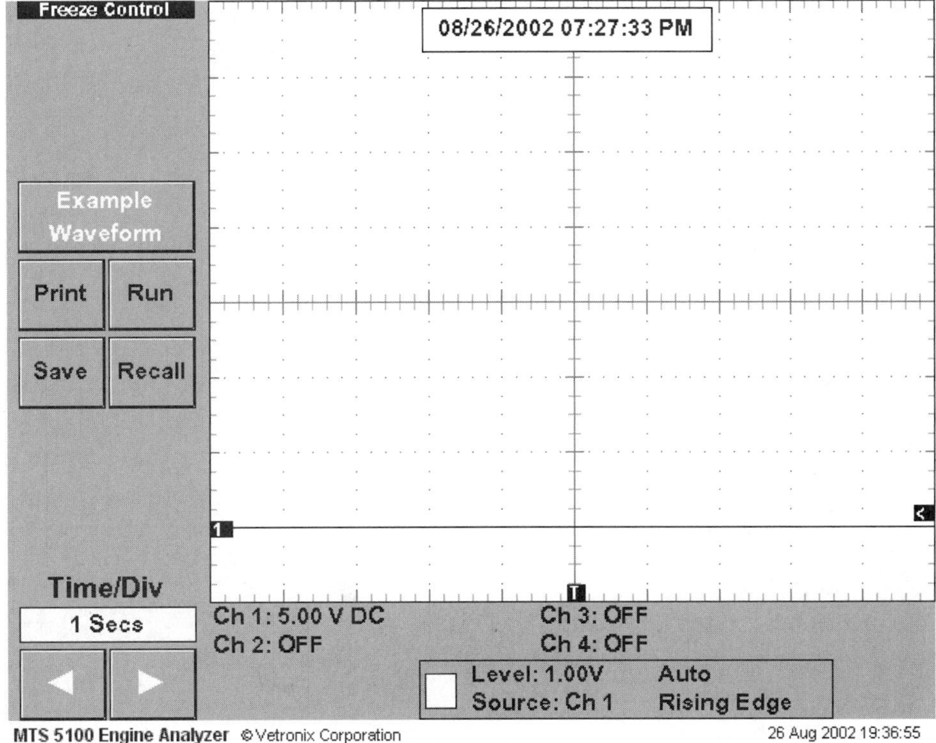

Figure 3. The zero-voltage line should be moved to one division from the bottom to measure DC voltage signals.

The DSO has the ability to go as high as 50.00 volts per division or as low as 0.050 volts per division. That is only 50 millivolts per division. As you scroll up and down, you will notice a pattern. Let us start with 1V. Hit the up arrow, and the scale changes to 2V. With another touch, we are at 5V. From 5V it will scale up to 10V, then 20V, and then 50V.

The pattern repeats itself: 1, 2, 5, over and over again until we run out of scales. The DSO pictured is a four-channel DSO, so we have the ability to have different voltages for each channel. This is helpful if we are trying to watch two, three, or four items that all have different voltages. Remember, no matter what DSO you have, the voltage will always be displayed in volts per division and there are usually eight voltage divisions up and down.

TIME SETTINGS

Now, let us look at the time scales. How long do you want to view the charging pattern? For 1 second total, with 10 divisions' worth of time across the screen, a 100 mSec/Div seems right, as **Figure 4** shows. Notice that the amount of time per division shows in the lower left corner of the screen next to the volts per division. Remember that a millisecond is one-thousandth of a second (0.001) and carries the small m as its abbreviation, so 100 milliseconds is one-tenth of a second (0.100 second). With 10 divisions each equal to one-tenth of a second, the total screen will equal 1 second. By touching the right arrow in the bottom left, we can increase the time per division all the way up to 20 seconds.

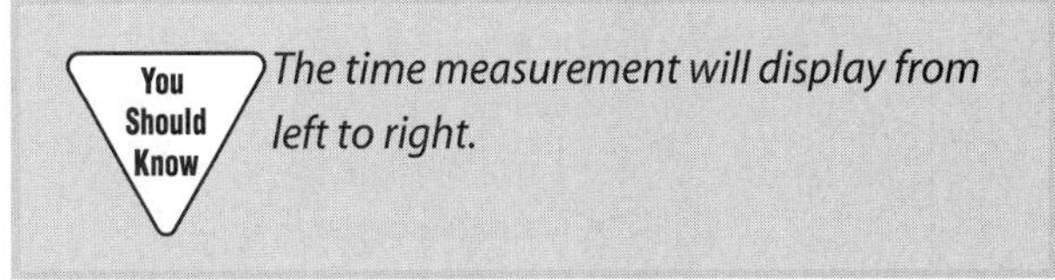

The time measurement will display from left to right.

At 20 sec/Div we would be able to view 200 seconds, or 3 minutes 20 seconds. On the other hand, by reducing the time per division, we will be able to capture time in increments as short as 20 microseconds. A microsecond is a millionth of a second. Within the automotive field, we generally find little application where we would need to measure a millionth of a second. Not all DSOs are capable of measuring a microsecond. However, most hand-held units can measure a millisecond per division. A millisecond is a thousandth of a second. Generally speaking, technicians should set the time per division so that they can see about two to four complete cycles of whatever they wish to look at.

The time setting carries over to all channels. We can have different voltages on each channel, but not different times.

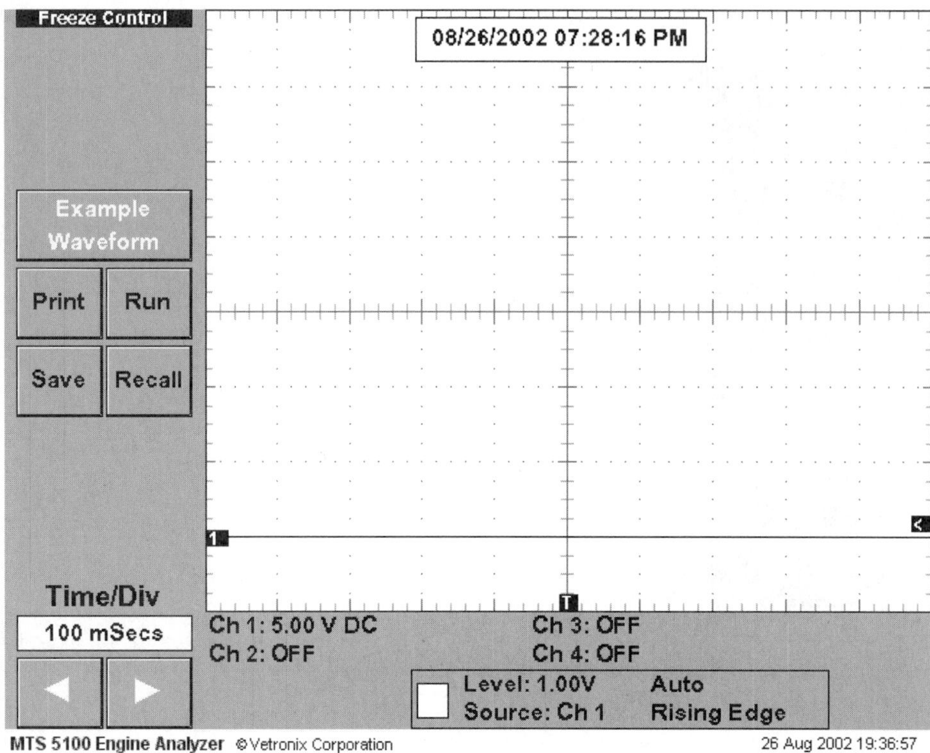

Figure 4. A DSO set for 100 m Sec per division will display 1 second total time.

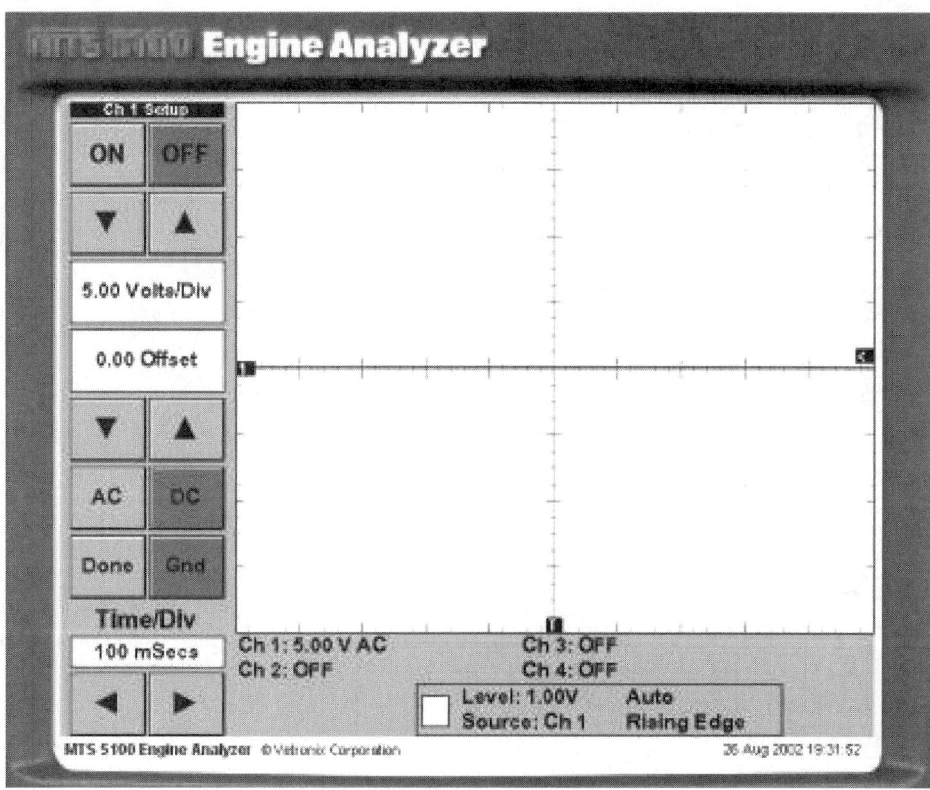

Figure 5. The DSO has been set to measure AC voltage, and the zero-volt line has been placed in the middle of the screen.

USING SIGNAL FINDER

Most DSOs will enable you to find a pattern. Under the channel controls menu is a box labeled signal finder. A touch of this button allows the DSO to analyze the voltages and the time frame and pick the volts per division and the time per division that are good for the screen.

The signal finder function will generally be able to display a pattern on the screen by analyzing the input for voltage and time fluctuations.

Once a pattern is displayed on the screen, you can adjust it to your liking. This auto-range button is a favorite because it saves you time. Many DSOs have auto range.

SETTING FOR AC OR DC VOLTAGE

The modern vehicle has both AC and DC signals that we must be capable of looking at. Setting the scope to a different type of voltage is simple. On the Ch 1 Setup menu on the left side of the screen is the AC/DC button **(Figure 5)**. We have changed to the AC scale, and the Ch 1 voltage in the lower left now shows 5.00 V AC. In addition, we have moved our zero up to the middle of the screen. We need to change to AC for some wheel speed sensors and some ignition components. Most of the vehicle, however, is DC. Remember to move the zero up from the bottom of the screen when you are checking an AC component so that you will have sufficient voltage capability below zero.

INPUT LEADS

The modern DSO has a variety of leads. Most have ignition capability, so there may be some ignition-specific leads. For the majority of inputs, we need to connect two leads to the scope: a positive or signal lead and a ground. Remember that everything that is displayed is a voltage above or below zero volts.

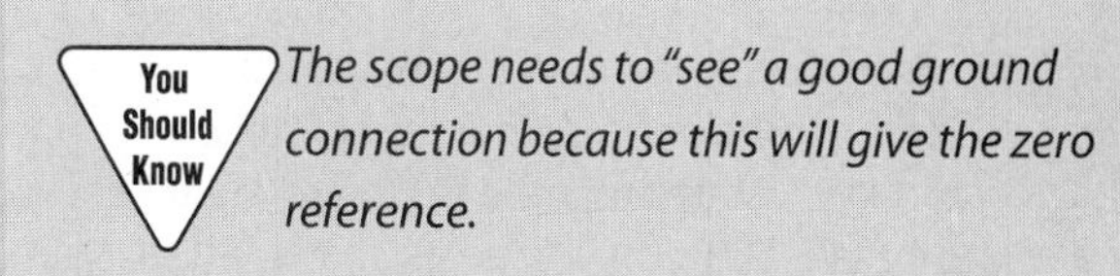

The scope needs to "see" a good ground connection because this will give the zero reference.

The DSO needs to have this zero-volt reference, and it will get it from the ground connection. Use a good, clean engine ground or the battery negative terminal. The other leads are to connect the other three channels for a total of four **(Figure 6)**. Additional connections on top of the DSO will be used for other functions.

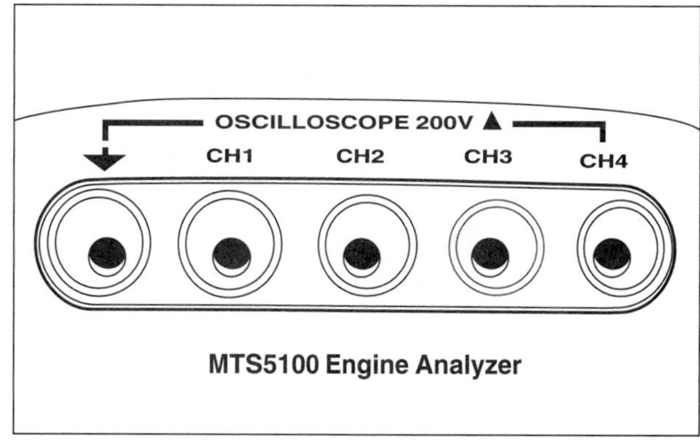

Figure 6. A four-channel DSO has four input connections plus a ground.

Summary

- A DSO gives the technician the ability to look at voltages over a period of time.
- A DSO usually has 10 divisions of time and eight divisions of voltage.
- The volts-per-division setting determines the scale of voltage.
- The time-per-division setting determines the scale of time.
- Normally, the zero line is placed near the bottom of the screen for DC measurements.
- Normally, the zero line is placed in the middle of the screen for AC measurements.
- A ground connection is necessary to give the scope reference.

Review Questions

1. Technician A states that a DSO set to 200 mV/Div and 200 mSec/Div will display a signal for a total of 2 seconds. Technician B states that a DSO set to 200 mV/Div and 200 mSec/Div will be able to display an 800-millivolt positive and an 800-millivolt negative signal. Who is correct?
   A. Technician A only
   B. Technician B only
   C. Both Technician A and Technician B
   D. Neither Technician A nor Technician B
2. Technician A states that the DSO is a voltmeter that is time based. Technician B states that the DSO can be used to view voltage changes over time. Who is correct?
   A. Technician A only
   B. Technician B only
   C. Both Technician A and Technician B
   D. Neither Technician A nor Technician B
3. The V/Div is the
   A. number of volts per division up and down the screen
   B. number of volts per division across the screen
   C. variable factor of time
   D. none of the above
4. The DSO will display volts
   A. up and down the screen
   B. across the screen
   C. from 0 to 25 volts only
   D. that have a positive polarity only
5. A 1 sec/Div setting will allow 1 minute of signal changes to be displayed.
   A. True
   B. False
6. A millisecond is a thousandth of a second.
   A. True
   B. False
7. A millivolt is a millionth of a volt.
   A. True
   B. False
8. A DSO should be set for AC or DC, depending on the expected input.
   A. True
   B. False
9. What is the advantage of a DSO over a DMM?
10. Why is the zero generally placed in the middle of the screen for AC measurements?

Chapter 17

DSO Trigger and Slope

Introduction

In Chapter 16, you found some of the benefits to using a DSO after setting volts and time per division. Getting a pattern on the screen is usually the most difficult operation when you first start using a DSO. The object of this chapter is to get that pattern on the screen in exactly the correct position. This is done with **trigger** and **slope** plus **trigger position**. We will guide you through the setting of all three. The DSO you are using might not be set up exactly the same as the one we are using, but the function of trigger and slope are common to most DSOs. The ability to set trigger position is not common, but we will explain that procedure so you will be ready to use almost any DSO.

ADVANTAGE OF USING TRIGGER AND SLOPE

In your toolbox will be some specialty tools that have only one function, such as an odd socket that works on only one specific ignition module. Without the tool, could you get the module off its mount? Perhaps, but how long would it take you? Trigger and slope are like that special socket. If they are used correctly and consistently, they will allow you to get the correct pattern in the right place on the screen. They will also allow you to turn the scope on when a specific event occurs, run the 10 divisions of time, and then freeze the pattern on the screen. Trigger can be an internal input that is taken from the pattern that will be on the screen or be an external input from a trigger probe or sync probe. We will deal with internal triggering first. Let us define the three internal functions.

1. Trigger is the specific voltage that the scope must see to begin tracing a pattern on the screen. For instance, if the voltage is 5 volts, the scope will not begin tracing anything on the screen until it sees 5 volts from the main input leads. If the leads are connected to the battery, where the voltage low is 10 volts and the voltage high is 14.5 volts, then the DSO will not turn on. The screen will remain blank because the input leads never "saw" the 5 volts that we set as the trigger. Trigger is always used in conjunction with slope.

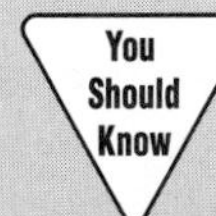

The trigger is the voltage, which will start the display. The slope is whether the voltage is rising or falling.

2. The slope is whether the trigger voltage is achieved rising or falling. For example, if we use a 12-volt trigger and a rising slope, the DSO will remain blank during the cranking of the engine when the battery voltage drops from 12.6 volts to approximately 10 volts. However, it will begin tracing a pattern when cranking is completed and the charging system begins to charge the battery.

   As the voltage rises past 12 volts, the scope will begin tracing the battery voltage pattern. The 12-volt trigger with a rising slope condition is met and the DSO begins tracing. If we have the DSO set for a single shot, it will give us 10 divisions of time on the screen and then freeze.
3. The trigger position is the spot on the screen where the pattern conditions of trigger and slope are met.

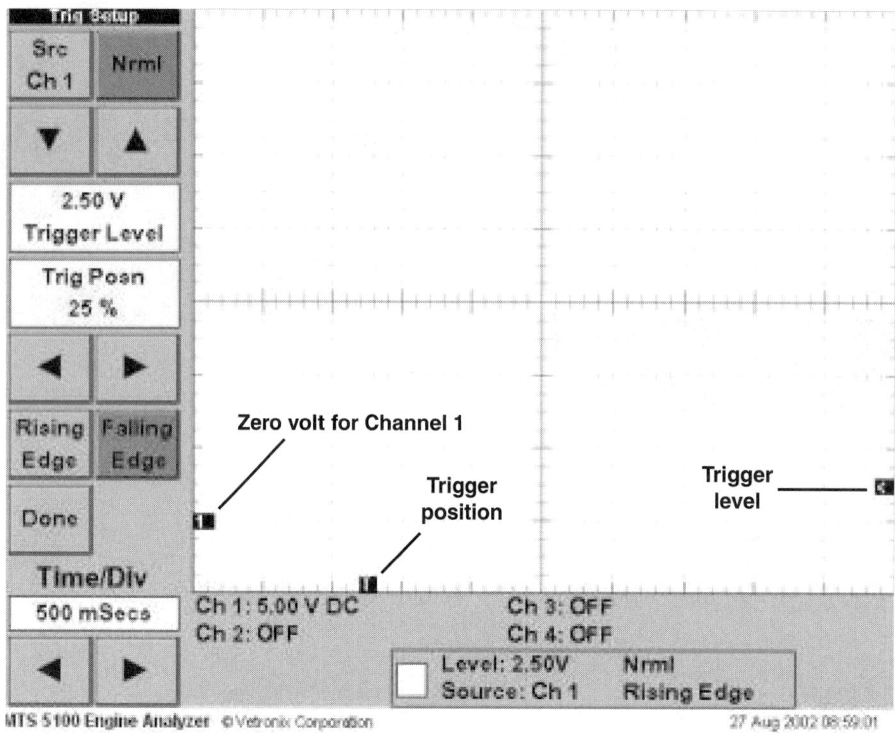

Figure 1. Trigger voltage and position are shown on the screen.

SETTING TRIGGER AND SLOPE

Let us look at the setting trigger screen and discuss the setting of these three functions. On the main DSO screen is the trigger setup button. After touching the button, the setup screen appears, as shown in **Figure 1**. The up and down arrows toward the top left of the screen will set the trigger level. The trigger voltage level will show on the screen as an arrow on the right edge of the screen. We are currently set to 2.5V, which is half a division because we are set for 5V per division. The trigger position is set using the right and left arrows on the left edge near the middle and will be represented by a "T" on the bottom edge of the screen. Currently, our trigger position is 25 percent of the distance from the left edge. Directly below the arrows is the slope function. Currently, the rising edge is highlighted. Let us put these together. When the channel 1 signal rises past 2.5 volts, the DSO will begin tracing, with the point of trigger being approximately 25 percent from the left edge. Notice that the information is printed for you along the bottom edge of the screen in a box, in addition to having the symbols printed on the DSO radicule (the 10 wide by eight high portion of the screen).

USING AN EXTERNAL TRIGGER

Most DSOs have the ability to use the incoming channel signal as the trigger. In addition, some DSOs have an external function, which allows the use of a **sync probe**. A type of sync probe is shown in **Figure 2** connected to a dedicated connection on the top of the analyzer.

Figure 3 shows the additional screens of information required to set the sync probe. We have highlighted the

Figure 2. A sync probe can be used as an external trigger.

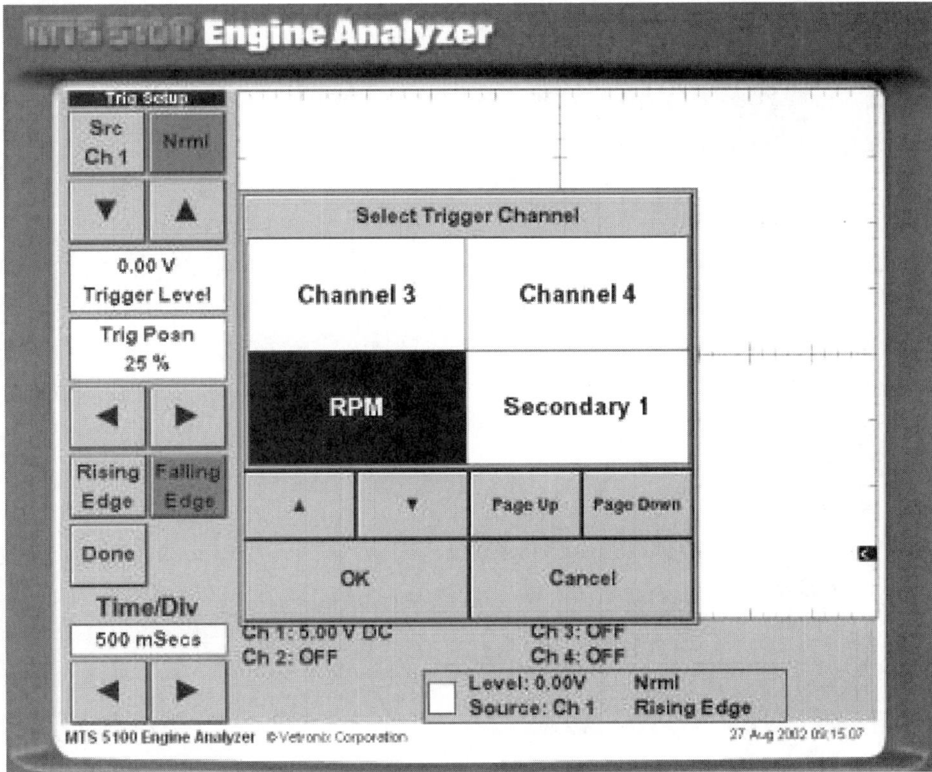

Figure 3. Information required to set the trigger.

RPM (revolutions per minute) button as the trigger channel. This means that the sync probe can be placed around the "source" of the triggering signal, such as a spark plug wire. This comes in handy when a repeating signal is present and you want to look at the signal when some other event occurs. For example, an ignition coil fires all cylinders, and you decide that you want to look at the coil while cylinder three is firing. If the sync probe is placed around plug wire three, the scope will start and display the pattern when the correct plug fires.

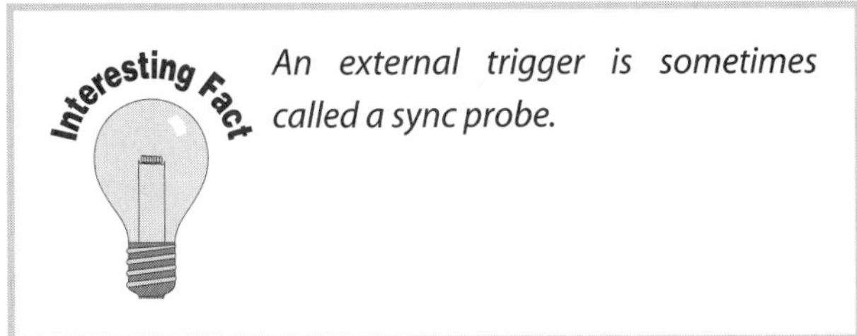

An external trigger is sometimes called a sync probe.

EXAMPLES

Let us use some examples and look at how we would set up the DSO to capture the event. We will start with a fuel injector. Although it is not the purpose of this section to discuss fuel injector diagnosis, we will learn how we can capture it on the screen. The first thing we must think about is how the injector functions. We realize that at this point you might not know, so we will fill in the necessary blanks. Power to the fuel injector comes through the ignition switch, while the ground is applied by the PCM, as diagrammed in **Figure 4**. We will place the DSO positive lead into the negative of the circuit because it will show the turning on and off of the injector. The positive side shows only power and will not be as valuable. When it closes, the injector will generate a spike usually of about 75–125 volts, so the DSO must be capable of capturing a voltage that high. Twenty volts per division should be enough. Injectors stay open for

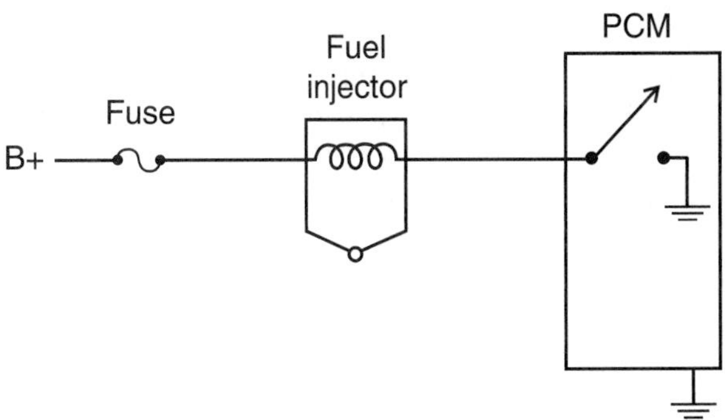

Figure 4. The PCM controls the ground side of the fuel injector.

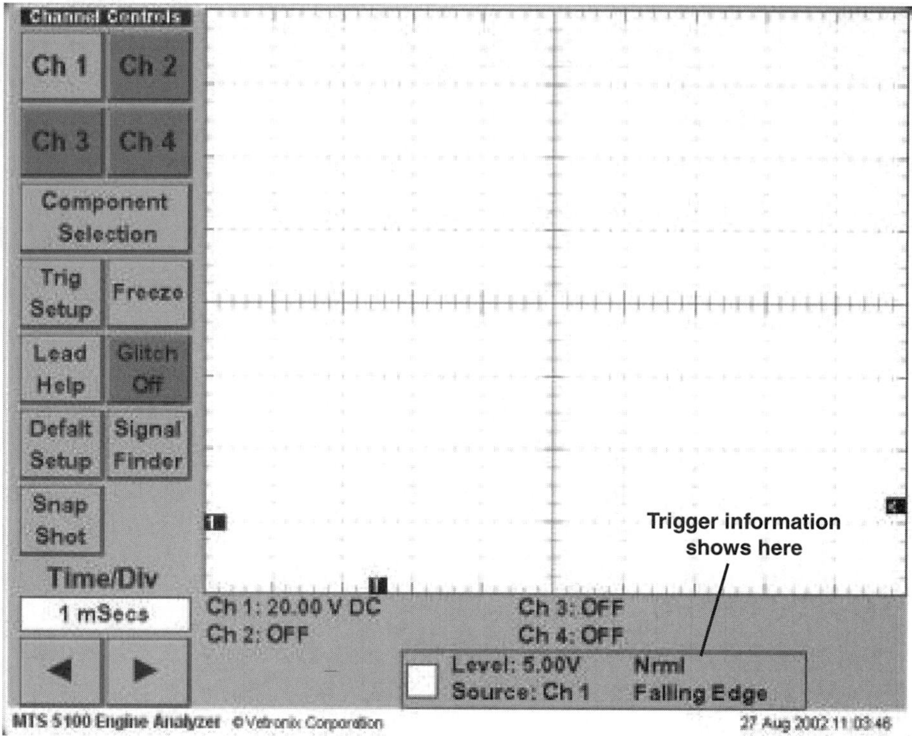

Figure 5. The screen shows a 5-volt trigger with a falling edge slope from a channel source.

approximately 2–5 milliseconds, so a time per division of 1 millisecond should work. Now let us look at trigger and slope. The ground side of the injector sees power first, and then it sees the ground supplied by the PCM, so voltage will drop from approximately 14 to 0 volts when it is injecting fuel. Any amount between 14 and 0 volts should be adequate for the trigger voltage. As for the slope, is it rising or falling? It goes from 14 volts down to 0 volts, so it is falling. The last question is, where do you want the pattern to be on the screen? Start with 25 percent and then adjust. It is important to note that 25 percent is the starting point, and adjustment might be required. **Figure 5** shows the setup. We have all the information we need: a 5.00-volt trigger with a falling edge (slope) and 20 volts per division at 1 millisecond per division with the trigger point at 25 percent. Connect it and see what happens. **Figure 6** is the screen capture from a fuel injector. In the first section, the screen shows power until the trigger point ("T" on bottom edge of screen). When the injector fires, the circuit goes to 0 volts because the PCM applied a ground for 3.5 milliseconds. A spike of 80 volts occurred just as the PCM turned off the injector. Remember that our goal of using trigger, slope, and trigger position was to get a pattern using the majority of the screen.

Let us consider one more example of using trigger and slope, but this time with an external rpm trigger. We connect the DSO to the ignition coil on an eight-cylinder vehicle with a distributor. This means that the ignition coil will fire eight times in a row. If we decide that we want to look at only cylinder three coil firing, we would connect the sync around that plug wire. **Figure 7** shows what the primary ignition for cylinder three looks like with an external rpm trigger. Notice that the bottom of the screen shows that the trigger source was rpm and that we used a rising edge slope. Our DSO settings included 50.00 V DC per division and 1 mSec per division. In the ignition Chapters 46–53, we look at how we analyze a primary pattern such as this. Before analyzing, you must be capable of getting the pattern on the screen, which has been the reason for introducing you to trigger, slope, and trigger position.

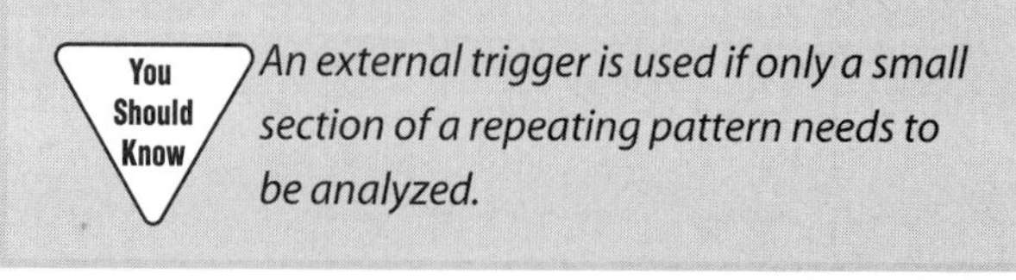

An external trigger is used if only a small section of a repeating pattern needs to be analyzed.

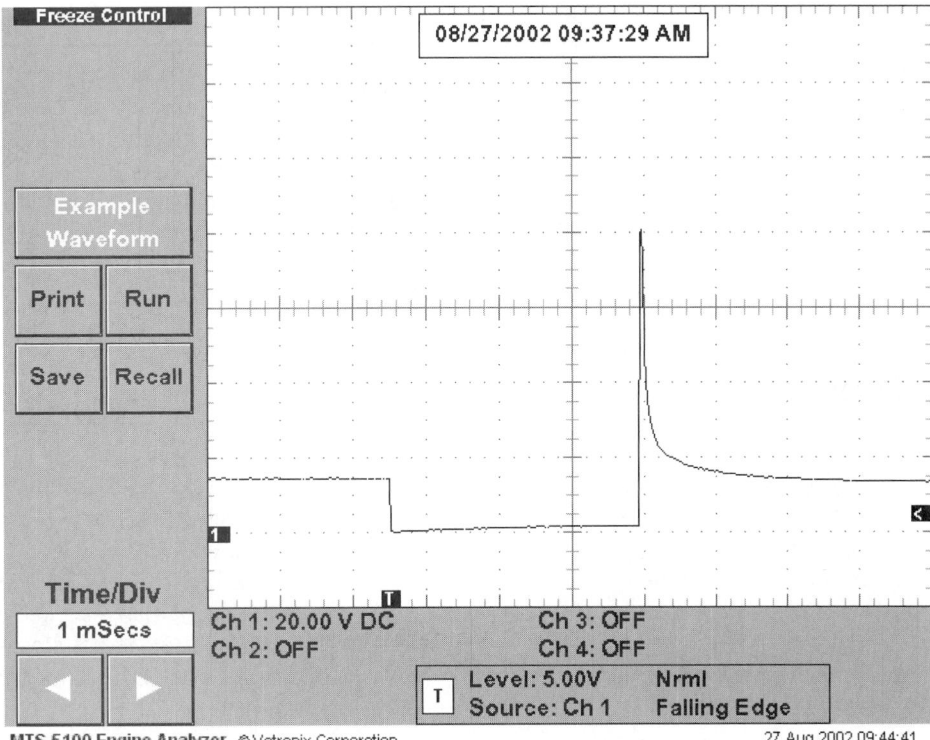

Figure 6. The presetting of trigger and slope places the fuel injection pattern where it is easiest to see.

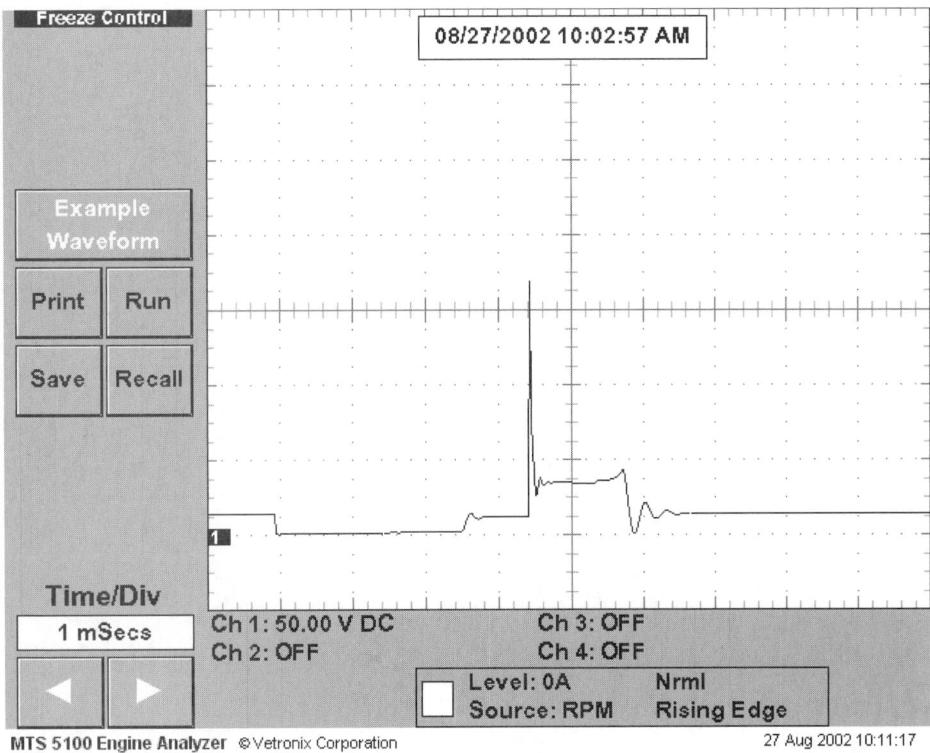

Figure 7. The external sync probe around the number three plug wire places the primary pattern for number three on the screen.

Summary

- Trigger is the voltage that will start the DSO display.
- Slope is whether the trigger voltage is rising or falling.
- Trigger position is the spot on the display where the pattern will be.
- The use of trigger, slope, and trigger position involves the placement of the pattern on the screen in the position that will allow the best analysis.
- A 5-volt trigger, falling slope, and 25 percent position means that the voltage will have to fall past 5 volts for the pattern to begin displaying 25 percent from the left edge of the screen.
- An external trigger can be used to place a specific pattern on the screen.

Review Questions

1. Trigger is the
   A. time per division of the DSO
   B. volts per division of the DSO
   C. voltage that will start the pattern
   D. position of the pattern on the display
2. The slope is
   A. whether the trigger voltage is rising or falling
   B. the position of the trigger on the screen
   C. external or internal
   D. none of the above
3. Technician A states that the use of trigger and slope determines the height of the pattern. Technician B states that the use of trigger and slope determines where on the screen the pattern will be. Who is correct?
   A. Technician A only
   B. Technician B only
   C. Both Technician A and Technician B
   D. Neither Technician A nor Technician B
4. A rising voltage signal that crosses 1 volt will be displayed on a DSO that has a trigger of 1 volt with a rising slope setting.
   A. True
   B. False
5. An external trigger is used to set the voltage level.
   A. True
   B. False

Reading and Interpreting a DSO Pattern

Introduction

Now that we know how to get a pattern on the screen, it is time to look at the most common types of patterns usually encountered by the technician. Generally, these fall into the categories of AC, DC, or pulsed. The pulsed signals are further divided into the categories of variable frequency, variable duty cycle, and variable pulse width. Although this may sound complicated, it really is not. It is, however, important information that the DSO will be showing you, and it is critical that you have a basic understanding of what you are looking at.

AC SIGNAL

You are probably familiar with AC signals and do not even realize it. In Chapter 49, we look at the pickup coil, or variable reluctance sensor, common to ignition systems. It generates a signal that is a form of an AC signal because it periodically changes its polarity or direction. An AC signal will have part of its signal above the zero line in the positive direction and part of its signal below zero in the negative direction **(Figure 1)**. Notice that the zero line, where the number 1 is, is located in the middle and that the DSO has been set for AC (Ch 1: 5.00 V AC). Follow along as we analyze this pattern. The voltage first begins in a positive direction and rises for slightly more than three divisions until it peaks. At this point, the voltage drops rapidly until it reaches and crosses the zero-volts line 1½ divisions later. The exact reverse of the pattern now takes place. The rapidly dropping voltage goes negative until it reaches about two divisions, where it bottoms out. At this point, the voltage begins to rise, crosses zero, and eventually gets to about 20 volts. At this point, the voltage again begins to drop, crosses zero in the negative direction, and the pattern repeats itself. This is an AC signal usually called a sine wave. What voltage was actually achieved? What is the scale or V/Div? The lower left of the screen shows 5.00 V AC. Remember that this means that each division, either negative or positive, equals 5.00 volts. Our pattern reached a positive peak of four full divisions, so:

5.00 volts × 4 divisions = +20.0 volts

Because the negative pattern was close to a mirror image of the positive pattern, it registered −15.0 volts. Looking at time, a full positive and negative cycle lasted for about two divisions, and again the lower left of the screen shows that each division is worth 5 mSec/Div, so:

5 mSec × 2 divisions = 10 mSec

The entire event took only 0.010 seconds. The number of complete AC cycles that occurs during 1 second is referred to as the frequency of the signal. If our example had occurred during 1 second, it would have had a frequency of 1 cycle per second. Frequency measurements are usually in hertz, which is the number of cycles per second. Hertz is usually abbreviated as Hz. How fast is a 60-hertz signal? If you said that it is 60 times a second, you are right. Remember that a complete cycle is necessary: both the positive and negative voltage waveforms must be present. Generally, a technician should set the time base to see a couple of complete cycles. Setting the time base higher will increase the number of cycles on the screen but will make your interpretation difficult because of the closeness of the waveforms. Adjust the time per division (Time/Div) so that one to two full cycles are on the screen, and your analysis will be easier. The exception to the one to two full cycles is if the signal is cylinder oriented. Our signal was from a four-cylinder pickup coil, so we set the time to display the full four patterns, each one corresponding to a cylinder.

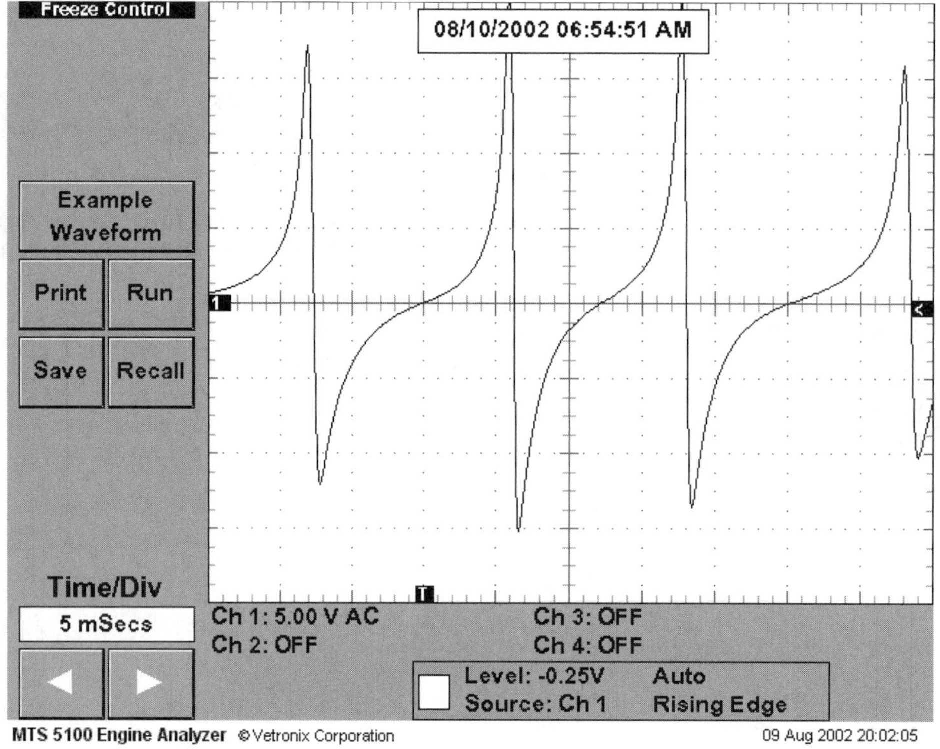

Figure 1. The AC signal from a pickup coil. The DSO has been set for AC and the zero placed in the middle.

DC SIGNAL

A DC signal is a voltage signal that contains only a positive or a negative polarity. Generally, positive voltages are encountered in the automotive electrical system. A useful feature of most DSOs is the ability to move the zero down so that a DC signal fills the screen. The move button will move the zero up or down as needed, as **Figure 2** shows. The zero line is where you see the number 1 appear on the left edge of the screen. Make sure that you have some space below the trace, as shown. If you place the zero at the bottom of the screen, you will not be able to see any negative pulses that might be present. Preset the scope with the ability to display some negative voltage, just in case.

For example, with the zero down on the second division line, you will have seven full divisions' worth of voltage to display. Pick a volts per division that allows for the greatest pattern height. This makes reading the signal easier and more accurate on your part. Let us set up the DSO for a battery voltage reading with the vehicle running. Typical charging voltage is approximately 14 volts. If we move the zero down to the top of the first division, we will have seven divisions' worth of voltage available for measurement. Fourteen volts with seven divisions would need a 2 V/Div setting, right? This would be the correct setting, but we must make sure that the entire pattern is displayed. Every peak must be visible. If all are not, increase your V/Div until the entire pattern is visible. In our example, the charging voltage was higher than 14, so we had to go up to 5 V/Div. You do not want to miss any voltage surges that might have an effect on PCM circuits.

Consider the time division across the screen. Your setting will depend on what you are looking for. We are interested in short glitches that might appear but are very fast, so the time per division is very fast: 100 microseconds. What is the level of charging voltage displayed in Figure 2? Count the divisions, and remember that each small line is 25 percent of a division. Our voltage looks like it is just about 13 volts and it is quite pure, meaning that it is flat with no ups or downs. If an intermittent problem is suspected, increase the time base so that the pattern will be present for a longer period of time, up to the maximum of the DSO. Some DSOs allow for capturing extended periods of time, even though they are displaying only 10. This gives you the ability to capture up to 20 minutes' worth of patterns and view them by moving the 20 divisions across the screen in a parade fashion. However, frequently the DC pattern will not be a constant flat line, but instead will show up in a pulsing format. The type of pulsing pattern can be characterized as frequency, duty cycle, or pulse width modified. Let us look at the these different forms of pulse patterns.

VARIABLE FREQUENCY

If a DC signal is pulsing on and off like that shown in **Figure 3** and the on and off times appear to be equal, the

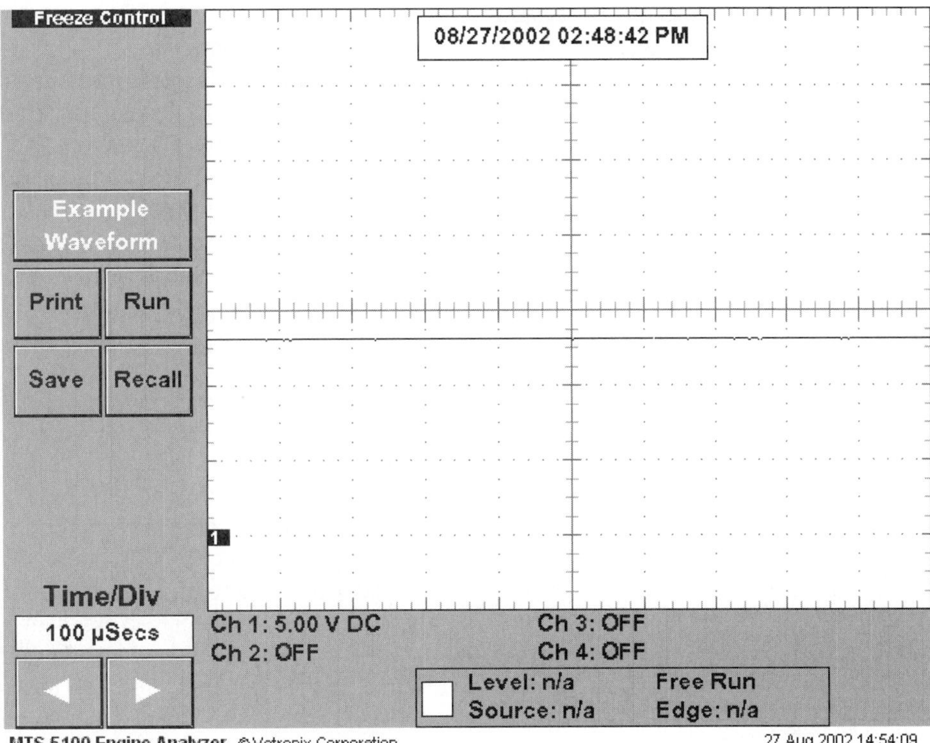

Figure 2. Battery voltage will be a straight DC line. Zero volt is indicated by the number 1 on the left edge of the screen.

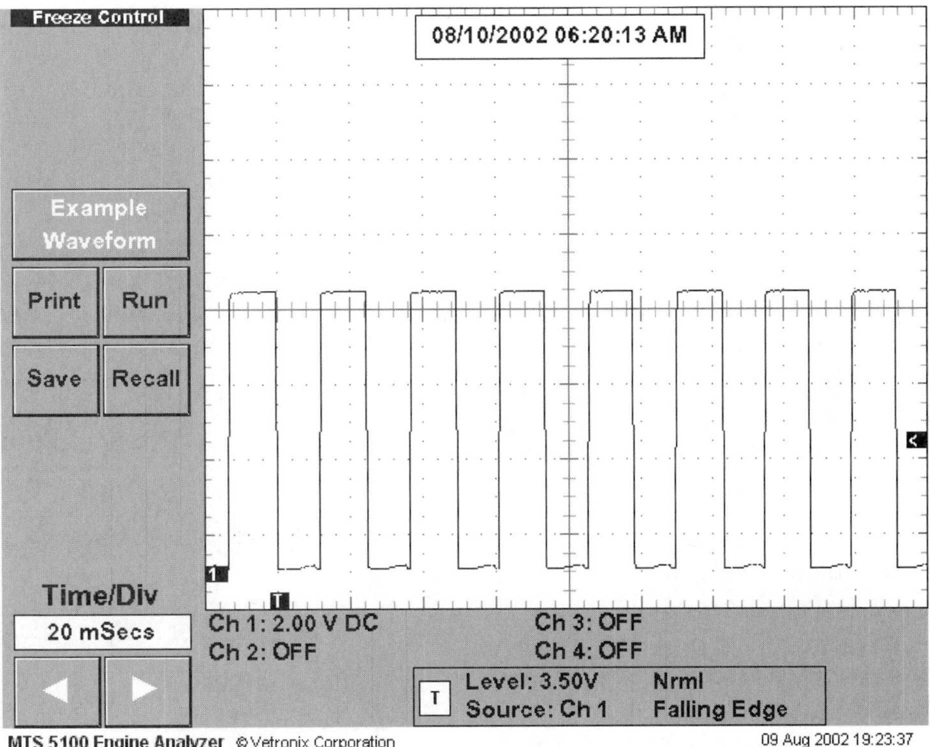

Figure 3. A pulsing DC voltage that is frequency oriented at 20 mSec per division, generating a 40-hertz signal.

signal is frequency oriented. Remember that we discussed the frequency of a signal when we looked at an AC signal. Frequency measured in hertz is the number of complete cycles per second. A frequency of 95 hertz means that the signal cycled 95 times every second. Keep in mind that this cycling might be an on/off signal or a voltage pulse that never reaches zero. The important component is that it actually cycles. Let us look at some examples that might make this concept easier to comprehend.

Look again at Figure 3. It is an example of a cycling signal that starts slightly above zero (where the number 1 is), rises over three divisions above zero, stays at this level for approximately ½ a division, and then drops back down the three divisions to zero. The signal stays at zero for ½ a division and then repeats itself. Notice that the signal repeats eight times across the screen. This is a frequency-oriented signal. It repeats over and over, and the rise time of the voltage equals the fall time. The equal rising and falling voltage and the equal time high and low are the keys to this being a frequency-oriented signal.

Figure 4 is also an example of a signal that has frequency orientation. The voltage rises from zero straight up, stays high for less than a division, and then falls back to zero, where it stays for less than a division. This signal is faster than that shown in Figure 3. How do you know? You have realized that the time base is the same, 20.0 milliseconds, and the number of full cycles is greater, so the frequency must be higher. Again, the characteristics of equal rise and fall voltages plus equal high and low time make this a frequency-oriented signal. You will find many examples of frequency-oriented signals on vehicles. Most of the **Hall effect sensors**, some **manifold absolute pressure (MAP) sensors**, and mass airflow sensors generate a frequency-oriented signal. The number of pulses or cycles per second (hertz) will be counted by a processor or module and be interpreted as speed, pressure, or mass. In this chapter we will look at some specific waveforms that are frequency oriented.

DUTY CYCLE PATTERNS

A duty cycle–oriented pattern is frequently encountered in automotive electronics. Specific examples include timing signals, mixture-control solenoids, some charging systems, and some idle air systems. If you keep a frequency-oriented pattern in mind, a duty cycle pattern is easy to understand. Duty cycle refers to the amount of on time during the length of one complete pattern or cycle. As the on time increases, the off time decreases. A charging system with a digital regulator is an excellent example to start with. Most regulators cycle on and off hundreds of times a second, so their frequency remains relatively the same. However, the on time is varied, based on required charging current. If more current is needed to run some additional accessories or charge the battery, the on time is increased and the off time is decreased. This is also referred to as **pulse width modulation (PWM)**. **Figure 5** shows a signal with approximately a 25 percent on time and a 75 percent off time. The top of the pattern is the off section, and the

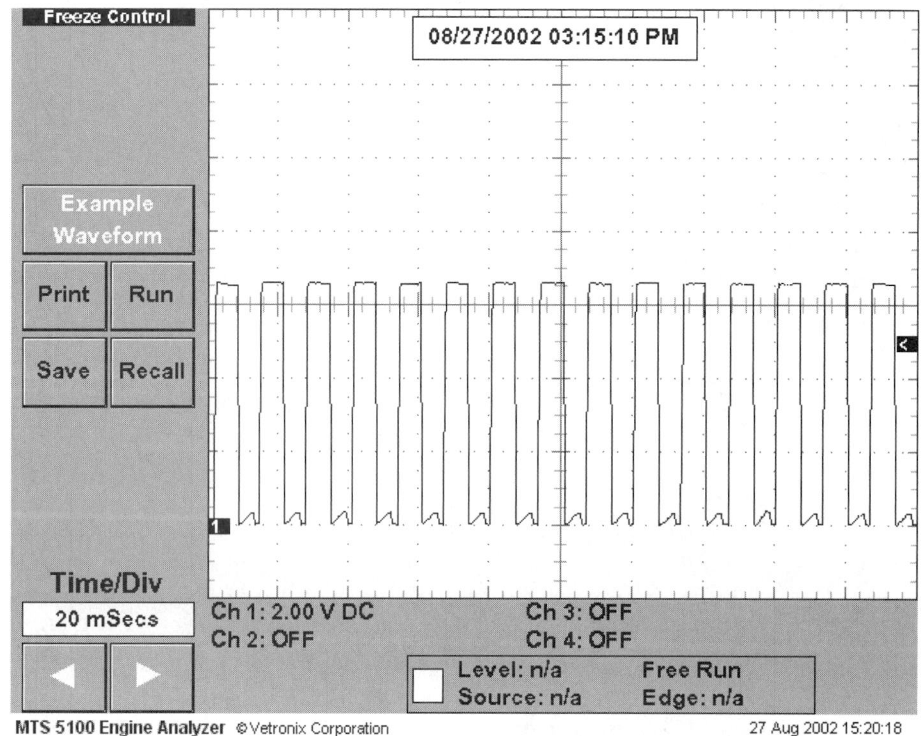

Figure 4. A 75-hertz DC signal.

bottom of the pattern is the on section. Notice that the complete on/off cycle lasts about 20–30 milliseconds. If the pulse width of the signal changes to 50 percent on and 50 percent off, the pattern would look like that shown in **Figure 6**. Notice that the complete on/off cycle again lasts about the same period of time. The key to understanding a

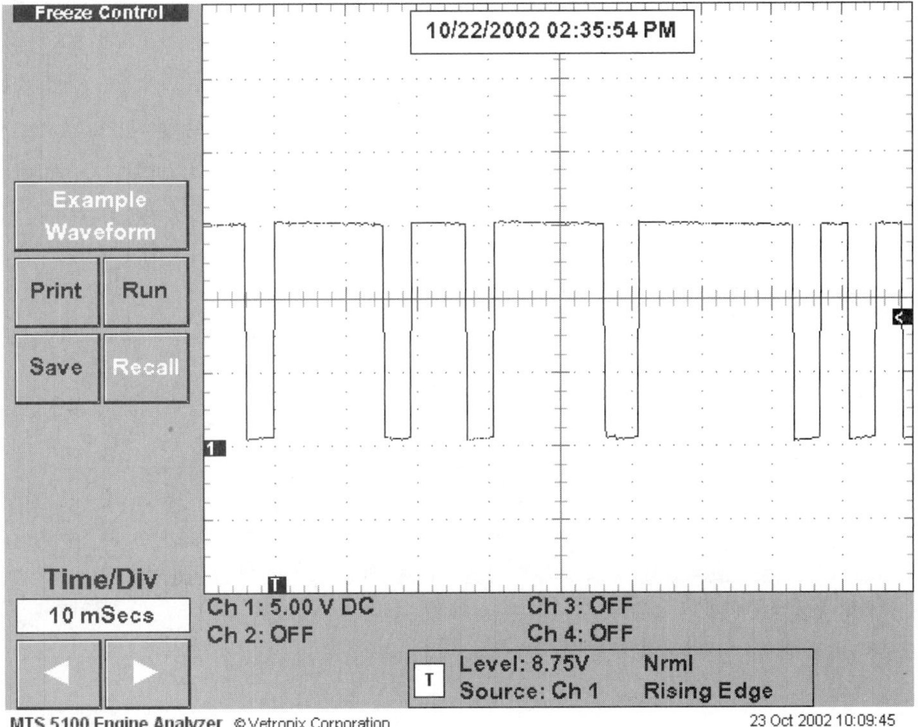

Figure 5. A 25 percent duty cycle pattern.

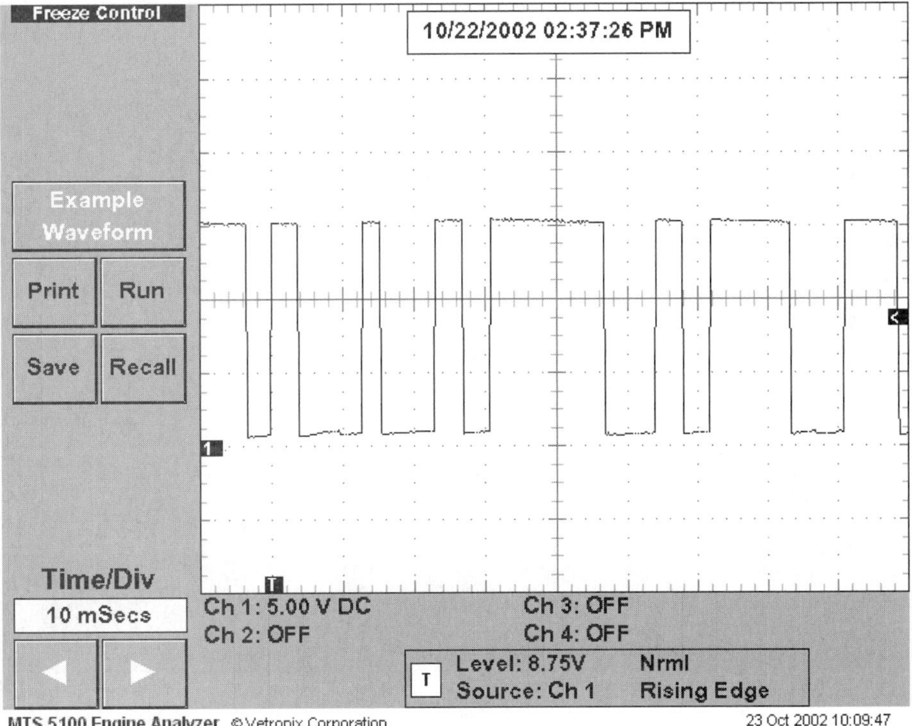

Figure 6. A 50 percent duty cycle pattern.

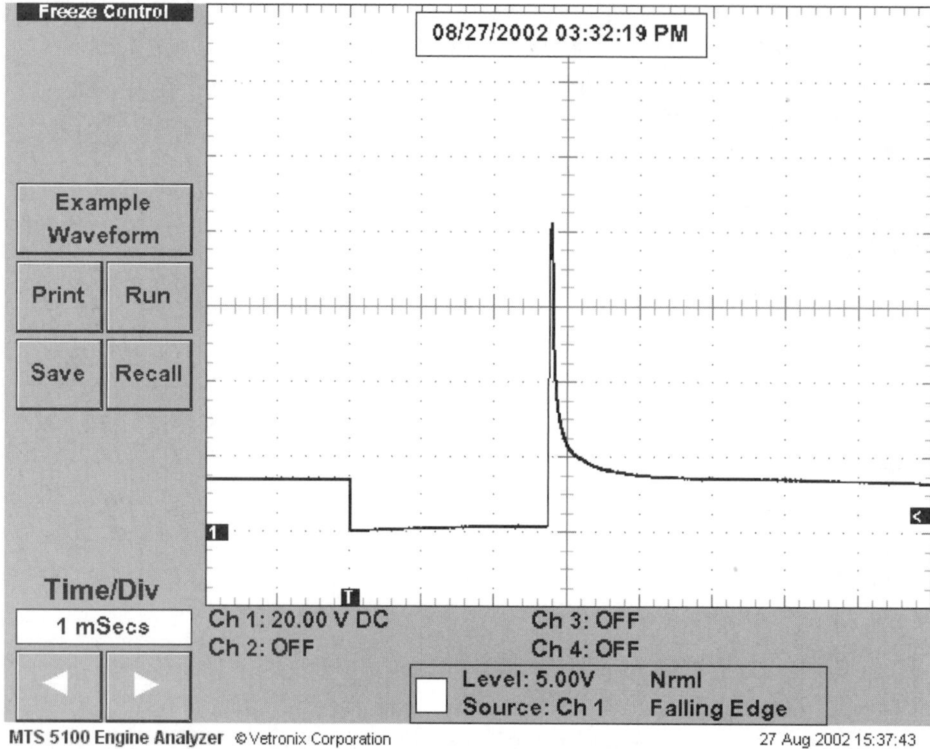

Figure 7. Fuel injectors are examples of PWM signals. This injector is open for 2.75 mSec.

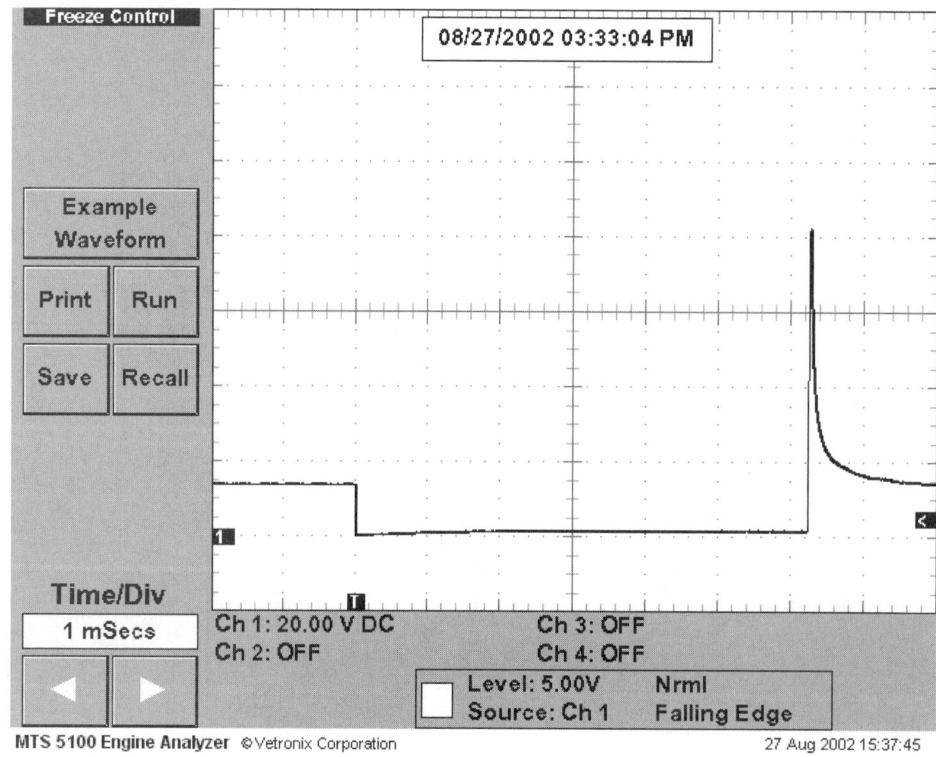

Figure 8. As load increases, pulse width increases to 6.25 mSec.

duty cycle or pulse width–modulated signal is in realizing that the number of pulses per second (the frequency) does not change. Only the relationship between the on and off times changes.

PULSE WIDTH

A specific example of a pulse width signal will change frequency under various conditions. A fuel injector's on time is usually measured in milliseconds and will vary with increasing and decreasing engine loads. Additional load will require additional fuel, so the injector on time will be increased. This is another way of saying that the pulse width will be increased. There is a difference between pulse width and duty cycle. Remember that duty cycle signals generally remain at a fixed frequency. If we continue with our fuel injector example, can the number of pulses per minute remain the same? It cannot. As the engine speed changes, the number of pulses must also change. Usually one pulse per rotation of the engine is used, so if the engine is running at 2000 rpm, we will need 2000 pulses per minute. If the engine speed is reduced to 900 rpm, the number of pulses must also be reduced. **Figure 7** shows a fuel injector pattern at idle. The injector is on during the down phase, in which the voltage descends to zero. We are injecting fuel for about 2¾ milliseconds. If this is a pulse width–modulated signal, then it must change under different conditions. **Figure 8** shows a pulse width signal recorded of the same vehicle under hard acceleration. Notice that the on time (pulse width) increased to more than 6 milliseconds. This substantial increase in pulse width represents the additional fuel required under load. There are other examples of a PWM signal on vehicles.

Summary

- An AC signal has a positive and a negative component.
- A DC signal is either all positive or all negative.
- Make sure that the basic settings of time and voltage are sufficient to have three to five patterns on the screen.
- A variable frequency signal changes the number of times it cycles in a given period of time.
- A variable frequency signal does not change the relative on/off times of the cycles and generally is fixed at 50 percent.
- A duty cycle signal varies the on/off time and usually keeps the frequency the same.
- A PWM signal varies the on time and usually varies the frequency.

Review Questions

1. A pulse width–modulated signal is one that will have
   A. a varying on/off time
   B. the same number of voltage changes per second
   C. a varying number of pulses per second
   D. none of the above
2. A variable frequency signal is one that will have
   A. a varying on/off time
   B. the same number of voltage changes per second
   C. a varying number of pulses per second
   D. none of the above
3. A fuel injector pattern is observed at idle rpm. Technician A states that this signal is an example of a pulse width–modulated signal. Technician B states that the on time should increase with additional engine load. Who is correct?
   A. Technician A only
   B. Technician B only
   C. Both Technician A and Technician B
   D. Neither Technician A nor Technician B
4. As the number of pulses per second increases, the ________ is increasing.
   A. duty cycle
   B. pulse width modulation
   C. AC
   D. frequency
5. As the on time of each pulse increases (without changing the number of pulses), the ________ is increasing.
   A. duty cycle
   B. pulse width modulation
   C. AC
   D. frequency
6. A changing pulse width indicates that the on time is changing.
   A. True
   B. False
7. More pulses per minute indicate a PWM signal.
   A. True
   B. False

8. An AC signal will have both negative and positive voltages.
   A. True
   B. False
9. A duty cycle signal changes frequency and on time.
   A. True
   B. False
10. What is the difference between a variable frequency signal and a duty cycle signal?
11. Why does a DC pattern usually have the zero line down toward the bottom of the screen but not at the bottom?

Using a Current Probe with a DSO

Introduction

The DSO by itself is a great tool, capable of giving the technician information that is useful in diagnosis and repair. However, realize that the DSO by itself will give you only a voltage pattern. Although much diagnosis can be accomplished with just a voltage waveform, having the ability to look closely at a current waveform increases the level of your diagnostic capabilities. A natural addition to the modern DSO is a current probe, which will generate a current waveform. With both voltage and current waveforms available, the technician has all the information required.

WHY USE A CURRENT PROBE?

The interrelationship among voltage, current, and resistance is Ohm's Law, and we have spent extensive time looking at it. Remember, if you know two of the three values, you can calculate the third. Adding a current waveform will give us that remaining value, leaving only resistance, which can be calculated using Ohm's Law.

If we know the current and the voltage, we can calculate the resistance under load of the circuit.

Another reason to consider the current probe is the ease of operation. Just clamp it around the wire, and it creates a waveform that indicates what is happening inside the wire. There is no need to backprobe a connector or disconnect a component. The current probe gives you current waveforms that are captured under load, which is far more accurate than using an ohmmeter. In addition, to use an ohmmeter, the component must be isolated from power and ground to protect the meter. That is not so with the current probe. The more you use it, the more you will rely on its information to help with diagnosis and repair.

SETTING AND CALIBRATING THE PROBE

Many different probes are on the market, but they all function basically the same. They all generate a voltage signal based on the level of current flowing through a wire in the center of the probe. **Figure 1** shows a common current probe.

You Should Know

Current probes have an internal battery and need to be zeroed each time they are used. In addition, if the probe is used under the hood, the heat of the engine requires that the probe be zeroed periodically.

The amount, or level, of the voltage signal is based on the output of the probe. Our probe generates a 100 millivolt signal for every amp of current flowing. This volt-

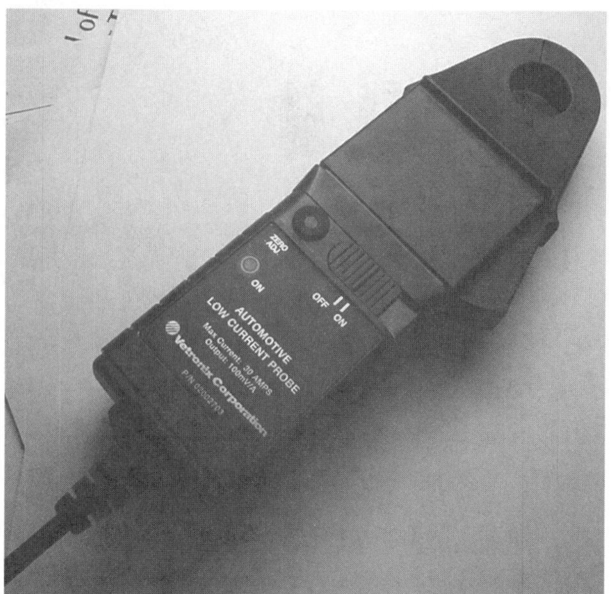

Figure 1. A current probe will generate a signal that is displayed on the DSO.

age can be read in two different ways. If we input the voltage signal directly into the DSOs, the pattern will show 100 millivolts for every amp flowing. However, some DSOs have the ability to recognize that a current probe is the input device and convert the signal to amperage that shows directly on the screen. It makes no difference to the accuracy or the interpretation of the signal.

All current probes have an internal battery and the ability to zero themselves. The small knob on the left side of our example is turned until zero volts or zero amps shows with the current probe not around a conductor.

SETTING UP THE DSO

Setting up the DSO is easy. Some DSOs have a current setting in the menu, as **Figure 2** shows. Click on the current waveforms, as Figure 2 shows, and the DSO will set itself up so that the screen looks like **Figure 3**. Notice that near the bottom left of the screen the Ch 1 notation is for 1.00 A DC. This means that each division is equal to 1.0 amp. The menu system did the work for you.

If your DSO is not menu driven and does not have a current waveform selection, set the volts per division to match the probe output as a starting point. Our probe generates a 100-millivolt signal for each amp it senses, so **Figure 4** shows that the Ch 1 setting is 0.10 V DC or, in other words, 100 millivolts per division.

Once the DSO is set, rotate the zero adjust until the current line is on zero. You are ready to proceed.

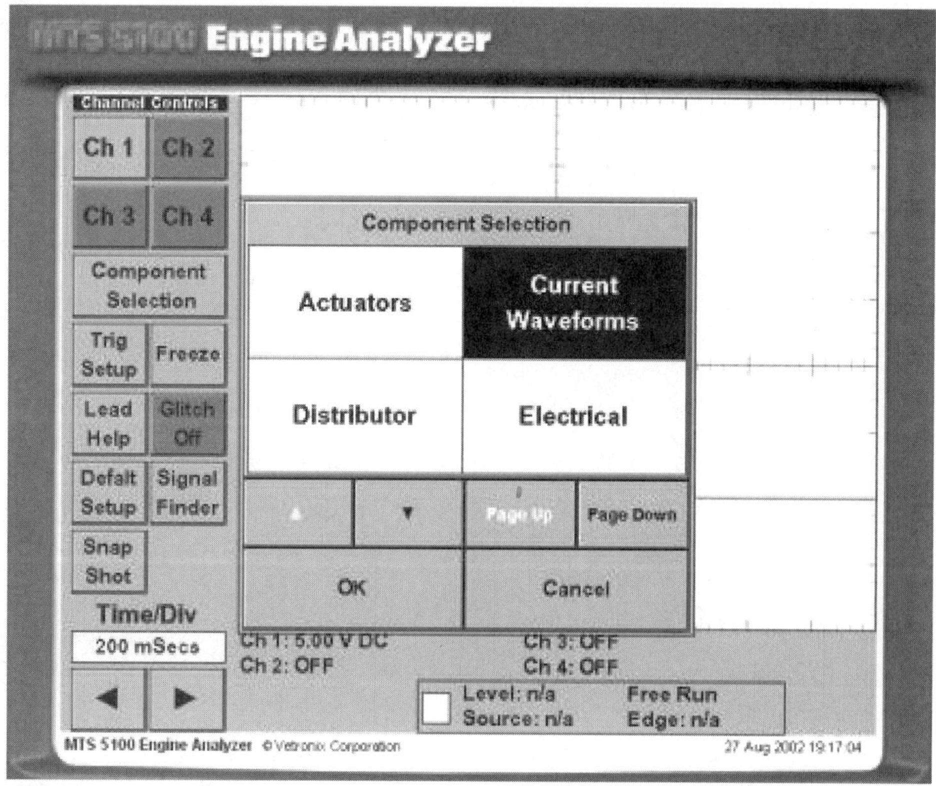

Figure 2. The component selection menu with current waveforms button highlighted.

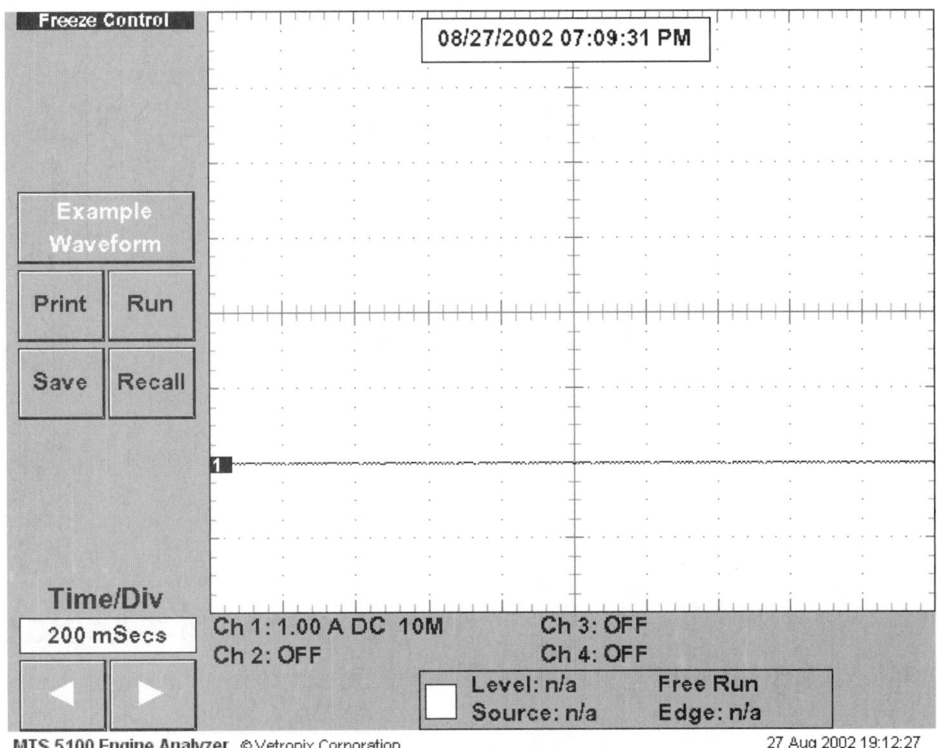

Figure 3. Ch 1 set for 1.00 A DC with no current flow trace is zeroed.

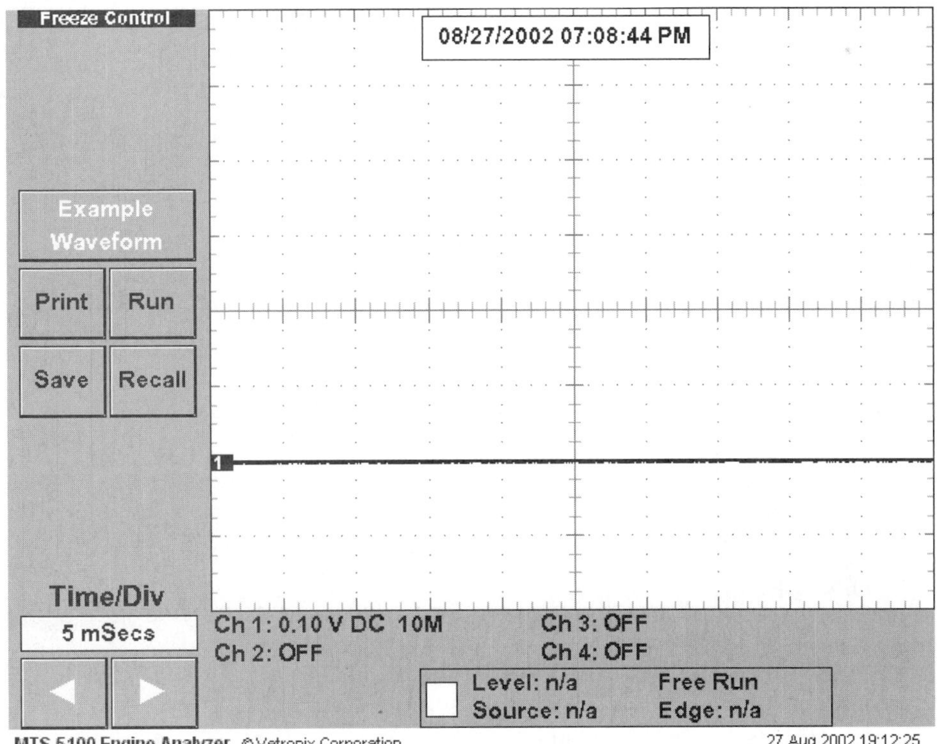

Figure 4. Ch 1 set to read 0.10 V DC from a 100 mV/amp probe.

CONNECTING THE PROBE INTO THE CIRCUIT

Connecting the probe into the circuit is probably the easiest thing to do. Find the circuit that either feeds current to the load or brings the ground to the load. It does not matter which side of the circuit you measure.

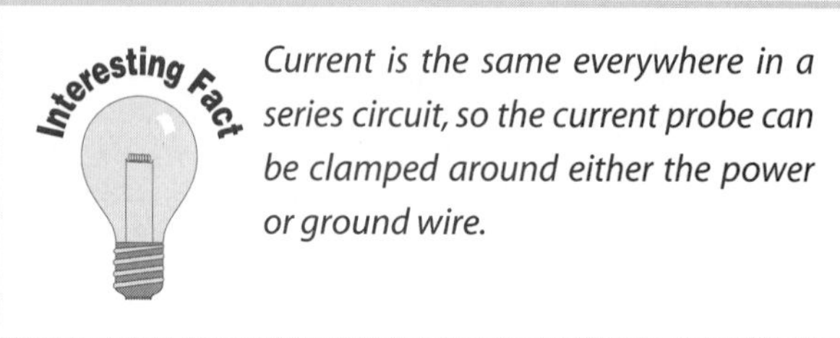

Current is the same everywhere in a series circuit, so the current probe can be clamped around either the power or ground wire.

The current is the same everywhere in a series circuit. Make sure the probe is completely closed around the wire, and make sure that you have only one wire. Multiple wires change the current values and give you wrong information. In addition, if you close the clamp around the feed wire and the ground wire to a load, the two signals will cancel each other out, giving you nothing on the screen. We are now ready to measure current flow.

MEASURING A NONPULSING LOAD

The value of the current probe is its ability to look at very fast pulses of current, which are impossible to look at with a DMM. However, let us start with something that is on but not pulsing. We will use the headlight system. We will place our probe around the battery cable and turn on the headlights. We should see a continuous indication of current flow. **Figure 5** shows the results of turning the headlights on with the probe around the battery cable. Notice that we had to change the scale because with 1 amp per division we were able to measure only 7 amps. With 5.00 amps per division, how much current did the headlight system draw? It drew approximately 17 amps. Remember that the number 1 on the left edge of the screen indicates where zero is. Also notice that we are in the free-run mode (lower trigger box). No trigger was necessary because the pattern was a constant 17 amps.

MEASURING A PULSING LOAD

Again, the value of the current probe is its ability to capture fast-moving events and freeze them on the screen for analysis. Let us use the same fuel injector that we have looked at using a voltage waveform. **Figure 6** shows the

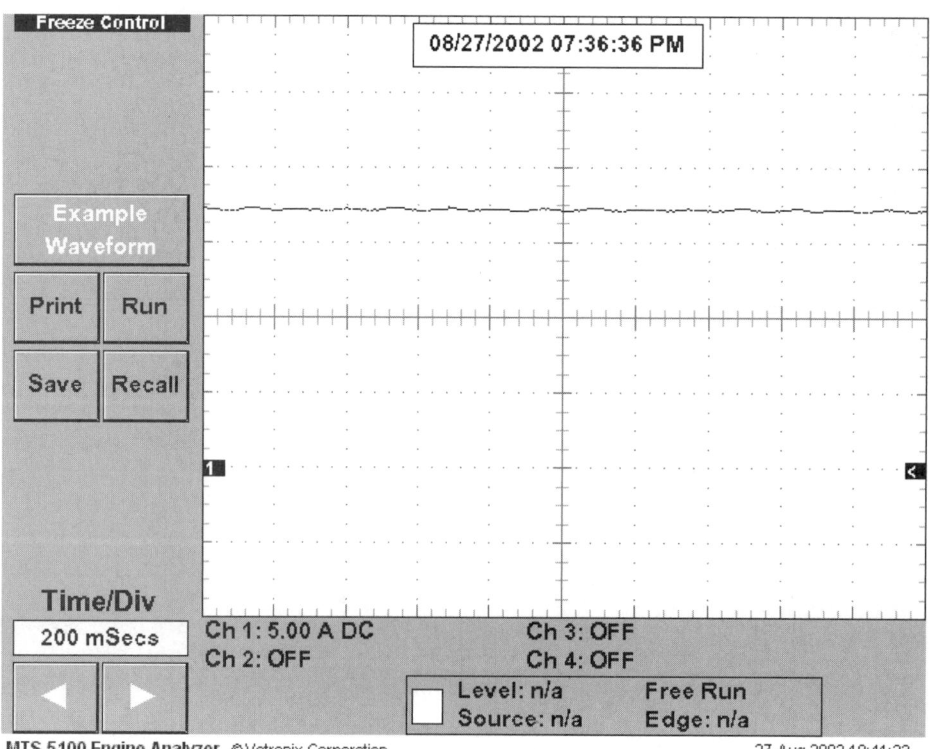

Figure 5. The steady flat line indicates that the headlight system is drawing 17 amps.

Figure 6. The current probe is placed around the fuel injector wire.

connection made at the injector. The probe is completely around one wire of the fuel injector. Either wire will give the same results, but with reverse polarity. If the pattern goes down, you have the polarity reversed. Turn the probe 180 degrees around, and the pattern will go up, which is the way we want it to go. **Figure 7** shows the pattern from the fuel injector at idle. The rise from zero amps occurs at the beginning of the third division. Current ramps up until it peaks at about 950 milliamps. Notice that we have changed the amps per division to enlarge the trace. After peak amperage is achieved, the injector shuts down when current is turned off in the middle of the screen. Do you notice the little blip in the middle of the fourth division? That is the **pintle bump**, or the place where the fuel injector opened. The specification for the pintle bump is that it should occur in the middle third of the current ramp. Ours just barely makes this specification. Additional information about how long the injector pulse lasted can also be obtained from the trace. With 1 millisecond per division, and three divisions of current from turn on until turn off, it looks like the pulse lasted 3 milliseconds.

Let us look at one more pulsing pattern. This time, we will use the ignition primary that we looked at with a voltage trace. Remember that we had to sync or trigger it to the number three cylinder, because there are eight patterns. We will do the same thing here. **Figure 8** shows the pattern for cylinder number three primary coil current. In Chapters 46–53, the ignition chapters, we will analyze it completely; however, let us look at what information the DSO with a current probe was able to give us. Primary current ramped up straight as an arrow until the maximum allowable current was achieved. At 2.0 amps per division and 3½ divisions, it looks like the coil was

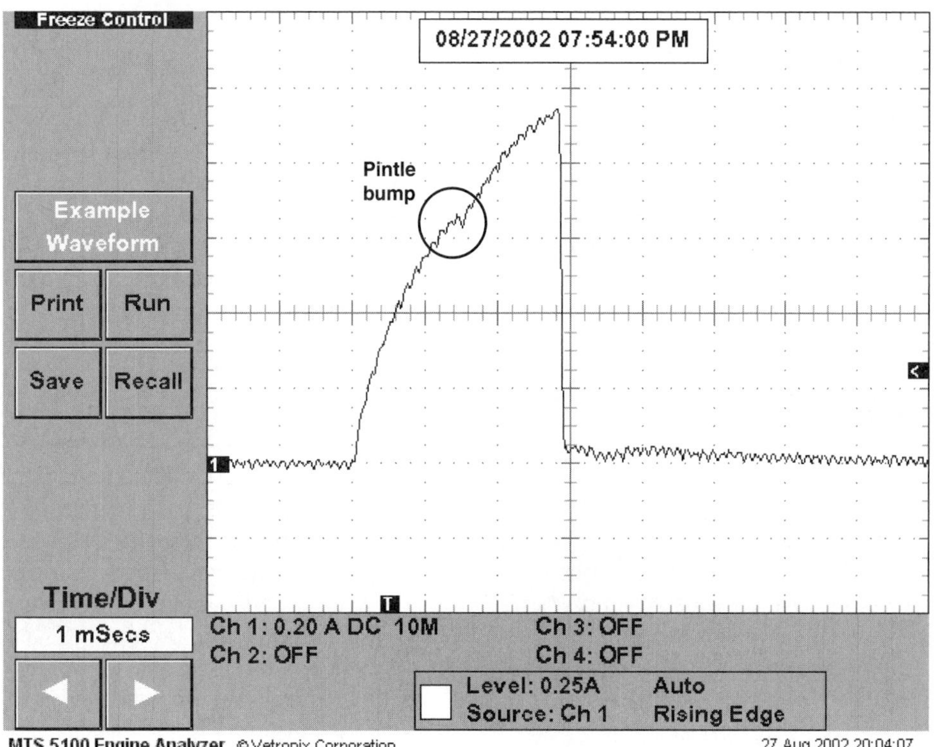

Figure 7. Fuel injector waveform showing a pintle bump in the fourth division.

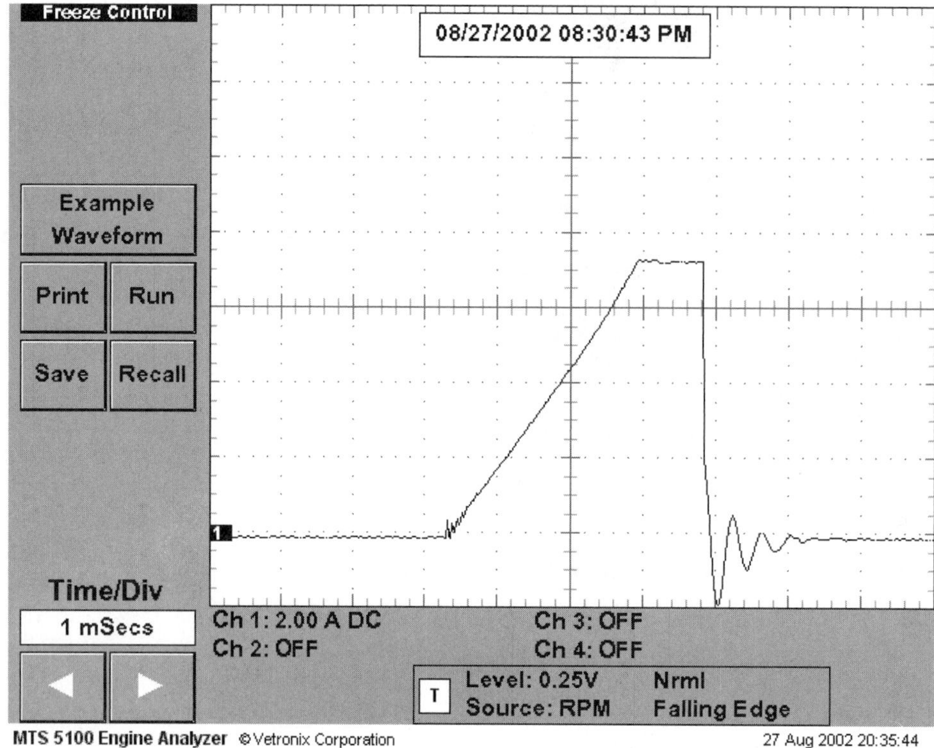

Figure 8. Primary ignition current flow showing a 7-amp maximum.

drawing approximately 7 amps, which is normal for this system. In addition, primary current was on for about 3.5 milliseconds from turn on until turn off. The pattern gives us additional information, which we will look at in Chapter 49.

Imagine the information that a dual trace of voltage and current at the same time would supply. That is one of the topics for Chapter 20, where we use the DSO dual trace capability.

Summary

- A current probe can be clamped anywhere in the circuit.
- Current is the same everywhere in a series circuit.
- Set the DSO volts per division to match the current probe's output.
- Zero the current probe before use.
- Trigger the pattern if necessary.
- The current probe can indicate when a fuel injector opens.
- The current probe can be used for pulsing or constant current flow.

Review Questions

1. A current probe has "100 mV per amp" printed on the front. Technician A states that this means that the probe will generate a 100-millivolt trace if 1 amp is flowing. Technician B states that the DSO might interpret the signal as one division equals 1 amp. Who is correct?
   A. Technician A only
   B. Technician B only
   C. Both Technician A and Technician B
   D. Neither Technician A nor Technician B
2. A current probe should be zeroed before to use to
   A. make sure the signal is accurate
   B. compensate for the battery voltage
   C. give a zero reference to the pattern
   D. all of the above
3. If the power and ground wires are both put inside the current probe, the DSO will show
   A. accurate information
   B. reverse polarity
   C. nothing, because the signals cancel each other out
   D. inaccurate information

4. A fuel injector has a bump on the current ramp. It indicates
   A. how much current the injector is drawing
   B. at what polarity the probe is located
   C. how long the injector remained open
   D. when the injector opened
5. Current is different everywhere in a series circuit.
   A. True
   B. False
6. Current probes rely on vehicle power.
   A. True
   B. False
7. Reduce the voltage (amperage) per division if the pattern is too small.
   A. True
   B. False
8. The on time of a circuit can be measured by looking at the current ramp.
   A. True
   B. False
9. What is the advantage in using current for diagnostic purposes?
10. Why do we not want to put the power and ground wires inside a current probe?

Using the DSO's Multiple-Trace Capability

Introduction

Now that you have seen some applications using the single-trace capability of the DSO, it is time to take a look at multiple traces. A DSO usually has the ability to display more than one trace. Two is the most common setup, but even four-trace DSOs are on the market. We will use the four-trace Vetronix 5100 Engine Analyzer but will rely on only two traces.

WHY USE MULTIPLE-TRACE CAPABILITY?

The question of why we should use multiple-trace capability is easy to answer. Frequently, on modern systems, there is an interrelationship between two or more components that we want to observe. The change in one is supposed to cause a change in the other. The easiest way to observe the change is to put both components together on the same screen and observe the relationship.

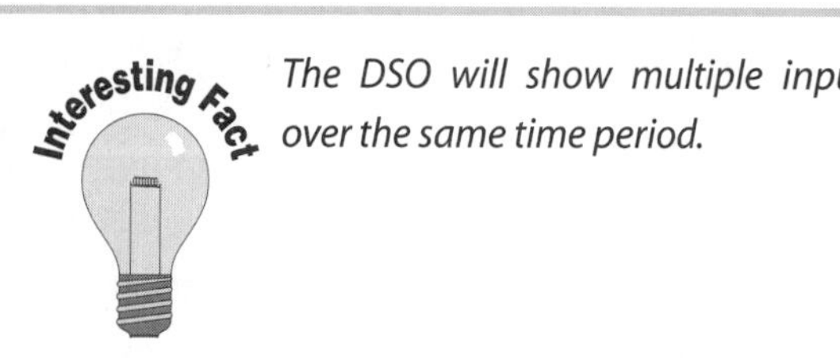

The DSO will show multiple inputs over the same time period.

Another reason for multiple-trace capability involves putting a voltage waveform and a current waveform for the same component on the screen together. If we know the voltage and the current, Ohm's Law will allow us to calculate the resistance of the circuit. The most frequent specification is a resistance one, and knowing what the real resistance is with current flowing is the best method. The DMM on ohms is limited in its ability to give accurate resistance readings because it does not push as much current as is normally flowing.

HOW TO SET AND ADJUST MULTIPLE TRACES

The DSO needs to recognize that you want to trace multiple inputs. Most DSOs default to a single trace when they are turned off and require resetting each time. The exception to this is the DSO that offers the ability to store some presets in memory. Let us go through the procedure with the same unit we have been using, the Vetronix 5100 Engine Analyzer. **Figure 1** shows how the DSO comes up when you turn the unit on. Notice in the upper left of the screen that only Ch 1 is highlighted. The other channels are turned off. This is verified below the radicule with the channel status. Channels 2, 3, and 4 are listed as off. If we touch the Ch 2 button, the setup screen appears and we can touch the on button, as **Figure 2** shows. Notice that the status information below the radicule has gone from off to 5.00 V DC for both Ch1 and Ch 2. Both channels are set for the same voltage by default. Notice also that there are now two numbers along the left edge of the radicule: number 1 is located in the middle of the left edge, and number 2 is one division from the bottom. The two channels are spread out but really only have half the screen each to display. This is a disadvantage to using multiple traces; the patterns will have to be smaller. By the time you have four traces on the screen, viewing detail becomes difficult. It is also possible to

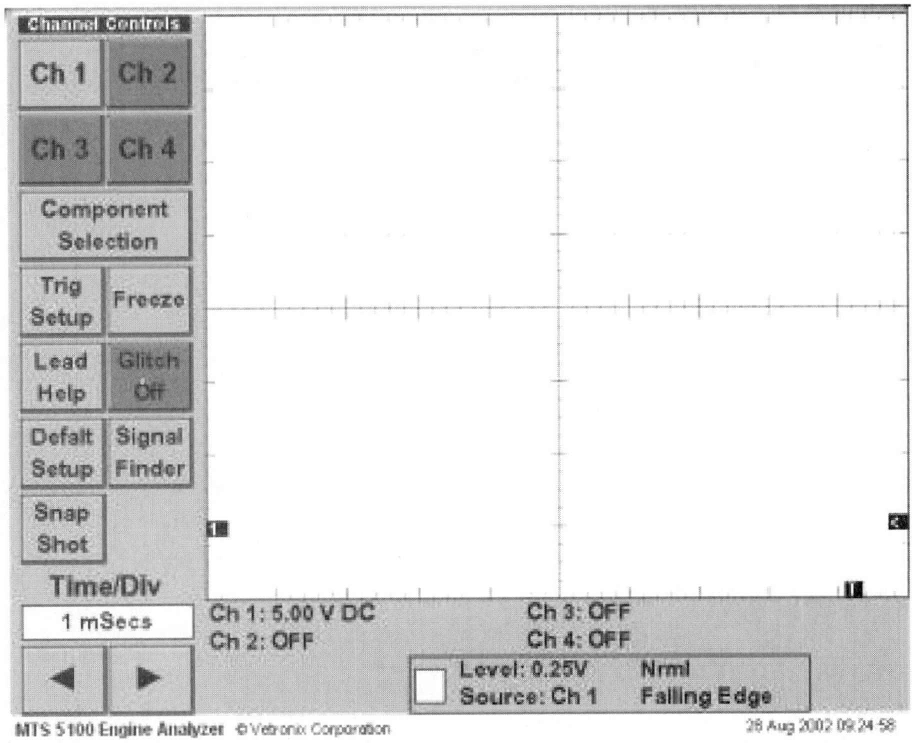

Figure 1. Channel 1 is on; channels 2, 3, and 4 are off.

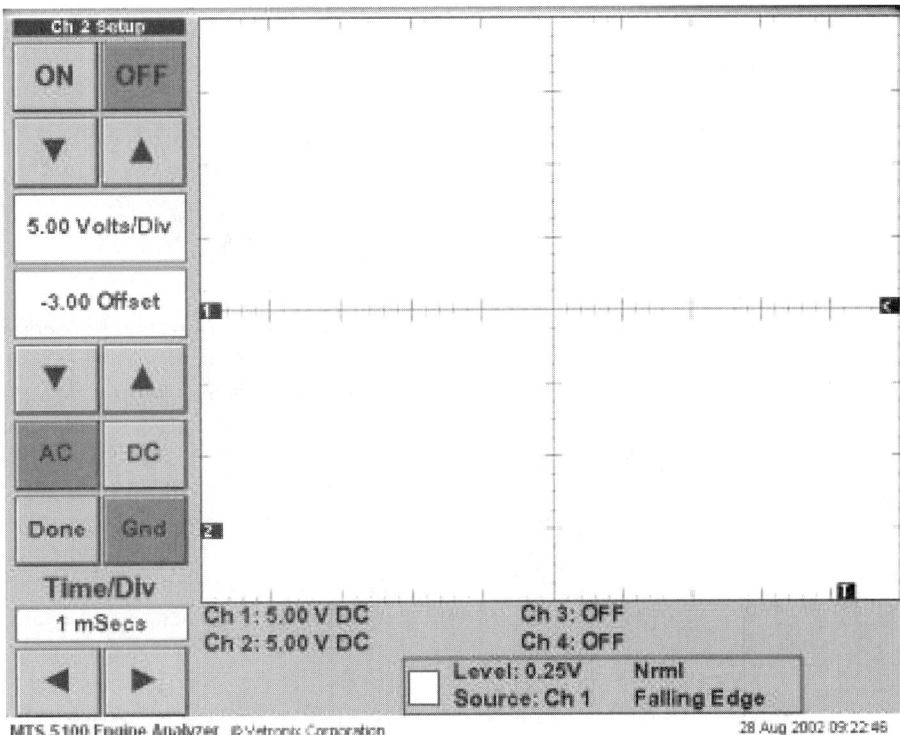

Figure 2. Setting up Ch 2 for 5.00 volts per division and 1 mSec per division. Ch 1 and Ch 2 share the screen.

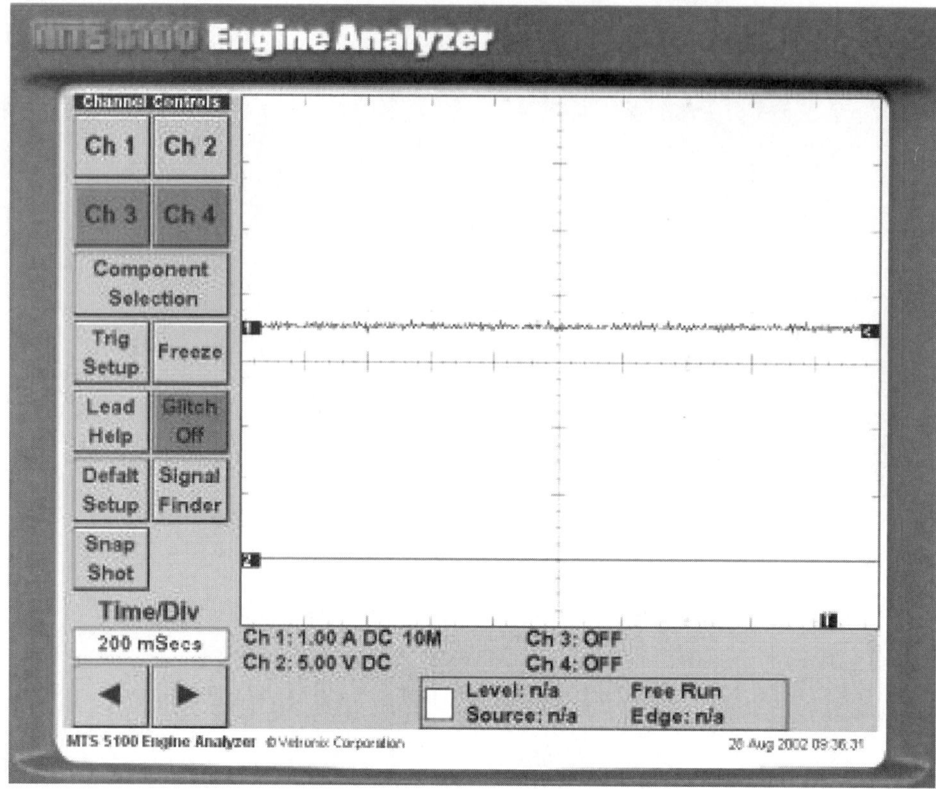

Figure 3. Ch 1 set for current probe input at 1.00 amp per division; Ch 2, at 5.00 volts per division.

overlay various patterns for comparison purposes. To accomplish this, we would move the different channels' zero position to the same position.

Dual-trace capability allows you to set each channel for a different voltage. You can even set one channel for current measurements and the other for voltage measurements with totally different scales, as shown in **Figure 3**. Most DSOs do not allow for different time per division settings, however. As an example, we have set up channel 1 for current and channel 2 for voltage. Channel 1 has 1.00 A DC per division, whereas channel 2 has 5.00 V DC per division. Notice that there is only one time per division, 200 mSec, listed. This is the time frame for both channels.

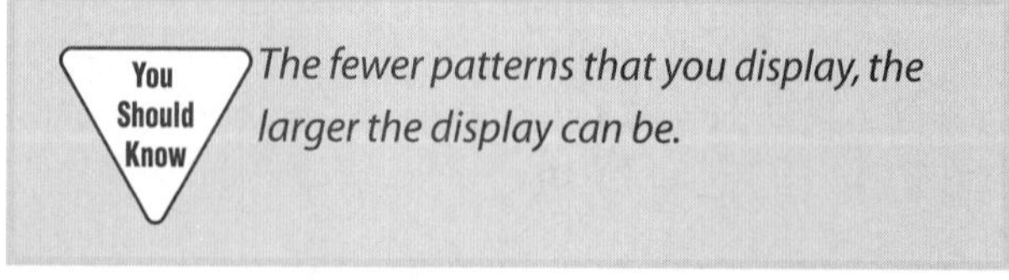

The fewer patterns that you display, the larger the display can be.

SETTING A TRIGGER

Setting or using a trigger for a dual trace is no more complicated than that for a single trace. First decide what the input for the trigger will be. Usually it can be Ch 1, Ch 2, or an external sync probe that we have used before. Do not forget that you must also specify rising or falling edge slope. **Figure 4** shows the setup for the trigger screen. The final information is listed in the box below the radicule. It shows that our source will be the rpm probe looking for a rising edge. 0A is listed as the level because any pulse that rises from zero up will be used as the trigger.

EXAMPLES

Let us set up and use the dual-trace capability to look at some common sensors. Recognize that the purpose of this is not necessarily to diagnose the inputs but to learn how to use the multiple-trace feature. Some of our examples are explained in detail throughout the text. When an engine starts, it needs certain inputs to recognize the position of the crankshaft and sometimes even which stroke the cylinder is on. On some General Motor engines, a sensor that generates two distinct patterns accomplishes this. The pattern shown in **Figure 5** comes off a crankshaft position sensor (CKP). It generates a Hall effect signal, which we study in Chapter 48. The crankshaft vibration dampener has two sets of slots or windows and vanes.

When a vane is in the sensor, the signal drops to just about zero. When a window is in the sensor, the signal rises to its input voltage. These two sensors are powered with 7

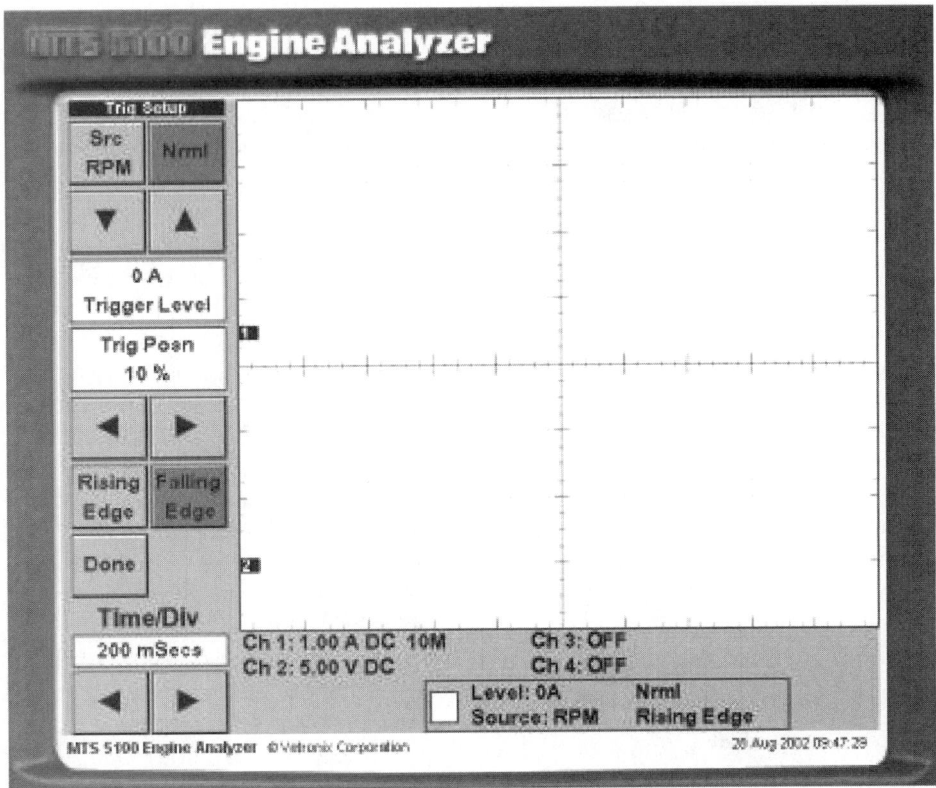

Figure 4. The settings for the trigger are found on a separate menu.

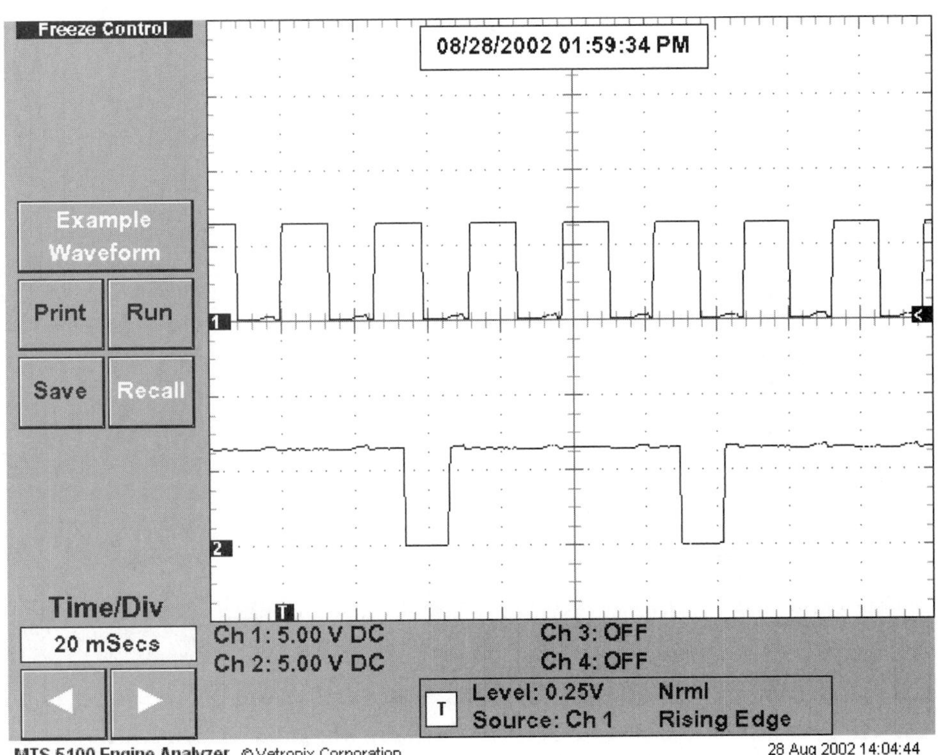

Figure 5. A 3X 1X CKP sensor.

volts. Channel 2 shows the 1X sensor's output. A window is inside the sensor once per crankshaft rotation. With two patterns on the screen, this crankshaft has rotated about three times. Channel 1 is displaying the other part of the sensor's output called the 3X output. The 3X will generate three pulses per rotation of the crankshaft. Trigger is being supplied by the channel 1 voltage waveform at a 0.25-volt level with a rising edge. The trigger point is where the "T" is located at the beginning of the second division. Why would we want to display both of these in a dual-trace mode? The same sensor generates both signals, and both are used to tell the ignition module and PCM the crankshaft position. Both signals are necessary for the vehicle to function normally. The on/off-type signal represents both sensors doing their jobs at the same time. Notice that the speed of the crankshaft is fast. At idle, this engine makes one complete rotation in about 70 milliseconds. Notice that the voltage turns, either up or down, are clean and crisp. If this engine has a problem, it does not lie within the CKP 1X 3X sensor.

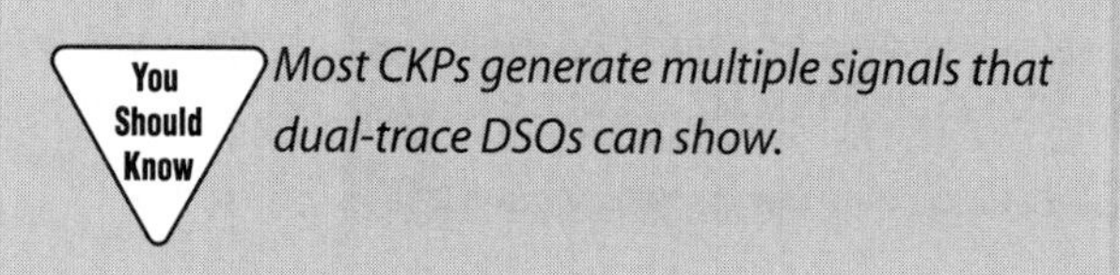

Most CKPs generate multiple signals that dual-trace DSOs can show.

Let us go back to some patterns that we looked at in Chapter 19, that of the ignition module and the fuel injector. You remember that we looked at each using a voltage waveform and later a current waveform. If we combine the two different patterns on the same screen, the amount of information presented is almost at the maximum. **Figure 6** shows off the dual-trace capability of the DSO. Channel 1 shows a current probe waveform from an ignition coil, whereas channel 2 displays the voltage waveform from the same ignition coil. The current probe on Channel 1 is supplying the trigger. Both occur during the same time frame. Think about what is happening and what the two patterns are showing you. The ignition module applies a ground to the coil primary. This is represented by the downward turn of the voltage trace (channel 2) at the end of the second division. If the voltage at the negative of the coil goes to zero, current should begin to flow. At the same time, at about the end of the second division, channel 1 shows a rapid increase, or ramping up, of current. When the ignition coil is fully saturated, at the end of the seventh division, current flow is limited by changing the condition at the negative of the coil. Our voltage waveform takes a turn up, and the current trace flat tops. How much current is flowing? Remember that our probe produces 100 millivolts for every amp, and the Ch 1 scale is 200 millivolts per division, so every division will equal 2 amps for a total of approximately

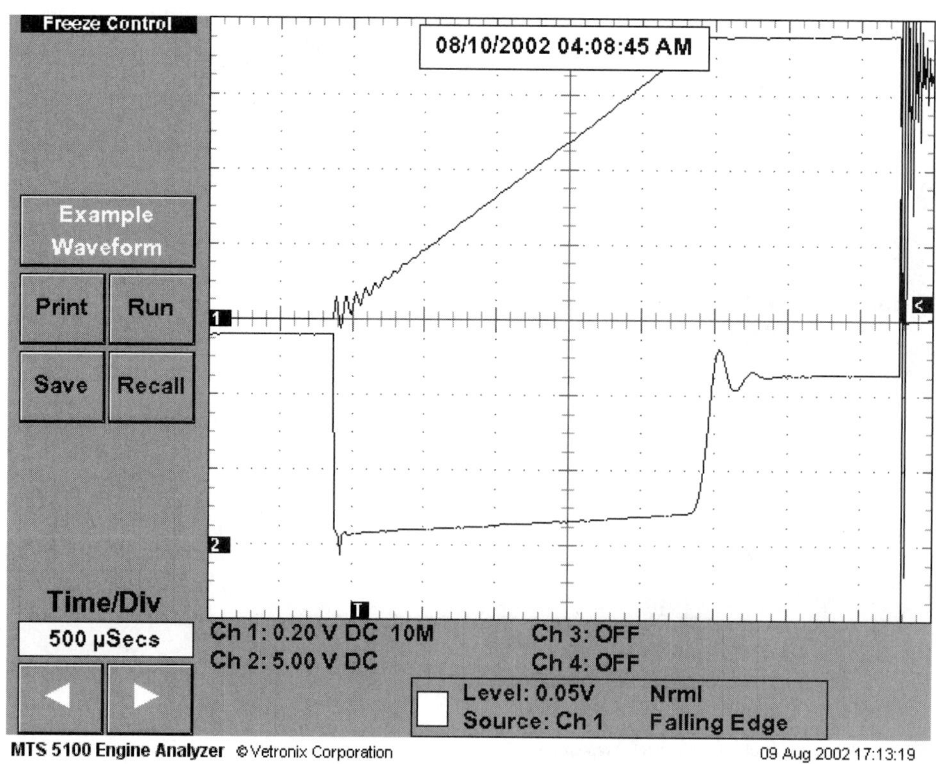

Figure 6. Primary ignition display with current on Ch 1 and voltage on Ch 2.

7.5 amps. Primary current flows until the last division, where the current trace drops and the voltage trace rises.

This is an excellent example of the ability to get all of the information that we need to diagnose this primary ignition system. With these two traces, we can see what the input voltage was, check the quality of the ground, see that current ramps up normally, and look at the module function. We discuss this more in Chapters 46–53, the ignition chapters. The dual trace gives us the ability to compare a voltage trace and a current trace for the same component during the same time period.

Our third example will be the fuel injector that we have seen with a voltage trace and a current trace, but not together. **Figure 7** shows a current ramp on channel 1 and a voltage trace on channel 2. Both are taken during the same time frame of the fuel injector firing. The current trace shows a lot of oscillations that we should look at. Notice that the oscillations are present even when there is no current flow at the beginning of the trace and at the end. There appears to be some source of interference to the current probe that is producing the oscillations. Because the oscillations are present even with no current flowing, we would not be concerned about them. The channel 1 input is supplying the trigger, which is the current trace. The pattern shows the power (voltage) to the injector before turning it on. When the ground is applied in the second division, current begins to flow. It continues until it peaks at about 0.75 amp. The scale is 0.5 amp per division and we have about 1½ divisions, so about 0.75 amp. How do we know if this is correct? Looking up this injector resistance in the manual shows that it is 11.8–12.6 ohms. Our voltage waveform shows 13 volts applied, so this becomes an Ohm's Law problem. What we are looking for is the range in current for this injector:

$$\frac{13 \text{ volts}}{11.8 \text{ ohms}} = 1.10 \text{ amps for the highest current}$$

and:

$$\frac{13 \text{ volts}}{12.6 \text{ ohms}} = 1.03 \text{ amps at the lowest}$$

This fuel injector is not drawing the correct current. Some additional factors can impact the level of current; however, the injector is the major one. Loose or corroded connections need to be checked before the replacement of the injector. Notice that we used information from both traces to diagnose this injector. A check of the other five injectors revealed that they were drawing 1 amp each, so it is time for a new injector.

Our last example is of a four-trace pattern **(Figure 8)**. You need to recognize that the application of four traces is limited, but here is a good example. A vehicle stops running

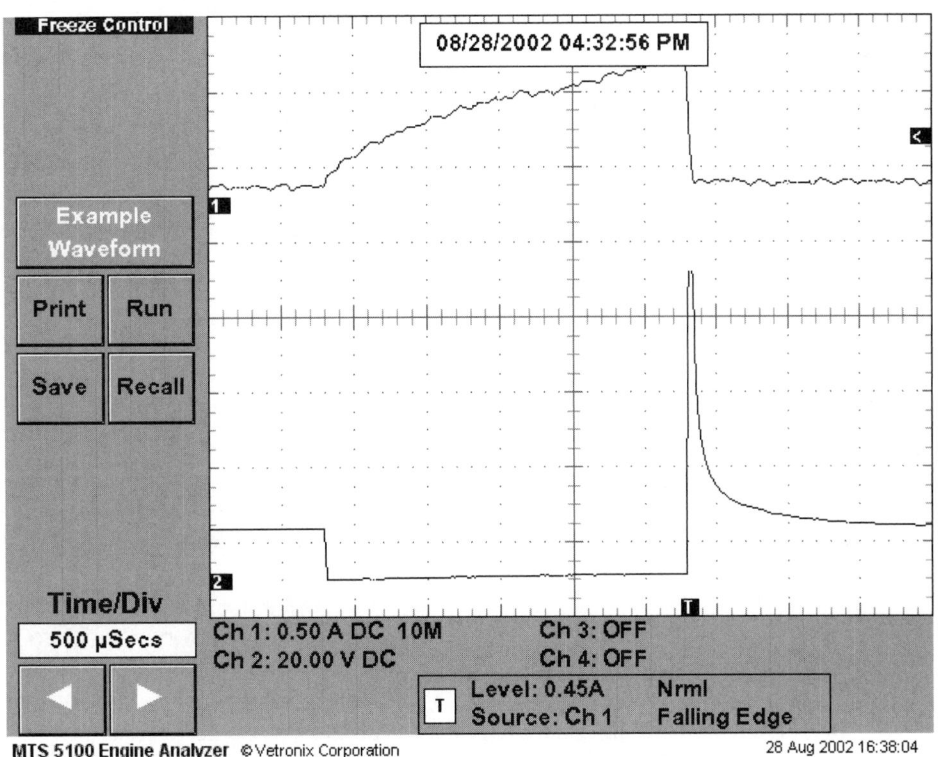

Figure 7. Fuel injector waveform with current on Ch 1 and voltage on Ch 2.

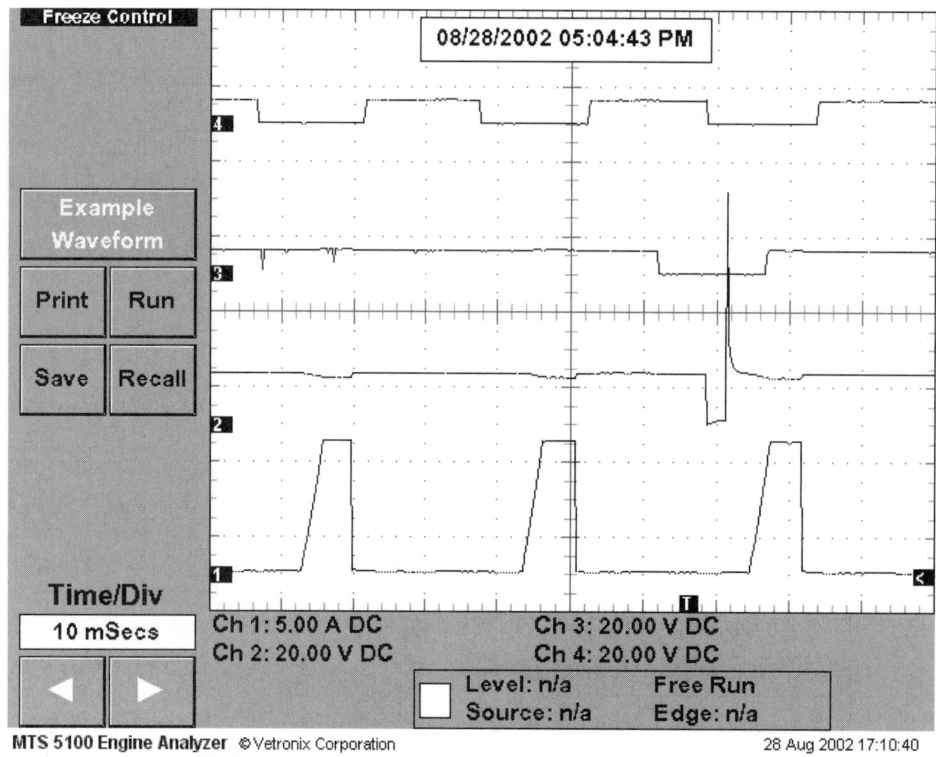

Figure 8. A four-trace DSO can display two inputs and two outputs at the same time.

for no apparent reason. Our technician connects the current probe to the ignition coil primary as the channel 1 input. Channel 2 should look familiar; it is a fuel injector. Channels 3 and 4 are of the 3X 1X sensors off the CKP. With the engine running, all channels display what they should. Now we just watch until the engine dies and figure out what is missing. Notice that a trigger has not been set. The DSO is in the free-run mode. The DSO gives us the opportunity to look into the circuits of the vehicle as long as we know what is considered normal. As you learn more about vehicles and the use of the DSO, you will get better at diagnostics. Remember that the DSO is not the simplest of tools. It will take practice and experience to make it a productive tool in your toolbox.

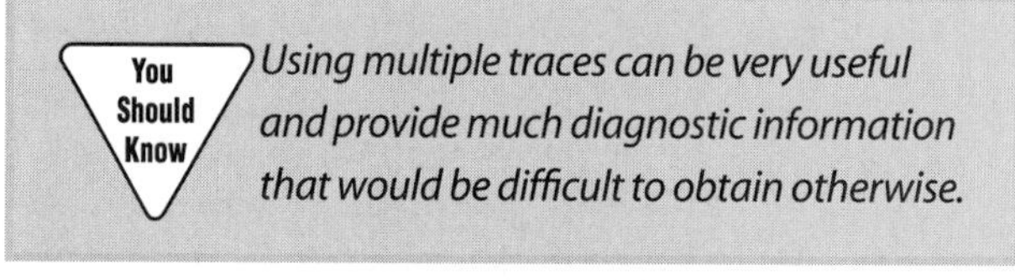

Using multiple traces can be very useful and provide much diagnostic information that would be difficult to obtain otherwise.

Summary

- Multiple-trace capability is a feature that is best used when two components react within the same time period.
- Each trace can have its own voltage setting.
- Each trace will have the same time per division.
- Dual tracing a fuel injector with current and voltage gives all of the information that the technician usually needs to diagnose and repair the system.
- Multiple traces can use the same trigger types that a single trace can use.
- The voltage trace and the current trace can give the technician enough information to use Ohm's Law and calculate the resistance.
- The more traces on the screen, the higher the voltage per division needs to be in order for the technician to be able to see the individual channels.

Review Questions

1. The voltage per division setting for each channel is
   A. the same
   B. different as needed
   C. used as the trigger
   D. the same as the time per division
2. Technician A states that a current waveform can be useful for determining the resistance under load of the component. Technician B states that a voltage waveform can be useful for determining the resistance under load of the component. Who is correct?
   A. Technician A only
   B. Technician B only
   C. Both Technician A and Technician B
   D. Neither Technician A nor Technician B
3. The trigger for a multiple trace DSO can come from
   A. channel 1
   B. channel 2
   C. the rpm probe
   D. all of the above
4. Using multiple-trace capability allows for
   A. different voltage settings for each channel
   B. current probe use
   C. the same time frame for all channels
   D. all of the above
5. Different voltage levels can be specified for each channel.
   A. True
   B. False
6. Different time values can be specified for each channel.
   A. True
   B. False
7. The position of each trace can be moved up or down.
   A. True
   B. False
8. Only an external sync can be used with multiple traces.
   A. True
   B. False
9. What is the main advantage to using multiple-trace capability?
10. What is the main disadvantage to using multiple-trace capability?

Electronic Fundamentals

SECTION OBJECTIVES

At the conclusion of this section, you should be able to:

- Understand the difference between electricity and electronics.
- Understand the wiring of a transistor as an amplifier.
- Understand the wiring of a transistor as a switch.
- Safely work around electronics.
- Test a thermistor.
- Test a TPS.
- Test inputs with a DMM.
- Test inputs with a DSO.
- Recognize the pattern from an AC sensor.
- Recognize the pattern from a Hall effect sensor.
- Recognize the pattern from a variable-frequency generator.
- Recognize the pattern from an optical sensor.
- Safely work around integrated circuits.
- Understand how an O_2S functions.
- Test an O_2S with propane.
- Determine if the vehicle is in fuel control with a DSO.
- Recognize common O_2S problems.

A vehicle will not be able to pass the simplest emission test if the oxygen sensor does not function correctly.

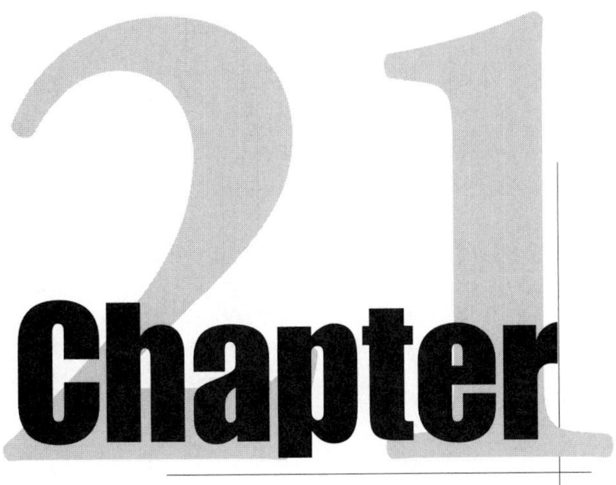

Chapter 21 Solid-State Devices

Introduction

The vehicle of today would be very different if it were not for electronics. For purposes of clarity, let us define electronics as the technology of controlling electricity. Electronics has become a special technology beyond electricity. Transistors, diodes, semiconductors, integrated circuits, solid-state devices, and microprocessors are all considered being part of the electronics era rather than just electrical devices. Within this chapter, we examine some of the building blocks to electronics and then look at some of the common electronic devices in use today on vehicles.

CONDUCTORS AND INSULATORS

Let us begin with a review of a conductor and an insulator (nonconductor). We defined a conductor as a material whose outer electron orbit, the valence orbit, contained few, loosely held electrons. A conductor has one or two electrons in the outer orbit, as **Figure 1** shows. These one or two electrons easily move to other atoms if an electrical pressure (voltage) is applied to them. We find copper and aluminum in common use in electrical systems. These are conductors that will carry the electrical current to the load, where it will do the work for us.

The electrical opposite to a conductor is a nonconductor. Nonconductors are most commonly referred to as insulators. Nonconductors have many electrons in their outer orbit, as **Figure 2** shows. The valence orbit (outer orbit) is usually full or nearly full in a nonconductor. It takes eight electrons to fill the outer orbit or shell. The electrons found in full valence orbits resist moving to other atoms when a voltage is applied to them. This characteristic is what makes them have insulating properties. Plastic and rubber, for instance, are common nonconductors, or insulators, found on vehicles.

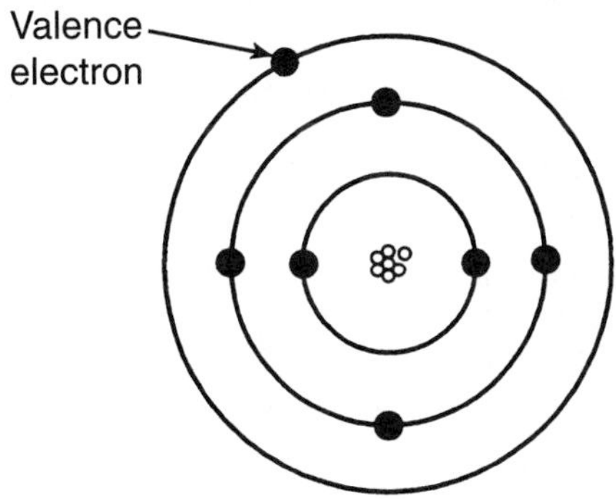

Figure 1. A good conductor will have one or two electrons in the outer orbit.

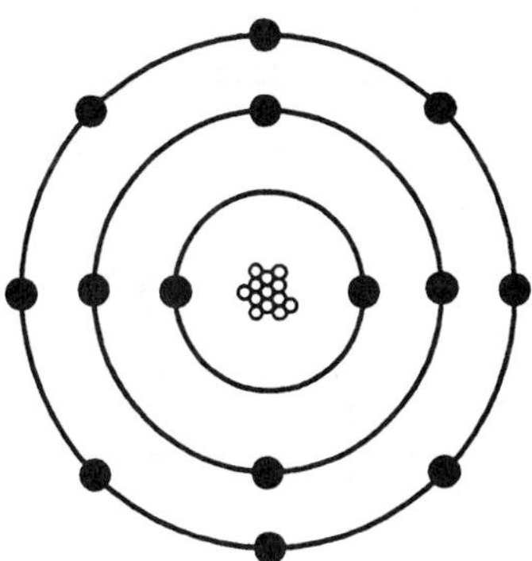

Figure 2. A good insulator will have a full outer orbit.

What if a valence orbit is not completely full and yet contains more than the usual one or two electrons common to conductors? Their outer orbits with more than a conductor's number of electrons and yet less than an insulator's number make them not quite a conductor and not quite a nonconductor. They are referred to as semiconductors and usually have four valence electrons in their outer orbit. They are neither good insulators nor good conductors. Energy will be necessary to make them conduct, but more energy than is necessary for a conductor and less than is required for an insulator. The principal elements that exhibit the semiconductor characteristics are germanium and silicon. The most commonly used is silicon because of the availability, resistance to heat, and cost of it. Virtually every beach covered with sand is a potential solid-state device comprised of semiconductors, because sand is silicon. To be of practical use, the silicon is treated or doped with impurities to give it certain electrical qualities. The treatment will make it either positive (P type) or negative (N type). Assembling various N types and various P types together gives us electronic devices able to control, switch, and amplify current.

THE DIODE

One of the simplest semiconductor devices in use on the vehicle is the **diode**. It was also the first application of solid-state electronics to find its way into the automotive industry **(Figure 3)**. The diode is simply a one-way check valve for electricity. It will allow current to flow in one direction only. Current trying to flow in the opposite direction will be blocked. There are many different types of diodes, but they all operate on the one-way check valve principle.

Figure 4 shows the electrical symbol for the diode. Notice that the left side of the diode (the side with the arrowhead) is identified with the notation anode. The flat-

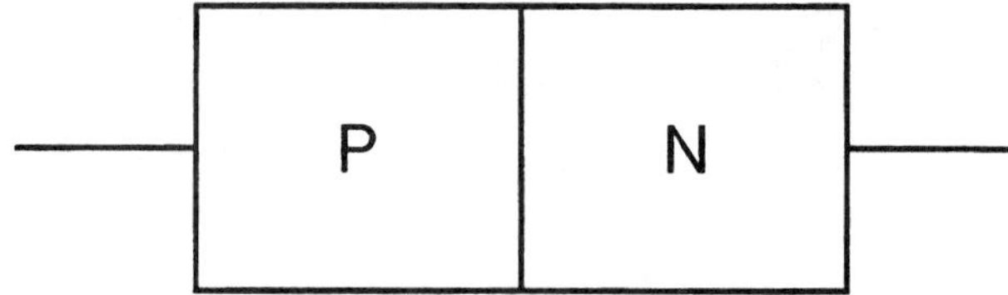

Figure 3. The diode is a one-way check valve for electricity.

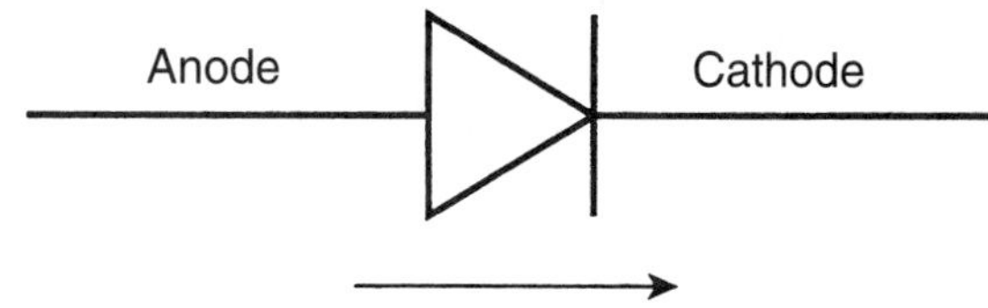

Figure 4. The schematic symbol for the diode.

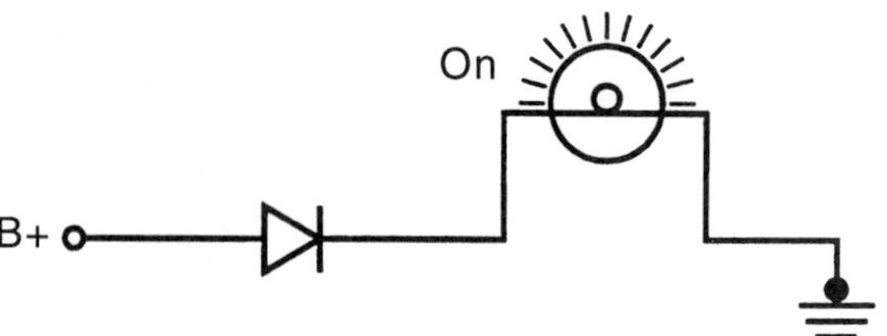

Figure 5. The diode conducts, and the light bulb is on.

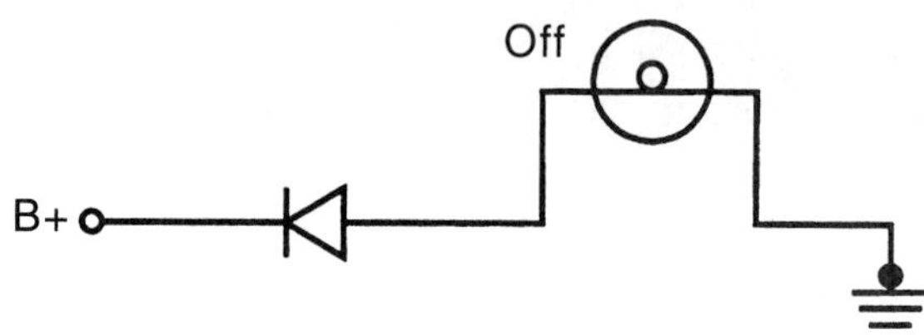

Figure 6. The diode blocks, and the light bulb is off.

line side of the diode is identified by the notation cathode. The arrow shows the direction that current will flow.

If we wire a diode in a circuit with a power source and a small light bulb, as shown in **Figure 5**, we can see the diode effect. With the anode toward the positive side of the circuit, the diode will conduct and the bulb will be on. Reversing the circuit, as shown in **Figure 6**, will turn the bulb off because the diode's cathode is toward the positive and will therefore not conduct.

Diodes are frequently referred to as two-terminal devices and are rated according to the job they must do. The amount of current that they can pass easily is called the **forward current rating**. Exceeding this current level will damage or destroy the diode, usually burning it open. An open diode cannot pass current in either direction. The diode will get very hot if large amounts of current are passed through it. Most diodes in automotive use today will have a forward voltage drop of about 0.7 volt. We previously discovered that if a voltage drop is present, then light, heat, or a magnetic field is being produced. Heat production in a diode destroys it, if the forward current rating is exceeded. For this reason, many diodes are mounted in finned heat sinks, which are generally made of aluminum **(Figure 7)**. The aluminum fins will dissipate the heat produced by the diodes in much the same manner as a radiator will remove unwanted engine heat. Removing unwanted heat will protect the diode. Airflow around the heat sink is important if the forward current rating is to be reached without damaging the diode. The second method of rating diodes is their **peak inverse voltage**. This is the voltage that the diode can withstand safely in its blocking mode. Typically, diodes range from 50 to 1000 volts or more. Exceeding the peak inverse voltage for even a split second can result in damage to the diode. Usually, diodes that have

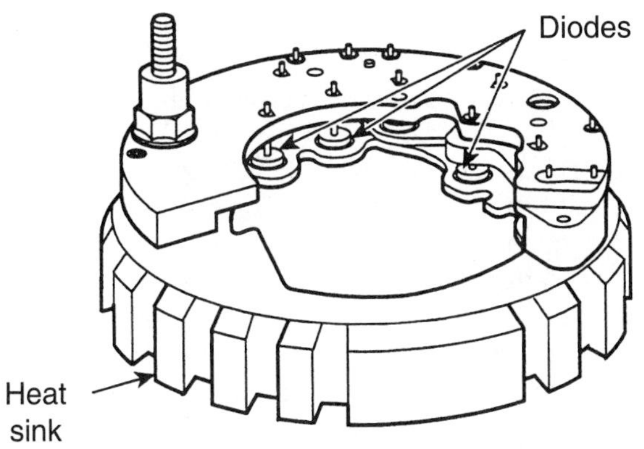

Figure 7. Diodes are usually mounted in a heat sink.

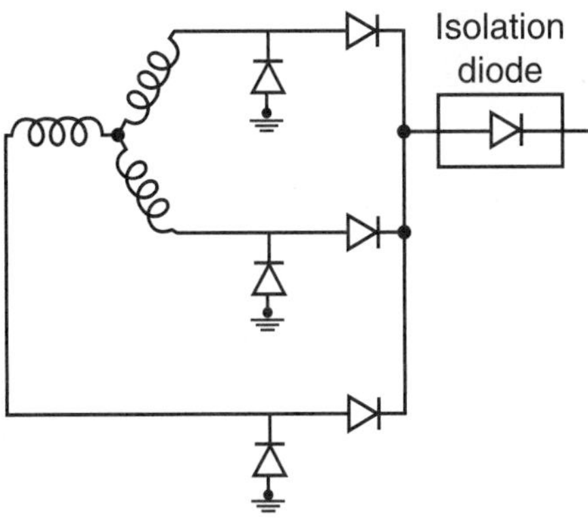

Figure 9. An isolation diode in a charging circuit.

been subjected to peak inverse voltages in excess of their rating will short circuit themselves. A short-circuited diode will pass current in both directions.

Diodes are found in three common applications in the vehicle of today. The first application is in rectification circuits, where they will take alternating current (current that flows first in one direction, then in the other direction) and change it into direct current. **Figure 8** shows the use of six diodes in a rectification circuit. (We see more of this in Chapter 43, Chapter 44, and Chapter 45.) This was most likely the first application of solid-state devices on a vehicle.

The second vehicle application is in isolation circuits. The blocking ability of the diode is used as a switch to prevent current from flowing where the manufacturer does not want it. **Figure 9** shows how isolation diodes are also used in some charging systems to totally isolate the battery from the alternator when the vehicle is off.

The third application of diodes is as a protection device for **integrated circuits (ICs)**. A diode is located near large magnetic coils such as the air-conditioning compressor clutch or where a relay coil is connected into a circuit that is IC controlled, as **Figure 10** shows.

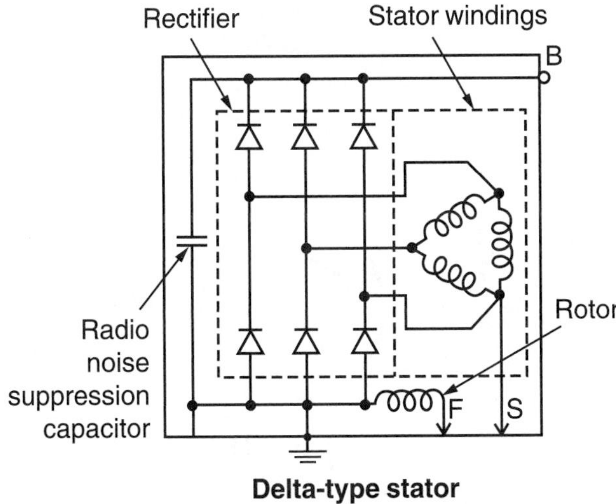

Figure 8. Diodes in a bridge rectify the AC into DC.

When current flows through a coil of wire, it will generate a strong magnetic field. Turning off this field will generate a high voltage spike that, if allowed to reach the IC, could destroy it. Our modern PCM has numerous ICs within it. Locating the diode as in the figure will prevent the spike from reaching back to the IC and yet will allow the IC to control the coil's magnetic field. The spike must not exceed the peak inverse voltage or the diode will conduct and destroy both the IC and itself.

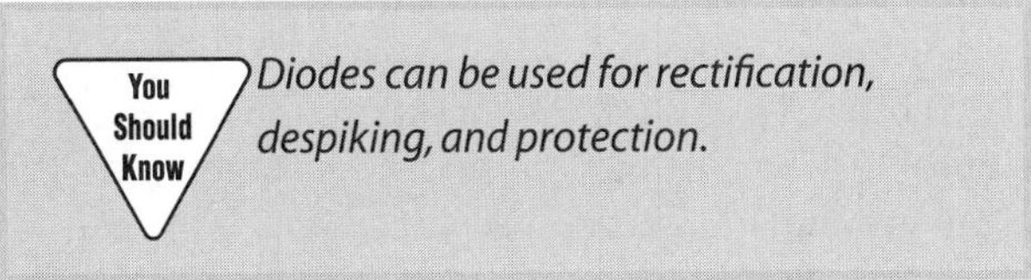

Diodes can be used for rectification, despiking, and protection.

Diodes used in this manner are called despiking or suppression diodes. When current is flowing through the coil (bottom diagram) in the normal direction, the diode is blocking. When the current to the device is turned off, a high-voltage spike will be generated with a reversed polarity. This reversed polarity causes the diode to conduct,

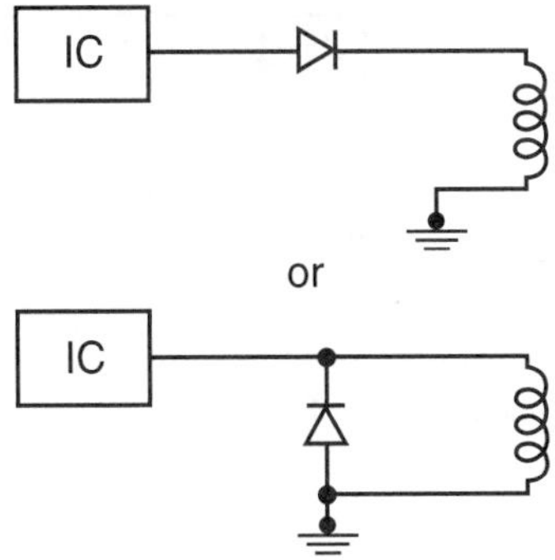

Figure 10. Diodes are used to protect other circuits.

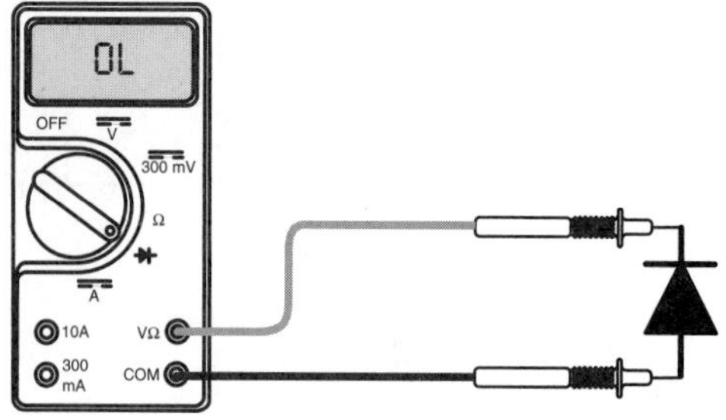

Figure 11. A DMM can be used to test a diode.

effectively giving the voltage spike a path to push current, other than to the IC.

We have found increased use of diodes in vehicles recently. The increased use of ICs in vehicle microprocessors has increased the use of protection diodes. Some vehicles use as many as 20 separate diodes to protect the expensive and important chips. The diode can be tested by using a DMM that has a diode test position. **Figure 11** shows our DMM testing a diode. The meter should show continuity in a forward direction and blocking (OL) in the opposite.

TRANSISTORS

Diodes are made by joining some P-type material and some N-type material, as we have seen. Taking this one step further puts us into the world of the **transistor**. The transistor is a semiconductor device made by joining together three layers of P- and N-type material.

The two basic styles of transistors are NPN and PNP. **Figure 12** diagrams the construction of both types. The diode previously discussed had two connections: one for the P material and one for the N material. Notice that the figure shows three connections available, one for each layer of positive or negative material. Each layer has its own name—emitter, collector, or base—and the entire component can be housed in a plastic or metal case. Size and construction details determine the application of the transistor. The automobile manufacturer decides which transistor will function correctly for the circuit being designed. You will not be responsible for the designing and wiring of transistorized circuits as a service technician; however, a general working knowledge can help you in the diagnosis of failed circuits. In addition, certain precautions are necessary when working on transistorized circuits, which we will also discuss.

Generally, the repair of a transistorized circuit involves the replacement of the entire circuit after diagnosis has pinpointed the failure. Individual repair or replacement of single transistors is usually not attempted in the field. In addition, the application of a single transistor in a circuit is not found frequently. Instead, a series of transistors, diodes, and ICs becomes a module with a single purpose. The ignition module is a good example of this. It will turn on and then off the ignition coil, based on inputs from other devices. It is not a single transistor all by itself.

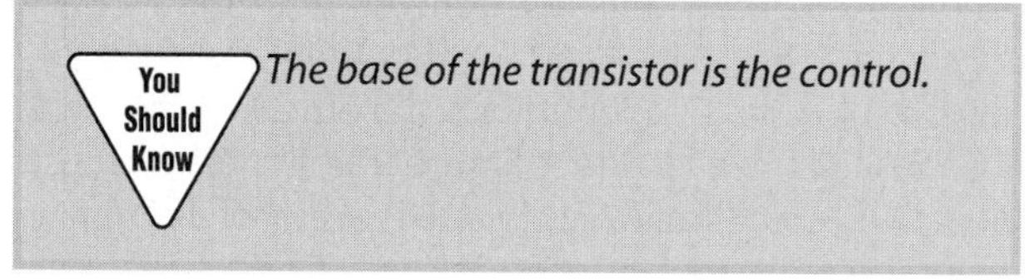

The base of the transistor is the control.

Figure 13 diagrams both NPN- and PNP-style transistors. Notice that an NPN-style transistor has both the base and the collector at positive potential, whereas a PNP-style

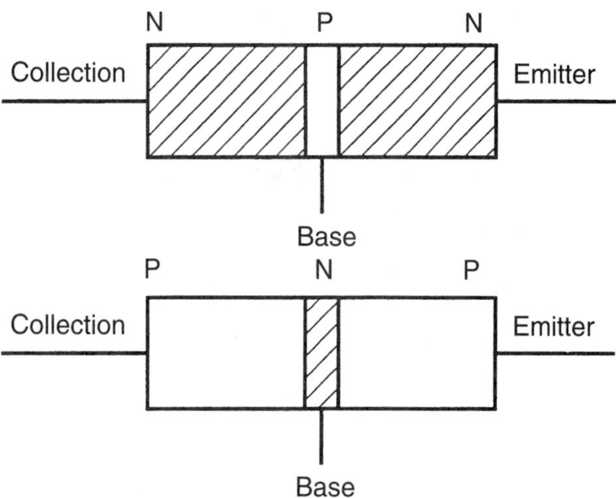

Figure 12. Transistors can be NPN or PNP.

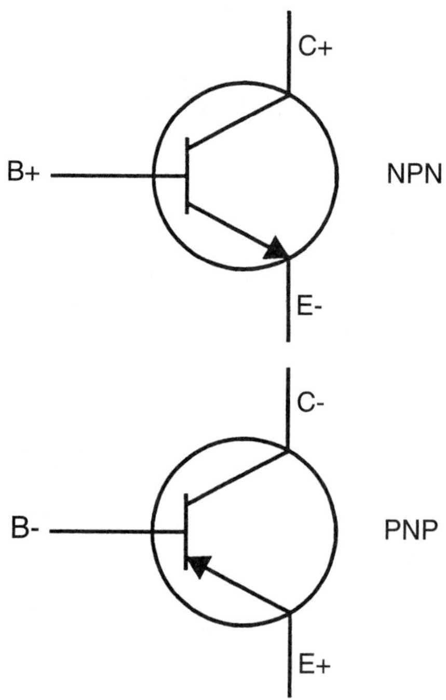

Figure 13. The schematic symbol for an NPN or a PNP transistor.

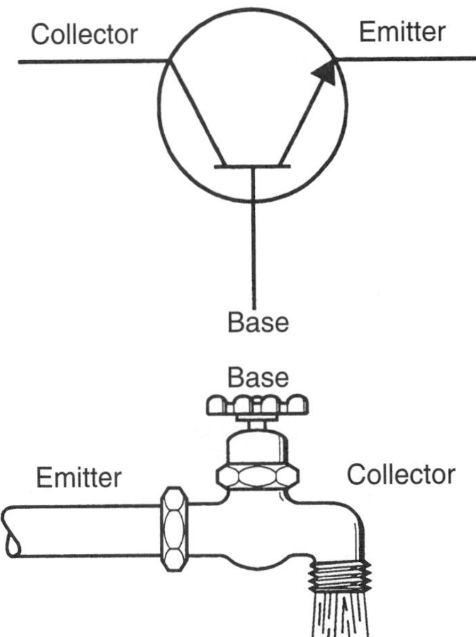

Figure 14. The transistor operates like an electrical faucet.

transistor has both the base and collector connected together at negative potential. The arrow drawn on the emitter always shows the direction of current flow (from positive to negative). A simple analogy is frequently made to water flow **(Figure 14)**. The emitter is the source (as in the water supply). The handle of the faucet is the equal to the base, whereas the end or opening where water will flow from is the equal of the collector. In a transistor, the collector is the output, whereas the emitter is the input. This leaves the base as the controller of current between the emitter and collector.

TRANSISTORS AS AMPLIFIERS

The transistor is a controlling device. It will generally be used to turn something on and off or it will be used as an amplifier. It should not be confused with a variable resistor or **thermistor**, which we examine in Chapter 22. The transistor does, however, follow all the rules of Ohm's Law. The simple diagram in **Figure 15** will give you an idea of how a transistor works as an amplifier. Notice that the voltmeter will measure the input voltage. The collector circuit has a small light bulb to ground. When current flows out the collector, the bulb will go on. We have connected a variable resistor between the base of the transistor and the positive potential. NPN transistors are positive base devices. The ammeter will measure the base current. As the variable resistor is turned, the bulb will change intensity. This is because the variable resistance changes the current flow to the base. As current flow is added to the base (less variable resistance), more current flows through the collector and emitter. If base current is turned off completely, what will the collector emitter circuit do? With no base current, the

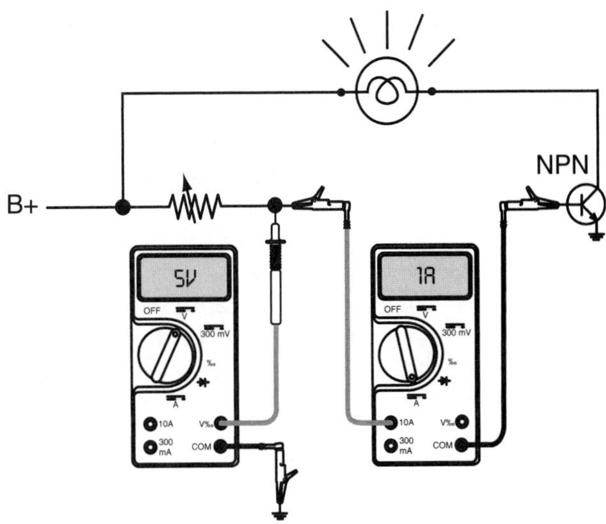

Figure 15. A transistor can be wired to amplify a small base signal.

transistor will shut down completely, and no collector emitter current will flow. In this manner, the base current determines the amount of current flowing through the collector emitter. When a transistor is wired as shown, it is being used as an amplifier. A small amount of base current change can cause a larger amount of collector emitter current change.

TRANSISTORS AS SWITCHES

Transistors are also used as switches, especially in electronic ignition and in the regulation of the charging system. Their advantage is that they can operate many thousands of times a second without difficulty, whereas a mechanical switch can develop problems at a couple hundred opening and closing cycles per second.

We examine the use of transistors in ignition systems in Chapters 49–56. **Figure 16** shows a transistor used as a switch rather than as an amplifier. An on/off signal or input to the base drives transistors. Our amplifier circuit had a vari-

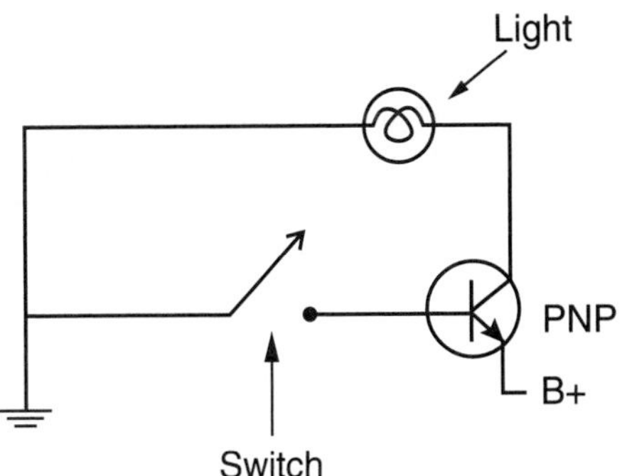

Figure 16. A transistor can be wired to act like a switch.

able signal or input to the base. Figure 16 shows a simple switching circuit that uses a transistor. When the base "sees" the positive current from the battery, it will conduct current through the collector and turn on the bulb. The resistor to the base limits the current that the base will see. In the example shown, the transistor can be switched on and off many thousand times a second for extended periods of time. This advantage to a transistorized application results in greater reliability over a relay or mechanical switch. There is no arcing inside a transistor, as is the case with relay contact points. We have seen increased use of transistorized relays in automotive applications in recent years. In addition, some mechanical relays have transistor control circuits in them.

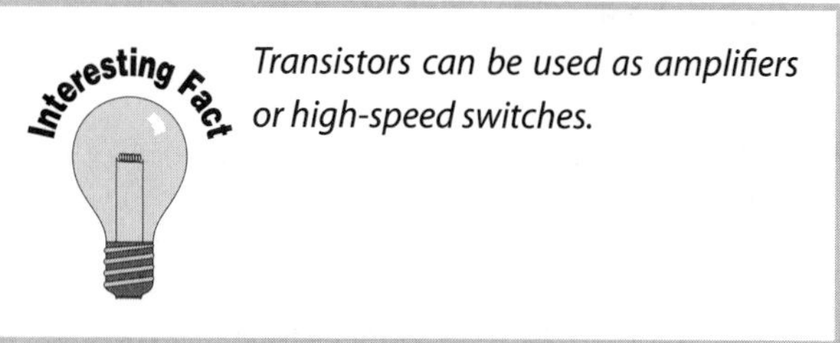

Transistors can be used as amplifiers or high-speed switches.

PRECAUTIONS NECESSARY WHEN WORKING ON ELECTRONICS

The precautions necessary when working around electronics are the same as those for diodes. As a matter of fact, the precautions listed at the end of this chapter are necessary for the entire modern vehicle. Making a polarity mistake or allowing an overvoltage condition to occur can cause extensive damage to most circuits. Usually if an electronic circuit works for a short period of time, it will function almost indefinitely. Failure of a circuit after a vehicle has been on the road for a number of miles can usually be traced to a problem outside of the semiconductors. Frequently, it can be traced to an overvoltage condition caused by the charging system or the use of too high a voltage during battery charging or jump starting. Always keep the following in mind:

1. Do not apply voltage directly to electronic devices unless the manufacturer's procedure calls for it.
2. Do not connect or disconnect an electronic device with the power on to the circuit.
3. Keep the battery voltage less than 20 volts during charging or jumping.
4. If a relay is added to the vehicle, use one that contains a despiking diode. This will prevent a relay surge from getting to an electronic device and damaging it. High-voltage surges will destroy chips.
5. Do not run the vehicle without the battery being connected.
6. Always make sure that the charging system is not producing excessive AC. AC destroys electronics that are designed for DC.

Summary

- The diode is the building block of electronics.
- A diode will block current in one polarity and pass current in the opposite polarity.
- An NPN transistor is usually used to switch negatively.
- A PNP transistor is usually used to switch positively.
- A transistor can be used to amplify a weak signal.
- The base is the controller of transistor current.
- The emitter usually has the load of the circuit.

Review Questions

1. A DMM in the diode test position is placed across a diode and indicates continuity in both directions. Technician A states that this indicates that the diode is blocking current in both directions. Technician B states that this indicates that the diode is open and should be replaced. Who is correct?
   A. Technician A only
   B. Technician B only
   C. Both Technician A and Technician B
   D. Neither Technician A nor Technician B
2. A despiking diode
   A. protects some solid-state circuitry
   B. cannot be tested with a DMM
   C. is not used on modern vehicles
   D. protects the load
3. Transistors (either NPN or PNP) have three leads identified as
   A. anode, cathode, gate
   B. neutral, positive, negative
   C. base, emitter, collector
   D. negative, positive, negative or positive, negative, positive

4. Diodes are designed with high resistance in one direction and low resistance in the other.
   A. True
   B. False
5. Most digital ohmmeters in use today can test diodes accurately.
   A. True
   B. False
6. A PNP transistor needs the base and collector at negative potential to have current come out of the emitter.
   A. True
   B. False
7. What are some of the advantages to the use of electronics?
8. What precautions are necessary when working on electronic circuits?

22 Electronic-Control Input Devices

Introduction

The age of electronic control of virtually everything has meant the increased use of input devices designed to give microprocessors information about temperatures, positions, pressures, etc. Virtually every system on the modern vehicle is processor controlled. In this chapter, we examine how some of the common sensors do their job electronically. In Chapter 23, we see how to test some of these inputs.

THERMISTORS

Thermistors are in use in most vehicles today. They are named because they are a temperature-sensitive resistor. To be correct, thermistors are not part of the semiconductor family. They are usually found connected to semiconductors or ICs, so we will look at their use here. Thermistors are used to detect various temperatures or changes in temperature. Their most frequent use involves the measurement of engine coolant temperature or inlet air temperature.

The first use of thermistors dates back to the mid sixties when electrical dashboard gauges first appeared. The thermistor was installed in the engine block with its end sitting in the engine coolant. As the coolant temperature changed, the resistance of the thermistor also changed. Ohm's Law at work says that, as the resistance of a circuit changes, the amperage will change also, assuming the voltage remains the same.

Figure 1 is a drawing of a common gauge circuit using a thermistor and an ammeter that has been calibrated in degrees Fahrenheit of temperature. As the temperature of the engine increases, the resistance of the thermistor decreases. This decreased series resistance increases the current flow through the ammeter and changes the needle's position. Rather than having the meter show amperage, the face is redesigned to show temperature. Many dashboard temperature gauges operate on this principle. The thermistor is called the **engine coolant temperature sensor (ECT)** and usually has at least one wire attached to it from the gauge. By installing it into the engine block, the sensor is grounded, completing the circuit. Many ECTs in use today have two wires connecting them into a circuit. Their operation is the same with the ground being supplied by the additional wire rather than the engine block. This style of thermistor is a negative coefficient type. This means that, as the temperature goes up, the resistance of the thermistor will go down.

When computers became responsible for air:fuel ratios and began to control ignition systems, the sensing of engine temperature and inlet air temperature became important. A cold engine requires a richer air:fuel ratio

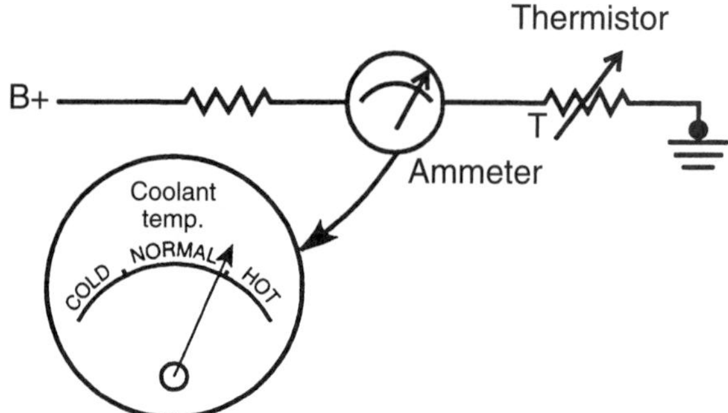

Figure 1. The thermistor controls the current flow through the meter.

IAT Sensor		
Temperature to Resistance Values (Approximate)		
°F	**°C**	**OHMS**
210	100	185
160	70	450
100	38	1,800
70	20	3,400
40	4	7,500
20	-7	13,500
0	-18	25,000
-40	-40	100,700

Figure 2. As the temperature decreases, the resistance increases.

(more gas), and a hot engine requires a leaner air:fuel ratio (less gas). The engineers turned to the thermistor as the device that would relay engine temperature to the computer. The engine computers in use cannot measure temperature directly. The input of engine temperature and air temperature information must be in a language that the computer will recognize. This language is electricity.

Notice the chart reproduced in **Figure 2**. It is a temperature-to-resistance chart for a negative-coefficient thermistor-style intake air temperature sensor (IAT).

Remember, negative coefficient means that, as the temperature increases, the resistance of the thermistor goes down. Notice that the thermistor is capable of inputting temperatures as cold as −40°F (below zero) and as hot as 210°F. This temperature range will become an input to the computer. "Input" is another way of saying information going into the computer. It is different from an output, which is the computer telling something to turn on or off. Everything associated with an on-board computer is either an input or an output. The thermistor signal is an input.

Most of the computer-controlled fuel systems in use today utilize air temperature as an input. Thermistors are easily installed and wired into the computer and will have their resistance changes seen as temperature changes. The use of multiple thermistors is increasing.

The computer circuits using thermistors usually send out a small signal to the sensor. The resistance of the sensor varies the current flow and voltage drop through the circuit. The computer reads this voltage drop and knows the temperature. The small signal sent out from the computer is usually referred to as the **reference voltage**. A reference voltage of 5 volts is common.

Figure 3 shows a wiring diagram that has the computer sending out the 5-volt open circuit reference to the intake air temperature sensor. Notice that there is a resistor in the 5-volt feed. This resistance value will vary from one manufacturer to another, but for our example, let us make it a 1500-ohm resistor. The IAT is also a resistance and you should remember that series resistance will cause the voltage to divide. The PCM monitors the voltage in between the fixed internal resistor of 1500 ohms and the variable IAT resistance. For example, an air temperature of 110°F would put the thermistor's approximate resistance as 1500 ohms. If we have 5 volts across two 1500-ohm resistors in series, the voltage between the resistors will be 2.5 volts, right? The voltage divides based on the resistance in a series circuit. In this way, the computer recognizes the 2.5 volts as an indication of 110°F intake air temperature. The computer will now use this information (input) to help it make correct and accurate decisions or outputs. Even though it is not the intent of this text to cover computer systems completely, the input devices on many of the vehicles are relatively easy to understand and diagnose using a voltmeter and an ohmmeter. Knowing the approximate engine temperature and the actual resistance of the ECT makes diagnosis a snap. For example (refer back to Figure 2), an intake air tem-

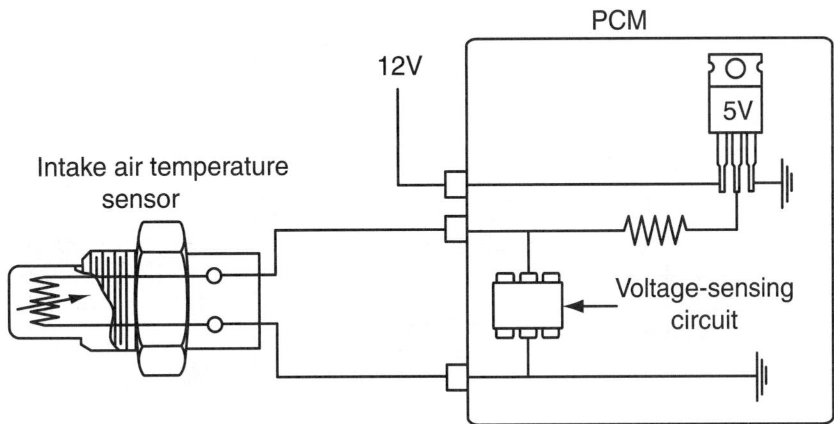

Figure 3. IAT circuit with 5 volts from the PCM.

perature of 100°F and an IAT resistance of 100,000 ohms would tell technicians that they are looking at a bad sensor. The actual air temperature and the temperature that the computer "sees" as an input is different by 140°F. An IAT resistance of 100,000 ohms should be a temperature of –40°F rather than the 100°F actual. In this manner, the technician can verify the information that the computer is accepting and determine if an input is accurate. Computer enthusiasts like to say "garbage in, garbage out," which translates to "wrong inputs will result in wrong outputs" and possibly driveability complaints. Thermistors are a frequently found input device any time a temperature reading is needed. This includes not only engine and air temperature, but transmission, transaxle, and air conditioning. It is not uncommon to have as many as 10 thermistors in use on today's vehicles.

POSITION SENSORS

Another computer input is the **throttle position sensor (TPS)**. As was the case with the coolant or intake air temperature, the computer must know what position the throttle is in to be able to determine air/fuel ratio, ignition functions, and various other outputs, which it is responsible for. The TPS is one of the inputs that virtually all computer systems use. The TPS uses the voltage division principle of a series circuit as its indicator of throttle position. As the throttle position changes, the TPS will vary the resistance of the circuit. Notice that reference voltage is again applied to the TPS **(Figure 4)**. Notice also that there is a circuit through the TPS to the PCM ground circuit. A very small amount of current is always flowing through the TPS. Because it is a resistance, the voltage will drop from 5 volts to the 0 volts of ground. The center terminal connects to a wiper that will touch the TPS resistor in a position that relates to the position of the throttle. For example, if the throttle is at 50 percent, the wiper will contact the resistor in the middle. If the voltage drops evenly across the resistance, then the voltage in the middle will be 2.5 volts. The 2.5 volts is interpreted by the PCM as a throttle position of 50 percent. The simplified chart in **Figure 5** shows that idle will return approximately an 0.5 volt signal; one-quarter throttle will be approximately 1.25 volts; half throttle will be 2.5 volts; and three-quarters throttle will be 3.75 volts. **Wide-open throttle (WOT)**, or 100 percent, will be approximately 4.5 volts. Notice that at the two extremes of idle and WOT, the voltage varied between 0.5 and 4.5 volts.

Throttle Angle	TP Voltage
0%	0.5V
25%	1.25V
50%	2.50V
75%	3.75V
100%	4.50V

Figure 5. As the throttle angle increases, the TP voltage increases.

Most TPSs operate on these principles. They can, however, operate in reverse: high voltage at idle and low voltage at WOT. They can also operate on a different reference voltage than that in this example, but in principle, they will be the same. Diagnosis of them is simplified if you understand what it is they are doing. Keep in mind that there is a mechanical connection between the throttle and the TPS. On a central fuel injection system (TBI), the TPS is mounted directly to the injection unit; on a multiport system (MFI), it is mounted to the throttle body unit **(Figure 6)**. In either case, as the throttle is moved, the wiper arm of the TPS is also moved. The three wires attached to it are the reference voltage, the return voltage, and the ground.

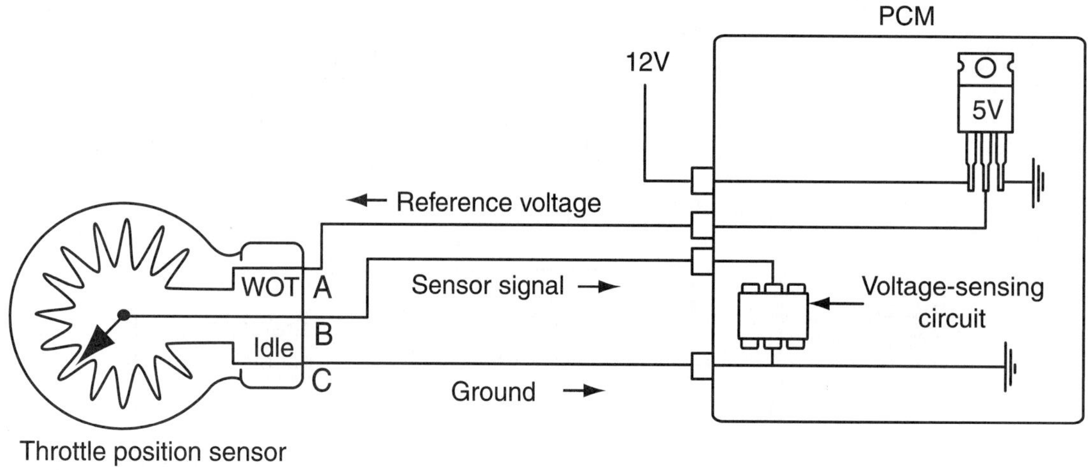

Figure 4. As the throttle moves, a wiper arm moves to send the voltage signal to the PCM.

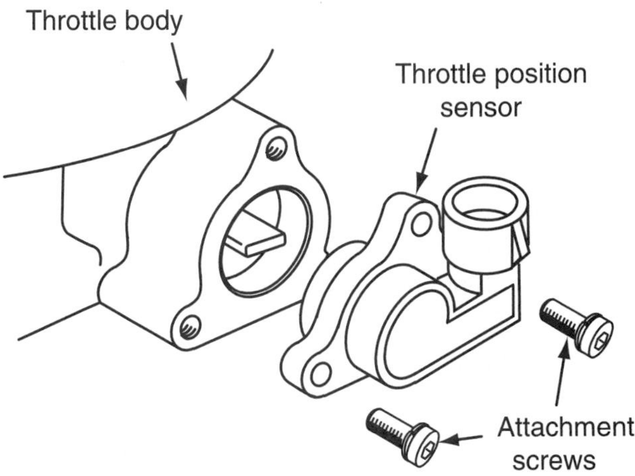

Figure 6. The TPS bolts to the throttle body assembly.

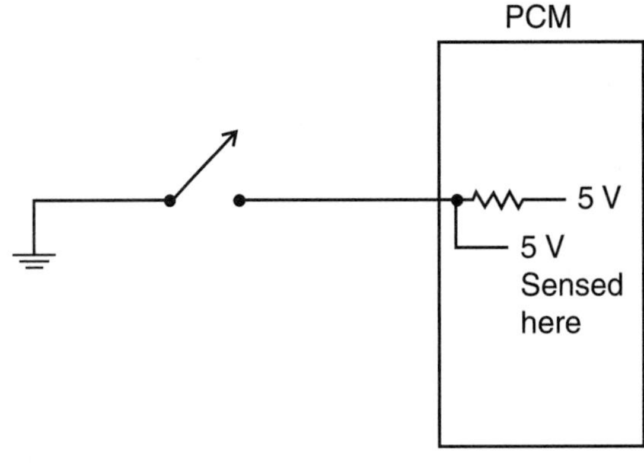

Figure 7. Five volts is sensed with the switch open.

THE SWITCH AS AN INPUT DEVICE

The switch that we looked at in Chapter 6, where we discussed circuits, can be used as an input device to an electronic module. For example, the manufacturer wants to know if a valve is open or closed and attaches a switch to the valve powering it from the PCM with a 5-volt feed that has a large amount of resistance, as shown in **Figure 7**. The resistance ensures that only a small amount of current will flow when the switch closes the path to ground. The PCM senses the voltage at the point just after the resistance. If the switch is open, the voltage will be 5 volts, and if the switch is closed (to ground), the voltage will be zero volts, as **Figure 8** shows. The PCM now has used the switch as an input device to indicate the position of the valve.

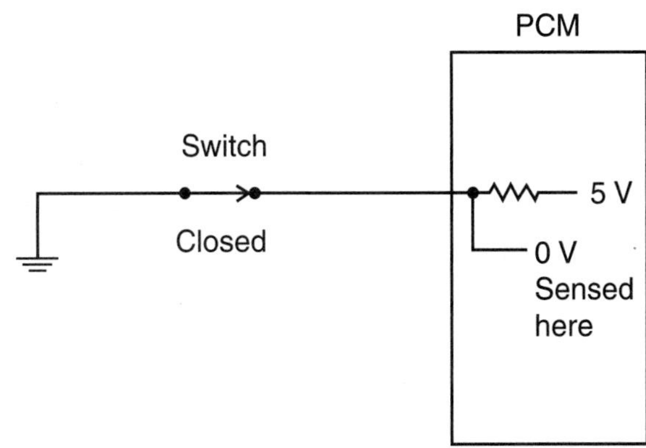

Figure 8. Zero volt is sensed with the switch closed.

Summary

- A thermistor is a resistor that has a changing resistance based on temperature.
- A negative coefficient thermistor will have an increasing resistance with decreasing temperature.
- Intake air temperature and engine coolant temperature are two examples of thermistors.
- An IAT relies on voltage division to indicate temperature to the PCM.
- A TPS will indicate the percent of throttle opening by a variable voltage return between 0.5 and 4.5 volts.
- A switch can be a PCM input with a 5-volt signal when the switch is open and zero volts when it is closed.

Review Questions

1. At one-half open throttle, the TPS return voltage on a 5-volt computer system is 0.5 volts. Technician A states that this is normal. Technician B states that this is not normal. Who is correct?
   A. Technician A only
   B. Technician B only
   C. Both Technician A and Technician B
   D. Neither Technician A nor Technician B
2. A negative-coefficient thermistor will have its total resistance
   A. go up as its temperature increases
   B. remain the same as its temperature increases
   C. go down as its temperature increases
   D. go down as its temperature decreases

3. Thermistors are frequently used as
   A. despiking units
   B. computer input devices
   C. computer output devices
   D. none of the above
4. Generally both an IAT and ECT are negative-coefficient thermistors.
   A. True
   B. False
5. Proper testing of a thermistor requires both a resistance reading and a temperature reading.
   A. True
   B. False
6. Thermistors are frequently used as computer output devices.
   A. True
   B. False
7. TPSs are variable resistors with two wires attached.
   A. True
   B. False
8. Reference voltage is the voltage signal that the PCM usually sends out to sensors.
   A. True
   B. False

Diagnosing and Servicing Electronic-Control Input Devices

Introduction

As we mentioned in Chapter 22, input devices fall into only a few categories. We looked at position sensors, thermistors, and switches. In this chapter, we look at simplified methods of testing these devices. In addition, we introduce to you a fundamental piece of test equipment, the scanner. Every technician's toolbox contains a scanner. It allows you to interface with the PCM and have the information displayed on screen about inputs and outputs.

USING A DMM TO TEST INPUTS

There are a variety of electronic input devices that are found on the modern vehicle. Many of these can be tested using the digital multimeter. The use of volts scales usually will give sufficient information for an accurate diagnosis.

Thermistors

Thermistors, you recall, are variable resistors whose resistance value will change with temperature. There are two methods to check the accuracy. The easiest is to measure the resistance after determining the temperature. If, for instance, the temperature of the coolant is 190°F and the resistance of the thermistor is 10,050 ohms, is the device good based on **Figure 1**? The chart indicates that at 190°F we should have approximately 240 ohms. The sensor is obviously not accurate. As a matter of fact, 10,000 ohms is closer to 32°F. The PCM will add the amount of fuel required at 32°F, which will be more than is required at 190°F.

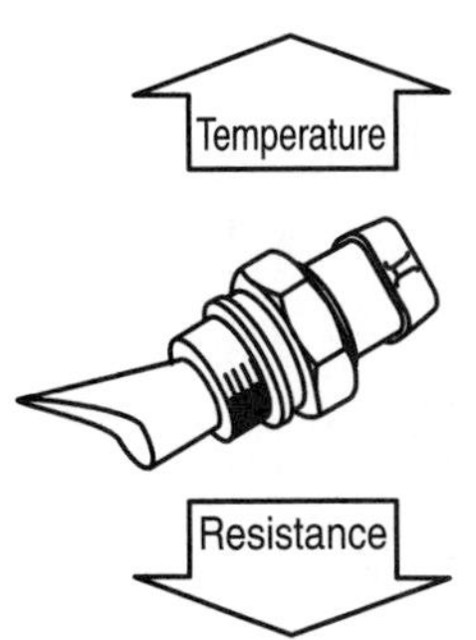

Thermistor temperature-to-resistance values		
°F	°C	Ohms
210	100	185
190	90	240
160	70	450
100	38	1,800
40	4	7,500
32	0	10,000
0	-18	25,000
-40	-40	100,700

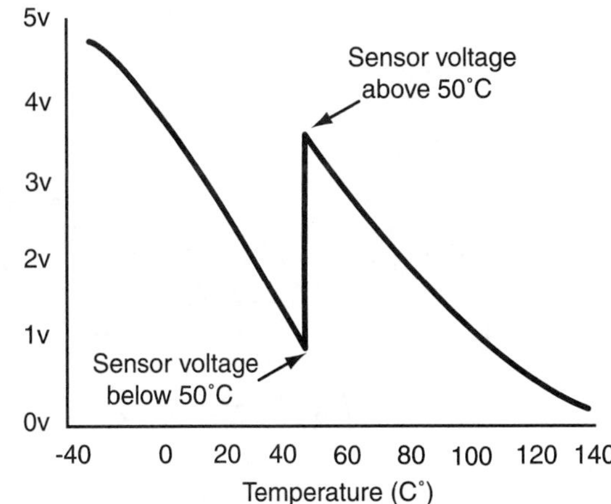

Figure 1. Temperature-to-resistance and temperature-to-voltage charts.

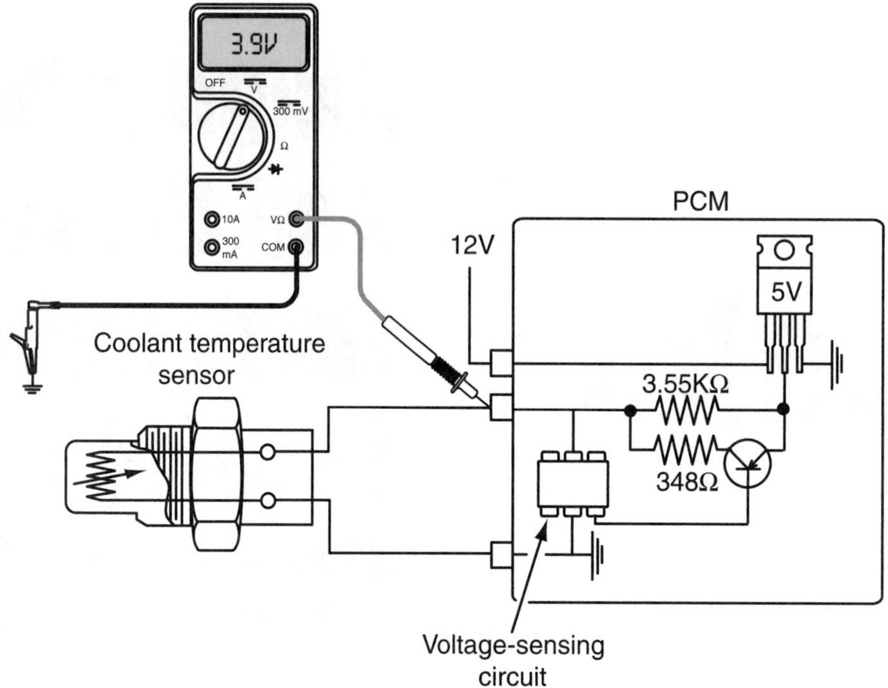

Figure 2. A 3.9-volt reading indicates a below-zero temperature.

Another method, and usually the preferred method, is to tap into the signal side of the circuit with a voltmeter as **Figure 2** shows. Our voltmeter reads 3.9 volts. This converts to a below-zero reading. What should the voltage be if the temperature of the engine is 60°C? The chart indicates that it should be approximately 3.0 volts.

Using the voltmeter to indicate the voltage that the PCM is seeing is more accurate than using an ohmmeter. The ohmmeter requires that the sensor be disconnected, whereas the DMM set on volts can be used with the sensor connected.

Position Sensors

The TPS uses a 5-volt feed and returns to the PCM an indication of position with voltage. **Figure 3** shows the circuit that we looked at in Chapter 22. If the DMM set on volts is placed between the sensor signal wire and ground, the display will indicate TPS voltage, which is an indication of position. If a slow sweep of the gas pedal does not show a slowly changing voltage upward, the sensor is not functioning. You will see that the DSO is a better choice for this sensor.

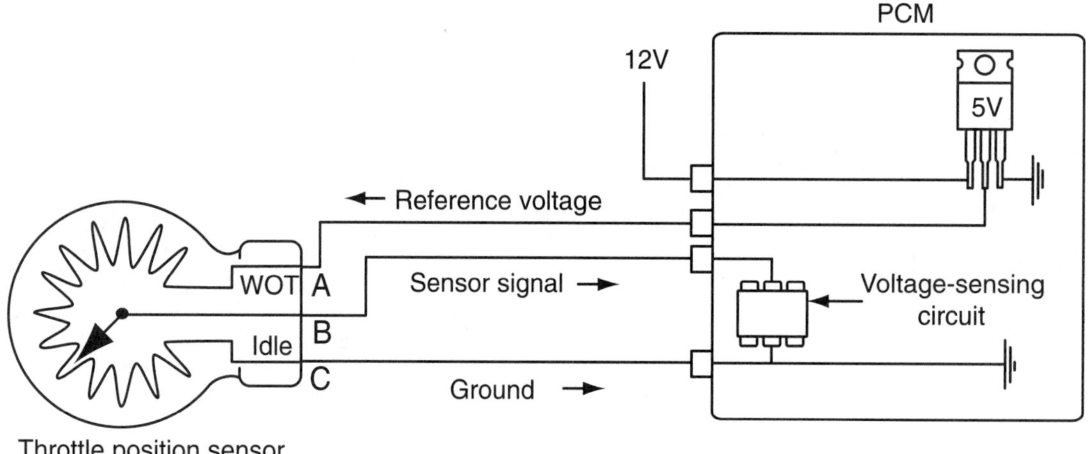

Figure 3. A typical TPS circuit.

Switches

There are two types of input switches in use today, and it is important to note which type you have before testing. **Figure 4** shows a pull-up switch. Pull-up is another way of saying the circuit will switch voltage or B+. With the voltmeter wired as shown, the display will read zero volts with the switch open and zero volts with it closed.

The opposite configuration is shown in **Figure 5**, which is of a pull-down switch circuit. What will the voltmeter read with the switch open? How about with the switch closed? Open it will read voltage and closed it will read zero volts (ground).

The open switch allows the battery source voltage to reach the meter. With the switch closed, the ground is applied and the voltmeter reads the ground voltage, which is zero volts.

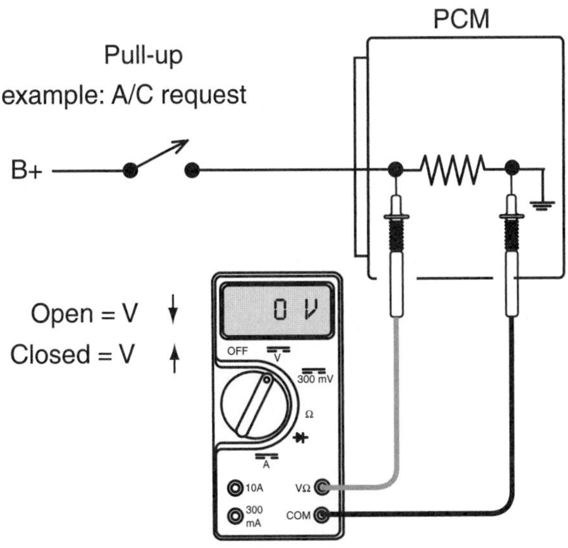

Figure 4. The voltmeter reads zero with the switch open and B+ with the switch closed.

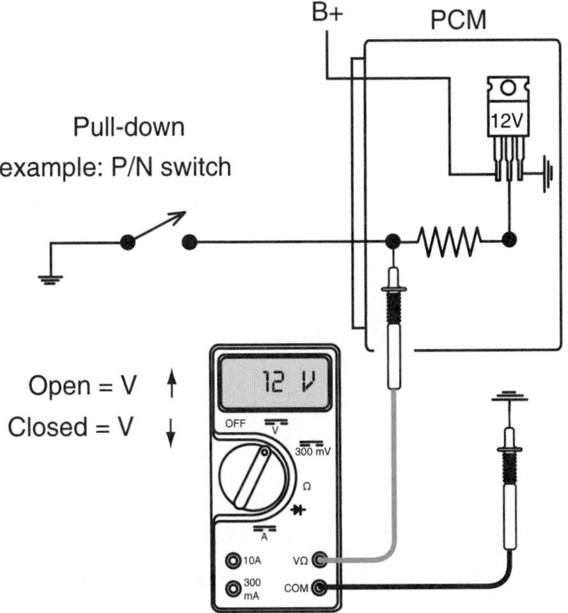

Figure 5. The voltmeter reads B+ with the switch open and zero with the switch closed.

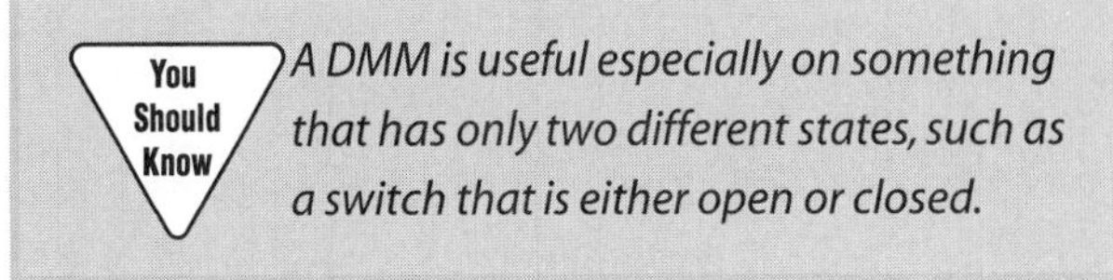

A DMM is useful especially on something that has only two different states, such as a switch that is either open or closed.

USING A DSO TO TEST INPUTS

Some examples of inputs lend themselves to the use of a DSO. Position sensors are one of them. First, we will capture a pattern from an ECT over a period of 500 seconds with a graphing multimeter. The graphing multimeter is a DSO-type device, but slower. We will connect the red lead to the ECT and the black lead to ground and start the engine cold. The voltage should change as the engine warms up. Notice from **Figure 6** that the ECT started at 1.5 volts and ended 500 seconds later at 1.5 volts. To test the accuracy of the signal, we would have to graph the temperature during the same time period. Remember the dual range thermistor circuit? This vehicle uses one. We started out at 1.5 volts, which is approximately 100°F. As the temperature of the engine went up, the sensor voltage went down until it reached the switch point of 122°F. The resistance value within the PCM changed at this point, and that is the straight up line in the second division that goes up to approximately 3.75 volts. As the warmup continues, the voltage continues to drop until the last division, when it rises up again.

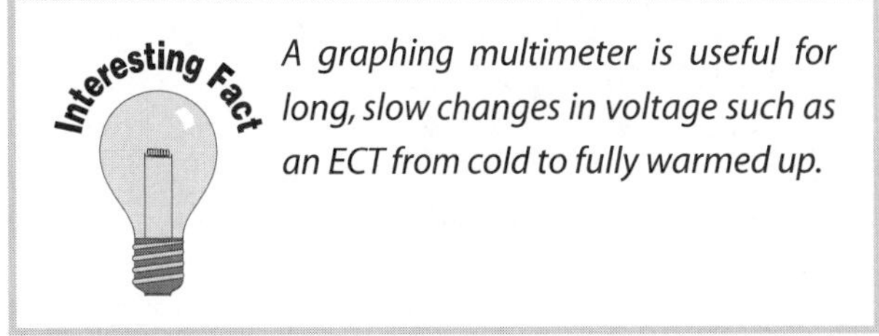

A graphing multimeter is useful for long, slow changes in voltage such as an ECT from cold to fully warmed up.

What do you think happened here? This is the temperature at which the engine thermostat opened and allowed cooler water into the engine. The ECT was able to read this decrease in coolant temperature and send the information to the PCM. Notice that the drop in voltage is smooth, with only two small voltage glitches in the fourth and fifth divi-

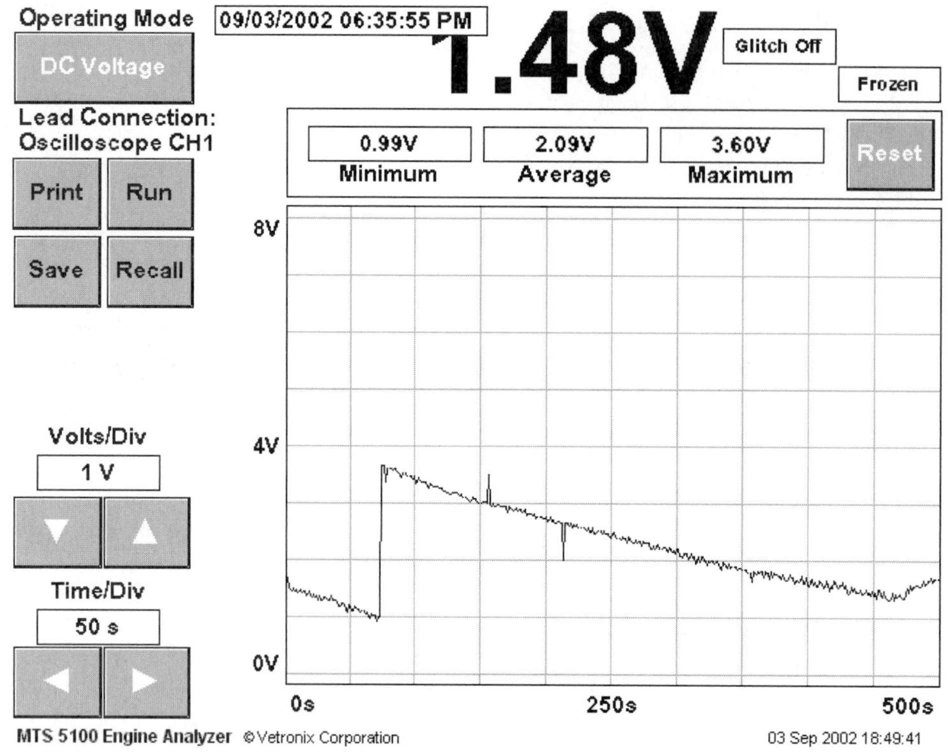

Figure 6. A voltage graph of the ECT during warmup.

sions. It is possible that a problem exists, but it looks more like interference from another source and need not be of concern.

Position Sensors

Position sensors are where the DSO pays for itself. After you talk to the customer and try to determine under what driving conditions the difficulty occurs, the DSO can capture a pattern off a TPS to match the conditions. **Figure 7** shows a normal TPS during a slow acceleration and deceleration. The DSO has been set for 1 volt per division and 200 milliseconds per division. We have a total on-screen time of 2 seconds to move the throttle from closed throttle to WOT. The key on–engine off voltage was slightly higher than 0.5 volts. The maximum voltage at WOT was 4.5 volts, and the ramp was smooth and glitch free. The engine was *not* running during this test. The key was on and the engine off.

Look at the TPS shown in **Figure 8**. Notice the dip in voltage in the fifth division. As the throttle was moved from closed to WOT, the voltage did not change smoothly as in the previous figure. There appears to be a problem in the TPS. The PCM will have difficulty figuring out where the throttle is if the gas pedal is about one-third open. At approximately the one-third position, the voltage dives back down to the almost closed level. Once past the position, it rises normally. This TPS should be replaced. Once the new one has been installed, the new one should be tested. This test is called a **TPS sweep**.

INTRODUCTION TO A SCANNER

When a PCM is busy controlling the engine systems, it takes the inputs and analyzes them before making a decision or output.

A scanner allows the technician to look inside the PCM and view what it sees as inputs. A single bit of data is displayed in a format that a scanner can interpret and is called a **PID**. A PID is a single piece of information. If we look at the TPS by itself, we are looking at the TPS PID. Many scanners are on the market, and it is not within this text to look at the different types of scanners available. We look at data that the PCM transmits to the scanner in an effort to diagnose the same three inputs: thermistors, position sensors, and switches.

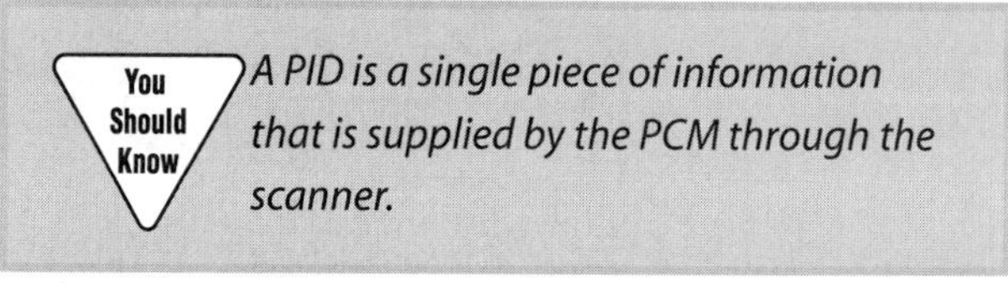

A PID is a single piece of information that is supplied by the PCM through the scanner.

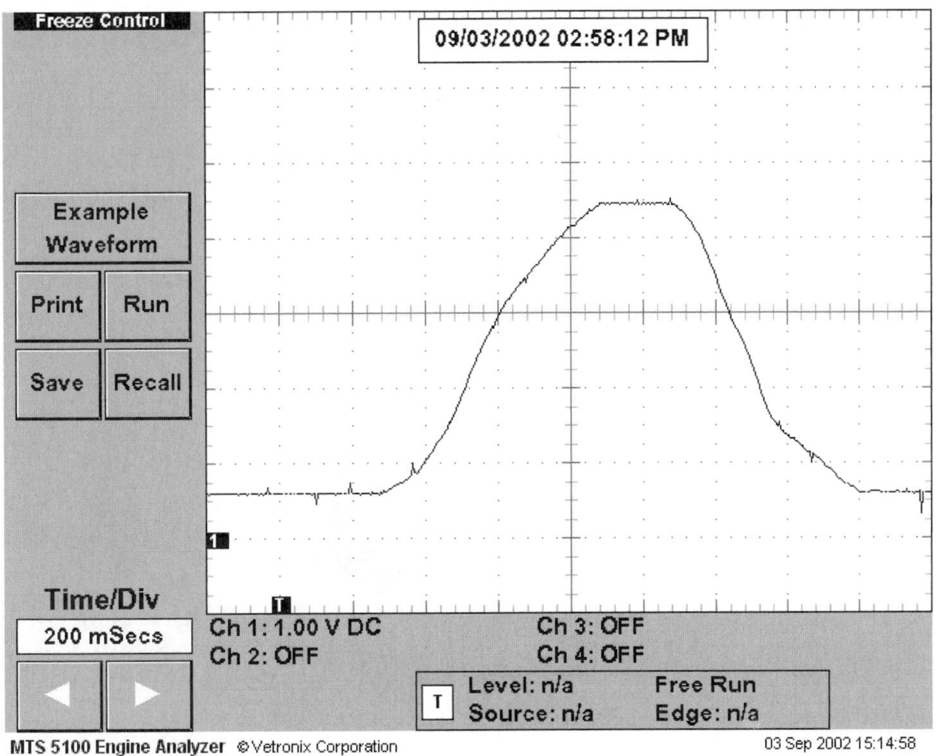

Figure 7. The voltage sweep from a functional TPS.

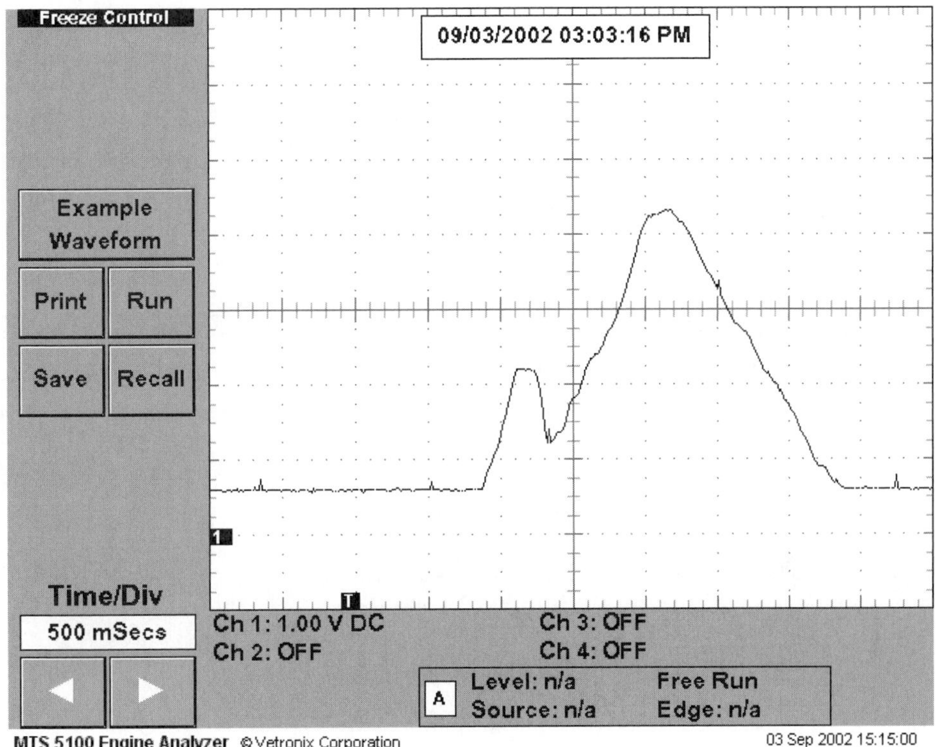

Figure 8. The voltage sweep from a faulty TPS.

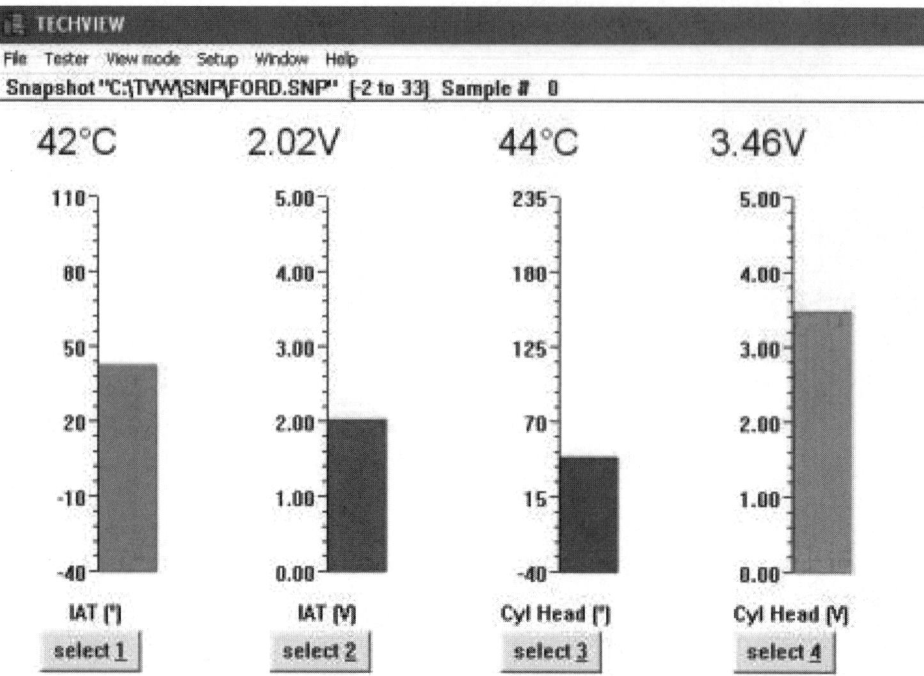

Figure 9. A scanner allows a bar graph display of four PIDs.

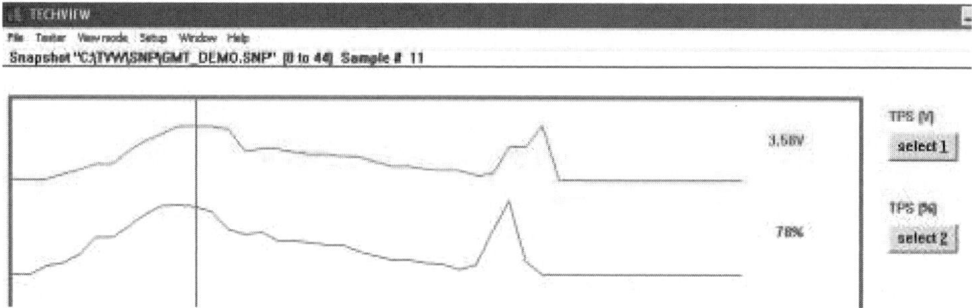

Figure 10. The scanner allows us to look at the TPS over a long period of time.

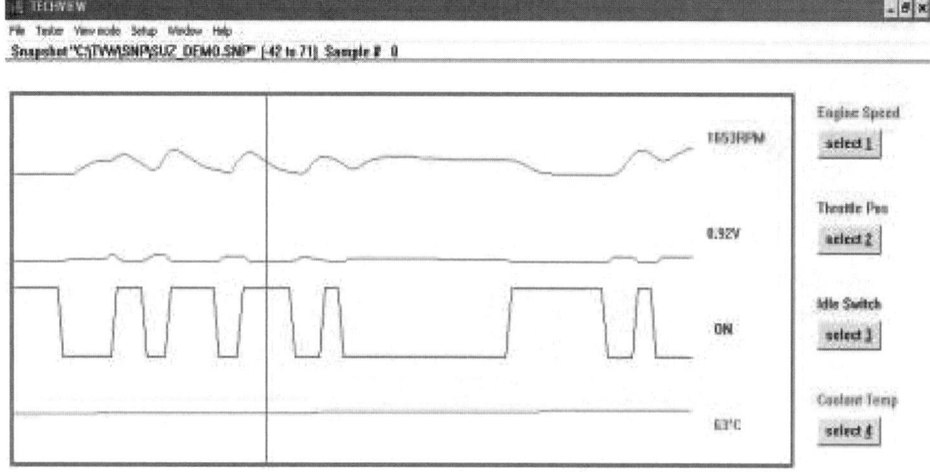

Figure 11. The scanner can show a switch as it opens and closes.

USING THE SCANNER TO TEST INPUTS

It is important to note that many of the input sensors found on the modern vehicle need to be tested by additional methods other than by using a DMM. If the PCM generates serial data that has sensor information, a scanner can be the tool of choice.

Thermistors

Figure 9 shows the scanner data for engine and air temperature in degrees Fahrenheit. If we start with a cold engine and let the engine warm up, the temperature will change directly on the screen. We can also display the data in graph form over a period of time. A passenger can also watch the data while the vehicle is driven until the problem is duplicated.

Throttle Position Sensors

Scanner data is very helpful when viewed over a period of time. If we move the throttle and watch the data change, we are doing a type of sweep.

A graph of the TPS position in percent and the TPS voltage is displayed in **Figure 10**. Notice that generally the voltage and the percent of throttle follow one another, as they should. Keep in mind that the data you are looking at is coming from the PCM and is its interpretation of what it sees. When we use the DSO, we are looking at the data directly. When we use the scanner, the data has been interpreted by the PCM. In theory, there should not be a difference; however, to be realistic there can be times when the two do not agree. This signifies a problem that will require further diagnosis and the use of the DSO.

A scanner will display changing values very well and allow you to freeze them for later inspection.

Switches

Switches are easily observed using a scanner. We choose the PID that signifies the switch, and the scanner will tell us if it is open or closed. **Figure 11** shows four PIDs at the same time. The top one is engine speed. The second one is TPS volts. The third one is an idle switch that closes when the throttle closes. It is a pull-up circuit, so the voltage goes up at each closing and drops when the switch opens, when the throttle starts to open. The fourth PID is engine temperature, which did not change during the course of the test. Is this a good switch? It looks good because it can be correlated to the throttle position. There is an instantaneous rise in voltage when the throttle is closed.

Summary

- A DMM can be used to easily look at inputs as long as they are not fast.
- We can use a voltmeter to look at thermistors as an input device.
- The voltage of a switch will change from B+ to zero volts as the switch opens and closes.
- The DSO is effective in looking at changing voltages such as those present in the TPS.
- A TPS sweep should reveal problems when observed on a DSO.
- A scanner allows the technician to look at individual PIDs.
- A PID is a single data stream from one device.
- The scanner will allow the technician to look at PIDs over a period of time, including during a test drive.
- The scanner will show the state of the switch and the history of its action.

Review Questions

1. Technician A states that the scanner can be used to observe a TPS sweep. Technician B states that the DSO can be used to observe a TPS sweep. Who is correct?
   A. Technician A only
   B. Technician B only
   C. Both Technician A and Technician B
   D. Neither Technician A nor Technician B
2. A DMM shows the same voltage during two different times of one warmup. This is because
   A. the sensor is defective
   B. the DMM should not be used to test a thermistor
   C. the sensor is of the dual-range variety
   D. none of the above

3. A scanner is set to display a switch that is opening and closing. The voltage should
   A. remain steady at B+
   B. remain steady at ground or zero volts
   C. pulse between zero volts and a negative value
   D. pulse between B+ and zero volts
4. A TPS sweep is done on a DSO. The pattern shows a rise and then a rapid fall in voltage at approximately one-quarter throttle. After one-quarter throttle, the voltage continues to rise to 4.5 volts. Technician A states that this might cause the PCM to cut back on fuel at the one-quarter throttle position. Technician B states that this is normal for a TPS pattern. Who is correct?
   A. Technician A only
   B. Technician B only
   C. Both Technician A and Technician B
   D. Neither Technician A nor Technician B
5. The preferred tool to look at the TPS is the
   A. DMM on ohms
   B. DMM on volts
   C. scanner
   D. DSO
6. As the engine coolant temperature increases, the sensor's voltage will rise.
   A. True
   B. False
7. A scanner can show a TPS PID in voltage or percent.
   A. True
   B. False

Chapter 24 Integrated Circuits As Input Devices

Introduction

A variety of ICs are on the modern vehicle. The PCM alone has hundreds of ICs. The diagnosis and repair of the PCM is not part of this text. Instead, we look at a few ICs that are frequently found as input devices to the PCM or the ignition module. We will examine some of them again when we go into the ignition system in depth in Chapters 49–53. Consider this to be an introduction.

IC PROTECTION

We have noted that the use of ICs is increasing in the automotive industry. This fact makes precautions all the more important. The following are some general rules or precautions that should be followed when working on ICs. You should note that they are the same precautions that were listed in the electronics section.

1. Do not apply voltage directly to the IC unless the manufacturer's procedure calls for it.
2. Do not connect or disconnect an IC with the power on to the circuit.
3. Keep the battery voltage less than 20 volts during charging or jumping.
4. If a relay is added to the vehicle, as in a rear window defogger, use one that contains a despiking diode. This will prevent a relay surge from getting to an IC and damaging it. High-voltage surges may destroy chips.
5. Do not run the vehicle without the battery being connected.
6. Always make sure that the charging system is not producing excessive AC. AC can damage sensitive computer circuits or at least change around some of the input information.

HALL EFFECT SENSORS

Probably the most common use for an IC found on most vehicles is a Hall effect sensor. They are used whenever the manufacturer wants to know the position of something, such as the crankshaft or camshaft. They are an IC that will output a signal that is a digital on/off voltage change. By counting the number of pulses, speed can also be calculated. An example of this type of input device is a **CKP**, or crankshaft position sensor. The CKP consists of a rotary vane cup and a Hall effect switch **(Figure 1)**. Voltage and a ground are applied to the Hall effect switch, which effectively turns it into a device that will be able to sense magnetic fields. A weak permanent magnet is located opposite the sensor with the vane cup in between.

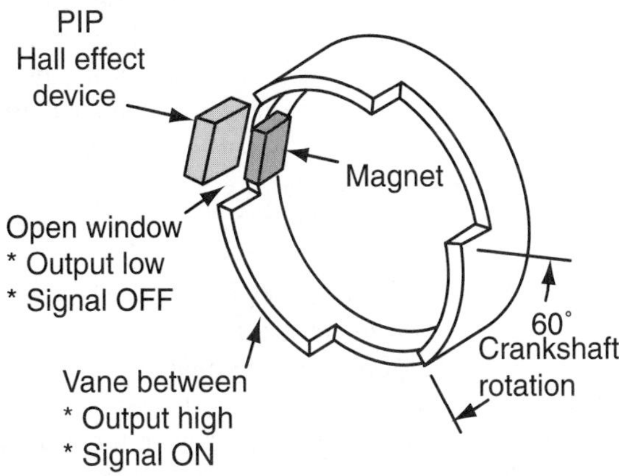

Figure 1. A simplified CKP.

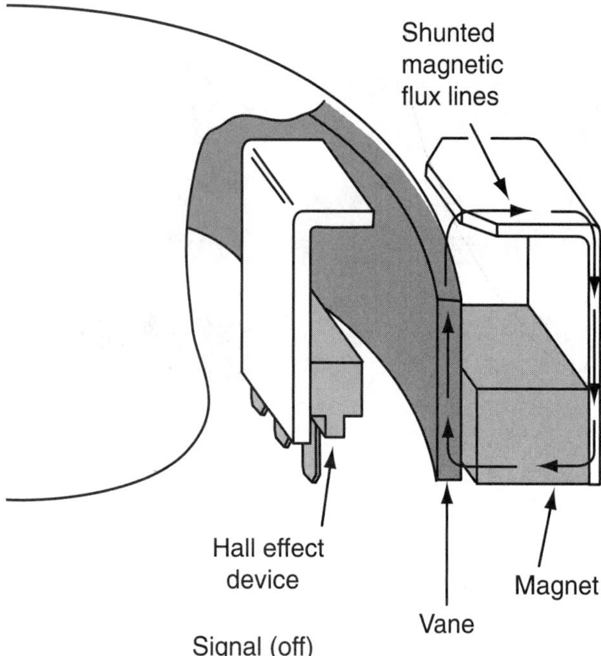

Figure 2. When the vane blocks the magnetic field, the voltage rises.

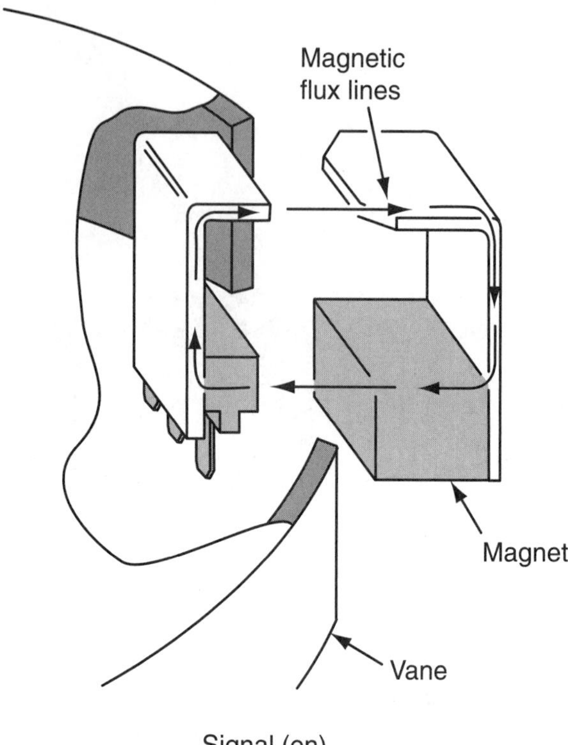

Figure 3. When the magnetic field is not blocked, the output voltage drops.

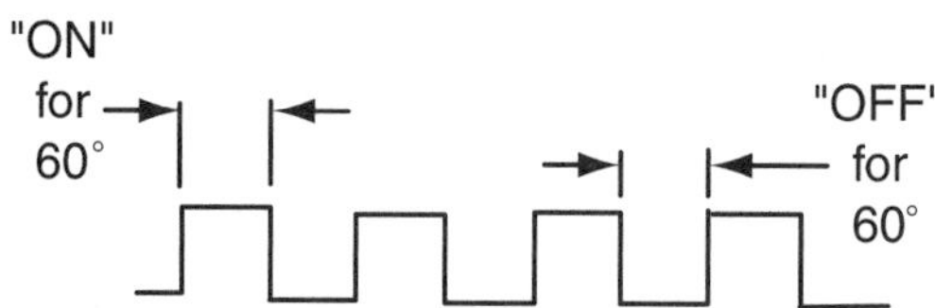

Figure 4. A four-cylinder CKP might generate four high-voltage and four low-voltage signals.

When the vane blocks the magnetic field from reaching the switch, the signal from the switch is on or high **(Figure 2)**. The output at this time is the input voltage. If the sensor is powered with 12 volts, then the output with the vane blocking the magnetic field is 12 volts. However, when a space is between the switch circuit and the magnet, the signal goes off or down **(Figure 3)**. The amplitude or voltage does not change in this type of sensor. Instead, the number of times the on/off cycle is repeated is changed. This is called a frequency-modulated signal. We will see many uses for this type of sensor on the modern vehicle. The signal from a CKP or any other sensor that uses a Hall effect switch is a digital on/off signal, which can be directly read by the processor. Its signal may be used to determine position, speed, or both. Keep in mind that the faster the rotary vane cup moves, the more frequent the output signal will be. A CKP for a four-cylinder distributor will generate four high- and four low-voltage signals per rotation **(Figure 4)**. This will translate into 4000 signals per minute at 2000 engine rpm (typical highway speeds). Two thousand engine rpm is 1000 distributor rpm. With four signals per distributor rotation:

4 × 1000 rpm = 4000 signals per minute delivered to the processor

The processor may be the **electronic control module (ECM)**, the PCM, or any other processor that needs information about speed and/or position. In addition, the manufacturer may add additional sensors to one magnet so that the output will be two different sets of high/low voltage changes. We will look at multiple sensors in the ignition chapters, Chapters 46–53.

The basic Hall effect sensor has three wires: power, ground, and a signal. However, if we add additional sensors, each will have its own signal wire.

VARIABLE-FREQUENCY GENERATORS

The manifold absolute pressure sensor, or MAP, is an example of an input signal device that uses a chip. There are two designs in frequent use today; both do the same job, that is, tell the computer-controlled fuel and ignition systems what the pressure is inside the intake manifold. Let us look at their designs and see what the IC, or chip, is doing.

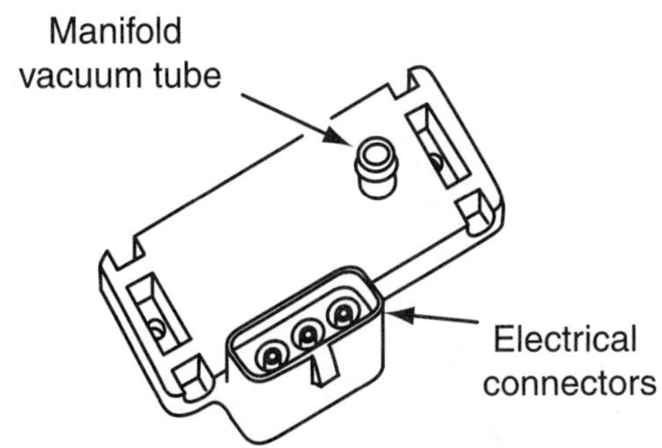

Figure 5. A General Motors MAP.

Figure 5 shows a common voltage division style of MAP circuit. It is calibrated to recognize pressures between 0 psi and 15 psi. The vacuum line attached to it goes to the intake manifold so that it sees the pressure in the manifold. As the throttle is opened and closed and the vehicle accelerates or cruises, the pressure inside the manifold will change. The computer needs to know this changing pressure if it is to change the air/fuel ratio (how much air and how much fuel). This pressure inside the manifold is close to normal atmospheric pressure because the throttle is not restricting the airflow. Obviously, the computer is an electrical device and cannot recognize pressure directly. The pressure must be converted to a computer-recognizable voltage signal by the chip and input into the computer.

Figure 6 shows the relationship between pressure and voltage that the IC should input to the computer. This chart is for a voltage style of MAP in common use among many manufacturers. When we idle or decelerate, the pressure in the manifold goes down. You have probably referred to this as a vacuum. It might go as low as 1 or 2 psi (which converts to 0.5 volts at the computer). Notice from the chart that 7 psi is 2.5 volts to the computer. In this way, WOT will be approximately 4.5 volts and deceleration will be approximately 0.5 volts. The MAP also can compensate for altitude and weather conditions.

Altitude compensation changes the MAP inputs to the computer. In this way, the computer can vary the air:fuel ratio as the vehicle climbs a mountain and the air gets thinner. Less air requires less fuel. The reduced air pressure changes the voltage signal to the computer by 0.1 volt for every 1000 feet, as the chart in **Figure 7** shows. We will see how to diagnose these input devices in Chapter 25. In the meantime, keep the purpose of this IC, or chip, in mind. It is an excellent example of ICs on the automobile.

Altitude	Volts
Below 1000	1.7–3.2
1000–2000	1.6–3.0
2000–3000	1.5–2.8
3000–4000	1.4–2.7
4000–5000	1.3–2.6
5000–6000	1.3–2.5
6000–7000	1.2–2.5
7000–8000	1.1–2.4
8000–9000	1.1–2.3
9000–10,000	1.0–2.2

Figure 7. Altitude compensation changes the MAP output.

Another type of MAP is shown in **Figure 8**. The information input to the computer will be the same, manifold pressure. The manner in which this IC functions is different, however. Instead of being a voltage divider, as was previously discussed, it is a frequency generator. Ford has used this type of sensor on some of their vehicles. Frequency refers to the number of voltage shifts during a period of time. For example, a frequency of 60 cycles, or hertz, is used in household electrical systems. This means that the voltage shifts or changes 60 times a second. This is its frequency.

Figure 9 shows the conversion of manifold pressure to frequency. The MAP in this circuit will vary or modulate the frequency of its output (input to the computer) as the pressure changes. The principle is the same, but you will see

Manifold Pressure (Vacuum) to Voltage Conversion		
Absolute Pressure	**Vacuum**	**Sensor Voltage**
14.7 psi	0"	4.5 V
10.5 psi	6.5"	3.75 V
7 psi	13"	2.5 V
3.5 psi	19.5"	1.25 V
0 psi	26"	0.5 V

Figure 6. Pressure/vacuum-to-sensor voltage.

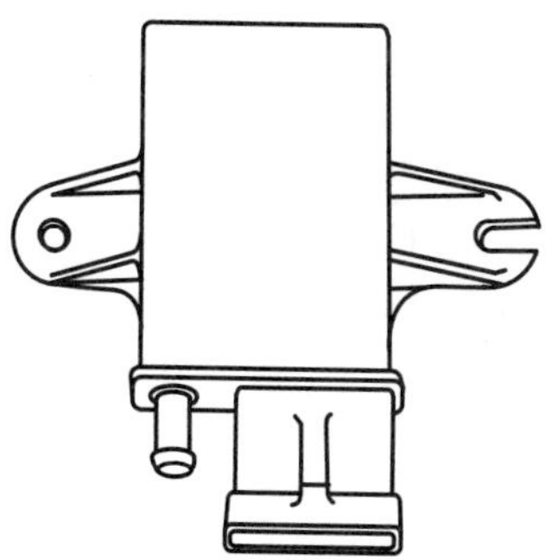

Figure 8. A Ford MAP.

Approximate Signal Frequencies of a Ford Map Sensor at Sea Level	
Manifold Vacuum	**Map Frequency**
0 in HG	160 Hz
13 in HG	125 Hz
26 in HG	90 Hz

Figure 9. Manifold vacuum changes will change the MAP output.

that the diagnosis procedure is different. Our first example can be diagnosed with a vacuum pump (to lower the pressure) and a voltmeter to note the pressure changes, whereas the second example will require the ability to measure frequency using either a scope or a DMM with frequency capability. As noted before, we will look at their diagnosis in Chapter 25.

PHOTO DIODE SENSORS

Position and speed are two important inputs to any PCM or ignition module. Systems are on the road that use the output from a circuit that uses a light-sensitive photoelectric cell and a light-emitting diode. **Figure 10** shows the inside of a distributor of a GM product. Notice that there are holes in a plate in two different rows. There are

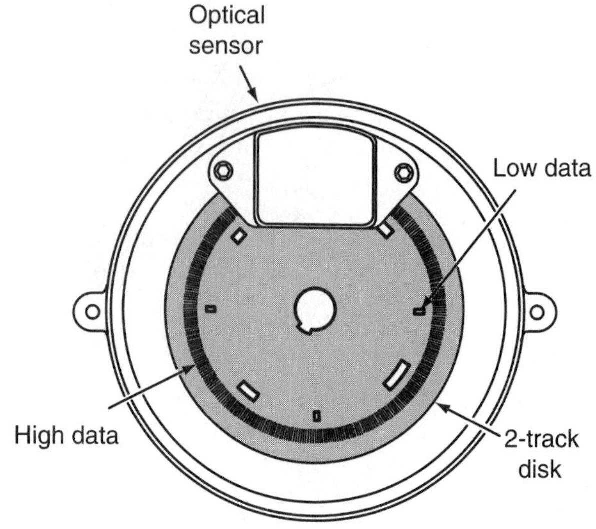

Figure 10. An optical sensor with two outputs.

360 holes or slots in the outer ring and eight slots in the inner row. Notice that the inner slots are different sizes. Their size will correspond to the cylinder number.

The sensor requires power and ground to light the diodes. When the diode's light is passing through a slot, the sensor "sees" the light and generates a voltage that will allow the IC to modulate the input voltage (usually 5 volts) up and down. The output from this sensor is similar to the Hall effect, an up and down voltage from 0 to 5 volts, which corresponds to engine speed for the outside slots or cylinder firing order for the inside slots **(Figure 11)**.

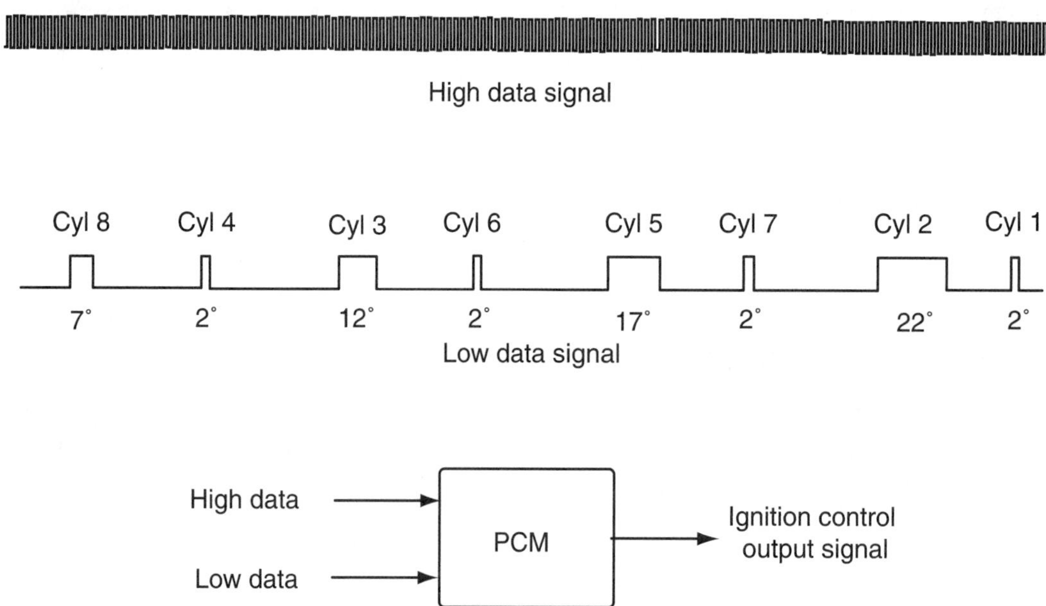

Figure 11. Two square wave signals from an optical distributor.

Summary

- Do not allow the charging voltage to go above 20 volts.
- Integrated circuit devices are frequently used to input information to the PCM or ignition module.
- Hall effect sensors generate a high signal when the magnetic field is blocked.
- Hall effect sensors generate a low signal when the magnetic field is not blocked.
- Many CKP sensors are Hall effect design.
- A Hall effect sensor generates a square wave signal whose frequency changes with speed.
- MAP sensors generate a voltage division signal or a frequency modulated signal.
- MAP sensors can change the air:fuel ratio based on altitude.
- Optical sensors generate a signal that looks like a Hall effect signal.
- Optical sensors usually have two rows of slots that will generate two different signals.

Review Questions

1. Manifold pressure is a ________-type input on a Ford product.
   A. inches of vacuum
   B. pulse width modulation
   C. frequency
   D. voltage
2. Battery voltage is jumped directly to the MAP sensor. Technician A states that this is an incorrect testing procedure and might destroy the IC inside the sensor. Technician B states that this procedure will not give any valid information. Who is correct?
   A. Technician A only
   B. Technician B only
   C. Both Technician A and Technician B
   D. Neither Technician A nor Technician B
3. An optical distributor is being discussed. Technician A states that the outside row of slots will generally have 360 slots with each one corresponding to a degree of distributor rotation. Technician B states that the inside row will generally have the same number of slots as cylinders with the signal being different for each cylinder. Who is correct?
   A. Technician A only
   B. Technician B only
   C. Both Technician A and Technician B
   D. Neither Technician A nor Technician B
4. The MAP signal for a GM vehicle is
   A. frequency modulated
   B. pulse high then low
   C. changing based on engine speed
   D. voltage from 0V–5V based on manifold pressure
5. Reference voltage is the voltage signal that the PCM usually sends out to sensors.
   A. True
   B. False
6. An ohmmeter cannot be used to test a MAP sensor.
   A. True
   B. False
7. MAP sensor signals vary with the pressure inside the manifold
   A. True
   B. False
8. Hall effect sensors are generally used to indicate position or speed.
   A. True
   B. False
9. Why is the CKP signal required by the PCM?
10. Why does the PCM need to know the altitude?

Chapter 25 Diagnosing and Servicing ICs

Introduction

In Chapter 24, we looked at how the various manufacturers used ICs as input devices, generally to the PCM. You need to recognize, however, that sometimes ICs are used as an input to ignition modules. We cover ICs in detail in Chapters 46–53, the ignition chapters. In this chapter, we study how to use modern test equipment to look at some selected inputs. We will rely primarily on the DSO as our tool of choice. The DSO offers us the most versatility and gives us the opportunity to observe the output of the sensor. Sometimes a scanner will give some valuable information because it shows us what the PCM thinks it has seen. Using the combination of the scanner and the DSO gives us all of the information that we should need to effectively repair the vehicle.

Do not forget that we should not disconnect an IC with power to it or jump power or ground to a terminal that we have not identified. ICs are not at all forgiving. Make a mistake, and the IC could be damaged. Then you would have two problems, the original one and the one you created, not a good situation to be in, especially considering the increasing cost of some of the vehicle components.

Always make sure you have the wiring diagram and the manufacturer's repair procedures in front of you before beginning the diagnosis. Both items may help keep you out of trouble.

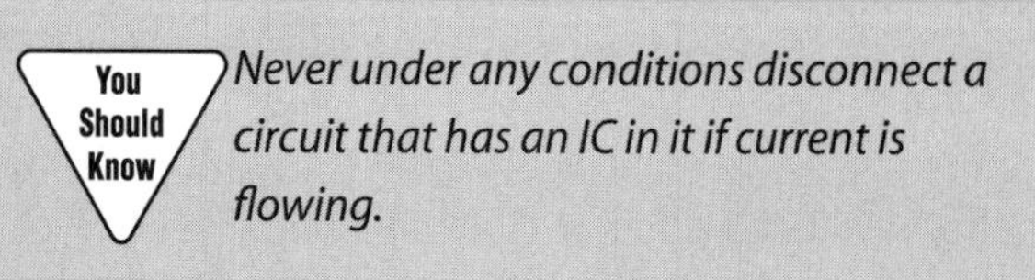

Never under any conditions disconnect a circuit that has an IC in it if current is flowing.

USING A DSO TO TEST INPUT SENSORS

When a technician needs to diagnose an input sensor that is an integrated circuit, the DSO becomes the perfect tool. The DSO will not draw a high enough current level to cause the circuit to malfunction and will still give the technician valuable information.

Hall Effect

Let us start with the Hall effect input sensor that we looked at in Chapter 24. Remember that it is a three-wire sensor with power and ground applied. The Hall effect sensors are frequently used in crankshaft and camshaft position devices. In addition, multiple Hall effect sensors might be in one housing, producing two patterns at the same time. We will look at multiple Hall effect sensors in Chapter 48. We will use a CKP as our example.

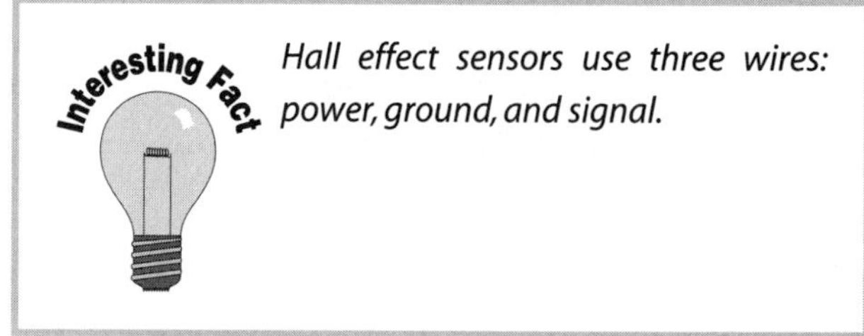

Hall effect sensors use three wires: power, ground, and signal.

Figure 1 shows a CKP installed on the front of a 3.4-liter GM product. The vibration dampener will have the windows and vanes. There are three conductors running to the sensor: power, ground, and signal. The wiring diagram will identify the colors of each. If we connect the DSO to the signal

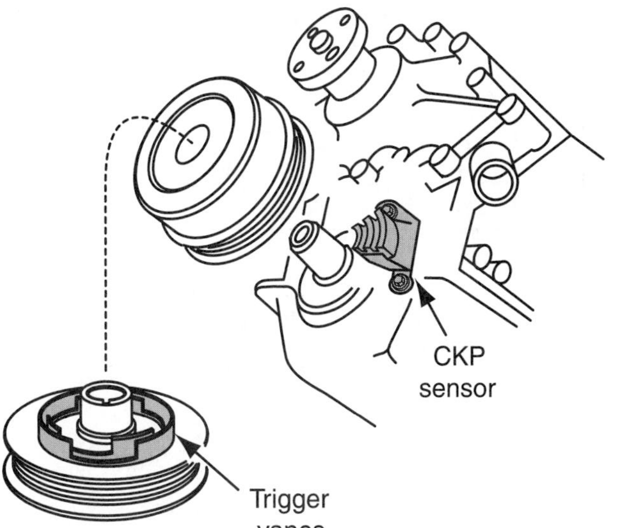

Figure 1. A CKP bolted to the front of an engine.

wire and either crank the engine over or run the engine, we should get a square wave pattern like that shown in **Figure 2**. Remember that when a window is in the sensor, the voltage should go down to zero volts. This allows us to look at the ground to the sensor in addition to the sensor's ability to recognize the window. Look at our example in Figure 2. The bottom of the square wave is right on the zero line (where the number 1 is). The sensor appears to have a good solid ground. When the vane is in the sensor, the voltage should rise up to about the same level as the input voltage. In our example, we are set for 5.00 volts per division and our top of the pattern is about 13 volts high. This sensor is fed with full battery voltage. Not all sensors are powered with full B+. The top line might be at 5 or 7 volts. The top line gives us a look at the power to the sensor. Return to our example. A good, solid, clean line at the top indicates a good power source. The up and down action of this sensor is a good example of how a good sensor's output should look.

MAP Sensors

Manifold absolute pressure sensors fall into two different categories. Let us look at variable voltage first. Most manufacturers use a MAP of this type. Notice from **Figure 3** that the MAP is fed with a +5 volt called reference on the GRY wire and has a ground applied to it on the BLK wire. The LT GRN wire brings the manifold pressure information to the PCM. There will be a vacuum line from the manifold to the MAP and there should be vacuum at the MAP with the engine running.

We should connect our DSO to the LT GRN signal wire. With the key on, we should be measuring atmospheric pressure and a relatively high voltage. With the engine running, the voltage will drop down and stay down because the engine vacuum has reduced the pressure in the manifold. If

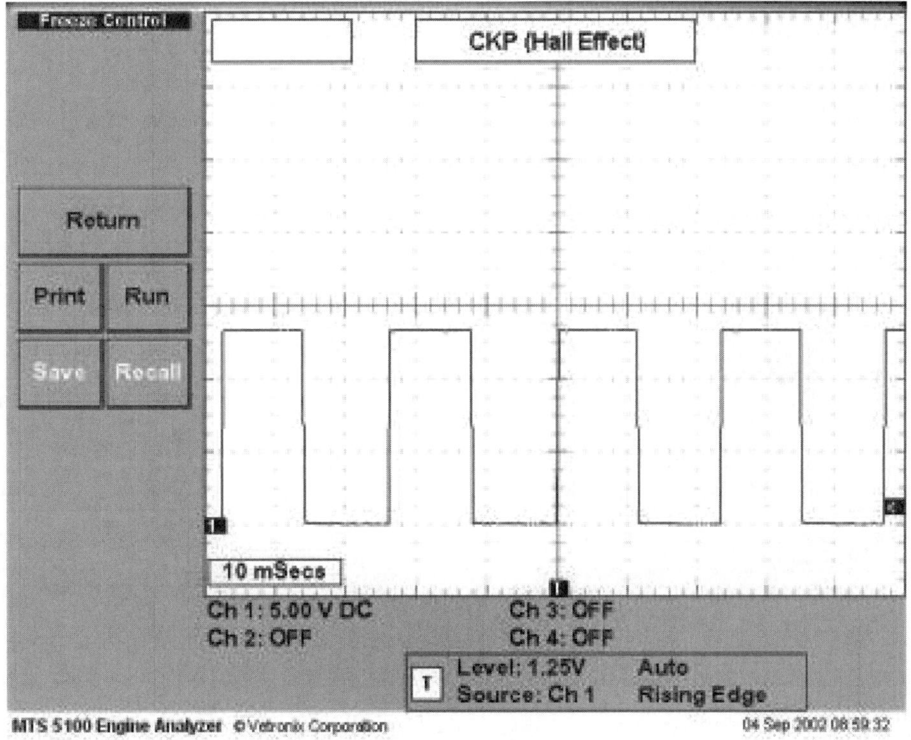

Figure 2. A Hall effect sensor will produce a square wave signal.

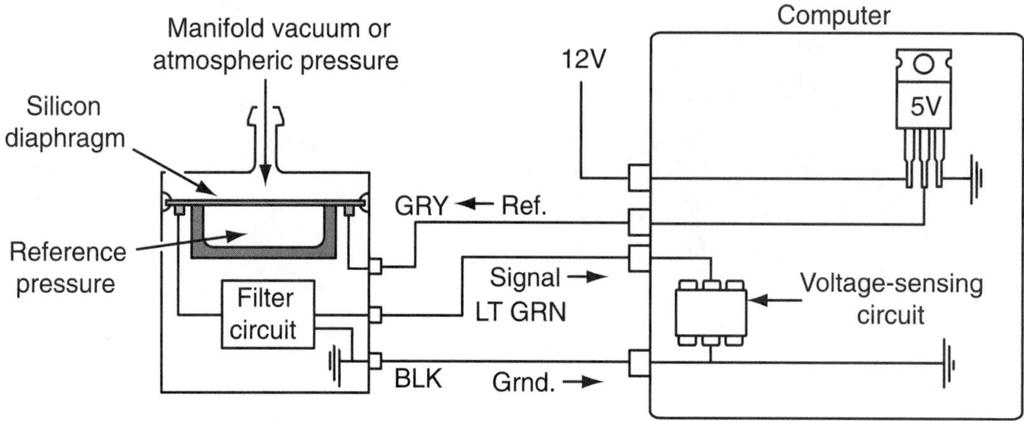

Figure 3. The MAP is powered with 5 volts from the PCM.

we accelerate rapidly, the MAP should respond with the opening of the throttle. This is a good place to use the dual-trace option of the DSO. **Figure 4** shows the TPS and the MAP on separate channels, key on–engine off. Channel 1 is the MAP and is reading just about 4.5 volts, whereas the channel 2 is the TPS and is reading about 0.6V. Let us start the engine and set the trigger for a single shot. The trace should start when we accelerate hard. We will open the throttle all the way to the wide-open position for a split second and then let off. **Figure 5** shows the relationship between the MAP and the TPS. Notice that they follow each other almost exactly. When we accelerate, the TPS goes to WOT (channel 2) and achieves a maximum voltage of approximately 4.5 volts. Manifold absolute pressure goes from an idle voltage (first and second divisions) of about 1.25 volts to a maximum of about 5 volts (in the middle of the pattern). As long as the throttle was at WOT, the MAP stayed high. As soon as the TPS dropped off because we let

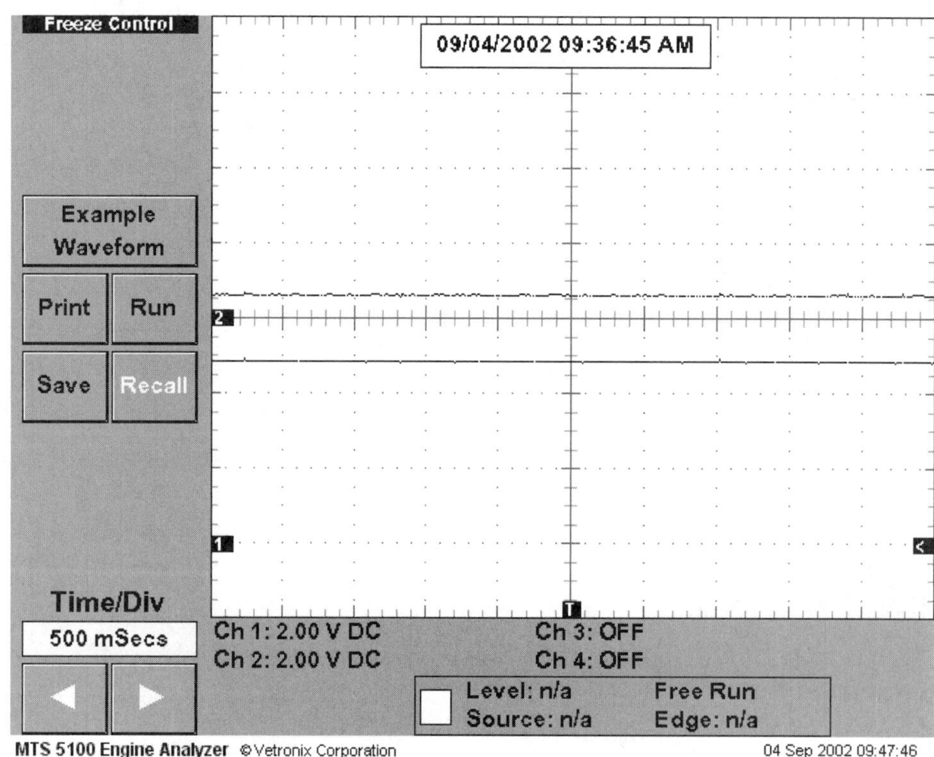

Figure 4. MAP output channel number 1; TPS output channel number 2 with KOEO.

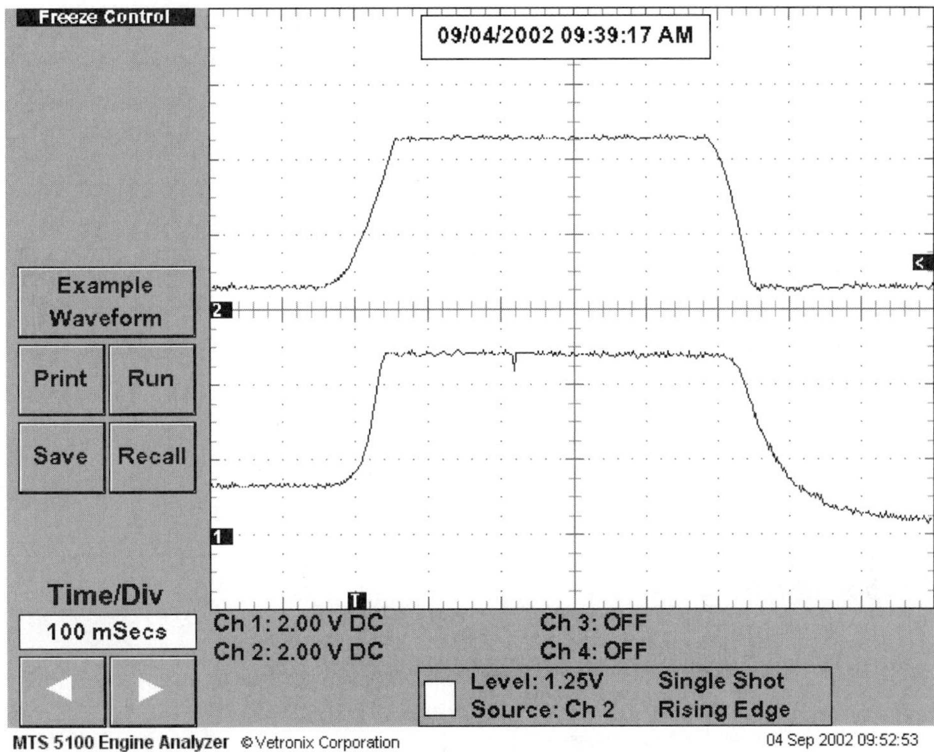

Figure 5. The TPS (number 2) and MAP (number 1) signals during a WOT acceleration.

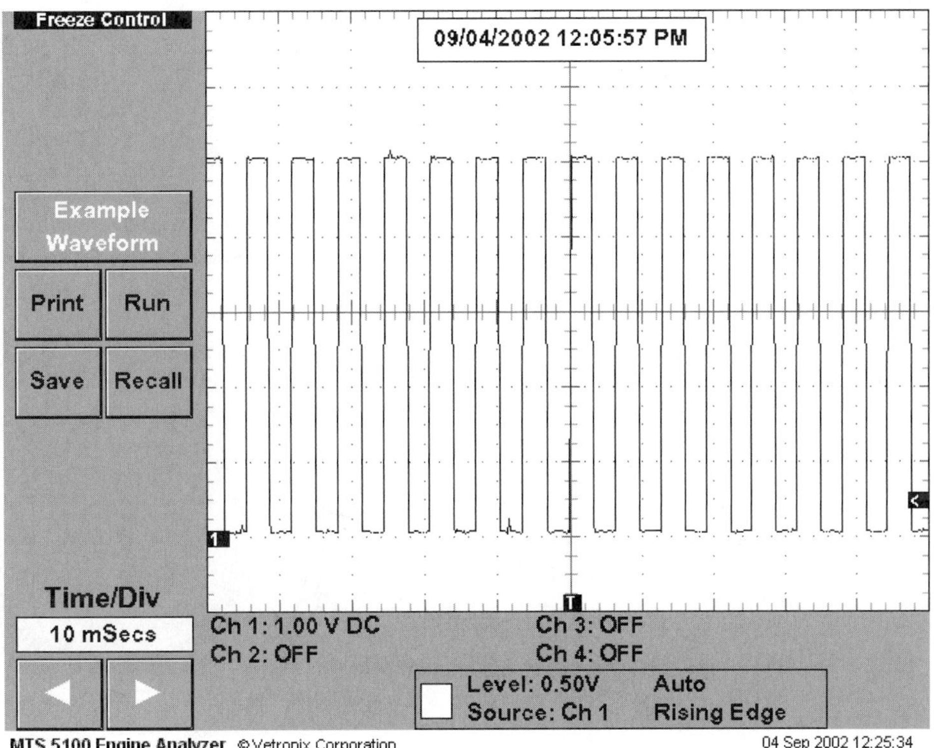

Figure 6. A Ford MAP output—KOEO.

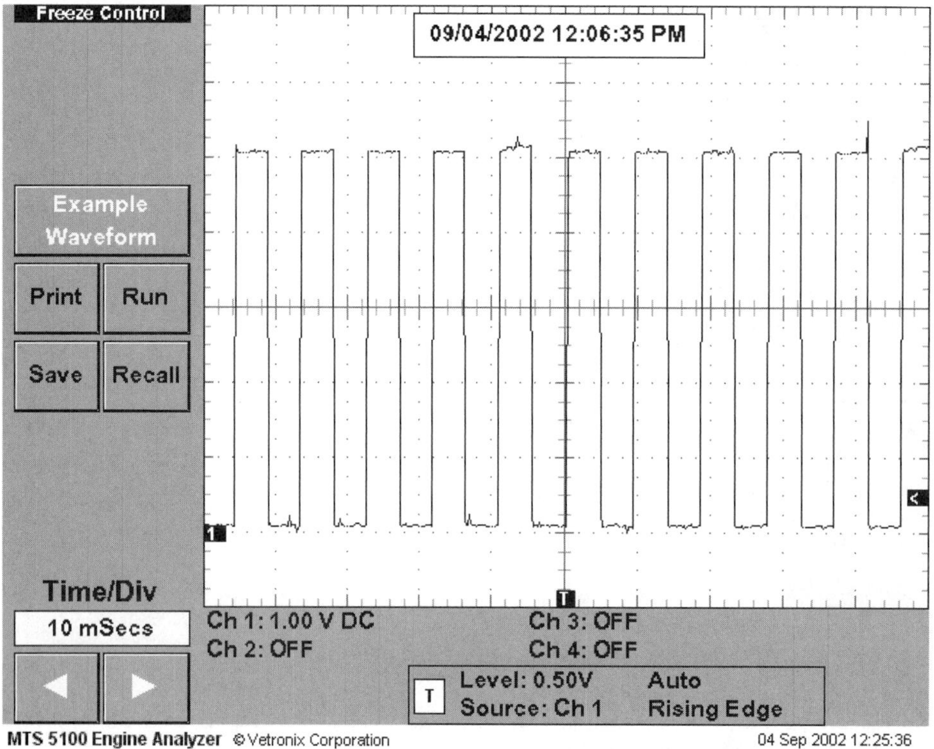

Figure 7. A Ford MAP output—KOER.

up on the throttle, MAP followed down as the engine decelerated. Vacuum reached a level lower than idle at the end of the pattern. Both of these components are doing their job.

The second type of MAP is a variable-frequency generator that we mentioned in Chapter 24. It will generate a square wave signal that will resemble the Hall effect sensor that we looked at before. We will look at two different patterns; one from key on—engine off (KOEO) and the second with key on—engine running (KOER). The MAP is fed with the same 5-volt reference and ground that the voltage style was. We should connect the DSO to the signal terminal and turn on the KOEO. **Figure 6** shows the results. Notice that the pattern is squared off with the bottom of the pattern at zero volts and the top at 5 volts. Also notice that the time of the top voltage and the time of the bottom voltage is the same. It is 50 percent up and 50 percent down. This relationship will not change. When we start the vehicle, the frequency will change. **Figure 7** shows fewer patterns during the same time frame, which means the frequency went down. The accuracy of the sensor is best tested using a vacuum pump and calculating the frequency or reading it off the scanner with the MAP PID chosen.

> **You Should Know** *Manifold absolute pressure sensors generate either a voltage signal or a frequency-modulated signal.*

Optical Sensors

The optical sensor that we discussed in Chapter 24 is another 5-volt–powered sensor **(Figure 8)**. The light shin-

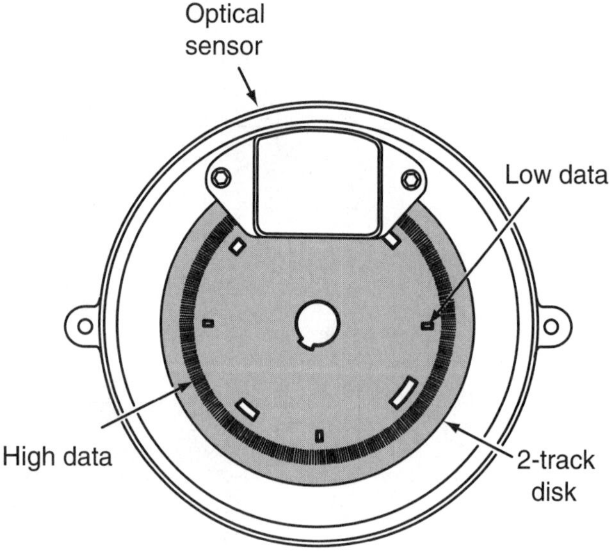

Figure 8. An optical sensor with two outputs.

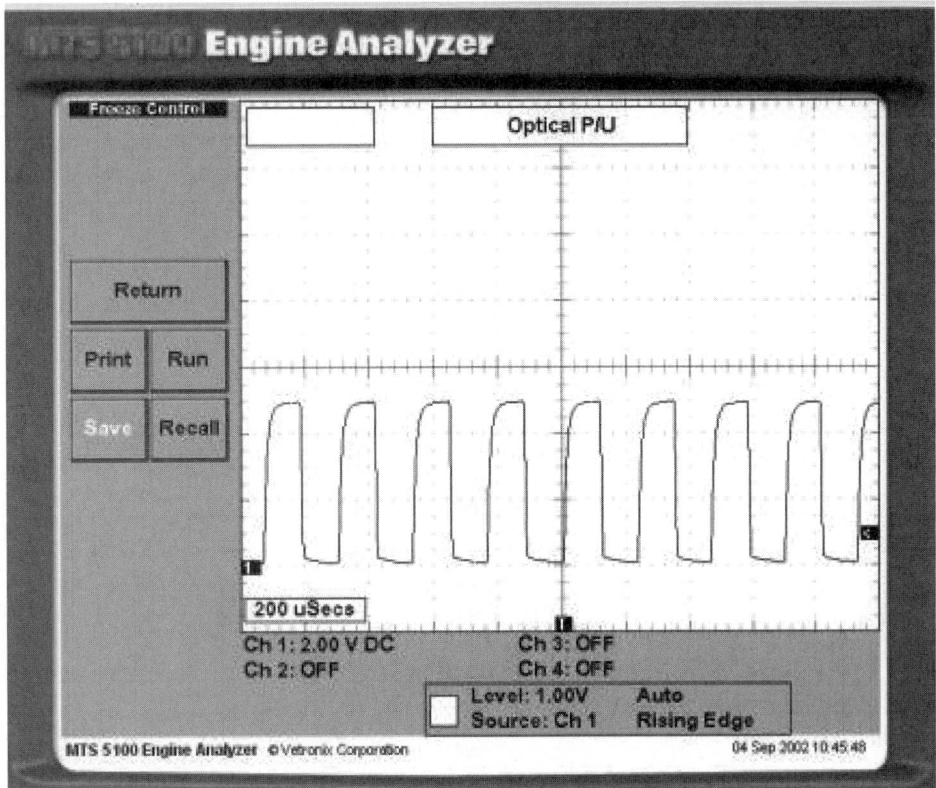

Figure 9. The output from an optical sensor.

ing through the slots will generate a zero-volt signal, and the light being blocked will generate the 5-volt signal. From this standpoint, the signal looks like the Hall effect sensor that we have looked at before.

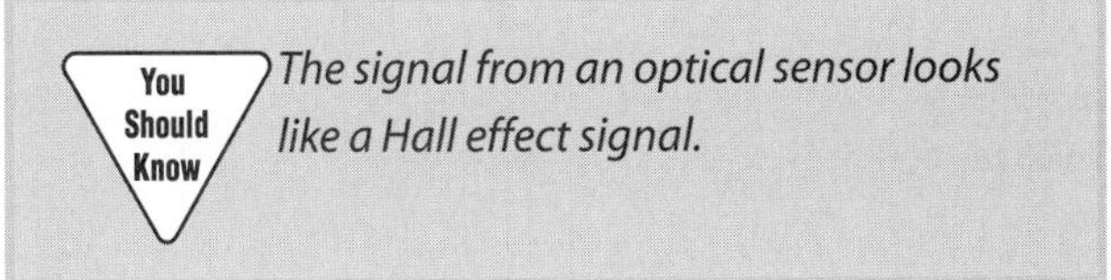

The signal from an optical sensor looks like a Hall effect signal.

Figure 9 shows a captured high data rate optical sensor. Notice that all of the waveform is uniform. The low data rate sensor would produce different width patterns. The bottom of the pattern is almost at the zero-volt level and the top is at 5 volts. The slightly rounded appearance to the upward section of the trace is normal. It is the downward turn of the trace that the PCM uses for determining rpm, and that edge is clean and sharp.

USING A DMM TO TEST INPUT SENSORS

You need to recognize that a DSO is preferred to just about any other tool for most input sensors. Many of the sensors are so fast that a DMM cannot follow them. Take, for example, the optical sensor that we just looked at. The high data rate signal is pulsed 360 times for every distributor rotation. If the vehicle is running down the road at 50 mph, the distributor might be turning at 1000 rpm (half engine speed). This means that 360,000 pulses are being produced every minute and 6,000 pulses per second. There is no way that any DMM could capture the signal, and if it could, our eyes could not follow it. The DSO can follow and trace signals at speeds even higher than the high data rate optical sensor. The exception is the frequency-generating style of MAP. Your DMM or your DSO might have a frequency position.

Figure 10 shows the frequency off a Ford MAP with KOEO. Notice that the frequency is displayed in the upper section of the screen. A graphing multimeter can be set up to graph the voltage changes. However, it is not as fast as the DSO and should not be used for fast sensor diagnosis. **Figure 11** shows the same sensor with the KOER. These

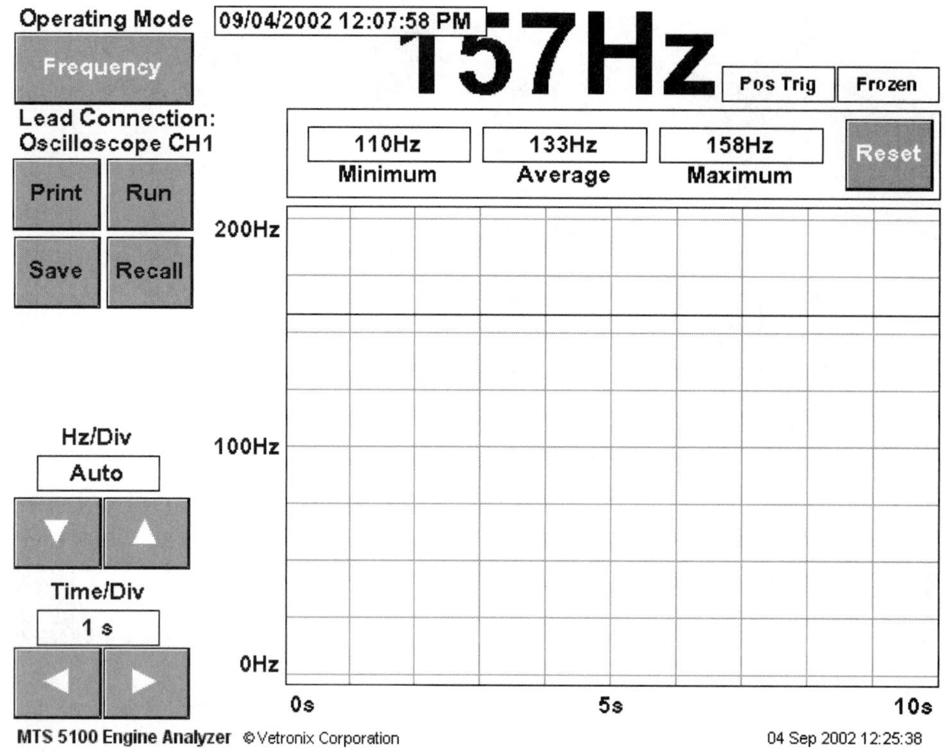

Figure 10. The frequency of a Ford MAP with KOEO.

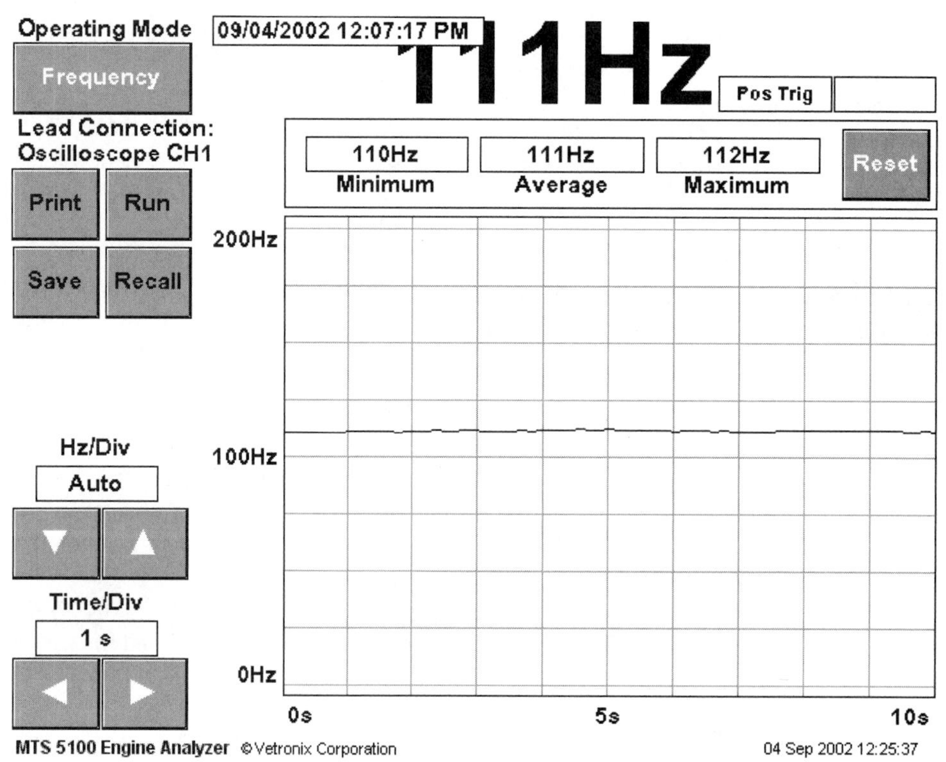

Figure 11. The frequency of a Ford MAP with KOER.

two frequencies can be compared with a specification graph such as that shown in **Figure 12**. There is no vacuum with KOEO (0 in Hg on chart), so the frequency should be high. With the engine running, a moderate level of vacuum will pull the frequency down.

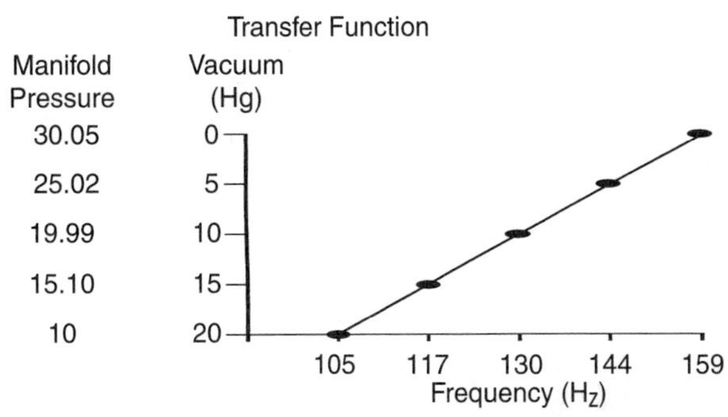

Figure 12. Pressure/vacuum versus frequency.

Summary

- A Hall effect sensor generates a pulsing DC signal that is speed oriented.
- The top of the Hall effect signal is usually the level of power feeding the sensor.
- The bottom of the Hall effect signal is the ground of the sensor.
- The Ford MAP generates a frequency-modulated DC signal.
- The Ford MAP has high frequency with the KOEO and lower frequency with the KOER.
- The MAP can generate a variable voltage that changes with manifold vacuum changes.
- Optical sensors generate a square wave DC signal that shows power and ground quality.
- A graphing multimeter can be used to indicate the frequency of a signal.

Review Questions

1. Disconnecting an IC with power on
   A. is acceptable practice because the IC is protected
   B. could cause damage
   C. is the easiest method to check for power
   D. can give a good indication if the circuit is live
2. A pattern from a Hall effect sensor is being observed. Technician A states that the top of the pattern shows the power applied to the IC. Technician B states that the bottom of the pattern shows the ground applied to the IC. Who is correct?
   A. Technician A only
   B. Technician B only
   C. Both Technician A and Technician B
   D. Neither Technician A nor Technician B
3. A frequency-modulated MAP will generate a(n)
   A. variable voltage signal
   B. AC signal
   C. variable frequency signal
   D. changing pulse width signal
4. The optical sensor's signal is similar to the signal from a
   A. magnetic generator
   B. voltage MAP
   C. switch
   D. Hall effect sensor
5. A Ford MAP is displayed on a DSO with KOEO. Technician A states that the signal will have the highest frequency under these conditions. Technician B states that there will not be any manifold vacuum applied to the sensor at this time. Who is correct?
   A. Technician A only
   B. Technician B only
   C. Both Technician A and Technician B
   D. Neither Technician A nor Technician B
6. A DMM set to measure volts is the best tool for testing a Ford MAP.
   A. True
   B. False
7. A graphing multimeter will display the high data rate of the optical sensor.
   A. True
   B. False
8. As the TPS voltage goes up, the MAP voltage should also go up.
   A. True
   B. False
9. The high data rate optical sensor is generally used to indicate rpm to the PCM.
   A. True
   B. False
10. Under what conditions could a graphing multimeter be used?
11. What will change on the DSO screen as the frequency of the signal changes? Why?

Chapter 26

Oxygen Sensors

Introduction

If you look on the wiring diagram of the modern vehicle, you will see that **oxygen sensors (O_2Ss)** are on the vast majority of them. It is not uncommon for the vehicle to have multiple O_2Ss. The advent of On-Board Diagnostics—second generation (OBDII), beginning with the 1996-model vehicles, saw additional O_2Ss placed after the catalytic converter. A V-type engine could have up to six O_2Ss. Their importance cannot be overemphasized. In this chapter, we look at the "front" O_2S, because it is the one that is running the engine. These O_2Ss are usually abbreviated as the S1 sensors **(Figure 1)**. This is not an OBDII text, so we will not look at the rear O_2S. In addition, the O_2S is within the engine control management system, and we are not going deeply into that subject. Why are we looking at the O_2S in a basic electricity/electronics text? Simply, the vehicle's various systems rely on the information from the O_2S. We are going to

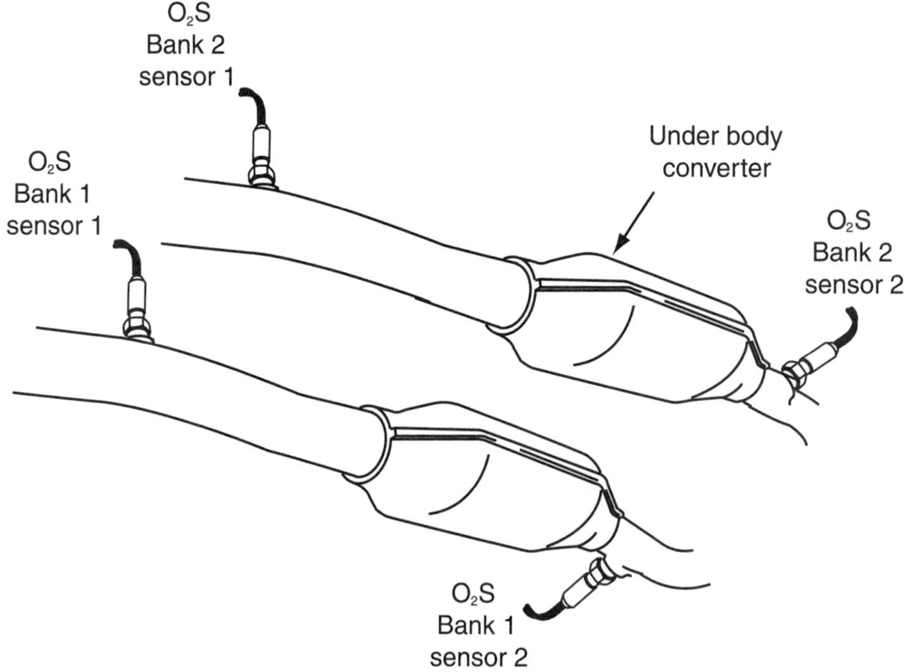

Figure 1. O_2S positions.

look at the ignition in depth, and you need to realize that the information from the O_2S is used to help determine timing advance. When tailpipe emissions became as important as they are today, manufacturers looked to a type of electronic sensor called an oxygen sensor.

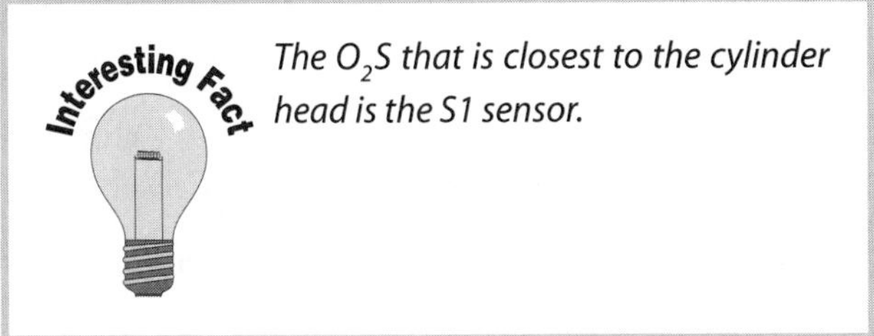

The O_2S that is closest to the cylinder head is the S1 sensor.

Controlling air:fuel ratios very tightly called for the processor (ECM or PCM) to have information about the effectiveness of the burn within the cylinder. An air:fuel ratio of 14.7 parts air to 1 part fuel by weight has been determined to be efficient for most engine operations and the catalytic converter. This air:fuel ratio is called a stoichiometric ratio. It was virtually impossible to achieve and hold without the use of electronics, especially the O_2S.

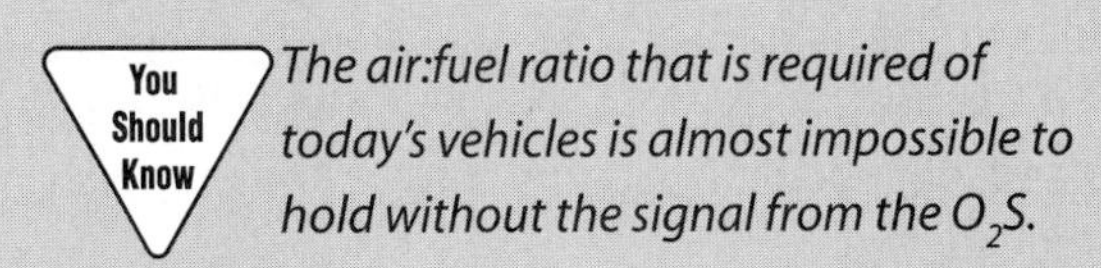

The air:fuel ratio that is required of today's vehicles is almost impossible to hold without the signal from the O_2S.

You will see when you get into testing the O_2S that the air:fuel ratio will float from slightly rich to slightly lean. This modulating, or changing, air:fuel ratio is one of the major factors that allow the vehicle's catalytic converter (CAT) to function. The CAT absorbs O_2 when the system is lean and uses this O_2 when the system is rich. The O_2 is used within the converter to convert the hydrocarbons and carbon monoxide into CO_2. You should refer to the fuel systems text within this series for more information on the CAT. Keep in mind, however, that the vehicle is designed with the modulating air:fuel ratio necessary for the CAT to function correctly. Vehicles will not pass an emission test without a functioning CAT, and a CAT cannot function without the air:fuel ratio modulating from rich to lean. The O_2S is responsible for this modulation. The O_2S signal is the most important signal from an emission standpoint, and that is one of the reasons that we decided to include it in this text.

HOW DOES AN O_2 SENSOR FUNCTION?

The most common O_2S is a zirconium dioxide–type voltage generator, which typically can generate a voltage between 0 and 1 volt. The sensor is composed of a hollow zirconium dioxide body closed at one end. The faces of the body are coated with thin films of platinum, which serves as electrodes **(Figure 2)**. Exhaust oxygen flows around the outside of the sensor and is exposed to one side of the sensor while outside ambient air is allowed to flow to the inside. With ambient oxygen in contact with one side of the sensor and exhaust gas oxygen exposed to the other, we

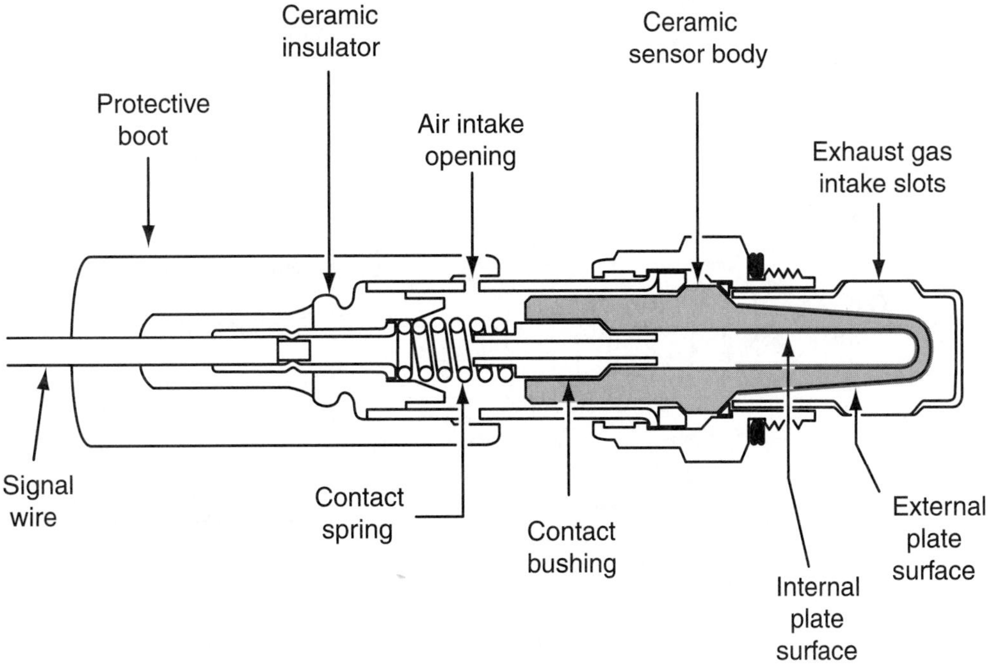

Figure 2. A single-wire O_2S.

have the making of a basic cell or battery. Remember that two dissimilar materials in an acid produce voltage. The dissimilarity in an O_2S comes from the two different levels of oxygen exposed to the two sides of the zirconium dioxide.

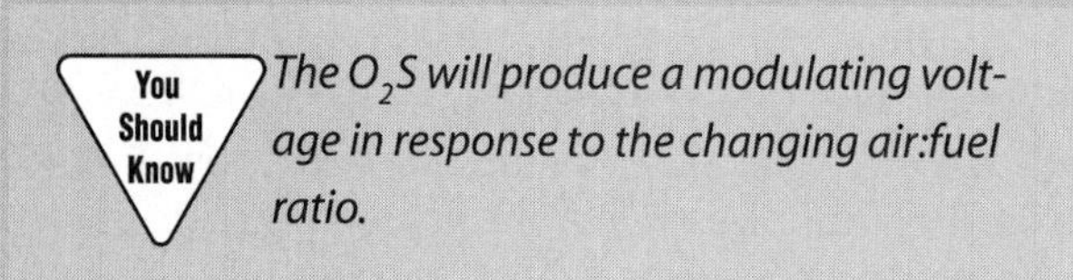

The O_2S will produce a modulating voltage in response to the changing air:fuel ratio.

The greater the difference, the greater the voltage generated. Air has approximately 20 percent oxygen before combining with fuel in an engine. The hydrocarbon from the fuel combines with the oxygen, reducing the 20 percent to usually 1–2 percent for a good running engine. At this level of oxygen, the sensor will generate a voltage of approximately 450 millivolts (0.450 volts). Most vehicles today have the PCM producing a "set-point" voltage of 450 millivolts, called the bias voltage, which is sent to the O_2S. Anything higher seen at the PCM is considered a rich exhaust, whereas anything lower is a lean exhaust. The PCM will take this information and convert it to an opposite command.

Remember that, if the cylinder burn was lean, it will leave behind in excess of 2 percent O_2, and the sensor will generate a low signal (generally lower than 0.450 volts). However, if a rich (more fuel) condition exists, more oxygen will be used to burn the fuel. This will leave behind less than 1 percent oxygen, and the signal from the sensor will go higher than 0.450 volts **(Figure 3)**. This high signal equals rich, low signal equals lean will be used by the PCM to determine if the air:fuel ratio is stoichiometric after the burn has taken place. In this way, it is a check on the burning within the cylinder. By holding the air:fuel ratio within the modulating (1–2 percent O_2), the emissions are controlled tightly. The O_2S is an extremely important emission sensor. Its output is a pulsing DC voltage between 0 and 1 volt **(Figure 4)**.

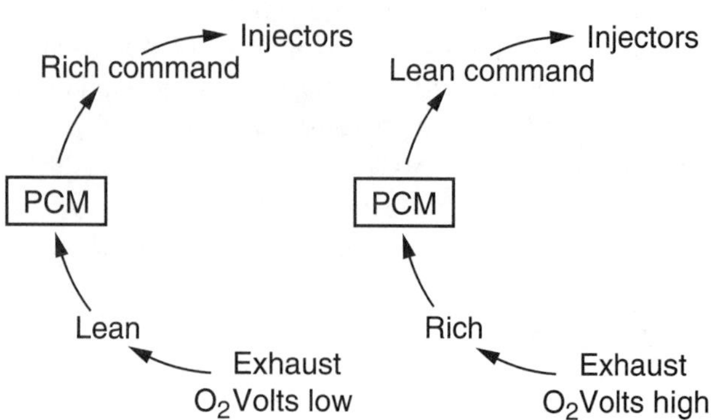

Figure 3. The O_2S signal is used in closed loop.

COMMON PROBLEMS

Problems related to the O_2S center around two areas. The first is insufficient activity, and the second is inaccuracies of the signal. In Chapter 27, we look at how to functionally test the O_2S. Let us look at the problems here. Insufficient activity is a common problem. Remember that the signal is supposed to modulate slightly between slightly rich to slightly lean.

If something has coated the sensor, such as silicone, lead, some additives, or antifreeze, the signal might slow down and not be as responsive as it should be. When this happens, it is likely that the signal will overshoot the mark. For example, if the system is going rich, but the voltage is

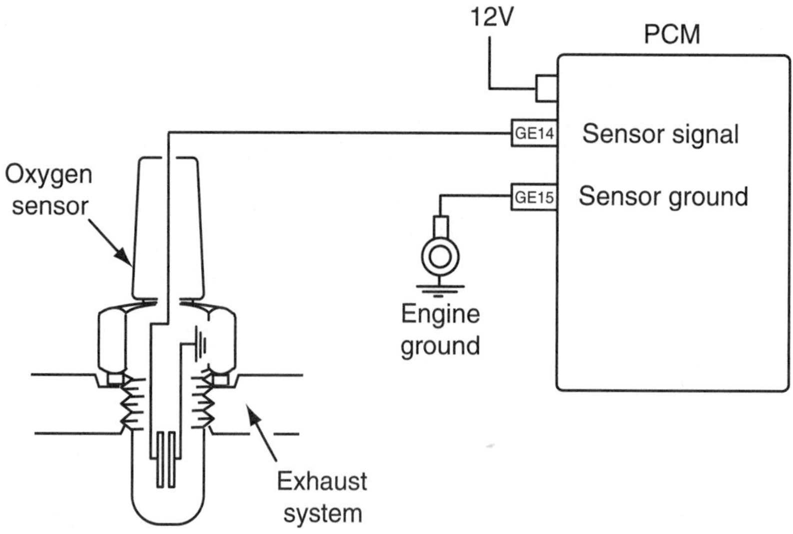

Figure 4. The signal from the O_2S goes directly to the PCM.

not rising as fast as the system change is taking place, the PCM is not aware of just how rich the system is and continues to add fuel until the O_2S signal finally catches up with the burn conditions. The signal delay has caused the system to get to a richer value than it should. The same thing happens on the lean side. The overshooting on the rich side causes additional CO to come out of the CAT. Overshooting on the lean side fills the CAT with O_2, but if the system continues to go leaner than it should, the possibility of a misfire increases. This lean misfire makes it impossible for the CAT to be effective, and if the CAT is ineffective, the vehicle emissions probably will increase above the required level.

> **Interesting Fact**
>
> *Anything that interferes with the ability of the O_2S to generate an accurate signal will change the air:fuel ratio.*

The second problem is the accuracy of the signal. When the air:fuel ratio is at a specific level, the O_2S must be capable of sending an accurate signal to the PCM **(Figure 5)**. The signal must represent what is occurring in the burn. Let us look at the two extremes.

If the signal is richer than what is really happening, what will the result be? Think your way through this. The signal is richer than reality. What will the PCM do with the signal? It will probably lean out the air:fuel ratio until the signal that it sees is within the parameters. If the PCM forces the system leaner than it should be, the possibility of a misfire increases. Misfires allow more emissions than the CAT can probably handle, especially **HC**, which is raw gas that never was ignited or burned.

Consider the opposite. If the signal is leaner than what is really happening, the PCM will add more fuel than is necessary. This will result in increased CO emissions. CO is produced if fuel is burned with too little oxygen or, in other words, the system was too rich for the amount of O_2. You can see that the accuracy of the O_2S signal is extremely important from an emission standpoint. In Chapter 27, we look at how we functionally test the O_2S.

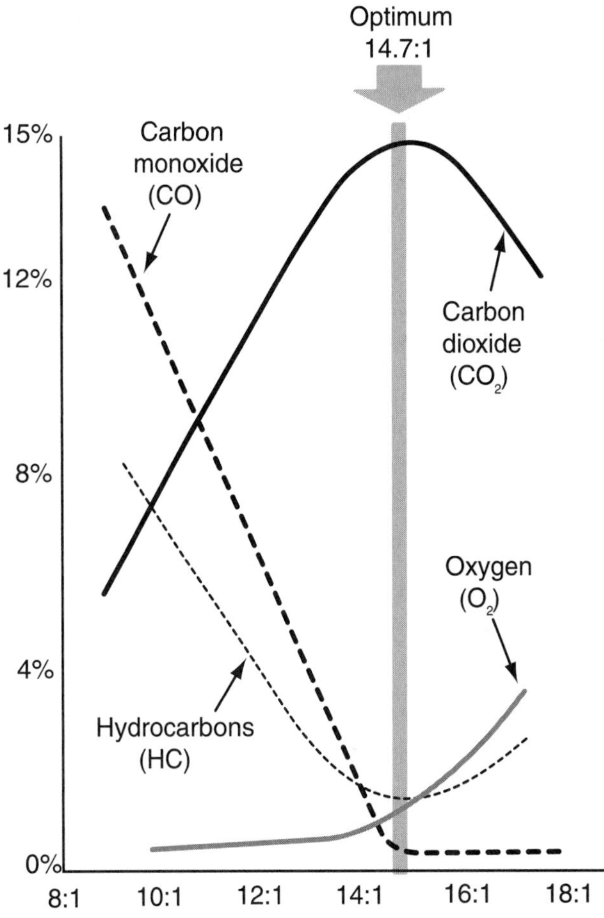

Figure 5. HC and CO are lowest at 14.7:1 air:fuel ratio.

Summary

- O_2Ss are found on all modern vehicles.
- The signal from the O_2S is used by the PCM to control air:fuel ratios.
- If the burn is lean, the signal from the O_2S will drop below 450 millivolts.
- If the burn is rich, the signal from the O_2S will rise above 450 millivolts.
- The catalytic converter functions because of the modulating air:fuel ratio.
- If the O_2S is not accurate on the rich side, the PCM will run the vehicle too lean.
- If the O_2S is not accurate on the lean side, the PCM will run the vehicle too rich.
- The O_2S must respond to changes quickly or the PCM will overshoot the correct voltage and air:fuel ratio.
- The most common type of O_2S is a zirconium dioxide–voltage generator.

Review Questions

1. The oxygen sensor will generate what type of signal?
   A. pulsing or floating DC
   B. square wave
   C. AC
   D. frequency modulated
2. A voltage in excess of 0.450 volts from the O_2S indicates a
   A. normally running engine
   B. rich condition
   C. lean condition
   D. faulty sensor
3. Technician A states that a signal from the O_2S higher than 450 millivolts should tell the PCM to add fuel. Technician B states that a signal from the O_2S lower than 450 millivolts indicates a lean condition. Who is correct?
   A. Technician A only
   B. Technician B only
   C. Both Technician A and Technician B
   D. Neither Technician A nor Technician B
4. A lean condition exists in the cylinder burn. The correct response from the PCM should be to
   A. add fuel
   B. take away fuel
   C. maintain the existing amount of fuel
   D. run the engine speed up until the system sees a rich condition
5. If the O_2S signal is richer than the air:fuel ratio, the PCM response might cause a(n)
   A. increase in CO
   B. decrease in HC
   C. increase in HC
   D. decrease in CO
6. Less than 0.450 volts from the O_2S indicates a rich exhaust.
   A. True
   B. False
7. A slow O_2S might allow the PCM to overshoot the ideal air:fuel ratio.
   A. True
   B. False
8. A lower-than-ideal signal should cause the PCM to add fuel.
   A. True
   B. False
9. Why are the manufacturers using O_2Ss on vehicles?
10. Explain the O_2S conditions that might cause an increase in HC to be produced.

Chapter 27 Diagnosing and Servicing Oxygen Sensors

Introduction

In Chapter 26, we stressed the importance of the O_2S signal. It has to be an accurate indication of what is occurring during the burn and must be fast enough to follow the changes that take place. The best method available to test the O_2S is the function test in which we inject propane and watch the results on the DSO. The purpose of this chapter is to go over this procedure. It is important that you practice the procedure. It will take some practice before you will become proficient in its use and in the interpretation of the results. Do not forget that the O_2S is one of the most important sensors on the vehicle, and its function must be up to standards or the PCM will make the wrong decision. The results will be an increase in the vehicle emissions, a reduction in mileage, or a driveability problem. It is also important to note that, beginning in 1996, vehicle manufacturers began testing the O_2S on the vehicle through OBD II. The testing procedure that we go over here is intended for a vehicle manufactured before 1996. It will work on vehicles built after 1996, but the OBD II system will periodically test the O_2S as the vehicle drives down the road.

USING PROPANE AND A DSO

The theory behind this test is simple. We will add or subtract propane and look at the results on the DSO. Propane is fuel and the engine will burn it. As it burns, the O_2S should see the burn and change the voltage. The voltage change will show on the DSO, and the information can then be used to determine if the O_2S is functioning according to the specifications.

Let us set up the DSO first. Do not forget that the basics of the DSO are included in Chapters 16–20. Do not hesitate to review them if anything does not make sense. The O_2S will generate a voltage generally between 0.00 and 0.999 volts. At no time should it go over 1.00 volt or be less than 0.00 volt. It is a DC signal that we want to be able to continue for many minutes on the screen. The best time setting appears to be 0.500 seconds or 500 milliseconds. The lower left of **Figure 1** shows our 500-millisecond setting. This will give us 5 seconds of total time across the screen. You will see why this is important in a little while. If we set the volts per division for 200 millivolts (0.20 volts DC), we will be able to see the entire O_2S. Place the zero line one division from the bottom, as our blank screen shows. We will not use a trigger, so the trigger box in the lower section of the figure shows "Free Run." These settings should allow us to get a good test pattern.

We need a method of controlling the flow of propane, and the enrichment tool shown in **Figure 2** does the job well. The knob controls the volume of propane, and the button on top turns on the flow. We should preset the flow to twice the number of cylinders. This is a V-6 engine that we will test, so:

$$6 \times 2 = 12$$

We will preset the flow to 12 **standard cubic feet per hour (SCFH)**.

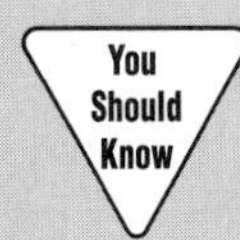

The propane should be injected as close to the intake as possible.

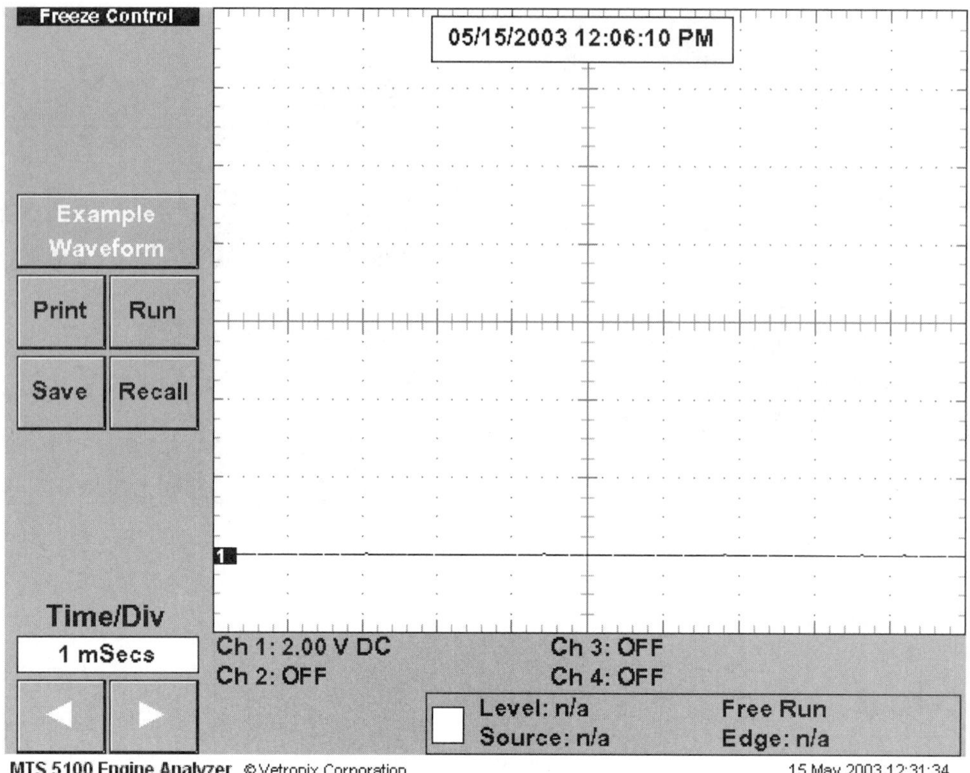

Figure 1. Ideal settings to test the O_2 are 500 mSec/Div and 200 mV/Div.

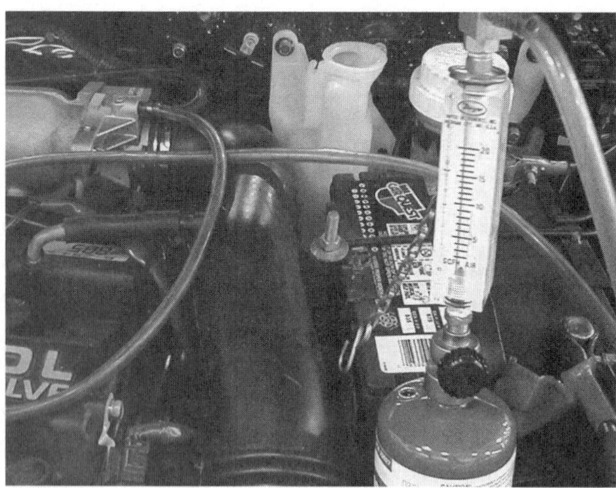

Figure 2. Propane is injected into the intake using an enrichment tool.

The flow will start and stop based on the position of the button on top. Pushing it down will turn the flow on, and releasing it will stop the flow. The propane should be injected as close to the intake as possible. Notice that in our example we have connected the propane hose to an intake port at the throttle body. In some applications, it is necessary only to place the hose in the air cleaner inlet; however, it needs to be as close to the intake as is possible for the test to function normally.

Connect the DSO to the O_2S and we will be ready to test. **Figure 3** shows the DSO connections made using the engine for ground and backprobing the O_2S connector for our input. This is a single-wire sensor that uses the engine block for its ground.

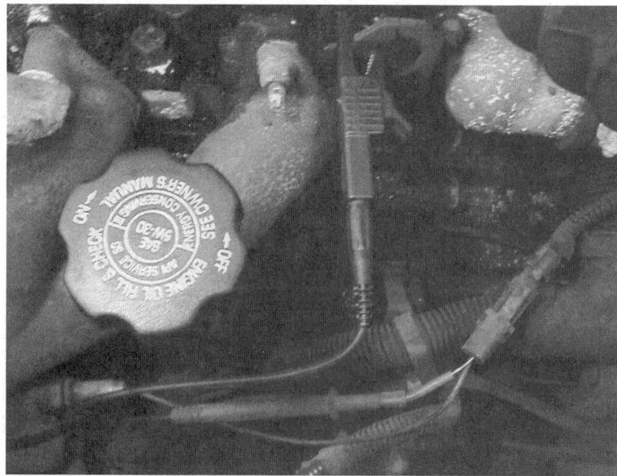

Figure 3. Ground the DSO close to the sensor if possible.

DETERMINING THE "RICH" VOLTAGE

There are three parts to testing an O_2S. We need to look at the rich maximum voltage, the lean minimum voltage, and the speed of the signal. We will do the rich first. With the engine fully warmed up and running at 2000–2500 rpm, you should have a pattern on the screen of the DSO. If you do not, go back and reset the basic settings. The most frequent difficulty involves either having the trigger on or having one of the basic settings wrong.

Turn on the propane by pressing the button down. The system should go rich and flatline the DSO voltage display on the screen. If the pattern did not flatline, increase the flow rate. The DSO must show a flatline because we want to look at the maximum voltage the O_2S can generate. **Figure 4** shows that our V-6 engine is generating an 840-millivolt signal, and it did flatline. The flatline ensures that this is the maximum voltage that the O_2S can generate. If the sensor is functioning correctly, the flatline should be in excess of 800 millivolts and less than 1.00 volt. This narrow window of 200 millivolts is required for accuracy during rich conditions.

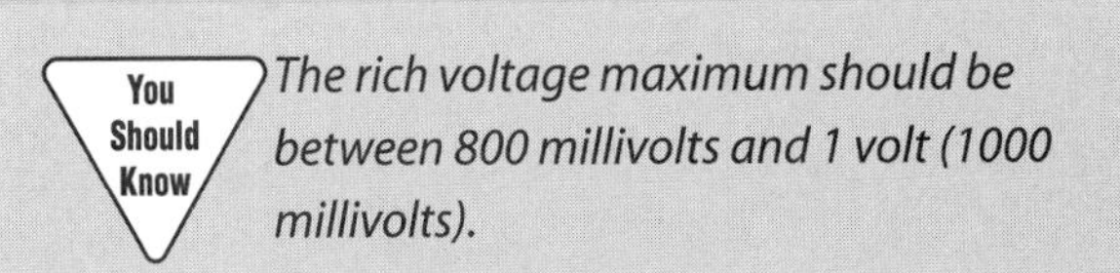

The rich voltage maximum should be between 800 millivolts and 1 volt (1000 millivolts).

Anything beyond this specification indicates that the O_2S cannot accurately indicate to the PCM the rich condition.

DETERMINING THE "LEAN" VOLTAGE

If adding propane caused a rich condition, then taking it away should cause a lean condition, right? The only trick to this is turning the propane off while the signal is rich. Generally 1 second with the propane on will cause the next second to be an indication of the lean voltage. Let us raise the engine speed back up to 2000–2500 rpm and push the propane button for 1 second. The pattern should flatline rich again, and then when we turn off the propane we should see another flatline lean. As was the case with the rich voltage, the pattern must flatline lean. If it will not flatline, there is an additional source of fuel. Pull off a large vacuum line to lean out the burn enough to get the flatline. The DSO must show the flatline lean or we are not looking at the minimum voltage that the O_2S will generate. **Figure 5** shows the results of turning off the propane. Notice that the pattern did flatline at 50 millivolts. The specification for the lean voltage is less than 175 millivolts and above zero volts. If the sensor's minimum voltage is not within this narrow band, the ability to detect a lean condition and accurately indicate it to the PCM is not possible.

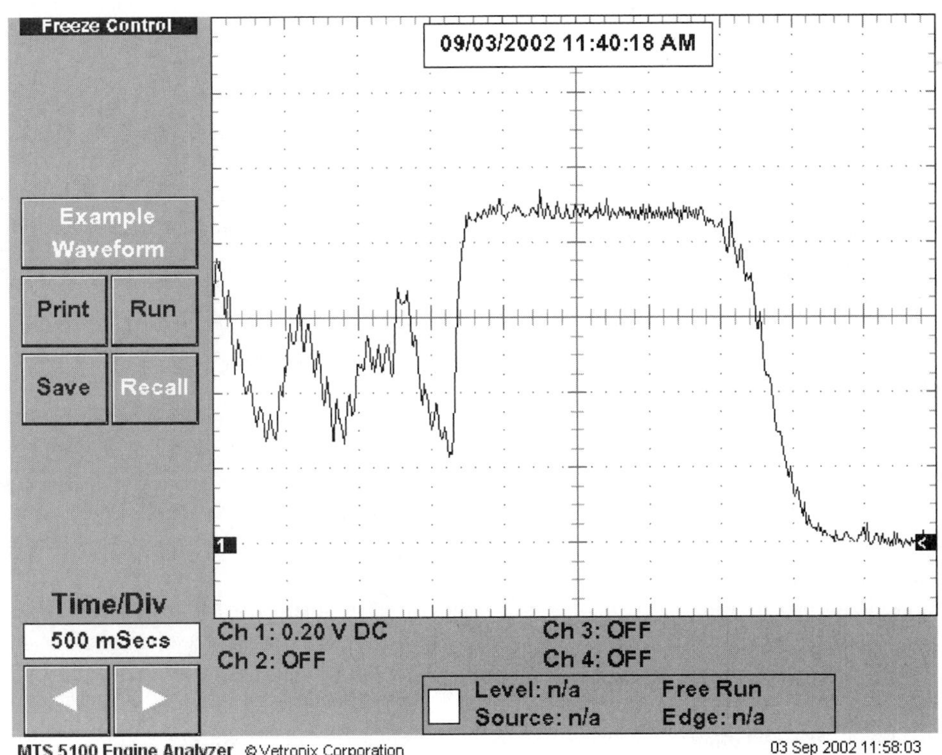

Figure 4. With propane on, the voltage flatlines rich.

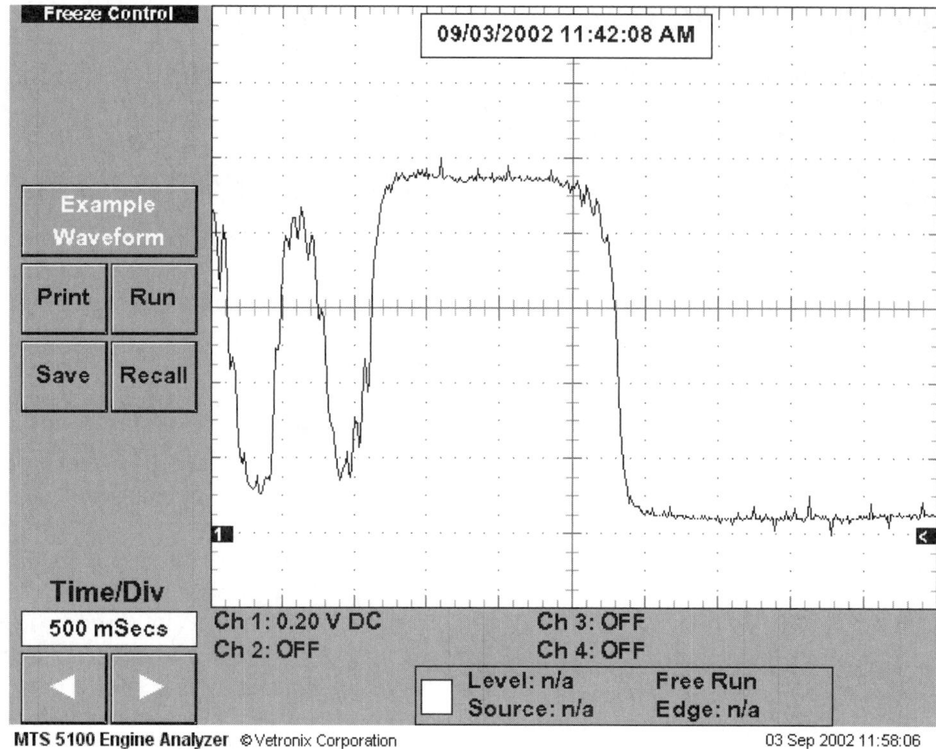

Figure 5. With propane off, the voltage flatlines lean.

The rich and lean voltages need to be an accurate indication of what actually occurred during the burn. The PCM will change the air:fuel ratio based on this information, and if it is not accurate, the changes will also not be accurate. If the rich and lean voltages are correct, we can look at the O_2S's speed.

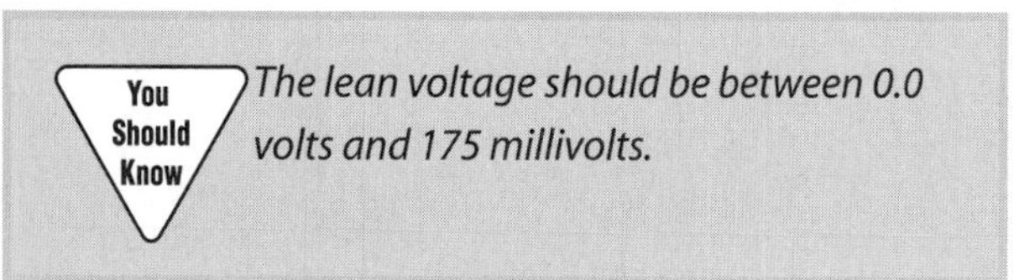

The lean voltage should be between 0.0 volts and 175 millivolts.

DETERMINING THE SPEED OF THE O_2 SENSOR

The accuracy and speed of the signal are both required outputs for the O_2S. Measuring the speed will again rely on the propane enrichment tool. Bring the engine speed back up to 2000–2500 rpm. Once the pattern on the DSO has stabilized, turn on the propane again for 1 second, off for 1 second, and then back on. The second on produces the pattern that will indicate the speed of the O_2S. We are interested in the amount of time it takes to go from lean to rich. The only way to get it lean is to force it rich first, so that explains why we turned the propane on, then off, and then back on again. If the O_2S is functioning quickly enough, it should have an almost instantaneous rise from lean to rich. We are especially concerned about the middle of the possible voltage range, so we look at the rise from 300 to 600 millivolts and want to see it change in less than 100 milliseconds. **Figure 6** shows the lean-to-rich transition, and it occurred in less than 100 milliseconds, so this sensor is good. It passed all three tests. The rich voltage, the lean voltage, and the speed are all within specification. It is possible to have all three tests on the same screen as shown.

DETERMINING IF THE VEHICLE IS IN FUEL CONTROL

Once we know that we have a functioning O_2S, we can determine if the vehicle is in fuel control. Being in fuel control means that the PCM is in charge of the air:fuel ratio and that no misfires are occurring. We can determine this from a single pattern. Bring the engine speed back up to 2000–2500 rpm. Watch the DSO O_2S pattern for a couple of minutes while everything stabilizes. No propane should be flowing, and the vehicle should be back in normal condition. If you had the air cleaner off, it should be put back on.

If the PCM is capable of holding the air:fuel ratio to specification, the average of the O_2S signal should be 450 millivolts. **Figure 7** shows the pattern from the V-6 that we

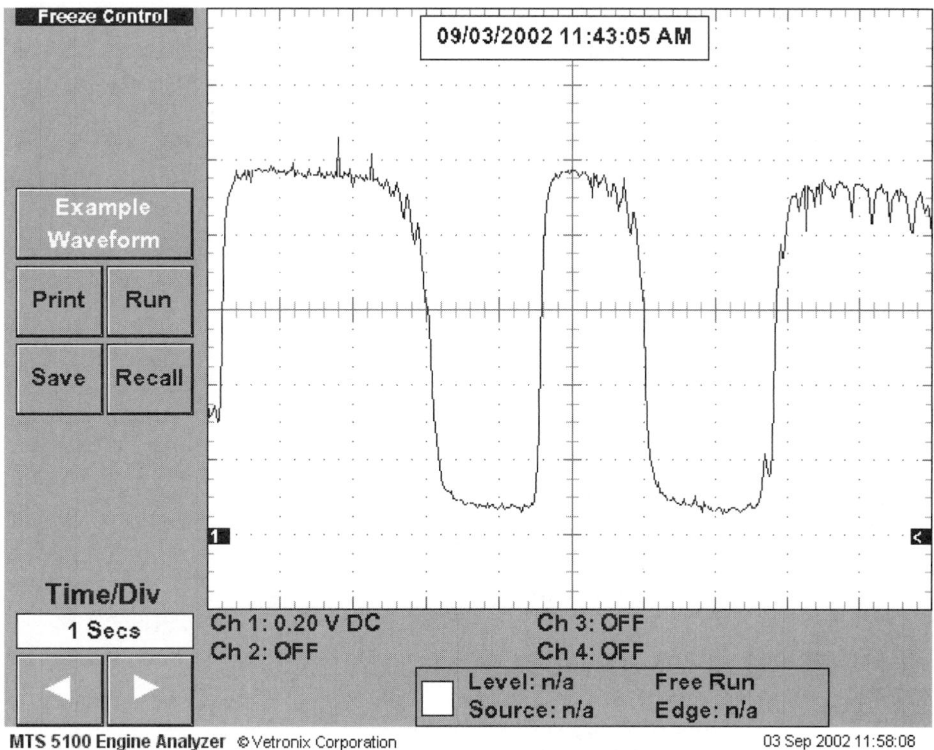

Figure 6. How fast the O_2S can react is important.

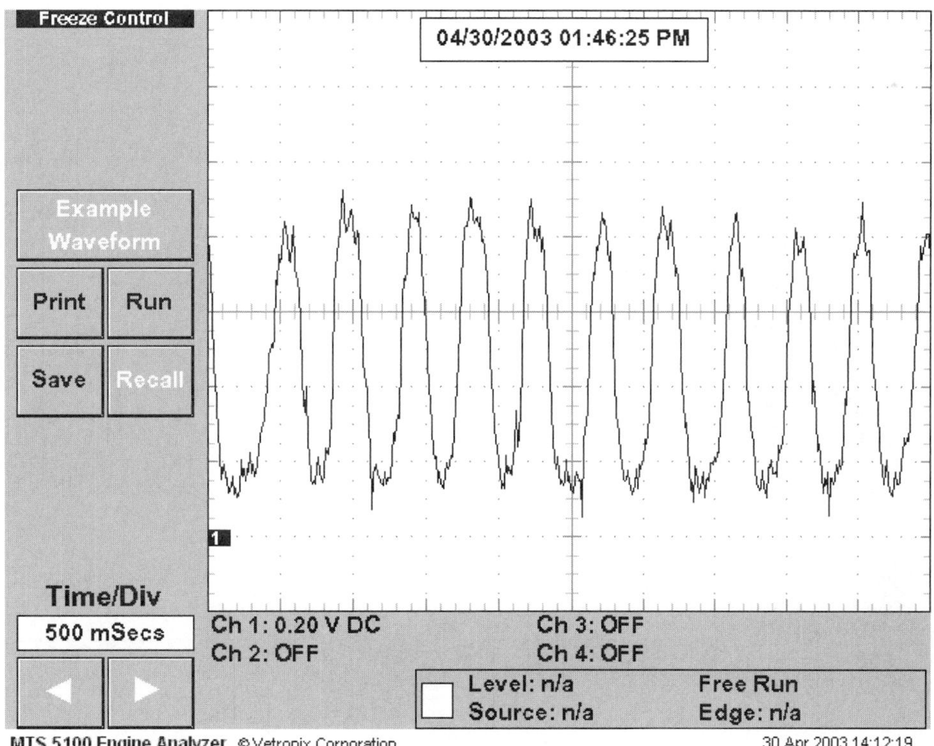

Figure 7. An engine in fuel control.

know has a functioning O_2S. Find 450 millivolts on the screen. The peaks above and the peaks below should be approximately half the distance. The average O_2S voltage should be 450 millivolts. Some DSOs will calculate this voltage average for you.

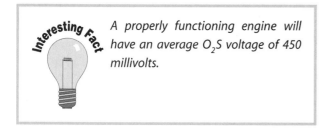

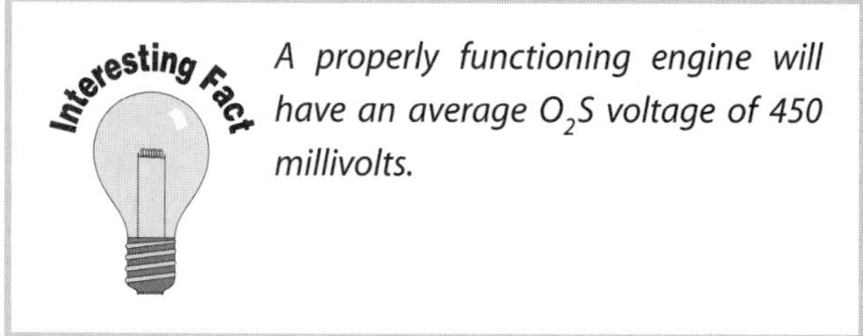

A properly functioning engine will have an average O_2S voltage of 450 millivolts.

The other factor in determining fuel control is to look at the number of times the pattern turns or shifts downward. The specification is for the number of times per second to be between 0.5 and 5.0 cycles. If we count the number of cycles on the screen in Figure 7, we see that we have approximately 11. If we divide by 5, because there are 5 seconds on the screen, we have a frequency slightly higher than 2. This is well within specification.

Figure 8 shows another vehicle that has a functioning O_2S. It has passed the rich, lean, and speed tests using propane. The vehicle is running at 2200 rpm and the pattern shows many short pulses. If you count each pulse and divide by 5 (seconds) the frequency is well above the 5 maximum. This vehicle is showing misfire. If the frequency of the O_2S is above 5 and the O_2S has been functionally tested, the vehicle is misfiring. There are almost 50 cycles showing on the 5 seconds of screen time. Dividing the 50 by 5 seconds gives us a frequency of 10, which is well above the maximum of 5. This vehicle had a bad fuel injector connector that did not allow the injector to function. This produced a completely dead cylinder.

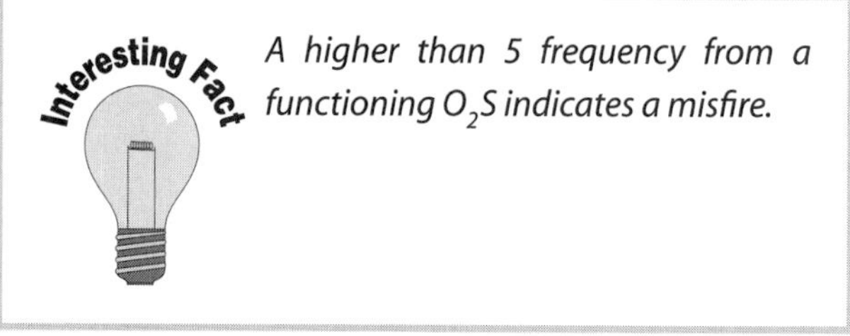

A higher than 5 frequency from a functioning O_2S indicates a misfire.

Each time the cylinder misfired, O_2 came out of the cylinder. The O_2S measured it and told the PCM that conditions were lean. The PCM's response was to add fuel. This vehicle failed the emission test for excessive CO. The extra fuel that the PCM added was more than was needed by the other cylinders.

Remember the sequence. First, test the function of the O_2S, measuring the rich voltage, the lean voltage, and the speed. If it passes, use its signal to determine fuel control.

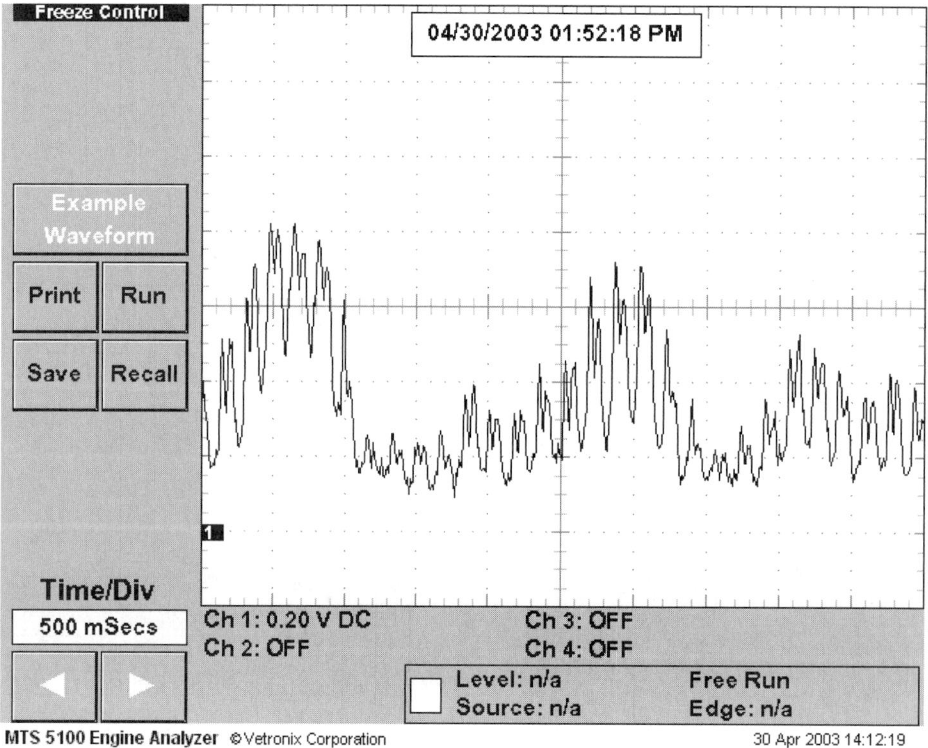

Figure 8. The DSO pattern shows significant misfire.

Summary

- Propane is added to the intake to test the rich voltage.
- Propane is turned off to test the lean voltage.
- A functioning O_2S should be able to go from lean to rich in less than 100 milliseconds.
- The normal frequency of an O_2S is between 0.5 and 5.
- Higher than normal frequency is an indication of a misfire.
- O_2Ss must be tested for accuracy and speed of the signal.

Review Questions

1. When propane is added to the intake of an engine running at 2000–2500 rpm, the pattern should
   A. flatline lean
   B. show a peak voltage
   C. flatline rich
   D. show a drop in voltage
2. When propane is taken away, the pattern should
   A. flatline lean
   B. show a peak voltage
   C. flatline rich
   D. show a drop in voltage
3. The specification for the rich maximum voltage is
   A. between zero volts and 175 millivolts
   B. slower than 100 milliseconds
   C. Faster than 100 milliseconds
   D. between 800 millivolts and 1.0 volt
4. The specification for the lean voltage is
   A. between zero volts and 175 millivolts
   B. slower than 100 milliseconds
   C. faster than 100 milliseconds
   D. between 800 millivolts and 1.0 volt
5. The specification for the speed of a functioning O_2S is
   A. between zero volts and 175 millivolts
   B. slower than 100 milliseconds
   C. faster than 100 milliseconds
   D. between 800 millivolts and 1.0 volt
6. A correctly functioning engine will have a frequency of____ cycles per second.
   A. 0.5–5
   B. more than 5
   C. less than 0.5
   D. 2000–2500
7. Misfire will cause the O_2S's frequency to be____cycles per second.
   A. 0.5–5
   B. more than 5
   C. less than 0.5
   D. 2000–2500
8. The rich voltage should be below 800 millivolts.
   A. True
   B. False
9. The lean voltage should be above 175 millivolts.
   A. True
   B. False
10. Excessive frequency (more than 5 cycles per second) indicates misfire.
    A. True
    B. False

Section 6

Wiring Diagrams

SECTION OBJECTIVES

At the conclusion of this section, you should be able to:

- Know the various symbols that are found on wiring diagrams.
- Know how to read location codes.
- Know how to read expanded codes.
- Use the wiring diagram and common-point diagnosis.
- Use the wiring diagram to find an open circuit.
- Use the wiring diagram to find a short circuit.
- Understand the precautions when working around printed circuit boards.

Wiring diagrams from vehicles made in the 1960s were generally printed on one page. The wiring diagram for modern vehicles may have as many as twenty pages.

Chapter 28 Wiring Diagram Symbols

Introduction

As the vehicle has progressed through the years, so has the complexity of its wiring. Following the maze of wires on the modern automobile is next to impossible today. The vehicle of today is wired in sections called looms or harnesses. These harnesses are prewired for a particular section of the vehicle, such as the dashboard or the engine compartment. The wires are all tied or bundled together and tightly taped. This makes the task of following a particular wire through the vehicle both time consuming and difficult, if not impossible. In addition, the wire might change color or go through a part of the vehicle that has nothing to do with the circuit being followed.

The wiring diagram is an essential item to repair electrical systems on the modern vehicle. The use of a wiring diagram allows you to follow the wire to some point where you can conveniently find it on the vehicle and test it. By eliminating the need to physically follow the wire throughout the vehicle, you will reduce your diagnosis time, resulting in reduced consumer cost.

The diagram for the Model T could be reduced down to billfold size and still show all the necessary information required to repair it **(Figure 1)**. By contrast, a wiring diagram for a modern vehicle can be several pages long with hundreds of wires, connectors, and components. Ford and other manufacturers are currently working to eliminate some of the maze of wiring through the use of communication multiplexing. Multiple processors are placed throughout the vehicle and communicate via a single conductor. The majority of the wires that run among sections of the vehicle are eliminated. Multiplexing promises to help sort out some of the wiring of today. Each year sees more models that are multiplexed.

Wiring diagrams usually come in one of two different styles: pictorial and schematic **(Figure 2)**. Pictorial diagrams show pictures of the components, whereas schematic diagrams use symbols to illustrate components. Sometimes manufacturers will use both styles on one diagram. Generally, the automobile manufacturers choose schematic diagrams because they are considered more informative, show internal electrical paths, and take up less space than pictorial diagrams. To be able to understand the typical

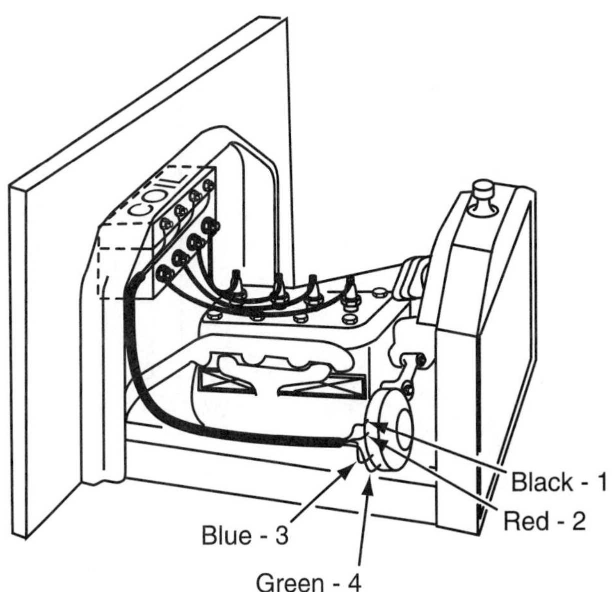

Figure 1. The entire Model T wiring diagram.

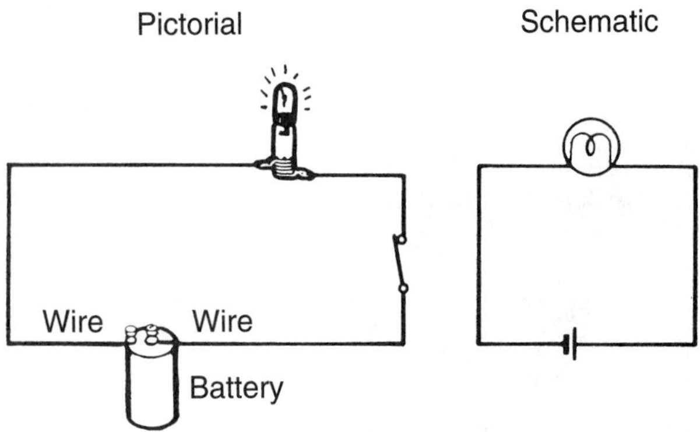

Figure 2. Pictorial and schematic diagrams.

schematic wiring diagram, you will need to know the most common symbols.

GROUNDS

Probably the most frequent symbol encountered is that of ground **(Figure 3)**. It can be represented in either case or remote form. **Figure 4** shows a section of a wiring diagram that has both remote and case grounds. Notice that the tail lamps have case grounds. The black dot right on the light indicates that the ground for the bulb will be completed when the bulb's socket is installed in the housing. The housing must be made of metal, which will conduct electricity. The side marker lamps and the license lamp, however, do not have the same black dot attached to the bulb. Instead, notice that they have a ground wire, which eventually connects to a ground. The circle and ground symbol indicates that the wire has a circle type islet connected to it, and a screw or bolt connects it to the metal of the frame.

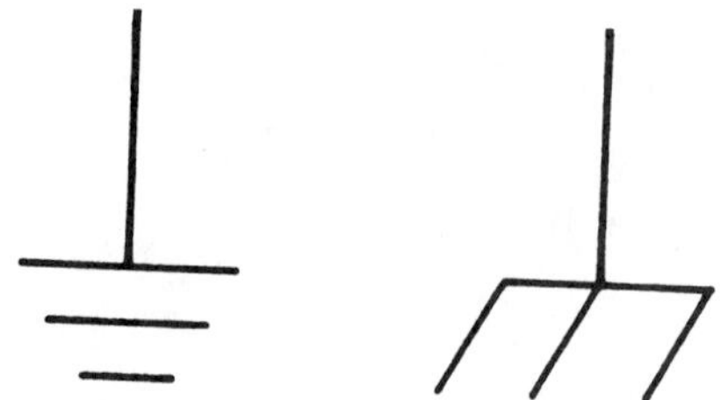

Figure 3. Common ground symbols.

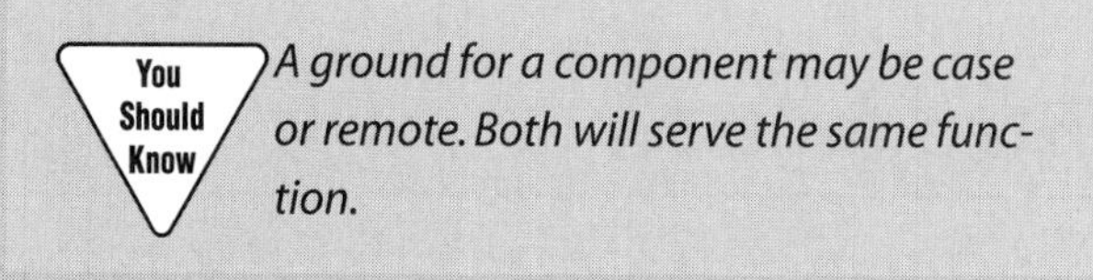

A ground for a component may be case or remote. Both will serve the same function.

In this case, remote means not attached. Notice also from the diagram that the side marker lamps are connected to the same ground, G-405. It is possible for a single ground to

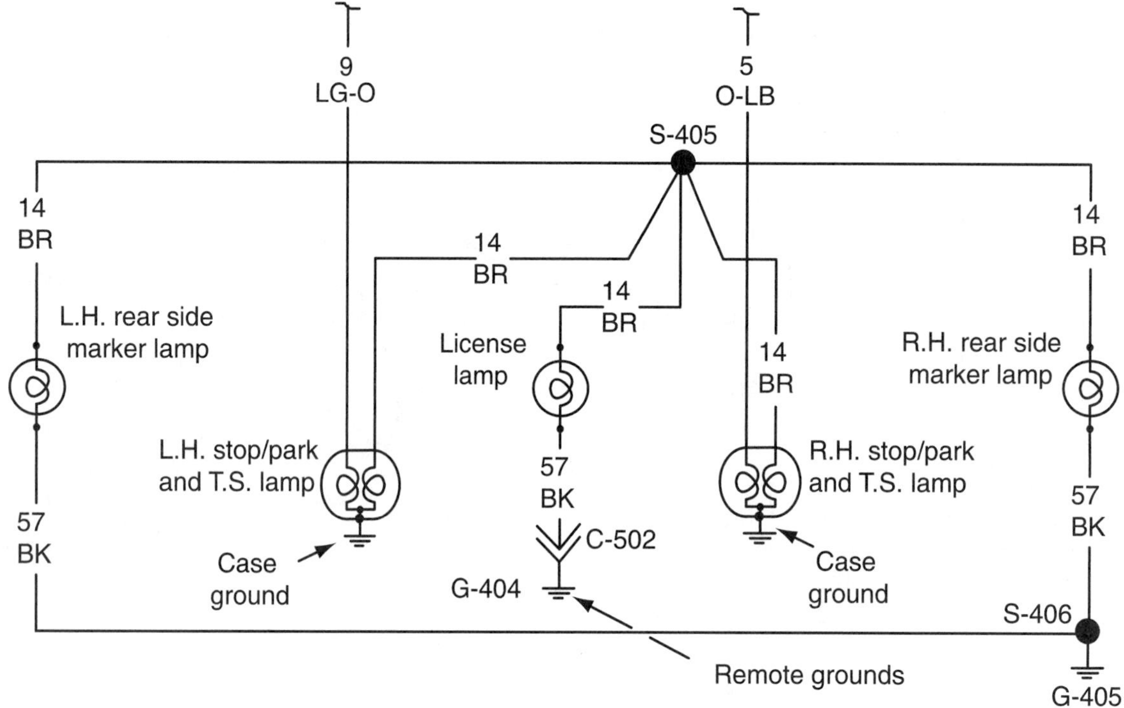

Figure 4. A diagram will show both case and remote grounds.

be the ground for more than one component. While we have this diagram to refer back to, let us discuss splices. A **splice** is a spot where two or more wires are connected. It is considered a **common point** or connection point. It is usually represented by a black dot. This section of the diagram shows two splices, one labeled S-406 near the RH rear side marker lamp and the other labeled S-405 near the top of the diagram. S-405 has five wires connected together, whereas S-406 has three. A wire can cross over another wire many times, but they are connected together only if they have the black dot indicating a splice. In addition, these splices all carry a number that we will look at later when we discuss location codes. Some of the more common splice drawings, including one of an **expanded splice**, are shown in **Figure 5**.

Remember that the object of any wiring schematic is to improve the ability of technicians to find their way around the vehicle. If the splice on the vehicle has more wires than can easily be shown on the diagram, an expanded drawing will be used with more than one splice drawn. To identify it as the same point or splice on the vehicle, both splices will have the same number. In the example, S-307 shows up at two splices with a wire labeled S-307 connecting the two together. The two splices and the wire connecting them together are the same splice drawn in two different spots to make the drawing easier to read.

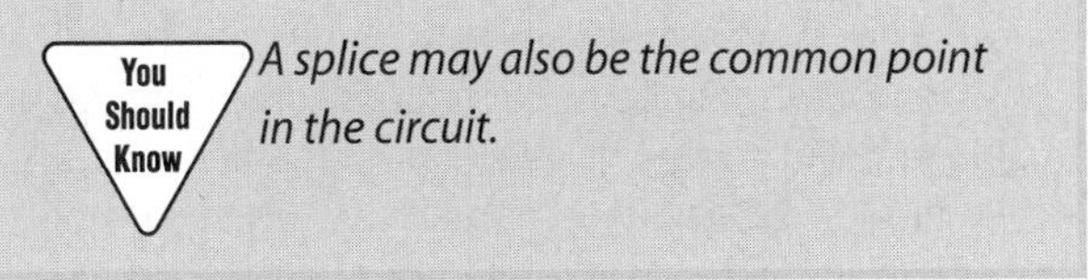

A splice may also be the common point in the circuit.

The S preceding the number indicates a splice. There can be only one splice on the entire vehicle with this number. If it appears more than one time, it is for drawing convenience and ease of diagnosis. There will only be one S-307 on the vehicle. A **connector** is the next item we discuss. A connector is a terminal where two wires are connected together **(Figure 6)**. The most common connector symbol found on diagrams is probably that which represents a **bullet connector**.

Refer back to the schematic in Figure 4 (grounds). In the middle of the drawing is C-502, which is a bullet connector

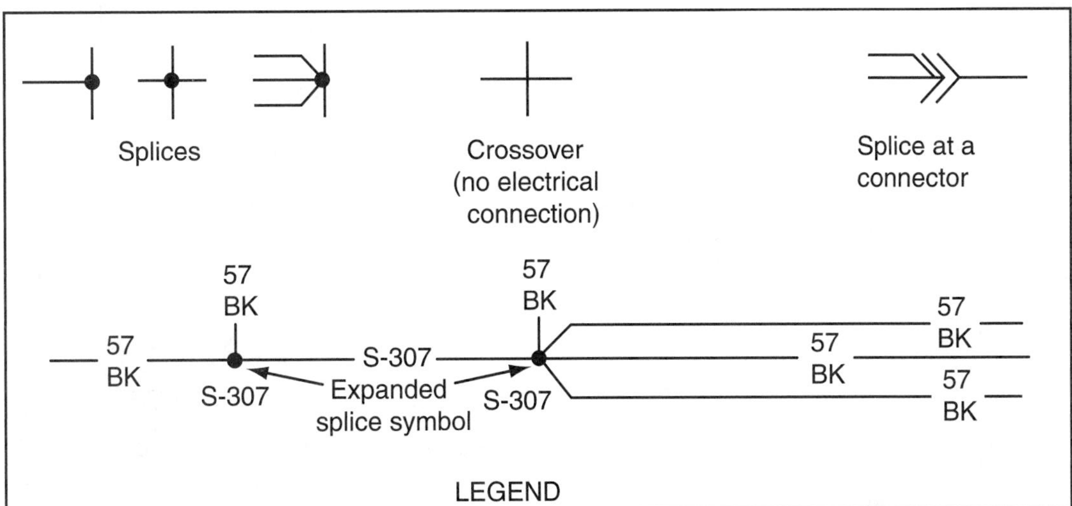

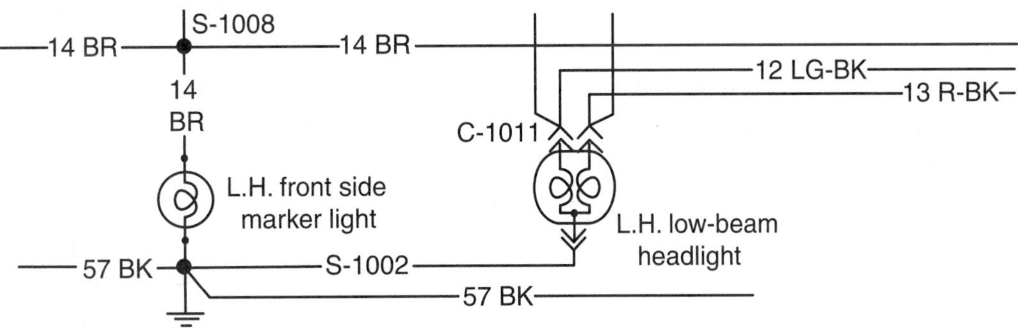

Figure 5. Different splice symbols.

Figure 6. Schematic symbol for a bullet connector.

just ahead of the license lamp. As is the case with splices, each connector is numbered and carries some designation. They are usually convenient places to test the circuit because they are easily disconnected. Sometimes the manufacturer might wish to connect more than one wire together in a group. A multiple-wire connector is used. Sometimes, when the connector is drawn schematically, the shape shown in **Figure 7** is used. Usually, both will not be present on the diagram. The dotted line through the three female wires and the dotted line through the male wires indicate that these three wires are in the same connector body all molded together. The closeness of the connections would probably lead you to figure this out, even if it did not have the dotted lines. However, sometimes the wires in a connector might have to be drawn far apart from one another. When this occurs, the diagram will use a technique similar to that of the expanded splice. Dotted lines and connector numbering will indicate that the wires are all in the same connector.

Let us look at another type of connector. **Figure 8** shows a headlight switch with its connector. Do not worry about the inside of the switch for now, but notice that the four wires coming into the top and the four wires coming into the bottom of the switch all carry the same connector number, C-707, and are drawn with a dotted line. This indicates that this molded connector, C-707, has eight wires in it and seven female terminals. (There are two wires connected to one terminal in the lower half.) Remember again that only one connector will be numbered C-707 on the vehicle and, even though it is drawn as two, its number indicates that it is one connector with eight wires attached to the headlamp switch.

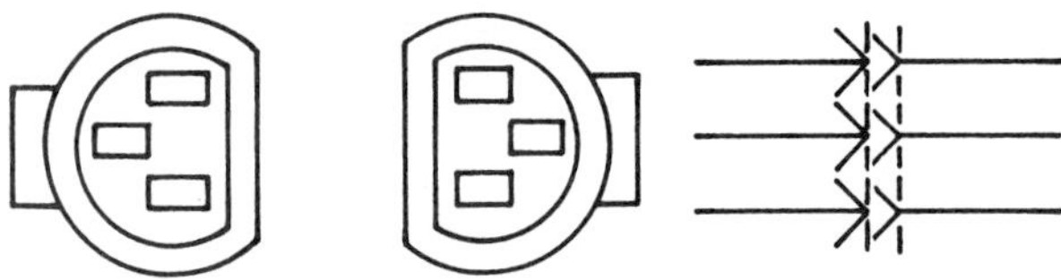

Figure 7. A pictorial diagram and a schematic diagram for a three-wire connector.

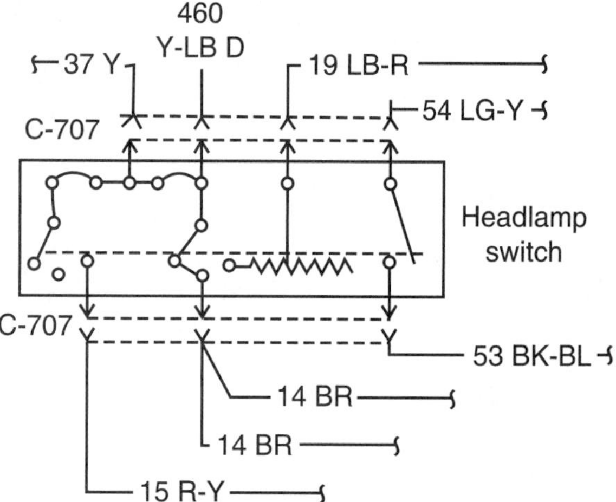

Figure 8. A common headlamp switch.

PROTECTION DEVICES

We have studied the three types of protection devices: fuses, circuit breakers, and fusible links. Now let us look at what they typically look like on a schematic diagram. **Figure 9** shows two different styles of fuses. On the left is the common symbol for a fuse that is mounted in a panel called the fuse box. The fuses are plug in and can be glass, ceramic, or blade variety. On the right is the same symbol

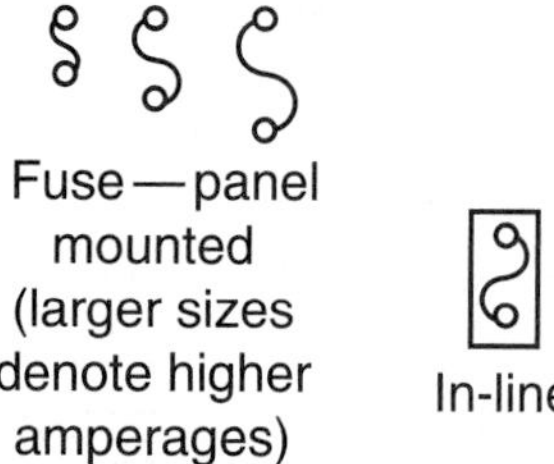

Figure 9. Different fuse schematics.

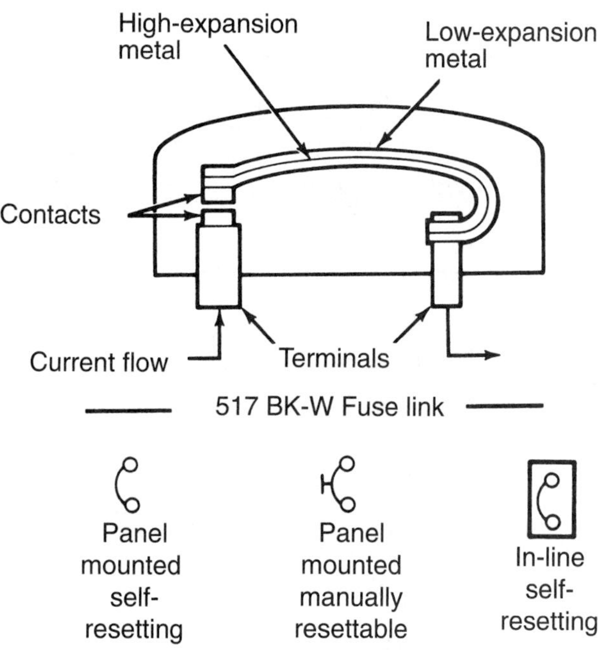

Figure 10. Circuit breakers are frequently used in motor circuits.

with a box drawn around it. The fuse is not in the fuse box. It is spliced into a circuit and is called an in-line fuse.

Circuit breakers are used in place of fuses in circuits that are prone to temporary overloads such as windshield wipers that might ice to the window in the winter. The temporary overload caused by the ice causes the circuit to draw additional current, which will overheat the metal inside the breaker and open the contacts. Once the breaker cools down, it will close again, restoring power to the circuit. Circuit breakers are represented in **Figure 10**. As was the case with fuses, a box drawn around the breaker symbol indicates that it is not in the fuse box but is located somewhere else. Fuse placement is sometimes shown as a graphic of the fuse panel **(Figure 11)**. In addition, a listing of what circuits are being protected is usually provided.

SWITCHES

The next group of schematic symbols to look at is switches. Switches control the on/off action or direct the flow of current through various circuits. The contacts inside the switch assembly carry the current when they are closed

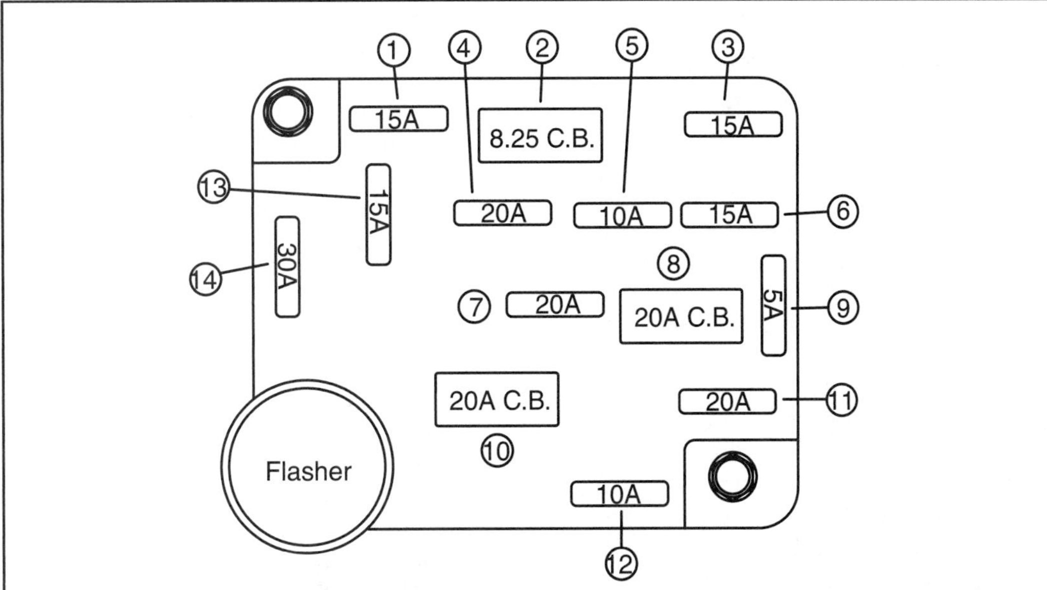

1. Stop lights, emergency warning system, speed control module, cornering light relays, and trailer tow relays
2. Windshield wiper/washer
3. Tail, park, license, coach, cluster illumination, side marker lights, and trailer tow relay
4. Trunk lid release, cornering lights, speed control, chime, heated backlite, and control A/C clutch
5. Electric heated mirror
6. Courtesy lights, clock feed, trunk light, miles-to-empty, ignition key warning chime, garage door opener, autolamp module, keyless entry, illuminated entry system, visor mirror light, and electric mirrors
7. Radio, power antenna, CB radio
8. Power seats, power door locks, keyless entry system, and door cigar lighters
9. Instrument panel lights and illuminated outside mirror
10. Power windows, sunroof relay, and power window relay
11. Horns and front and instrument panel cigar lighters
12. Warning lights, seat belt chimes, throttle-solenoid positioner, autolamps system, and low fuel module
13. Turn signal lights, back-up lights, trailer tow relays, keyless entry module, illuminated entry module, and cornering lights
14. ATC blower motor

Figure 11. The wiring diagram might contain fuse information.

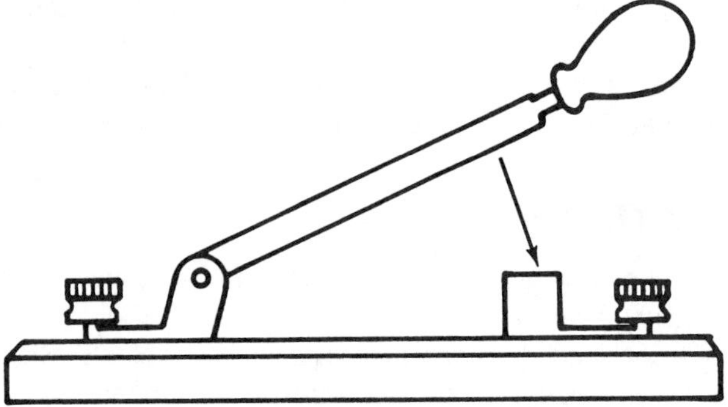

Figure 12. An SPST switch.

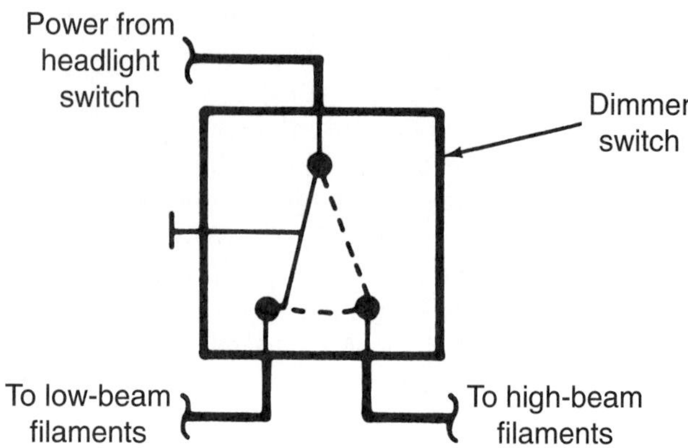

Figure 14. An SPDT headlight dimmer switch.

and open the circuit when they open. Another term for opening is to "break" the circuit, whereas closing is considered "making" the circuit **(Figure 12)**. You will frequently hear the words "pole" and "throw" applied to a switch. This is a holdover from the days of the knife switch. In this type of switch, the conductors were attached to poles, and the knife could be thrown to close the contacts. The knife switch would be considered a single-pole, single-throw switch. Simplified, this means that there is one input and one output. Pole refers to the number of inputs, whereas throw refers to the number of outputs. Therefore, a switch with one input and one output is referred to as **single-pole, single-throw (SPST)**. We find frequent use of additional styles of switches on the automobile: **single-pole, double-throw (SPDT)**; **double-pole, double-throw (DPDT)**; and **multiple-pole, multiple-throw (MPMT)**. Switch contacts are also designed according to whether they are **normally open (NO)** or **normally closed (NC)**. The momentary contact switch in **Figure 13** is an example of a switch drawn NO. The contacts will not allow current through in their normally open position. They are, therefore, normally open.

Usually the wiring diagram will have the switch drawn in the position it will be in with the vehicle off, and this is defined as "normal" **(Figure 14)**. A headlight dimmer switch is an example of an SPDT. It has one input and two outputs (one output for the low-beam headlights and one for the high beams). In addition, notice that one set of contacts will be drawn normally open, whereas the other will be drawn normally closed.

Figure 15 is a combination **transmission range switch (TR)** and back-up light switch. It is an example of an MPMT. It has two inputs and the possibility of six outputs or positions. Two movable wipers move together. The dotted line between the poles usually indicates that the poles are "ganged," or connected together, mechanically. Based on the drawing, when will power be able to flow through the switch to the starter relay? The jumper between P and N indicates that current will be able to flow through when the switch is in the P (park) or N (neutral) position. When will the back-up lights go on? You can figure out from the drawing that they will be able to go on only when the switch is in R, or reverse, position. Ganged switches are frequently used in applications where the motion of the switch is needed to control more than one circuit. In this

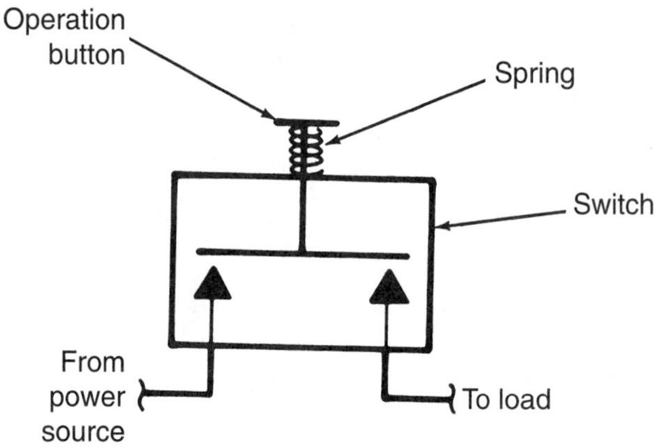

Figure 13. A momentary contact switch.

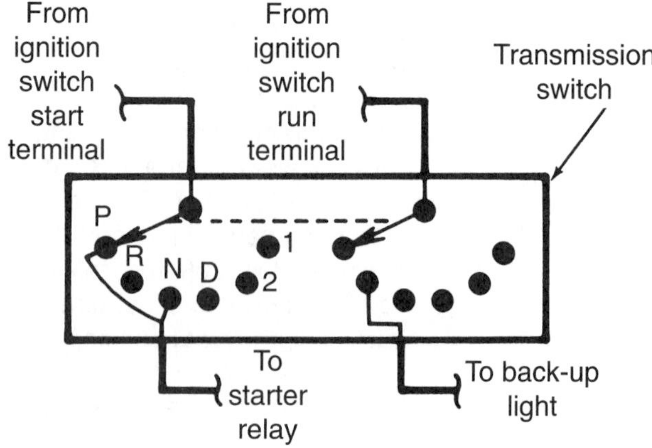

Figure 15. The dotted line between switch wipers indicates that both wipers move together.

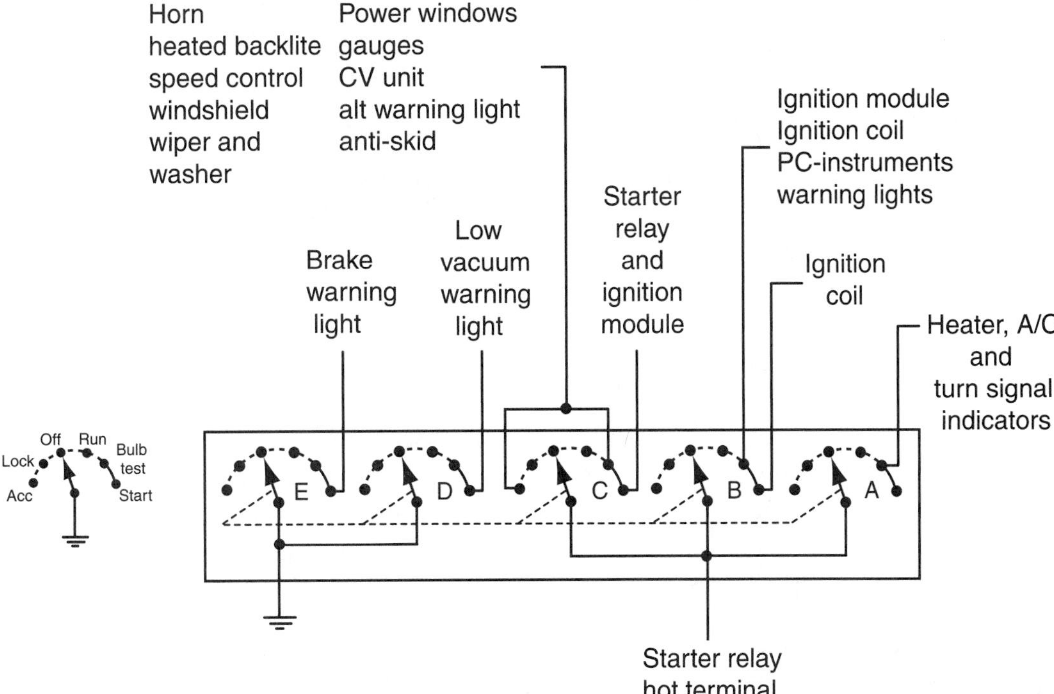

Figure 16. The ignition switch is usually the most complicated switch on the vehicle.

example, the movement of the gear shift allows the starter to be engaged in park and neutral and the back-up light to be on in reverse; thus, two functions operate off the one switch.

Figure 16 is another example of a ganged switch, the ignition switch. It is probably the most complicated switch used on most vehicles. Our example has five wipers ganged together. The dotted line extending to all wipers indicates that they will all move together. In addition, the legend printed in the lower left-hand corner indicates the five different positions that the switch can be in.

All wipers will move together and are currently shown in the L, or lock, position. This is the main switch for the entire vehicle and has many wires connected to it that perform many different functions. For example, the two wipers on the left are going to be utilized to turn the warning lights on during cranking to test the bulbs. The ground available on the two wipers (points E and D) will be applied to the two circuits and will turn the bulb on. This is the test of the circuit that most warning lights go through during cranking. The other three wipers all have B+ applied to them, and they will switch it over to different circuits that require power. Let us test your understanding of the wipers. What positions will allow the turn signals to function? The last wiper has power to this circuit when it is in the R, or run, position. This makes sense. We want the turn signals to be functional only when the engine is running. How about the power windows? Wiper C shows a connection to the circuit that will feed the power windows. Notice the jumper between positions A and R. This means that the power windows, and anything else on the circuit, will be operational in both the accessory and run positions. You should be capable of following the current path through this MPMT. You may encounter a situation in which the contacts are not functioning as they should. Your understanding and ability to interpret the diagram will allow you to test the switch with a meter or test light.

SOLENOIDS

Frequently switches, especially the momentary contact type, will be used to control electrical solenoids. Electric door locks, hood latches, and trunk latches can be solenoid operated. **Figure 17** is a diagram of a solenoid-operated luggage compartment latch. Let us use all the information to date and follow the flow of current until it reaches ground. B+ is available through wire 296 to the deck lid release switch at connector C-1603. The momentary contact switch allows B+ through this SPST switch and out C-1603. C-1603 must therefore be a molded two-wire connector. When the contacts close, B+ flows through three additional connectors—C-1604, C-1605, and C-1606—and reaches the solenoid coil. Current flowing through the coil creates a magnetic field and pulls the latch open. The solenoid is grounded with a case ground to the metal surrounding it.

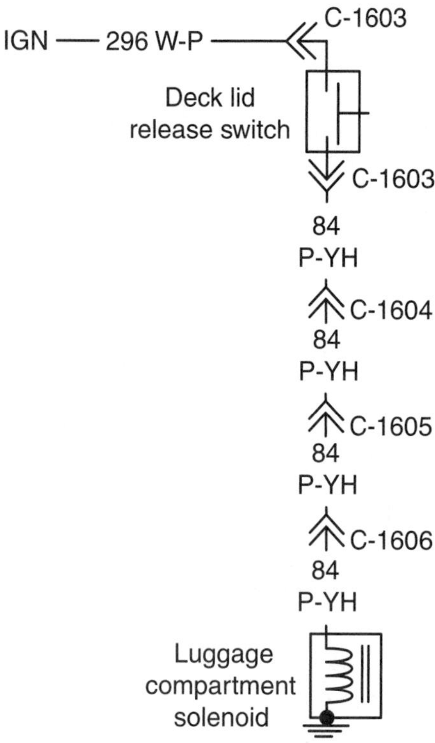

Figure 17. Solenoid-operated latch schematic.

RELAYS

The use of relays continues to increase. You remember that a relay is a device that uses low current to control high current. **Figure 18** is a diagram of a common Bosch-type relay found on many vehicles. The numbering of the terminals is relatively consistent throughout the industry. The low current, which is usually called the control current, flows through terminals 85 and 86, where it will generate a magnetic field. The field will close the contacts numbered 30 and 87. Once there is continuity through the contacts, the device will turn on.

Generally these relays are wired so that power for the device is available on terminal 30 and the load is wired to terminal 87. Some of the relays will also have a diode across the coil terminals to eliminate or control the spike that will occur in the coil when current is turned off.

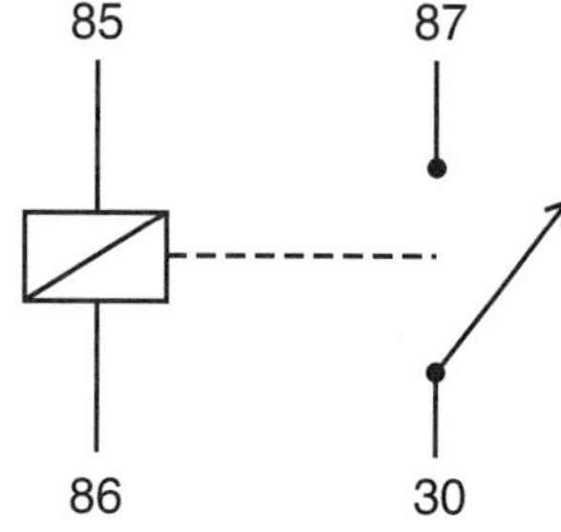

Figure 18. The schematic diagram for a common relay.

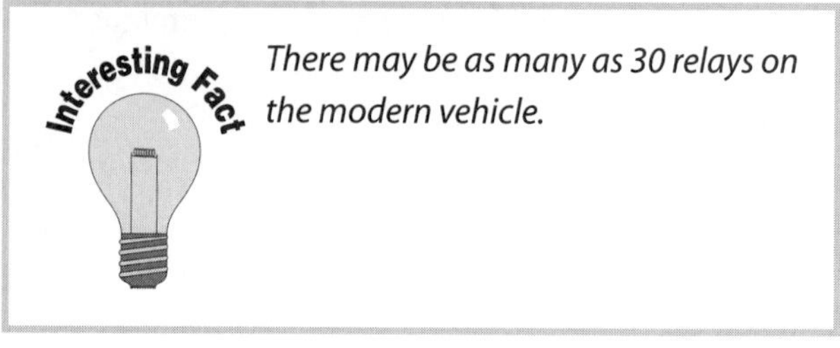

There may be as many as 30 relays on the modern vehicle.

RESISTORS

The next symbols that we will look at are those of the various types of resistors in use in the vehicle's electrical system. We have used the basic resistor symbol **(Figure 19)** in our discussion of circuits. You should realize that there are numerous applications using fixed resistors. For example, blower motor speeds or windshield wiper speeds are frequently controlled by using a resistor in series. Most automotive wiring diagrams will have the specified ohms of resistance printed near the resistor for reference purposes. Sometimes the need arises for a variable resistance, such as a dimmer on the dashboard lights. Increasing the circuit resistance dims the lights, whereas decreasing the resistance brightens the lights. This is usually done with a rheostat **(Figure 20)**. Notice the arrow drawn through the center of the resistor. This is what indicates that it is a variable resistance. Unlike the fixed resistor, it will usually not have a value printed near it on the schematic diagram. The increased use of computers in vehicles has seen the introduction of an additional form of the resistor: the **potentiometer**. This device is used as an input signal for the computer **(Figure 21)**. Most TPSs are potentiometers. They are drawn on schematic diagrams as shown. The three connections coming off the sensor go to the computer and

Figure 19. A fixed resistance.

Figure 20. A variable resistance.

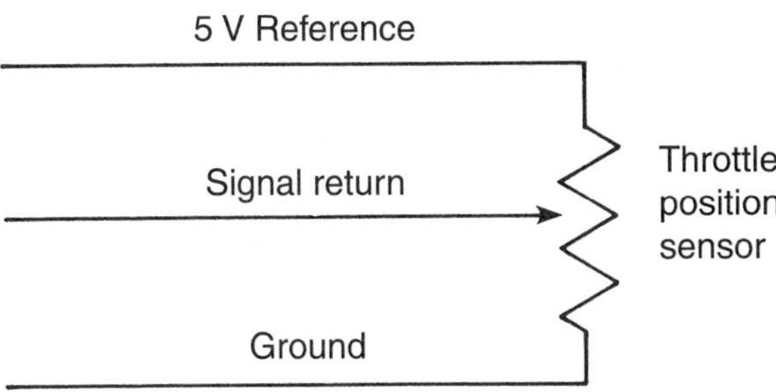

Figure 21. A potentiometer sends a voltage signal to indicate position.

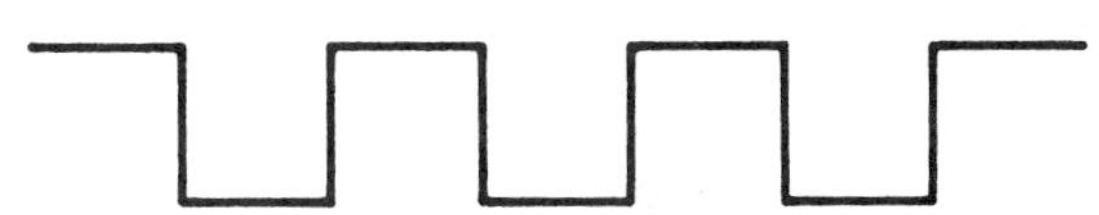

Figure 22. The schematic symbol for a heater.

will indirectly tell it the throttle-opening angle. From a drawing standpoint, the potentiometer is similar to a rheostat and it does change resistance. The major difference comes in the number of wires. The potentiometer will have three wires, whereas the straight variable resistor will usually only have two.

The last form of resistance commonly found on the vehicle is the heating resistor. The symbol is shown in **Figure 22**. The squared-off resistor symbol will be found for the rear window defogger on the schematic diagram. It is a resistance heater that will develop sufficient heat when current is passed through it to melt the snow off the window.

MOTORS

There are many electric motors in use on today's vehicle. The more common ones are shown schematically in **Figure 23**. The differences among them are: how they are grounded, either case or remote; whether or not they are reversible; and if they are multispeed. The schematic for the single-speed bidirectional motor would be found for power windows, antennas, or seats, examples in which the customer reverses the motor direction to control the windows and antenna. This motor usually has two wires coming in and no ground. Reversing the current flow through the motor windings causes the motor to reverse its direction. The single-speed unidirectional (one-direction) motor shown in the top left is the type found on engine cooling fans and heater blower motors. Notice that only one wire is coming in, and the unit is case grounded. The third motor shown is similar to the style used in windshield wipers, two-speed unidirectional—one direction but multispeeds and either case or remote grounded.

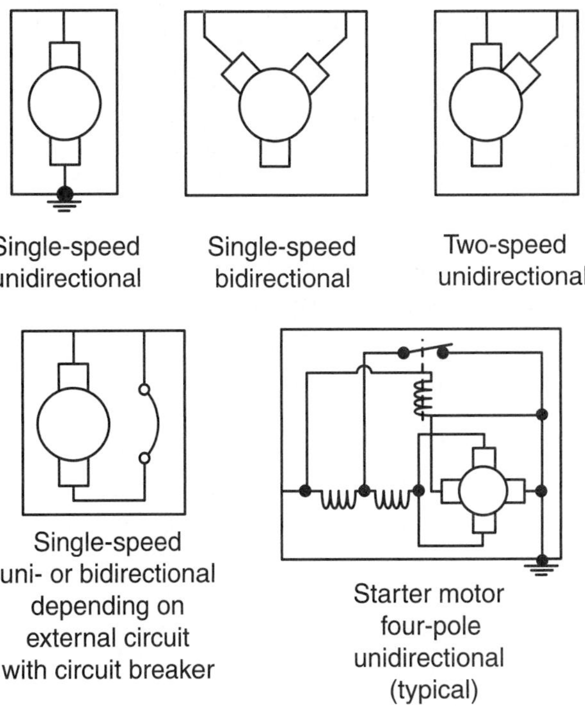

Figure 23. Motor schematic types.

Many motors have a circuit breaker installed internally, as the schematic in **Figure 24** indicates. The circuit breaker will protect the motor if the mechanism is locked up and cannot move. The one shown can be automatically reset once the motor is able to turn again.

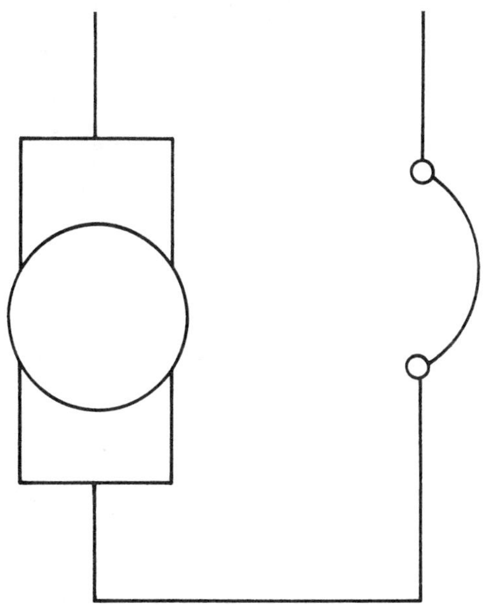

Figure 24. The circuit breaker protects the reversible motor.

LOCATION CODES

Now that we have looked at some of the common schematic symbols and have a basic idea of how to trace through a wiring diagram, look at how one would actually follow a circuit on the vehicle. Remember that it is virtually impossible to follow a single wire through the vehicle. It might change color many times and go through sections of the vehicle that are totally unrelated to the circuit being followed. This is the main reason why the technician must rely on the wiring diagram. It will give you information that will be invaluable when trying to follow the circuit on the vehicle.

Location codes are sometimes printed directly on the diagram, as **Figure 25** shows, or they can be found in another section of the information. In either case, they are important to the technician because they give information about where to find a connector, a ground, or a relay. Countless hours of diagnosis are saved through the use of the location code.

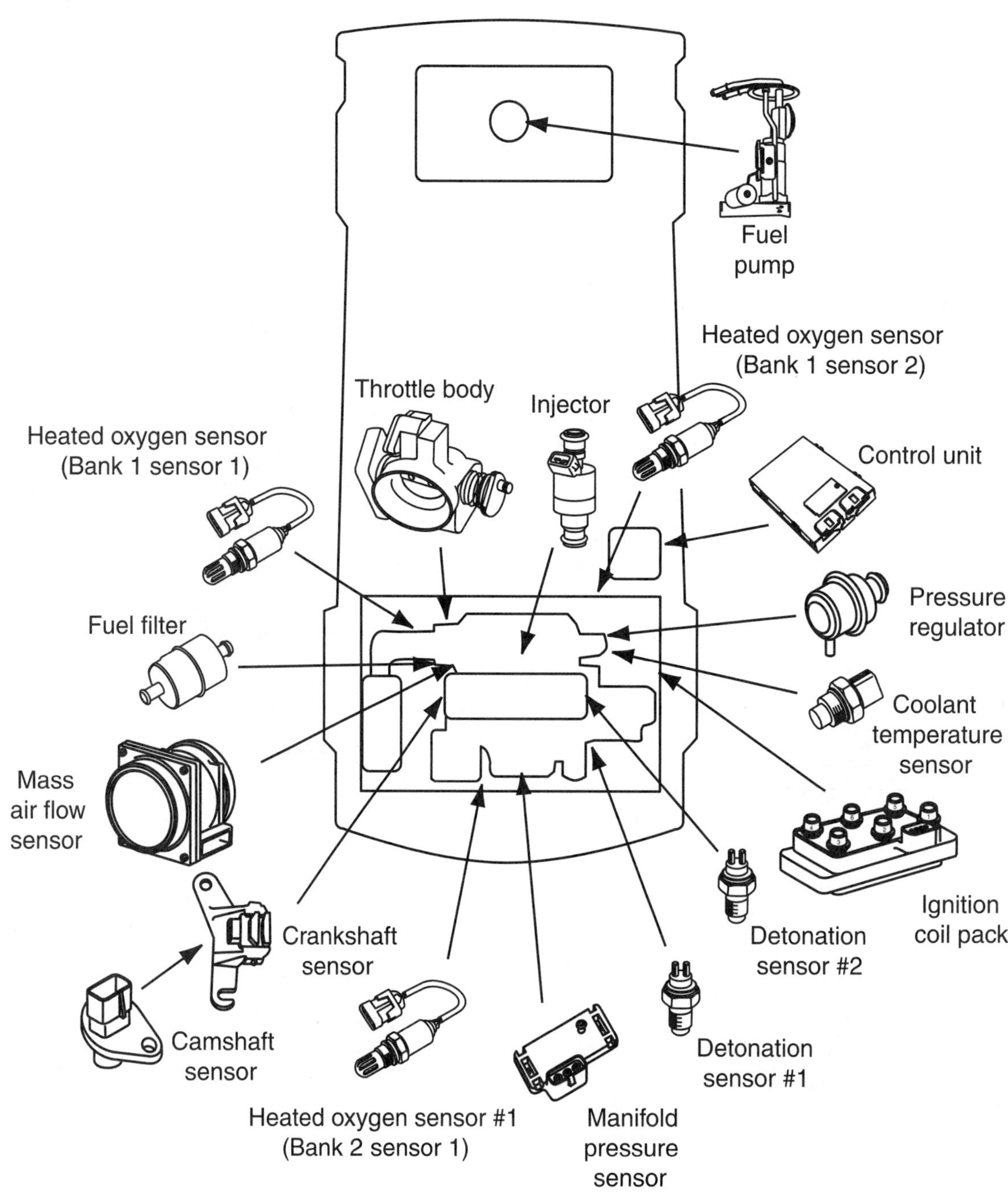

Figure 25. Frequently, location codes are part of the wiring diagram.

Summary

- Wiring diagrams may be drawn in pictorial or schematic style.
- Grounds may be either remote or case.
- A case ground is one in which the component is grounded by attaching it to the chassis.
- A remote ground will have a wire connecting it to a ground screw.
- A common point is a connection for two or more wires and is represented by a dot on the diagram.
- An expanded splice or connector is a drawing style that allows for easier reading of the diagram.
- A dotted line between two or more connections indicates that the connections are all in one housing.
- Fuses and circuit breakers have unique schematic symbols.
- A box drawn around a fuse or breaker usually indicates that it is not in the fuse panel.
- Switches can be SPST, SPDT, or MPMT.
- A dotted line between contacts indicates that they move together.
- A dotted line inside a relay indicates that the contacts will close magnetically.
- The coil inside the relay will have control current flowing through it.

Review Questions

1. Technician A states that a common point is represented by a black dot on the wiring diagram. Technician B states that a connector can be a common point. Who is correct?
   A. Technician A only
   B. Technician B only
   C. Both Technician A and Technician B
   D. Neither Technician A nor Technician B
2. A switch is drawn on the diagram with NC printed beside it. This indicates that the switch will be
   A. normally closed with the vehicle running
   B. open with the vehicle not running
   C. closed with the vehicle off
   D. open with the vehicle running
3. A dotted line between connectors indicates that the
   A. connectors are electrically connected
   B. connectors are physically connected
   C. connectors are remote
   D. None of the above
4. A remote ground is where
   A. a ground connection is made at some point other than at the component
   B. the ground is completed by placing the component into a metal housing
   C. two or more grounds are connected
   D. a splice takes place
5. Technician A states that two connectors with the same number are physically the same connector. Technician B states that the two connectors are drawn apart so that the drawing is easier to read. Who is correct?
   A. Technician A only
   B. Technician B only
   C. Both Technician A and Technician B
   D. Neither Technician A nor Technician B
6. A resistor drawn with an arrow through it is a thermistor.
   A. True
   B. False
7. Dotted lines between switch wipers indicate that all wipers will move together.
   A. True
   B. False
8. A case ground is a remote ground.
   A. True
   B. False
9. A common point is a point where two or more wires are connected.
   A. True
   B. False
10. What are the major differences between a pictorial diagram and a schematic diagram?
11. Why is a circuit breaker sometimes shown wired in series with a motor?

Chapter 29 Using the Wiring Diagram As a Service Tool

Introduction

We hope that by now you are convinced about the advantages of having a wiring diagram in front of you as you begin circuit diagnosis. As the vehicle has evolved and more electrical accessories are added, the circuits become both more involved and more difficult to diagnose. In addition, having modules take over certain functions has also complicated the situation. The point of this chapter is to have you examine the wiring diagram for the vehicle that you are working on before opening the toolbox. The diagram should be thought of as a map that will show you how to get to a specific destination. If you began driving across the country without a road map, you would get lost frequently. The map allows you to decide in advance what roads to take. Add the Internet for a road construction update and you have enough information to travel intelligently. That is the purpose of the three techniques we discuss in this short chapter—getting the diagnosis accomplished intelligently.

COMMON-POINT DIAGNOSTICS

Many technicians refer to common-point diagnosis as a tool or trick of the trade. It refers to the use of the wiring diagram to prediagnose, especially an open circuit. When the manufacturers design the vehicle, they electrically tie different circuits together in a series-parallel arrangement.

Figure 1 is a simple example of series-parallel circuits showing that the fuse is series wired to the three parallel loads. The common point in the circuit is just after the fuse where the three loads branch off. If the vehicle has a load A, which will not function in this circuit, the technician would

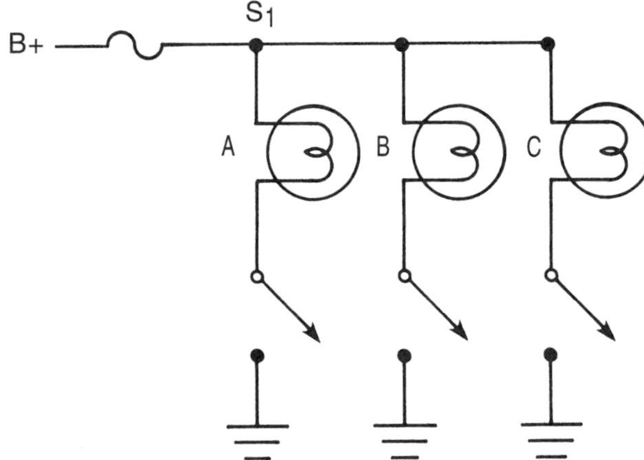

Figure 1. A series-wired fuse for three parallel loads.

take advantage of the information that the wiring diagram has supplied. Two additional loads are common-point wired (B and C) to the fuse at S_1 (splice 1). By being alert to this simple fact, technicians can save themselves valuable diagnosis time. If loads B and C both function, the open must be beyond common-point S_1. If none of the loads function, the open is probably before the common point. The wiring diagram has just saved us from having to test at least half of the circuit, and if the half we do not have to test is hidden behind panels, we have substantially reduced our diagnosis time.

USING THE WIRING DIAGRAM TO FIND AN OPEN CIRCUIT

Let us look at another example and use common-point diagnosis to decrease our diagnosis time in a lighting circuit. The customer complaint is no back-up lights. Where do we

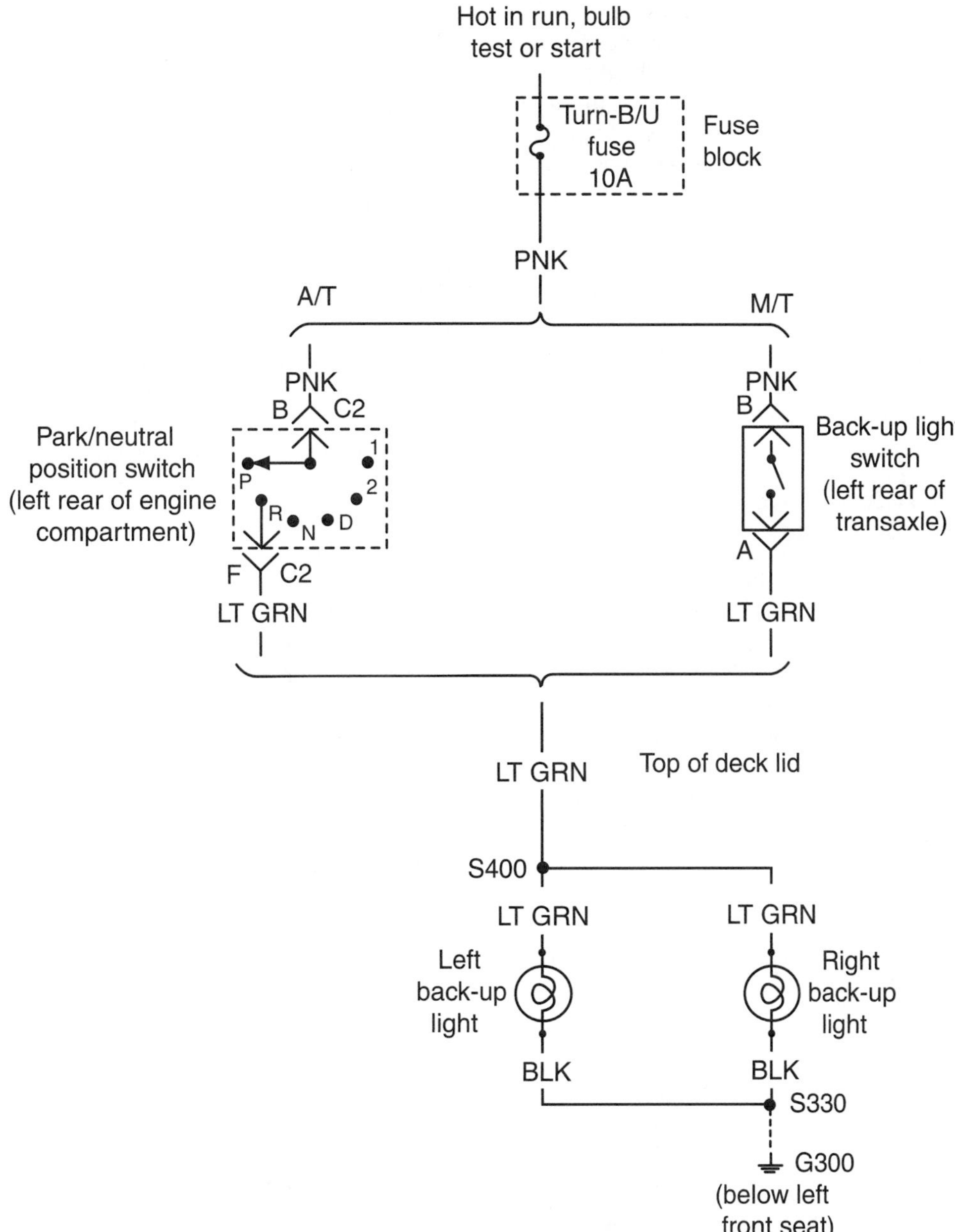

Figure 2. A back-up light circuit complete with location codes.

start? **Figure 2** shows a vehicle back-up lighting circuit. Our first consideration is to see how the back-up lights are wired. Power for the circuit comes from the 10-amp turn-B/U fuse in the fuse block. The power to the fuse must come from the ignition switch because the notation of "hot in run" is above the fuse. From the fuse, the power goes to the park/neutral position switch for an automatic transmission vehicle and then through the body harness, where it common points off to the two bulbs. The circuit ends at the ground below the left front seat. Our goal is to use the information from the diagram to help diagnose the circuit. Follow this logical use of the diagram. Try out the turn signals. If they work, the problem is after the fuse because the fuse powers both circuits. If neither works, begin the diagnosis at the fuse. Let us assume the turn signals work. There is no need to test at the fuse because we know it must be working. The next most logical test position is the common point at the top of the rear deck lid. This is a very accessible point. We test for power at the common point. No power points us toward the front of the vehicle, whereas power points us to the rear. Two simple tests have directed us to the correct section of the vehicle and eliminated the need to test the entire circuit. Let us go back to the common point and suppose that we have found it to be hot. The park/neutral position switch must be functional, and all of the wiring up to this point is good. It is time to pull the bulb and test for power and ground in the connector. No power indicates an open circuit between the bulb connector and the common point (that was hot). Power with the test light connected to a good ground indicates that a good circuit exists up to the bulb. It is unlikely that both bulbs are burned out, but that is easy enough to test. If the bulbs are good, we have a ground problem. Splice 330 is a common point for the ground circuit. A test light connected to power should light if this is a good ground, right? If not, we need to go under the left (driver's) front seat where the ground is connected. This circuit is not that complicated, but it is a good example of the procedure that you will go through to diagnose an open-circuited back-up light circuit.

USING THE WIRING DIAGRAM TO FIND A SHORT CIRCUIT

Let us try another example, but this time the customer complaint is a blown number 4 fuse **(Figure 3)**. Obviously, we would not want to crawl around this extensive circuit because it covers almost the entire vehicle. The time involved to locate the short circuit would be hours. Instead, common-point diagnosis can be used to identify the most likely location of the short before electrical testing begins. Remember, short circuits decrease the overall circuit resistance and generally blow the circuit protector, which in this case is fuse number 4.

To isolate the circuit, we would first turn off all loads on fuse 4: make sure the air conditioning (AC) is off, the turn signal is in a center position, and the gear selector is in park with the vehicle off. This should eliminate all current flow through the circuit. Next, we would plug in another fuse. Remember that our goal here is to isolate the short to as small a section of the entire circuit as is possible. Turn on the ignition; did the fuse blow? If it did, where is the short? The common point after the fuse up to the switches is the only part of the circuit that is live. We would now know that our work will be in the dashboard area after the fuse and before the individual circuit switches. We would now separate the circuit at the connectors after the common point and before the switches to further isolate the circuit. If the fuse did not blow when we turned on the ignition switch, we know that our short is after the common point and the switches. By now you probably have figured out that the next step, if the fuse has not blown, would be to turn on each circuit separately until the fuse burns out. The short would be in the section of the circuit that, when energized, caused the fuse to burn out. Diagnosis like this can save you a tremendous amount of time. If, for example, the vehicle described above came in with the complaint of inoperative AC and you did not at least walk through the circuit to determine if additional loads were on the same fuse, you might spend unnecessary time looking for a short in the AC circuit when, in fact, you might have a short in the back-up light assembly. Until the gear selector is placed in reverse and power runs to the back-up light assembly, the short would not appear. The customer complaint of inoperative AC or inoperative turn signals ignores the additional circuit on the fuse. Most customers are unaware of inoperative back-up lights because they cannot see them. They recognize only the circuits that have a direct bearing on themselves. Being aware of additional circuits on a fuse and using common-point diagnosis will pay the cost of the wiring diagrams on the first job.

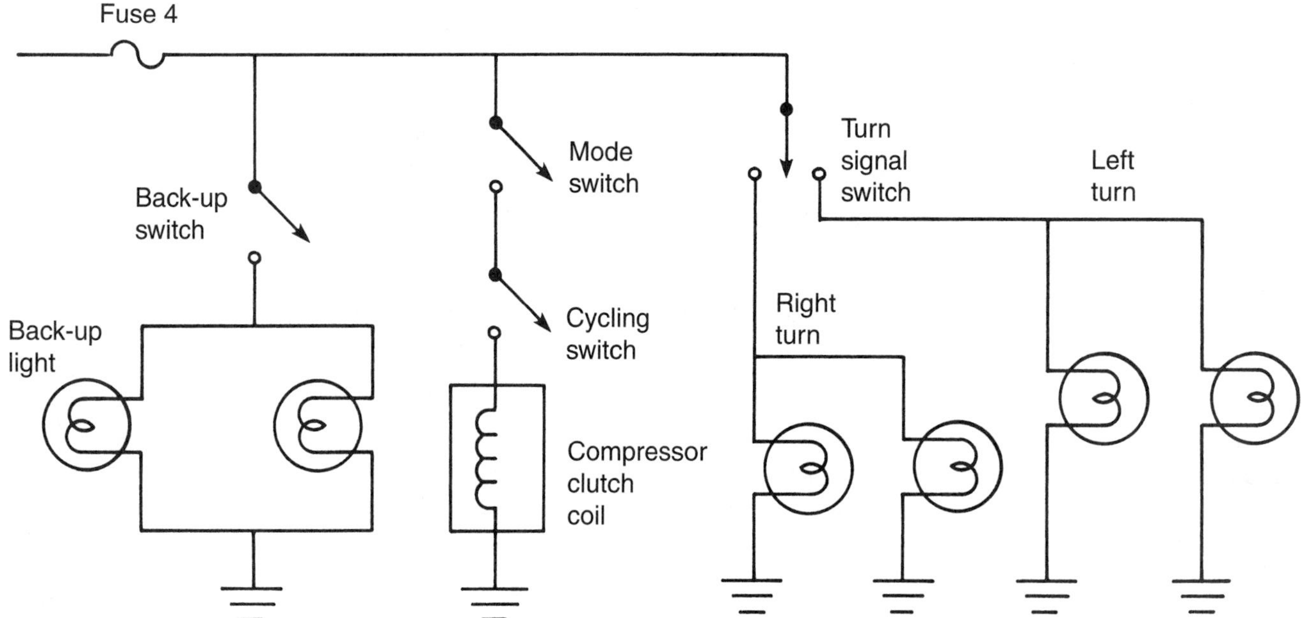

Figure 3. Use the circuit switches to isolate the section with the short.

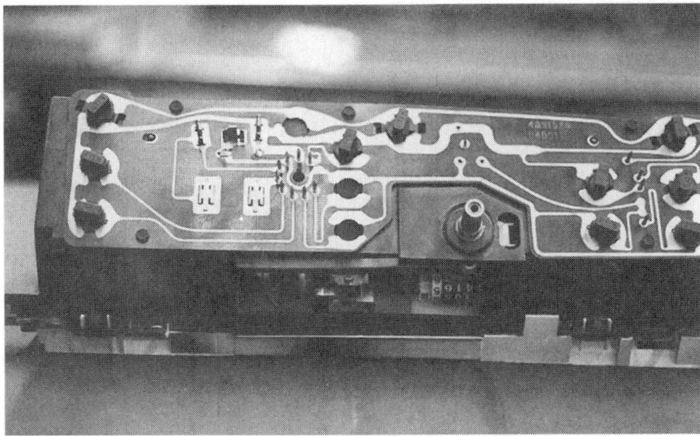

Figure 4. A printed circuit board in place on the back of an instrument cluster.

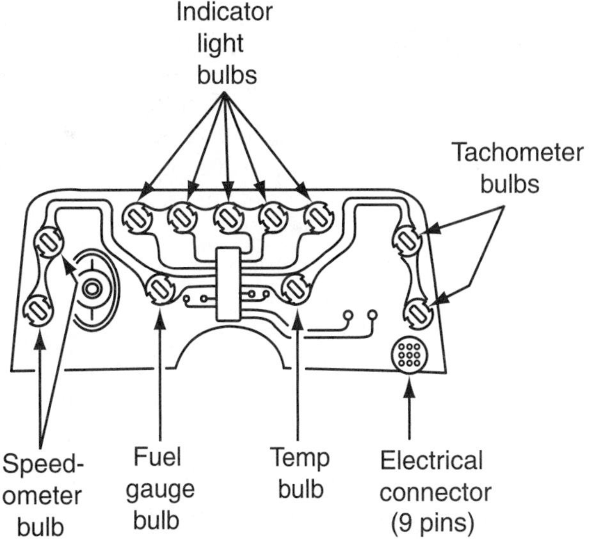

Figure 5. The schematic diagram of the printed circuit.

PRINTED CIRCUITS

The last thing we look at in this chapter is the use of printed circuit boards that are on many dash assemblies. **Figure 4** is a picture of a printed circuit board from the back of a dashboard. It is composed of a film of plastic with small strips of copper. The current for various sections of the dash will run through these strips of copper. Lights, gauges, and indicators will plug directly into the printed circuit board, and one harness connector will bring both B+, ground, and the gauge signals to all the circuits. Printed circuit boards became popular as more manufacturers began to use plastic dashboards. The majority of the wiring diagrams show the layout of the board. It is possible to trace through the circuit as you would with any other wiring diagram. The connector for the entire dash shows as a circle on the right side with nine terminals **(Figure 5)**.

Precautions When Dealing with a Printed Circuit Board

There are some precautions that should be observed when working on a printed circuit.

1. Never touch the surface of the board. Dirt, salts, and acids on your fingers can etch the surface and set up a resistive condition. It is possible to knock out an entire section of the dash with a fingerprint.
2. The printed circuit board is damaged easily because it is thin. Be careful when plugging in bulbs, so as not to tear the surface. Replacement circuit boards are usually available, but the inconvenience and time necessary to obtain a replacement make a mistake costly. Handle circuit boards with care.

One of the advantages to the use of printed circuits is their high degree of reliability. In addition, they allow the dash to be pulled back and worked on with the entire circuit laid out and functional. Because both B+ and ground come through the connector, the board will function even if it is not completely installed. This allows you to use your voltmeter or test light on the circuit with it functioning normally.

Summary

- The wiring diagram should be thought of as the road map into the circuit.
- The diagram shows us how the manufacturer has connected the maze of parallel circuits.
- The wiring diagram shows us the common points in the circuit that will become our test points.
- If two or more circuits are off the same fuse and only one is not functioning, the fuse must be good.
- Open-circuit diagnosis involves looking for power along the path that the diagram shows.
- Short-circuit diagnosis involves isolating all of the circuits on a fuse and individually turning them on until the fuse burns.
- Printed circuit boards are frequently used in dash circuits.

Review Questions

1. Two circuits are powered off the same fuse but have individual grounds. One circuit works; the other does not. Technician A states that the problem could be ahead of the fuse. Technician B states that the ground might be faulty. Who is correct?
   A. Technician A only
   B. Technician B only
   C. Both Technician A and Technician B
   D. Neither Technician A nor Technician B
2. A splice
   A. is a common point
   B. could be on the power or the ground side
   C. connects two or more wires together
   D. all of the above
3. A printed circuit board is being tested. The technician tests for the ground connection and finds it to be open. This could cause
   A. the fuel gauge to not function
   B. the dash light to not function
   C. the entire dash to not function
   D. no problem at all
4. The fuse burns out when a switch is turned on. Technician A states that this means that the short to ground is after the switch. Technician B states that the common point just after the fuse could be shorted to ground. Who is correct?
   A. Technician A only
   B. Technician B only
   C. Both Technician A and Technician B
   D. Neither Technician A nor Technician B
5. A common point is where two or more wires are connected.
   A. True
   B. False
6. Common points can be only on the power side.
   A. True
   B. False
7. If the fuse blows when the switch is closed, the short must be after the switch.
   A. True
   B. False
8. Printed circuit boards should not be touched.
   A. True
   B. False
9. Briefly describe the procedure for finding an open circuit.
10. Briefly describe the procedure for finding a short circuit

Section 7

Batteries

SECTION OBJECTIVES

At the conclusion of this section, you should be able to:

- Recognize how batteries are rated.
- Understand the basic construction of a battery.
- Do a load test.
- Test for state of charge using a hydrometer.
- Test for state of charge using an optical refractometer.
- Test for state of charge using open-circuit voltage.
- Do a three-minute charge test.
- Understand the causes of sulfation.
- Jump-start a dead battery.
- Charge a battery.

Battery service accounts for a large percentage of service by the time a vehicle is around three years old. Few batteries last much beyond four years.

Chapter 30 Automotive Batteries

Introduction

An automotive battery is a device that produces electricity by means of chemical action. Technically, to be considered a battery, it must be composed of two or more cells, each cell being independent and containing all the chemicals and elements necessary to produce electricity. The term "battery" means groups. The modern 12.6-volt battery that we use in vehicles is composed of six cells. Over the years, the requirements of additional current have been supplied by the charging system; however, the entire system will not work if the battery does not do its job effectively. What is that job? To start the vehicle. Our modern vehicle battery must supply the current demands of the starting, fuel, and ignition system or the vehicle will not run. As we look at the battery in detail, keep in mind its purpose. Nothing works unless the battery does its job.

TYPES OF CELLS

Two common types of cells are available today: primary and secondary. Their difference is in the ability of a secondary cell to be recharged. Primary cells cannot effectively be recharged and are the common A, C, D, and so on, cells that we use in flashlights and radios. When their ability to supply current at the correct voltage is gone, they must be replaced. Let us look at this primary cell closer. A working knowledge of how it works will help your understanding of a secondary cell.

Figure 1 shows a drawing of a carbon-zinc primary cell. The case of the cell is negative and is made of zinc, and the center terminal is positive and is a carbon rod. Between the carbon rod and the zinc case is a paste called the **electrolyte**. It is the chemical action of the electrolyte with the carbon and zinc that produces the 1.5 volts and small amount of current the cell has. As you run the flashlight or radio, the electrolyte wears away and the light gets dimmer and dimmer until the cell is dead. Using a battery charger on a zinc-carbon cell will never recharge the cell back to its full potential. This is why they are called primary cells.

There are many variations of the zinc-carbon cell. Among these are the alkaline, mercury, and lithium, among others. The chemical action within these cells is slightly different and results in a greater current capacity when compared with carbon-zinc. Generally, their voltage also remains higher throughout their use.

CONNECTING CELLS

Have you ever noticed that cells are installed in flashlights with the positive of one cell in contact with the negative of another cell? Have you wondered about the 9-volt battery used in many transistor radios? Both of these are examples of series wiring of cells together. **Figure 2** shows a pictorial diagram of a common two-cell flashlight. By wiring the cells in series (positive to negative), we are able to increase the voltage. Two cells in series produce 3 volts:

$$0.15 \times 2 = 3 \text{ volts}$$

If you think four cells in series produce 6 volts, you are correct:

$$1.5 \text{ volts} \times 4 = 6 \text{ volts}$$

Now look at **Figure 3**. A 9-volt battery has six cells wired in series and stacked on top of one another:

$$1.5 \times 6 = 9 \text{ volts}$$

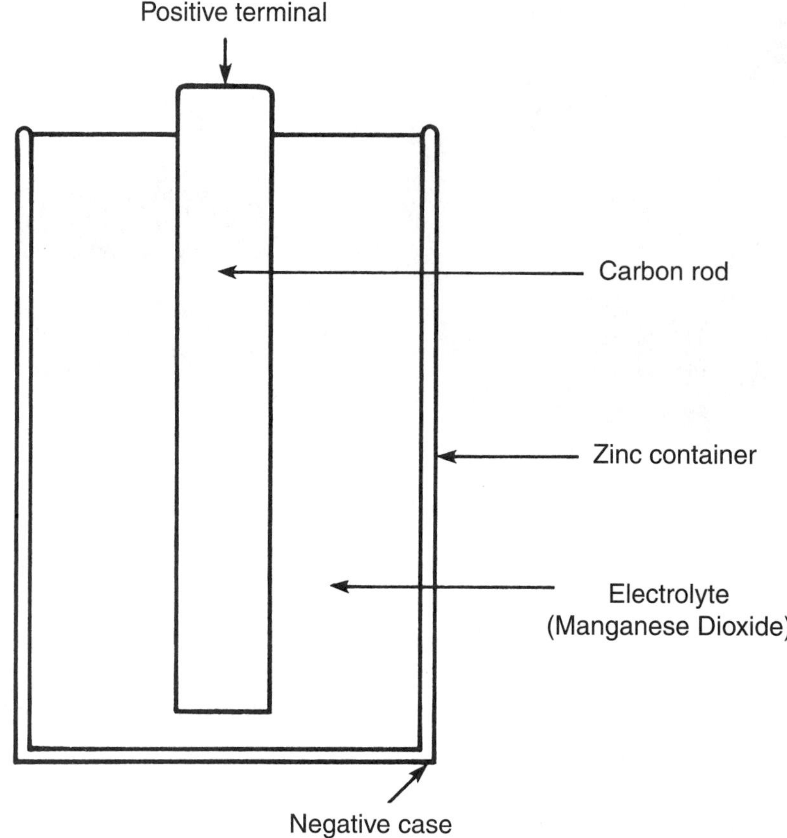

Figure 1. A primary cell.

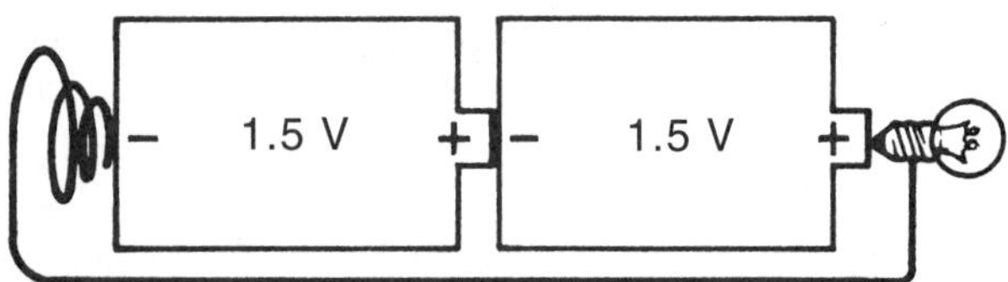

Figure 2. A two-cell flashlight operates on 3 volts.

The conclusion here is that wiring of cells in series increases the voltage.

By now you have probably figured out what happens if we wire cells in parallel (positive to positive and negative to negative). The voltage will remain the same, but the ability of the cell to produce current will be doubled. If we wire a 1.5-volt cell to one cell and turn it on, the bulb will remain on for a period of time. If we wire two cells in parallel, the amount of time the bulb will stay on will be doubled. The conclusion here is that wiring of cells in parallel increases the available current while keeping the voltage the same.

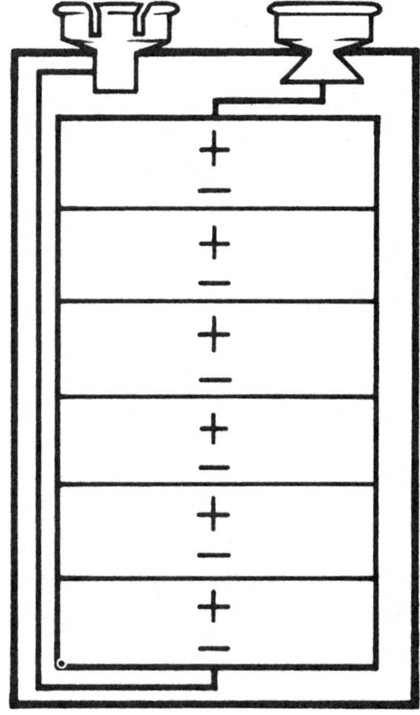

Figure 3. Six cells wired in series equal 9 volts.

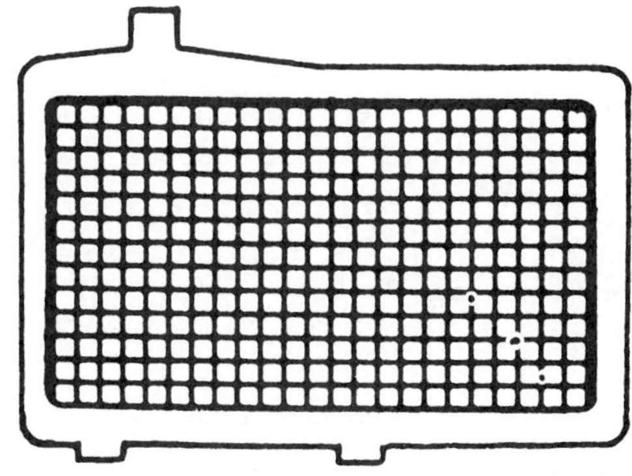

Figure 4. A typical battery grid.

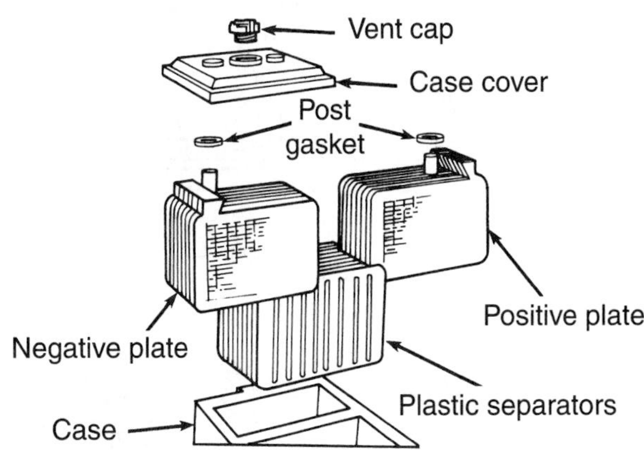

Figure 5. Components of a secondary (wet) cell.

AUTOMOTIVE BATTERY CELLS

Now that we have an understanding of how a primary cell functions, let us turn our attention to the rechargeable cells that make up our automotive storage batteries. Rechargeability is the key difference that makes this type of secondary cell ideally suited for automotive use. Each morning the amperage necessary to crank and start the vehicle would quickly kill a primary cell. But, with a secondary cell, the charging system replaces the amperage used and brings the battery back up to full potential.

Let us take the top off a typical battery and look at the cell. A cell is composed of positive plates, negative plates, separators, and electrolyte. A plate, either positive or negative, starts out with a grid similar to a window screen. **Figure 4** shows a typical grid. Grids are usually made of lead alloys, either antimony or calcium. The active material (lead dioxide for positive, sponge lead for negative) is pressed into the grid in paste form. Notice that the two plates are chemically different, but both are forms of lead. Do you remember our primary cell? It also was composed of two different materials, zinc and carbon. By taking a series of plates and wiring them together by a strap, we form either positive- or negative-plate groups. These plates are insulated from each other by separators made of plastic or glass. Older batteries were separated by wood. **Figure 5** shows the assembled element that will be placed into the battery case. This element will sit in its own individual container, isolated from the other cells with a weak solution of sulfuric acid and water called the electrolyte.

Remember our discussion of series and parallel wiring of cells. Within the individual cell we have just assembled, all of the positive and all of the negative plates are wired together in series. This increases the amperage capacity of the cell, while the cell voltage remains the same (2.1 volts). The more active the material contained within the element, the greater the amperage capacity of the cell. The six cells each will produce 2.1 volts. They are wired in series to produce the 12.6 volts we need. Notice **Figure 6**. It shows the connectors that connect the positive plates of cell one to the negative plates of cell two. Cell two's positive plates are connected to the negative plates of cell three, and so on. This series wiring of each cell to the other cells produces the required 12.6 volts from a fully charged battery. Notice also from Figure 6 that the element sits down in the case on element rests, which raise it up slightly from the bottom. These spaces form the sediment chamber where active material that flakes off the plates can collect without shorting out the cell.

Most batteries are produced with the plates sitting the length of the cell. There are, however, some batteries in which smaller plates are placed the width of the cell, as diagrammed in **Figure 7**. This type of cell will typically have many more plates, with each one being smaller. They are wired with all the positives and negatives together. The key to understanding battery capacity is in realizing that the style of plate has little to do with function. The amount of active material and the amount of electrolyte are what will determine the current capacity of any battery.

SIZES AND RATINGS

When we look at automotive-type batteries, we have to realize that there are different sizes of batteries out in the field. Usually, their size is related to the job that they must do. A large V-8 requires a greater starting amperage than a 1.6 liter, 4-cylinder engine. Usually, the size of a battery is related to the amperage necessary to crank the engine. Automotive manufacturers will usually install larger batteries in vehicles with larger engines. Electrical accessories are also sometimes taken into consideration when figuring the required size of the battery. We will need to know the actual size and the recommended size of the battery when we test in Chapter 31. Basically, there are three common methods of rating both United States and foreign batteries.

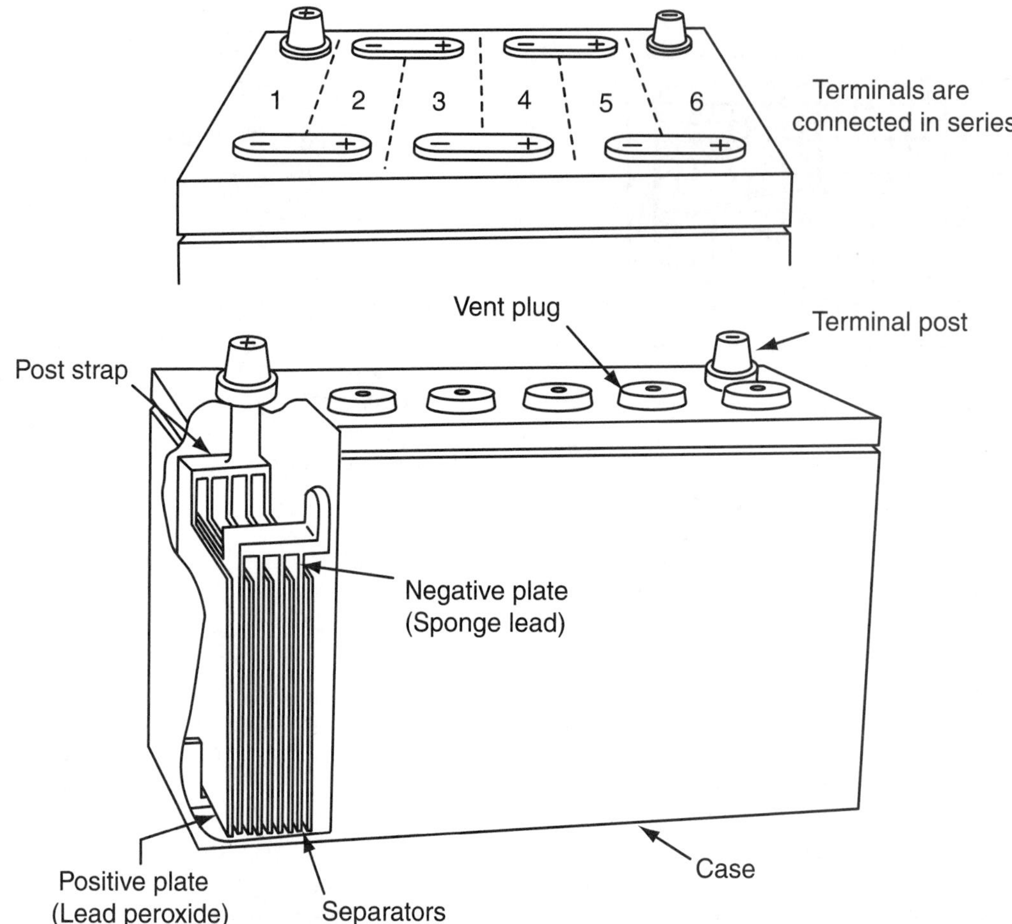

Figure 6. Six secondary cells wired in series equal 12.6 volts.

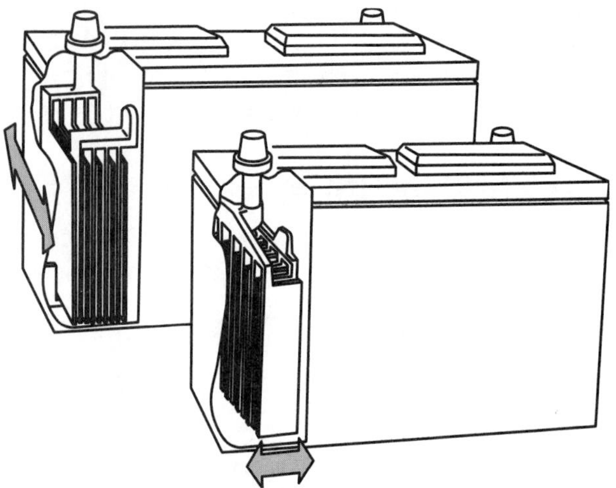

Figure 7. Plates can be placed in either direction.

1. *Ampere Hour.* Probably the oldest and not very common method of rating batteries is ampere hour, or A/H. It is also referred to as the 20-hour method. Simply stated, it tells the amperage that a battery can produce until its terminal voltage falls to 10.50 volts. For example, if a battery can be discharged for 20 hours at a rate of 2.5 amps before its terminal voltage reaches 10.50 volts, it would be rated at 50 ampere hour:

$$2.5 \times 20 = 50 \text{ ampere hour}$$

In theory, a 50–ampere hour battery can produce a current level that when multiplied by time equals the number 50:

$$25 \times 2$$

or

$$10 \times 5$$

and so on. This low rate of discharge does not necessarily indicate the cranking capacity of the battery, but larger ampere hour batteries will deliver more current. Typically 60– to 80–ampere hour batteries are used by the automotive manufacturers as original equipment. From a practical standpoint, the ability to produce current for 20 hours does not necessarily indicate the ability to crank an engine during a 20°F below zero morning. It is for this reason that cold cranking amps have replaced ampere hours.

2. *Cold Cranking Amps.* Cold cranking amperage is an indication of the battery's ability to supply current when

cold. It is more difficult to supply current cold for short spans of time. The rating is done at 0°F and is an indication of the amperage the battery can supply for 30 seconds while maintaining a voltage of 1.2 volts or more per cell (7.2 volts total). Cold cranking amps (CCAs) usually fall in the range of 500–850 CCA, for passenger vehicles. Replacement batteries go as high as 1200 CCAs. A battery rated in CCAs has usually been manufactured with an increased number of thinner plates. The thinner plates will deliver larger amounts of current for shorter periods of time. Realistically, if it takes more than 30 seconds of cranking to get the vehicle started, there is something wrong. As a technician, put your time into finding out why the vehicle does not start quickly. Most modern fuel-injected vehicles will start within 5 seconds of cranking.

> You Should Know *Never replace a battery with one that has a lower rating. Always look up the size that the original equipment manufacturer (OEM) installed and assume this to be the smallest replacement battery.*

3. *Reserve Capacity.* The last rating we discuss is reserve capacity, and it is defined as the ability of a battery to maintain vehicle load with the charging system inoperative. This minimum would probably include lights, ignition, wipers, and heater and be expressed in minutes. The reserve capacity is the minutes a new fully charged battery can be discharged at a rate of 25 amperes. A rating of 120 minutes means that the battery could be discharged at a rate of 25 amps for 2 hours (120 minutes). Although this rating is not popular with vehicle manufacturers, it still remains, especially in the aftermarket. Another type of battery is called a deep-cycle battery and it is rated in reserve capacity. Deep-cycle batteries have a lower number of thick plates. The thick plates allow for a lower rate of discharge for extended periods of time. Boats using an electric trolling motor or recreational vehicles with 12-volt loads are good examples of the use of a deep-cycle battery.

> Interesting Fact *Some vehicles have a battery rated in CCAs plus a deep-cycle battery on board. The CCA battery is the starting battery and the deep-cycle battery is the recreational battery supplying current to, for instance, the interior of the motor home.*

MAINTENANCE-FREE BATTERIES

If we look at the typical lead-acid battery, we would realize that the charging of it produces both hydrogen and oxygen in gas form. In the mid seventies, the battery manufacturers began replacing the antimony in battery grids with calcium. The reason for this change was that a grid made of antimony would gas under most charging situations. Even a slow charge would produce the gases through the electrolysis of the water. To maintain the appropriate amount of electrolyte meant that the technician or consumer had to replace the hydrogen and oxygen with water (H_2O). It was normal for a lead-antimony battery to use an ounce or more of water per cell per month. By reducing the antimony and replacing it with calcium, this water usage was reduced to approximately 10 percent of what it was. The term "maintenance-free" was attached to the calcium grid battery because under "normal" conditions it would not require the addition of water in each cell during its life **(Figure 8)**.

Normal should be defined at this point. A correctly functioning charging system with a voltage generally lower than 15 volts for a warm engine and battery is normal. Realize that if a maintenance-free battery is overcharged (more than 15 volts) for extended periods of time, it will gas just as the lead-antimony battery did. If the manufacturer has sealed the cell tops to the point where you cannot add water, this gassing will eventually ruin the battery. Once the cell plates are not covered with electrolyte, they will begin to dry out and will be easily broken off from vibration. The life of the battery will be greatly diminished by this loss of water. At the same time, the concentration of electrolyte will be increased. It is impor-

Figure 8. A maintenance-free sealed battery.

Figure 9. An OCV of 12.6 volts indicates a fully charged battery.

tant to note here that not all maintenance-free batteries are completely sealed. If the cell tops are removable, distilled or demineralized water should be added, if needed, just as it would be added for a lead-antimony cell. Never add water that is not pure. Distilled is preferred, especially if tap water has minerals suspended in it. Even calcium grid batteries will gas slightly under normal conditions. They will usually not require the addition of water during their lifetime, however.

Maintenance-free batteries with removable cell covers can have their specific gravity tested with a hydrometer. We do, however, find many nonremovable cell covers in use today. These batteries are tested in the same manner (see Chapter 31), except that their state of charge is estimated based on their **open-circuit voltage (OCV)**. OCV is read by putting a voltmeter across the battery with no current flowing **(Figure 9)**. All electrical accessories that draw current must be turned off. In addition, if the battery has been recently charged, a slight load (20 amps) must be applied for about a minute to drain off the surface charge before the open-circuit voltage is noted. The OCV can be compared with specific gravity based on **Table 1**. In this manner, the open-circuit voltage can give us an indication of the battery's state of charge with a reasonable degree of accuracy.

Charge	Specific Gravity	12V Bat	6V Bat
100%	1.265	12.6	6.3
75%	1.225	12.4	6.2
50%	1.190	12.2	6.1
25%	1.155	12.0	6.0
Discharged	1.120	11.9	5.9

Table 1. OCV and specific gravity.

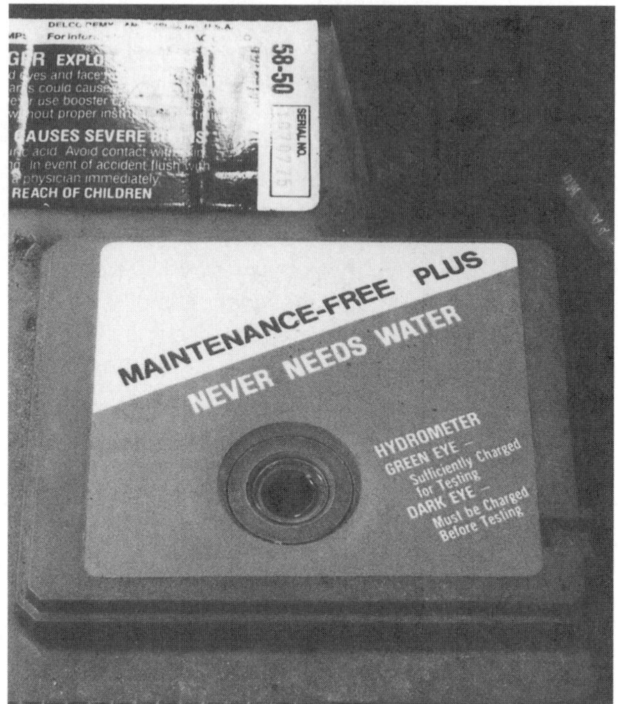

Figure 10. A battery with a charge indicator.

Some manufacturers are building a hydrometer into their cells before they seal the cell covers. **Figure 10** shows a battery with a charge indicator. This indicator is accurate for the one cell it is in and should not be relied on to give total battery condition. If the one cell is charged but the open-circuit voltage is low, that would be an indication that other cells must not be up to full charge. Do not assume that the charging eye gives total battery information. Its information should be added to an OCV measurement and other battery tests, which we examine in Chapter 31.

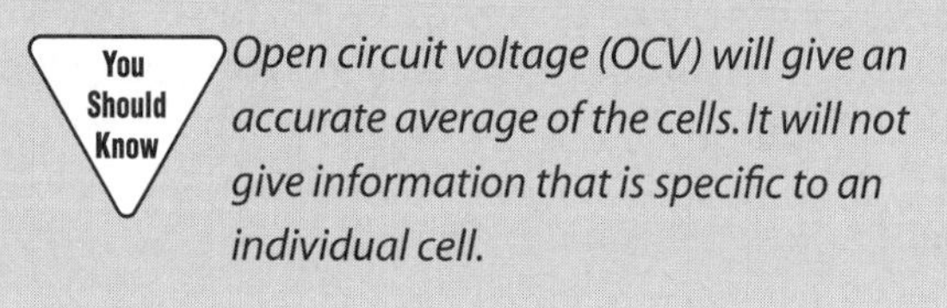

Open circuit voltage (OCV) will give an accurate average of the cells. It will not give information that is specific to an individual cell.

OXYGEN-RECOMBINATION BATTERIES

A recent addition to the battery family is the oxygen-recombination battery **(Figure 11)**. Consisting of a completely sealed unit with no liquid acid, it has the advantage that it can be mounted in any position, even upside down. The reason why liquid acid is not necessary

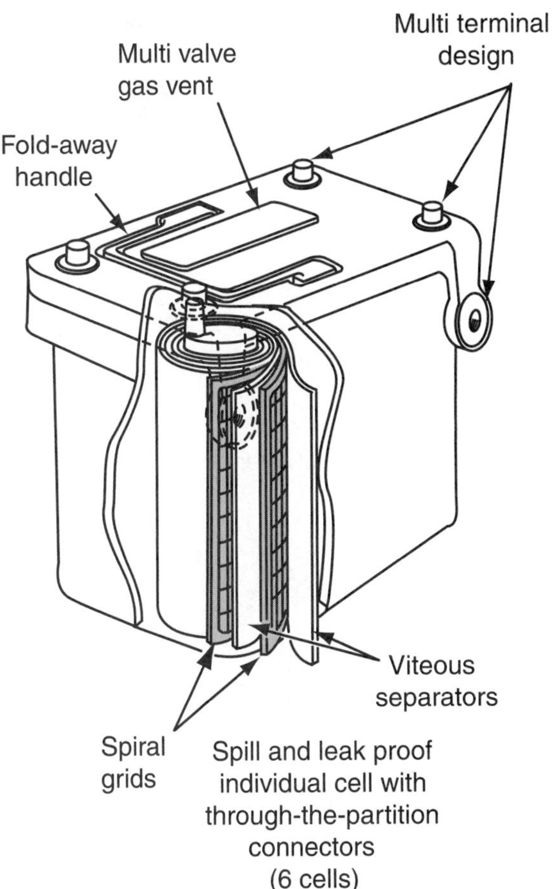

Figure 11. An oxygen-recombination battery with no free acid.

is because of the normal charging cycle. In a typical battery, even one of the calcium variety, once the plates have accepted all the current they need to become fully charged, any additional charging will split the water by electrolysis. The oxygen formed at the surface of the positive plate rises to the top and is vented. The same thing is true for the hydrogen produced at the negative plate. The oxygen-recombination design begins to function when the battery approaches full charge. The oxygen that is generated at the positive plate is sandwiched between compressed glassy separators saturated with electrolyte. This bubbling oxygen moves across the cell rather than upward. When the oxygen contacts the negative electrode, the oxygen becomes chemically absorbed with the lead as oxides. Now, as the charging current arrives at the negative plate, the electrons become occupied in releasing this bound-up oxygen back to the electrolyte rather than producing hydrogen gas. The effect is that the negative electrode never reaches a state of full charge and, therefore, does not generate any hydrogen. With no hydrogen being produced and the oxygen being released into the electrolyte, there is not the need to add water.

Oxygen recombination has been practiced for years in nickel-cadmium batteries. Until recently, it was not successfully applied to batteries in automobiles. The glass separators that act like blotting paper have made the oxygen-recombination method practical for automotive batteries. In all likelihood, this method will become more popular in the future. You will have to rely on the open-circuit voltage to determine the state of charge. Keep in mind that the vehicle's charging system voltage must not exceed 15 volts for any period of time on this type of battery or it will be permanently damaged. Exceeding 15 volts (with a warm battery) produces more oxygen than can be recombined. Without a reserve of electrolyte in liquid form to supply the O_2, the battery's electrolyte is destroyed.

BATTERY SAFETY

The following are safety rules that must be observed when working around or on batteries.

1. Always remove the negative battery cable before doing electrical work on the vehicle. It is imperative that the battery cable be disconnected when removing starting or charging components. It is also important that you realize that disconnecting the battery will eliminate learned strategies, some alarm functions, radio presets, clocks, etc. These will require resetting at the completion of the repair.
2. Never do anything that would cause a spark or arc near the battery. The hydrogen that the battery might be releasing is explosive.
3. Always wear eye protection.
4. Wash your hands after you have handled a battery.
5. Rinse off the vehicle finish if battery acid has been spilled on it.
6. Do not attempt to jump start a frozen battery.

Summary

- Wiring individual cell plates in parallel will increase the cell's ability to produce current.
- Wiring separate cells in series will increase the battery's voltage.
- Primary cells cannot be effectively recharged.
- Secondary cells can be recharged.
- The most common battery capacity rating in use is CCAs.

- The negative battery cable should be removed before doing most underhood work.
- The CCA rating is a 0°F rating.
- Oxygen-recombination batteries have the advantage that they can be mounted in any position.
- Oxygen-recombination batteries cannot be charged at a voltage higher than 15 volts.
- Maintenance-free batteries may have a built-in hydrometer in one cell.

Review Questions

1. Technician A says that a primary cell can be recharged with a battery charger. Technician B says that a secondary cell produces a higher voltage than a primary cell. Who is right?
   A. Technician A only
   B. Technician B only
   C. Both Technician A and Technician B
   D. Neither Technician A nor Technician B
2. Six secondary cells are wired in parallel. Technician A says that their combined voltage will be 2.1 volts. Technician B says that their combined amperage capacity will be six times that of a single cell. Who is right?
   A. Technician A only
   B. Technician B only
   C. Both Technician A and Technician B
   D. Neither Technician A nor Technician B
3. Six secondary cells are wired in series. Technician A says that their combined voltage will be 12.6 volts. Technician B says that their combined amperage will be six times that of a single cell. Who is right?
   A. Technician A only
   B. Technician B only
   C. Both Technician A and Technician B
   D. Neither Technician A nor Technician B
4. A low-maintenance battery usually has grids made of
   A. calcium
   B. antimony
   C. glass
   D. lead dioxide
5. A 50–ampere hour battery is capable of delivering 10 amps for five hours before it is dead.
   A. True
   B. False
6. Engine size is the usual determiner of battery size.
   A. True
   B. False
7. Charging system voltage should not exceed 15 volts for a warm battery (80°F).
   A. True
   B. False
8. Oxygen-recombination batteries frequently require water.
   A. True
   B. False
9. Distilled or demineralized water is the best to add if a cell requires water.
   A. True
   B. False
10. An open-circuit voltage in the 12.6-volt range indicates an overcharged battery.
    A. True
    B. False
11. A 650-CCA battery is specified as original equipment. A replacement battery should be 375 CCA at minimum.
    A. True
    B. False
12. A 120-minute reserve capacity battery can deliver 25 amps for 12 hours before it is dead.
    A. True
    B. False

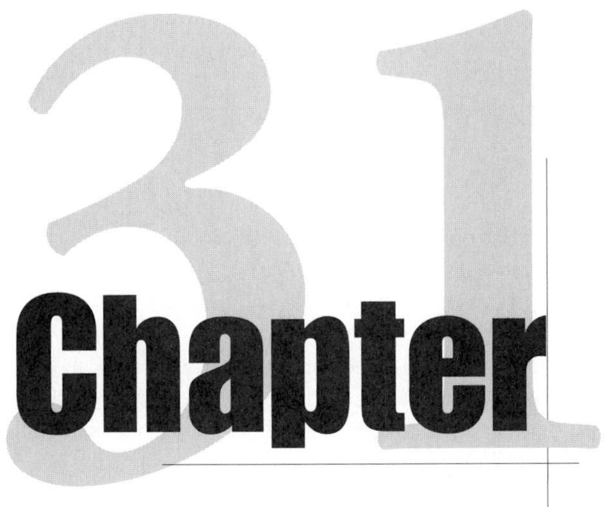

Diagnosing Batteries

Introduction

Now that we have spent some time studying batteries and have a good understanding of how a battery is constructed and how it works, it is time to get down to the important vehicle tests that are commonly done. The ability to be able to predict failure and determine if a battery is the cause of difficulty is the single goal of this chapter. The battery is often blamed for other system problems. It is probably one of the most replaced components in the vehicle, and its cost makes it all the more important that our diagnosis is accurate.

Before we get into testing, we have to discuss and understand the reasons why we test. If the reasons are clear, the testing will be all the clearer. We will also be better equipped to discuss with our customers what we are doing and why we are doing it. Communication with the customer is important if we are to be successful. The best diagnosticians must still have the ability to explain to customers what they are doing. We sometimes have to talk customers into what is best for their vehicles—not necessarily the least expensive, but the best. We take into consideration how the vehicle is used, the weather it will be subjected to, and the age of the vehicle before we make any recommendations. Do not forget to communicate with your customers.

WHY SHOULD WE TEST BATTERIES?

Why do we test batteries? Simply stated, we usually test for one or more of three reasons. The first is to isolate a problem to one component. The second is for preventive maintenance purposes. The third is to predict failure. Let us look at each one of these reasons briefly.

1. *To Isolate a Problem to One Component.* The age of replacing parts on the vehicle until it is repaired is over. It is important to realize that we must make every effort to isolate the customer's problem to a single component. If the battery is bad, replace it. If the starter is bad, replace it, and so on. Do not replace the entire system because the vehicle did not start. Isolate each area until you pinpoint the difficulty. A battery can hide or mask other areas or components that might need repair. The starting system is a great example of this. A battery that does not have the ability to keep its voltage up during cranking will force the starter into drawing more current. As you look at starting current and see that it is too high, you might be tempted to replace the starter only to find that the new one cranks at the same amperage. The battery in this example has masked the problem and caused our guess to be wrong. We never want to guess if we can help it. By isolating the battery, we are testing it independent of the vehicle. When we have finished, we are confident that the rest of the vehicle has not given us improper or false information. Never use the vehicle's starting system to test the battery. If you do, you will be one of those technicians who is frequently replacing both the starter and the battery and hoping that the customer complaint is cured.

2. *Preventive Maintenance.* By testing and cleaning corrosion off battery cables and posts, we are ensuring that the current will have a low-resistance path and will be delivered where it will do the work. Preventive maintenance also means that we will do everything in our power to ensure that the vehicle will start. Does the battery appear to have enough power to make it through the winter? Is there something we can do to prevent breakdown? If we can prevent costly breakdowns, and quite possibly towing bills, our customers will have greater faith in our diagnostic abilities and in us.

3. *Predicting Failure.* Predicting failure goes hand in hand with preventive maintenance. If by our testing we can predict that the battery will not be capable of starting the vehicle during the winter months, we are predicting failure. Explaining this to most customers will make them realize that now is the time for replacing the battery, not one morning when they are late for work, it is cold outside, and their vehicle's battery is dead. Certain tests can be used to determine if the battery will soon develop problems. The object of predicting failure and preventive maintenance is to alert the customer before the problem occurs. Do not wait for the vehicle to not start before you suggest to the customer that testing of the starting and charging system is appropriate. Encourage a semiannual battery starting and charging checkup with some preventive maintenance. This will prevent most problems that normally occur.

A REVIEW OF GENERAL BATTERY SAFETY

Whenever we work around a battery, we must always keep three things in mind: acid, amps, and sparks. Let us be more specific. Batteries are filled with a solution of sulfuric acid. If this solution comes in contact with your eyes, they could be permanently damaged. Skin burns, eaten away clothes, and damaged vehicle finishes are also possible. Many batteries are semisealed nowadays, and some people believe that this eliminates the acid danger. Nothing could be further from the truth. It is true that gluing the top cell coverings down will help keep the acid in if you tip the battery over. But realistically, tipping the battery over is not a common problem. Always wear safety glasses with side shields, as illustrated in **Figure 1**. These will prevent acid from accidentally being splashed into your eyes during battery servicing. Immediately wash battery acid off your skin or the vehicle finish with lots of water. A mild soap will prevent the acid from irritating your skin. Always wash your hands after handling a battery, as a small amount of acid might have remained on the battery and on your hands. Hands with a little acid on them will usually not be irritated. Our eyes will not be as lucky, however. Rubbing even a little acid residue into the sensitive eye can cause damage. If you accidentally get acid into your eyes, you must rinse them out and seek medical attention immediately. Many shops are purchasing eyewash fountains such as that shown in **Figure 2**. The small expense involved can prevent your blindness.

This battery that we are servicing has tremendous potential to deliver very high amperage. Many batteries are capable of delivering 1000 amps or more into a dead short. If we put something across the two posts with little or no resistance, such as a wrench, sparks fly and damage to sensitive solid-state circuits could result. It is possible to short out a battery by accident and burn out a component that was on during the short. Always disconnect the negative cable (ground) before doing anything major under the hood of the vehicle. This precaution is especially important to observe if you are working on electrical components or items near the battery. This will ensure that damage will not be done if a short occurs. The battery cable is not fused and can cause extensive damage and possibly an underhood fire. Remember that flowing amperage will generate light, magnetic fields, and heat. Under dead short conditions, the amount of heat can be extreme.

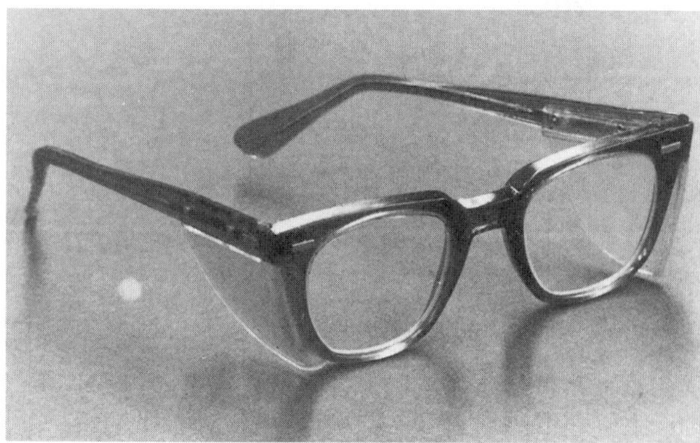

Figure 1. Always wear safety glasses with side shields.

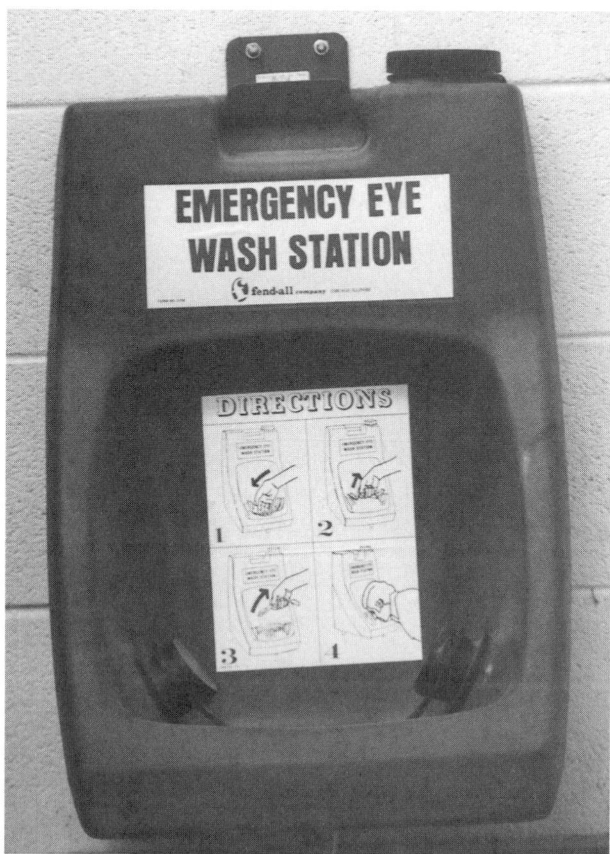

Figure 2. Know where the emergency eyewash station is.

Do not forget that a charging battery can give off hydrogen and oxygen, which are explosive. Frequently, batteries explode right in the face of the person working on them. Never do anything that might cause a spark or flame near the top of a battery. This includes connecting or disconnecting a battery charger with the charger on. A common mistake is to charge a battery and then immediately attach the vehicle's cables with something on in the car. If, for example, the doors are open or the ignition switch is on, current will flow as soon as the cable is touching. This will cause a small arc that could ignite the hydrogen and oxygen still sitting around the top of the battery. The exploding battery will shower acid and pieces of plastic for quite a distance. The least of your worries is the customer who might have been watching the whole episode and quickly decided that you do not know what you are doing. The worst of your worries could be permanent blindness. Doing the job right in the first place will prevent problems.

You Should Know

Always disconnect the ground cable before doing major work on the vehicle **(Figure 3).** *Make sure that no sparks or arcs occur near the battery, especially if it is being charged, and always wear safety glasses.*

TESTING THE BATTERY STATE OF CHARGE WITH A HYDROMETER

In Chapter 30, we discussed a cell's specific gravity and defined it simply as the thickness of the electrolyte. Anything thicker than water has a specific gravity higher than 1.000, and anything thinner than water has a specific gravity lower than 1.000. We can use specific gravity to tell us the state of charge of a cell. A reading of 1.275 (usually read "twelve seventy-five") indicates that the electrolyte in the tested cell is 0.275 times thicker than water. The thickness comes from the sulfuric acid in the electrolyte. As the battery discharges, the sulfuric acid is chemically absorbed into the plates, leaving behind a greater concentration of water. This will lower the specific gravity of the cell. The opposite is also true: as the battery is charged, the sulfuric acid is driven (chemically) out of the plates and into the electrolyte, thickening it and raising its specific gravity. Interpreting the specific gravity of the cells in a battery gives us two bits of important information. First, it tells us whether or not the cell is charged. Second, if we compare all six-cell readings (12.6-volt battery), we can determine the overall battery condition. Think about the importance of this information. Knowing whether or not the battery is charged is important to us. Further testing of a discharged battery does not prove anything and could result in our misdiagnosing a problem. Most batteries that are not charged will fail a load test, tempting us to install a new battery. Two days later, the customer is back with the same problem. If by contrast the battery is charged, we can expect it to pass a load test and will replace it if it does not. We have also received additional information about the charging system. A fully charged battery *usually* indicates a reasonably charging system. Comparison of all of the cells' state of charge or specific gravity gives us the second important bit of information.

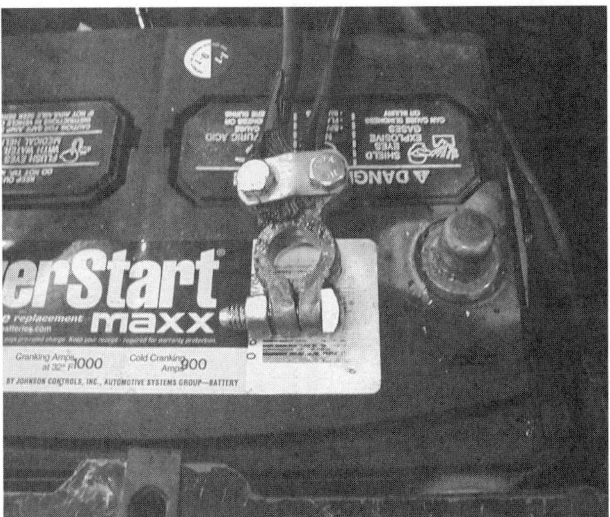

Figure 3. Disconnect the negative battery cable before working.

When the specific gravity of one cell is quite a bit different from that of the other five cells, it might have a tendency to pull the other cells down to its lower level. Remember, the cells in a battery are electrically connected and, therefore, one cell that is not charged as well will pull the other cells down to its level as it attempts to balance. Eventually, all cells will be equally discharged. This can occur overnight, so the customer will have a dead battery in the morning. Battery manufacturers like to have less than a 0.025 difference from one cell to the others. The greater the difference, the greater the likelihood of self-discharge. By the time a battery reaches a difference of 0.075, it will most likely self-discharge during a short enough period of time to cause the customer some inconvenience. Use common sense when faced with different specific gravity readings. The greater the difference, the more you should encourage your customer to invest in a new battery. Keep in mind, though, that not all batteries will begin to self-discharge at a greater than 0.025-point spread. Some batteries with a large point spread last a long time, yet some with very little difference fail the very next day. Explain the options to your customers. If they are hesitant, ask them to come back in a couple of weeks so that you can retest the batteries. If the specific gravity point spread is getting greater, you will not have as much difficulty convincing them to purchase a new battery.

Self-discharge of a battery overnight can often be traced back to a battery with a large point spread. Putting an ammeter on the battery and seeing no current draw with a vehicle that has a dead battery each morning usually indicates a self-discharge condition from cells whose specific gravity point spread is too great.

0.025 or less	Normal
0.025–0.050	Some batteries will discharge.
0.050–0.075	Many batteries will discharge.
0.075–0.100	Most batteries will discharge.
0.100+	All batteries will discharge.

Notice the relative and somewhat vague terms: some, many, and most. You will never be able to predict exactly which batteries will self-discharge. Look for the point spread and inform your customer. As the point spread gets greater, so do the odds that battery failure is approaching. Encourage your customer to purchase the battery before total failure takes place.

Look at **Figure 4**. It is a picture of a common type of hydrometer. Notice the float inside the barrel. The liquid electrolyte will float it up and allow us to read the liquid level against the scale printed on it. It sounds complicated but really is not. If you look at the liquid line on the float and it reads 1.250, what does this mean? It basically means that this cell is 75 percent charged or another way of saying it is one-quarter discharged. **Table 1** will give you the percent charged from the specific gravity. Do not forget that the reading 1.250 is usually referred to as twelve fifty. Recording the specific gravity of each cell will give us the information we want and allow us to make an intelligent recommendation to customers about their vehicles.

Figure 4. A common temperature-compensated hydrometer.

Charge	Specific Gravity	12V Bat	6V Bat
100%	1.265	12.6	6.3
75%	1.225	12.4	6.2
50%	1.190	12.2	6.1
25%	1.155	12.0	6.0
0%	1.120	11.9	5.9

Table 1. OCV and/or specific gravity versus state of charge.

Most of the time when we test batteries with a hydrometer, we are especially interested in whether the battery is charged or discharged. We must realize that the hydrometer will not give us accurate information if the temperature of the electrolyte is either warmer or colder than 80°F. For this reason, most hydrometers are temperature compensated. As the electrolyte gets colder, it will thicken, and we will have to subtract some points to get an accurate reading. For every 10°F, the specific gravity will change by 0.004 of 4 points. Refer to **Figure 5**. What will the actual specific

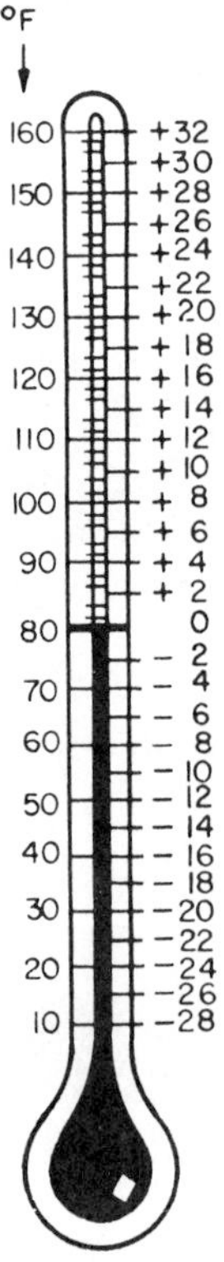

Figure 5. Temperature compensation based on the specific gravity temperature.

Specifications

Battery Size ______________ Engine Cubic Inches ______________

Hydrometer Readings

Cell Number	1	2	3	4	5	6	
Negative Post							Positive Post

Results (Check one.)

_____ All cells are within 25 points and are charged (continued testing).
_____ All cells are equal and are discharged (charge battery).
_____ Cell readings vary more than 25 points (skip load test and advise customer to replace battery.)

OR

Open Circuit Voltage Test

Load battery to 20 amps for about 1 minute before testing OCV.

______________ volts—open circuit (no load)

Refer to the chart and convert either the hydrometer reading or the OCV over to state of charge.

Charge	Specific Gravity	OCV (12.6)	OCV (6.3)
100%	1.265	12.6	6.3
75%	1.225	12.4	6.2
50%	1.190	12.2	6.1
25%	1.155	12.0	6.0
Discharged	1.100	11.9	5.9

Battery Load Test (for batteries at least 75% charged)

Set tester to the following:
1. Tester to STARTING
2. Variable LOAD off
3. Voltage—INTERNAL—higher than battery voltage

Attach the BST as follows:
1. Place large cables of tester clamp over the battery clamps, observing that polarity of the battery and tester are the same.
2. Attach inductive pick-up clamp around either of the BST cables *not* battery cables.

Determine the correct load:

Battery Amp/Hour × 3 = Load or
Cold Cranking Amps ÷ 2 = Load

NOTE: Compare the specific battery to the one actually installed. Load to the LARGER of the two.

Apply the calculated load for 15 seconds, observe voltmeter, and turn load OFF.

_____ GOOD—battery voltage remained above 9.6 volts during load test
_____ BAD—battery voltage dropped below 9.6 volts during load test (perform 3-minute charge test to check for sulfation)

3-Minute Charge Test (only for batteries that fail the load test or are less than 75% charged)

1. Disconnect the ground battery cable.
2. Connect a voltmeter across the battery terminals (observe polarity).
3. Connect battery charger (observe polarity).
4. Turn the charger on to a setting around 40 amps.
5. Maintain 40 amp rate for 3 minutes while observing voltmeter.

_____ GOOD—voltmeter reads less than 15.5 (battery is not sulfated).
_____ BAD—voltmeter reads more than 15.5 (battery is sulfated and should be replaced).

Figure 6. A battery testing worksheet.

gravity of the cell be if you get a 1.220 reading at 50°F? You should have said 1.208 because 50°F calls for a compensation of −12.

$$1.220 - 0.012 = 1.208$$

It is simple. It is also important because we are basing further testing and diagnosis on actual specific gravity readings. They should be temperature compensated. The lab worksheet shown in **Figure 6** should give you all the information you need to effectively analyze a battery with a hydrometer. Testing batteries with hydrometers is an invaluable source of information, assuming that the top caps are not glued down. We cover testing the batteries that we cannot use a hydrometer on later in this chapter.

TESTING THE BATTERY STATE OF CHARGE WITH AN OPTICAL REFRACTOMETER

The **optical refractometer** is an extremely useful device. It is frequently used to determine the condition and concentration of antifreeze in the cooling system. It can also be used to give specific gravity information about the electrolyte. Using a refractometer will give you the same information that the hydrometer gave you. It will, however, be more accurate and easier to use than the hydrometer. Another advantage to the refractometer is its ability to generate specific gravity information from a couple of drops of electrolyte as opposed to the hydrometer that requires a full barrel of electrolyte. If the battery cell caps can be removed, the refractometer is the quickest and most accurate method of determining the state of charge from the specific gravity. Remove a couple of drops of the first cell's electrolyte, as shown in **Figure 7**. Place it on the glass slide of the refractometer and hold the lens up to your eye. The light will refract through the electrolyte and display the specific gravity of the cell in the window, as **Figure 8** shows. Wipe the glass slide clean and repeat for the other cells. An optical refractometer is faster and much more accurate than the hydrometer. The only disadvantage is its cost. You can buy five top-of-the-line hydrometers for the cost of an optical refractometer.

TESTING THE BATTERY STATE OF CHARGE USING OPEN-CIRCUIT VOLTAGE

If the hydrometer readings are impossible to obtain because the battery does not have removable caps, perform an open-circuit voltage test. An open-circuit voltage test is read with no load on the battery and then converted to a state of charge. If the battery has been recently charged, the surface charge on the plates must be drained off. This is done by placing an approximately 20-amp load on the battery for about one minute. Without draining off the surface charge, the battery will appear to be at a higher charge than it actually is. The conversion from OCV to state of charge is reproduced in Table 1. It is also on the worksheet in Figure 6. This is by far the simplest state of charge test, but it does have some problems. When the information it yields is compared to the hydrometer or optical refractometer, the first thing you should realize is that it is an average and is not cell specific. A single cell that is beginning to fail will be easily located and diagnosed using specific gravity readings. It takes a very bad cell to begin to have an impact on the OCV readings. Although OCV is not the first choice, it may be the only choice. If the battery manufacturer has glued the cell covers down, we will not be able to get individual cell specific gravity readings, and OCV is the *only* choice. It will give you enough information to determine the overall state of charge. Do not forget that state-of-charge information is required before doing a load test.

Figure 7. Place some electrolyte on the slide.

LOAD TESTING

Now that you realize how and why we use a hydrometer, optical refractometer, or OCV, we can move on to the most important battery tester: the load tester. This tester has been around for a long time, and its use is extremely important. Remember that battery size is directly related to the number of cubic inches of displacement. The ability to crank the engine is the most important consideration. The battery needs to be able to supply sufficient amperage to crank the engine fast enough to get it started. Cranking speed is important on many electronic fuel-injected vehicles. Without sufficient speed, the computer will not turn on the fuel metering system and the engine will not start. High cranking speed also helps to prevent flooding of the engine. We will be measuring cranking speed in Chapters 33–42, the starting systems section.

Let us go back to the battery. A good supply of current with the voltage remaining high (typically above 9.6 volts) is necessary, so this battery we are using must be able to keep its voltage above 9.6 volts while it is delivering amperage. Testing the battery's ability to do this is called load testing.

Do not confuse load testing with cranking the engine over. Load testing involves pulling a specific amperage out of the battery. It is another example of independent testing rather than relying on the vehicle starting system to do the test. Look at **Figure 9**. It is a picture of a common style of load tester. It is referred to as a **volts-amps tester**, or **VAT**.

With this device, we will be able to test the battery in addition to the starting system and eventually the charging system. It has a large carbon pile included with the amme-

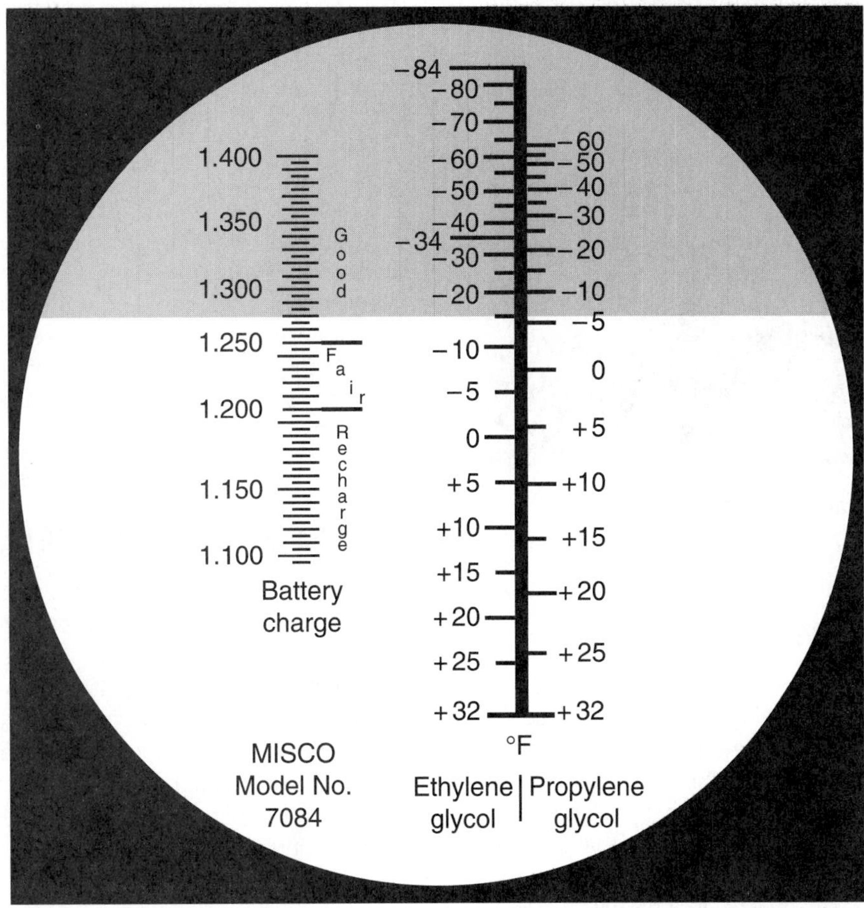

Figure 8. The window shows the state of charge.

Figure 9. A common volts-amps load tester.

ter and the voltmeter. The knob in the center of the tester will allow us to vary the pressure on the carbon pile. The more pressure we apply, the less resistance the carbon will offer to the flow of current. When we attach it to the battery, as illustrated in **Figure 10**, the carbon will draw current out of the battery, producing heat. The ammeter will read the amount of current flowing. This then gives us an adjustable load for battery testing. This device is indispensable because it allows us to test each battery to its own level. The smaller the battery, the lower the load. The larger the battery, the more the load. Applying the load to the battery for 15 seconds and watching that the battery voltage remains above 9.6 volts is the key here.

Load testers are normally connected across the battery to be tested, observing that the polarity of the meter is the same as the polarity of the VAT. In addition, the voltmeter must be on a range that is higher than the battery voltage. Our meter will have to be set on the 18-volt range. Once connected, the voltmeter should read battery voltage before we apply the load. The ammeter should read 0 (zero), indicating that no current is flowing through the load. If it does not read 0 amps, it should be adjusted or zeroed.

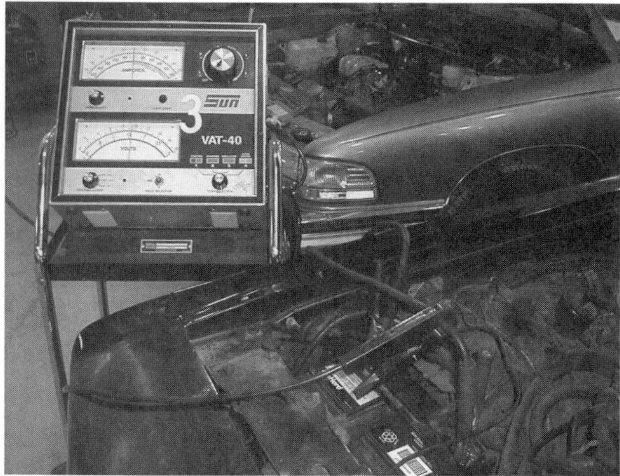

Figure 10. A VAT ready to do a battery test.

Procedure for Determining Battery Load

Now comes the actual loading of the battery. To determine the load, we first determine the size of the battery the manufacturer installed in the vehicle. This number will be our reference and the minimum battery that should be in the vehicle. At no time should you assume that the battery currently installed is the correct one; compare and make sure. The actual load that we will apply will be determined by looking at the size of the battery installed and the manufacturer's original equipment. We load to the higher or larger battery. For example, let us assume that the manufacturer had originally installed an 800-CCA battery in the vehicle. The owner has since installed a 650-CCA battery. We would load it to the 800-CCA level, not the 650-CCA level. If, on the other hand, the owner has installed a 1000-CCA battery, we would load down to the 1000-CCA level. Always load down to the vehicle manufacturer's battery level as the minimum. Typically, the manufacturer installs the minimum required size battery. We try to convince consumers, however, to install larger batteries when the original equipment fails. Load down to the larger of the two choices.

The load applied will be determined by the size of the battery as follows. The two most common methods of rating batteries are used in the following examples. If the battery is rated in CCAs, divide the rating by 2 to get the load. An 800-CCA battery will be loaded down to 400 amps for 15 seconds with the battery voltage remaining above 9.6 volts during the load.

$$800 \div 2 = 400 \text{ amps}$$

If the battery is rated in ampere hour to determine the load to be applied, multiply the ampere hour by 3. In other words, a 100–ampere hour battery will be loaded down to 300 amps for 15 seconds with the battery voltage remaining above 9.6 volts during the load.

$$100 \times 3 = 300 \text{ amps}$$

We do the same procedure with a battery of any size: Three times the ampere hour or half the cold cranking amps.

The worksheet reproduced in Figure 6 shows that we first determine the specific gravity cell by cell. If the readings show that the battery is reasonably charged (75 percent or more), we load test. If the battery is sealed, we rely on OCV to determine if the battery is 75 percent or more charged. By looking at the results of the hydrometer or OCV and the load test, we have a good idea of the condition of the battery. With the results, we can make an intelligent recommendation to the customer.

THREE-MINUTE BATTERY TEST

The load test that we have just discussed is the industry standard. If the battery can pass the load test, the odds are it will be capable of cranking the engine under most conditions. This, of course, assumes that the battery is adequately sized for both the engine size and the outside temperature. As a battery ages, a deterioration called **sulfation** might occur. Sulfation is a chemical action within the battery that interferes with the ability of the cells to deliver current and accept a charge. It is frequently found in batteries that have been sitting unused for extended periods of time.

The three-minute charge test is a reasonably accurate method for diagnosing a sulfated battery **(Figure 11)**. Usually, it is done if the battery is suspected of being sulfated or has failed the load test. When a battery fails the load test, it is not necessarily the battery's fault. It is possible that the battery has not been receiving an adequate charge from the vehicle's charging system over an extended period of time, leading to sulfation. The three-minute charge test

Figure 11. Setup for a three-minute charge test.

determines the battery's ability to accept a charge and is easily accomplished.

Connect the shop battery charger to the vehicle's battery after first removing the ground cable from the battery. On-board computers for fuel and ignition control are sensitive to higher-than-normal voltages. Your battery charger might raise the system voltage above the level that the computer likes. Your customers will not be happy if your battery testing burns out their computers. This is why we disconnect the battery ground cable first. Put a voltmeter across the terminals. Do not forget that the meter is polarity sensitive. Turn on the charger to a setting of approximately 40 amps. Maintain this rate of charge for three minutes. Immediately after the 3 minutes, observe the voltmeter with the charger still charging. If the voltmeter reads less than 15.5 volts, the battery is not sulfated. If the voltmeter reads more than 15.5 volts (7.75 volts for a 6-volt battery), the internal resistance of the battery is too high as a result of sulfation or poor internal connections to accept a full charge. The battery should be replaced. No further testing is necessary. The reason for doing the three-minute charge test should be reviewed again. This test will pick out a sulfated battery that might be the result of an undercharge condition on the vehicle. A complete charging system test should be performed on the vehicle. The new battery might develop the same sulfation if the cause of the undercharge condition is not eliminated.

When you start with a state of charge test and finish with the load test, you have all of the information that you need to condemn the battery. Recognize, however, that indirectly you are also getting information on the condition of the charging system and the general condition of the electrical system. A fully charged battery should be capable of passing a load test. The combination of the hydrometer, optical refractometer or OCV, and the load test yields more than enough information on which to base a decision about the battery's future. Remember that a load test on a discharged battery proves nothing. Something as simple as a loose fan belt can cause the charging system to never fully charge the battery. This battery will probably fail the load test. Replacing it will cure the no-start that brought the vehicle into your shop but will cause the customer to not be pleased with your diagnosis when the *new* battery runs down and causes another no-start. Using the three-minute test is not a substitute for charging a weak battery before load testing it. Also keep in mind that specific gravity can be useful in predicting a future self-discharging battery. A large point spread is almost always associated with a battery that runs itself down. Low open-circuit voltage readings after the battery has been sitting a couple of hours are also indicative of a weak cell. The weak cell has a tendency to pull the rest of the cells down to its level. You might be forced into charging a battery, disconnecting it, and letting it sit overnight. When you check it in the morning, if the battery is dead, replace it. It is self-discharging. The procedure might not sound too convenient, but it certainly is conclusive.

BATTERY CCA TESTING

In recent years, some of the automotive equipment manufacturers have been able to develop methods of testing the available capacity of a battery through the use of a computer. The majority of the testing systems apply a calculated load to the battery for a specific period of time while the voltage is graphically plotted. The resulting graph of the dropping voltage is compared with those in the computer's memory. The graphs in memory are the actual voltage changes that specific CCA batteries will experience. In this way, the voltage of the tested battery compared with the stored voltage graphs allows the computer to accurately test the available CCA. This test is an example of computer technology being applied today to help the technician. The ability to be able to tell the customer how many CCAs are available allows the technician to help guide the repair and predict failure. Typically, the CCAs of most batteries will drop as the battery ages, even though a standard load test will be passed. Now the technician can have additional information available to help predict failure rather than waiting until the failure has already taken place.

Summary

- Batteries should be tested periodically because they have a limited life.
- Specific gravity is an important predictor of battery life.
- Specific gravity readings should not vary excessively cell to cell.
- Either a hydrometer or an optical refractometer can give cell-by-cell specific gravity readings.
- Battery state of charge can be determined by looking at specific gravity or OCV.
- OCV testing is usually done only on sealed batteries.
- Batteries that are 75 percent or more charged may be load tested.
- Load testing involves drawing a specific amount of current for 15 seconds and watching the battery voltage.
- A three-minute charge test may be done on batteries that fail the load test.
- A three-minute charge will identify if a battery is sulfated.

Review Questions

1. Technician A says that a 12.6-volt OCV reading indicates that a battery needs recharging. Technician B says that a 1.265 hydrometer reading indicates that a battery needs recharging. Who is correct?
   A. Technician A only
   B. Technician B only
   C. Both Technician A and Technician B
   D. Neither Technician A nor Technician B
2. A 450-CCA battery is being load tested. The correct load will be
   A. 1350 amps
   B. 450 amps
   C. 225 amps
   D. 990 amps.
3. Technician A says that during a load test the battery voltage must not fall below 9.6 volts. Technician B says that the load applied should be three times the ampere hour rating. Who is correct?
   A. Technician A only
   B. Technician B only
   C. Both Technician A and Technician B
   D. Neither Technician A nor Technician B
4. Hydrometer readings are between 1.260 and 1.275. Technician A says that a load test is necessary to determine the rest of the battery information. Technician B says that a load test is not necessary. Who is correct?
   A. Technician A only
   B. Technician B only
   C. Both Technician A and Technician B
   D. Neither Technician A nor Technician B
5. Hydrometer readings of 1.200 to 1.220 are taken. Technician A says to load test and then do a three-minute charge test. Technician B says to recharge and then load test. Who is correct?
   A. Technician A only
   B. Technician B only
   C. Both Technician A and Technician B
   D. Neither Technician A nor Technician B
6. Technician A says that the three-minute charge test can help identify a sulfated battery. Technician B says that a sulfated battery will usually fail a load test. Who is correct?
   A. Technician A only
   B. Technician B only
   C. Both Technician A and Technician B
   D. Neither Technician A nor Technician B
7. A three-minute charge test is being done on a battery that has failed a load test. Technician A says to set the charger to 40 amps. Technician B says that the voltage should stay below 15.5 volts. Who is correct?
   A. Technician A only
   B. Technician B only
   C. Both Technician A and Technician B
   D. Neither Technician A nor Technician B
8. Negative battery cables are removed first
   A. to help prevent open circuits
   B. only on side terminal batteries
   C. to help prevent short circuits to ground
   D. all of the above
9. A load of 325 amps for a 650-CCA battery is the correct test load.
   A. True
   B. False
10. During load testing, 10.6 volts indicates the need for a new battery.
    A. True
    B. False
11. A three-minute charge test can be used instead of a hydrometer to test a cell's state of charge.
    A. True
    B. False
12. Having to add water to a low-maintenance battery can be an indication of a higher-than-specified charging voltage.
    A. True
    B. False
13. Acid is added to a battery that has a low electrolyte level.
    A. True
    B. False
14. Fast charging a battery (higher than 15 volts) will increase its shelf life.
    A. True
    B. False
15. Sulfation is increased if a battery is left in a discharged state for extended periods of time.
    A. True
    B. False
16. Charging batteries can give off hydrogen and oxygen, which are explosive.
    A. True
    B. False

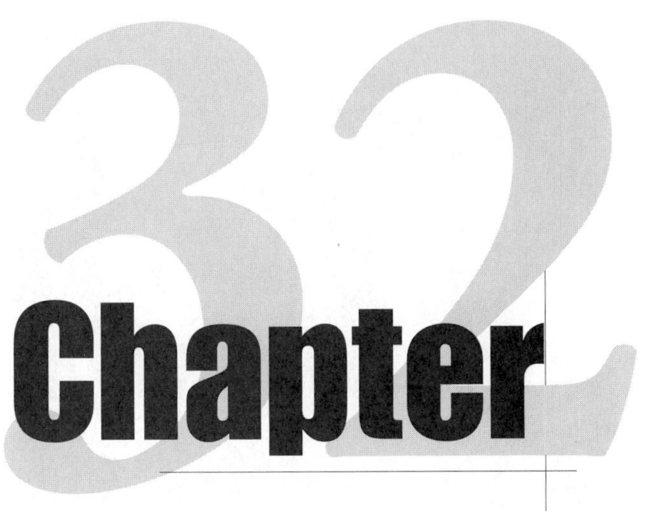

Chapter 32 Servicing Batteries

Introduction

Battery servicing is rather limited. Most problems found will result in the technician installing a replacement battery. As a result, we look at only three items within this chapter: routine maintenance, charging a dead battery, and jump-starting a vehicle. Little else is needed to keep the vehicle starting battery in good shape. There is a direct correlation between battery life and the charging system, but we will save that for Chapter 45.

BATTERY MAINTENANCE

Cleaning the battery is probably one of the simplest and yet most important things that a technician can do to increase the life of a customer's battery. We should clean not only the cables but also the top. Notice the voltmeter reading in **Figure 1**. It shows a difference in pressure across the top of the battery. Current must be slowly draining across the top of the battery. Years ago, the tops of most batteries were made of a soft tar-like substance. As the battery charged and discharged and the gases were vented from the cells, the tar absorbed chemicals and dirt that were conductive. This allowed small amounts of current to flow across the top of the battery and gradually discharge the cells. All batteries currently being produced have hard plastic tops covering the cells. This design greatly reduces the current drain across a dirty top, but it does not prevent it completely. For this reason, we should clean and neutralize any acid on the top of the battery. Make a paste of baking soda and water, and brush it on the top of the battery. After it bubbles and fizzes for awhile, rinse it off with large amounts of water. This procedure will ensure that current will not gradually drain off across the top. Our goal is to keep the battery fully charged once the vehicle is turned off, especially if the vehicle will not be run for a while. Winter storage of a vehicle is a good example of the necessity of a clean top. If during a month or more of storage a small drain occurs across the top of the battery, it is likely that the battery will sulfate and possibly be ruined by spring.

Battery tops were originally made of a tar-like substance that greatly increased the self-discharge rate of the battery.

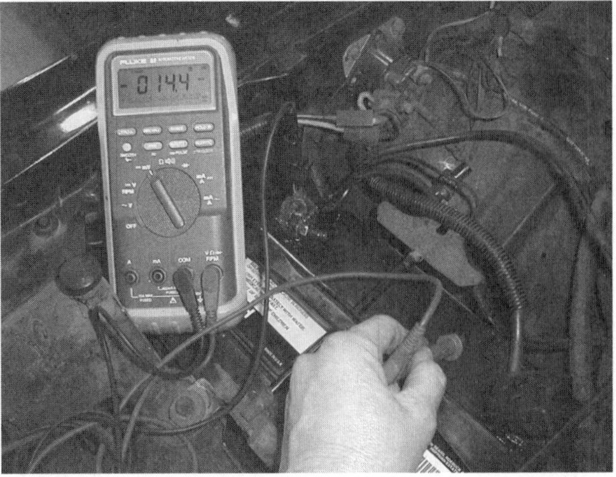

Figure 1. Current may flow across a dirty battery top.

Cable ends need more frequent care. Their connection is extremely important. You will see, in Section 8, Starting Systems, that any corrosion or resistance between the battery and the cable end will increase the amperage that will be needed to crank the engine. This increased amperage is harder on the entire system, including the battery. For this reason, we should remove the cable ends from the posts and clean the electrical contact surfaces as often as is necessary to ensure that the connection is tight and corrosion free. We usually remove the negative cable first. The reason for this is simple. When we use a wrench on battery bolts, the wrench will be at the posts' electrical potential. With the whole area around the battery usually being ground potential, it makes more sense to remove the ground cable first. If we accidentally come in contact with the frame or body with our wrench, no current will flow. If, however, we are removing the positive first and we touch the surrounding metal, sparks fly. Neither the battery nor an expensive wrench will like to have a couple of hundred amps flowing through it. Once the negative cable has been disconnected, the battery is now isolated from the vehicle and, if the wrench comes in contact with the ground, current cannot flow. Always remove the negative cable first and you will save yourself from exploding the battery, embarrassing yourself in front of a customer, and spending money replacing tools that you have welded.

Once the cable is off the post, the post and the cable end can be cleaned with a battery brush, as shown in **Figure 2** and **Figure 3**. Once the cable and post are clean, use a little of that baking soda paste to finish the job. It will neutralize any remaining acid and help to prevent corrosion. If the battery that you are working on has side terminals, they should be removed, cleaned, and replaced. Just because you cannot see the corrosion forming under the protective collar does not mean it is not taking place.

Figure 2. A cleaner brush will clean the post.

Figure 3. A cleaner brush will clean the cable end.

Figure 4 shows the area under the cable end. A special brush is used for a side terminal battery.

In most cases when we find a badly corroded cable end, it can be cleaned and reinstalled with a new battery bolt and be as good as new. However, we sometimes find that, through repeated removing and replacing, the lead in the cable end cracks or stretches to the point where it can no longer be tightened around the post. At this point, you have two options—either to replace the entire cable or to replace the cable end. Cables do not last forever, so if a voltage drop test indicates excessive resistance inside the cable, do not hesitate to replace it. Starting system voltage drop testing is outlined in Section 8, Starting Systems. If the voltage drop across the cable is acceptable and there is sufficient length to the cable, replacing the end is a good solution to the problem, as long as it is done correctly. First, cut back the insulation until no corrosion is present. You will have to cut off all the corroded section of the cable. Clean, corrosion-free copper is necessary if the repair is to last.

Figure 4. Cleaning a side terminal battery.

Spread the cable strands out to inspect them, making sure that no corrosion exists as shown in **Figure 5**. Corrosion that remains will spread to the rest of the cable end in a short time, necessitating another repair. Once you are assured that the strands are clean, twist them together. Solder the end together to make a good connection **(Figure 6)**. A propane torch can be used to get the end hot enough to melt the solder. Once it is hot, allow the solder to flow up the cable about an inch. This soldering of the end is not always done out in the field. However, if done, the new cable end will be as good as the original equipment. Put a new cable end in place. Do not forget any small wires that were originally part of the end, and tighten the two small bolts down until the cable end is tightly held in place on the cable and the job is done correctly.

If the cable end does not have to be replaced, it should be removed, cleaned, and reinstalled. The use of a battery cable remover is recommended because it will push straight down on the post rather than applying pressure to the side. Side pressure from a screwdriver or wrench prying under the cable end might break off the post and ruin an otherwise good battery. Once the cable ends have been removed, clean both the post and inside cable clamp until they are shiny and bright. Once the battery top, the post, and the cable clamp are clean, reinstall and tighten down the clamp bolt. If you have trouble fitting the cable end over the post, spread the clamp slightly. Just spread it enough to allow it to slip easily over the post. It is also a good trade practice to have extra battery clamp bolts on hand because you will find, frequently, that they cannot be reused because of corrosion. Replacing them as necessary ensures a tight connection. All the current that the starter will draw must come through these connections. They *must* be tight and corrosion free.

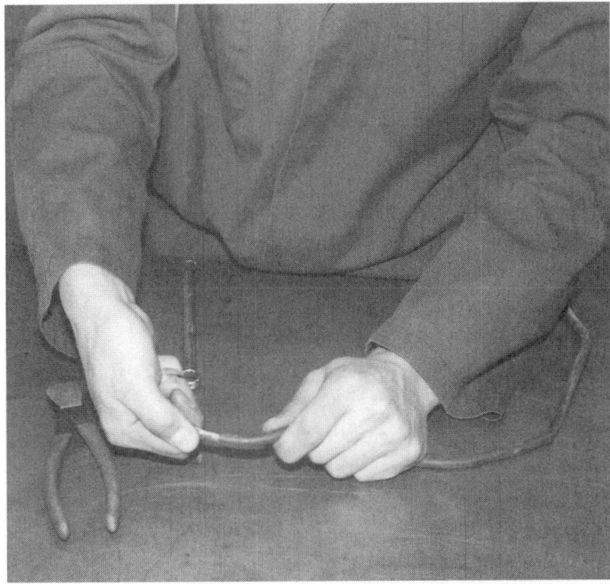

Figure 5. Spread the cable strands to inspect them for corrosion.

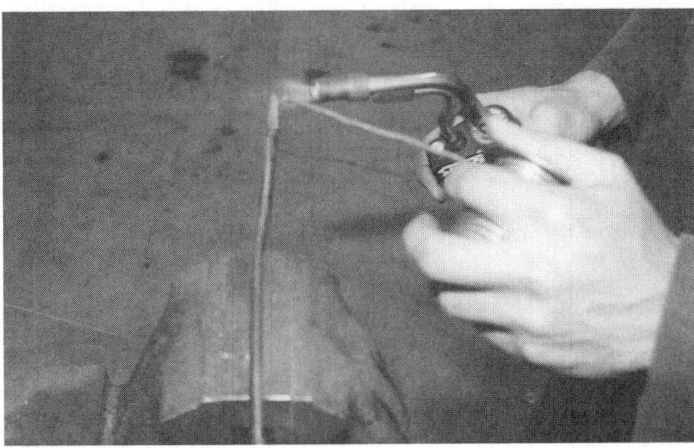

Figure 6. Solder the cable end.

Side terminal batteries are a little less likely to develop corrosion but still must be cleaned periodically. Follow the same procedure previously outlined with one exception: always use a torque wrench on the cable end bolt. Overtightening will strip out the threads in the battery, ruining it. The bolt threads into a lead strap that is easily stripped out. Be careful. A torque of 60–90 inch-pounds is recommended. Notice that is inch-pounds, *not* foot-pounds. If a torque wrench is not available, hand tighten the bolt and then tighten about one-quarter of a turn more. With clean threads on the bolt, the one-quarter turn past hand tight will get you close to the recommended torque.

Filling batteries with distilled water is a simple task that should not be overlooked. Remember that by-products of charging most batteries are hydrogen and oxygen. Water, or H_2O, replaces the liquid lost through electrolysis. Usually, we fill cells to the "split ring." This means that water is added until it just touches the bottom of the fill hole. Overfilling will cause the electrolyte to bubble out during charging.

Whenever we have to add water to a modern battery, we should be asking ourselves, why? Why does the battery need water? Has it been in service for a long period of time? Has it been overcharged by either a battery charger or the vehicle's charging system? We should realize that under most normal conditions batteries of the calcium-lead variety require little water during their lifetime. Usually the water placed in the battery during manufacture will be sufficient for years, unless an overcharge condition exists. The old lead-antimony batteries used, by comparison, large quantities of water. Adding water to a battery cannot be accomplished with a sealed battery. If the top is not sealed and the cells are a bit low on water, by all means refill them with distilled or demineralized water. Then look for the cause of the battery being low on water. Remember that overcharging is the most common cause.

Batteries are heavy and awkward to carry. Do not attempt to carry one if you have a bad back. Also, do not carry one against your body because this might leave

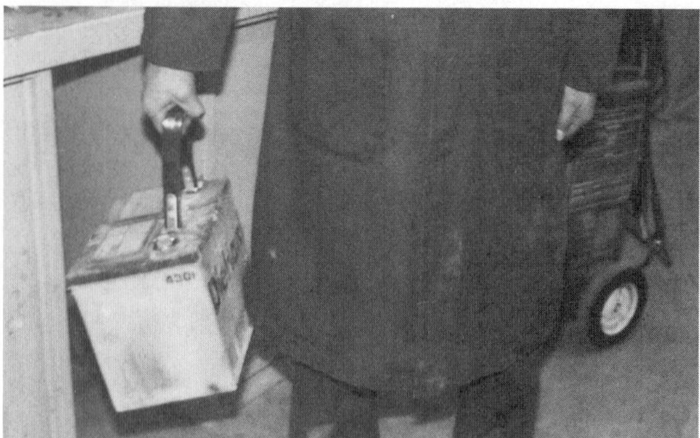

Figure 7. Use a battery carrier or strap.

behind some acid on your clothes, ruining them. Use a battery carrier or battery strap, as shown in **Figure 7**.

CHARGING A DEAD BATTERY

If an open circuit voltage test or a hydrometer indicates that a battery needs to be recharged, connect the charger positive to positive and negative to negative. It is always a good idea to remove the negative battery cable from the post before charging. This will isolate the vehicle from the charger and prevent the charger from damaging any microprocessor or electronic components. Battery chargers should be checked frequently to make sure that they are not producing too high a voltage while they charge the battery. Many battery manufacturers recommend that no higher than 15 volts be applied during charging of a warm battery. If the charging voltage is approximately 15 volts and the vehicle has been disconnected, you need not worry about damaging either the battery or the vehicle systems during charging. The use of a slow charger (4–6 amps) is the easiest on the battery but not always the most practical solution to the customer who needs the vehicle *now*. Customers want their vehicles as quickly as possible.

There are two different methods to determine charging rate. The easiest method is to attach a voltmeter across the battery while it is being charged. The charging rate is low enough if the voltage does not exceed 15 volts (assuming that the battery is at room temperature). If the voltage exceeds 15 volts, simply reduce the charging rate until the meter is just below 15 volts. Keeping the voltage at approximately 15 volts during charging will ensure the quickest charge for the customer and will also be the safest charge rate for the battery. The second method involves a fast charge followed by a slow charge (less than 5 amps). Many shop chargers have timers for fast charging that automatically reduce the charging rate to a slow charge once the time period (usually one hour) has passed. Do not overcharge a battery just because the customer is in a hurry. If you smell rotten eggs around a charging battery, it is being charged too fast. Reduce the charging rate to eliminate the gassing.

Figure 8. Sparks near the top of a charging battery caused an explosion.

More small chargers are becoming "smart chargers," which means that they monitor the battery charging voltage and automatically reduce the rate as the voltage increases. No matter what method you use to charge the battery, keep in mind that as the battery is charging it will produce hydrogen and oxygen in small quantities. Hydrogen and oxygen are highly explosive. Avoid any sparks or arc near the top of a charging battery or the battery might explode, as **Figure 8** shows.

> **You Should Know** *Hydrogen and oxygen are explosive. Do not disconnect the charging cables with the charger on or you might blow the battery up.* Always wear safety glasses.

JUMP-STARTING A BATTERY

Jumping a dead battery from a live, charged battery is frequently done. Many times customers leave their lights on or crank their vehicles until the batteries are discharged to the point where recharges or jump-starts are necessary. The procedure is not difficult, but it is imperative that it be done correctly or damage will occur.

Make sure that the two vehicles are not touching. Connect the positive jumper cable to the charged battery's positive post, then to the discharged battery's positive post.

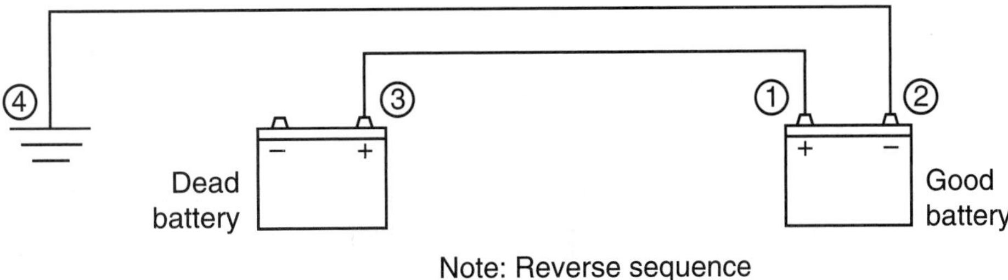

Figure 9. Jump-start sequence.

Connect the negative jumper cable to the charged battery's negative post. Connect the other end to the engine block, as far away from the battery as possible. Make the final connection on the discharged battery's side of the circuit. This final connection will produce a spark. This spark should not occur anywhere near the charged battery because the spark might cause the battery to explode. Once the vehicle is started, disconnect the negative cable from the block immediately. **Figure 9** shows this procedure. Under no circumstances should you make your final connection or your first disconnection at the charging battery. Jump-starting batteries to get a vehicle started should not be attempted if the battery has been discharged overnight in subfreezing temperatures. A discharged battery freezes at 19°F above zero. Trying to jump-start a frozen battery will sometimes result in the battery case exploding. The hydrogen and oxygen cannot get through the ice at the top of the cells, and the resulting pressure can blow the side out of the case. Usually freezing the battery has ruined it anyway, and so it makes little sense to try to jump-start it.

In this age of microprocessors and electronic gadgetry on vehicles, you must be aware of system polarity and system voltage. Reversing the polarity or using a starting unit whose voltage is too high could do thousands of dollars of damage to the vehicle systems. Under no circumstances should you jump-start a computerized vehicle using anything except another vehicle. If you decide to charge a battery using another vehicle, disconnect the negative cable to isolate the vehicle systems from the charging battery. Many manufacturers void the vehicle warranty if it was jump-started using anything other than another vehicle. We must be constantly aware of polarity and voltage when we work around modern vehicles.

Summary

- Cleaning the top of a battery helps to reduce or eliminate discharging.
- Cleaning cable ends and posts will eliminate voltage drops, which will increase cranking current.
- New bolts should be used when reinstalling battery cable ends.
- Side terminals should be torqued to 60–90 inch-pounds.
- Only distilled or demineralized water should be added to a battery.
- During charging, try to maintain a voltage of approximately 15 volts.
- Reduce the charging current if the battery voltage exceeds 15 volts.
- Never cause a spark or arc near the top of the battery. It could cause the battery to explode.
- When jump-starting a vehicle, connect the dead battery's negative cable last.
- When jump-starting, always wire the two batteries in parallel, never in series.
- Never attempt to jump-start or charge a frozen battery.

Review Questions

1. Battery tops are cleaned periodically to
   A. look nice
   B. prevent sulfation
   C. prevent slow discharge across the top
   D. eliminate excessive resistance that will increase starting amperage draw
2. Negative battery cables are removed first
   A. to help prevent open circuits
   B. only on side terminal batteries
   C. to help prevent short circuits to ground
   D. all of the above

3. A battery is being jump-started from a running vehicle. Technician A states that the negative cable of the running vehicle should be connected last. Technician B states that the positive cable of the dead vehicle should be disconnected first. Who is correct?
   A. Technician A only
   B. Technician B only
   C. Both Technician A and Technician B
   D. Neither Technician A nor Technician B
4. Jumper cables are connected to the vehicle's dead battery first.
   A. True
   B. False
5. Having to add water to a low-maintenance battery can be an indication of a higher-than-specified charging voltage.
   A. True
   B. False
6. Acid is added to a battery that has a low electrolyte level.
   A. True
   B. False
7. Fast charging a battery (higher than 15 volts) will increase its shelf life.
   A. True
   B. False

Starting Systems

SECTION OBJECTIVES

At the conclusion of this section, you should be able to:

- Understand how magnetism and motion are related.
- Understand how a motor is wired to deliver starting power.
- Recognize how the various types of starters function.
- Recognize the differences in starter design.
- Diagnose starting problems involving starter current.
- Understand the relationship between starter cranking speed and starting current.
- Measure voltage drops in the starting system.
- Trace the starting circuit on a wiring diagram.
- Understand and be capable of diagnosing the TR switch.
- Use the 12-volt test light to diagnose starting control difficulties.
- Measure cranking speed.
- Remove and replace a starter assembly.
- Measure pinion clearance and adjust it if necessary.

Interesting Fact

The starting system is frequently referred to as an educated short circuit because of the extreme levels of current that it handles. There is no circuit on the vehicle that draws as much current and does as much work.

Starting Systems

Introduction

Up to this point, we have spent considerable time studying electricity without looking at any specific electrical system. It is now time to put many of the principles discussed into practice as we analyze the starting system. Many technicians think of it as the "first" system, because it is the system that must function correctly before any other system is called upon to do its job. Correctly functioning ignition, charging, and fuel systems would not be of much value if the starting system could not do its job first. We discuss some motor principles involving electromagnets and then discuss how we put magnetism to work in cranking the engine. We end with discussing common systems. The lab manual will allow you the opportunity to test the starting system on the vehicle. Diagnosis and repair of current starting systems is not difficult, but it is important. Most consumers take the system for granted because they are used to the instant response they get when the key switch is turned to the start position. High reliability is achieved with minimum maintenance and diagnosis done at the professional technician's level. Keep the principles in mind and the rest is easy.

A TYPICAL STARTING CIRCUIT

Let us look at a typical starting or cranking circuit briefly before we zero in on the particulars. Refer to **Figure 1**. When the consumer turns the key switch, a small amount of current flows from the battery through the TR switch or the **clutch switch (CS)** over to the starter solenoid. This current is called **control current** and is used to pull a gear into mesh with the engine's **flywheel** and close a large switch between the battery and the starting motor. Current now flows through the motor and cranks the engine. Releasing the key switch stops the flow of control current and allows a large spring to pull the gear out of mesh, thus allowing the engine to spin or run under its own power. This sequence is repeated each time the key switch is turned. We will look at each component of this system in turn. As we do this, keep in mind the underlying design principle. Starting systems are designed to deliver great amounts of horsepower for short periods of time. They are designed to handle very large amounts of current. Some technicians refer to them as "educated short circuits." Their resistance is so low that they are, for all practical purposes, a dead short to ground. Connections, cables, switches, solenoids, and batteries must be able to handle as many as 1000 amps under adverse conditions. Understanding how this works starts with an understanding of magnetism.

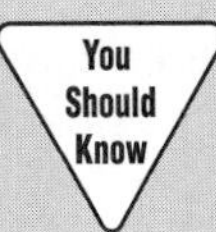

The typical starting system should be capable of cranking the engine over at a minimum of 200 rpm.

MAGNETISM

One of the most familiar and yet misunderstood principles is magnetism. A magnet is a material that will attract iron and steel, plus a few additional materials. It will not attract plastic, aluminum, wood, and paper, among others. The magnetic attraction is stronger at the two ends, or poles. A magnet that is allowed to spin freely will align itself north and

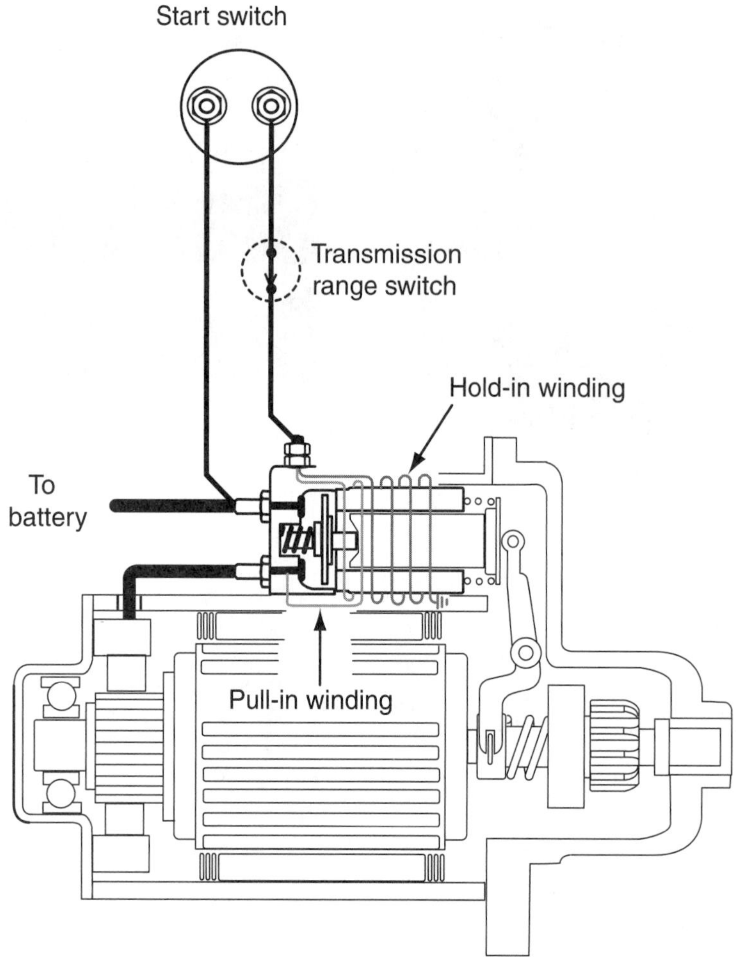

Figure 1. The basic cranking circuit.

south. The end facing north is called the north pole, and the end facing south is called the south pole. A hiker's compass points to the north because the compass needle is a small magnet. As a child, you perhaps played with two magnets and discovered that like poles (two north [N] poles or two south [S] poles) would repel or push each other away and unlike poles (one N pole and one S pole) would attract each other. These principles are best illustrated in **Figure 2**, which shows the magnetic fields that surround a magnet. You can easily duplicate these fields by placing a piece of paper over a magnet and sprinkling iron filings over the paper. The filings will align themselves with the force fields of the magnet. Putting unlike poles together allows the fields to join, whereas putting like poles together forces the fields to split, as shown in **Figure 3**. This magnetic field attraction and repulsion are what will turn our engines over.

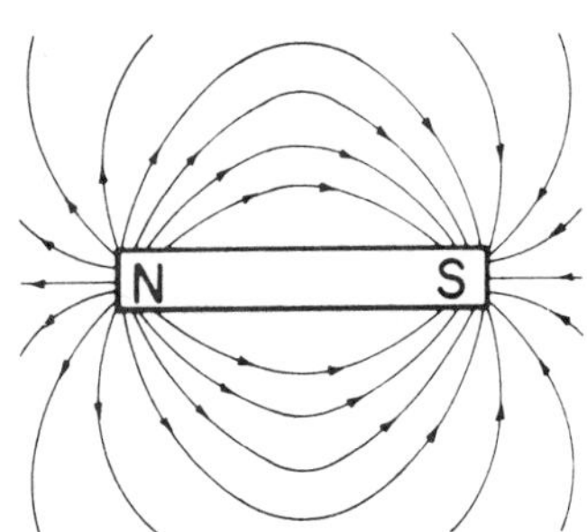

Figure 2. Magnetic fields surround a magnet.

ELECTROMAGNETISM

The strength of a magnetic field can be greatly increased through the use of electricity. A magnetic field is always around a conductor that is carrying current. The inductive current probe pickup on our battery tester was able to convert this magnetic field into an ammeter reading. Any time current is flowing through a conductor, a magnetic field will surround it. With this principle in mind, we can make a simple, but powerful, electromagnet. Look at **Figure 4**. It shows the weak magnetic field surrounding a single conductor. In other words, this conductor has a

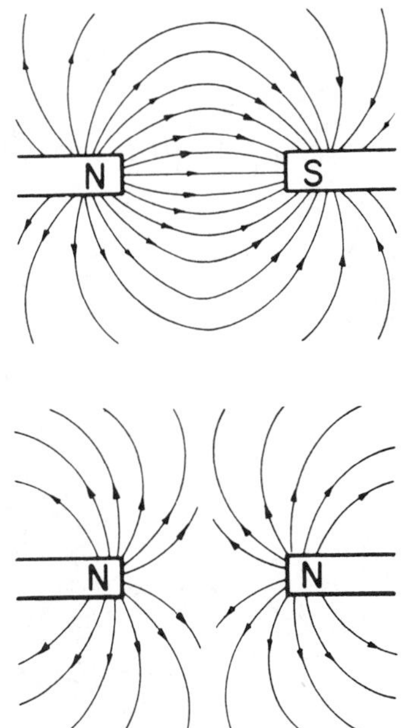

Figure 3. Like poles attract; unlike poles repel.

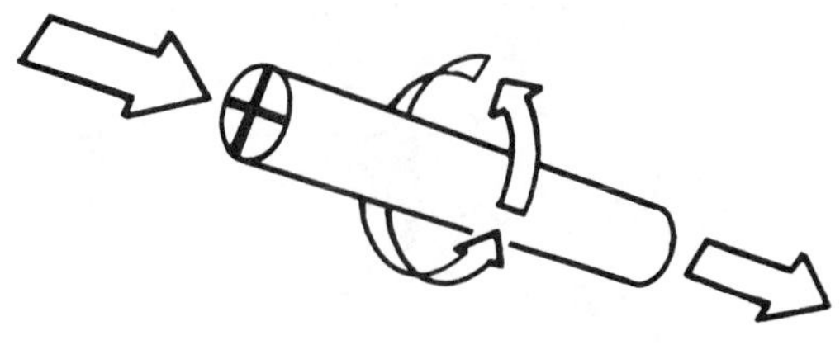

Figure 4. A weak magnetic field surrounds a current-carrying conductor.

low flux density. Flux density is the concentration of the lines of force. If we place two conductors side by side, as shown in **Figure 5**, we will double the flux density and double the magnetic field. In a similar manner, increasing the amount of current flowing through the conductors will also increase the strength of the field. As we progress through this section of the text, keep in mind two magnetic field principles:

1. The strength of a magnetic field is related to the current flowing through the conductor. Increasing the current flow increases the strength of the field (flux density).
2. Adding more current-carrying conductors will also increase the strength of the magnetic field (flux density).

Many starting systems operate on both principles. Increasing the amount of current and increasing the number of conductors will greatly increase the strength of the magnetic field.

Let us look at these principles in detail. If we wish to increase the strength of a magnetic field, we can form a single wire into a loop, as shown in **Figure 6**. Notice that looping the wire, as shown, will double the field strength (flux density) where the wire is running parallel to itself. The direction of current flow is the same: from left to right in both sections of the wire. Magnetic lines of force cannot cross one another. The iron filings proved this, so the magnetic lines of force at the bottom of the loop will double in density. As we add more loops, the fields from each loop will join together and increase the flux density. **Figure 7** shows how these lines of force will join and attract each other. This attraction is so strong that insulators must be placed between the conductors to prevent them from

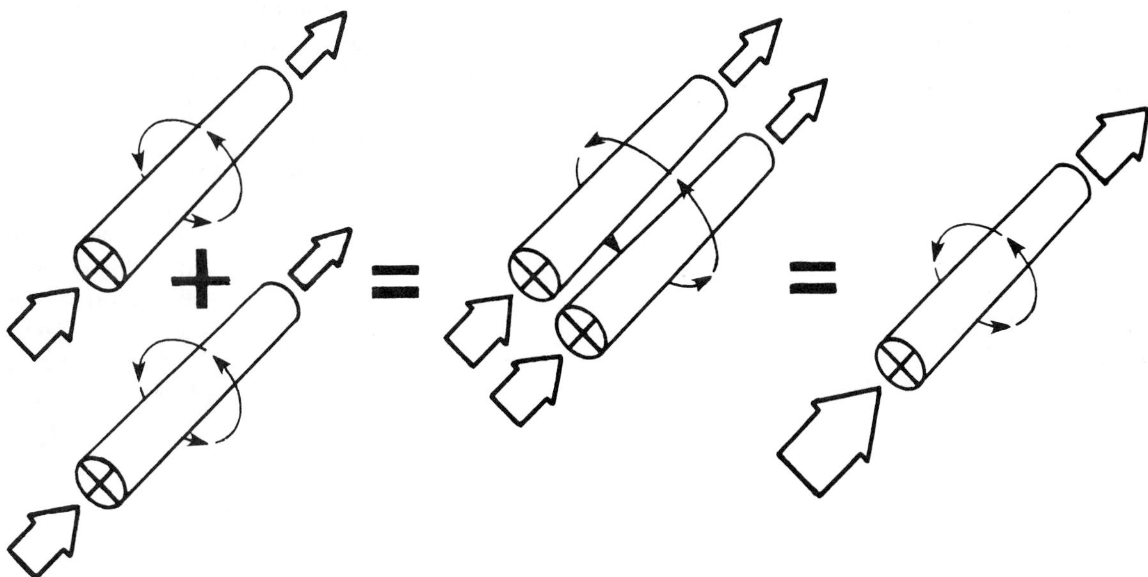

Figure 5. Increasing the flux density.

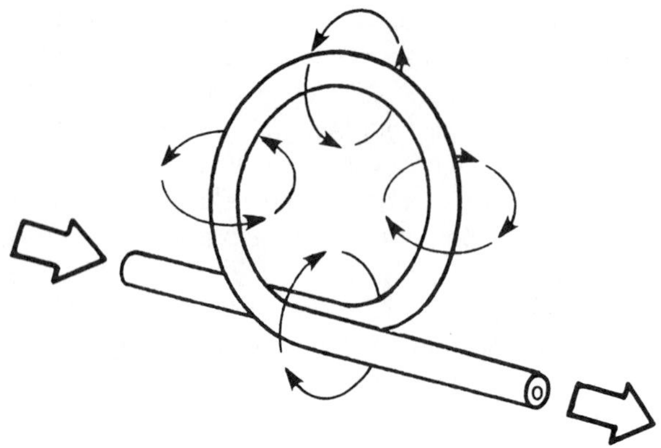

Figure 6. Flux density is increased by coiling the conductor.

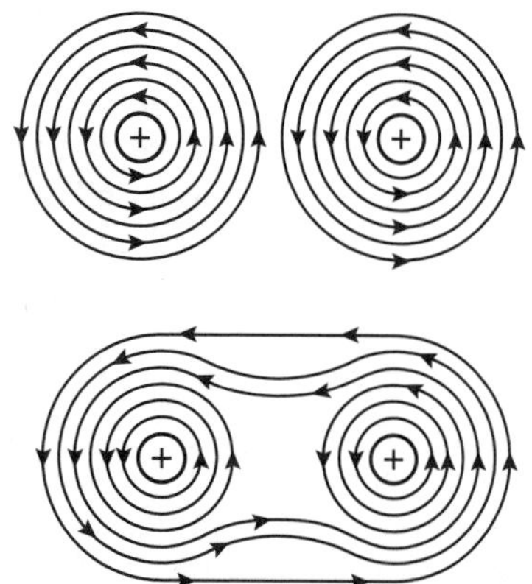

Figure 7. Magnetic lines of force join together.

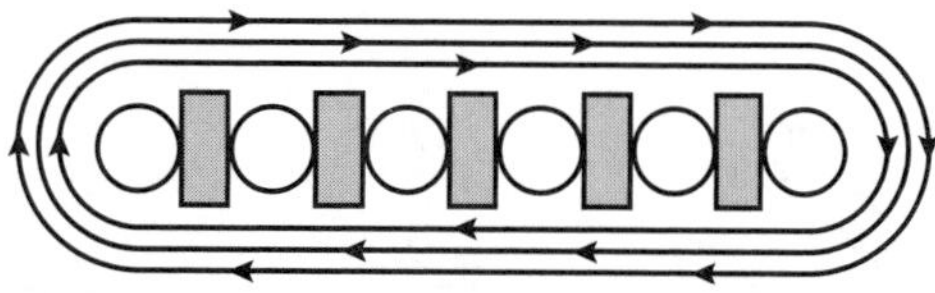

Figure 8. Insulators are used to separate current-carrying conductors.

shorting out **(Figure 8)**. The strength and thickness of the insulation will depend on the amount of voltage applied to the circuit. You should remember this from our discussion of conductors and insulators. Current determines the size of the conductor, and voltage determines the amount of insulation required. The more conductors with current flowing in the same direction, the greater the flux density and the stronger the magnetic field.

Let us examine this same principle with current flowing in opposite directions, as **Figure 9** shows. Notice that the magnetic lines of force have a tendency to move apart because they have their force fields repel each other.

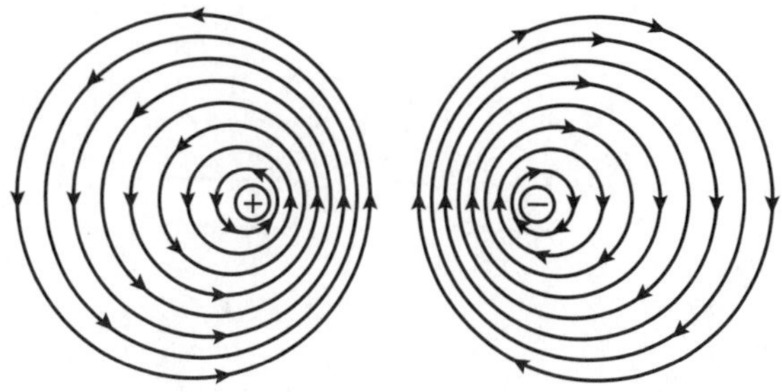

Figure 9. Fields repelling because of reverse polarity.

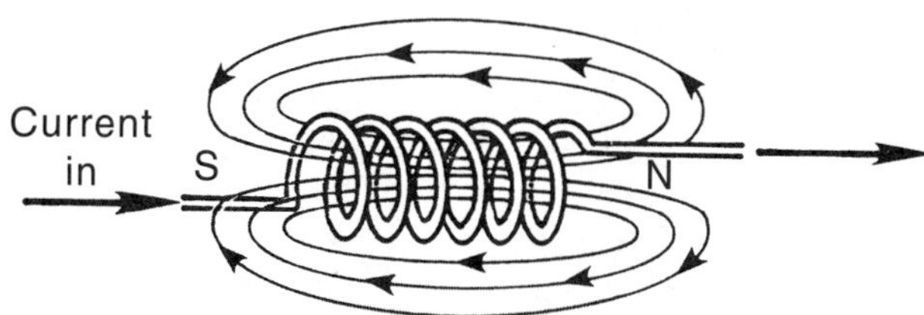

Figure 10. Coiling greatly increases the magnetic field.

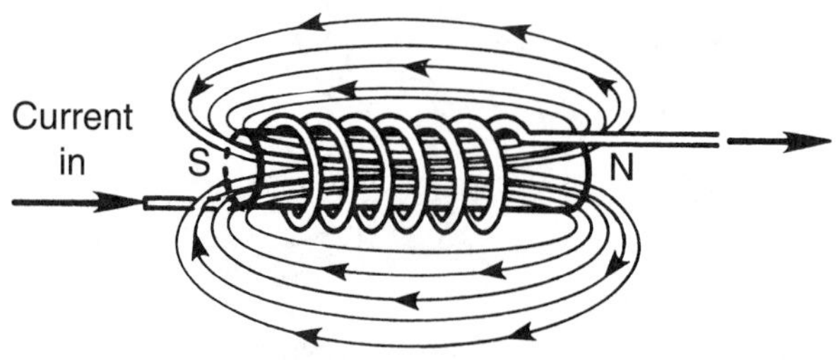

Figure 11. An iron core concentrates lines of force.

Remember that like fields will repel one another. Current flowing in the same direction sets up fields that will join together, whereas current flowing in opposite directions will set up fields that will repel each other. These principles are fundamental to your understanding of electromagnetism in the vehicle. By winding the current-carrying conductor into a coil, we will greatly increase the flux density, as shown in **Figure 10**. One additional magnetic principle can be applied to increase the strength of the field: an iron core can be placed in the center of the coil. Iron will conduct or concentrate the lines of force easier than air. The iron frame used in electromagnets allows a place to assemble the windings and greatly increases the strength of the field, as shown in **Figure 11**.

MOTOR PRINCIPLES

Examine **Figure 12**. It shows a horseshoe magnet with lines of force from the N pole to the S pole. This is its field. If a current-carrying conductor is placed within this field, we

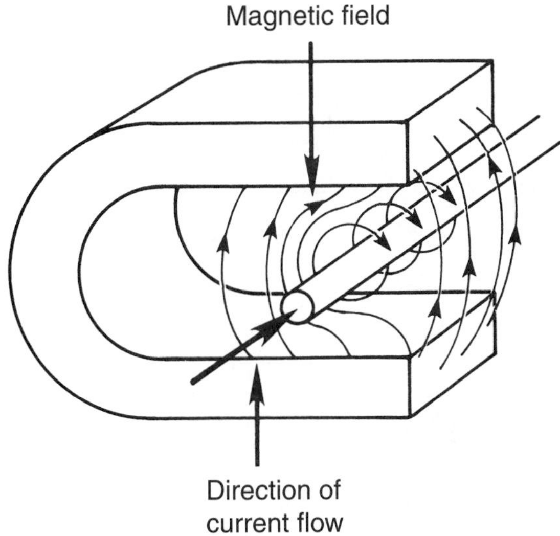

Figure 12. A current-carrying conductor will move when placed in a magnetic field.

will have two force fields. On the left side of the conductor, the lines of force are in the same direction and will therefore concentrate the flux density. On the right side of the conductor, the lines of force are opposite those of the horseshoe magnet and will therefore result in a weaker field. With a strong, heavily concentrated field on one side and a weak, loosely concentrated field on the other, the conductor will tend to move from the strong field toward the weak field, from left to right, as illustrated in **Figure 13**.

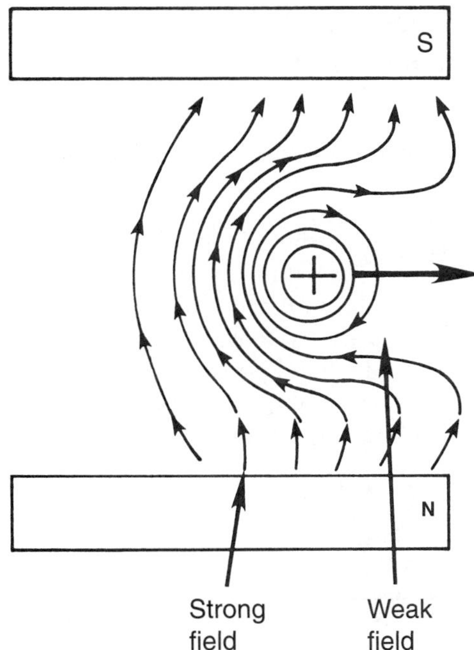

Figure 13. The conductor moves from the strong to the weak field.

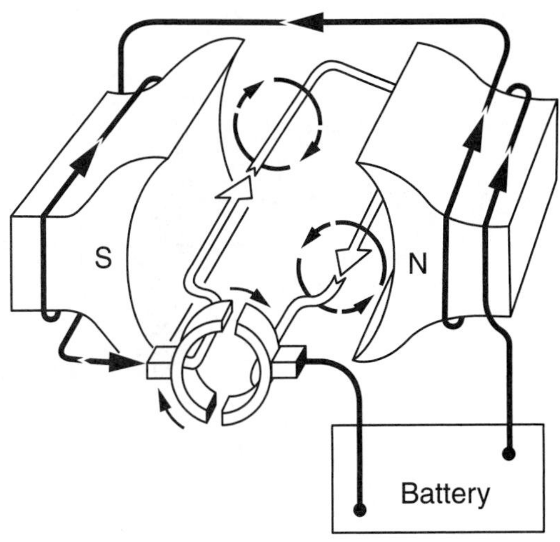

Figure 14. A simple electromagnetic motor.

The amount of force against the conductor is dependent on the strength of the magnet and on the amount of current flowing through the conductor. This movement is summarized by the statement that a current-carrying conductor will tend to move if placed in a magnetic field. Refer to **Figure 14**. It shows a simple electromagnetic style of motor. Trace the current as it flows from the battery around the two iron pole pieces. The direction of the windings of the left pole place it at a south polarity and the right side at a north polarity. Current continues through a half circle called the **commutator**, a loop of wire, and then out the other commutator. Current will now set up a magnetic field around the loop of wire that will interact with the north and south fields and put a turning or rotational force on the loop, as shown in **Figure 15**. This force will cause the loop to turn. When the loop turns half a turn, the position of the commutator bars will be reversed, and the direction of current flow through the loop will also be reversed. This will put the same clockwise rotational force on the loop as it had previously. This action will continue until the current is turned off. This two-pole, single-loop electric motor has the capability to develop a small amount of power and spin.

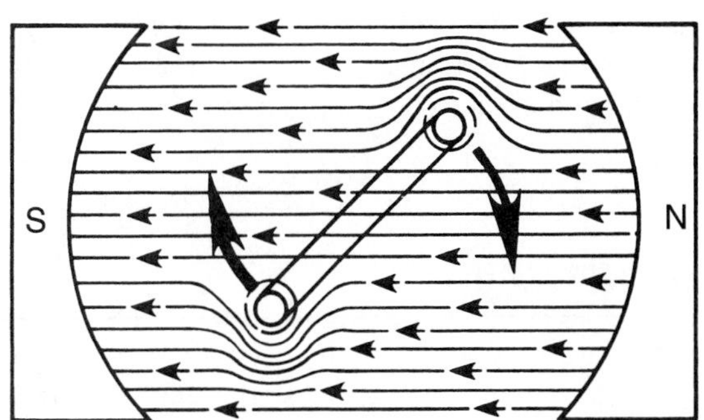

Figure 15. The current forces the armature to rotate.

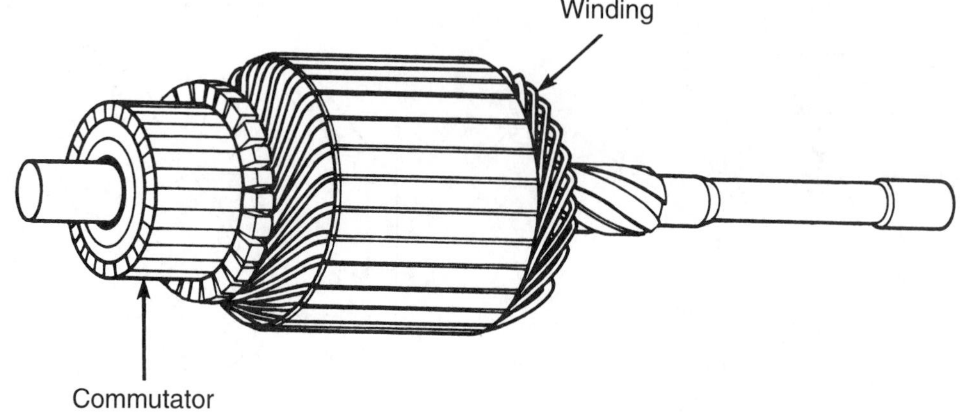

Figure 16. An armature.

Bringing the current into the loop of wire is accomplished with the commutator and a set of brushes. This arrangement allows the current to flow into something that is moving (the loop), through something that is stationary (the brushes). Each end of the loop will have a commutator bar soldered into it.

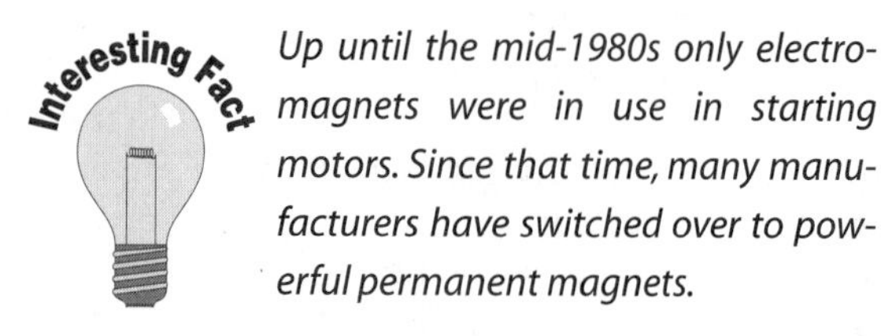

Up until the mid-1980s only electromagnets were in use in starting motors. Since that time, many manufacturers have switched over to powerful permanent magnets.

Our single-loop motor would not develop sufficient horsepower or **torque** (twisting effort) to turn over the engine. The power it can develop must be increased to be of value to the starting system. Simply increasing the number of loops and adding additional pole pieces will do this. Most electromagnetic starters use four pole pieces and many loops of wire in one assembly, called an armature. **Figure 16** shows a common armature. **Figure 17** shows a common four-pole case and the interaction of the magnetic fields. There are many different methods of wiring the four poles and the armature together. The difference is dependent on engine speed, torque required, and battery size, among others. Technicians need not concern themselves with the design of the motor because choices will probably not be available at the time of repair. The various worldwide manufacturers determine the design of the motor necessary for their application. Replacement in the field with a new, rebuilt, or remanufactured starter is the only option available to most technicians.

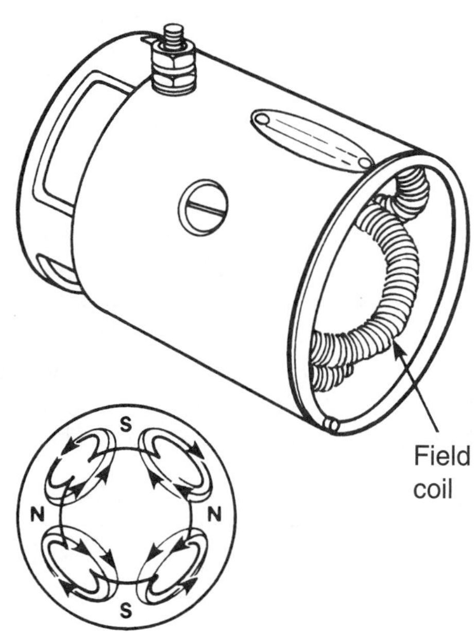

Figure 17. Four-pole magnetic field interaction.

Summary

- The starting system is the "first" system to function.
- Control current comes to the starter solenoid from the ignition switch.
- A magnet that is allowed to turn freely will align itself north and south.
- Two like poles will repel each other.

- Two unlike poles will attract each other.
- Flux density is the concentration of magnetic lines of force.
- A current-carrying conductor that is placed in a magnetic field will tend to move.
- Increasing the amount of current flowing through a conductor will increase its flux density.
- The voltage applied determines the insulation required.
- The current flowing determines the size of the conductor.
- A commutator and brushes allow for the transfer of current between a moving armature and a stationary field coil.
- Most electromagnetic starters use multiple poles in their case circuit.

Review Questions

1. A current-carrying conductor will tend to do what when placed in a magnetic field?
   A. move
   B. remain stationary
   C. move toward the north pole
   D. move toward the south pole
2. Technician A states that coiling a wire around a soft iron core will increase the strength of its magnetic field. Technician B states that increasing the amount of current flowing through a conductor will increase the strength of its magnetic field. Who is correct?
   A. Technician A only
   B. Technician B only
   C. Both Technician A and Technician B
   D. Neither Technician A nor Technican B
3. Flux density is the
   A. number of conductors in a coil of wire
   B. strength of the magnetic field
   C. amount of current flowing through a conductor
   D. none of the above
4. A current-carrying conductor will tend to move when placed in a magnetic field.
   A. True
   B. False
5. Coiling a current-carrying conductor around a soft iron core will decrease its flux density.
   A. True
   B. False

Solenoid Shift Starters

Introduction

The most popular starter design is the **solenoid shift**. It is used, with some variation, by virtually every automotive manufacturer. Whether the motor is an electromagnetic style or a permanent magnetic style, if there is a solenoid mounted on the motor it is considered a solenoid shift design. The name really says it all. The solenoid will shift the motor into the flywheel and apply current to the motor windings to crank the vehicle.

SOLENOIDS

Previously, we mentioned that an additional item would be necessary to shift the pinion into mesh with the flywheel. This item is the solenoid. It will have the job of pushing the drive pinion into mesh when the key switch is in the start position. Refer to **Figure 1**, which shows a solenoid mounted on top of a starter motor. Notice the shift fork. It will pivot on the pivot pin and push the starter drive mechanism (overrunning clutch) into mesh when the plunger is drawn into the solenoid windings. Current from the key switch will energize the solenoid windings and make it a strong electromagnet. This magnetic field will attract or pull the plunger into the hollow windings. Electricity running through the windings will keep the overrunning clutch in mesh with the ring gear teeth.

The first job or function of the solenoid is to engage the starter drive with the engine. The second function is to electrically connect the battery to the starter motor. This is the relay function. The solenoid will act as a switch to energize the motor once the drive pinion is engaged. This sequencing of activity is extremely important. If the motor is energized before the pinion is engaged, damage to both the flywheel and the pinion results. The plunger contact disc located at the back of the solenoid will be pushed against two contacts by the solenoid plunger. One of the contacts is B+ and the other is the motor feed (M) or starter input. Some older solenoids may also have an additional terminal that becomes B+ with the solenoid energized. This is called the **ignition bypass terminal**. It is used to supply current to the ignition system during cranking. Its use has gradually tapered off because most styles of electronic ignition do not require a bypass circuit. We cover the bypass circuit in Chapter 48.

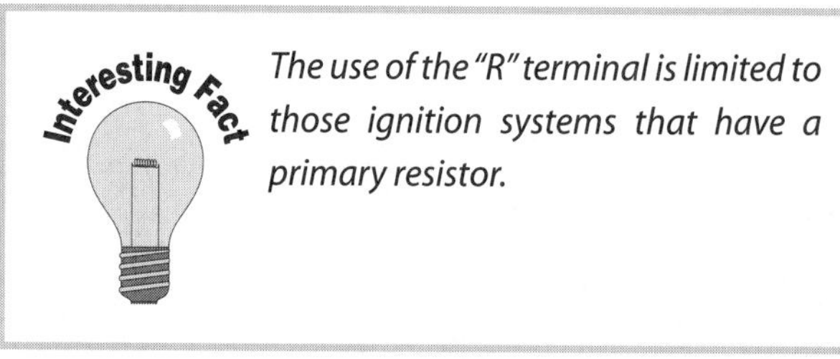

The use of the "R" terminal is limited to those ignition systems that have a primary resistor.

Most solenoids use two windings. They are called the **pull-in** and the **hold-in windings**. Their functions are identified by their names. How they work is interesting and a simple example of parallel loads. Refer to **Figure 2**, which shows a schematic of the two windings. Notice that the pull-in winding starts at the S terminal (start) coming from the ignition switch and ends or grounds at the M, or motor, terminal. This terminal will look like a ground because of the extremely low resistance of the stopped starter motor. Earlier in this chapter, we looked at how the motor functions. With the armature stopped, the windings will offer no resis-

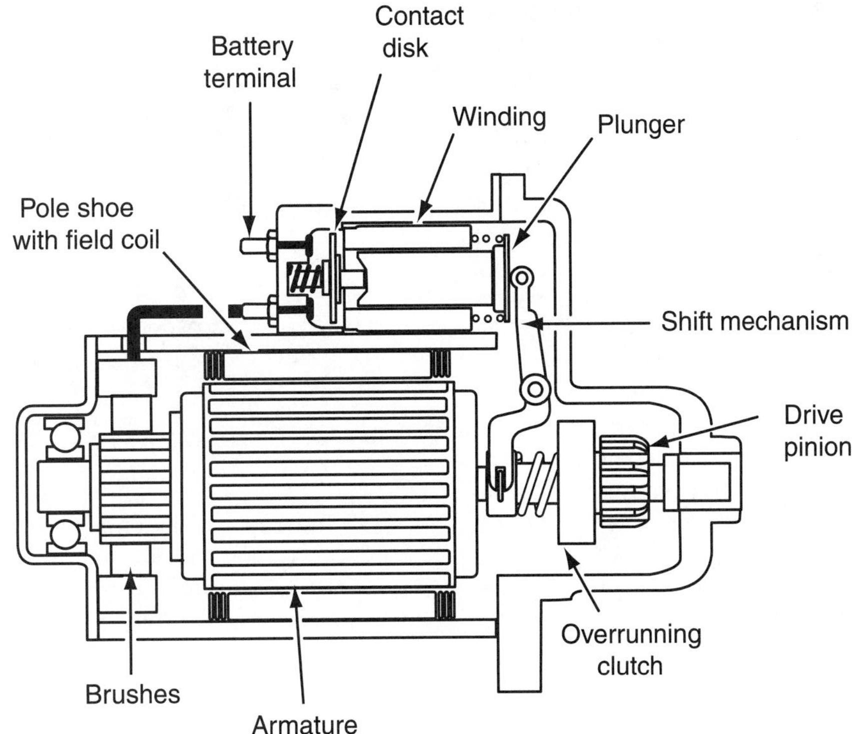

Figure 1. A typical solenoid shift starter.

tance to low amounts of current. When the S terminal is energized with B+, the ground for the winding will be supplied by the starter motor and current will flow. By contrast the hold-in winding has its own ground, usually soldered directly to the case of the solenoid. It does not rely on the starter motor circuit for its path to ground. When B+ is delivered to the S terminal, both windings will be energized in parallel and a strong magnetic field will be produced. This is because both windings are drawing current, sometimes as high as 50 amps. Once the plunger spring is compressed and the plunger is pulled in, a lot less current will be necessary to hold it in place. The contact disc will now be at B+ potential because it will have been pushed against the battery terminal and the M terminal. Notice that the end of the pull-in winding is also attached to the M terminal. The M terminal now becomes B+, when the S terminal received its B+ from the ignition switch. This places the same potential (B+) at both ends of the pull-in winding. With the same pressure, or potential, at both ends, all current flow through the winding stops. This design then allows the current, which the pull-in winding was drawing, to become available for use by the starter motor. The hold-in winding continues to draw current because the two ends of its coil have B+ and ground.

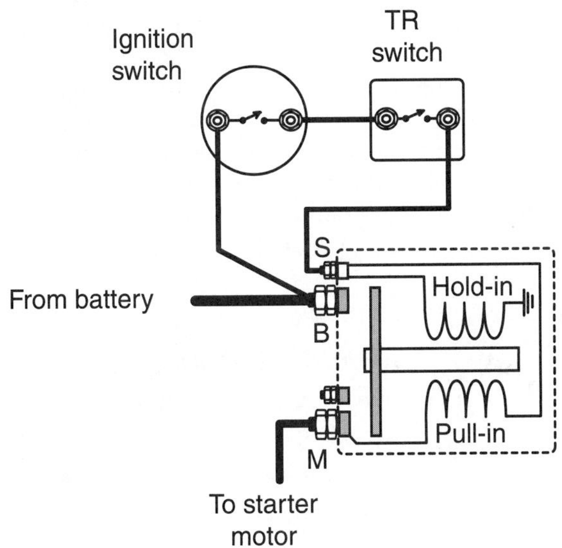

Figure 2. A solenoid shift schematic.

ARMATURE

Once the starter solenoid energizes the starter, current flow through the motor will develop the high torque and horsepower required to crank over the engine. Let us review the flow of current through the typical electromechanical starter. Current has only one job to perform. It must develop magnetic fields that will pull and push the armature. Most starters that have electromagnet-type field coils will have four coils wired in a series-parallel configuration. This will allow two of the coils to be north polarity and two of the coils to be south polarity. Remember that the field coils do not rotate. They are stationary because they are bolted to the motor case. In addition to developing the north and south field coils, electricity will flow through the stationary brushes and into the armature. The armature is supported on both ends by bushings or bearings so that it can rotate freely. The brushes deliver current to the motor through copper connections called commutator bars. There are two bars for each loop of wire in the armature and multiple loops per commutator bar. Indirectly, all armature windings are connected together at the commutator. The many loops of wire are insulated from one another and from the armature shaft. Current flow through the armature runs from a brush through a commutator bar, through the armature winding, and out another brush. Usually two of the four brushes are connected to ground and are, therefore, at a negative potential, whereas two of the brushes are at a positive potential. In this way, the armature will have four sets of rotating fields, two per armature winding.

Do you remember the action of the solenoid? The first function of the solenoid is to move the starter drive mechanism into mesh with the flywheel. The business end of the armature is where we usually find the drive mechanism. The end of the armature shaft is threaded with heavy, strong threads so that the drive will rotate slightly as the solenoid pushes it into the flywheel. This allows the teeth to easily mesh without "grinding."

STARTER DRIVE MECHANISM OR OVERRUNNING CLUTCH

The **overrunning clutch** is a roller-type clutch that transmits the torque of the armature in one direction only. In this manner, it functions in preventing the engine from spinning the starter armature. Refer to **Figure 3**. Notice the spring-loaded rollers between the drive pinion and the roller retainer. The torque of the armature will be transferred to the drive pinion through these rollers. The clutch housing is connected to the armature and will spin at armature speed. This action will drive the spring-loaded rollers into the small ends of the tapered slots, where they will wedge against the pinion.

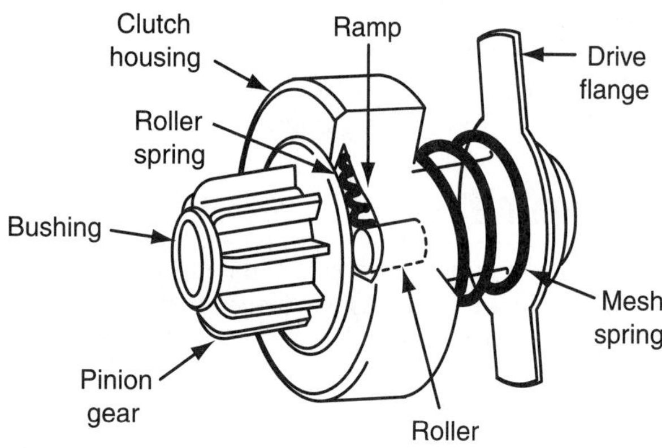

Figure 3. An overrunning clutch.

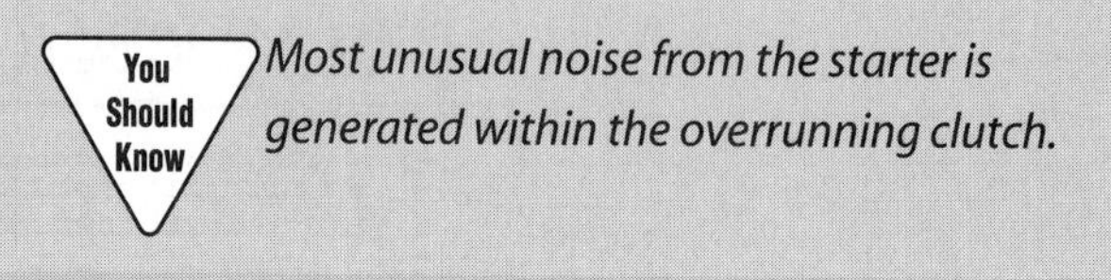

Most unusual noise from the starter is generated within the overrunning clutch.

The rollers will lock the pinion onto the armature and allow the motor to turn the flywheel. When the engine starts, the flywheel ring gear begins to drive the pinion faster than the armature. This spinning action releases the rollers and allows the pinion to spin faster than the armature. The name says it all; the unit allows the pinion to "overrun" the armature.

CURRENT FLOW THROUGH THE STARTER

Let us trace through the starting circuit and follow the flow of current. With the gearshift lever in either park or neutral, the ignition switch is turned to start. This brings B+ over to the S terminal of the solenoid. Both the pull-in and hold-in windings energize and develop the strong magnetic field necessary to pull the plunger in against spring tension and push the starter drive into mesh with the flywheel. The plunger disc comes in contact with both the battery terminal and the motor terminal. This puts B+ on both sides of the pull-in winding at the same time that it energizes the motor. The plunger disc brings current through the motor where the field coils and armature are energized. The engine cranks by the spinning armature driving the starter pinion, which is meshed with the flywheel. Once the engine starts, the pinion overruns the armature until the ignition switch is released and the plunger spring pulls the drive out of mesh.

A correctly functioning starting system should crank the engine over around 200 rpm or greater. Anything substantially less might cause a no-start condition.

Summary

- Solenoids have two functions: to move the starter drive into mesh with the flywheel and to connect the starter motor to the battery.
- The pull-in winding of the solenoid draws high current as it pulls the solenoid plunger and the starter drive.
- The hold-in winding of the solenoid draws low current as it holds the plunger and starter drive in mesh.
- Most starter motors have four field coils.
- Most starter motors have four brushes that bring current through the armature.
- The overrunning clutch style of starter drive allows the armature to spin the engine but prevents the engine from spinning the armature.

Review Questions

1. Technician A states that the solenoid will move the pinion into mesh with the flywheel. Technician B states that the solenoid will switch the current on to the starter motor. Who is correct?
   A. Technician A only
   B. Technician B only
   C. Both Technician A and Technician B
   D. Neither Technician A nor Technician B
2. The component that is used to bring current into the armature is/are the
   A. overrunning clutch
   B. brushes
   C. field coils
   D. solenoid
3. The purpose of the pull in winding within the solenoid is to
   A. pull the plunger into the solenoid
   B. move the starter drive into mesh with the flywheel
   C. allow the overrunning clutch pinion to mesh with the flywheel
   D. all of the above
4. The hold-in winding will
   A. switch current into the armature
   B. move the drive pinion into mesh with the flywheel
   C. hold the plunger in the solenoid
   D. turn on the current to the armature
5. Technician A states that the purpose of the overrunning clutch is to protect the flywheel from spinning too fast. Technician B states that the overrunning clutch allows the armature to spin the flywheel but does not allow the engine to spin the armature. Who is correct?
   A. Technician A only
   B. Technician B only
   C. Both Technician A and Technician B
   D. Neither Technician A nor Technician B

Diagnosing and Servicing Solenoid Shift Starting Systems

Introduction

In Chapter 34, we discussed the operation of the solenoid shift starter. It is now time to look at the entire system as it is commonly found in vehicles of today. It is interesting to note that, of all the current systems on vehicles today, the starting system has experienced the fewest changes. Onboard electronics, digital control, and integrated circuits have had a tremendous impact on most vehicle systems. The starting system has not seen these changes. It remains basically the same as it was in the 1930s with the exception of the introduction of the permanent magnet motor. The starting system should be considered one of the basic systems that all technicians must know. Employers expect their employees to be able to diagnose starting system problems quickly, get them repaired, and get the vehicle back on the road.

In this chapter, we look at a detailed on-the-vehicle test of the solenoid shift–style starting system for both a no-crank condition and a preventive maintenance diagnosis condition.

REVIEW OF THE VOLTS-AMPS TESTER

Let us begin our discussion with a review of the VAT that we used in battery testing. Refer to **Figure 1** and **Figure 2**, which show two common VAT load–type testers. Both have an ammeter capable of measuring 500 amps on the high scale and approximately 100 amps on the low scale.

Figure 1. A VAT.

Figure 2. A VAT-type tester.

Battery Worksheet

Specifications

Battery Size ______________ Engine Cubic Inches ______________

Hydrometer Readings

Cell Number	1	2	3	4	5	6	
Negative Post							Positive Post

Results (Check one.)

_____ All cells are within 25 points and are charged (continued testing).
_____ All cells are equal and are discharged (charge battery).
_____ Cell readings vary more than 25 points (skip load test and advise customer to replace battery.)

OR

Open Circuit Voltage Test

Load battery to 20 amps for about 1 minute before testing OCV.

______________ volts—open circuit (no load)

Refer to the chart and convert either the hydrometer reading or the OCV over to state of charge.

Charge	Specific Gravity	OCV (12.6)	OCV (6.3)
100%	1.265	12.6	6.3
75%	1.225	12.4	6.2
50%	1.190	12.2	6.1
25%	1.155	12.0	6.0
Discharged	1.100	11.9	5.9

Battery Load Test (for batteries at least 75% charged)

Set tester to the following:
1. Tester to STARTING
2. Variable LOAD off
3. Voltage—INTERNAL—higher than battery voltage

Attach the BST as follows:
1. Place large cables of tester clamp over the battery clamps, observing that polarity of the battery and tester are the same.
2. Attach inductive pick-up clamp around either of the BST cables *not* battery cables.

Determine the correct load:

Battery Amp/Hour × 3 = Load or
Cold Cranking Amps ÷ 2 = Load

NOTE: Compare the specific battery to the one actually installed. Load to the LARGER of the two.

Apply the calculated load for 15 seconds, observe voltmeter, and turn load OFF.

_____ GOOD—battery voltage remained above 9.6 volts during load test
_____ BAD—battery voltage dropped below 9.6 volts during load test (perform 3-minute charge test to check for sulfation)

3-Minute Charge Test (only for batteries that fail the load test)

1. Disconnect the ground battery cable.
2. Connect a voltmeter across the battery terminals (observe polarity).
3. Connect battery charger (observe polarity).
4. Turn the charger on to a setting around 40 amps.
5. Maintain 40 amp rate for 3 minutes while observing voltmeter.

_____ GOOD—voltmeter reads less than 15.5 (recharge slowly and reload test).
_____ BAD—voltmeter reads more than 15.5 (battery is sulfated and should be replaced).

Figure 3. A battery test worksheet.

In addition, both have voltmeters with two or more scales and a carbon pile that can load a battery or system down. This is the basic battery-starting-charging system tester. The information it will supply is necessary in our diagnosis. The VAT is a fundamental piece of diagnostic gear that most shops have. Many manufacturers offer it in their tester line.

The first part of this chapter discusses a testing series that a technician would do on a vehicle that does crank. The reason for doing it might be preventive maintenance, perhaps before a season change such as before a harsh northern winter, or an intermittent cranking problem. This series of tests can be done on any vehicle. However, it should be noted that the specifications supplied from various manufacturers might be slightly different from the general guidelines printed. We must also realize at this point that a full battery diagnosis should be completed before looking at the starting system. The reasons for this are based on the interrelationship between battery performance and starting performance. Starting problems can be caused by the battery, just as battery problems can be caused by the starting system. The battery must have the ability to keep its voltage above 9.6 volts and deliver the current necessary to crank the engine. For example, a starting system will draw twice the current if the battery voltage drops to half. To illustrate this example requires a review of wattage:

$$\text{voltage} \times \text{amperage} = \text{wattage}$$

It is the wattage that will do the work of cranking the engine. Many V-8s crank at approximately 2000 watts. At 10 volts applied to the starter, 200 amps will flow because:

$$10 \text{ volts} \times 200 \text{ amps} = 2000 \text{ watts}$$

However, if the battery cannot keep its voltage up, the starter will draw more current. More current drops the voltage lower, increasing the current, which will drop the voltage, and so on. If 7.5 volts are delivered to the starter, it will draw 266 amps.

$$7.5 \text{ volts} \times 266 \text{ amps} = 1995 \text{ watts}$$

From a practical standpoint, it is important to note that the system may be incapable of delivering extremely high amperage and the engine will just not crank. However, you can see how important it will be to deliver amperage at a reasonable voltage level. This is why we stressed the importance of correct battery size in previous chapters. Larger batteries will work less and, in theory, last longer in your customer's vehicle. Always install batteries that are at least equal to that installed by the original equipment manufacturer. **Figure 3** is a reprint of the battery tests we outlined in Chapter 32. Do not hesitate to review them thoroughly. They must be accomplished before you begin to look at the starting system components. If the battery fails the load test and is fully charged, it must be replaced before going on. A battery that is weak will throw off your starting system testing.

CRANKING CURRENT

With a fully charged, performance-tested battery installed, we can begin testing the starting system. We will want to crank the engine without starting it, so we should disable the ignition or fuel system. If the vehicle has a coil wire, either jumper the coil wire directly to ground at the distributor cap end **(Figure 4)**, or remove the battery feed from the ignition system **(Figure 5)**. Both of these methods will allow the engine to crank without starting. Under no circumstances should you crank the engine with the coil wire disconnected and open circuited because this could stress the coil and **ignition module** and might damage them. Most vehicles produced today do not have coil wires, and it is impractical to disconnect each plug wire. In this example, the fuel injectors might be disconnected or an input sensor that will shut down fuel or injection might be disconnected. On some vehicles, a single fuse might shut down all of the injectors. Each vehicle is slightly different and your job is to figure out what will be necessary on the vehicle you are working on. Consult your instructor or another experienced technician if you need help.

Figure 4. Ground the coil wire.

Figure 5. Remove the power from the ignition coil.

> **You Should Know** *The listed amperages are considered to be the maximum. It is all right if the system cranks over at a lower current as long as normal cranking speed is observed.*

The **inductive probe** of the VAT was around either the positive or negative cable of the tester for battery testing **(Figure 6)**. It will now have to be moved to the vehicle battery cable so that the VAT will record what the engine draws rather than what the VAT draws. The load should be off. After the installation of the inductive probe, the meters should read battery voltage and zero amperage (unless there is a load on in the vehicle such as a light or the ignition switch). Turn on the headlights. The ammeter should move into the negative. If so, the cables are correctly installed. Turn off the headlights. If there is a zero adjust for the ammeter, adjust it until the needle rests or sits at 0 (zero). We are now ready to run a cranking current test. Either move the meter where you can observe it or ask someone else to crank the engine for about five seconds. Record both the voltmeter and the ammeter readings. Correctly functioning systems will crank at 9.6 volts or higher and will draw amperage relative to the cubic inches of the engine. Most V-8s will draw approximately 200 amps; six cylinders, approximately 150 amps; and four cylinders, approximately 125 amps. Remember that these numbers are just guidelines and are the maximum amperages at normal engine operating temperature. If manufacturers' specifications are available, use them.

Interpreting the numbers is not difficult but requires an understanding of the wattage principle that we discussed previously. If the specification on a particular vehicle were 9.6 volts and 200 amps and our results showed 300 amps at 6.8 volts, we would conclude that more testing is necessary to pinpoint the problem. A voltmeter reading lower than specifications or an ammeter reading higher than specifications is sufficient cause for detailed testing even if the vehicle does start. Preventive maintenance, if done correctly, will avoid future no-start conditions. Ammeter readings around or lower than specifications and voltmeter readings higher than the specifications are considered good, but detailed testing should be completed if an intermittent problem is present. The detailed testing will attempt to pinpoint bad components. At this point, we should realize that our results were taken at the battery and that these voltmeter readings might not be exactly representative of the actual voltage delivered to the starter. We frequently find voltage drops or losses through cables, connections, relays, or solenoids that diminish the delivered voltage to the starter. These voltage drops force the starter into drawing more current (wattage at work again). The detailed tests we will now discuss will allow us to look at the individual components that can have an adverse effect on the starting system.

Figure 6. Inductive pickup around the battery cable.

CRANKING SPEED

The speed at which an engine spins is important. It takes fuel, compression, and ignition to get an engine running. Compression and fuel will be dependent on cranking speed. The slower an engine cranks, the more compression loss there is. This loss will make it harder to draw in the air: fuel mixture and ignite it. General specifications call for approximately 200 rpm minimum for most starting systems. This speed can be measured easily with a tachometer. Most technicians do not measure them, however. A technician's experience and trained ear can be an accurate indicator of correct speed. After you are a few years in the field, your ear will be tuned to accurately evaluate the engine speed also. In the meantime, use a tachometer if there is a doubt in your mind that the engine is cranking over fast enough.

There is one set of conditions where we usually ignore the detailed tests and immediately pull the starter. This is the combination of a slower-than-normal cranking speed and a lower-than-normal cranking amperage. This set of conditions can occur only from internal resistance within the starter motor. Do you remember the relationship of speed to current draw?

Slow		High
Armature	=	Cranking
Speed		Amperage

When an engine cranks slowly, we should expect to find high, not low, amperage. Remember this combination. Lower-than-normal cranking speed and lower-than-normal amperage can be caused only by a problem inside the starter. Higher-than-normal amperage and slower-than-normal cranking speed usually require some detailed testing of the B+ and ground path. The theory behind detail voltage drop testing is simple. Voltage is lost or dropped when current flows through resistance. This is Ohm's Law at work

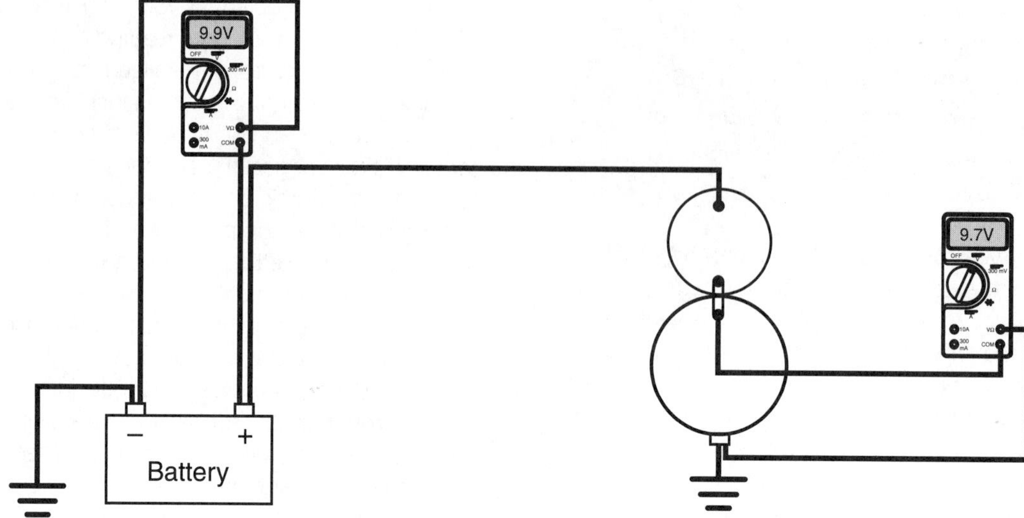

Figure 7. Battery voltage versus starter voltage during cranking.

again just as we have seen in the past. Most manufacturers design their starting systems with cables, connections, relays, and solenoids that have very little, if any, resistance to the flow of starting current. Typically, it is less than 0.2 volt on each side of the circuit (B+ and ground). This then means that the voltage across the starter input to the starter ground should be within 0.4 volt of battery voltage. **Figure 7** illustrates this principle. What we must do in the case of excessive current draw or in the case of an intermittent starting problem is to ensure that the voltage across the starter remains in the 0.4-volt or less area when compared with battery voltage. For example, if the battery voltage during cranking were 9.8 volts and the voltage delivered to the starter were 9.5 volts, we would know that the resistance of the circuit had caused a 0.3 volt drop. In most cases, this would be acceptable. If, however, the battery voltage were 9.8 volts and the starter voltage were 8.8 volts, we would know that a problem exists that will cause the starter to draw more current than is necessary. Do not forget that the current will go up if the voltage goes down (wattage again). Our next step will be to locate the resistance that is causing the 1-volt drop.

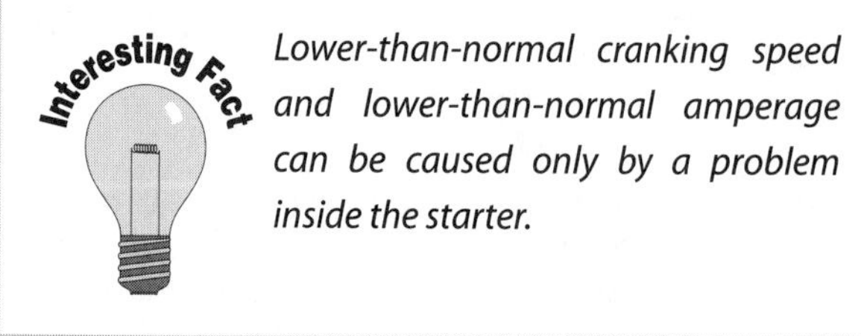

Lower-than-normal cranking speed and lower-than-normal amperage can be caused only by a problem inside the starter.

MEASURING VOLTAGE DROPS

Here is where we will use the lowest scale of the voltmeter, typically 2 to 4 volts. Keep in mind that current flowing through resistance is the cause of voltage drops. A resistance will have a different pressure or voltage on each side. In Chapter 9, we measured the voltage drops across the resistance of a load. We will do much the same thing now while the engine is cranking. The voltmeter will be placed across either the B+ or the ground circuit, as shown diagrammed in **Figure 8**. This setup allows us to pinpoint the circuit half that has the excessive resistance. Once the circuit half is identified, it becomes a simple task to divide the circuit into small pieces and place the voltmeter leads across the pieces, looking for voltage drops with the engine cranking. **Figure 9** illustrates this. The specifications for the B+ or ground side of the circuit is 0.2 volt maximum with normal current flowing. If higher-than-normal current is flowing, adjust the specification upward. If, for instance, 50 percent more current is flowing, the specification will be adjusted upward by 50 percent also to 0.3 volt. We will be able to locate our problem when the voltmeter reads the voltage drop. As you do detailed voltage drop testing, remember that the positive voltmeter lead will be connected into the part of the circuit that is most positive and the negative lead into the most negative. For preventive maintenance checks, both the positive and negative paths should be tested, and anything in excess of the 0.2-volt B+ or 0.2-volt ground specification should be repaired or replaced. If a connection appears to be where the resistance is, it should be cleaned, retightened, and retested. Corrosion and/or loose connections will cause voltage drops and increased current draw. The only part of the circuit that is designed to have over the 0.2-volt drop is the starter motor. Because the solenoid is mounted onto the starter, it is tested along with the rest of the B+ path cables. It is considered to be part of the total allowable 0.2-volt drop.

By doing the detailed voltage drop testing outlined here, you will be able to pinpoint the resistance. Repairing it

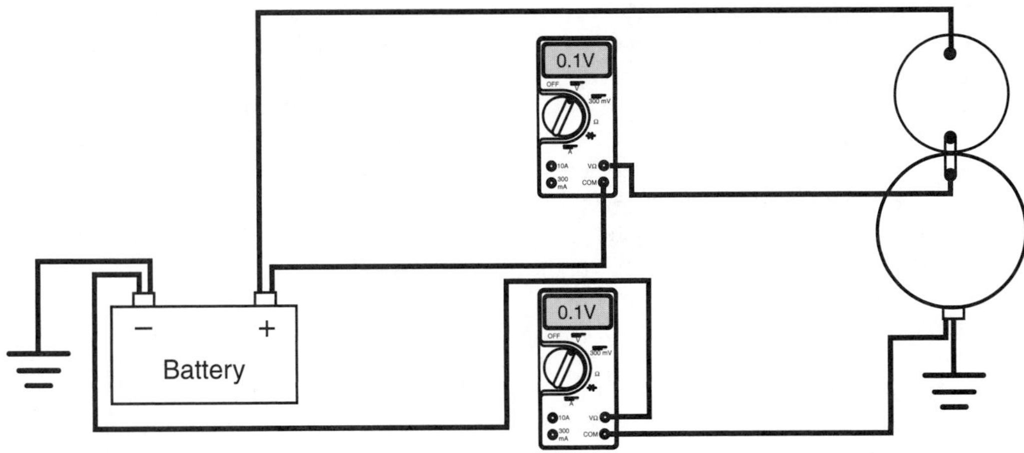

Figure 8. Measuring B+ or ground voltage drops.

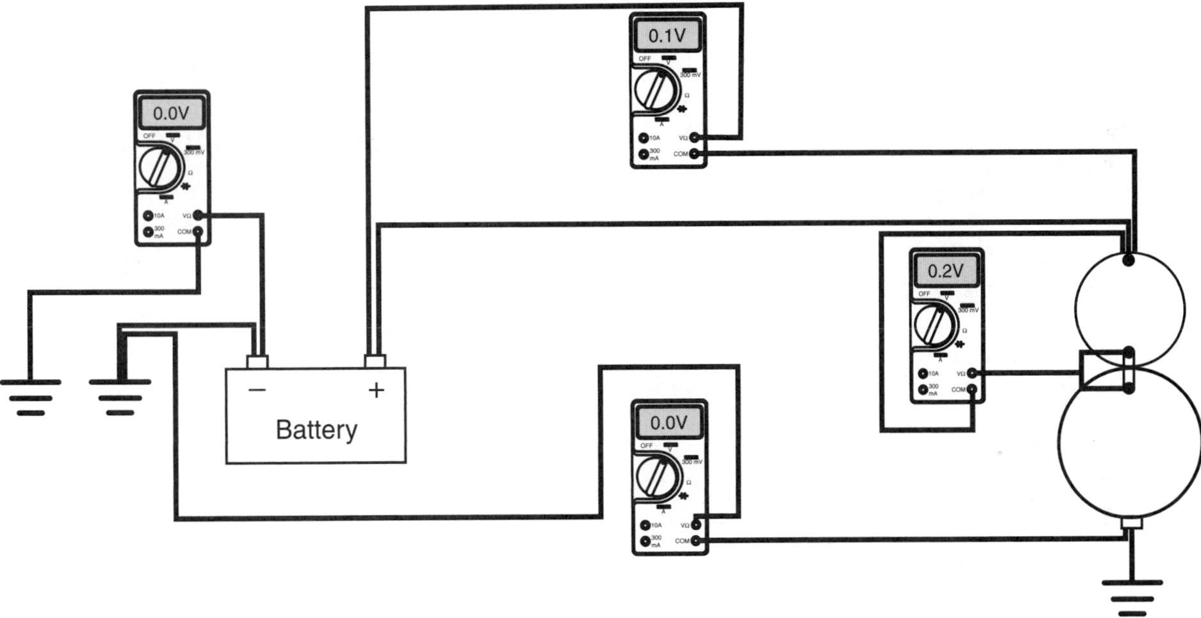

Figure 9. Divide the circuit into small pieces to find the voltage drop.

is simply a matter of replacing the cable, solenoid, or whatever had the voltage drop. Voltage drop testing should be done at least once a year to pick out a resistive component before it causes a no-start condition. Most customers appreciate preventive maintenance designed to prevent no-start conditions. After the voltage drops have been eliminated or at least reduced to an acceptable level, cranking amperage is used to condemn a starter motor. Higher-than-normal amperage without excessive voltage drops is an indication of starter problems or an overly tight engine. The usual procedure, at this point, is to pull the starter off the vehicle and disassemble it. The technician will look for obvious causes of excessive amperage draw, such as worn bushings or shorted windings. The key concept here is to not automatically pull the starter every time excessive amperage shows up. Eliminate voltage drops and test batteries first. Pulling the starter should be the last, not the first, step you take. Keep in mind that a tight engine will increase the amperage draw of the system also.

PROCEDURE FOR DETERMINING OPEN CIRCUIT IN NO-CRANK CONDITION

The previous preventive maintenance approach to starting systems works well if the vehicle does, in fact, crank. Voltage drop testing is important and should be done frequently. However, if the vehicle will not crank a different approach to the situation is necessary. Most no-crank situations are the result of open circuits and can be diagnosed easily with a 12-volt test light. Let us look at the three most common types of starting systems and go through a no-crank situation in each.

The first step you always take in any no-crank situation is to test the battery. Open-circuit voltage or hydrometer readings and then a load test will tell the story. A dead or very weak battery can cause a no-crank situation. Battery testing *must* be the first approach in all types of systems. Once we are assured that the battery has the ability to supply current at a reasonable voltage level, we can begin our search for what will usually be an open circuit.

Let us look at the procedure using the solenoid shift starting system **(Figure 10)**. Notice the M terminal on the solenoid. This is the end of the circuit. If a 12-volt test light lights or a voltmeter shows voltage when the ignition key is in the start position, the B+ feed to the motor is all right and we would move on to a ground circuit test. B+ to the motor is the object of the connections, solenoid, and battery. Keep in mind that the test light must be brightly lit. A dim light indicates excessive resistance within the circuit. Do not hesitate to use a voltmeter to measure the light bulb's voltage if you have difficulty recognizing voltage levels by looking at bulb brightness. In excess of 9.6 volts and, in actuality, approximately 12.6 volts will usually be delivered to the M, or motor, terminal of the starter. We now know that we have one of the three necessary factors to crank the engine: B+. The two unknowns at this time are the starter and the ground return path.

Let us look at the ground return path next. This is a simple test just like our first one was. Place the 12-volt test light's ground connection on the starter body (clean off the grease first) and the probe on the "known hot" M terminal. Turn the key to the start position. If the light lights, the ground return path is all right, as shown in **Figure 11**. This would leave only the starter as our open circuit component, and at this point we would pull it, bench test it, and then replace it.

The trick to starting diagnosis is to approach each section at the starter motor end and verify that B+ is present and a good ground return path is available before pulling the motor. Too often the starter is pulled and replaced when it is not the open component.

Let us go back to the B+ test and assume that the light did not light. Simply backtrack up the current path toward the battery until you get the test light to light. You will have to test the two sections separately. First, test the heavy current battery cable circuit, as shown in **Figure 12**. The light should light with the key off (this is just like being on the battery positive post) and should remain lit when the key is placed in the start position. If the light does not stay lit, repair the cable or end connections.

DIAGNOSING UNUSUAL STARTING NOISE

Most starter noise is related to either armature bearing/bushing wear or a faulty starter drive. The most common of these is the starter drive. Let us look at the bearing wear first. The ratio between armature speed and engine cranking speed is related to the number of teeth on the flywheel and the number of teeth on the starter drive. Many systems fall

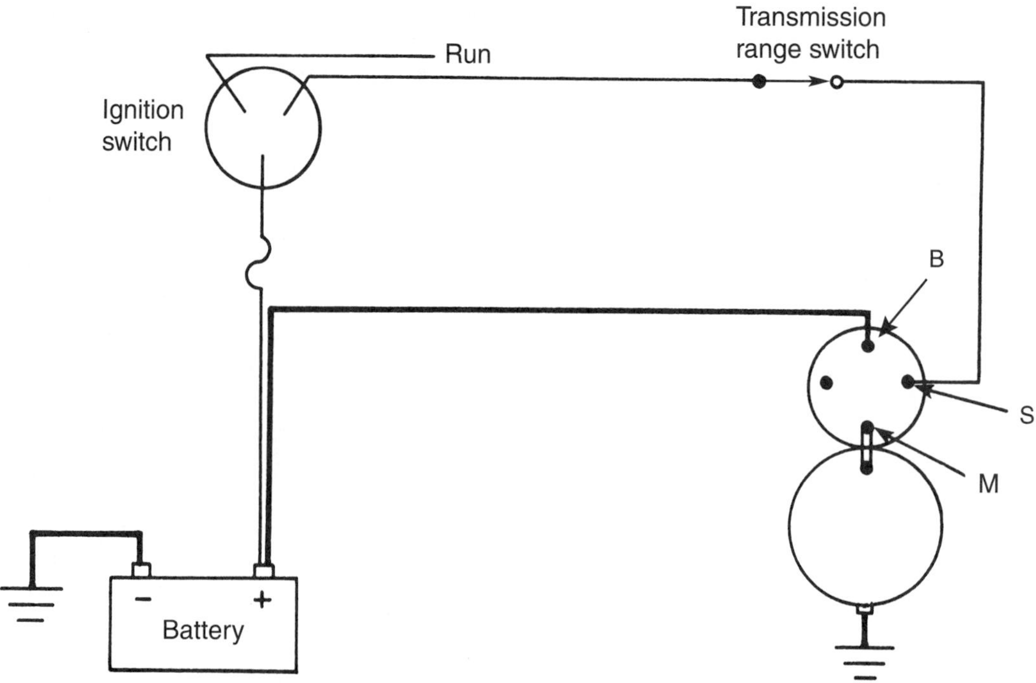

Figure 10. A simplified solenoid shift system.

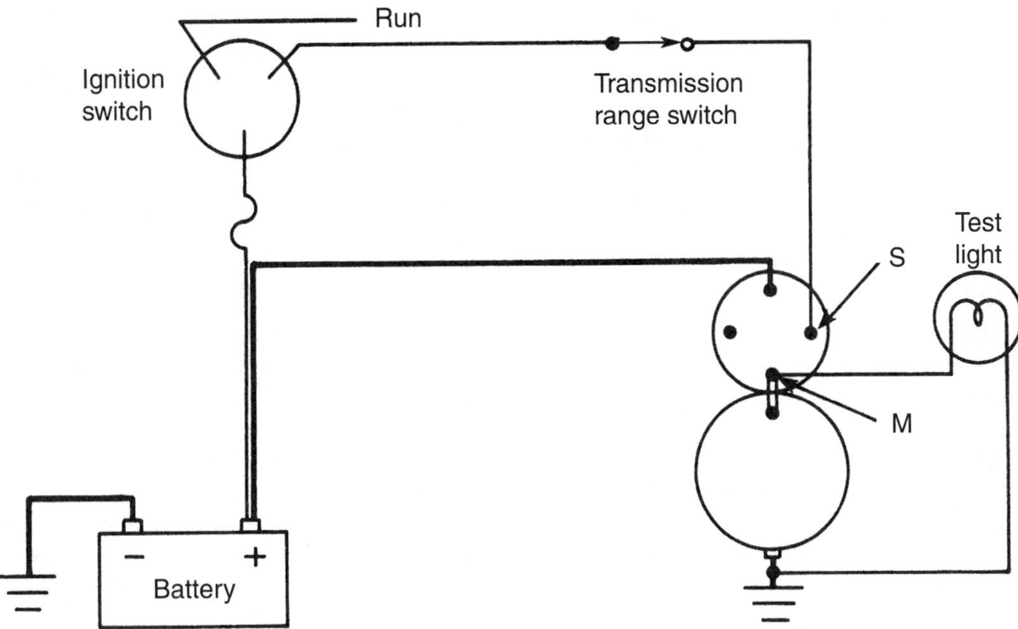

Figure 11. Testing for power.

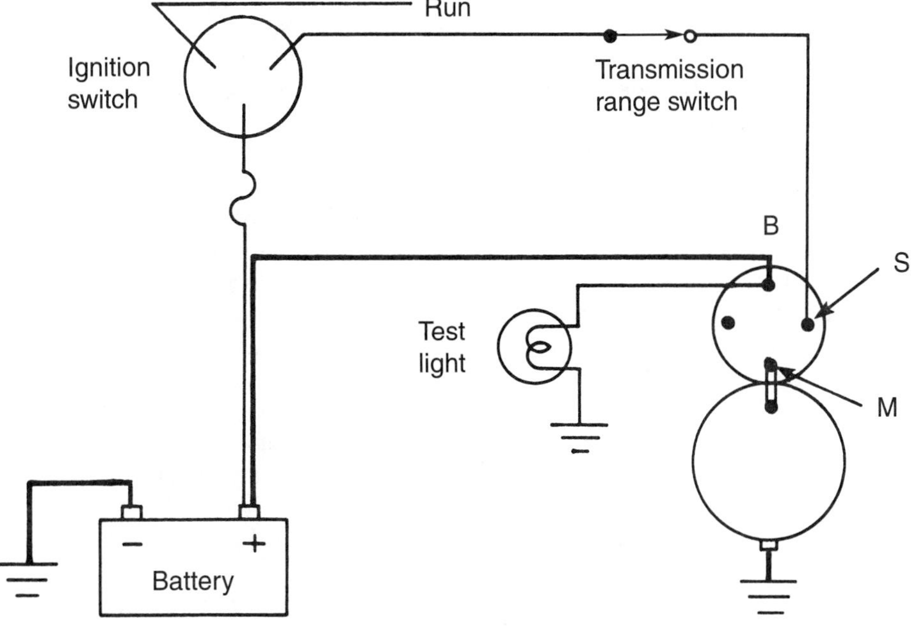

Figure 12. Testing for power to the B terminal.

in the 20:1 ratio. This means that the armature will be turning 20 times faster than the flywheel. Remember that normal cranking speed is approximately 200 rpm. This will place armature speed at approximately 4000 rpm. In addition, many solenoid shift systems in use today have permanent magnet motors with gear reduction that will add an additional 3:1 ratio. This will place armature speed at approximately 12,000 rpm. The bushings or bearings are capable of handling this high speed for many years of cranking. However, when they begin to show wear, they might develop a whining noise as the engine is cranked. Most of the time the noise will become excessive long before a no-start condition is encountered. The noise will be heard even after the key is released as the armature slows down and stops. It is this no-crank noise that will distinguish bearing noise from the more common starter drive noise.

When the armature starter drive wears out, it will begin to slip while cranking the engine. The tremendous torque necessary for cranking the engine is delivered through the starter drive. The rollers must lock down and connect the armature to the drive pinion or the engine will not turn over. If there is a slight slippage through the rollers, the drive mechanism will make a grinding noise as they spin within the collar. Eventually, the rollers will wear sufficiently that they will not lock down. This allows the armature to spin independent of the pinion, causing a no-start condition. Releasing the ignition key and trying to start the engine again might cause the rollers to lock temporarily. Once the starter drive begins to wear, it is only a matter of time before a no-start condition will occur. As a technician, make sure you realize that starter drive noise is indicative of a future no-start condition. Inform your customer that preventive maintenance means a replacement drive or more likely a replacement starter assembly.

Summary

- Wattage is found by multiplying voltage times current (volts × amps = watts).
- Most starting systems will draw less than 2000 watts.
- Battery testing should always be performed before testing the starting system.
- During cranking, the battery voltage should remain above 9.6 volts.
- Normal cranking speed is 200 rpm or higher.
- Lower-than-normal cranking current with lower-than-normal cranking speed can be caused only by internal resistance within the starter motor.
- B+ or ground path voltage drops should each be less than 0.2 volt with the engine cranking.
- A voltmeter wired in parallel with the solenoid will show the voltage drop while the engine is cranking.
- A 12-volt test light is used for no-crank testing.
- Most starter noise comes from the starter drive slipping.

Review Questions

1. Technician A states that slowing down the starter armature will result in increased amperage draw. Technician B states that speeding up the armature will result in decreased amperage draw. Who is correct?
   A. Technician A only
   B. Technician B only
   C. Both Technician A and Technician B
   D. Neither Technician A nor Technician B
2. A 0.7-volt drop is observed across the starter solenoid with the engine cranking. What should be done to the system?
   A. Nothing—it is acceptable.
   B. Install a larger battery.
   C. Replace the solenoid.
   D. none of the above
3. If an engine requires 2000 watts to crank, how many amps will the system draw if the battery voltage is 5 volts?
   A. 400 amps
   B. 1000 amps
   C. 200 amps
   D. Not enough information is given.
4. A starter is very noisy. Technician A states that the starter drive is probably the cause of the excessive noise. Technician B states that excessive current draw could cause the excessive noise. Who is correct?
   A. Technician A only
   B. Technician B only
   C. Both Technician A and Technician B
   D. Neither Technician A nor Technician B
5. A starter amperage draw test is done. The results are less than specification cranking speed and lower than normal amperage. What should be done next?
   A. Remove the starter—it is bad.
   B. Run voltage drop testing of B+ and ground.
   C. Test the battery.
   D. Test for tight engine.
6. The usual specification for voltage drop across the B+ circuit up to the starter is
   A. 0.1 volt
   B. 0.2 volt
   C. 0.3 volt
   D. 0.4 volt
7. Slower-than-normal cranking speed will result in lower-than-normal amperage.
   A. True
   B. False
8. Most noise from starters can be traced to starter drives.
   A. True
   B. False
9. Excessive voltage drops will increase starting amperage draw.
   A. True
   B. False
10. Slower-than-normal speed and lower-than-normal cranking amperage are the result of starter internal problems.
    A. True
    B. False

Positive Engagement Starters

Introduction

The solenoid shift starter is by far the most used style worldwide. However, since 1960, Ford has used a modified design called the positive engagement type on many of its applications. It is used with a starting relay and a different method of engaging the starter drive. Most of the positive engagement starters do not use a solenoid. The combination of a starting relay and a shorting switch inside the starter performs the functions of the solenoid.

ARMATURE

The armature within the positive engagement starter is virtually the same as the one in the solenoid shift starter. It consists of a series of windings insulated from the shaft with commutator bars soldered onto each winding **(Figure 1)**. On the end of the armature is an overrunning clutch–style starter drive, much the same as the solenoid shift type **(Figure 2)**. It is in the control of the starter drive that the system differs. Positive engagement starters use a shorting switch.

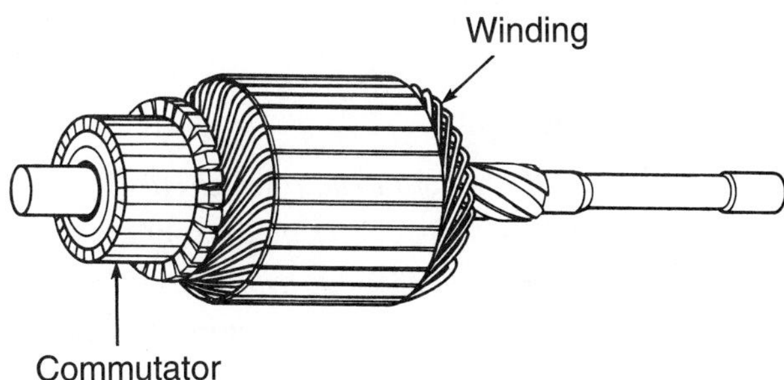

Figure 1. A starter armature.

THE SHORTING SWITCH

The simple schematic shown in **Figure 3** illustrates the principle of the **shorting switch**. The heavy contacts located between the two field coils are normally closed (NC) and will short the first coil directly to ground until they open. Opening them will allow current to flow through the rest of the starter. Remember the sequencing that the solenoid accomplished? It first engaged the drive and then energized the motor. The grounding contacts will accom-

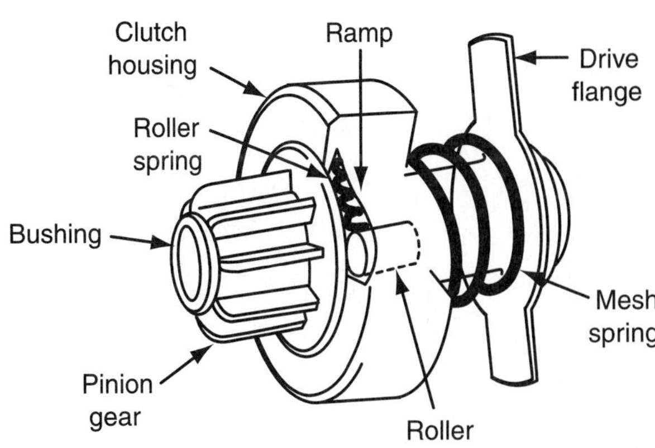

Figure 2. A starter drive (overrunning clutch).

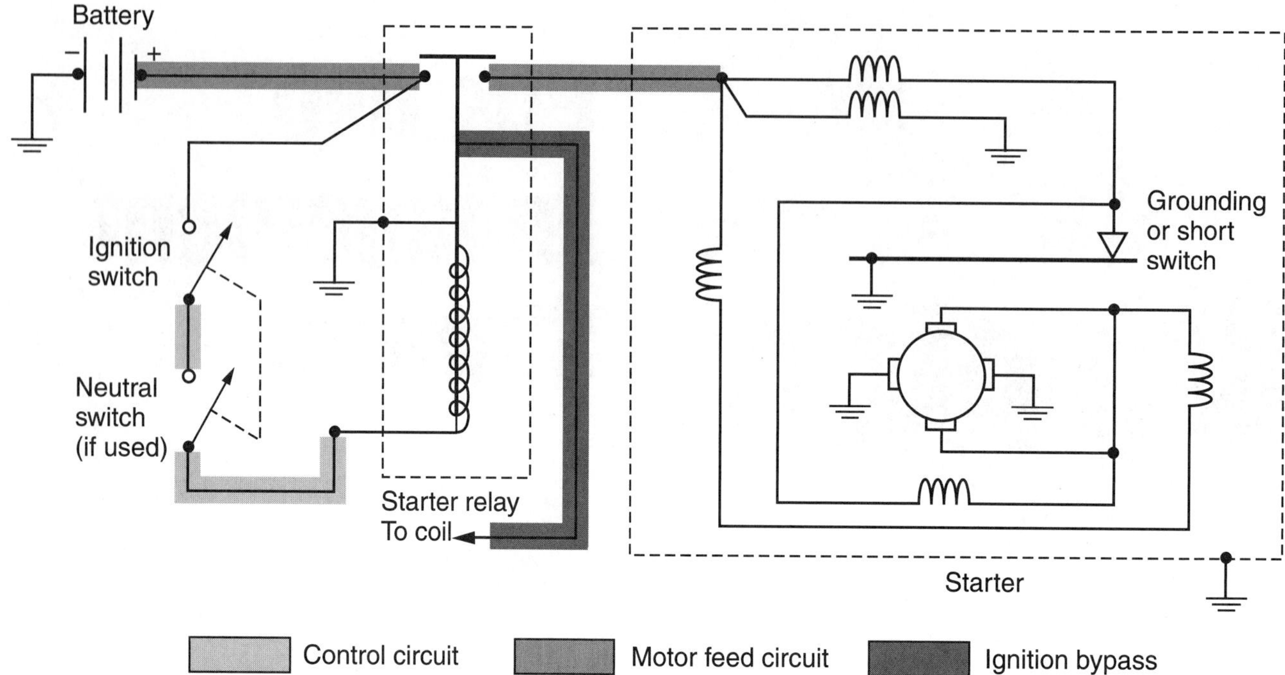

Figure 3. A positive engagement wiring diagram.

plish the same thing. Refer to **Figure 4** and **Figure 5** together now and follow the action. Current arrives at the motor input from the starting relay (key switch in start position) and energizes the first open field coil. The switch is closed, so most of the current goes through the first coil only and then to ground. The other path available is longer because it involves two field coils and the armature. Remember that electricity takes the path of least resistance, or in this case, one field coil and ground. As much as 400 amps will flow through this one coil to ground and produce a tremendous magnetic field. This field will pull down on the movable pole shoe, forcing the drive pinion into mesh with the flywheel. During this movement, the armature does not have any current flowing through it and is therefore stopped. On the end of the movable pole shoe is a tab that will open the grounding contacts once the drive is in mesh. As the contacts open, the current now flows through the rest of the motor and the engine cranks. Many of these motors also use a smaller winding connected in parallel to the drive coil to assist in the holding down of the movable pole shoe. This prevents the spring from disengaging the drive if battery voltage drops down low during cranking.

Figure 4. Positive engagement wiring using a starter relay.

If the grounding switch does not open, the starter motor cannot turn because all of the available current is flowing through ground rather than through the armature.

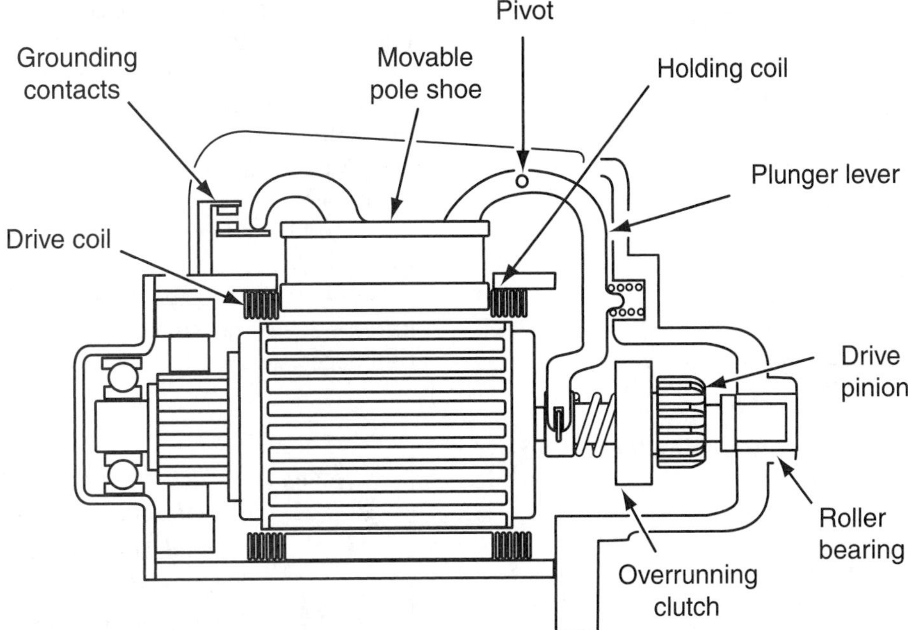

Figure 5. A positive engagement starter.

In this manner, the starter armature is prevented from turning until the starter drive is engaged with the flywheel. The initial current that pulled the starter drive into mesh is also used to open the switch to ground.

CURRENT FLOW THROUGH THE STARTING SYSTEM

Let us follow the current through the positive engagement starter **(Figure 6)**. Notice that the battery cable does not connect directly onto the starter as in the solenoid shift. Instead the B+ cable connects to the **starter relay**.

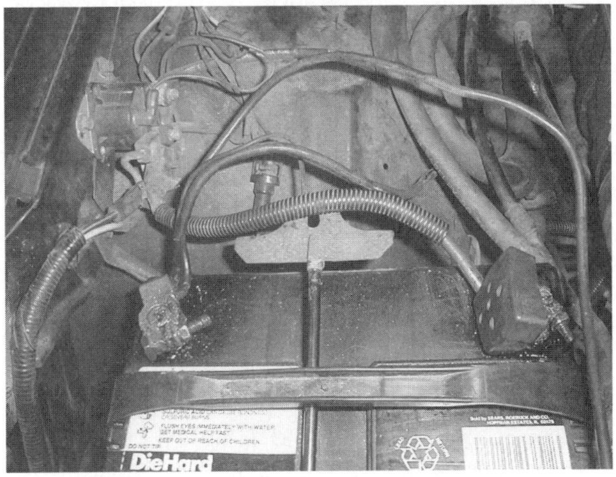

Figure 6. Battery cables to and from a relay.

The output of the relay becomes the input of the motor. Notice that this means that this style has an additional battery cable that will have to be diagnosed in case of difficulties. We will look at the control side of the system in Chapter 41. For our purposes now, let us assume that the relay has closed, bringing current to the motor. The shift arm with its movable pole shoe is up because of spring tension, and the drive is not in mesh with the flywheel. The ground switch is closed, applying a ground after the first open pole. As current begins to flow, the strong magnetic field pulls down on the movable pole shoe. This moves the overrunning clutch into mesh and opens the switch. Once the switch is open, current will flow through the other electromagnets and the armature, allowing the engine to crank.

We have analyzed the current path through the typical motor and have seen that the current will develop magnetic fields that will cause the armature to turn. It is time to discuss the relationship between speed, torque, and current draw. This is a good time to discuss something called **counterelectromotive force**, or **CEMF**. We have all the conditions necessary for the generation of a voltage: a wire loop spinning in a magnetic field. In Chapter 43, Chapter 44, and Chapter 45, we discuss charging systems and you will see that a conductor, a magnetic field, and motion between the two are all that is necessary for the generation of electricity. This is called **induced voltage** and will be used to charge our vehicle battery with the charging system. However, we will also have an induced voltage by the action of the armature spinning in the magnetic field of the pole pieces. This induced voltage will be opposite to the battery voltage that is pushing

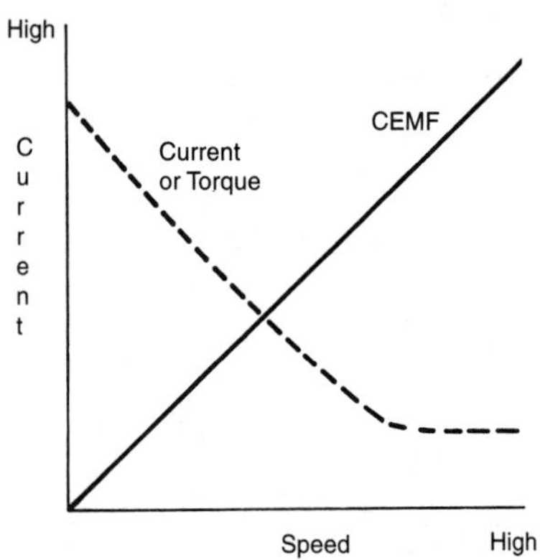

Figure 7. The faster the armature spins, the less current it will draw.

the current through the starter motor. The net result is less current flow through the starter as it spins faster. The faster the armature spins, the more induced voltage is generated. Remember that this voltage is opposite, or counter, to the battery voltage and will reduce the motor's current draw. **Figure 7** shows the relationship between speed and CEMF. Notice that CEMF is at zero if the motor armature speed is at zero. This will allow the maximum current flow through the motor if the armature is stopped. Torque or twisting effort will also be highest with the motor armature stopped. As the armature spins faster and faster, the current and torque will be reduced proportionately. Keep this in mind as you begin your diagnosis of starting difficulties. Slow the armature down, and the current draw will automatically increase. Speed the armature up, and the current draw will be reduced.

When a vehicle has oil that is too thick for the outside temperature, the additional drag that the lubricating oil puts on the engine results in increased cranking amperage.

Summary

- The armature and starter drive for a positive engagement starter is virtually the same as in a solenoid shift starter.
- Positive engagement–style starting systems have a starting relay and an additional battery cable.
- The large switch inside the starter motor connects the open field coil to ground when it is closed.
- The large switch inside the starter motor allows the current to flow through the field coils and armature when they are open.
- Current for the motor comes through the starting relay.
- CEMF is highest at high starter speeds.
- As CEMF is increased, current flow is reduced.

Review Questions

1. Technician A states that the solenoid moves the starter drive into mesh with the flywheel. Technician B states that the shift fork is connected to the solenoid. Who is correct?
   A. Technician A only
   B. Technician B only
   C. Both Technician A and Technician B
   D. Neither Technician A nor Technician B
2. Current flowing through the sliding pole electromagnet will
   A. move the starter drive into mesh with the flywheel
   B. close the shorting switch
   C. spin the armature
   D. none of the above
3. Technician A states that current flowing through a starter generates CEMF. Technician B states that the faster a motor spins, the lower its current draw will be. Who is correct?
   A. Technician A only
   B. Technician B only
   C. Both Technician A and Technician B
   D. Neither Technician A nor Technician B
4. Current for the armature flows through the shorting switch.
   A. True
   B. False
5. The starting relay must close before any current can flow to the motor.
   A. True
   B. False
6. A positive engagement system has two positive battery cables to bring current to the starter motor.
   A. True
   B. False

Chapter 37 Diagnosing and Servicing Positive Engagement Starting Systems

Introduction

In Chapter 35 we looked at the diagnosis of solenoid shift starting systems. There is little difference between the solenoid shift style and the positive engagement style from a diagnosis standpoint. We test for current draw and engine speed in the same manner, using an inductive pickup on a combination VAT. **Figure 1** and **Figure 2** show the same kind of VAT that we used in Chapter 35. The major diagnostic difference comes in voltage drop testing.

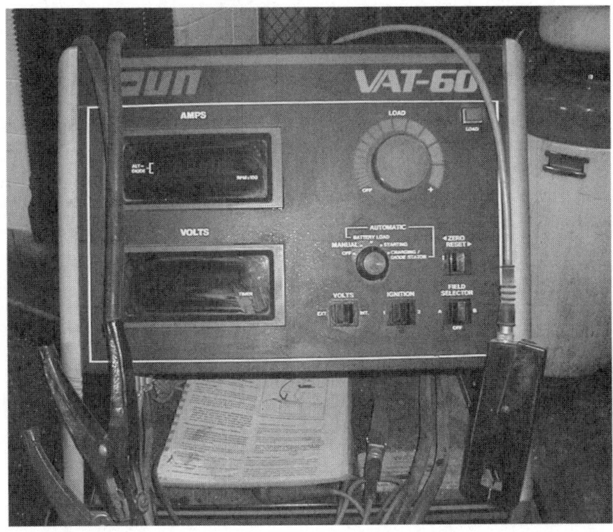

Figure 1. A VAT.

MEASURING CRANKING SPEED

Let us set up the test just like we did for a solenoid shift system. First make sure that the vehicle will not start by removing power from the ignition system **(Figure 3)** or by pulling and grounding the distributor end of the coil wire, as shown in **Figure 4**. This will allow the engine to crank over without starting. Place the inductive pickup from the ammeter around either battery cable and connect the voltmeter leads to the battery terminals. Connect up a tachometer to measure engine speed, and we have all of the connections that we need for initial diagnosis. Crank

Figure 2. A VAT.

Figure 3. Remove power to the ignition system.

Figure 4. Ground coil wire.

the engine over for a long enough time that engine speed is stabilized and record cranking current, voltage, and revolutions per minute. Do you remember the specifications? A positive engagement system will crank over just like the solenoid shift system. A 200-rpm minimum cranking speed is typical, with battery voltage staying above 9.6 volts. Current will be based on the size of the engine. The V-8s will draw approximately 200 amps or less. The V-6s will draw approximately 175 amps or less; and four-cylinder engines will draw approximately 150 amps or less. Few systems will draw in excess of 200 amps.

MEASURING VOLTAGE DROPS

Measurement of voltage drops during cranking will be similar to that of solenoid shift systems with the exception that there is an additional battery cable. Refer to **Figure 5**. Notice that the starter relay is mounted off the starter motor. The relay will supply current to the motor when it is energized. The first voltage that is important is the motor voltage. Place the voltmeter across the motor with the positive voltmeter lead on the M terminal and the negative lead on the starter case. Make sure that the ground is connected to a clean surface **(Figure 6)**. If you have two voltmeters, connect the other one across the battery terminals. With the engine cranking, the difference between the two meters will indicate the circuit's voltage drop. Do you remember the specification? It is 0.2 volt for each side or 0.4 volt for the whole circuit. If the difference is less than 0.4 volts between the two meters, you are finished checking for excessive voltage drops and the system passes. However, if the difference exceeds 0.4 volt, it is necessary to isolate the individual pieces of the circuit, as shown in **Figure 7**. Remember that the voltmeter is mea-

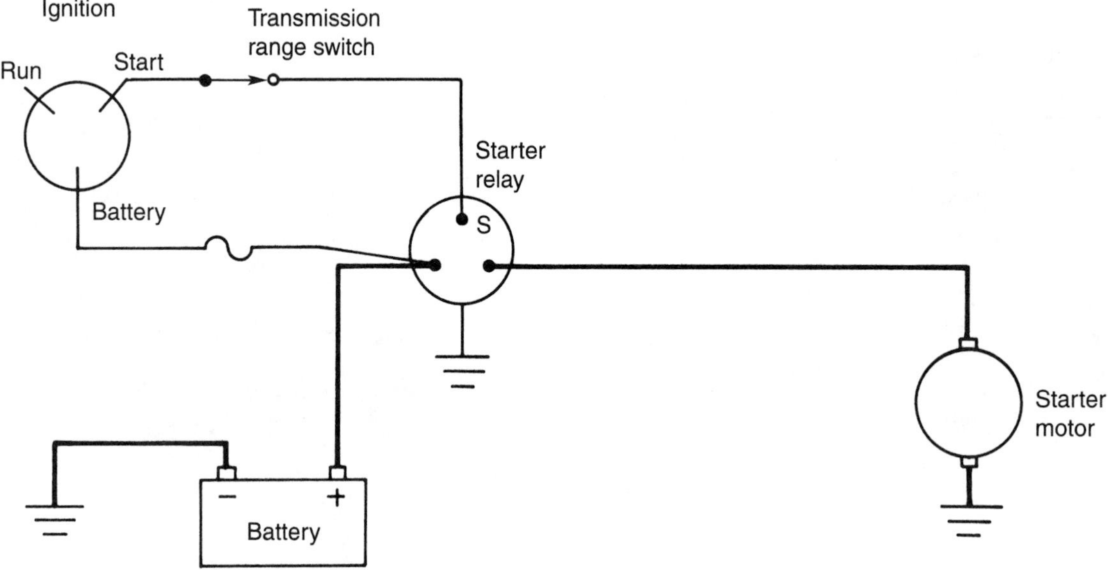

Figure 5. A typical positive engagement system.

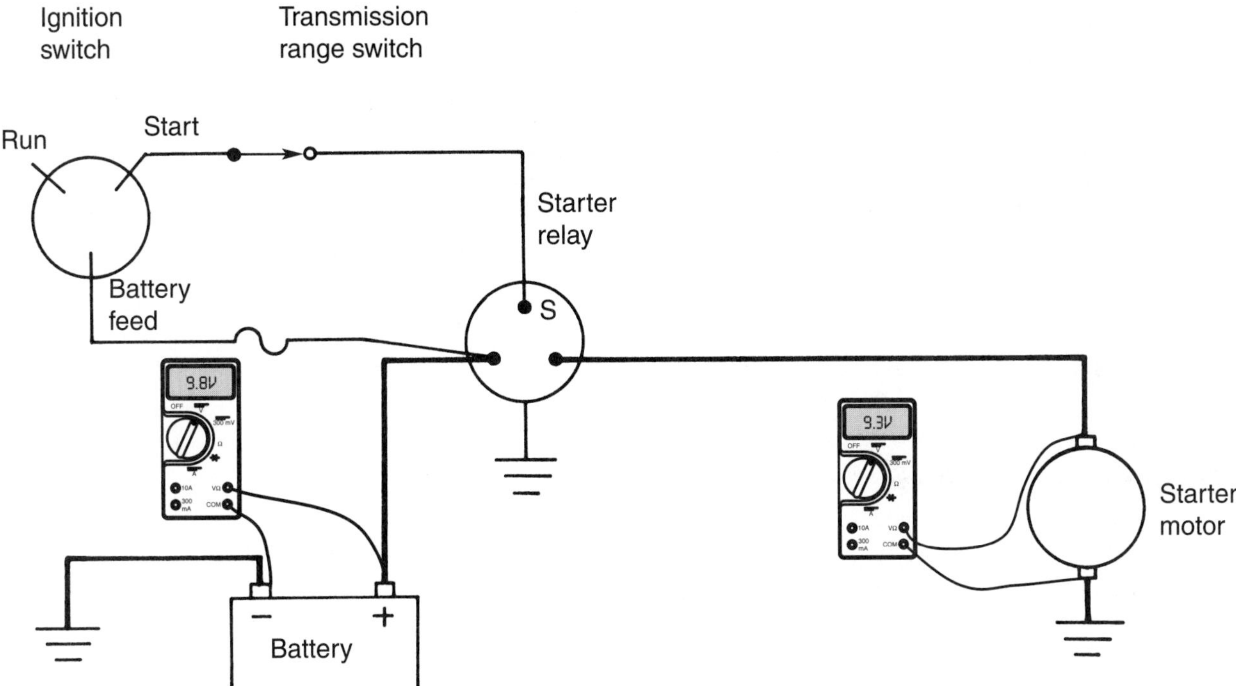

Figure 6. Starter voltage versus battery voltage.

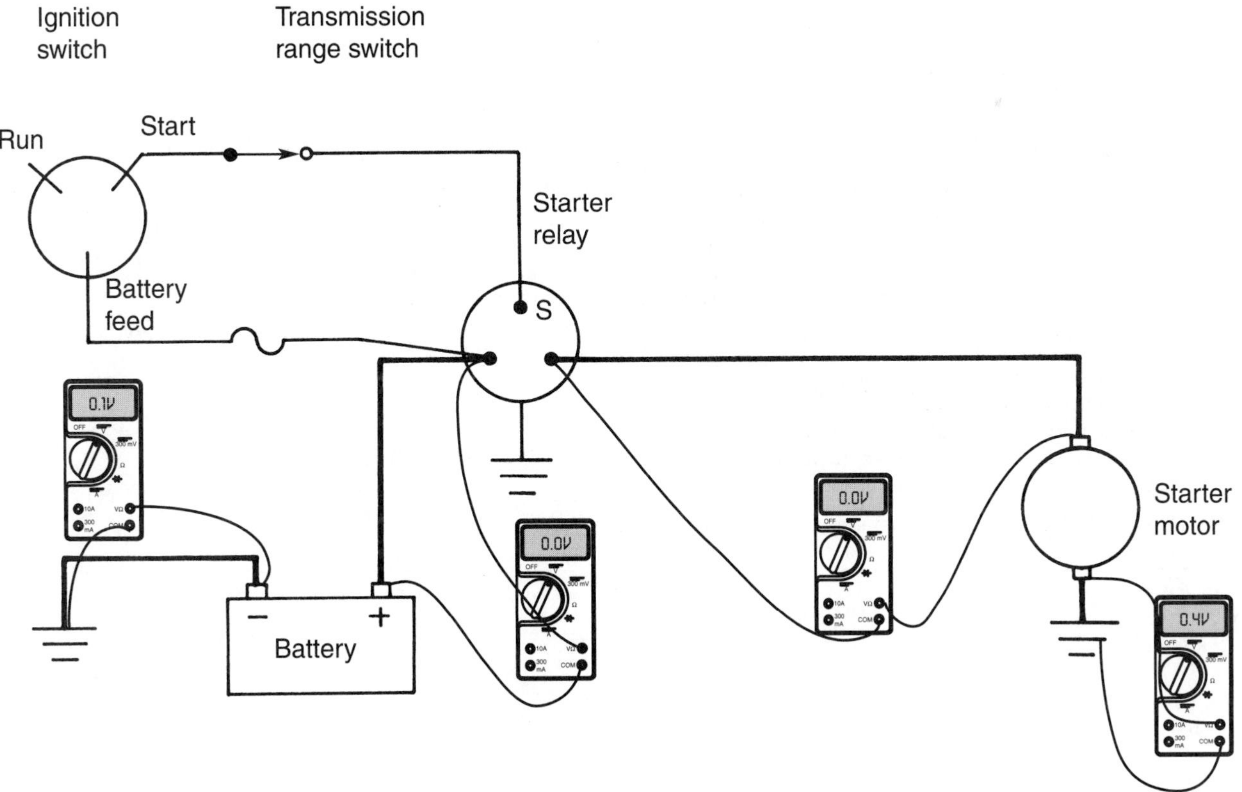

Figure 7. Isolate to pinpoint the voltage drop.

Starter Amperage Draw

1. Place the inductive pick-up clamp around either the positive or negative battery cable. Note: If the battery cable has a small wire attached at the post clamp, it should be placed inside the inductive pickup.
2. Make sure the engine will not start by either removing B+ from the coil (electronic ignition) or by pulling and grounding the coil wire at the distributor cap. Do not allow the ignition system to run open circuit.
3. Crank the car over just long enough to stabilize the amp meter needle; observe and record volt and amp meter.

Voltage ______________ (while cranking)
Amperage ______________ (while cranking)

____ Good—Voltage above 9.6 V
____ Amperage—200 A V8
175 A 6 cylinder
150 A 4 cylinder

NOTE: Amperage readings are approximate and will vary somewhat from one manufacturer to another.

Starter Circuit Voltage Drop

Ground Path

1. Use external voltage setting and the external voltmeter leads.
2. Place the positive voltmeter lead on the starter body (ground).
3. Place the negative voltmeter on the negative *post* of the battery.
4. Crank the car over.

Record Your Reading

____ Less than 0.2 volt reading while cranking; continue with positive voltage drop test
____ More than 0.2 volt reading indicates excessive resistance in the ground return path. Isolate each section of the return path and record the voltage drop with the voltmeter leads placed as follows:

____ Bat negative post to bat clamp
____ Bat clamp to cable ground lug
____ Ground lug to engine block
____ Engine block to starter motor

NOTE: The bad section of the circuit will have the voltage drop.

Positive Path Voltage Drop

1. Hook the negative voltmeter lead on the starter input terminal (after solenoid).
2. Hook the positive voltmeter lead to the positive battery post.
3. Crank the engine over and read the voltmeter. Record below.
   ____ Less than 0.2 volt while cranking reading; continue with charging system testing.
   ____ More than 0.2 volt reading indicates excessive resistance in the B+ path.

Isolate each section of the path and record below:

____ V positive post to positive clamp
____ V positive clamp to cable lug
____ V solenoid input to solenoid output
____ V solenoid output to cable lug (if solenoid is not mounted on starter)
____ V cable lug to starter motor input

NOTE: The bad or excessive resistance part of the circuit will have the voltage drop.

Figure 8. An on-car starting test worksheet.

suring voltage drops that are the result of excessive resistance. Any section that has excessive resistance will show an excessive voltage drop. All readings are taken with the engine cranking.

Figure 8 shows a summary of all the on-car starting tests, including voltage drop testing.

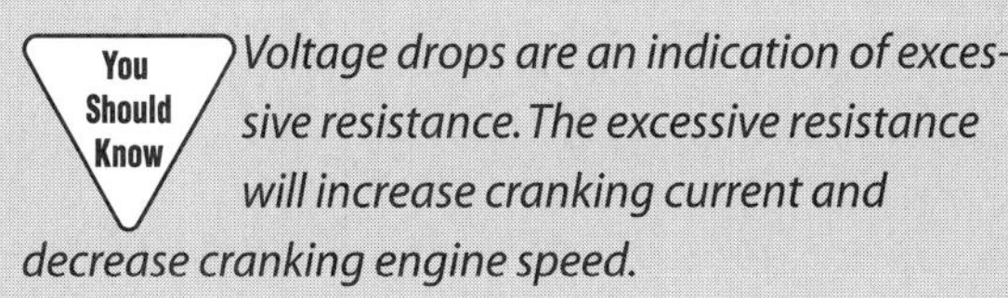

Voltage drops are an indication of excessive resistance. The excessive resistance will increase cranking current and decrease cranking engine speed.

DIAGNOSING UNUSUAL STARTER NOISE

There really is no difference between noise that is generated on a solenoid shift system or a positive engagement system, with one exception. The exception is the condition of the grounding or shorting switch not opening. If the movable pole shoe is pulled down by the magnetic field but the switch does not open, the system will not crank the engine over. There is insufficient field with only one pole in operation to spin the armature with enough force to crank the engine. There will be in excess of 400 amps flowing through the one winding, which will set up a loud hum or buzzing noise. This is the only noise that is different from a solenoid shift starter. Bearing noise or the grinding noise from the starter drive slipping remains the same in both styles of starter motors.

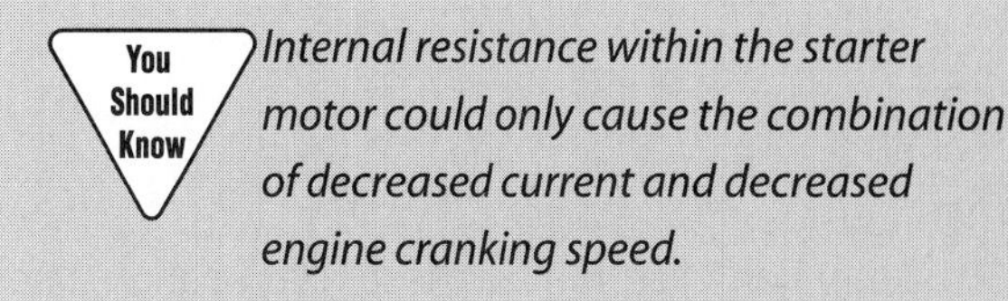

Internal resistance within the starter motor could only cause the combination of decreased current and decreased engine cranking speed.

DIAGNOSING EXCESSIVE CURRENT OR REDUCED CRANKING SPEED

Excessive current and reduced cranking speed generally go hand in hand. Whatever has caused the motor to slow down will increase the current flow. Remember the CEMF. As the armature slows down, it will automatically draw more current unless there is excessive resistance within the motor. Internal resistance will slow down the armature and decrease the current. We discussed this in Chapter 34, the solenoid shift chapter.

Decreased engine cranking speed with increased current has the following causes:

1. A battery that cannot deliver adequate current.
2. An engine that is "tight," that is, possibly having thick oil, excessive compression, blown head gaskets, etc.
3. Voltage drops in the heavy current cables, connections, or relay.
4. A starter motor that is dragging because of bearing problems.

If you start with a battery test, do a cranking current with rpm test, and end with voltage drop testing, the cause of the increased current with decreased cranking speed will be apparent.

Summary

- Specifications for a solenoid shift and a positive engagement starter are the same.
- A 0.2-volt drop is allowed on either the B+ or the B– side of the starting circuit.
- Excessive voltage drops will increase current flow.
- The grounding or shorting switch must open for current to be available to the motor.
- The starting relay accomplishes one of the solenoid's functions, that of turning on the heavy current flow to the motor.
- The movable pole shoe will engage the starter drive to the flywheel.
- A loud hum in conjunction with excessive current and no cranking can be caused from a shorting switch that does not open.

Review Questions

1. Technician A states that excessive current with no cranking might be caused from a closed shorting or grounding switch. Technician B states that an open shorting or grounding switch will allow current flow through the windings of the starter motor. Who is correct?
   A. Technician A only
   B. Technician B only
   C. Both Technician A and Technician B
   D. Neither Technician A nor Technician B
2. Decreased current flow with decreased cranking engine speed can be caused from
   A. thick oil
   B. a voltage drop across the starter relay
   C. internal starter motor resistance
   D. high compression

3. A V-8 should crank over at_____rpm.
   A. 250
   B. 200
   C. 175
   D. 150
4. A battery has failed the load test. Technician A states that this might cause reduced current flow. Technician B states that this might cause decreased cranking current. Who is correct?
   A. Technician A only
   B. Technician B only
   C. Both Technician A and Technician B
   D. Neither Technician A nor Technician B
5. The grounding switch must be closed before turning the ignition key to the start position.
   A. True
   B. False
6. The starting relay is considered all right if it has less than a 0.4-volt drop with starting current flowing.
   A. True
   B. False

Chapter 38 Gear-Reduction Starters

Introduction

The use of gear reduction in starting systems has grown steadily until today it has become the most common type of starter motor. It is important to note that the gear-reduction system is within the starter motor. External controls, relays, solenoids, etc., remain the same. Within this chapter we look at the most common type of gear reduction in use today, that of the permanent magnet style. We also look at a selected older electromagnetic gear-reduction starter.

ARMATURE

The major difference between a permanent magnet starter and an electromagnet starter is the armature. **Figure 1** shows the typical starter drive directly attached to the starter armature. We have looked at this in Chapter 33. Notice the difference in **Figure 2**. There is a set of gears between the starter armature and the starter drive mechanism. Also notice that there are no electromagnets. Electrically, this starter motor is simpler because it does not require current for field coils. Current is delivered directly

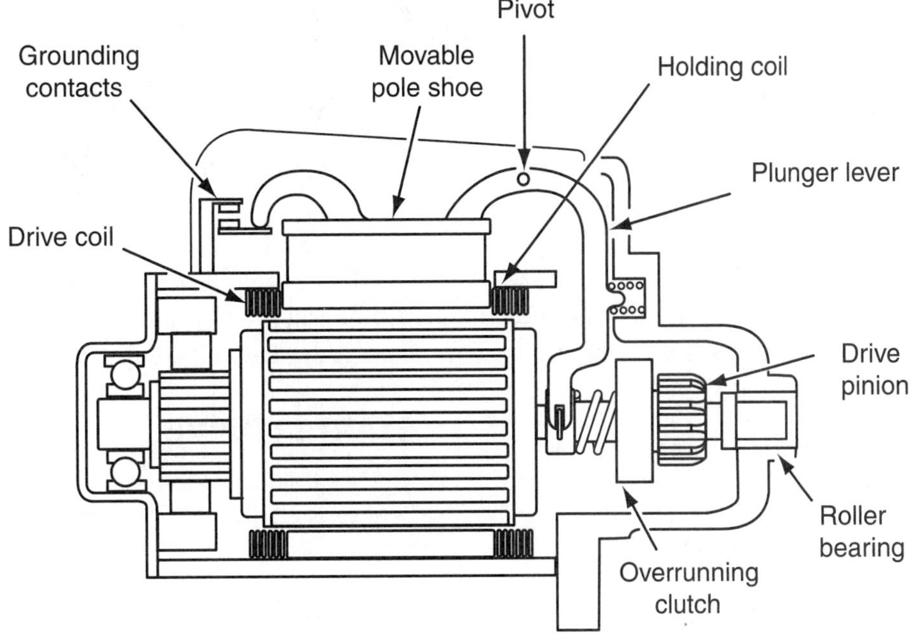

Figure 1. A starter drive connected directly on shaft.

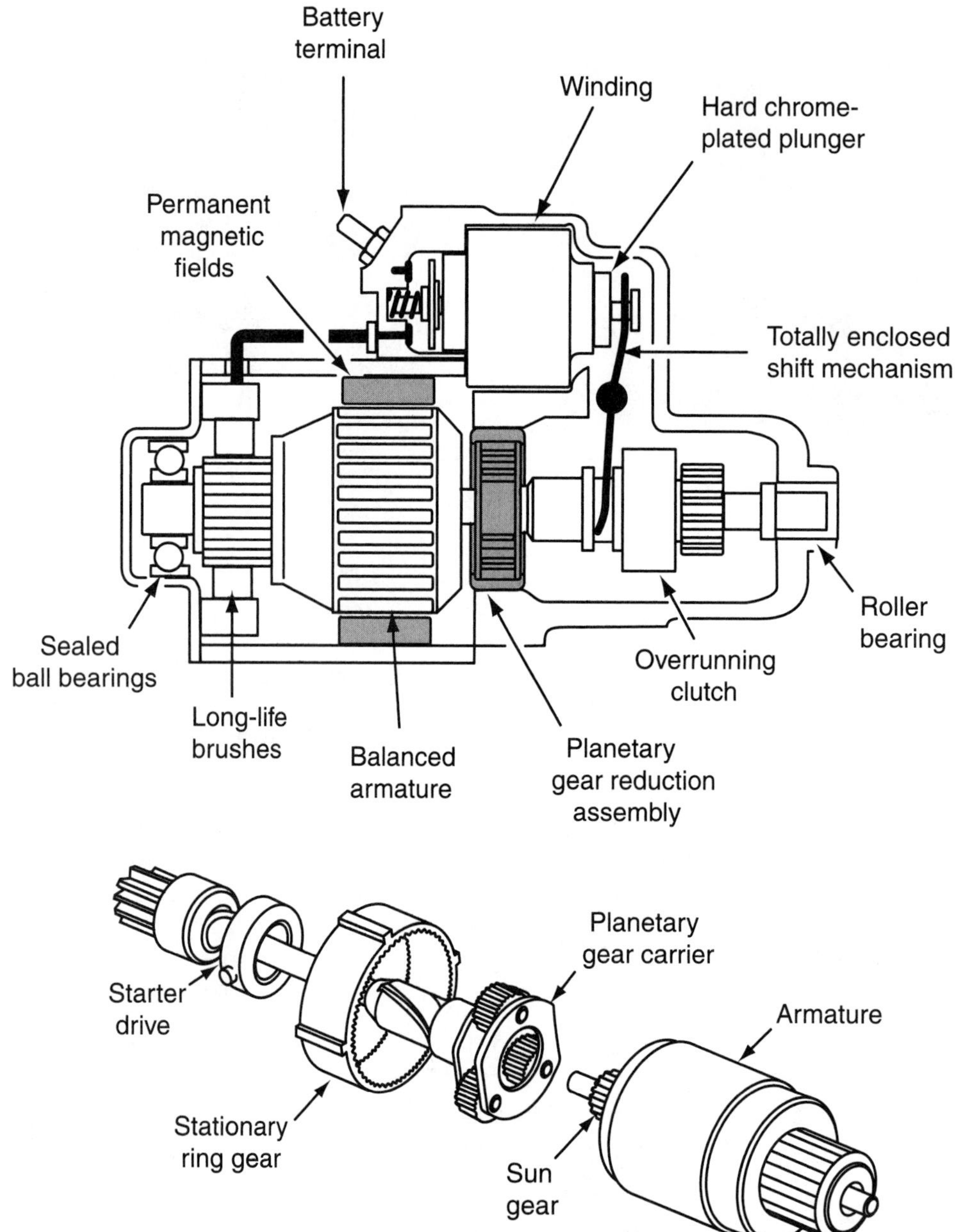

Figure 2. A permanent magnet gear-reduction starter.

to the armature through the commutator and brushes. With the exception of no electromagnets for the fields, this unit will function exactly as do the other styles that we have looked at. This style has been increasingly used because production costs are greatly reduced. Maintenance and testing procedures are the same as for other designs. Notice the use of a **planetary gear**–style gear-reduction assembly on the front of the armature. This allows the armature to spin at greater speeds to increase the torque delivered. In addition, notice the decreased size of the starter body. Reducing the overall size while keeping the torque high improves the adaptability of the unit in certain vehicles in which space is at a premium. The unit still requires the use of a starting solenoid.

STARTER DRIVE

The ability of the starter to turn over the engine via the flywheel is still required. The connection between the armature and the starter drive will be slightly different when compared with the common solenoid shift electromagnetic style. The drive, however, will be the same. An overrunning-style clutch drive is shown in **Figure 3**. It has the same two functions as before: that of allowing the arma-

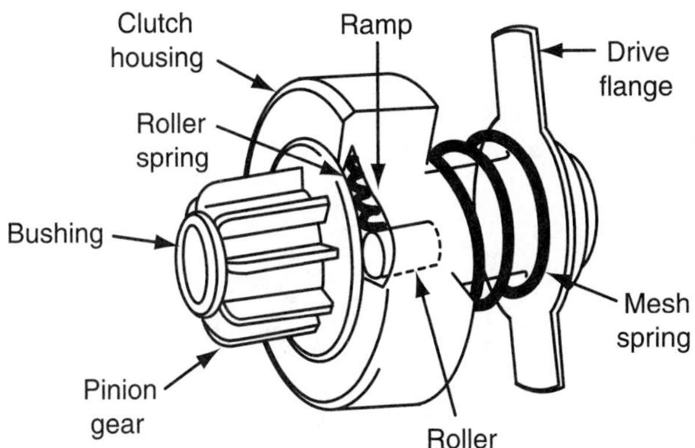

Figure 3. An overrunning clutch.

ture to spin the flywheel but not letting the flywheel spin the armature. We discussed this one-way action in Chapter 35, Chapter 36, and Chapter 37. There will be a ratio of teeth on the drive to teeth on the flywheel of approximately 17–20:1. This ratio will greatly increase the torque available.

TRANSMISSION

Gear-reduction starters use a transmission to increase torque. The gears in this transmission allow the armature to rotate about three times for each time the starter drive rotates. Two different styles of transmissions are in use: sliding gears and planetary gears. Sliding gears were common in Chrysler vehicles and vehicles that used Nippondenso starters that were electromagnetic. The use of sliding gears has been greatly diminished recently; however, many vehicles on the road still have them. Refer to **Figure 4** as we look at how they function. Notice that the armature does not directly spin the starter drive. It is connected to the **reduction gear**, which in turn spins on a pinion shaft. Splined onto the pinion shaft are the overrunning clutch and drive pinion. A shift fork is controlled by the solenoid and will move the overrunning clutch into mesh with the flywheel just before power is applied to the motor. Figure 4 is of a Chrysler starter motor. The Nippondenso starter is similar in function. Typically, this type of arrangement results in a 3.5:1 gear reduction. This means that the armature will spin 3.5 times for each rotation of the starter drive.

Beginning in the late 1980s, most of the manufacturers began replacing the electromagnetic-style starter with the permanent magnet style. Virtually all of the permanent magnet motors use gear reduction to increase the torque available to the engine. This increased torque is especially useful under extreme cold or extreme heat conditions. The majority of the gear-reduction, permanent magnet starters

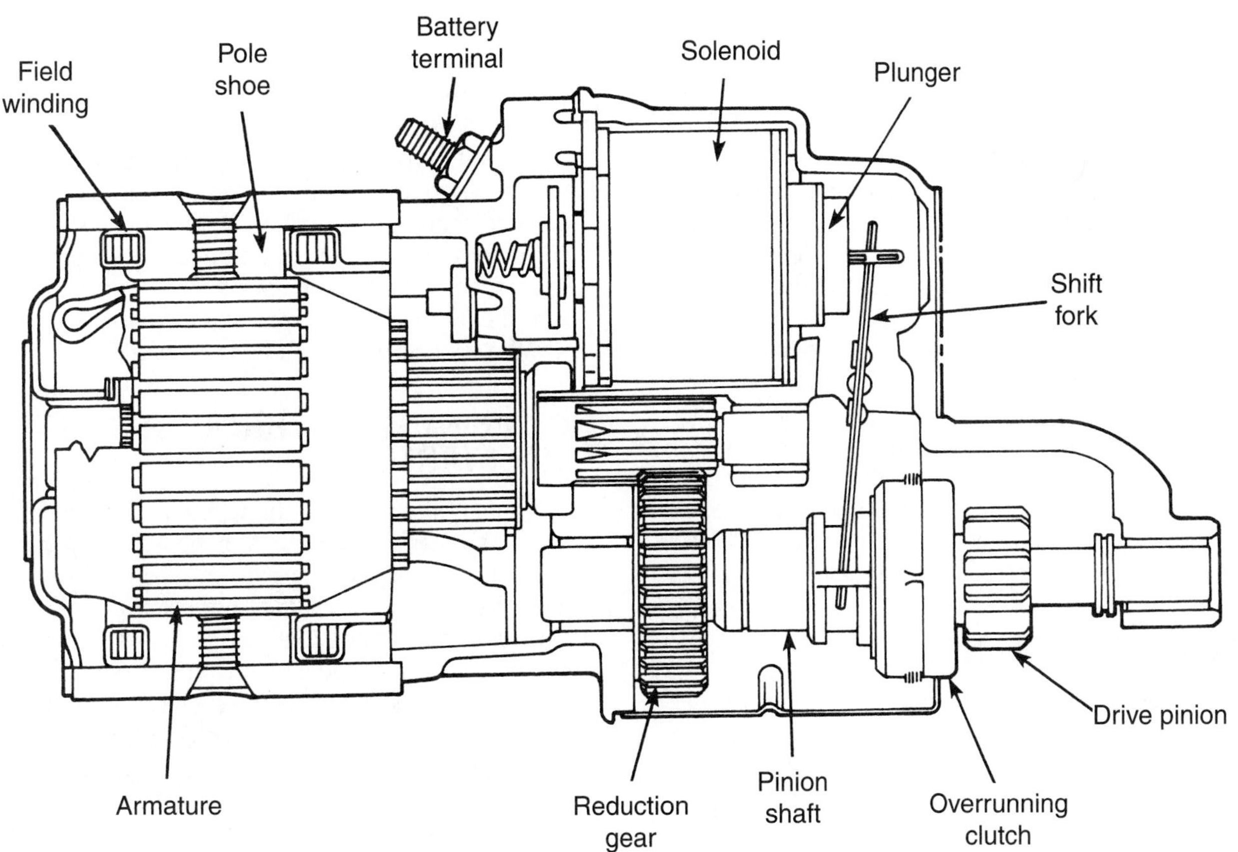

Figure 4. A gear-reduction starter with sliding gear.

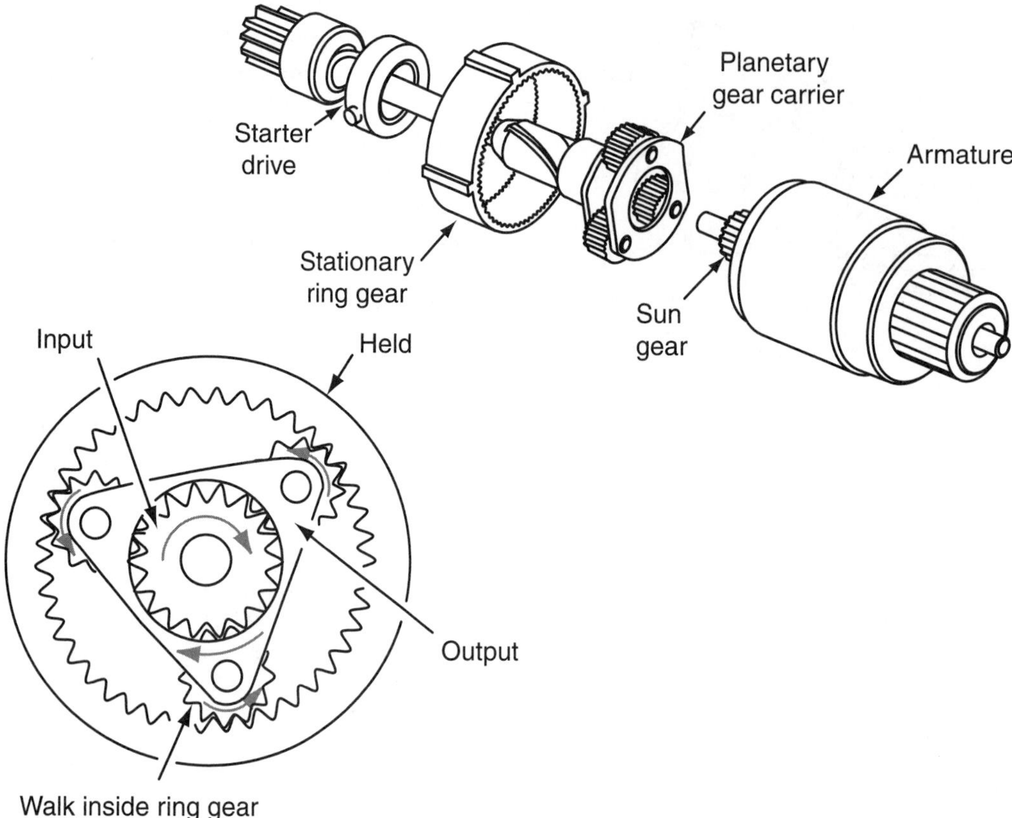

Figure 5. A planetary gearset.

use a planetary gear system to obtain the required gear reduction. Refer to **Figure 5** as we look at how a planetary gearset functions. On the end of the armature is a small gear called the **sun gear**. It will spin whenever the armature is spinning because it is part of the armature shaft. It will spin inside the planetary gear carrier. The three small gears in the carrier are called the planet gears. The planet gears will spin anytime the armature is spinning. As they turn they will "walk" around the inside of the stationary **ring gear**. It is this walking that moves the planetary gear carrier and the overrunning clutch shaft and gives us the gear reduction that is required. Again the overrunning clutch is moved into mesh with the flywheel by the solenoid. This style of starter is the most common in use today. It is still considered a solenoid shift starter. To be completely correct, it would be called a solenoid shift, permanent magnet, gear-reduction starter.

USE OF THE SOLENOID

We have seen how the solenoid continues to be used in gear reduction starters as a primary control. It still functions as a timing device controlling when current is allowed to flow through the starter. This timing is extremely important. The life of the starter pinion would be greatly decreased if we started the armature spinning and then jammed it into the flywheel gear. The acceptable sequence will be to move the drive into mesh with the flywheel and then apply power to the motor. This is easily done by the contacts at the back of the solenoid **(Figure 6)**. The plunger has to be completely inside the solenoid to move the plunger contact disk against the motor feed terminal. Power for the motor will come through the contact disk. The solenoid plunger will have the starter drive in mesh with the flywheel before moving the contact disk. In this way the solenoid will "time" the starting action.

CURRENT FLOW THROUGH THE STARTER

Current flow through the typical permanent magnet motor is greatly simplified. With no electromagnets to power, the armature remains the only "powered" section of the motor. Usually two sets of brushes are riding on the armature commutator. Each set has a B+ brush and a ground brush. The two ground brushes are connected directly to the metal case of the starter. They will become ground when the motor is bolted onto the engine and the negative of the battery is connected. The two positive brushes usually join together in a parallel connection that is the main feed into the motor (the M terminal). The M terminal is usually connected to the output of the solenoid. In this way when the solenoid contact disk connects the bat-

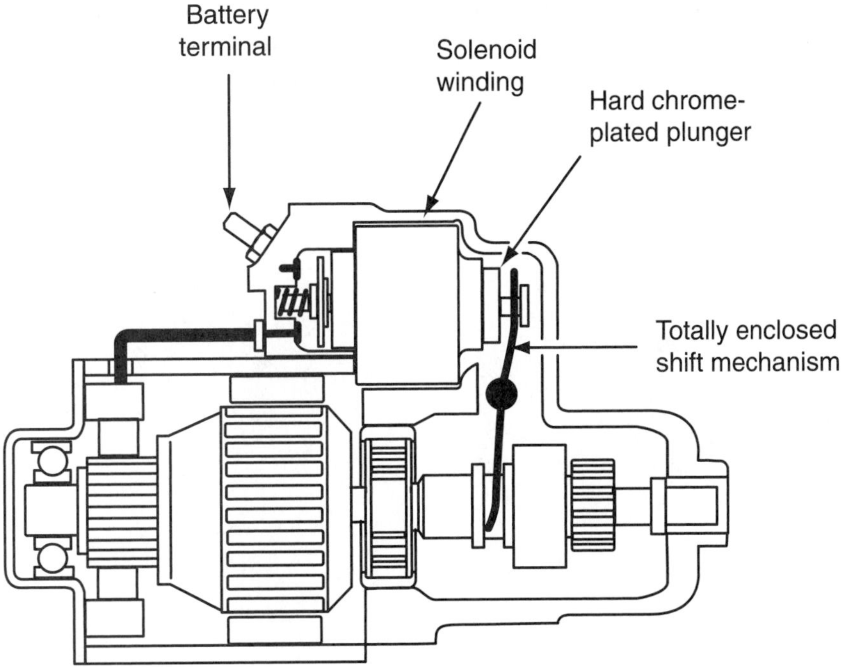

Figure 6. The solenoid is mounted directly to the starter motor.

tery to the M terminal, power will be available to the motor's armature through the B+ brushes. With battery + and battery – available to the armature, it will begin to spin. As it spins, it will in turn spin the starter drive through the gear-reduction transmission. It is important to note that the typical repair does not usually separate the starter from the solenoid. They are usually considered the starting unit and are replaced together.

Summary

- The armature of a gear-reduction starter does not directly drive the overrunning clutch.
- Most gear-reduction starters are also permanent magnet style.
- Permanent magnets replace the electromagnets found in older starters.
- Gear reduction is accomplished through the use of a sliding gear or a planetary gearset.
- The overrunning clutch style of starter drive is the most common.
- Most gear-reduction starters use a solenoid.
- The solenoid times power to the starter after the drive is meshed with the flywheel.
- Generally, there are two sets of brushes in a permanent magnet starter.

Review Questions

1. The use of gear-reduction starters has increased primarily because of
   A. increased use of permanent magnets
   B. the use of larger engines
   C. the need for increased starting speeds (cranking speed)
   D. the simplicity of the unit
2. Technician A states that gear-reduction starters allow the armature to spin faster, thereby reducing the amount of current draw. Technician B states that gear-reduction starters spin the engine slightly slower than a direct-drive starter. Who is correct?
   A. Technician A
   B. Technician B
   C. Both Technician A and Technician B
   D. Neither Technician A nor Technician B

3. Gear-reduction starters use an overrunning clutch style of starter drive.
   A. True
   B. False
4. Gear-reduction starters generally spin the engine over faster than direct-drive units.
   A. True
   B. False
5. The most common type of gear-reduction starter uses a planetary gearset.
   A. True
   B. False

Chapter 39 Diagnosing Gear-Reduction Starters

Introduction

The diagnosis of a gear-reduction starter is not that much different from that of a solenoid shift or positive engagement starter. There are some minor specific differences that we cover within this chapter. Our starting procedure will be the same as for other starting systems, beginning with a fully charged good battery and following through to current testing, voltage-drop testing, and looking for the causes of difficulties.

CRANKING CURRENT

When it comes to the diagnosis of starting systems, it is important to realize that the battery should be considered part of the system and must be tested first to determine its ability to supply the current demands of the system. Remember from Chapter 31 that a less than fully charged battery or an undersized battery will increase starting current and decrease cranking speed. Always do a full battery test before attempting to diagnose starting difficulties. Refer to the battery diagnosis chart printed in Chapter 31.

Cranking current is a helpful diagnosis tool. It and cranking speed are primary pieces of information that will help in either preventive maintenance or no-start diagnosis. The engine has to be able to crank over long enough to get stable readings, so ignition and fuel should be disabled. Do not forget that unburned fuel that enters the catalytic converter could cause it to overheat once the engine starts. Disconnect the injectors **(Figure 1)** or the primary input device that the processor will be using to fire the injectors. In **Figure 2**, the main ignition module connector is disconnected. This will interrupt the crankshaft signal coming to the ignition mod-

Figure 1. Disconnect the fuel injectors before testing.

Figure 2. Disconnect the ignition system before testing.

1250 W	⟷	1750 W	⟷	2000 W
4 cyl	⟷	V-6	⟷	V-8

Figure 3. Cranking watts versus engine size.

ule and the processor, effectively shutting down both ignition and fuel. With both shut down, you will be able to crank the engine over for a long enough time to get an accurate current reading. Make sure that all of the current necessary to crank the engine over will be measured by the inductive pickup. The upper limits for the engine are expressed in watts: 2000 watts for a V-8, 1750 watts for a V-6, and 1250 watts for a four-cylinder engine. Remember that watts is the voltage multiplied by current **(Figure 3)**. During cranking, the battery voltage should not drop much below 10 volts. If it drops below 10 volts, there is a problem. In theory, you have already tested the battery, and it should not be part of the problem. If you have not tested the battery, do it now; we have discussed how it can affect the amount of cranking current. With a correctly sized, fully charged, functional battery, the cranking current should be under the upper limits of 200 amps for a V-8, 175 amps for a V-6, and 125 amps for a four-cylinder engine, with battery voltage staying at approximately 10 volts during cranking. If the wattage or cranking current is too high, we need to realize that we have either an engine problem, a starter problem, or excessive external voltage drops.

DIAGNOSING EXCESSIVE CRANKING CURRENT

Excessive cranking current is one of the most frequent problems encountered in starting systems. Once we eliminate the battery as a possible cause, we are left with an engine problem, a starter problem, or excessive voltage drops. With the engine cranking over, listen to the noises generated. Do they sound normal? Frequently, a faulty starter will be noisy. If things sound normal, we need to move on to testing the voltage drops of the system. Remember that the voltage delivered to the motor helps to determine the cranking current. Higher voltage allows the motor to spin faster and draw less current. If a voltage drop is present, the delivered voltage to the motor will be reduced, the motor will spin slower, and the current will increase. If this sounds familiar, it is because we have looked at it in the other starter-diagnosing chapters. The only difference is that a gear-reduction starter will generally draw less than a non–gear reduction starter. This makes slight voltage drops less important. When the sum of all of the voltage drops exceeds around approximately 0.5 volt, we have a problem that needs to be addressed. Here is where we will use the lowest scale of the voltmeter, typically 2–4 volts. Keep in mind that current flowing through resistance is the cause of voltage drops. A resistance will have a different pressure or voltage on each side. In Chapter 5, we measured the voltage drops across the resistance of a load. We will do much the same thing now while the engine is cranking.

The voltmeter will be placed across either the B+ or the ground circuit, as shown in **Figure 4**. This setup allows us to pinpoint the circuit half that has the excessive resistance. Once the circuit half is identified, it becomes a simple task to divide the circuit into small pieces and place the voltmeter leads across the pieces, looking for voltage drops with the engine cranking. **Figure 5** illustrates this. The specification for the B+ or ground side of the circuit is 0.2 volt maximum with normal current flowing. If higher-than-normal current is flowing, adjust the specification upward. If, for instance, 50 percent more current is flowing, the specification will be adjusted upward by 50 percent also to 0.3 volt. We will be able to locate our problem when the voltmeter reads the voltage-drop. As you do detailed voltage-drop testing, remember that the positive voltmeter lead will be connected into the part of the circuit that is most positive and the negative lead into the most negative. For preventive

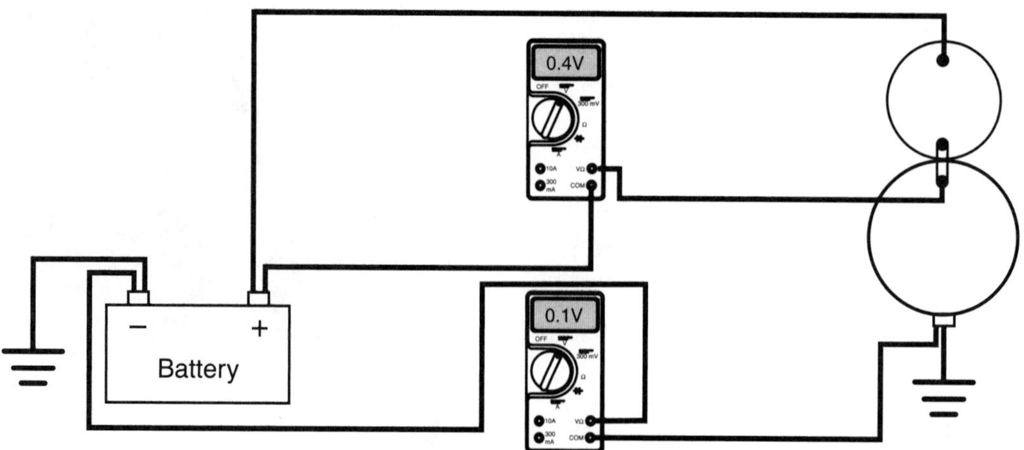

Figure 4. Measuring voltage drops.

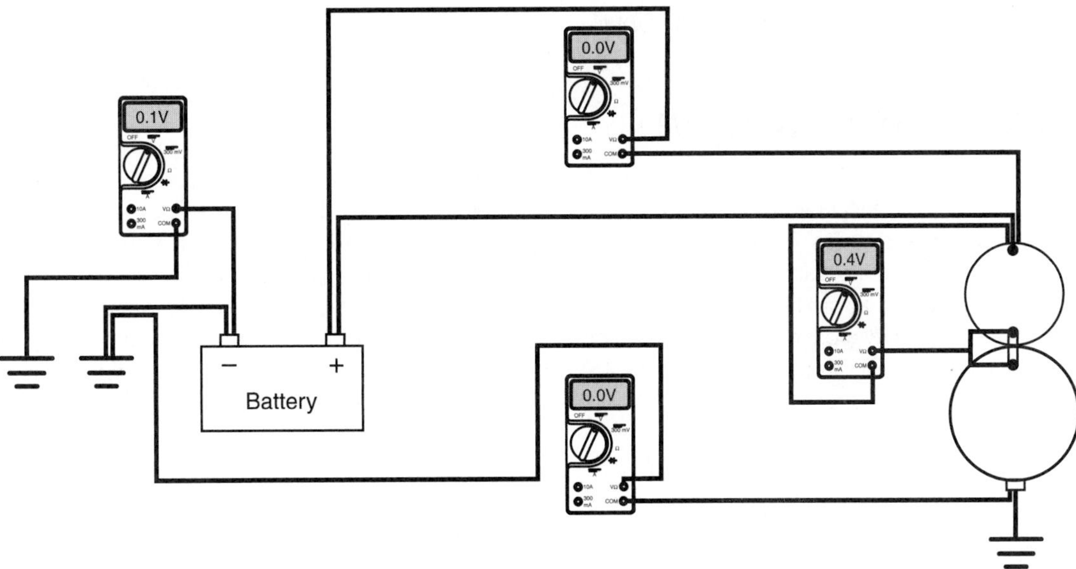

Figure 5. Isolate the circuit to find specific voltage drops.

maintenance checks, both the positive and negative paths should be tested, and anything in excess of the 0.2-volt B+ or 0.2-volt ground specification should be repaired or replaced. If a connection appears to be where the resistance is, it should be cleaned, retightened, and retested. Corrosion and/or loose connections will cause voltage drops and increased current draw. The only part of the circuit that is designed to have more than the 0.2-volt drop is the starter motor. Because the solenoid is mounted onto the starter, it is tested along with the rest of the B+ path cables. It is considered to be part of the total allowable 0.2-volt drop.

By doing the detailed voltage-drop testing outlined here, you will be able to pinpoint the resistance. Repairing it is simply a matter of replacing the cable, solenoid, or whatever had the voltage drop. Voltage-drop testing should be done at least once a year to pick out a resistive component before it causes a no-start condition. Most customers appreciate preventive maintenance designed to prevent no-start conditions. After the voltage drops have been eliminated or at least reduced to an acceptable level, cranking amperage is used to condemn a starter motor. Higher-than-normal amperage without excessive voltage drops is an indication of starter problems or an overly tight engine. The usual procedure, at this point, is to pull the starter off the vehicle and disassemble it. The technician will look for obvious causes of excessive amperage draw, such as worn bushings or shorted windings. The key concept here is to not automatically pull the starter every time excessive amperage shows up. Eliminate voltage drops and test batteries first. Pulling the starter should be the last, not the first, step you do. Keep in mind also that a tight engine will increase the amperage draw of the system. The tight engine is usually diagnosed by first checking all other causes of excessive starting current. Once the other causes are eliminated, it will leave the tight engine as the only option left. Obviously, the tight engine is also going to be the most expensive for the consumer because it will usually necessitate rebuilding or replacing the engine. Frequently, the engine will also give the consumer some warning with strange noises, such as knocking. One cause of excessive cranking current caused by the ignition system might be mistaken for a tight engine. This cause is overadvanced ignition timing. It is a possibility on vehicles with adjustable timing that the timing has either slipped or been set too advanced. This will increase the amount of current necessary to crank the engine over because the starter is fighting the advanced timing. We look at timing in Chapter 49. The easiest way to eliminate timing from the possible causes of excessive current is to make sure that

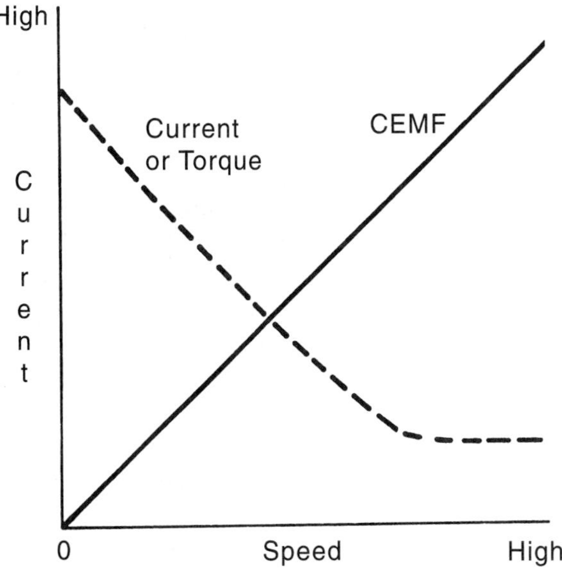

Figure 6. The relationship between speed and current.

the ignition system is disconnected when you measure cranking amperage.

DIAGNOSING REDUCED CRANKING SPEED

There is little difference between diagnosing excessive current and reduced cranking speed. The reason for this is that cranking speed will determine cranking current. It is virtually impossible to separate one from the other. If the armature is turning slowly, it will draw more current. The speed of the armature is dependent on the amount of delivered voltage. The relationship between current and speed is shown in **Figure 6**. Usually when you experience increased current draw you will have decreased speed. This diagnosis then becomes the same. Test the battery, then look for voltage drops, then look at the starter, and leave the engine last.

There is one exception to the higher current and reduced-speed diagnosis method. If the engine cranks slowly (substantially less than 200 rpm) and the cranking current is lower than normal, you need to replace the starter motor. Excessive internal resistance is the only cause of reduced speed and low current. Once the current is flowing through the inside of the motor, it will react like any other resistive load. If additional resistance is present, the current will be reduced and less work will be accomplished

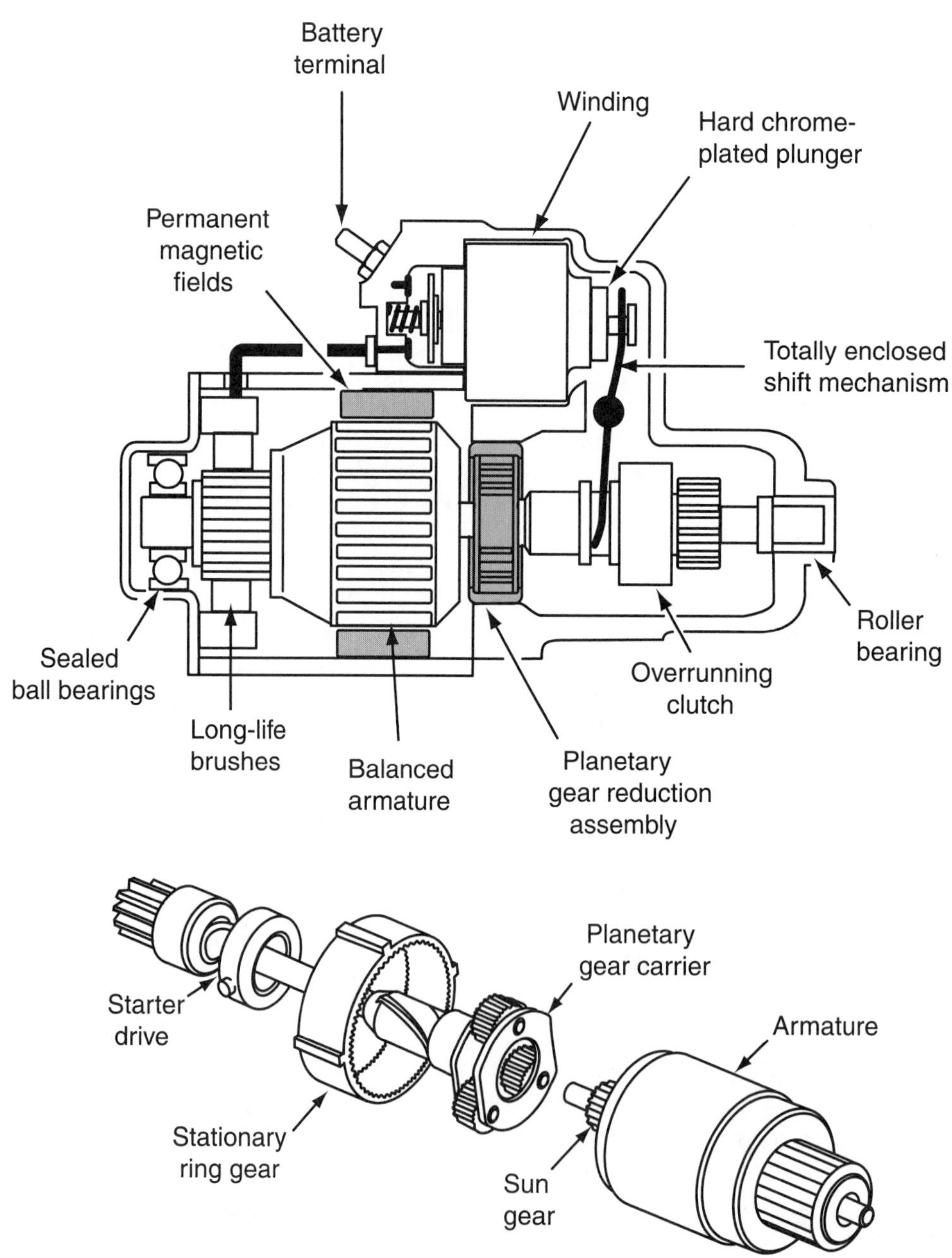

Figure 7. The gear-reduction unit can be the source of noise.

(reduced speed). Make sure you realize that internal resistance reduces cranking speed and reduces cranking current, whereas external resistance reduces cranking speed and increases cranking current. Taking note of the difference could save you a lot of diagnosis time in the future.

DIAGNOSING UNUSUAL STARTING NOISE

Starting motors are like any other motor. They contain bearings, brushes, a commutator, and an armature, which can generate slight noise. However, the starter motor has the additional possibilities of the starter drive slipping **(Figure 7)** or a problem with the gear-reduction unit. The drive and the gear-reduction unit will generate the same type of noise and are difficult to separate without taking the motor assembly apart. Realistically, few shops change out starter drives, and no one replaces gear reduction units, with the exception of starter remanufacturers/rebuilders. With the starter removed from the vehicle, an inspection of the drive gear **(Figure 8)** might reveal damaged teeth. Also, at this time, manually turn the engine over and inspect the teeth of the flywheel or flex plate ring gear **(Figure 9)**. Damage to either might generate the objectionable noise while cranking. If the ring gear teeth are in good shape, it is a good chance that the starter needs to be replaced. Make sure that you inspect the entire ring gear, not just the small section exposed by the removal of the starter.

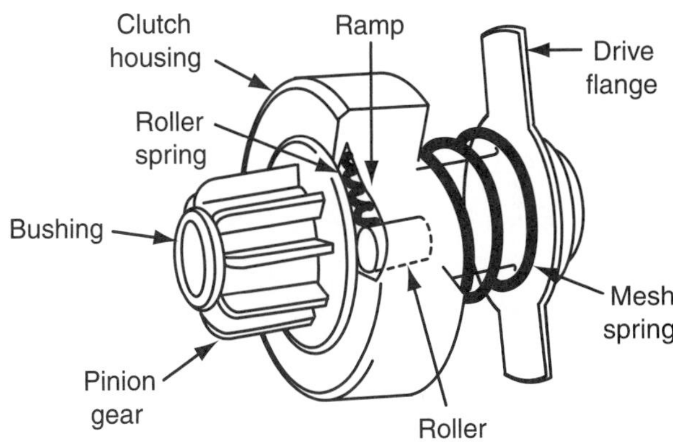

Figure 8. Starter drives can generate noise.

Figure 9. Inspect the teeth of the flywheel or flex plate for damage.

Summary

- A full battery test should always precede starting system testing.
- A partially charged battery will increase cranking current and decrease cranking speed.
- To effectively test the cranking system, the ignition and fuel systems must be disconnected so that the engine can crank and not start.
- Batteries should deliver current to the starter without their voltage dropping below 9.6 volts.
- The maximum voltage drop allowed on either the B+ or B– side of the circuit is 0.2 volt.
- Reduced cranking speed will automatically increase the cranking current.
- Overly advanced timing can increase cranking current and decrease cranking speed.
- Decreased current and decreased cranking speed can be caused only by internal (inside the motor) resistance.

Review Questions

1. There is a 0.8-volt drop on the ground side of the starting circuit. Technician A states that this might cause increased cranking current. Technician B states that this might cause reduced cranking speed. Who is correct?
   A. Technician A only
   B. Technician B only
   C. Both Technician A and Technician B
   D. Neither Technician A nor Technician B
2. A battery fails the load test. What should be done before starting testing?
   A. Nothing; continue with the starting testing.
   B. The battery cables should be cleaned.
   C. The battery should be replaced.
   D. A three-minute charge test should be run.

3. A gear-reduction starter cranks the engine over at 150 rpm. This is
   A. OK for any starter
   B. too fast
   C. too slow
   D. OK for this starter
4. A tight engine will result in reduced cranking current and reduced cranking speed.
   A. True
   B. False
5. What are the causes of excessive current during starting?
6. What are the causes of reduced cranking speed?

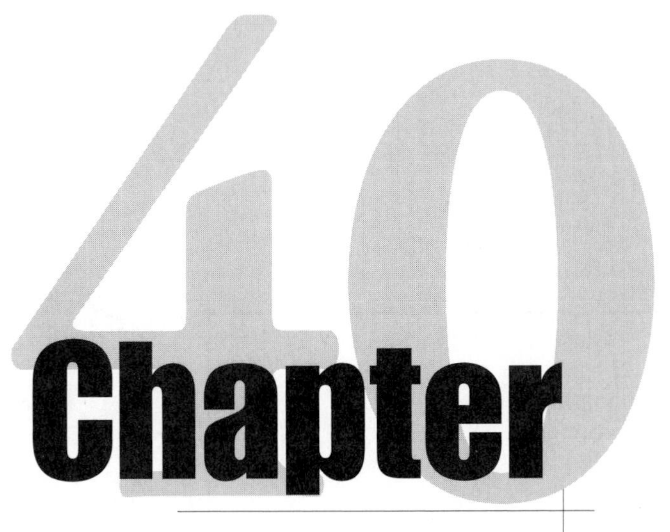

Chapter 40 Starter Controls

Introduction

Getting the engine to crank over is the focus of this section of this book. Each manufacturer uses a slightly different method and different components. The object of all of the controls is simple and straightforward, though: crank the engine only when certain conditions are present.

STARTING RELAY

We have looked at the solenoid already, and it is a type of relay. However, some of the manufacturers use a relay that when closed will allow current to flow to the solenoid S terminal or directly to the starting motor. **Figure 1** is a photo of a common Ford starting relay. It is usually mounted to the fender close to the battery and has three terminals. **Figure 2** identifies the basics of the circuit. The B terminal is directly connected to the positive terminal of the battery and uses a red wire (RED). The S terminal is connected to the starter motor solenoid and uses a yellow wire with a blue stripe (YEL/LT BLU), and the S terminal comes from the ignition switch through the TR and uses a red wire with a light blue stripe (RED/LT BLU). The TR is a normally closed switch that will open if the gear shift is in anything except neutral or park. As a safety feature, this prevents the starting system from functioning with the vehicle in gear. When the ignition switch is turned through the run position and into the start position, current flows through the switch, the TR, and down to the starter relay winding. Notice that there is a black (BLK) ground wire that grounds the relay to the center rear of the engine. The magnetic field will pull the relay plunger in and close the relay contact switch. With the switch closed, current will flow to the solenoid and starter motor. The relay allows more control over the system and follows the basic function of controlling high current with low current. A high resistance or open condition in the relay coil could produce a no-start. If there is insufficient magnetic field to close the contacts, the starter motor solenoid does not see B+ from the battery and the engine never cranks. There are different uses for starter relays, but the Ford setup described here is common. General Motors, on the other hand, uses a small relay, called the crank relay. Notice from the wiring diagram in **Figure 3** that the system is similar to that of the Ford because neither passes the current that cranks the vehicle over. The ignition switch directly supplies current to the relay coil, whereas the ground for the coil will be

Figure 1. A Ford starting relay.

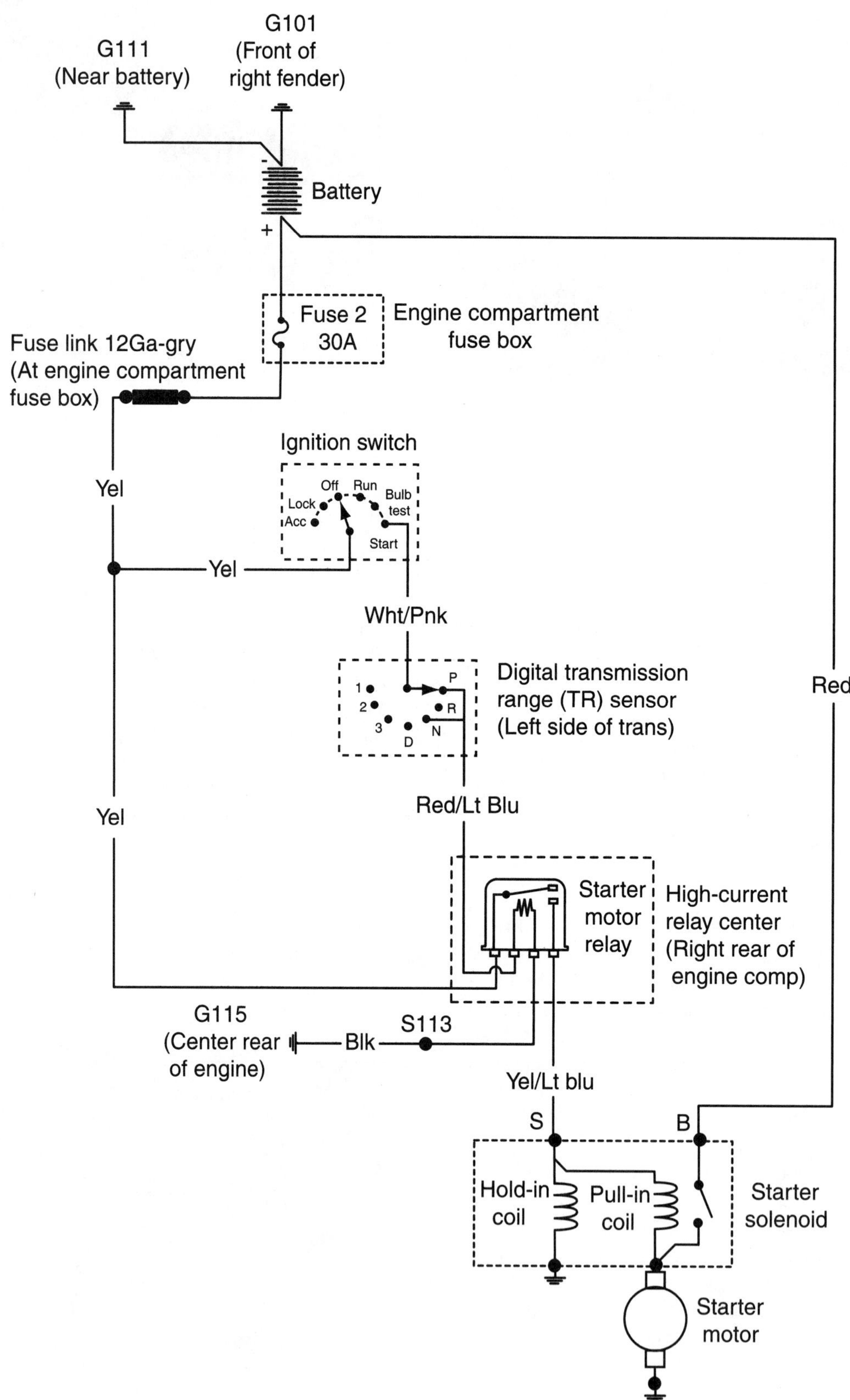

Figure 2. A Mercury starting circuit.

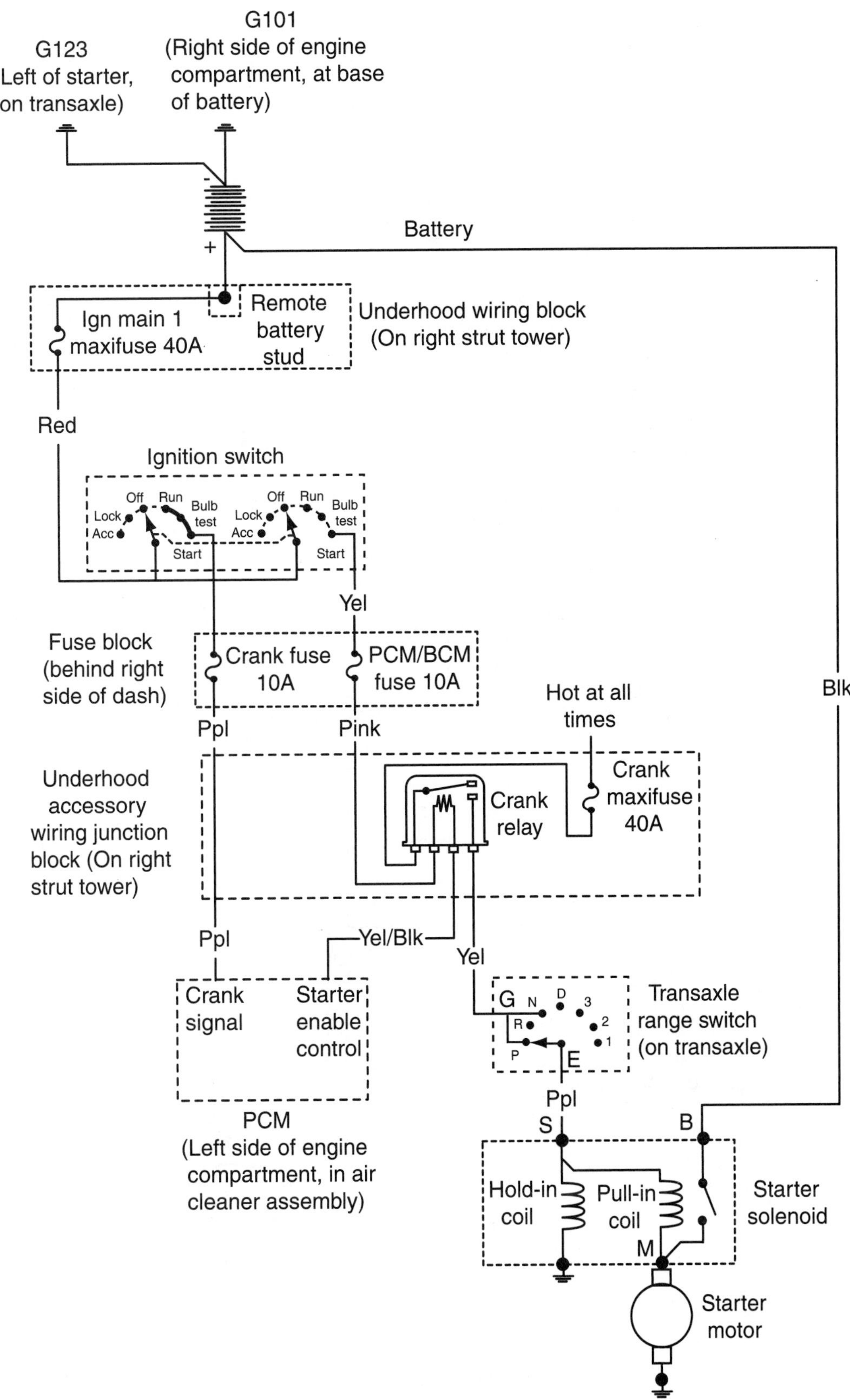

Figure 3. A Buick starting circuit.

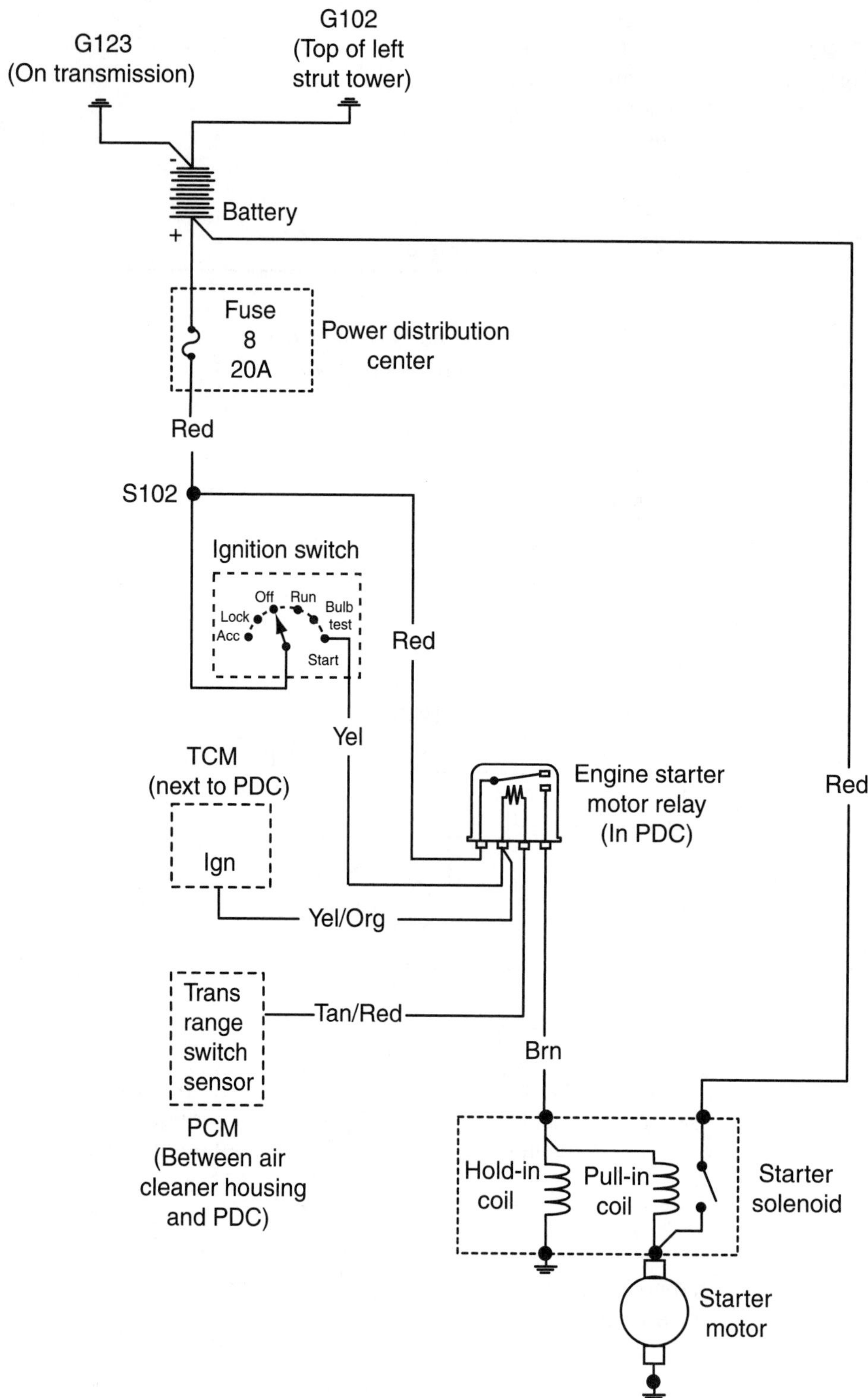

Figure 4. A Chrysler starting circuit.

supplied by the PCM through a circuit called **starter enable control**. This circuit allows the PCM to have some control over starting. In this way, if there has been some type of failure that will prevent the vehicle from starting, it might not crank over. The starter will not be "enabled" and the vehicle will not crank. Power from the relay goes directly through the TR and then to the solenoid—a variation, at least as far as the TR is shown in **Figure 4**, which shows that from a Chrysler vehicle. The only difference from the General Motors version is the lack of a TR switch. The control of the circuit is again achieved through the PCM. The PCM has the TR as an input, so it knows the transmission gear. If the vehicle is in gear, the PCM removes the ground from the engine starting motor relay and the vehicle does not crank.

USE OF A TRANSMISSION RANGE SWITCH

Vehicles must provide some control over the starting circuit to prevent the engine from cranking and starting if the transmission is in gear. Without this control, it is likely that the vehicle could cause a serious accident. Just how the TR is used varies by manufacturer, but let us look at the switch in general terms. First of all, it is a one position normally closed, other positions open switch. This means that only the one position that the gear switch is in will be closed and all other terminals will be open. We have a wiper that moves through the positions and directs the continuity to each position as the gear shift moves.

The TR is a safety feature of the vehicle. It should never be jumped or bypassed because the vehicle could start in gear and cause an accident.

Refer to **Figure 5**, which shows the TR from a Ford product. Power comes from the ignition switch when it is placed in the start position. The white wire with a pink stripe at terminal 12 brings power into the TR. The arrow inside the TR shows the position of the switch. It is currently in the park position. This will allow B+ at terminal 10, which leads down to the starter relay. What would happen if the switch were in the R position (reverse)? There is no continuity through the TR in this position. The switch is open and the engine cannot crank. The TR can be placed anywhere in the starting circuit. Refer to **Figure 6**. Notice that the TR is placed after the crank relay in the load circuit to the solenoid. These circuits are examples of the use of a TR in the starting circuit. There are many variations, but they all operate on the same principle: a series switch that will prevent the vehicle from cranking if the vehicle is in gear. If the vehicle has a manual transmission, the TR becomes a CS, which is normally open and closes when the driver depresses the clutch. The CS functions the same as the TR.

THEFT-CONTROL STARTER CIRCUITS

Many vehicles come from the factory with theft control. It is frequently part of the alarm system. **Figure 7** shows a Lumina starting circuit. Power for the solenoid comes through a theft-deterrent relay, which is under the dash. The theft-deterrent module, which is also under the dash, controls this relay. The module needs to see that the owner has turned off the alarm and theft system by using either the key or the electronic unlock button or it will not apply the ground to the relay. Without the ground, the vehicle will not crank over. Notice that the TR is placed in the relay control coil circuit. This is the simplest of theft-deterrent systems.

Another form of theft deterrent is shown in **Figure 8**. Notice in the right center that the ignition key is shown with a resistor pellet. When a resistance-pellet ignition key is inserted into the ignition key lock cylinder, the white wire with a black stripe (WHT/BLK) and the purple wire with a white stripe (PPL/WHT) will send the resistive value to the body control module. If the value is correct and matches the preset value within the module, it will apply the ground to the starter relay via the yellow wire with a black stripe (YEL/BLK) (pin C11).

General Motors' systems will shut the vehicle starting system off for a predetermined time if the wrong ignition key is used.

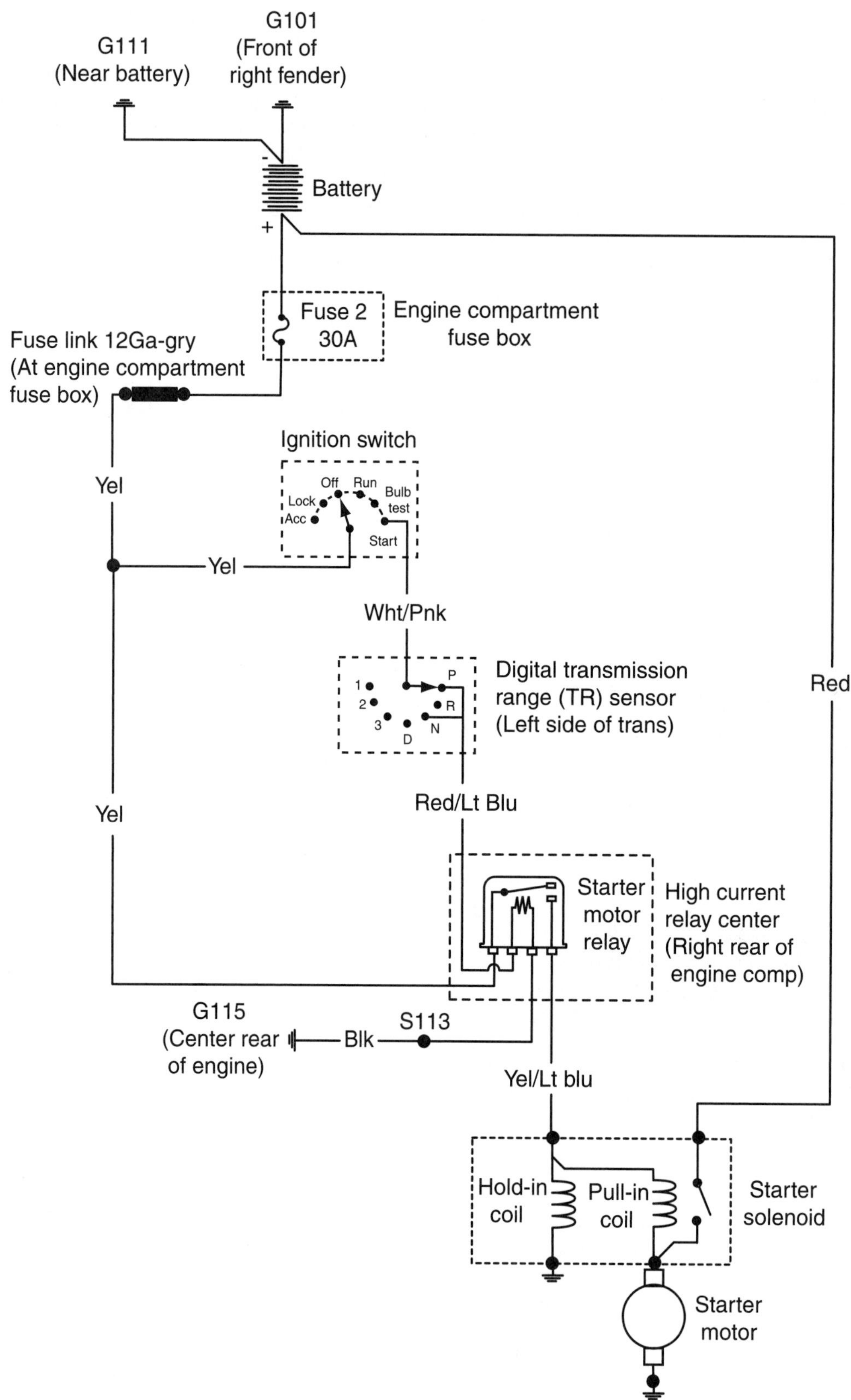

Figure 5. Mercury starter controls.

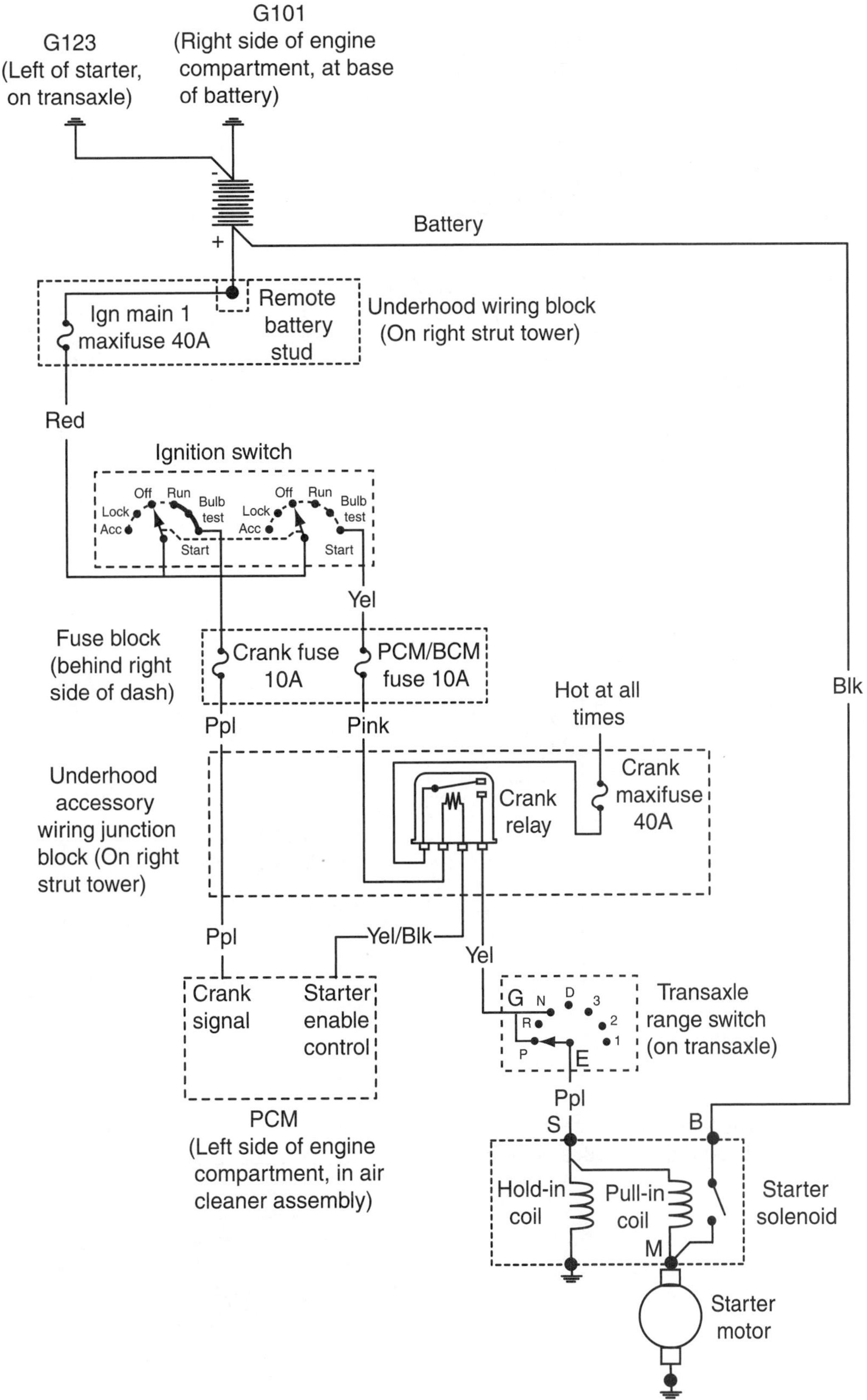

Figure 6. Buick starting controls.

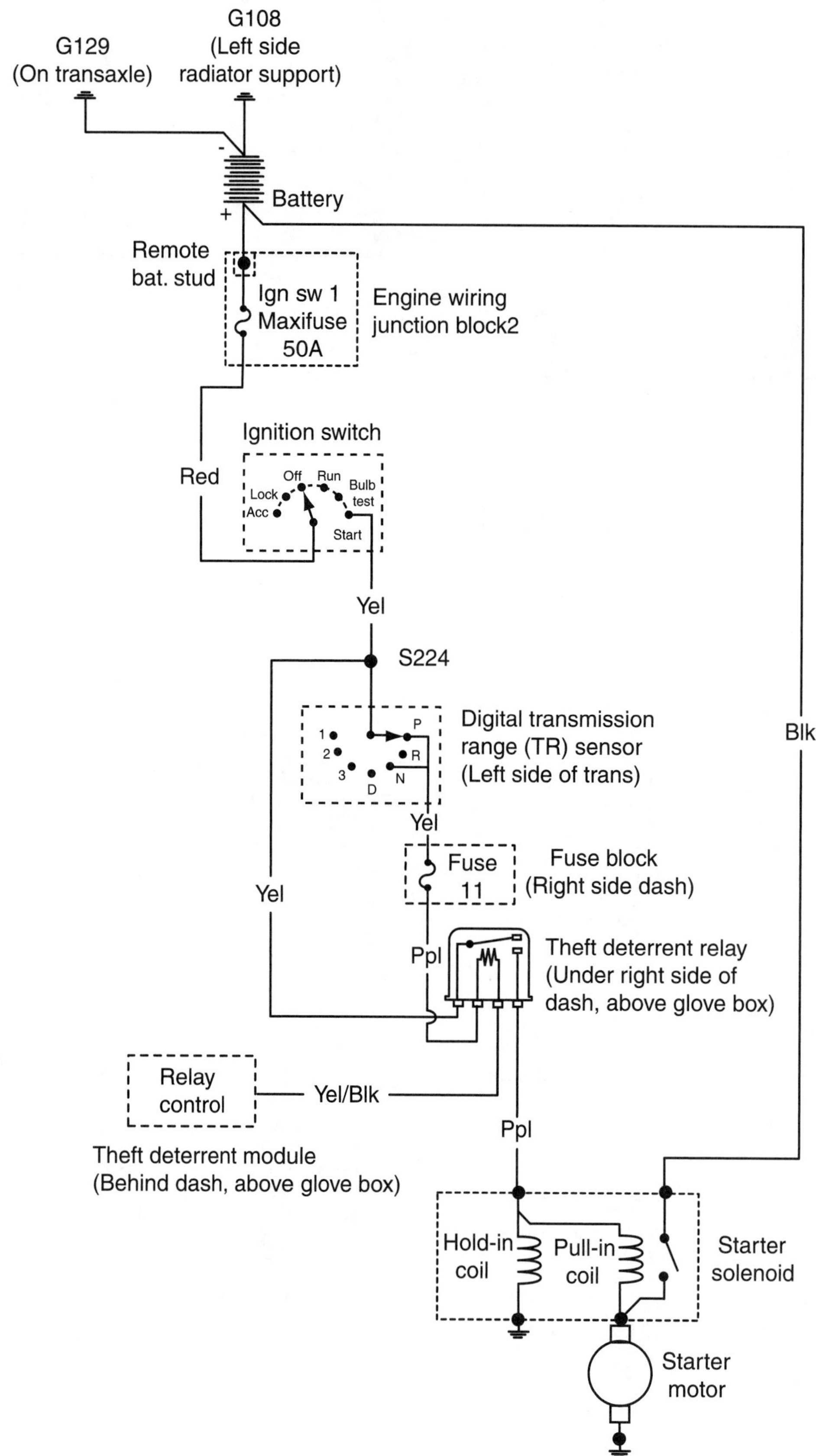

Figure 7. A Chevrolet starting circuit with theft deterrent module.

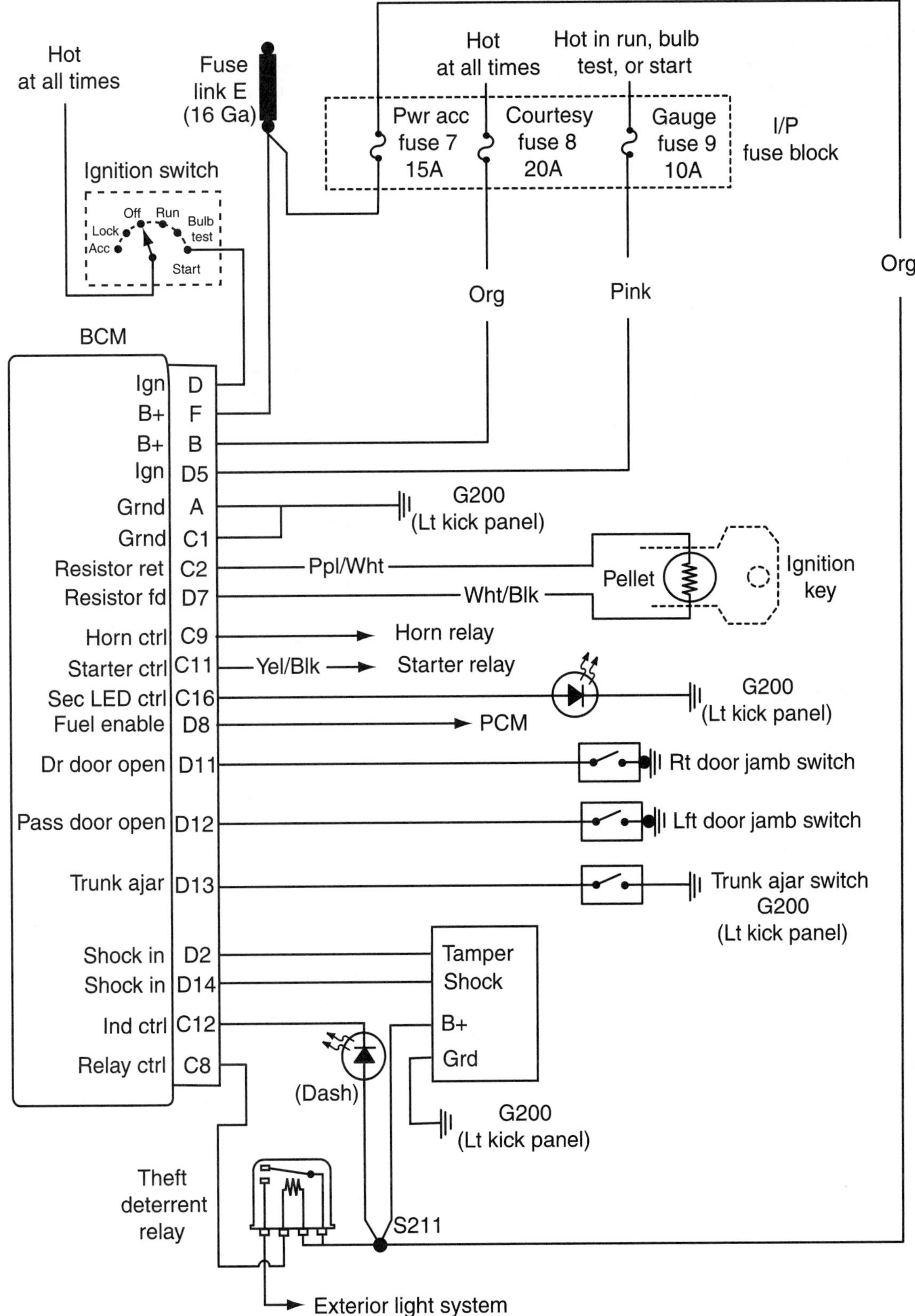

Figure 8. A Pontiac antitheft circuit.

Summary

- The TR switch is a part of the control circuit of the starter.
- The TR will be closed in the park and neutral positions (transmission gear).
- Theft deterrent is normally applied to a starting relay control coil.
- A TR can be placed anywhere in the starter motor control circuit.
- Starter relays are a frequent part of starter control circuits.
- The **clutch position switch (CP)** replaced the TR on a manual transmission vehicle.
- **Body control modules (BCM)** are frequently used to deter theft.

Review Questions

1. Technician A states that the TR is found in the power circuit to the solenoid. Technician B states that the TR is found in the power circuit to the control relay. Who is correct?
   A. Technician A only
   B. Technician B only
   C. Both Technician A and Technician B
   D. Neither Technician A nor Technician B
2. The TR is normally
   A. closed in drive
   B. open in neutral
   C. closed in park
   D. open in park
3. The CP is a normally open switch controlled by the clutch pedal.
   A. True
   B. False
4. If the resistor within the ignition key is the wrong resistance for the vehicle, the antitheft system will prevent the vehicle from cranking.
   A. True
   B. False
5. Why would a manufacturer choose to install a starter control relay?

Chapter 41

Diagnosing Starting Controls

Introduction

The approach to diagnosing a vehicle that will not turn over is different from that of a vehicle which cranks slowly. Generally, the only tools we need are a 12-volt test light and sometimes a DMM. Before looking at the procedure for a no-crank condition, review battery testing. It is possible that a no-crank condition could be caused by a faulty battery or one that is discharged. Always run a battery load test to determine the condition of the battery before any starting testing. Recognize that the procedure for a vehicle that will crank involves measuring cranking speed, cranking current, and voltage drops. Frequently, when a vehicle will not crank we diagnose the probable open circuit and then run the speed, current, and voltage drop tests. We must first get the engine to crank, which is the subject of this chapter.

USING A 12-VOLT TEST LIGHT

Much of the vehicle electronics that we see today require the use of a DSO, as we have discussed previously. The starting system is, for the most part, a resistive circuit with little electronics. This will allow us to use a 12-volt test light as a primary diagnosis tool. The exception is the side of the circuit that uses any module for control. Theft-deterrent or antitheft circuits should not be tested with a 12-volt test light. The light will draw some current, which is helpful for nonelectronic circuits but could damage electronic circuits.

An advantage to using a 12-volt test light on nonelectronic circuits is that the light will draw sufficient current to load the circuit down. Voltage drops occur when current flows through resistance. If there is an excessive resistance condition, the light will be unable to draw enough current to make the test light glow. A "no light" condition indicates either an open circuit or a high-resistance condition. The light allows us to "see" the high-resistance condition, as shown in **Figure 1**. However, if a digital voltmeter is used on the same circuit, it will not indicate the high-resistance condition. Why not? A typical voltmeter has 10,000,000 ohms of input impedance and will therefore draw almost no current. The meter placed at the point of the high resistance or open circuit will not draw sufficient current to cause the voltage drop. It could read 12.4 or 12.5 volts when placed in the circuit, as illustrated in **Figure 2**. For this reason, never use a digital voltmeter on an open circuit or one that you suspect has high resistance.

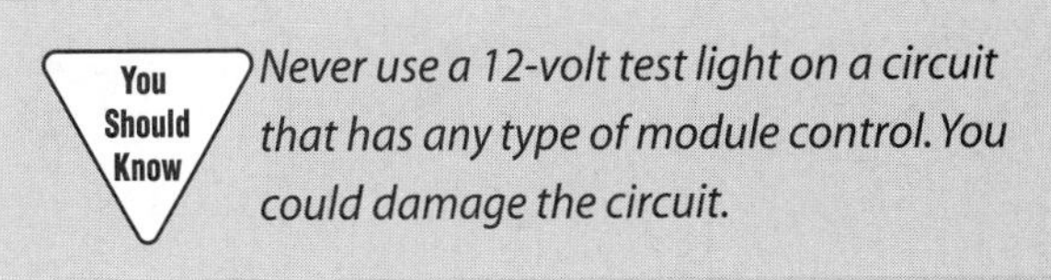

You Should Know
Never use a 12-volt test light on a circuit that has any type of module control. You could damage the circuit.

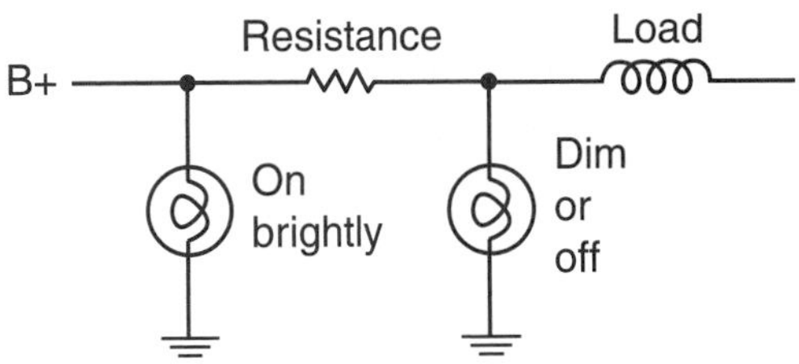

Figure 1. The test light "sees" the resistance of the open circuit.

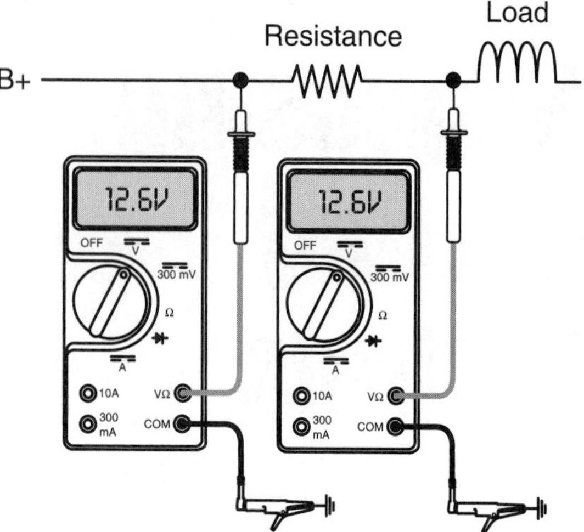

Figure 2. The voltmeter does not "see" the resistance of the open circuit.

The exception to the use of the 12-volt test light on starting circuits is one in which a module is being used in the circuit. The BCM found in **Figure 3** supplies a ground to the starter relay. For this reason, we would use a DMM set to volts and placed on the yellow wire with a black stripe (YEL/BLK). With your voltmeter's positive lead placed on the battery positive terminal and the negative lead placed on the ground wire for the starter relay, the voltmeter will read battery voltage if the lead is being grounded by the BCM. If for some reason the wire is open or the BCM is not applying the ground, the voltmeter will read zero volts.

NO-START TESTING

No-crank or no-start testing should always begin with an analysis of just how the circuit is supposed to work. Refer to the wiring diagram and "walk" through how the circuit gets the engine to crank. Coloring the positive part of the circuit red and the negative part of the circuit green will be helpful in getting it to make sense. Let us use **Figure 4** and walk through the circuit beginning at the ignition switch. Notice that two switches are ganged together (as indicated by the dotted line between wipers). These switches will move together, and we need to look at both. The left one receives power from terminal D5 (a red wire) and switches it through terminal D1, through a 10-amp fuse, and down to the PCM as the crank signal. This positive pulse to the PCM will signify that the ignition key is in the start position and "wake up" the PCM's circuits. The other ignition switch will bring power (B+) to the relay coil of the crank relay. The ground for the crank relay will come from the PCM starter enable control circuit. If these circuits are functional, the relay should close. This should bring power through the 40-amp crank maxifuse, the relay, the TR switch, and finally down to the starter solenoid at the S terminal. If the solenoid functions, power should be switched from the B terminal to the M terminal and the engine should crank. In addition, the solenoid and starter are case grounded to the engine block. Does this make sense? This information is important to understand before diagnosing a no-crank condition.

With the information about which terminal is positive and which is negative, we can now diagnose this no-crank condition. Begin with the battery test to ensure that the battery has sufficient power to crank the engine over. If it

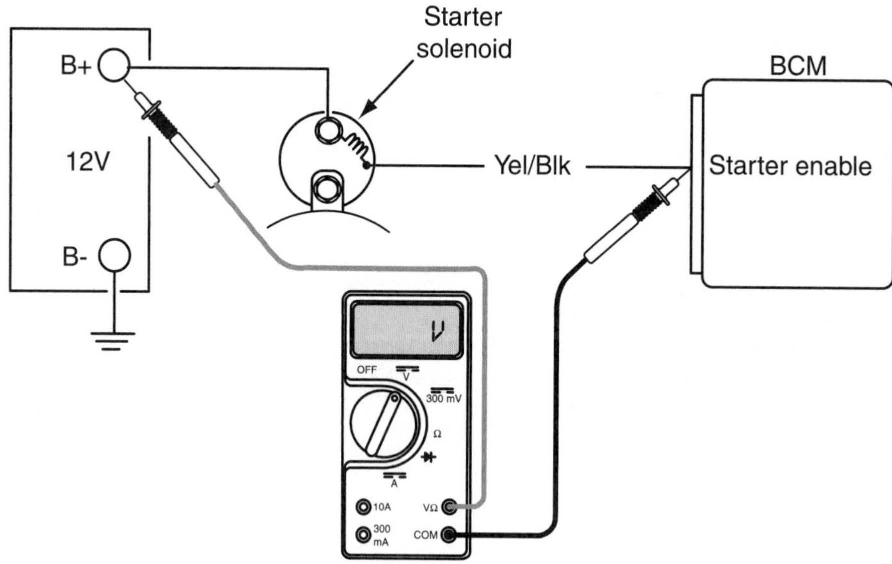

Figure 3. Pontiac starting controls.

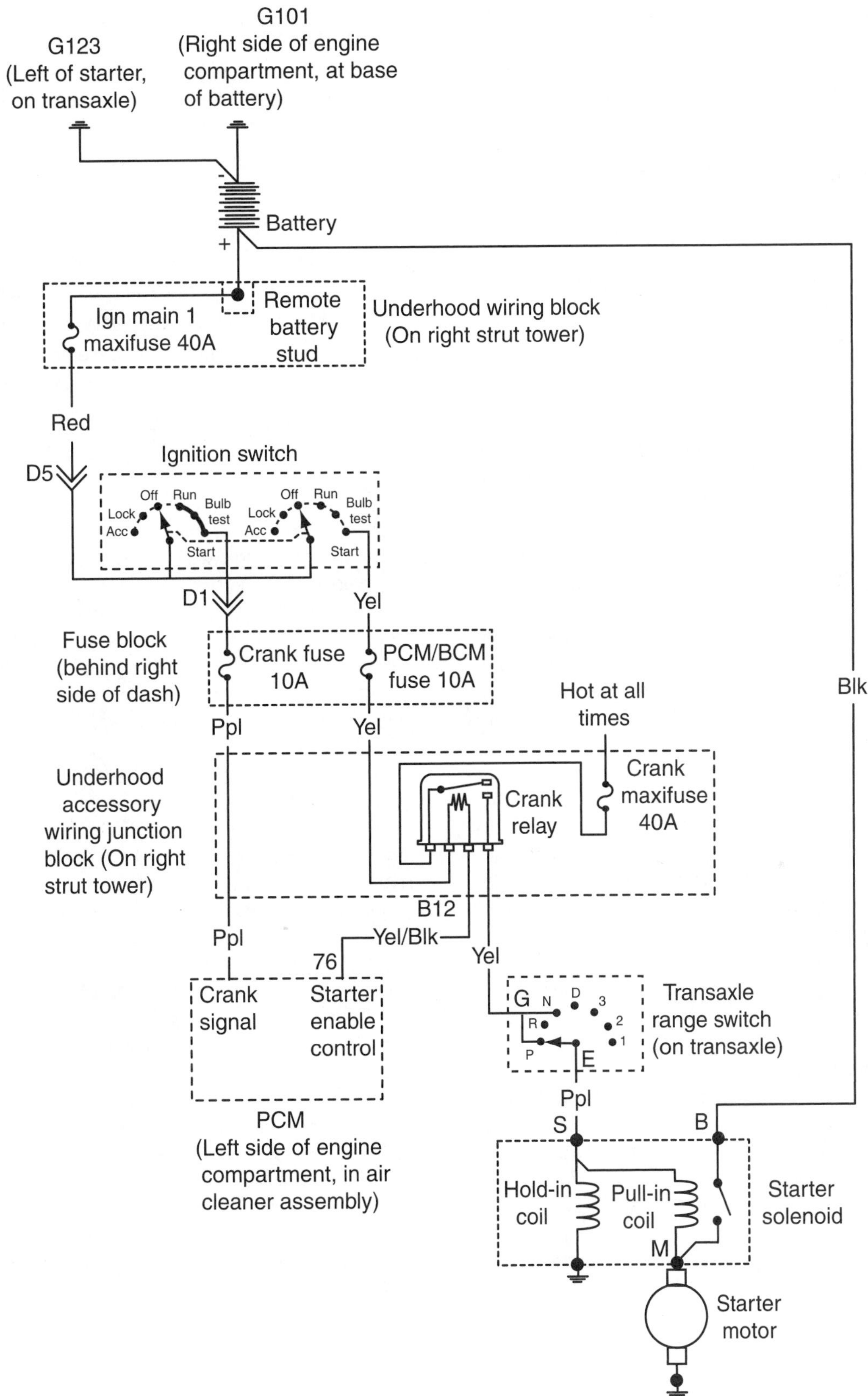

Figure 4. Buick starting controls.

passes the load test, it should have enough power to crank the engine over. Do not forget the basics: are the battery cables connected tightly? Make sure you visually inspect the circuit. Frequently, the high resistance or open circuit might be visible. Connect your 12-volt test light to a good engine ground or directly to the battery negative terminal. Test the light by touching it to the positive terminal of the battery. It should glow brightly, indicating that it is functional. Starting at the load, the starter, we begin looking for power with the key in the start position at the M terminal. With each test, you have two choices: power (light on) or no power (light off). The two choices indicate where our high resistance or open circuit is. If the light lights at the M terminal, the control circuit is doing all that it can and the motor needs to be replaced. If you have no light, you have no power and an indication of a problem further back into the control circuit. Let us assume you have no power. Where do you go next? You should have figured out that the next point of testing is the S terminal. Again, you have two choices. Power here indicates that the solenoid is bad; no power indicates that we need to keep looking. Follow the S terminal to the TR and test for power at the output terminal E (purple wire). Again, you have two choices. Power here indicates a broken wire between the E and S terminals; no power means keep on going. Place your test light on the input to the TR—terminal G, a yellow wire. Turn the ignition key to start again and observe the light. Power here indicates a faulty TR, and no power indicates that you need to continue testing. The procedure is repeated for each section or connection of the circuit. Do not forget that this circuit has two parallel sections, the solenoid control circuit and the crank relay circuit, and both legs must be functioning correctly for the engine to crank. Diagnosis is simple and straightforward, working your way from the load to the source looking for power. The bad component, open wire, or high resistance is always between the section that had the light off and the one that had the light on. The B+ section of both circuits is by far the longest and includes the most connections that need to be tested.

Let us consider the ground circuit beginning at the relay connection B12 (YEL/BLK) running down to the starter enable control within the PCM. This circuit must be treated differently for two reasons. First, it is a ground, not power; and second, it is supplied by the PCM. Remember that we really should not test anything around a module with a test light. There is a possibility that the light might draw more current than the module can supply safely. Break out that multimeter and look for a ground signal on pin 76 of the PCM with the key in the start position. Connect the positive lead to power at the battery and the negative lead to pin 76. If pin 76 is a ground, the voltmeter will read 12.6 volts. If it is not a ground, pin 76 will read zero volts.

Using red for positive and green for negative will highlight the two sections of the wiring diagram. This will be helpful during diagnosis because it tells the technician what to expect at a particular terminal.

Let us diagnose the Mercury Grand Marquis in the same manner **(Figure 5)**. This is probably one of the simplest no–crank over systems to test. First, walk through the circuit and color it if necessary. Case grounds at the starter supply the grounds for the motor and solenoid and a BLK wire goes to G115 for the starter relay. G115 is located at the center rear of the engine. Ask yourself which wires are positive and which wires are negative. With a no-crank condition, we would start at the YEL/LT BLU wire at the solenoid, turn the key to the start position, and look for power. Power indicates a starter problem; no power indicates a problem toward the relay. Go to the YEL/LT BLU wire at the relay. With the key in the start position, the wire should have power. Does it? If yes, you have a broken wire; if no, we must continue testing. Does the yellow wire coming into the relay have power? If yes, the relay has the power to function. If no, fuse 2, the 30-amp fuse, is a likely candidate and the next place to test. If power is available to the relay but not available through the relay, we must look at the control side. The RED/LT BLU wire should bring power from the TR, and the BLK wire should apply a ground. Test the ground using your test light to draw power. Connect the clip to B+ and the pin to the BLK wire. If it lights, it is an OK ground. No light indicates that a repair of the ground is in order. We can use a test light here because there is no module involved. With a good ground and power on the RED/LT BLU wire, we know that we have a faulty relay. If there is no power, work your way toward the battery until you find power.

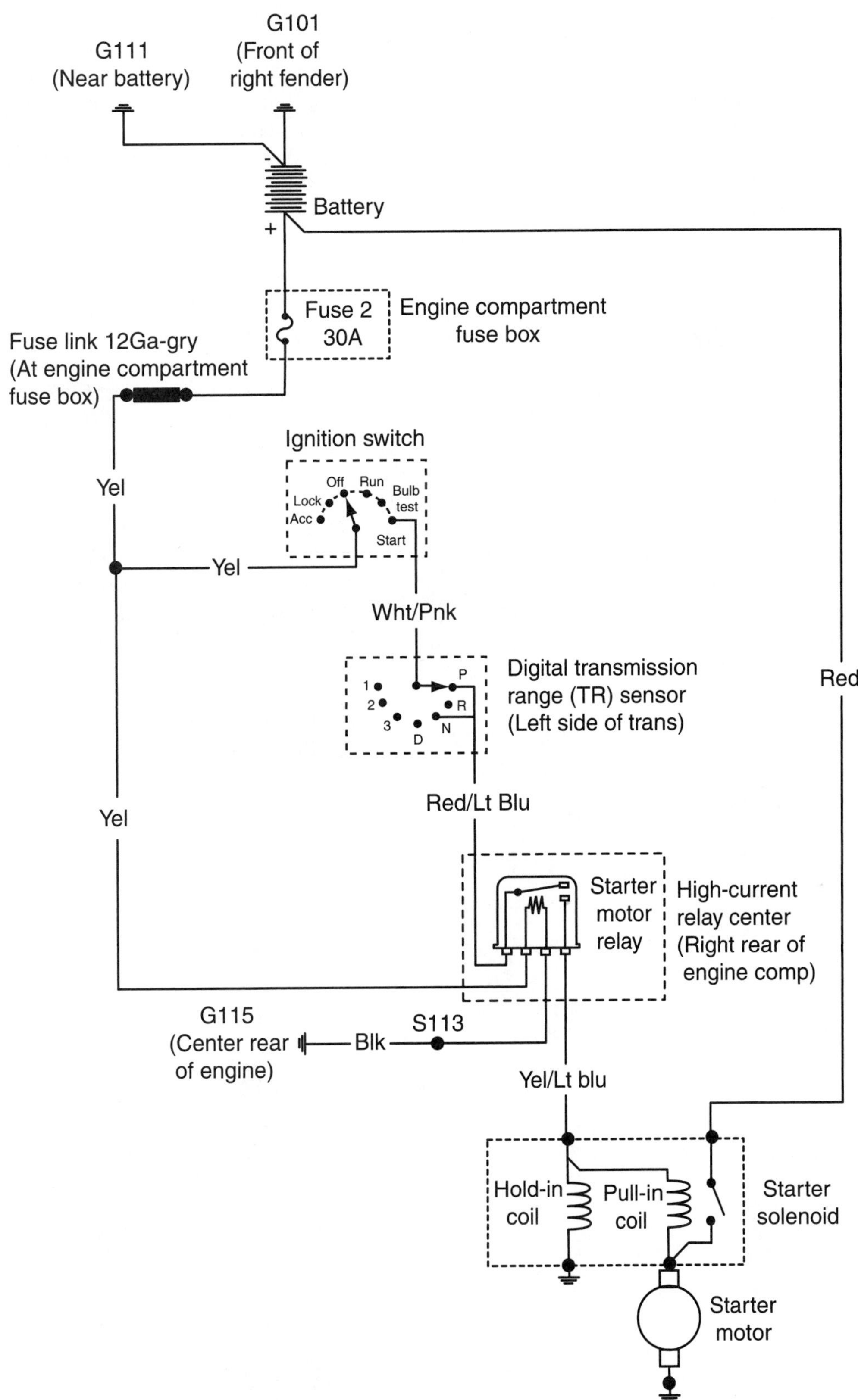

Figure 5. Mercury starting controls.

Summary

- Use a 12-volt test light for most no–crank over tests.
- Do not use a 12-volt test light if there is a module in the circuit being tested.
- A battery load test should precede any no-crank testing.
- PCMs are involved with certain starting systems, usually for alarm control.
- A test light is used to look for power with the key in the start position.
- A DMM is used to find a ground by connecting it to power.

Review Questions

1. An open TR will be identified by having the test light
   A. be on at the input lead and on at the output lead
   B. be on at the input lead and off at the output lead
   C. be off at the input lead and on at the output lead
   D. be off at the input lead and off at the output lead
2. The test light is placed on the M terminal and ground. When the key is turned to the start position, the light lights. Technician A states that this indicates that the circuit from the battery to the S terminal must be open. Technician B states that this indicates that the starter motor needs to be replaced. Who is correct?
   A. Technician A only
   B. Technician B only
   C. Both Technician A and Technician B
   D. Neither Technician A nor Technician B
3. Technician A states that most TR switches are normally open. Technician B states that most CP switches are normally closed. Who is correct?
   A. Technician A only
   B. Technician B only
   C. Both Technician A and Technician B
   D. Neither Technician A nor Technician B
4. Power is available to the load side and the control side of the relay. A ground is available to the control coil. The load contact will not close. What is wrong with the circuit?
   A. There is a bad ground connection.
   B. The ignition switch may be faulty.
   C. The starter solenoid is faulty.
   D. The relay is faulty.
5. Why do we use a DMM to test a circuit that has a module in it?

Chapter 42 Starting System Servicing

Introduction

Over the years, starting system servicing has changed. Few repair centers tear apart a starter and do any internal repair. The labor rates dictate that replacement of the starter is a more cost-effective approach for most vehicle owners. In this chapter, we look at common servicing techniques and go over some of the basics of removing and replacing the starter assembly.

CHECKING CRANKING SPEED

Cranking the engine over is the reason for the starting circuit. The speed at which an engine spins is very important. It takes fuel, compression, and ignition to get an engine running. Compression and fuel will be dependent on cranking speed. The slower an engine cranks, the more compression loss there is. This loss will make it harder to draw in the air:fuel mixture and ignite it. General specifications call for approximately 175–200 rpm for a good starter. This speed can be measured easily with a tachometer. Most technicians do not measure them, however. A technician's experience and trained ear can be an accurate indicator of correct speed. A few years in the field and your ear will be "tuned" also. In the meantime, use a tachometer if there is a doubt in your mind that the engine is cranking over fast enough.

Additionally, some of the vehicles will generate cranking rpm along the serial data line from the PCM. This makes the information available with a scanner. Some DSOs can also indicate rpm through the use of the tachometer or sync lead. Place the lead around a plug wire, as **Figure 1** shows, and cranking rpm will be displayed, as **Figure 2** shows. Our 3.4-liter General Motors engine is cranking over at 196 rpm, which is good. It is important that you realize that to be able to capture rpm the DSO or tachometer needs to know the type of ignition system. We cover the different styles in Chapters 46–53, but notice that **Figure 3** is asking the technician what type of ignition is on the vehi-

> **Interesting Fact**
>
> *Most engines should crank over at least 175 rpm for the fuel system and ignition systems to function correctly.*

Figure 1. To measure cranking rpm, the sync probe is placed around a plug wire.

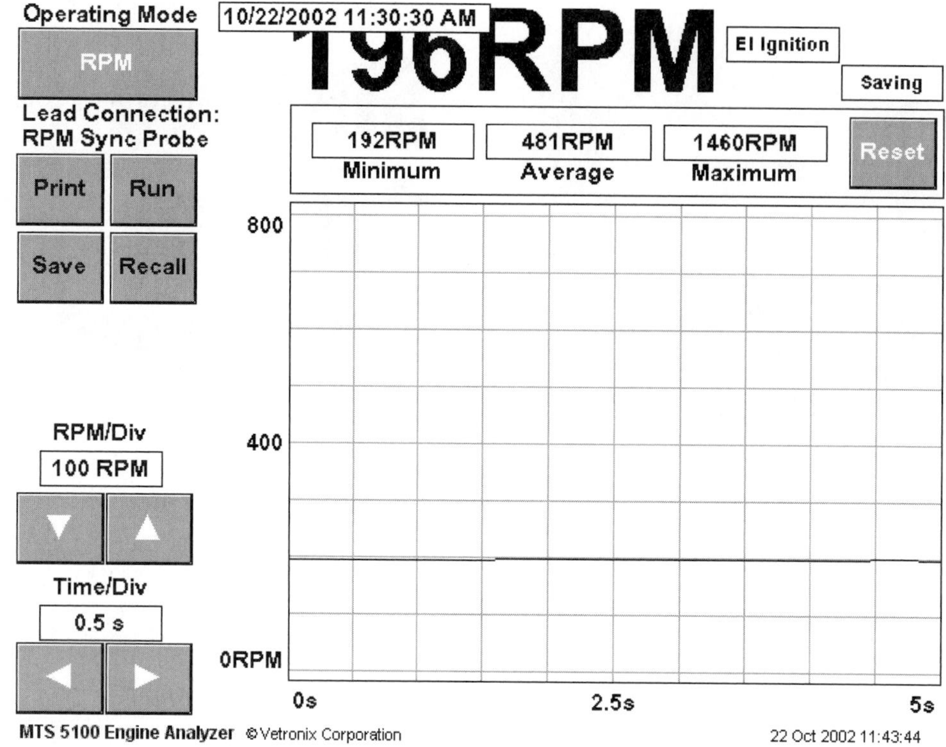

Figure 2. Cranking rpm displayed on a DSO.

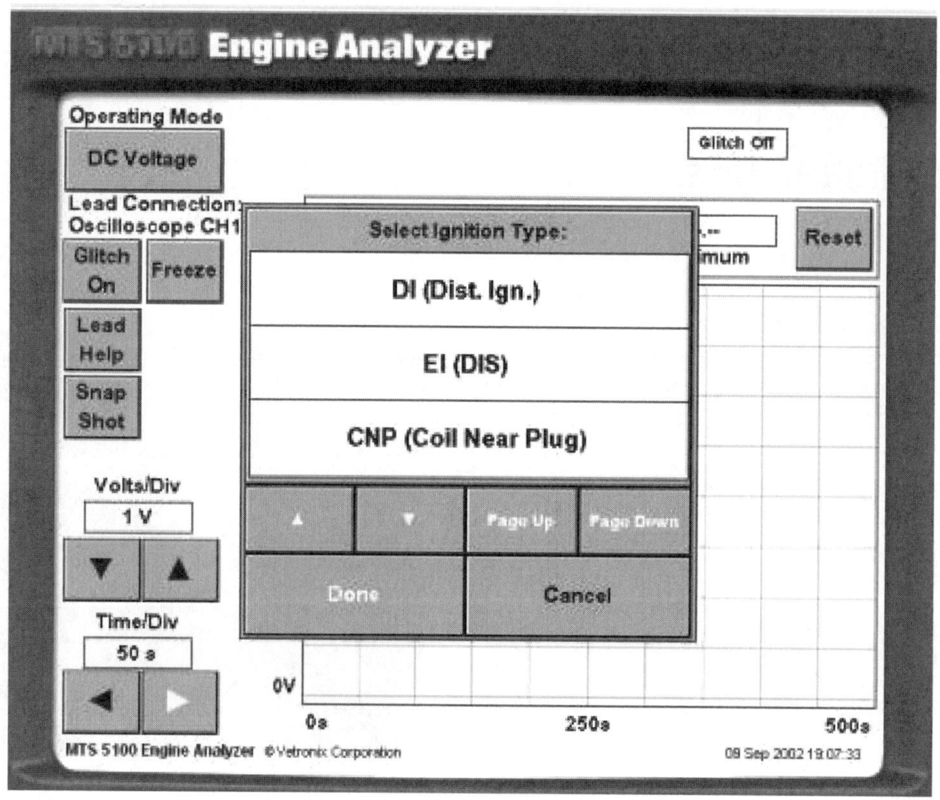

Figure 3. The DSO must know the type of ignition to determine cranking rpm correctly.

cle. Does it have a distributor? If so, it is a **DI** ignition. Is it distributorless? If so, it is an **EI** system. Or does it have a single coil located near each plug? If so, it is a coil near plug (CNP). Once you indicate to the DSO the type of ignition, it can easily calculate cranking rpm.

Is 196 rpm adequate? Yes, it is. This engine has a permanent magnet gear-reduction starter and should start with this cranking rpm.

REMOVING AND REPLACING A STARTER

Removing and replacing a starter or **R&R** as it is commonly referred to, is a frequent task done in most shops.

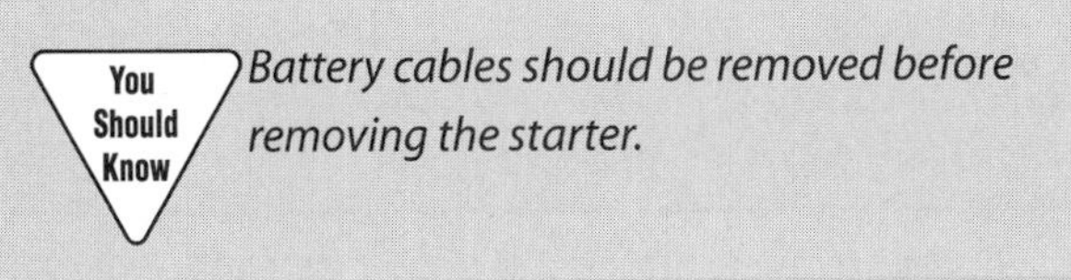

Battery cables should be removed before removing the starter.

We have looked at the why in previous chapters, so now let us look at the how. Keep in mind that each vehicle will have some differences, but this procedure will work for most. **Figure 4** shows a starter in the engine.

1. Disconnect the negative battery cable. This will turn off the system, in case the battery cable should come in contact with metal while you are removing it. Do not forget that the battery cable is hot, or live. It is a direct connection to the battery. When you disconnect it, keep in mind that relearn procedures might be necessary for the PCM, radio, or other modules with memories.
2. Raise the vehicle off the ground and support it on stands if necessary. Few starters can be removed with the vehicle on the ground.
3. Remove any covers or shields that are in the way. Frequently, water shields are placed so that the starter will not get the full force of water splashed off the tires.
4. Remove the cable and wires from the solenoid, if so equipped. Sometimes the starter has to be partially removed before you can gain access to the solenoid.
5. Remove the bolts that attach the starter to the engine and remove the starter.

Always make sure that you check to make sure the replacement starter is exactly the same as the one you are replacing. Reverse the procedure and recheck current draw and rpm. Always recheck the original reason why you are replacing the starter assembly.

One last consideration is to not let the weight of the starter hang from the solenoid wires. If you do, it is likely that you will damage either the wires or the insulation, and that will present problems in the future. Also, use a torque wrench to tighten all bolts to specification.

Figure 4. The starter motor installed in a V-type engine.

CHECKING AND ADJUSTING PINION DEPTH

Most starting systems in use today do not have the ability to adjust the pinion depth; however, you need to recognize that if the pinion is too tight the starter will bind, especially when it is hot. Damage might occur to the drive pinion gear if there is too much space between the flywheel teeth and the gear. **Figure 5** shows what the pinion looks like when it is meshed with the flywheel.

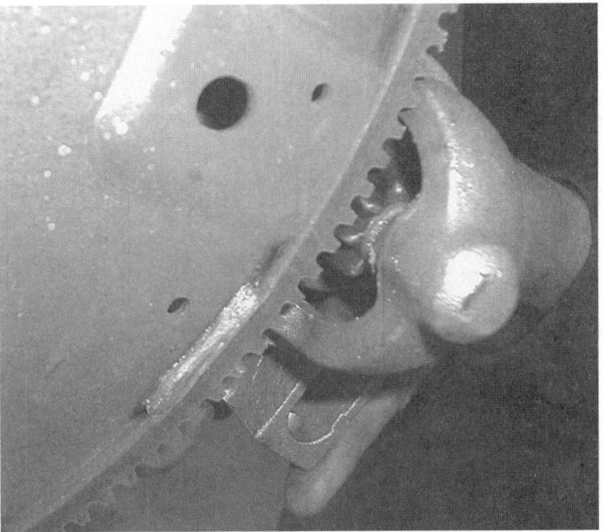

Figure 5. The drive pinion must mesh correctly with the flywheel teeth.

Some manufacturers specify a procedure that involves placing a round feeler gauge between the flywheel teeth and the pinion teeth. To be able to accomplish this, the solenoid is usually removed, the drive pulled into mesh, and the air gap measured, as shown in **Figure 6**. You need to realize that if the depth is incorrect the starter needs to be replaced. A few starters can use shims to move the drive away from the flywheel, but most manufacturers make no provision for adjusting pinion depth.

Figure 6. Using a feeler gauge to check the pinion-to-flywheel fit.

Summary

- Normal cranking speed is approximately 200 rpm.
- Slower-than-normal cranking speed requires a complete starting system checkout.
- To determine rpm, the tachometer or DSO needs to know what type of ignition system the vehicle has.
- Disconnect the battery cable before removing a starter.
- Do not let the starter hang from the connecting cable and wires.
- Some vehicles will require a relearn procedure when the battery is reconnected.
- Pinion depth is important to cranking speed and current.
- Pinion depth can be checked on most systems, but few allow for it to be adjusted.

Review Questions

1. Technician A states that cranking rpm should be in excess of 200 rpm. Technician B states that RPM is read off the fuel system sensor. Who is correct?
   A. Technician A only
   B. Technician B only
   C. Both Technician A and Technician B
   D. Neither Technician A nor Technician B
2. Pinion depth determines
   A. cranking current
   B. cranking rpm
   C. noise
   D. all of the above
3. After removing the battery cable for starter R & R, it may be necessary to
   A. do a solenoid R & R
   B. reset the charging system
   C. charge the battery
   D. perform the relearn process for the PCM
4. Starter pinion depth is adjustable on all starting systems.
   A. True
   B. False
5. Cranking speed should be in excess of 200 rpm on most systems.
   A. True
   B. False
6. Starter R & R requires that the battery cable be removed.
   A. True
   B. False
7. Explain how to remove a starter.
8. Why should the battery cable be removed before starter R & R?

Section 9

Charging Systems

SECTION OBJECTIVES

At the conclusion of this section, you should be able to:

- Understand how the charging system functions.
- Understand the function of each part of the system.
- Understand how the diode rectifier bridge functions.
- Check alternator output and compare it with specifications.
- Full field an alternator if applicable.
- Measure voltage drops on the charging system.
- Measure with a DMM the level of AC being produced by the alternator.
- Measure with a DSO the level of AC being produced by the alternator.
- Test the charging system's voltage regulation.
- Test the pulse width modulation (PWM) from the digital regulator with a DSO.
- Safely diagnose and repair the charging system.

Frequently, the charging system is the cause of other system failures. It should always be tested if any electronic component has failed.

Chapter 43 Charging System Overview

Introduction

Now that we have the vehicle started with the battery and starting system, we must restore the battery to a full state of charge and get ready for the next startup. In addition, vehicle systems and any accessories that are being used by the vehicle operator must have current available to them. These two functions, keeping the battery fully charged and supplying current to keep the vehicle running, are the job of the charging system. In this chapter, we look at how this system functions effectively. We also look at how we regulate or control the charging system. In Chapter 44, we look at the various types of field circuits, and in Chapter 45, we look at diagnosing and servicing the charging system on the vehicle.

The charging system has evolved right along with the rest of the electrical system. Starting with big bulky generators that generally could not effectively handle the high current demands of the modern vehicle and ending with computer regulation of the charging system of today, this evolution has kept this system in pace with the rest of technology. We do not cover generators or older alternators within this text because their use has gradually diminished to the point where you will probably never encounter one in the trade. The purpose of the charging system is to convert the mechanical motion of the engine to electrical power to recharge the battery and run the vehicle. The same principles were in use in generators as those in the modern charging system. This principle states that a voltage will be produced if motion between a conductor and a magnetic field occurs. This principle is exhibited in **Figure 1**. The amount of voltage developed will be dependent on the strength of the field and the speed of the motion. With this basic principle in mind, let us look at the modern charging system beginning with the alternator.

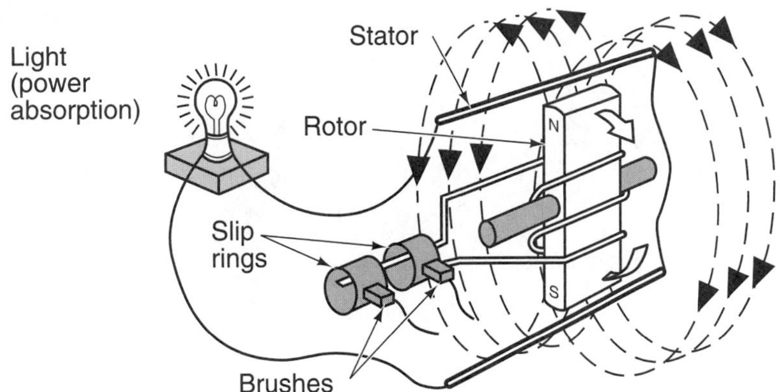

Figure 1. Voltage is produced if motion between a conductor and a magnetic field occurs.

The charging system can be divided into three major components: the battery, the alternator, and the regulator. We have spent sufficient time discussing the battery, so we will not say too much about it here. We will, however, restate that the battery can be responsible for many charging system problems, just as it can be for starting difficulties. The charging system cannot function correctly if the battery will not hold a charge or has limited capacity. In the future chapters, we again emphasize to test the battery first. Our discussion of the rest of the system within this chapter will assume that a correct-capacity, well functioning, fully charged battery is in the vehicle.

THE ALTERNATOR

Let us look at the alternator. It is the device that turns the mechanical motion of the engine into the electrical power needed **(Figure 2)**. Basically, it consists of three major components: the rotor, the stator, and the rectifier bridge. These three components are housed in two end frames. As we look at each component, keep in mind that the basic principle of electrical generation is motion between magnetic lines of force, and conductors will induce electrical pressure (voltage). In alternators, the rotor will supply the magnetic lines of force and the engine spinning the rotor will supply the motion.

Rotors

The object of the rotor will be to supply a spinning magnetic field. Examine **Figure 3**. It shows a rotor assembly that will also have a cooling fan and a pulley attached. The rotor is a coil of wire sandwiched between metal rotor halves. When current is passed through this coil of wire, a magnetic field will be produced. The strength of this field will be dependent on the amount of current flowing through it. This current is called **field current**, and by controlling its amount, we will control the output of the alternator. If a large amount of field current is flowing through the rotor, alternator output will be high. If a small amount of field current is flowing, alternator output will be low. It should also be realized now that most charging systems do not vary the amount of field current but instead vary the length of time that the field current flows. As an example, if 4 amps is flowing through the rotor winding for 50 percent of the time, it will be the equivalent of 2 amps continuously. The end result is the same, whether you vary the amount of current flowing or the amount of time current flows. Either way will give control over the output of the alternator.

Notice that the rotor coil is between two sets of interlacing fingers that will take on the polarity (north or south) of the side of the coil that they touch. In this way, we will spin a magnetic field that will first be north and then south polarity. Have you wondered how we can

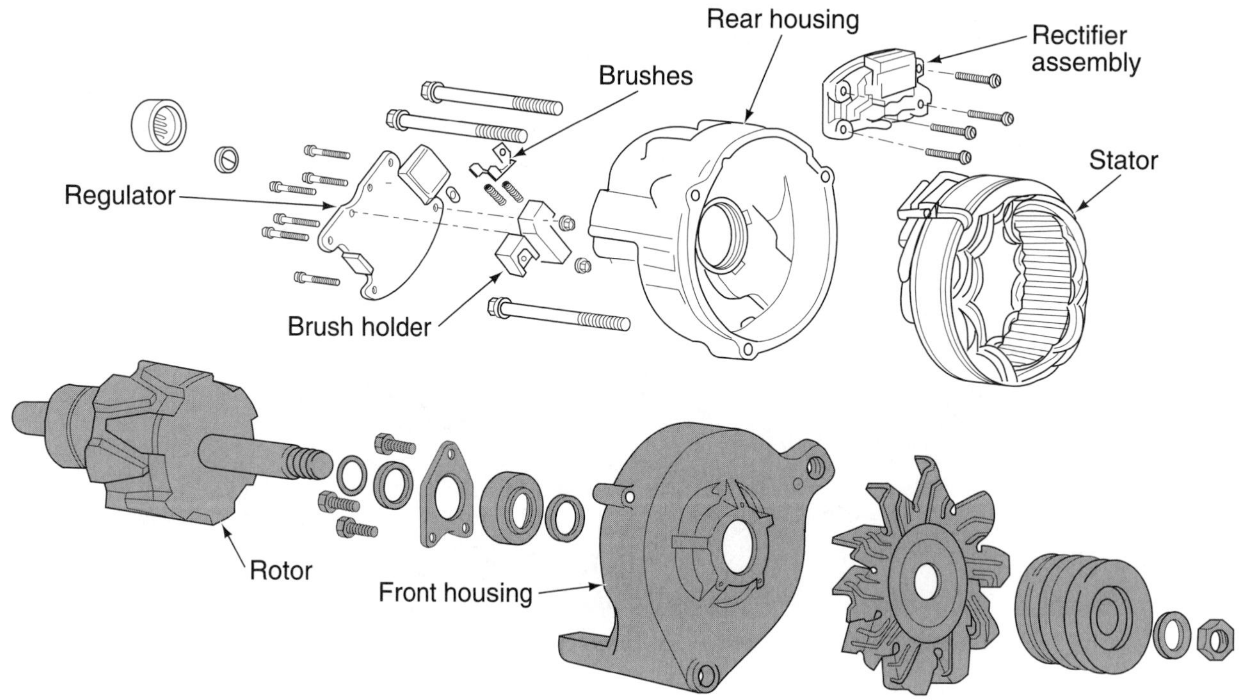

Figure 2. Components of an alternator.

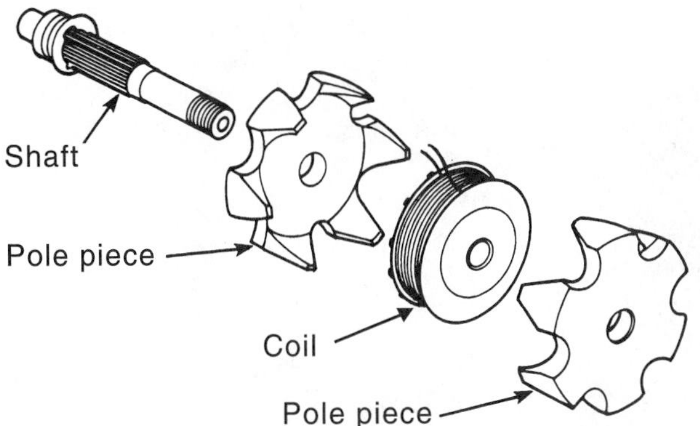

Figure 3. The rotor will supply a spinning magnetic field.

bring current into something that is spinning? Notice **Figure 4**. It shows how the wires from the coil are attached to two smooth rings that are insulated from the rotor shaft. These are called **slip rings**, and they function similar to an armature's commutator. The stationary brush passes field current into a slip ring. From here, field current flows through the field coil, back to the other slip ring, and then out the stationary brush. This gives us a method of getting an adjustable or controllable magnetic field that is spinning. The rotor is supported by bearings on each end, which will allow it to spin at speeds in excess of 15,000 rpm.

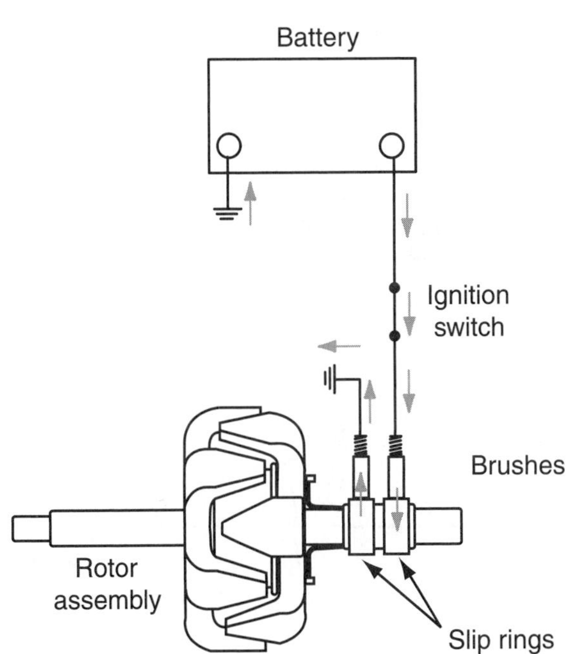

Figure 4. Field current flows through the brushes and slip rings.

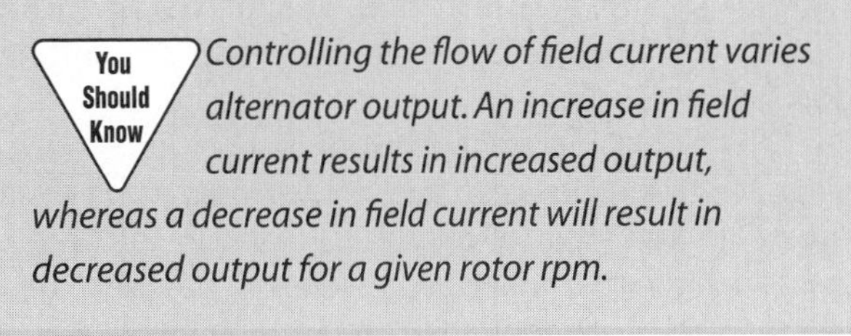

Controlling the flow of field current varies alternator output. An increase in field current results in increased output, whereas a decrease in field current will result in decreased output for a given rotor rpm.

The Stator

Remember that generation of electricity involves motion between magnetic lines of force and conductors. Our rotor was able to supply both the motion and the variable magnetic field, so only the conductors are left in the generation of electricity. Remember the magnetic field moving near the wire in Figure 1. The wire was having its valence electrons pressurized by the lines of force. Electricity is the movement of the valence electrons. **Figure 5** shows the simplest one-wire stator. Notice that the loop of wire has its two ends attached to a load (A and B on the illustration). This setup shows that if a conductor is energized by a north polarity magnetic field, the electrons will be pressurized in one direction, and if a south polarity field energizes them, they will be pressurized in the opposite direction. Electrons (or current) that move in one direction, stop, and then move in the opposite direction are called alternating current, or AC. Remember that

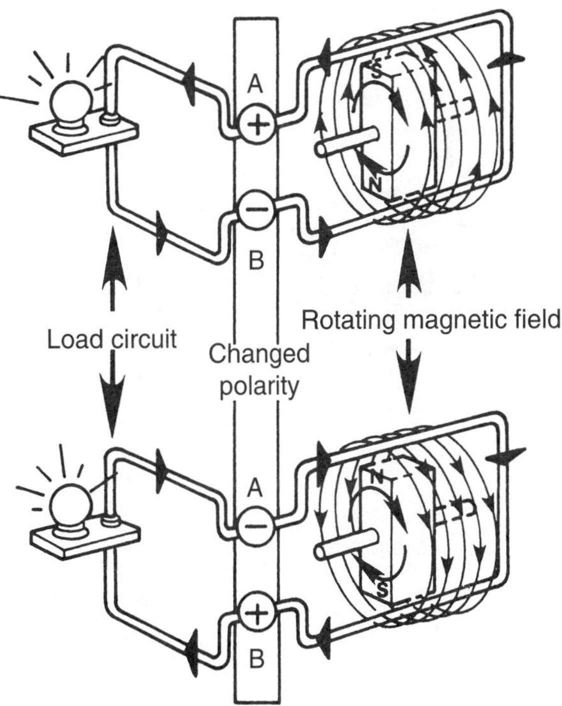

Figure 5. AC will be generated because of the spinning N and S magnetic field.

the rotor is spinning with interlacing fingers of north and south polarity. This will generate AC in our simple one-loop stator.

The amount of current generated in one single loop of wire is too small to be practical. Vehicles today may require more than 100 amps of current to maintain the battery charge and run the required accessories. To be able to generate this much current, the manufacturers wind lots of wire into an iron-laminated core or frame. Most alternators use three windings to generate the required amperage. They are placed in slightly different positions so their electrical pulses will be staggered and are wired in either a delta or a wye configuration, as seen in **Figure 6**, and called the stator. The delta winding received its name because its shape resembles the Greek capital letter delta. Some older smaller alternators used to use windings in a wye configuration. The wye winding receives its name because its drawing resembles the letter "Y". Notice the difference in the windings. Usually, lower-output older alternators used wye windings, whereas higher output modern alternators use delta windings. A delta winding produces more current because the windings will take on a type of parallel circuit, whereas a wye winding will be more like a series circuit.

The rotor fits inside the **stator**, as seen in **Figure 7**, with a small air gap. This allows the magnetic field of the rotor to energize all of the stator winding at the same time so the generation of electricity can be quite high if needed.

Before we move on to diodes, let us look at AC and examine something called the sine wave. Alternating current produces first a positive pulse and then a negative pulse. This can be represented on an oscilloscope **(Figure 8)**. Notice that the complete waveforms start at zero, go positive, and then drop to zero again before turning negative. The angle and polarity of the field coil fingers are what cause this sine wave in the stator. When the north pole magnetic field cuts across the stator wire, it generates a positive voltage within the wire. When the south polarity magnetic field cuts across the stator wire, a negative voltage is induced in the wire. A single loop of wire energized by a single north then a south pole magnetic field results in a single-phase voltage. Remember that there are three stator windings overlapped. This will produce the overlapping sine wave shown in **Figure 9**. This voltage, because it was produced by three windings, will be called **three-phase voltage**. An important point to remember is that AC voltage and current come off the stator. AC will not charge a DC battery, however. It will have to be changed or rectified into DC before we can efficiently use it.

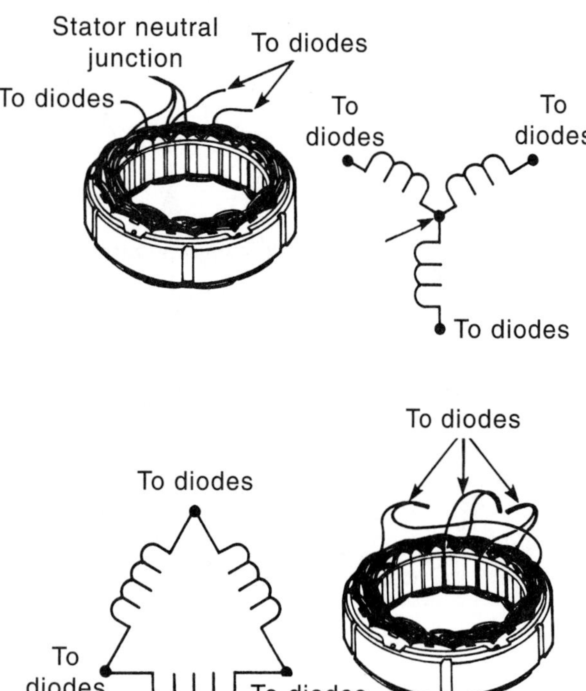

Figure 6. Delta- and wye-wound stators.

Figure 7. The rotor fits inside the stator.

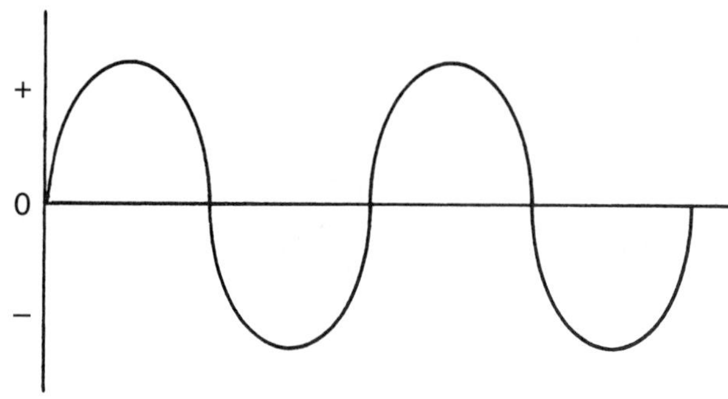

Figure 8. Each stator winding will generate a sine wave (AC).

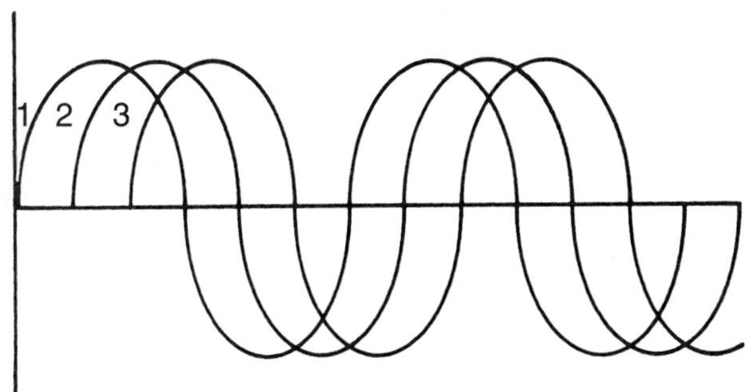

Figure 9. Three-phase AC voltage generated in a typical stator.

The Diode Rectifier Bridge

Let us review where we are right now. Field current is flowing through the rotor, generating a magnetic field. The engine is turning the rotor with a fan belt or serpentine belt, thus giving us a spinning magnetic field. The magnetic lines of force are cutting across the windings of the stator, generating AC in the windings of the stator. If this brief overview does not make sense, do not hesitate to review the last few pages covering the generation of AC.

As we mentioned before, AC will not charge a battery because it is a DC source of current. To charge a battery, we need a DC source at higher-than-battery voltage level. The AC coming out of the stator will have to be rectified by a set of diodes.

We discussed diodes in Chapter 21, and found that they were an electrical one-way check valve that would allow current to flow through them in only one direction. When the current reverses itself as it does in AC, the diode blocks and no current flows. With this in mind, look at **Figure 10**. It shows that if we take AC and run it through a diode, the diode will block off the negative pulse and produce the scope pattern shown. What would the pattern look like if the diode were reversed? You should have been able to figure out that it would look exactly the same, except that it would be blocked during the positive pulse and allow the negative pulse to flow through it **(Figure 11)**. Because the single diode was able to block only half of the pulses (either the positive or the negative) and was able to pass half of the pulses, we would call this **half-wave rectification**. We have taken the AC and changed it into a pulsing DC, but we are wasting the other half of the AC wave. To be able to have full wave rectification, we need to add more diodes, as shown in **Figure 12**. Notice that in either direction we get DC out of the four diodes. Follow the arrows through the coil on the left. When current is from bottom to top, diode 1 is blocking because it sees

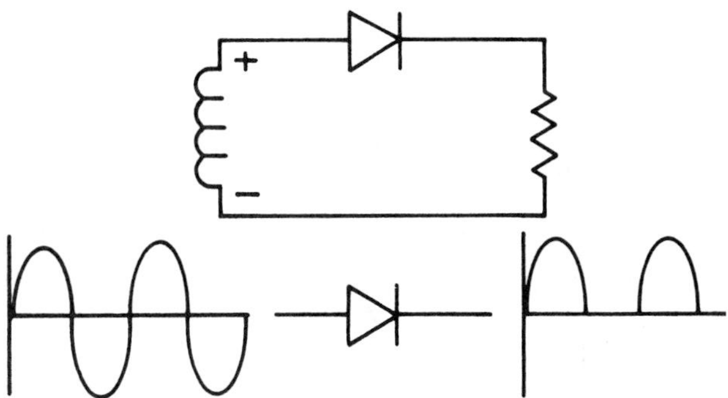

Figure 10. A single diode blocking the negative pulse.

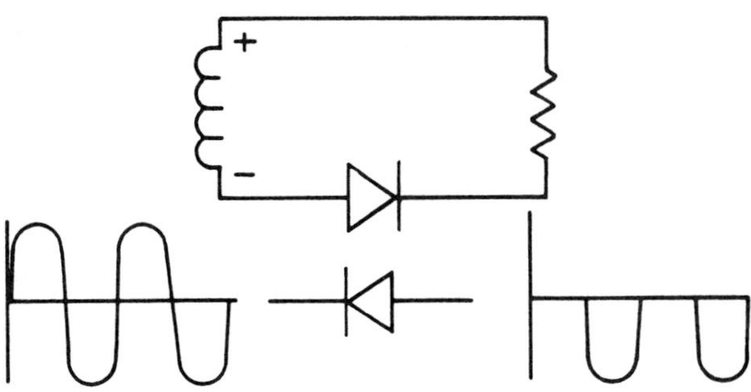

Figure 11. A single diode blocking the positive pulse.

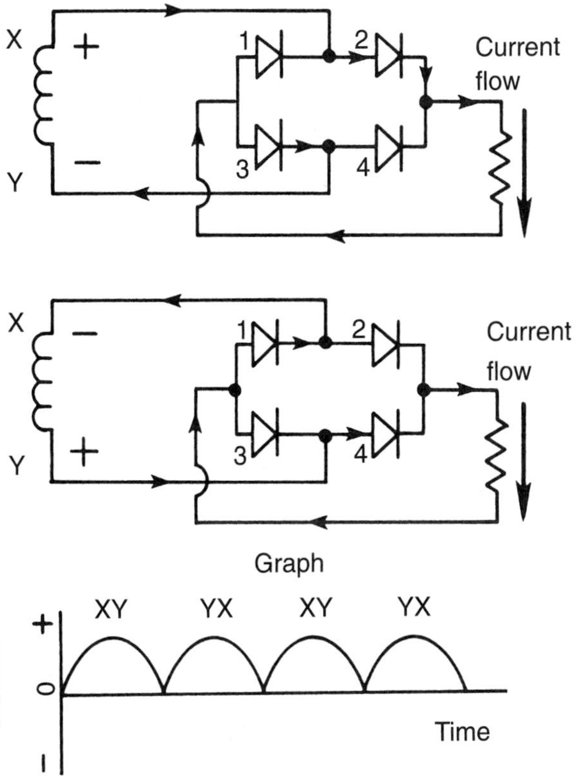

Figure 12. Four diodes are required for full-wave rectification.

pressure. Diode 2 is passing current out to the load. Diode 3 is passing current from the load, and diode 4 is blocking. When we reverse the direction of current through our coil of wire, all diode action is reversed so that diodes 1 and 4 are passing current, while diodes 2 and 3 are blocking current. The end result of both directions through the coil was the same. In this example, AC changes into DC and produces a pattern like that shown. This is called full-wave rectification.

Do you remember our stator winding? Let us examine diode action through six diodes and three stator windings. **Figure 13** shows a Y-wound stator with each winding attached to two diodes. Each pair of diodes has one grounded, or negative, diode and one insulated, or positive, diode. Notice also that the center of the Y contains a common point for all windings. This is called the stator neutral junction and can have a connection attached. At any time during the rotor movement, two windings will be in series, while the third coil will be neutral and doing nothing. Follow the arrows through the two windings and you will realize that the rectification of the AC into DC will be the same as before. As the rotor revolves, it will energize a different set of windings and in different directions. The end result will be the same, however. Current in any direction through two windings in series will produce the DC required by the battery.

Wiring the stator and diodes into the more common delta pattern does not change around the diode action. You can see from **Figure 14** that the major difference is that two windings are in series with each other and in par-

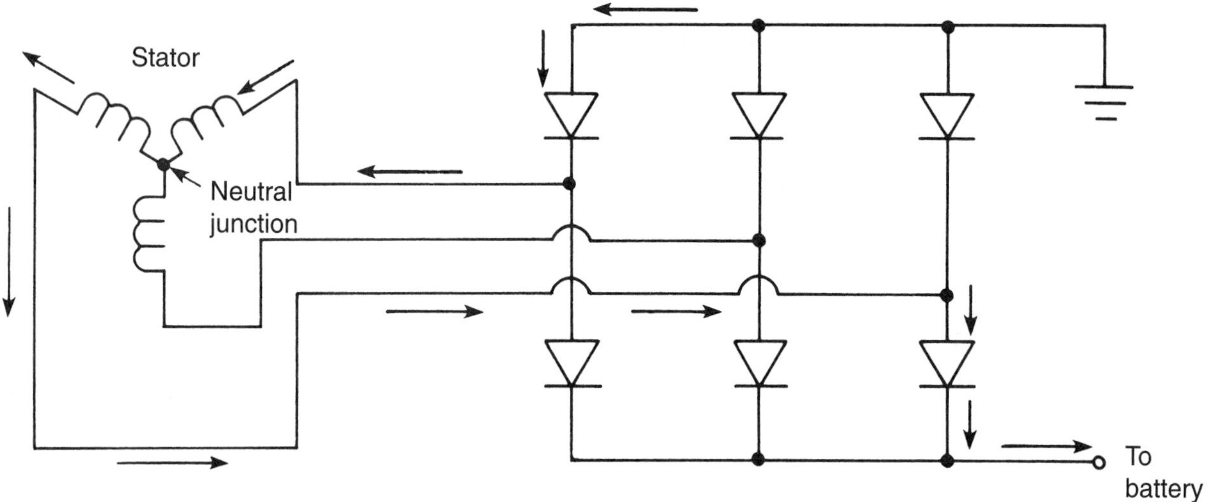

Figure 13. A Y-wound stator wired to the six diodes of the rectification bridge.

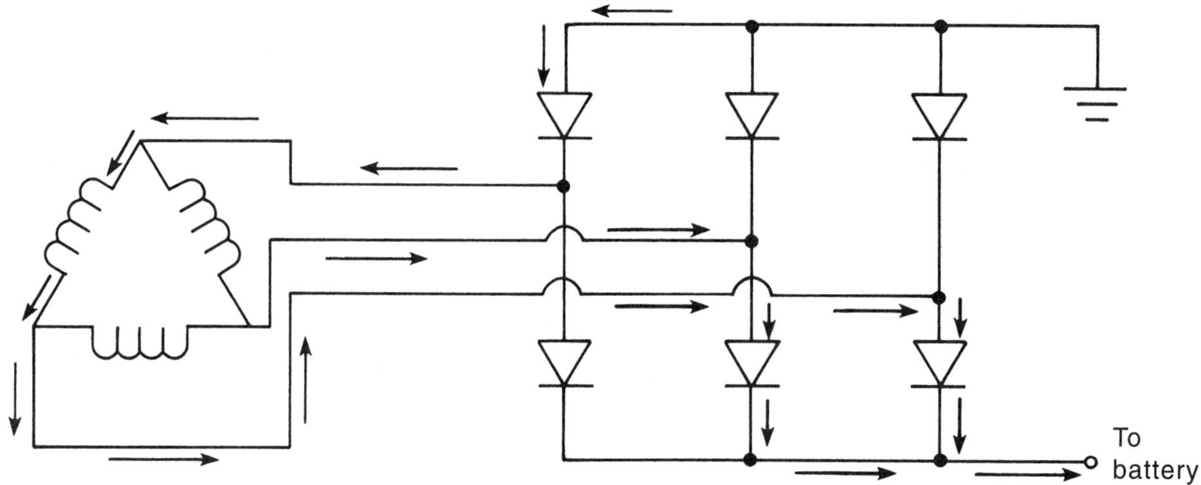

Figure 14. A delta-wound stator wired to the six diodes of the rectification bridge.

Figure 15. Diodes are normally in a vented assembly.

allel with the third winding. This accounts for the increased current that is available from a delta-wound alternator. The parallel paths will allow more current to flow through the diodes. The action of the diodes is the same, however.

Most alternators have their diodes in a bridge assembly, as shown in **Figure 15**. They must be electrically connected to the stator and either ground or positive and mechanically connected to something that will dissipate the heat they will generate.

End Frames

The end frames of the alternator provide the support for the rotor spinning over 15,000 rpm. Usually, a roller or needle bearing is used in one end and a ball bearing in the other. They ordinarily have no method of being lubricated because they are packed with lubricant and sealed by the manufacturer. In addition, the rear end frame usually provides the insulated through connectors necessary to bring field current to the rotor, B+ out, and any other stator, diode, and rotor connections. Different alternators found in the field will have different connections for various functions. Maintenance manuals will detail exactly what the different connections are for.

You Should Know

Alternators are generally made of aluminum and are therefore not to be pried on. Any excess force could easily bend the case, making an otherwise electrically good alternator worthless.

The alternator's case might have imprinted information about the type, style, or amperage output of the unit. Early Ford alternators were color coded, as can be seen in **Figure 16**. It can save the technician quite a bit of time when diagnosing charging system difficulties, if the size of the alternator is known.

Color Coding for Amperage Output	
Amperage	**Color**
38 amps	Purple
42 amps	Orange
55 amps	Red
61 amps	Green
65 amps	Black
70 amps	Black
85 amps	Red
90 amps	Red

Figure 16. Color coding indicates the rated output.

Before attempting any diagnosis of the charging system, it is important to note the rated output of the alternator that the manufacturer installed at the factory and the output of the one currently on the vehicle. If replacement becomes necessary, the physical size of the alternator must also be the same as the original.

REGULATORS

Now that we have a functioning alternator, we need to realize that the system must be regulated to be efficient. Regulation of the charging system is extremely important for a variety of reasons. The first is the most obvious: keeping the battery fully charged. If the charging current is lower than the amount of current needed to run the vehicle, the battery will gradually be run down. We must try to have the charging system supply enough current to both run the vehicle, including the accessories, and charge the battery. Once the battery is fully charged, the charging rate must decrease to a low enough level so as not to overcharge the battery. In addition, the advent of electronic accessories on the vehicle has made the charging voltage level critically important. Onboard computers and other digital equipment can be damaged if the charging system is allowed to raise up the system voltage too high. The modern charging system relies on the **regulator** to keep the battery fully charged, but not overcharged, run the vehicle electrically, and protect the system from an overvoltage condition. You can see just how important the regulation of the charging system is.

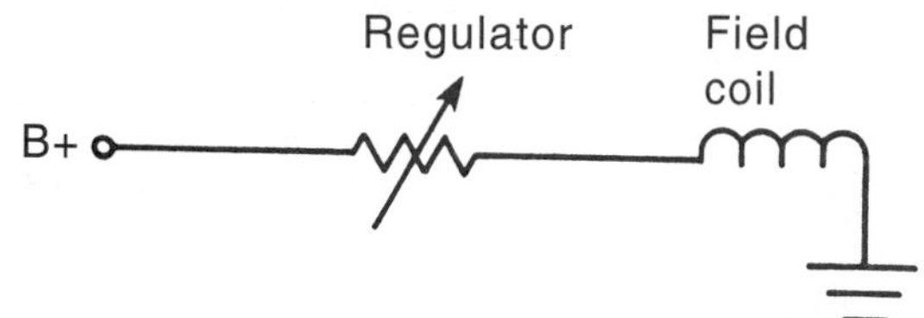

Figure 17. Regulators control field current.

Regulation is achieved by varying the amount of time that field current is flowing through the rotor. If field current on-time is high, output will be high. If field current on-time is low, output will also be low. The simplest way to think of the regulator is to imagine it as a variable resistor in series with the field coil **(Figure 17)**. If the resistance is high, field current will be low, and if the resistance is low, field current will be high. By controlling the amount of resistance in series with the field coil, we would be able to control field current and alternator output. Almost no regulator functions exactly as do variable resistors, but the principle is the same as using pulse-width modulation (PWM) to control field current based on charging system demands or needs. If the system voltage is 14 volts and the PWM is 50 percent, then the effect will be the same as 7 volts directly to the rotor. **Figure 18** shows the DSO pattern of a 50 percent PWM. A 10 percent PWM would be 1.4 volts; a 20 percent PWM would be 2.8 volts and so on. To be an effective regulator, two inputs are required to determine charging operation: voltage and temperature.

First let us look at voltage. Charging voltage is critically important if we are to expect a fully charged battery. If the charging system is functioning at 11.5 volts, the battery will never come all the way up on charge. To ensure full charge, most regulators are set for a system voltage between 13.5 and 14.5 volts. To simplify this concept, look at **Figure 19**. This block diagram shows what would happen to the vehicle system if the voltage were at 12.6 volts. First, the entire vehicle would be operating at 12.6 volts, and in all likelihood current would be coming out of the battery instead of going into it. By raising the voltage of the alternator above the voltage of the battery, as diagrammed in **Figure 20**, the alternator now becomes the source of current for the vehicle systems, and current will now flow into the battery, charging it. By keeping the charging voltage slightly higher than the 12.6 volts of the battery, we ensure that current will flow into, rather than out of, the battery. The regulator must have system voltage as an input if it is to regulate the charging system. This is sometimes referred to as **sensing voltage**, because the regulator is "sensing" system voltage and interpreting it to determine the need for charging current. Low sensing voltage (less than regulator setting) will cause an increase in charging current by increasing field current PWM. Higher sensing voltage will cause a corresponding decrease in field current on time and system output. The

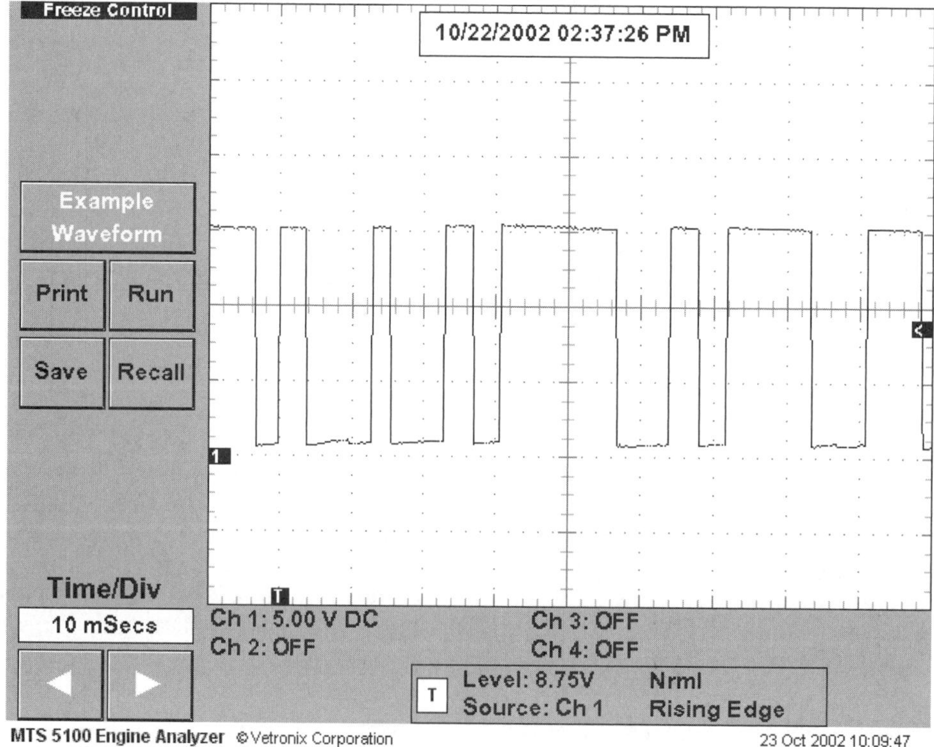

Figure 18. A DSO shows 50% PWM.

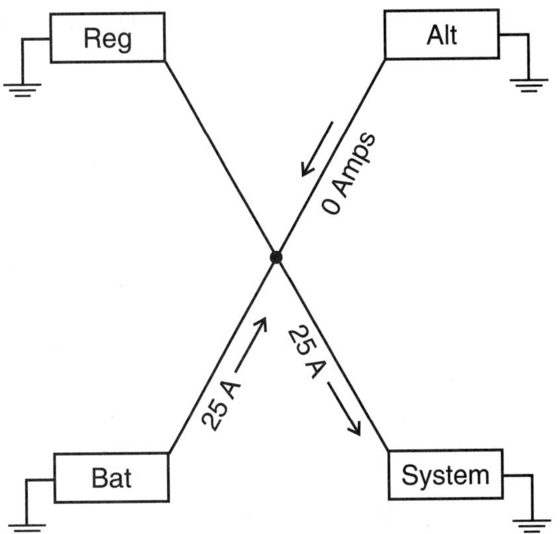

Figure 19. If the charging system does not produce the required 25 amps, the battery will have to supply it.

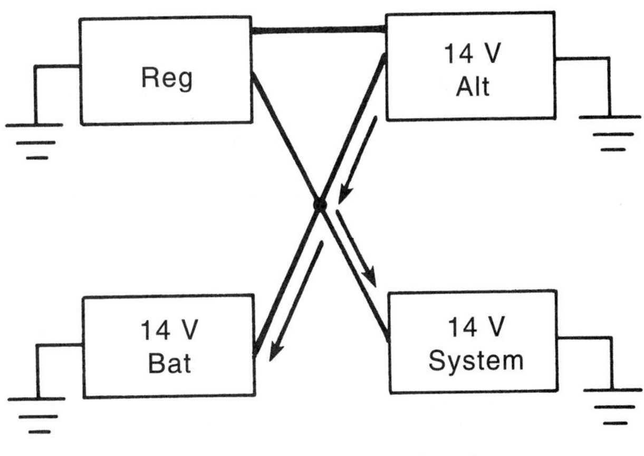

Figure 20. Raising charging voltage up to 14 volts charges the battery.

regulator is therefore responding to changes in system voltage by increasing or decreasing the amount of time that field current flows. For example, you are running down the road with no accessories on and a fully charged battery. The regulator senses a high system voltage because the battery is fully charged and reduces the charging current until it is at a level to run the ignition system and trickle charge (2–4 amps) the battery. If you turn on the headlights, you will begin a sequence of events like this. First, the additional draw will drop or load the battery voltage down. This reduced voltage will be sensed by the regulator through the sensing circuit. It will, in turn, increase the PWM of the field circuit, turning field current on for a longer period of time. This increase in field current on time will produce a stronger magnetic field and increase the alternator output in an effort to keep the system voltage within the specified range. This will all happen within a split second. Turning off accessories causes the opposite to occur. System voltage rises and the regulator cuts back on field current and alternator output.

The second input to the regulator is temperature. All regulators are temperature compensated as **Table 1** shows. Notice that as the temperature goes down, the voltage regulator setting increases. This is to compensate for the battery being less willing to accept a charge if it is cold. The regulator will therefore raise up the system voltage until it is at a level that the battery will readily accept. Bringing the battery up on charge quickly is the goal here, even in very cold climates.

	Volts	
Temperature	**Minimum**	**Maximum**
20°F	14.3	15.3
80°F	13.8	14.4
140°F	13.3	14.0
Over 140°F	Less than 13.8	—

Table 1. Temperature changing the charging voltage.

Summary

- The charging system has two functions: to deliver sufficient current to run the vehicle and to charge the battery.
- A voltage is produced when motion occurs between a magnetic field and conductors.
- The purpose of the rotor is to produce a rotating magnetic field.
- The purpose of the stator is to generate charging current.
- The purpose of the rectifier is to change the AC from the stator into DC that will be the alternator's output.
- A rectifier bridge uses four diodes at a time to change AC into DC.
- Most regulators will use PWM of field current to control alternator output.
- Slip rings and brushes bring field current into the spinning rotor.

- Full-wave rectification is achieved through the use of a rectifier bridge, which will usually have six diodes.
- Charging voltage is adjusted based on the temperature of the battery.
- Sensing voltage is the input into the regulator that is used to determine required output.

Review Questions

1. Alternator output at the battery terminal of the alternator is
   A. DC
   B. AC
   C. three-phase unrectified
   D. none of the above
2. The magnetic field current of an alternator is carried in the
   A. stator
   B. rotor
   C. housing
   D. diodes
3. An alternator stator with three windings usually has ________ output diodes.
   A. 3
   B. 4
   C. 6
   D. 8
4. Technician A says that the alternator cannot produce alternating current until the stator windings are energized by battery current. Technician B says that the alternator cannot produce alternating current until battery current flows through the field coil of the rotor. Who is correct?
   A. Technician A only
   B. Technician B only
   C. Both Technician A and Technician B
   D. Neither Technician A nor Technician B
5. The output of an alternator is created in the
   A. stator
   B. rotor
   C. housing
   D. brushes
6. Technician A says that the drive pulley is attached to the rotor. Technician B says that the magnetic field cuts across the stator conductors. Who is correct?
   A. Technician A only
   B. Technician B only
   C. Both Technician A and Technician B
   D. Neither Technician A nor Technician B
7. Increasing the regulator field PWM means that
   A. more output current is desired
   B. the rotor is turning too slow
   C. less output current is desired
   D. B+ sensing voltage is lower than specified
8. The alternator field circuit is composed of
   A. battery, stator, and diodes
   B. battery, regulator, and stator
   C. battery, rotor, and regulator
   D. regulator, stator, and rotor
9. The magnetic field is carried by the stator.
   A. True
   B. False
10. The field current flows through the stator.
    A. True
    B. False
11. Brushes and slip rings carry the full alternator output.
    A. True
    B. False
12. Field current for the alternator comes from the battery.
    A. True
    B. False
13. Alternator output decreases as field current PWM increases.
    A. True
    B. False
14. Alternator maximum available output increases as rotor speed increases.
    A. True
    B. False
15. Larger alternators generally have delta-wound stators.
    A. True
    B. False
16. Sensing voltage is the voltage that tells the regulator just how much output current is necessary.
    A. True
    B. False

Chapter 44

Field Circuits

Introduction

Regulators fall into different categories of operation and also different styles. Within the automotive industry, we find two styles currently in use: digital and computerized. The digital or computerized regulator is exclusively used on current United States and foreign vehicles. In this chapter, we look at regulator placement and identify types of field circuits. We also look at the use of computer regulation and charging indicators. Frequently, manufacturers will identify in their service literature the type of field circuit. It will be helpful if you can interpret their identification system before you begin to service the various units on the vehicle.

THE A CIRCUIT

The first field circuit to look at is the A circuit. **Figure 1** is a block diagram of a typical A field circuit. Notice that the regulator is on the ground side of the charging system and that B+ for the field coil is picked up inside the alternator. Remember that the PWM switch can be placed anywhere in a series circuit and it will have the same effect: that of controlling current flow. Having the PWM switch of the regulator on the ground side of the field coil will allow the control of field current just as easily as having the switch on the B+ side. An A circuit places the regulator *After* the field coil.

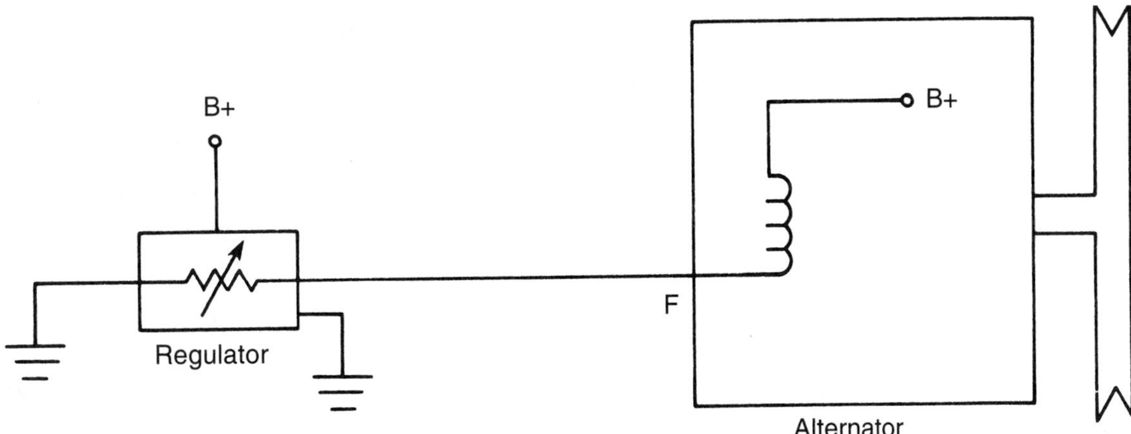

Figure 1. An A field circuit.

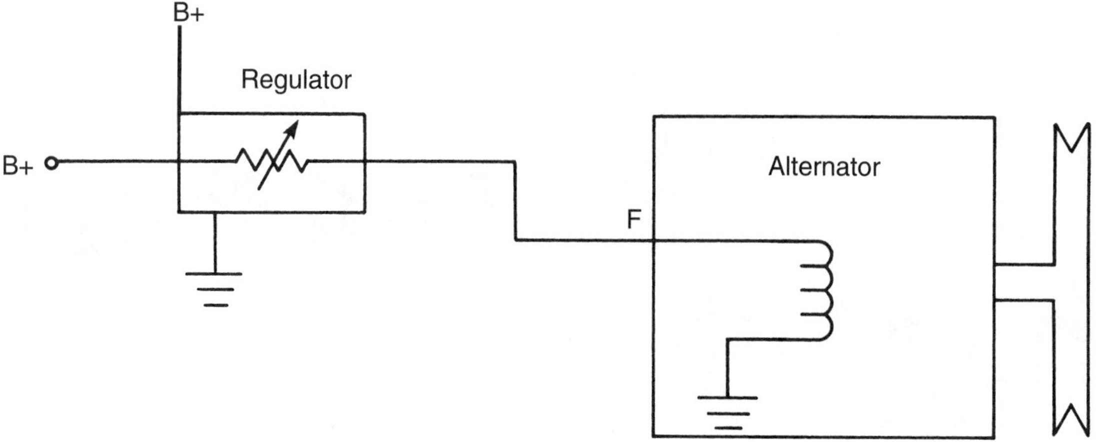

Figure 2. A B circuit.

THE B CIRCUIT

You have probably figured out that a B circuit has the regulator's PWM switch between the B+ feed and the field coil as diagrammed in **Figure 2**. The same results will occur, however, in either case. Notice that the field coil is grounded inside the alternator. A B circuit places the regulator *B*efore the field coil.

> **You Should Know** *Usually 13.5 volts is the beginning level at which regulation of the charging system becomes necessary. Lower than 13.5 volts should signal the regulator to allow full field current to flow through the field coil.*

ISOLATED FIELD CIRCUIT

Isolated field alternators pick up B+ and ground externally. They are usually easily identified because they have two field wires attached to the outside of the alternator case. Isolated is another way of saying insulated, or not connected. Isolated field alternators can have their regulator placement on either the ground (A circuit) side or on the B+ (B circuit) side. **Figure 3** block diagrams both an isolated A and an isolated B circuit. It will be helpful if you under-

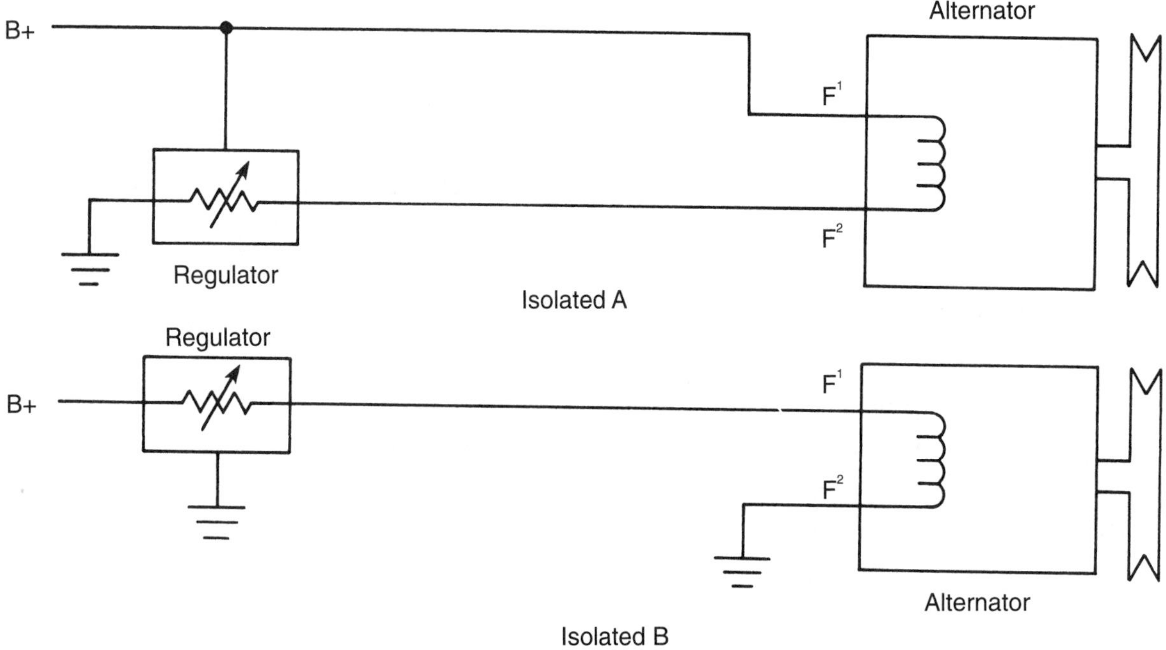

Figure 3. Isolated A and B field circuits.

stand the different types of field circuits before you begin to diagnose charging systems.

ELECTRONIC REGULATORS

Let us turn our attention to what is probably the most common regulator on the road today: the electronic regulator. It can be mounted externally or, in most cases, inside the alternator. There are some instances in which it will be bolted to the outside of the alternator case. It can be transistorized, but is most often an integrated circuit design. Long life without maintenance, close voltage limiting, and inexpensive production are its main advantages. With no moving parts and contact points to burn from arcing, the unit will virtually last the life of the vehicle. The heart of the electronic regulator lies in a component that we have not studied called the **zener diode**. Much like the diodes in the alternator, it has a significant difference, however: the ability to conduct in the reverse direction without damage. The zener diode used in regulators is doped so that it will conduct in reverse once a predetermined voltage has been achieved. You have probably guessed that this voltage level will be between 13.5 and 14.5 volts. **Figure 4** shows a typical transistorized regulator circuit in simplified form. The key to the circuit is the zener diode, which has B+ on the anode (the side with the arrow). When the voltage of the vehicle system rises above the specification, the diode will conduct and permit current to flow to the base of transistor 1. This turns the transistor on, which will in turn switch off transistor 2. Transistor 2 is in control of field current for the alternator. With it off, no field current can flow, thus shutting off the alternator until the voltage level drops below specification. Transistor 2 is acting like a switch turning field current on then off as the system voltage rises and falls above and below the specified voltage. This occurs many times a second and cannot be measured with a standard voltmeter. The thermistor in the upper left of the diagram will give the temperature voltage change necessary to keep the battery charged in cold weather. Some older regulators had contact points with surfaces to pit and wear. This caused the voltage levels to change over time. With the use of electronics, and no moving parts, the zener diode voltage will become the regulated voltage of the alternator for the life of the regulator and require no adjustment.

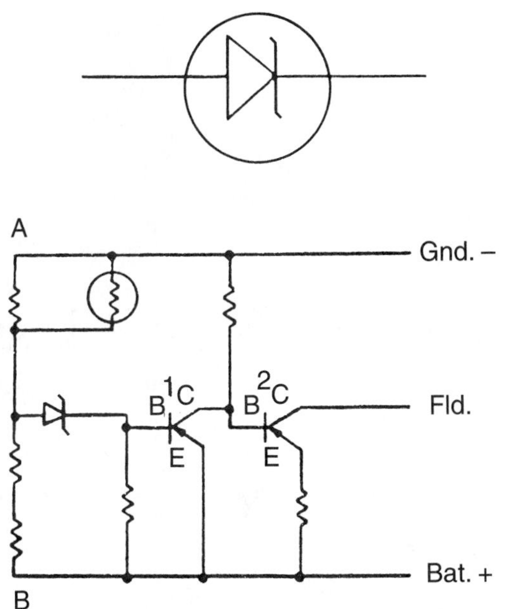

Figure 4. The heart of an electronic regulator is the zener diode.

This on then off again operation occurring many times a second is an example of PWM, which is used in many applications on today's vehicles and started with regulators and charging systems in the 1970s. Now, with computer control of many systems, PWM is the accepted method of control. It is easy to understand. Let us use a 100-amp alternator as an example. If the demands of the vehicle's electrical system and the battery requirements are 50 amps, the regulator will turn the alternator's field on for approximately half (50 percent) of the time, then off for 50 percent. This will occur rapidly. This 50 percent on then 50 percent off will produce approximately 50 amps out of a 100-amp alternator. This is shown in **Figure 5**. If the system's require-

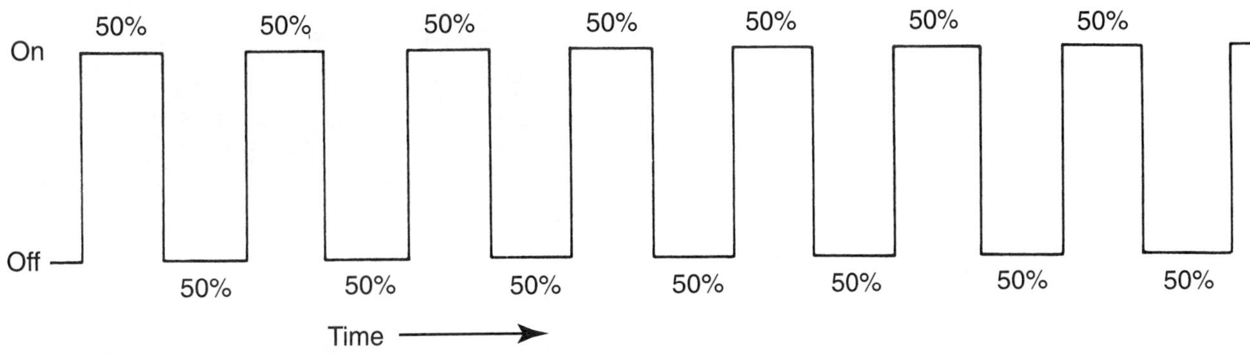

Figure 5. Fifty percent PWM.

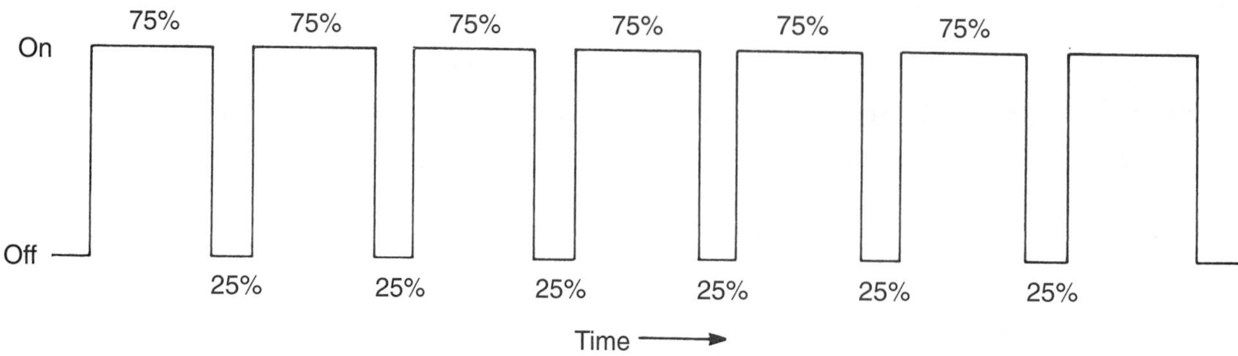

Figure 6. Seventy-five percent PWM.

ments change so that 75 amps are needed, the regulator will increase the on time of the field to 75 percent and decrease the off time to 25 percent. This is shown in **Figure 6**. There are many uses of PWM on the modern vehicle. The extremely fast processors that are in use in today's vehicles will allow PWM of many thousands of times a second. Regulators are usually sealed, with no means of setting voltage. They are serviced by complete replacement, which we will cover in Chapter 45. Remember that the function of the regulator is to control field current to keep the system voltage within the specification. The advantage to the electronic regulator is the sharp cutoff voltage. The regulator will be capable of controlling system voltage to as little as plus or minus 0.5 volt. Older mechanical regulators normally had typical ranges of 13.5 to 14.5 volts or more. You will realize in Chapter 45 that testing system voltage is easy.

Figure 7. Ford's alternator with the regulator mounted to the back of it.

EXTERNAL REGULATORS

Figure 7 shows an external regulator mounted to the back of an alternator, whereas **Figure 8** shows the regulator mounted to the fender of a vehicle. In either case, the regulator is considered external because it is not inside the alternator. An advantage of an external regulator is that it may be replaced separately from the alternator. The main disadvantage is the need for wires to connect the alternator to the regulator. Let us look at the Ford IAR charging system and analyze regulator function. Refer to **Figure 9**, a diagram in which you can see that field current comes directly from the battery through terminal A. The regulator is screwed down to the brush assembly with two screws, A and F. These two screws have continuity to the brush terminals on either side of the field. We will use these connections in Chapter 45. The I or indicator terminal comes from the instrument cluster and is ignition switched. Within the regulator is a circuit that will ground the I terminal if the alternator is not producing any current. Grounding this terminal will turn on the indicator lamp in the instrument cluster. Notice also that there are two B+ terminals to split the output until it reaches the outside of the alternator. Before the wiring reaches the starter relay battery terminal, the two conductors will join together. The Ford IAR is truly a unique charging system because of the regulator being external but part of the alternator.

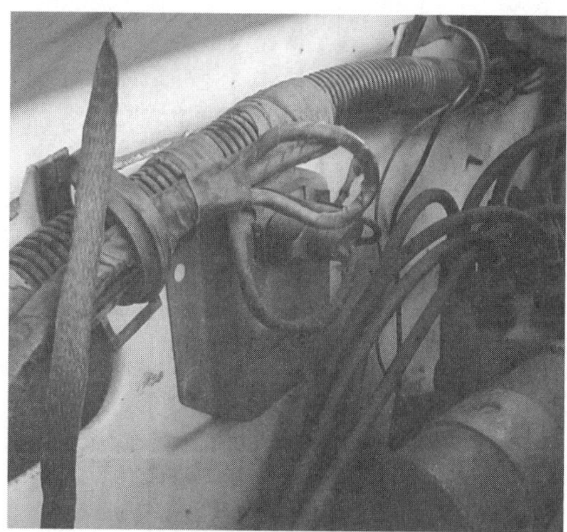

Figure 8. External regulators can be mounted almost anywhere.

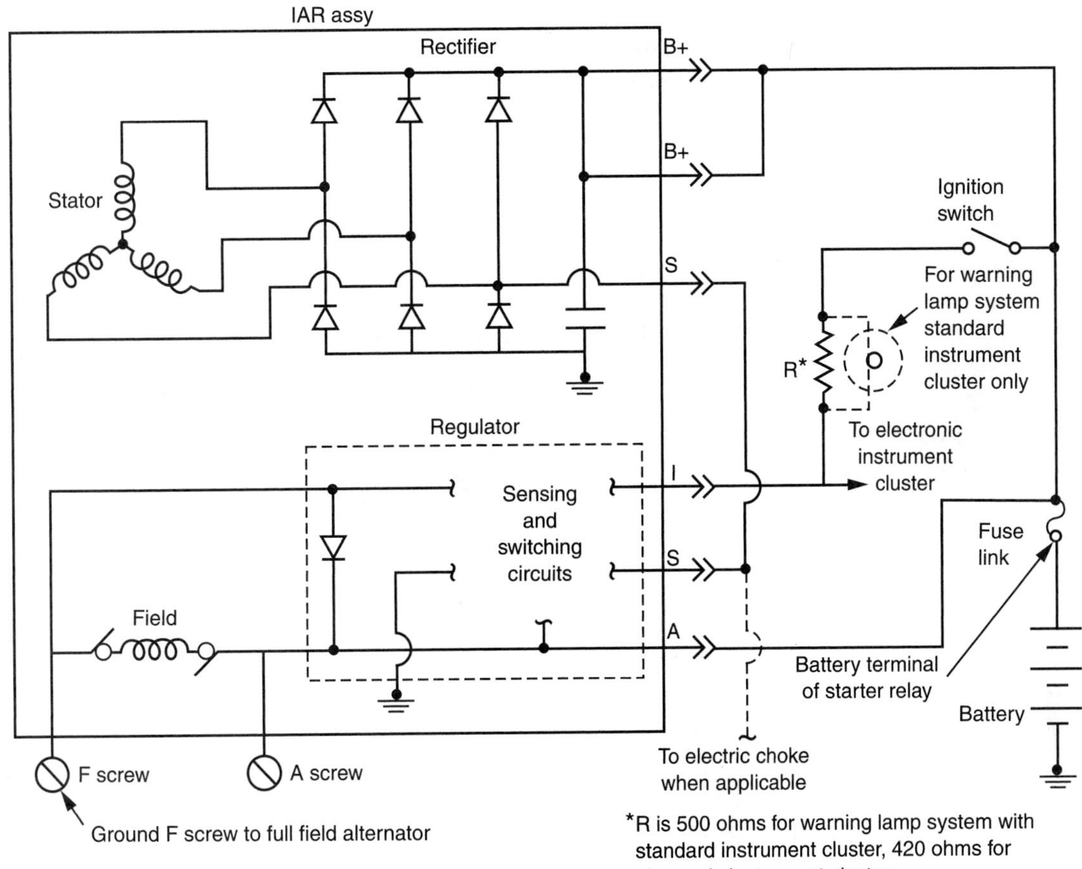

Figure 9. Ford charging system.

INTERNAL ELECTRONIC REGULATORS

The next group of regulators to look at is exactly the same electrically as our last group. The difference is in their mounting. Putting the regulator into the alternator is a logical choice because it eliminates the interconnecting wires and greatly simplifies the vehicle-building process. You should realize that most internally regulated alternators have no provision to separately test the regulator. If either the regulator or the alternator were to malfunction, you would normally replace the entire unit. Let us go through the operation of the charging system with an internal electronic regulator. Follow along on **Figure 10**, which is from a General Motors CS alternator. The single fusible link connects the alternator to the battery. The other two connections are for the tachometer if the vehicle uses a diesel engine and the charge indicator in the instrument cluster. When the key is turned on (KOEO), the BRN wire at the alternator will "look" like a ground because the regulator inside does not see any alternator output. With power on the top of the bulb and a ground on the bottom, the light goes on. As soon as the engine starts (KOER), the regulator will apply the charge voltage to the bottom of the bulb. With battery voltage on top and charge voltage on the bottom, the bulb will go off. If the alternator is functional, both voltages should be the same. All regulation of the charging field current is done internally with the current initially supplied from the fusible link. Once the alternator starts charging, field current will come from the output.

COMPUTER REGULATION

On many vehicles of today, the regulator function has been taken over by the vehicle PCM. The operation is exactly the same as the integrated circuit electronic regulator we have studied, and because the vehicle PCM is the regulator, it is considered to be external. The field circuit will be controlled using PWM to keep the system voltage within the specified limits. As the computer has gradually taken over various electrical operations on the vehicle, it is an obvious advantage to eliminate the regulator and turn its function over to the computer. The circuit is just as easy to diagnose as those we have previously studied. There is an obvious difference, however. Technicians, who are not the good diagnosticians they should be, probably are used to replacing the regulator any time a vehicle comes in with a no-charge condition. This same procedure, when applied to a computer-regulated vehicle, will involve the

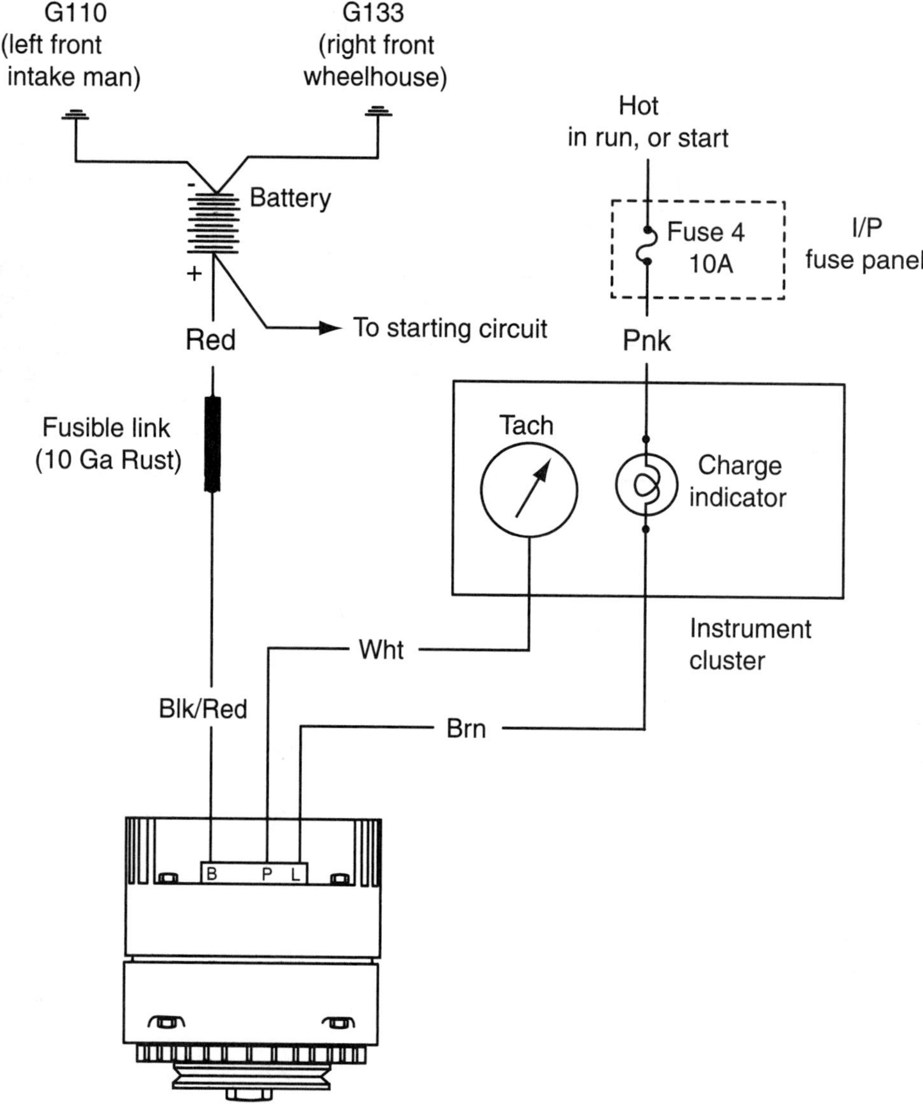

Figure 10. Delco CS alternator wiring diagram.

replacement of an expensive major component. The importance of diagnosing the exact problem and repairing it rather than replacing components until the vehicle charges cannot be overstressed. **Figure 11** shows a current computer regulator circuit from a Chrysler vehicle. There are only three wires on the alternator. The first is a large black cable with a gray stripe (BLK/GRY) leading from the alternator through a fuse link and up to the red lead in the power distribution center. The red wire goes to the battery. This wire is the charge wire from the alternator to the battery. The dark green wire with an orange stripe (DK GRN/ORG) is power to the field, and the dark green (DK GRN) is the ground wire for the field that leads to the PCM. Notice that power for the field comes from the load side of the **automatic shutdown relay (ASD)** inside the power distribution center and that the PCM supplies the ground for the relay coil at the ASD relay control terminal. Once the engine begins to crank, the ASD is closed by the PCM, bringing power to the alternator field. The PCM field driver supplies the ground for the field. A **driver** is a term used to describe a switching circuit. The PCM will now pulse-width modulate the ground side of the field circuit based on sensing voltage. This is an example of an isolated field A circuit alternator that is computer regulated.

The other system we look at is a General Motors PCM-controlled charging system diagrammed in **Figure 12**. General Motors uses two different types of PCM-controlled

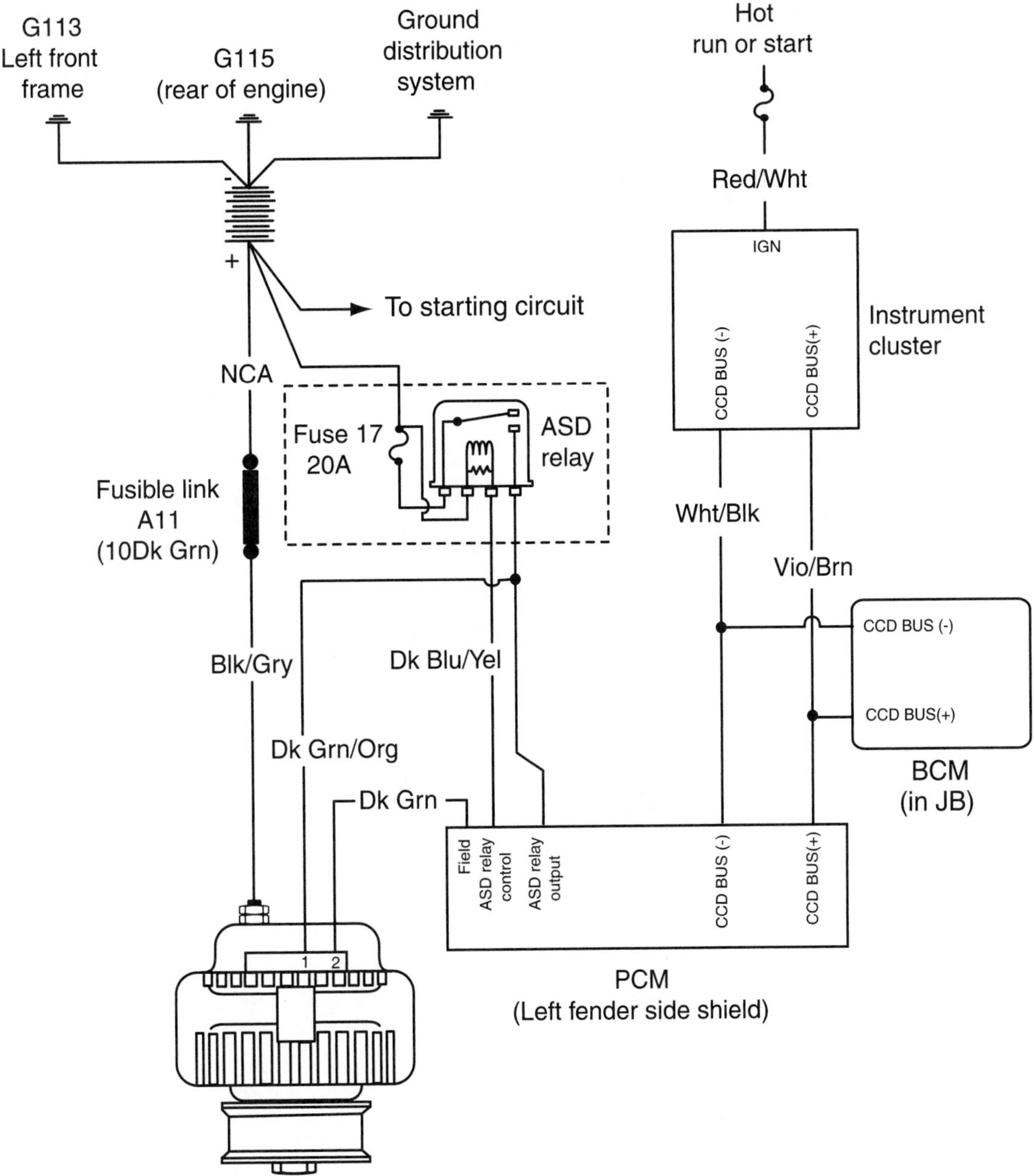

Figure 11. A Chrysler PCM-controlled charging system.

systems. One is similar to the Chrysler that we just looked at. We will look at the system that is different from the Chrysler in that the PCM will supply power to the field. Notice that power comes from the fuse block through the instrument cluster and into the PCM with a dark green (DK GRN) wire. Power out of the PCM is pulse-width modulated to the alternator on the red wire. The other end of the field is grounded inside the alternator. The other red cable brings charging current out of the alternator through a fusible link to the battery.

Either of these two examples can be full fielded, a process by which the technician bypasses the PCM and takes over the control of the alternator. This testing process is covered in Chapter 45.

CHARGING INDICATORS

In vehicle charging, indicators usually take one of three available forms: the indicator lamp, an ammeter, or a voltmeter. We have looked at regulator circuits, which

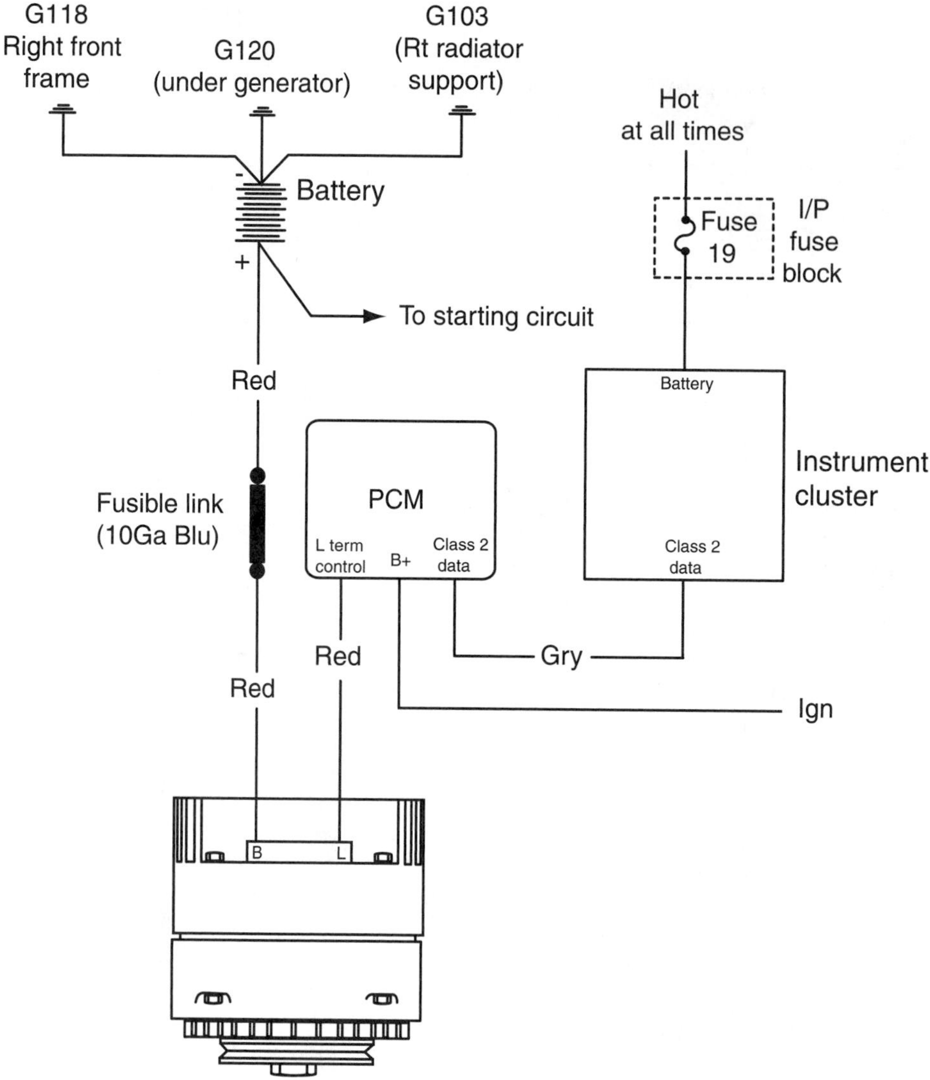

Figure 12. A two-wire General Motors PCM-controlled charging system.

have used an indicator lamp, and you have realized that the lamp will go out when equal voltage is applied to both sides of the circuit or when the regulator removes the ground from the indicator circuit. **Figure 13** shows that one side of the bulb (charge indicator) is connected to B+ coming from the fuse panel that is live, or hot, when the ignition switch is in the run position. The other side is connected to alternator output that will become B+ once some output is present. It is important to note that the typical indicator lamp circuit does not recognize how much current is being developed or what system voltage is. It will be off if there is any amount of current being produced and, therefore, should not be relied on to give much diagnostic information regarding system performance. It is, however, a good indicator for the consumer because it will usually glow or be on brightly if there is no charging current. A KOEO condition should cause the light to be on, whereas a KOER condition should cause it to be off.

The ammeter or charge indicator mounted on the dash gives the consumer much additional information over that of the indicator lamp. Many consumers, however, do not know what to do with the information. Correctly functioning, it will show a high charging rate with a recently started vehicle **(Figure 14)**. Once the battery comes up on charge, it will gradually taper down until it is

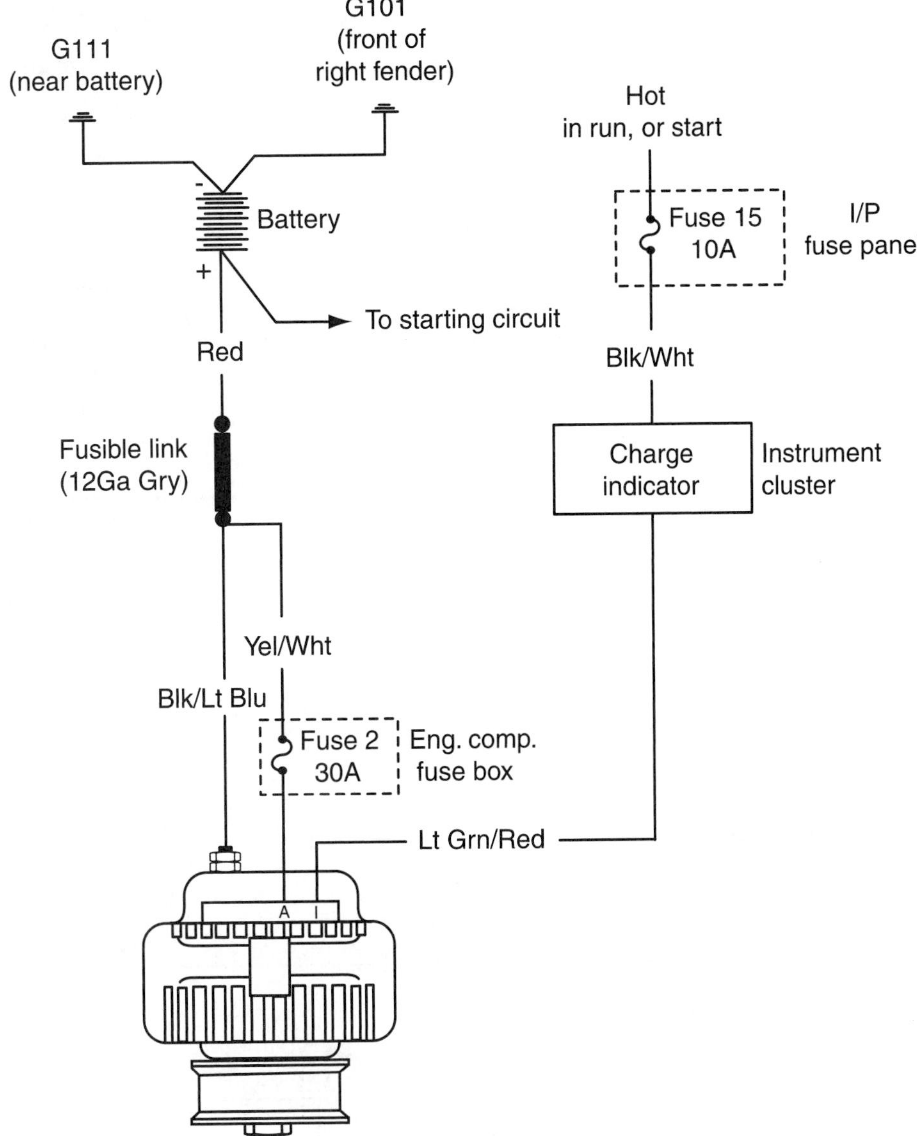

Figure 13. A charge indicator in instrument cluster.

only slightly higher than zero. Occasionally, it will read discharge. A discharge condition will be present on some, mostly older vehicles, at idle with a lot of accessories on. The condition should change once the vehicle speed goes above idle. The wiring for most ammeters places them between the battery and a junction where the B+ for all systems is picked up. Ammeters are more costly to manufacture and install than voltmeters, so their use has gradually tapered off. Ammeters can be useful to knowledgeable technicians because ammeters can indicate battery, alternator, or regulator difficulties before a no-start condition exists.

Voltmeters mounted on the dashboard are also found on many vehicles. Like ammeters they can be useful to the knowledgeable consumer or confusing and ignored by others. Usually they are fed off the ignition switch so that they measure system voltage with the key on. They can be helpful in picking out bad regulators, alternators, or batteries. **Figure 15** shows a typical in-dash voltmeter setup. When the vehicle is first started, the voltmeter will climb until the regulator reaches its set point. This is the voltage level that the entire vehicle is running on. Only idle conditions with virtually all of the accessories on will cause it to drop from its set point, but it should never be below the 12.6 volts of the battery.

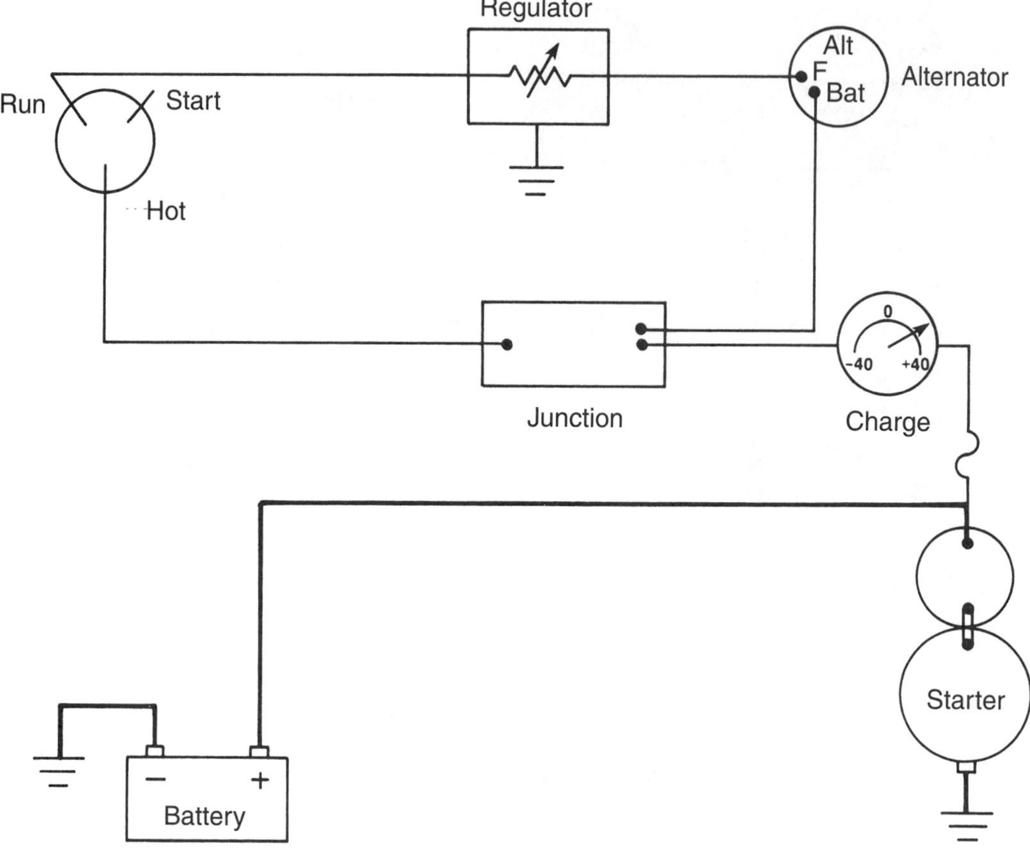

Figure 14. An ammeter wired to indicate charging current.

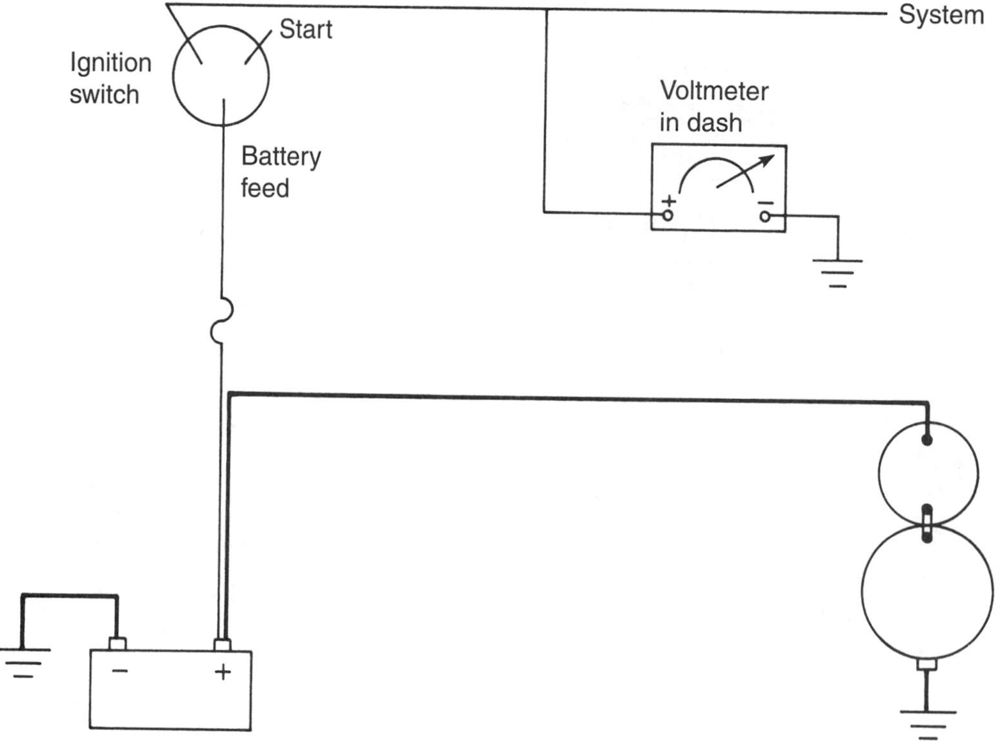

Figure 15. Voltmeters greatly simplify instrument cluster wiring.

Summary

- An A field circuit has the regulator on the ground side.
- A B field circuit has the regulator on the B+ side.
- Regulators control the amount of time that field current flows (PWM).
- The heart of most regulators is the zener diode, which will conduct or block based on the voltage.
- Pulse-width modulation is the process of turning the field on for a specific period of time to control alternator output.
- Fifty percent PWM should produce approximately 50 percent of the rated alternator output.
- One hundred percent PWM should produce approximately 100 percent of the rated alternator output.
- Regulators can be mounted external to the alternator or inside (internal to) the alternator.
- Some external regulators are mounted to the outside of the alternator.
- Computers (PCMs) can be used to control alternator field current as either an A or a B circuit.
- The "driver" inside the PCM turns the field current on and off, thereby controlling the output.
- Voltmeters, ammeters, or indicator lights are placed on the instrument cluster to inform the driver of charging difficulties.

Review Questions

1. Technician A states that most regulators' pulse-width modulate the field current to control output current. Technician B states that PCMs are used on some vehicles to control field current. Who is correct?
   A. Technician A
   B. Technician B
   C. Both Technician A and Technician B
   D. Neither Technician A nor Technician B
2. An 80-amp alternator has its field circuit pulse-width modulated to a 75 percent level. This should allow the alternator to produce
   A. 20 amps
   B. 60 amps
   C. 80 amps
   D. not enough information is given
3. Sensing voltage is the voltage applied to the
   A. alternator
   B. indicator lamp circuit
   C. voltmeter circuit
   D. regulator
4. An A circuit field is being regulated by the PCM. The PCM will supply a pulsing ________ to the field.
   A. ground
   B. power
   C. 5 volts
   D. either ground or power
5. One hundred percent PWM is the same as a full-fielded condition.
   A. True
   B. False
6. PCMs can be on either the ground or the power side of the field.
   A. True
   B. False
7. Sensing voltage is used for ammeter control.
   A. True
   B. False
8. Regulators directly measure alternator output.
   A. True
   B. False
9. How does PWM control alternator output?
10. Why have vehicle PCMs become the regulators on many vehicles?

Diagnosing and Servicing the Charging System

Introduction

Now that we have taken a look at the various components that make up the typical charging system, it is time to begin our discussion of charging system problems and service. As reliable as the modern system is, it is not perfect, and over time it may experience failure. Like the starting system, it cannot be looked at alone. The battery must be considered part of the system. Diagnosis of the battery must precede any evaluation of the charging system. A complete battery and starting check before charging system diagnosis should be done. The most perfect charging system will not be able to charge a faulty battery correctly. Sometimes technicians are tempted to shortcut the testing procedures and perhaps bypass a step or two. Eventually, these shortcuts produce a misdiagnosis, and the technician has a customer who is not very happy. For this reason, it is extremely important that a complete diagnosis of what appears to be a charging system difficulty should precede any parts replacement. An example of the importance of this could be a no-start with a three-year-old battery. The customer calls for a new battery. As the technician, you should start with a state-of-charge test and then do a load test before condemning the battery. If you skip the state-of-charge test and perform the load test and the battery fails, replacement of the battery might not cure the problem. In this example, let us say that the charging system is allowing the alternator to not fully charge the battery under heavy electrical load. In a few days, your customer is back with a dead battery, or worse yet, someone else discovers and solves the no-start problem. What do you think your customer's (or perhaps ex-customer's) opinion of your mechanical skill will be? Shortcutting correct procedure may result in misdiagnosis. Even though time is valuable, take the necessary time to completely evaluate the situation.

THE STATE OF CHARGE

Beginning with the battery, first consider the state of charge. Indirectly this will give us information regarding the charging system. A fully charged battery probably has a charging system that is functioning. Do not assume the opposite, however. A partially discharged battery does not necessarily indicate a charging problem. The battery could have an internal drain or the vehicle could be drawing excessive current when the engine is off. If the battery passes the state-of-charge test (80 percent or better), a load test should be performed to determine the capacity. Failure of the load test should lead you right into the three-minute charge test to test for sulfation. These three tests are again reproduced for you in **Figure 1**. If you have forgotten any of them, do not hesitate to review Chapters 30–32. Remember also that sulfation of a battery will occur if the battery is never brought up to full charge. It is possible for a charging system difficulty to produce a premature sulfation condition. Replacing a sulfated battery with a new one will cure only the immediate vehicle difficulty and not cure the cause. Weak, sulfated, or self-discharging batteries should be replaced before you try to diagnose the charging system. Substitute a known good battery for the testing procedure if you have to, but do not proceed until the battery is good or your results might not be accurate.

Specifications

Battery Size ______________ Engine Cubic Inches ______________

Hydrometer Readings

Cell Number	1	2	3	4	5	6	
Negative Post							Positive Post

Results (Check one.)

_____ All cells are within 25 points and are charged (continued testing).
_____ All cells are equal and are discharged (charge battery).
_____ Cell readings vary more than 25 points (skip load test and advise customer to replace battery.)

OR

Open Circuit Voltage Test

Load battery to 20 amps for about 1 minute before testing OCV.

______________ volts—open circuit (no load)

Refer to the chart and convert either the hydrometer reading or the OCV over to state of charge.

Charge	Specific Gravity	OCV (12.6)	OCV (6.3)
100%	1.265	12.6	6.3
75%	1.225	12.4	6.2
50%	1.190	12.2	6.1
25%	1.155	12.0	6.0
Discharged	1.100	11.9	5.9

Battery Load Test (for batteries at least 75% charged)

Set tester to the following:
1. Tester to STARTING
2. Variable LOAD off
3. Voltage—INTERNAL—higher than battery voltage

Attach the BST as follows:
1. Place large cables of tester clamp over the battery clamps, observing that polarity of the battery and tester are the same.
2. Attach inductive pick-up clamp around either of the BST cables *not* battery cables.

Determine the correct load:

Battery Amp/Hour × 3 = Load or
Cold Cranking Amps ÷ 2 = Load

NOTE: Compare the specific battery to the one actually installed. Load to the LARGER of the two.

Apply the calculated load for 15 seconds, observe voltmeter, and turn load OFF.

_____ GOOD—battery voltage remained above 9.6 volts during load test
_____ BAD—battery voltage dropped below 9.6 volts during load test (perform 3-minute charge test to check for sulfation)

3-Minute Charge Test (only for batteries that fail the load test)

1. Disconnect the ground battery cable.
2. Connect a voltmeter across the battery terminals (observe polarity).
3. Connect battery charger (observe polarity).
4. Turn the charger on to a setting around 40 amps.
5. Maintain 40 amp rate for 3 minutes while observing voltmeter.

_____ GOOD—voltmeter reads less than 15.5 (recharge slowly and reload test).
_____ BAD—voltmeter reads more than 15.5 (battery is sulfated and should be replaced).

Figure 1. A battery testing worksheet.

CHARGING SYSTEM TESTING

Once we are assured that the battery in the vehicle is functioning correctly, we can continue with test procedures. We will now look at alternator maximum output and see if it matches the specification for the vehicle. Keep in mind that the specifications listed in various manuals might list a couple of different alternators for the vehicle. When in doubt, look on the alternator case. Most manufacturers will print the rated output or color code it to indicate size. Usually, manufacturers will increase the size of the alternator as they add more current-drawing accessories. Make note of the specifications because you will refer back to them while you complete the testing procedure.

The VAT should be hooked up to the vehicle as you had it for starting testing: large cables over the battery posts and the current inductive pickup around the vehicle's battery cable. It is also possible to put the pick-up around the alternator output cable. Do not forget to include any small wires that are attached to the battery cable inside the inductive pickup, because this might be a charging wire. This is illustrated in **Figure 2**. Make sure that you are close to the battery with the inductive pickup or it might sense a magnetic field from some component under the hood and give you a false reading. Remember also that during starting, the meter range must be set high enough so as not to damage the meter movement. Once the vehicle is running, the range can be lowered for a more accurate reading (usually approximately 100 amps). Also, at this time, make sure all accessories on the vehicle are off.

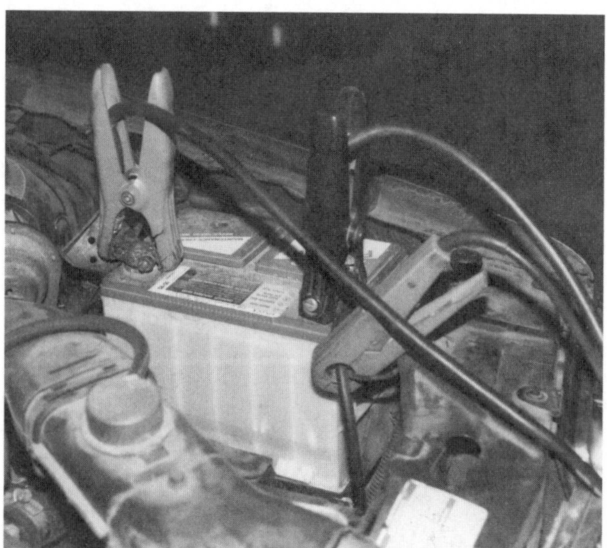

Figure 2. VAT connections for charging testing.

TESTING ALTERNATOR MAXIMUM OUTPUT

One of the most important considerations to keep in mind when looking at charging systems is the question, does the alternator have the ability to produce its rated output? Most alternators will not be capable of producing their maximum output at idle. Approximately 2,000 engine rpm will be necessary. Use a tachometer and try to block the throttle open to 2,000 rpm while you perform an output test. The regulator will have to **full field** the alternator for us to be able to read the maximum output. Full field is the way we describe having the regulator increase the PWM of the field to 100 percent, or full-field current. The easiest method for us to force this full-field operation is to use the carbon pile load to reduce battery voltage. If you remember our discussion of regulators in Chapter 44, reduced voltage to the regulator is the sensing signal for more alternator output. Usually, the load will have to reduce system voltage to 12–13 volts before full fielding will occur and alternator maximum output is achieved. Our ammeter will now register the current that is going into the battery. Record this level, turn off the load, reduce the vehicle speed to idle, and turn the engine off. If you clamped the ammeter around the alternator output cable, the number you recorded is the maximum alternator output. If you clamped it around the battery cable, you will have to add the current draw that the vehicle required to keep running. The ignition system and any full-time accessories were drawing current along with the field coil in the alternator. The current they were using was produced by the alternator but never delivered to the battery for our inductive pickup to register. Typically, it takes approximately 10–15 amps to run the modern vehicle. Add this to the amps reading taken at the battery. The number we now have should be very close (within 2–3 amps) or within 10 percent of the rated output of the alternator. If it is, we now know that the alternator can produce its rated output and that the regulator does have the ability to full field the alternator when it sees reduced system voltage.

If the total current available does not match the specification, we have identified a problem and will have to isolate the alternator, regulator, and interconnecting wires to determine just what is wrong. Recognize that if the alternator uses an internal regulator, trade standards usually will not allow you to "split" the alternator to replace the regulator. The combination alternator and regulator are considered to be a unit and are usually replaced as a single component. However, if the alternator uses an externally mounted regulator, the two components are considered separate and either may be replaced. Note that computer-controlled alternators are considered to be externally regulated. Your diagnosis should allow you to determine whether the alternator or the regulator is faulty. It is possi-

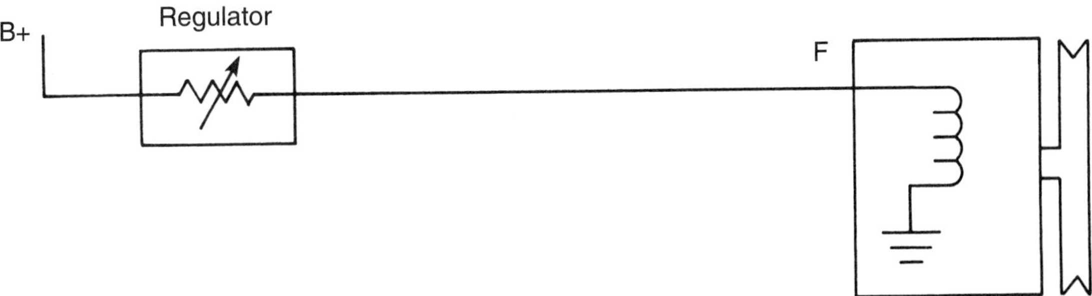

Figure 3. A B-style field circuit.

ble that both may have to be replaced, and replacing both the alternator and the regulator will no doubt cure the problem but may leave you open to criticism regarding your diagnosis and possible over repair. Make sure of your diagnosis so that the repair will cure the charging problem. Isolation of the various components is best done by the process of full fielding the alternator with the regulator disconnected from the circuit.

FULL FIELDING THE ALTERNATOR

If we supply full-field current to the alternator and still get reduced output, we have proved that the cause must be inside the alternator. If, however, full output is received when we full field the alternator, we know that the regulator must have been the cause of the problem. Let us take it step by step with an externally regulated alternator of the B circuit variety. You should remember that a B circuit indicates that the regulator is between the rotor and B+ and that the rotor is grounded inside the alternator. **Figure 3** shows a block diagram of this style.

Procedure for Full Fielding the Alternator

First, we disconnect the regulator with the ignition key off.

> **You Should Know** *Disconnecting or reconnecting an electronic component with current flowing should never be done. You might be the cause of it burning out.*

A good rule to follow is never disconnect or reconnect anything with the key on. By disconnecting the regulator, we have lost our source of B+ for the alternator's field circuit. Next, we start the engine and set it to the same approximate 2000 rpm that we used before. Load down the battery with the carbon pile and then apply full B+ directly from the battery to the field terminal of the regulator connector, as shown in **Figure 4**. If you are doubtful as to which connection is the field lead, consult the wiring diagram. Pin location and wire color should make the identification easy. By applying B+ to the field directly, we are bypassing the regulator circuit completely. The load can now be turned until the voltage is about the same as it was when we were asking the regulator to full field the alternator. This is important because we will want to compare the amperage available with the regulator in full-field operation and with the regulator bypassed. The amperage should be compared only if the voltage is the same. In addition, a full-fielded alternator with the regulator bypassed will raise the system voltage quickly. There is the risk that the unregulated voltage could climb high enough to cause some damage to a sensitive electronic component. Do not let the system voltage rise above typical regulator setting

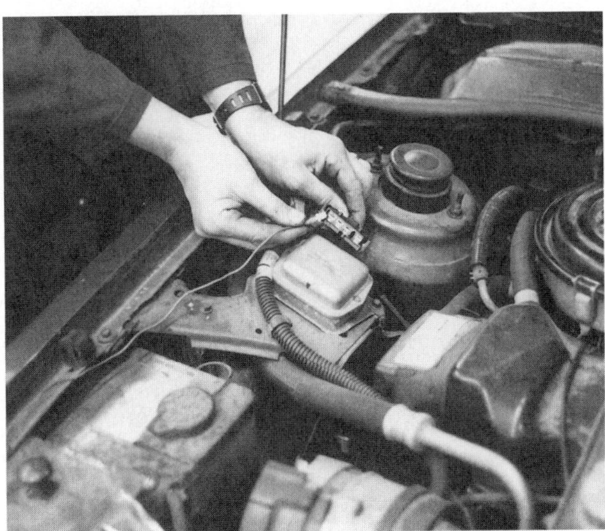

Figure 4. Full fielding an external regulator.

(approximately 15 volts) and you will not run the risk of too high a voltage. If, after you full field the alternator, it produces full-specified output, the regulator is at fault. But if after full fielding it produces less than rated or the same as when the regulator was in control, the alternator is at fault.

What would you do if the following situation presented itself?

Specification is 85 amps
Regulator in control 51 amps at 12.8 volts
Bypassed full-fielded 83 amps at 12.8 volts

With the above information, you would concentrate your efforts at the regulator. It apparently does not have the ability to full field the alternator. Field current must be reduced as it flows through the regulator. Regulator resistance in series with the field coil is the most common cause.

What would you do with the following situation?

Specification is 85 amps
Regulator in control 71 amps at 12.8 volts
Bypassed full-fielded 71 amps at 12.8 volts

Figure 5. A serpentine belt tension tester.

This situation is the opposite set of circumstances. The regulator is doing all it can. It must be supplying full B+ to the rotor because the output is the same as when we supply B+ directly from the battery. It looks like the alternator is at fault here. This condition will require us to perform one additional step before we condemn the alternator. This step will be to full field the alternator directly at the field terminal of the alternator. By doing this, we will be eliminating the wiring from the regulator to the alternator as a possible cause. Excessive resistance along this wire could cause the previous situation. If after directly full fielding the amperage is the same, you can assume that the alternator or drive belt is faulty and needs to be serviced or replaced.

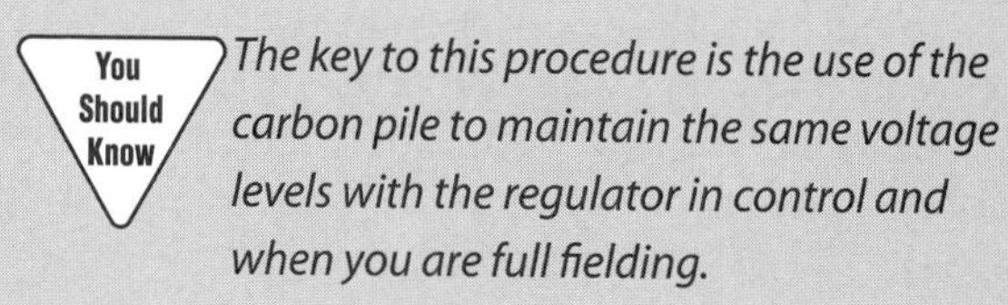

The key to this procedure is the use of the carbon pile to maintain the same voltage levels with the regulator in control and when you are full fielding.

In addition, do not disconnect or reconnect the regulator with the engine on.

Now is a good time to mention the use of the belt tension gauge. If the drive belt is loose, the rotor might slip and alternator output reduce to the level of our second example. More than one alternator has been replaced when the real reason for the reduced output was a loose drive belt. **Figure 5** shows a common style of serpentine belt tension tester measuring the tension of this V-8 belt. The belt is placed between the guides, and the tension is read directly off the scale. **Figure 6** shows the block diagram of an externally regulated A circuit alternator. Full

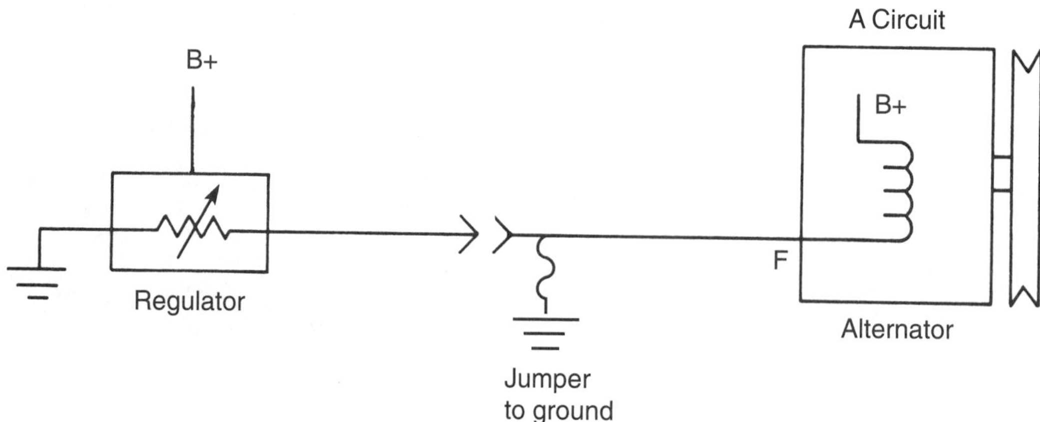

Figure 6. An A circuit is full fielded by applying a ground.

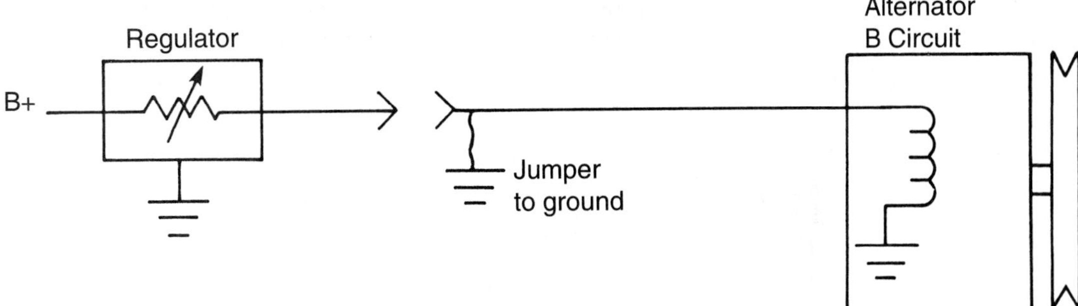

Figure 7. No current flows if a ground is applied to a B circuit.

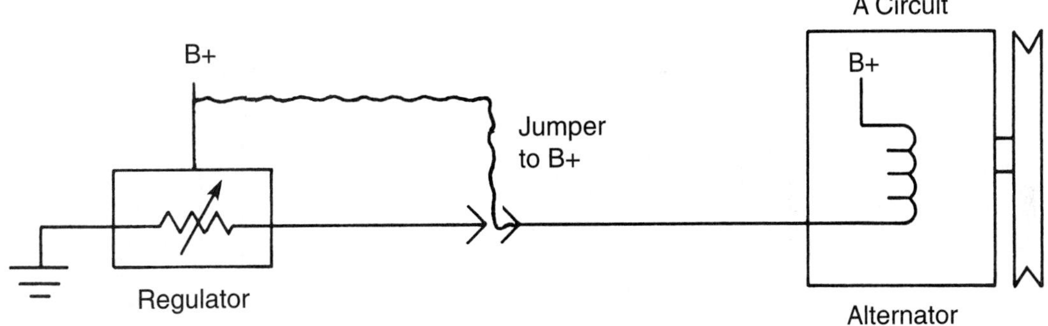

Figure 8. No current flows if B+ is applied to an A circuit.

fielding it will be as easy as in our B circuit example. The only difference will be in that we will supply a ground to the field terminal rather than B+. The same principles apply to the results. If you are not sure whether you are working on an A or a B circuit, try both B+ and ground. You will not do any damage if you apply the wrong test procedure in this case. A B circuit has one brush grounded in the alternator and requires B+ for correct full-fielding procedure. If you apply a ground to the F terminal thinking it to be an A circuit, your circuit looks like that diagrammed in **Figure 7**. No current flows because you have a ground on both sides of the field coil. Similarly, B+ to an A circuit looks like that illustrated in **Figure 8** and results in no current flow. Always try both B+ and ground if you have not identified the type of field circuit.

Full Fielding an Isolated Field

The full-fielding procedure previously outlined will work well for either an A or a B circuit but must be modified very slightly for an isolated field. To review, an isolated field can be either an A or B circuit but will pick up B+ and ground outside of the alternator. Isolated field alternators have two field leads rather than the one usually found on other styles. **Figure 9** shows a common wiring diagram for an isolated field PCM-controlled Chrysler alternator. Notice that it is an A circuit because the PCM regulator is on the ground side of the field coil. B+ for the circuit comes off the ASD relay and enters the alternator with a DK GRN/ORG field connection. The circuit continues through the rotor and out the DK GRN field connection over to the PCM where the regulator function is. The PCM receives its sensing voltage from the ignition switch B+. Full fielding requires that we apply full B+ and ground to the rotor. On isolated field alternators, you will have to jumper one lead to B+ and jumper the other field terminal to ground, if the terminal configuration allows it. By doing both, you will be assured that you have truly separated the alternator from the rest of the circuit. If the terminal configuration will not allow you to jumper at the alternator, the next best choice is to apply the ground at the PCM. You must keep in mind, however, that if you jumper the PCM field terminal to ground (with the PCM wire disconnected) and the alternator produces reduced output, you still could have a good alternator. If reduced B+ is available for the field coil, reduced output will be the end result. Always verify that both B+ and ground are available when full fielding an isolated field alternator if possible.

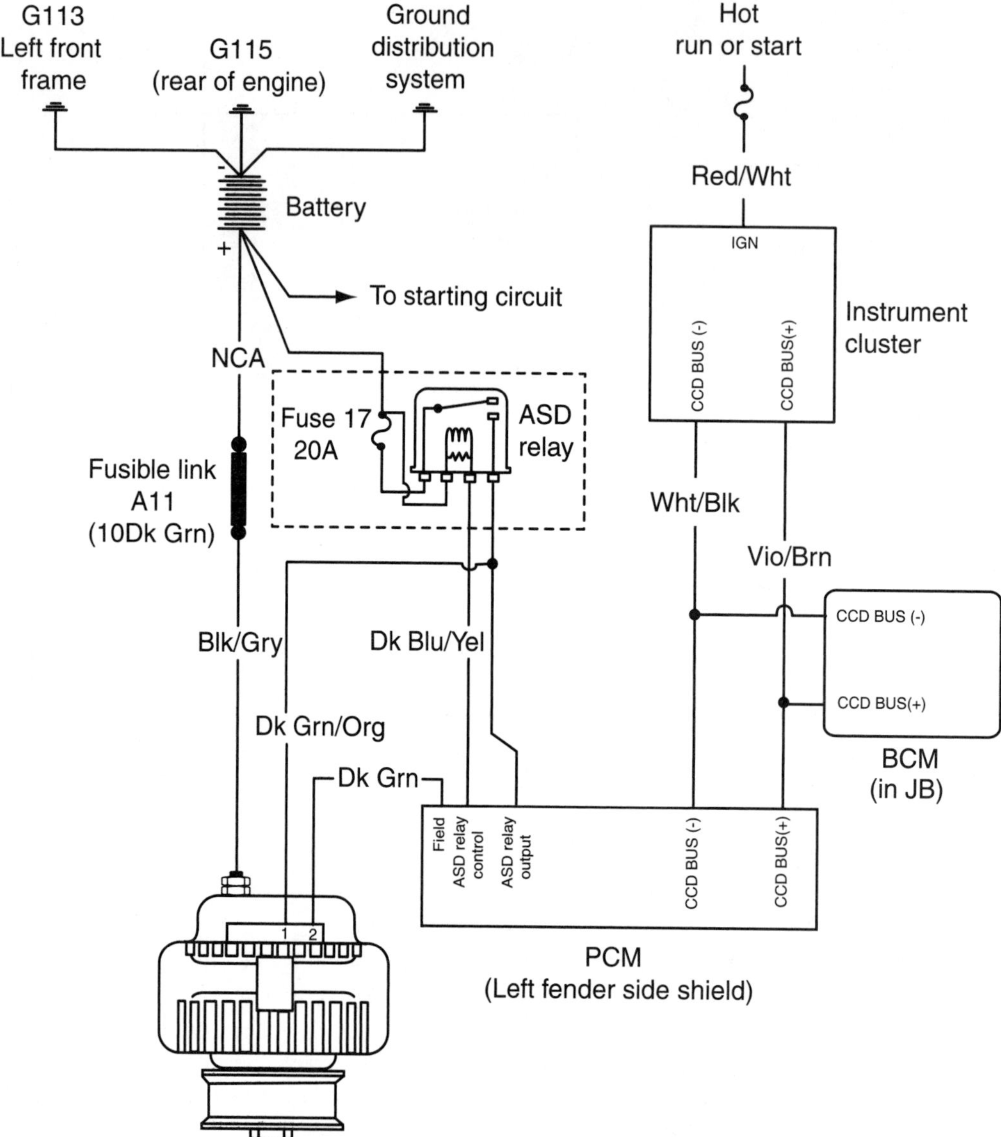

Figure 9. A wiring diagram for an isolated field PCM-controlled Chrysler alternator.

In the mid eighties, Ford moved the regulator on some of their vehicles to the outside of the alternator **(Figure 10)**. This unit has external connecting wires for the regulator. Field current B+ comes from a connection ahead of the ignition switch. The regulator supplies the ground for the circuit. Notice the F terminal on the bottom of the diagram. This is the terminal that you will ground when you wish to full field the alternator. The F terminal is usually protected by a cover but is easily accessible for testing purposes. **Figure 11** is a photo of this type of Ford alternator and regulator. Notice that the regulator has been removed and that the brush assembly is an integral part of the assembly. This regulator and brush assembly is easy to replace out in the field. One of the mounting screws is the insulated F terminal, which you will ground if you want to full field this alternator.

TESTING THE CHARGING SYSTEM VOLTAGE REGULATION

As we have stated previously, regulators have two major functions. The first is to have the ability to control the output. You have seen how this function is tested and compared with the bypassed regulator output to determine regulator condition. Regulators are replaced if they cannot full field or control the alternator output. The second function of the regulator is to keep the system voltage within a predetermined range, typically, 13.5–14.5 volts at normal operating temperature. As the temperature goes down, the regulator will raise the operating voltage to help charge the cold battery. The next process we should check on our charging system is its operating voltage. This is simply accomplished with the same volt-amp meter connections.

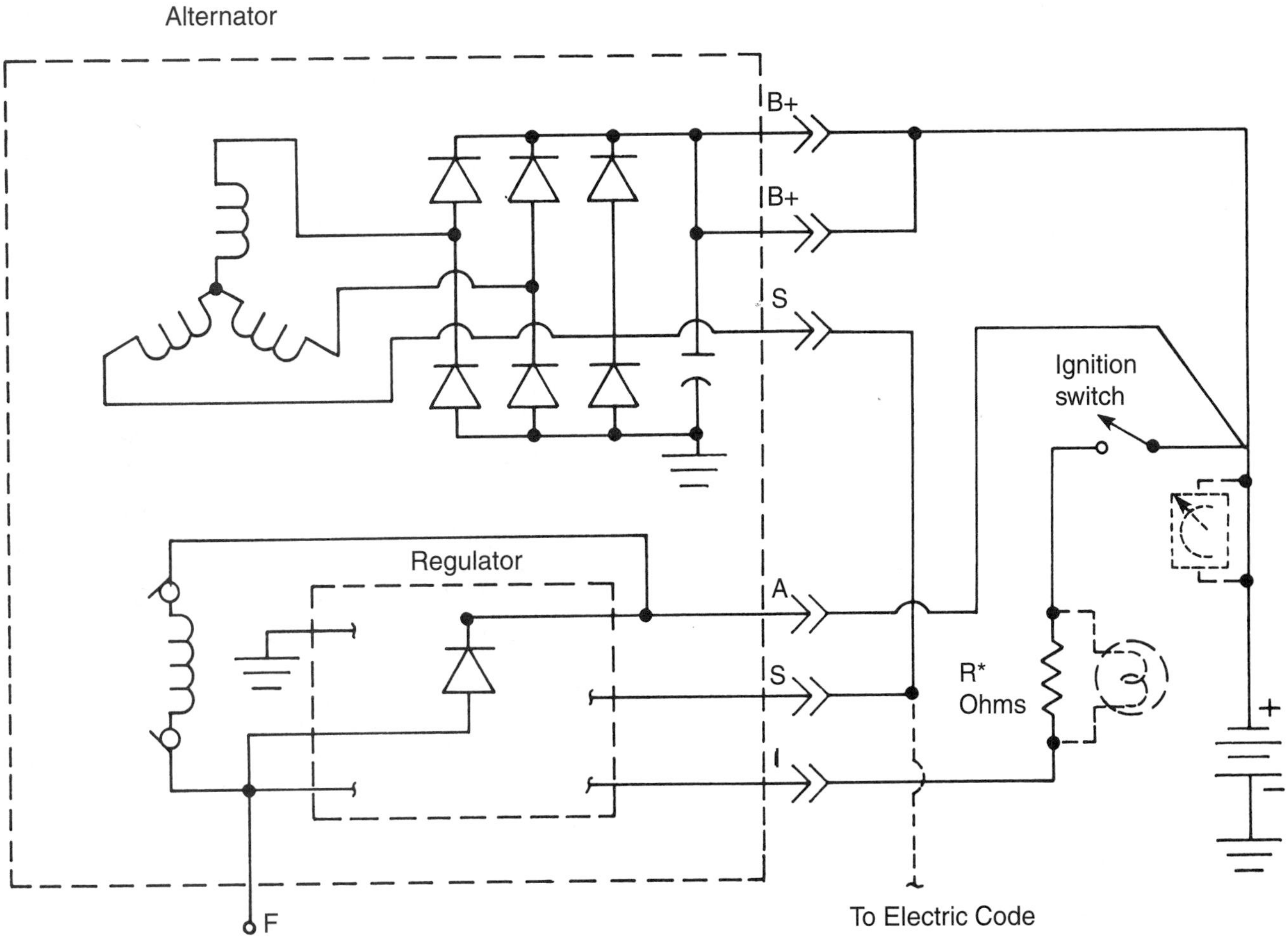

Alternator
B+
B+
S
Ignition switch
Regulator
A
S
R*
Ohms
I
+
−
F
To Electric Code

Figure 10. Ground the F terminal to full field the alternator.

Figure 11. A regulator removed from a Ford alternator.

As the battery comes up on charge, you will see a gradual decrease in the charging current. This decrease should occur within the specified operating voltage. Even with a fully charged battery and a charging rate of a couple of amps, the system voltage should not go above the regulator setting.

If the voltage setting is too low, the battery will not receive sufficient current to become fully charged and will gradually sulfate. If in testing regulator voltage it falls lower than the manufacturer's specification, the regulator will have to be replaced. Years ago, regulators were adjustable. Modern regulators, however, cannot be readjusted to bring their voltage within specifications and are replaced as a unit.

A higher-than-specification voltage level might not be the regulator's fault, however. **Figure 12** shows how a 2-volt drop on the ground side of the regulator will raise the system voltage by 2 volts. Excessive resistance within the sensing circuit will reduce the actual voltage that the regulator sees as system operating voltage. It will operate the alternator at a higher-than-normal output until it reaches the specified voltage that it sees. Any resistance causing voltage drop will add to the operating voltage. Sensing voltage must be accurate or the operating voltage will not be accurate. If you are testing a vehicle and find a higher-than-normal operating voltage, test the ground of the reg-

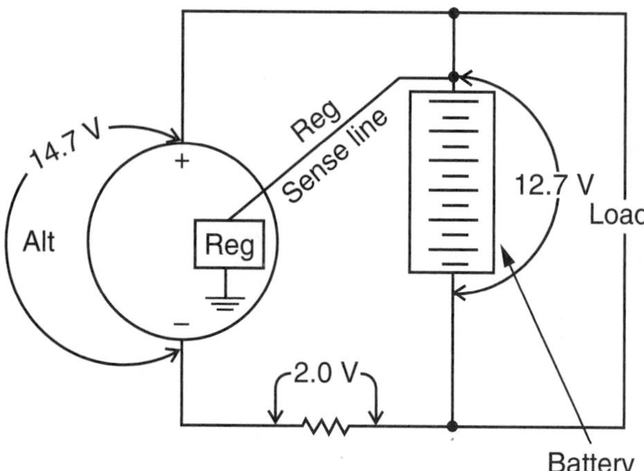

Figure 12. A 2-volt resistive drop will raise charging voltage.

ulator with a sensitive voltmeter between the case of the regulator and the ground of the battery. With the vehicle running, there should not be any difference between the two ground points and the meter should read zero. **Figure 13** shows this procedure. If the ground side is functioning correctly, use your voltmeter and test the sensing voltage input to the regulator as shown in **Figure 14**. The meter should read the same voltage as battery voltage. With a good ground and a good input voltage, the regulator should be capable of holding the voltage within specifications. If not, it should be replaced.

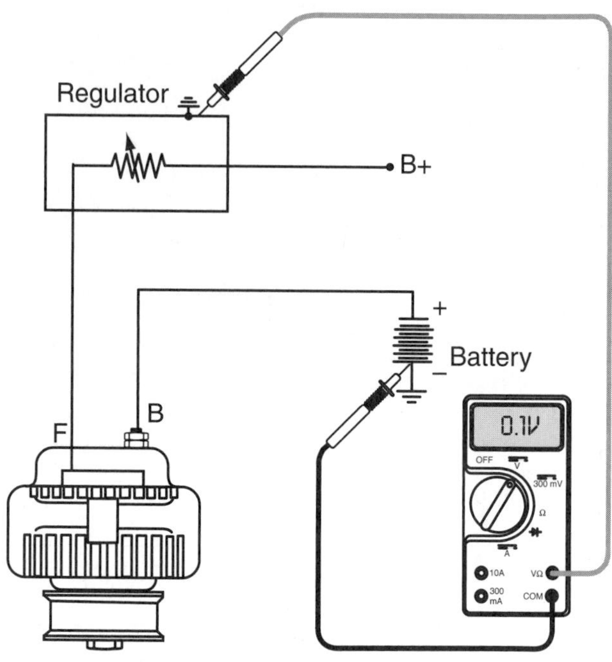

Figure 13. Checking the regulator ground.

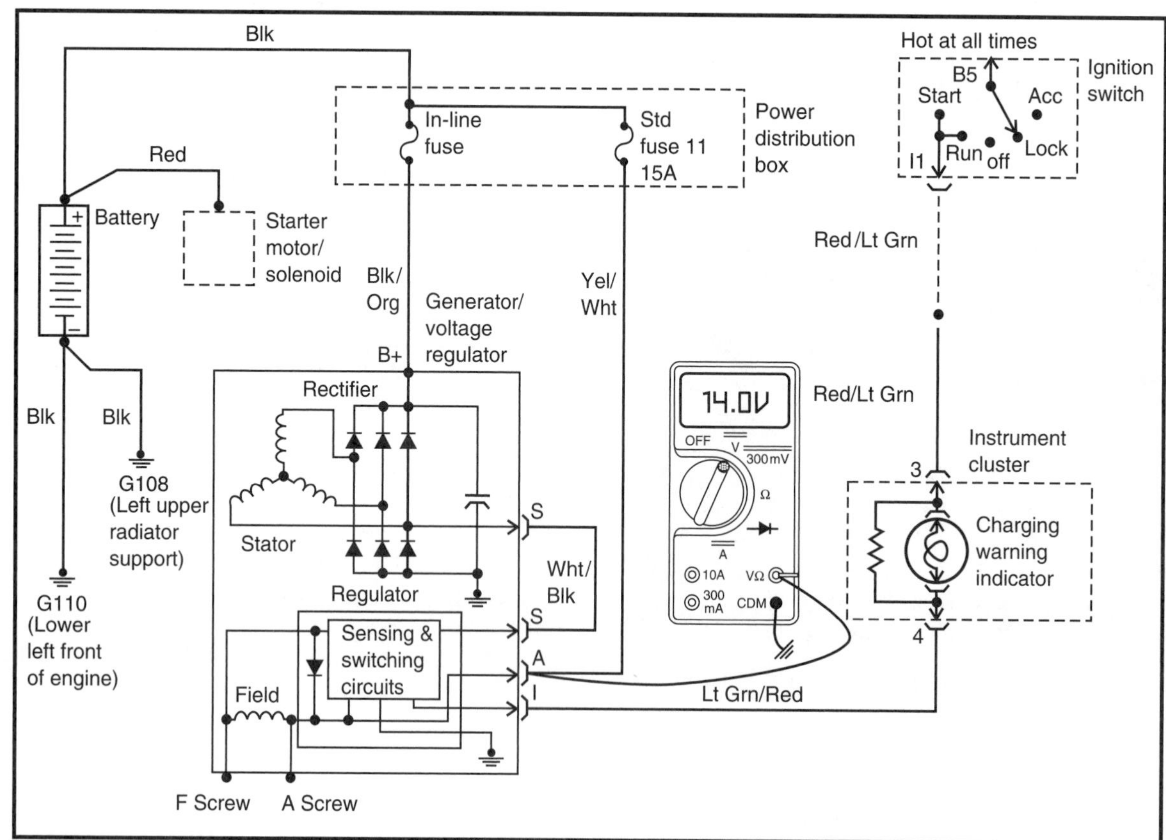

Figure 14. Checking sensing voltage.

CHARGING VOLTAGE DROPS

Testing for voltage drops is usually not routinely done on a vehicle but should be done any time a component is replaced or a stubborn charging problem cannot be identified. Simplified, it is the process to determine if the three major components, battery, regulator, and alternator, are all functioning at the same potential. We have seen why this is important for the regulator-sensing circuit because any resistance will raise the charging voltage above the specification. Refer to **Figure 15**, which is a diagram of an internally regulated CS alternator used by General Motors. Notice the use of a fuse link off the battery. Charging current to both run the vehicle and charge the battery must come through this link. It is a primary protection device that will have a slight voltage drop across with normal amperage. With increased amperage comes increased voltage drop and heat generation. Remember that a fuse link is two number sizes smaller in size than the wire it protects. This is so that it will burn before the wire or system it protects. By measuring the difference in potential at the alternator output terminal and at the battery positive post, we can look at the fuse link and assess its condition. **Figure 16** shows the voltmeter connections for this test. It is necessary to have a small amount of amperage actually flowing through the circuit. Usually approximately 10 amps is specified for a maximum voltage drop of 0.3 volt with an indicator lamp or 0.7 volt with an in-dash ammeter. The circuit necessary for an ammeter is longer and will automatically have a greater voltage drop.

We must look at the ground path as well as the B+ path. With the 10 amps flowing as in the previous test, the voltmeter is moved over to the ground path. One voltmeter lead is on the case of the alternator, and the other is on the negative post of the battery. There are not any fuse links on the ground return path, and so we expect to have negligi-

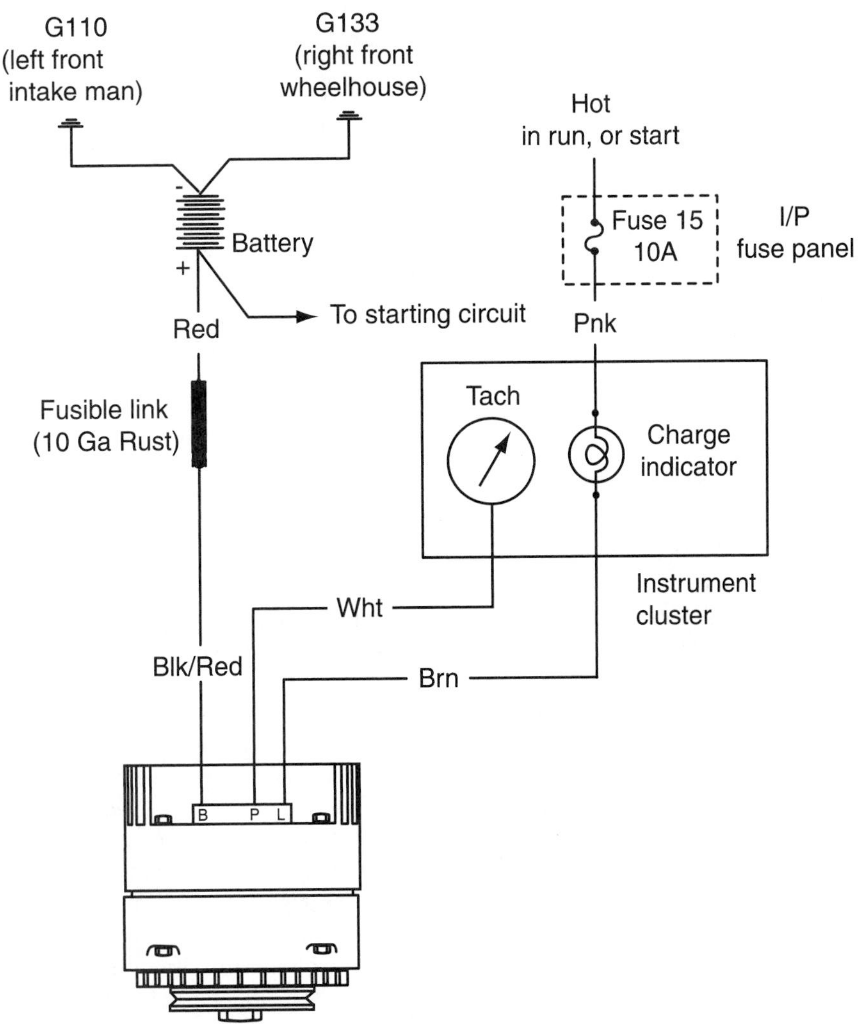

Figure 15. Charging current flows through the fusible link (10 Ga Rust).

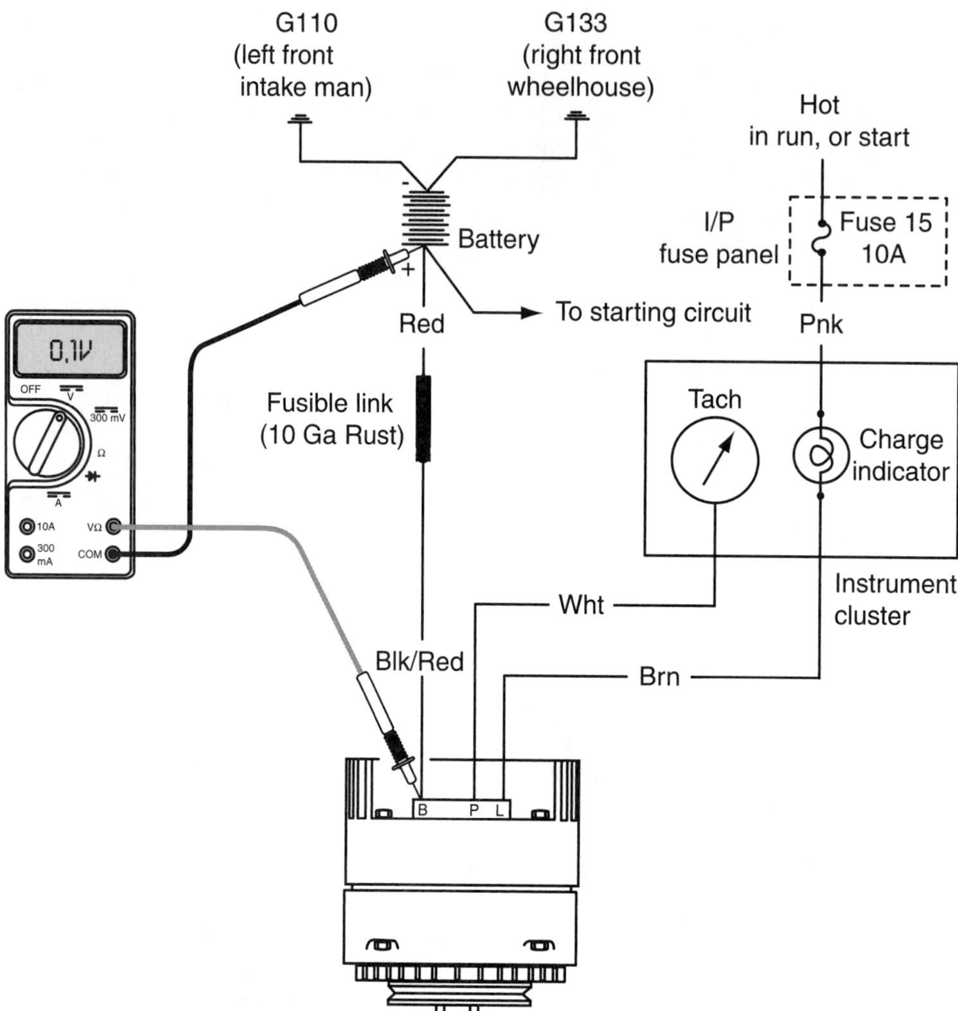

Figure 16. Measuring the voltage drop of the charge wire.

ble voltage drops. Less than 0.1 volt will be acceptable. **Figure 17** shows a drawing of the ground return path voltmeter connections. Once the alternator ground is verified, we will move the voltmeter lead from the alternator to the case of the regulator if it is externally mounted. With approximately 20 amps flowing, there should be no voltmeter reading. On internally regulated alternators, we skip this test because we cannot easily get to the regulator ground connection inside the alternator.

The B+ path and ground path voltage drop tests are all necessary because the charging system will be capable of functioning correctly only if all the major components are at the same potential. You remember that voltage drops are present whenever current flows through resistance. Usually, this resistance does not appear overnight. It has taken years to corrode or burn. Testing for the resistance before it becomes an open circuit and the car dies is why this test is so important. It is preventive maintenance oriented and when done correctly will usually identify a problem before it becomes a no-start and tow.

REVIEW OF ALTERNATOR/ REGULATOR TESTING

Let us look at where we have been so far in this chapter. We have discussed a series of tests to identify charging system difficulties. We began with a full series of battery tests because battery condition is fundamental to a properly functioning charging system. Next, we moved to the alternator and looked at belt tension and maximum output, first with the regulator in command, and then with the alternator full fielded. This was done to pick out a bad alternator or regulator. We then checked out the voltage setting of the regulator. Rounding out the series was a check of the battery, alternator and regulator, B+ voltage, and ground

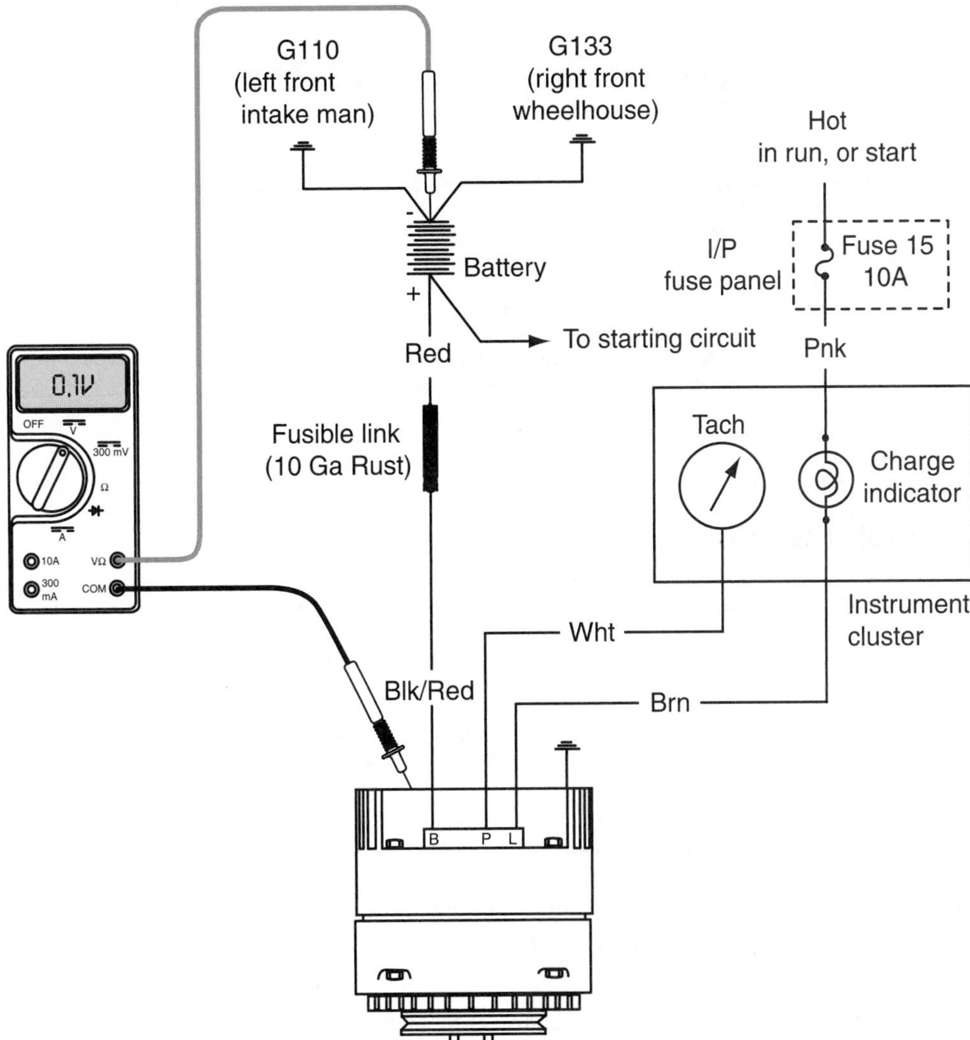

Figure 17. Measuring ground voltage drops.

potential, looking for excessive voltage drops that could have an effect on the system's performance. This series of tests is outlined for you in **Figure 18**.

CHARGING LIGHT DIAGNOSIS

The use of the in-dash charging light is rather limited from the technician's standpoint, but this is a good time to note its use. Let me emphasize that it is intended to be a warning light for the consumer rather than for the technician; however, basic information can be interpreted off the light. It is especially important to discuss its operation with your customer and note if it has not been functioning in the normal manner. Let us define normal operation. With the key on and the engine off, the light should be on. Most vehicle charging systems sense that field current is flowing and use it to turn the bulb on with the key on. The high resistance of the bulb, when compared to the field coil, puts the largest voltage drop across the bulb, resulting in it being on brightly. This high resistance also allows little field current to flow, just enough to begin operation once the vehicle is started. Once started, the light should go out and remain off during all operating conditions. Failure of the

The charging light can be used as an indication that the field has continuity. If the bulb is on with the KOEO it usually indicates that some field current is flowing.

Charging Worksheet

Charging System-Alternator Output

1. With tester attached to vehicle's battery as in starter draw, reconnect the ignition system and start the vehicle. Place fast idle cam to raise idle to 2,000 rpm approximately.
2. Change selector switch over to charging (or an amperage setting around 100 amps.)
3. Rotate the load control until the highest amp meter reading is obtained. Record here: ____________ amps @ ____________ voltage. Turn load off.
4. Turn vehicle off and disconnect the field lead from the regulator or the alternator. Note if A or B circuit.
5. Attach jumper wire or the SUN field lead to the F terminal.
6. Restart the vehicle, raise idle to 2000 rpm and full field the alternator.
7. Rotate the load control to obtain the same voltage as in Step 3. Record here: ____________ amps @ ____________ voltage. Turn load off.
8. Compare reading obtained in Steps 3 and 7. They should both be the same and to specs. If not
   a. the regulator is faulty, or
   b. voltage drop exists between the alternator, regulator, and/or battery.

Voltage Regulator Setting

1. With the regulator connected, start the vehicle; idle speed.
2. Raise the engine speed to a fast idle. The voltmeter should start to climb. Maintain fast idle with *no* load until voltage peaks.
3. Record the highest voltage: ____________ volts. The highest voltage read should match the specs. Any difference indicates
   a. the regulator is faulty, or
   b. a voltage drop exists between the alternator, regulator, and/or battery.
   Note: Position #3 on a VAT or a 1/4 ohm position can be used to simulate a fully charged battery for this test.

Charging System Voltage Drop

1. With the charging system hooked up, start the vehicle and raise the speed up to a fast idle.
2. Vary the load until the alternator's output is around 20 amps.
3. Maintain the load and use a voltmeter to measure the B+ drop between the output terminal of the alternator (positve voltmeter lead) and the battery positive post (negative voltmeter lead). Record below:

   ____________ volts

   Note: Charging systems with warning lights should have less than 0.3 volt drop.
   Charging systems with ammeters should have less than 0.7 volt drop.
4. Place the negative lead of the voltmeter on the alternator case and the positive meter lead on the regulator case. Record below:

   ____________ volts

   Note: Any reading more than 0.3 volt indicates excessive resistance. Most resistance problems are traced to improper, loose, or corroded terminals and connections.
5. Place the negative lead of the voltmeter on the alternator case and the positive meter leads on the battery negative post. Record below:

   ____________ volts

   Note: Any reading more than 0.3 volt indicates excessive resistance. Most resistance problems are traced to improper, loose, or corroded terminals and connections.

Figure 18. Basic charging system tests.

bulb to light with the key on and the engine off indicates a problem in the field circuit. Failure of the bulb to go out once the engine is running indicates a lack of charging current. In either case, a complete performance test of the entire system beginning with the battery and ending with voltage drop testing is indicated.

The dimly lighted indicator lamp is harder to diagnose. Start with a complete charging system test and determine if there are any components at fault. If output is correct and regulator voltage is correct, it usually involves excessive resistance, causing a voltage drop somewhere in the circuit. Look at **Figure 19**. It shows the typical indicator lamp cir-

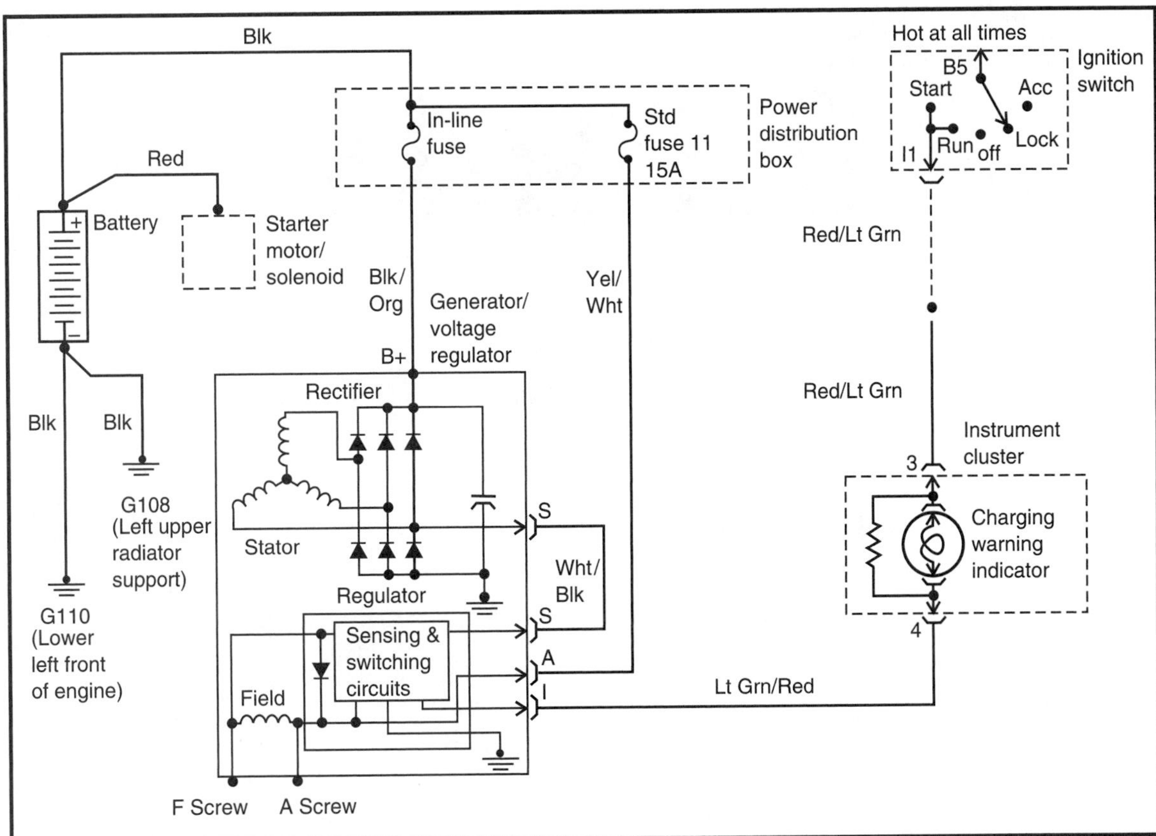

Figure 19. Charging indicator wiring.

cuit for an externally regulated alternator. If with this vehicle we experience a glowing alternator light, we will have to assume that the voltage on the two sides of the bulb is different. One side of the bulb is fed off the ignition switch, whereas the other is fed off the alternator. Getting out the voltmeter and testing the circuit voltage at both points and then working your way closer to the bulb until the difference is found is usually the recommended procedure. When the voltage drops down, you have found the spot where the resistance is located. Some alternator indicator lamps operate off a ground supplied by the voltage regulator. With this type, excessive resistance can cause the bulb to be dim, but it is the regulator that is responsible for turning it on in a no-charge current condition.

TESTING FOR EXCESSIVE AC USING A DMM AND A DSO

The diodes are responsible for rectifying the AC from the stator into the DC needed by the vehicle. If a diode is breaking down, it might allow some AC to get out of the alternator. This AC will not affect most of the components on the vehicle. Some electronic circuitry could be damaged by it, however. As a result, it is a good idea to run a diode pattern check whenever failure of an electronic component has taken place. Testing is simple to perform and is best done with a DSO. A DMM will give similar information if it is capable of an RMS reading.

Let us use the DMM first. Connect the positive lead to the alternator output or B+ terminal. Connect the negative lead to the case of the alternator. Start the vehicle and turn on enough accessories so that approximately 10–20 amps is flowing from the alternator **(Figure 20)**. With the DMM on AC, take a measurement. Each manufacturer specifies the maximum amount of AC that is acceptable. Usually this number is in the 0.250-volt AC (250 millivolts) range. This is an extremely important test especially after you have replaced an electronic component, such as an ignition module. You need to ask yourself the question: why did the module fail? One of the possible answers is excessive AC coming out of the alternator. After the repair, check the level of AC and if it is excessive (greater than 250 millivolts), the alternator needs to be replaced. The reading that you get by using a DMM is an average of the three legs of the stator and the rectifier bridge. It does not show you if one of the legs is beginning to have a problem, because it is an average. To look at each leg of the stator and rectifier bridge you need to use a DSO.

The connection for the DSO is exactly the same: positive to the alternator output, negative to the alternator case

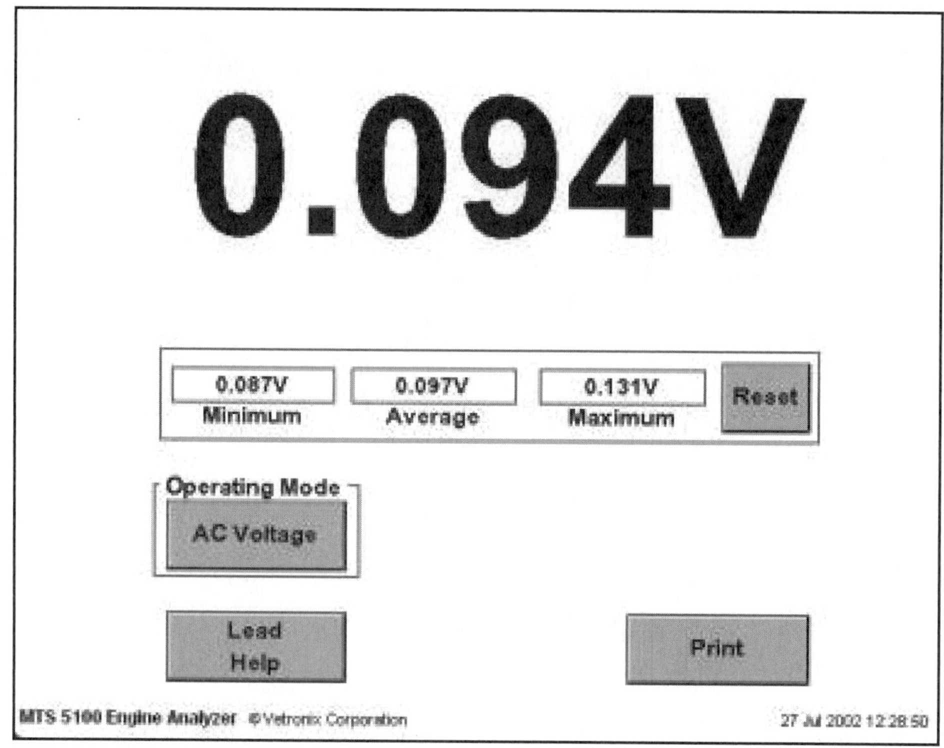

Figure 20. A DMM testing for AC voltage.

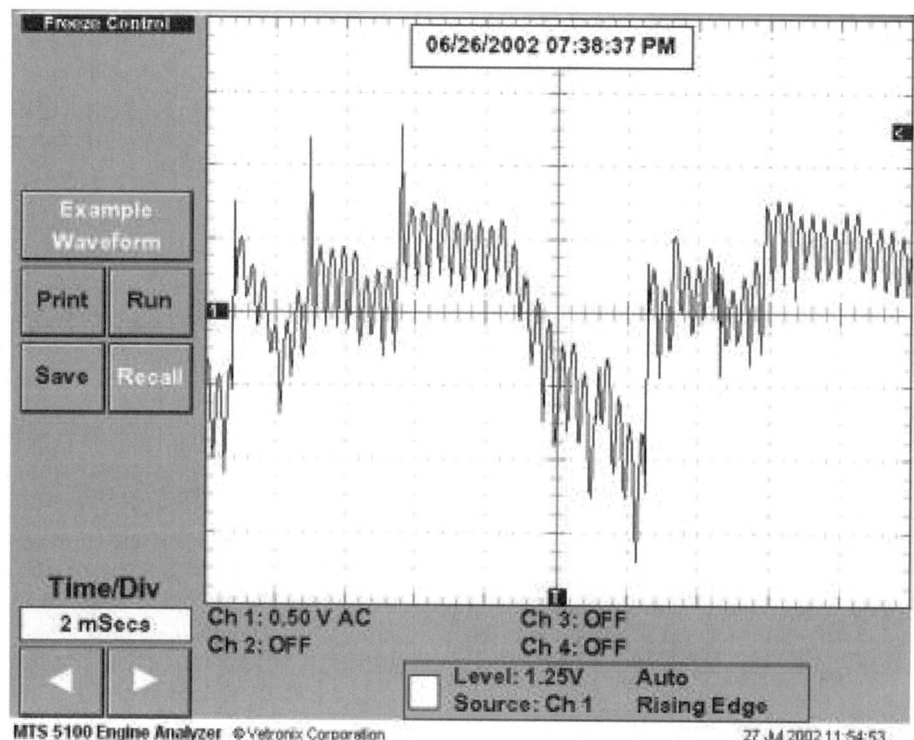

Figure 21. Excessive AC shown on a DSO.

ground. AC couple the scope, start the vehicle, and turn on some accessories. Vary the time and voltage setting until you get a clear pattern as in **Figure 21**. Notice that the scope has been set to 0.50 volt AC per division. Each pulse you see is an AC pattern from the alternator. The average pulse is about one division or 500 millivolts AC peak to peak. This is twice the allowable level and could easily cause damage. As a matter of fact, this alternator pattern is off a vehicle that came into the shop with a no-start condition. The diagnosis revealed a failed ignition module. After the module was replaced, an AC check of the alternator was made and revealed the bad alternator. **Figure 22** is the pattern off the new alternator. Notice that the level of AC is greatly reduced and below the 250-millivolt maximum allowable specification. If the alternator had not been replaced, it is possible that the new ignition module might have failed from excessive AC. It is also important to note that a badly sulfated battery can cause excessive AC out of the alternator.

You Should Know *The maximum level of AC that should come out of an alternator is 250 millivolts.*

In addition, there are some general rules to be observed when working on charging systems. They are:

1. Never run the vehicle with the battery disconnected. It helps to stabilize the voltage and absorbs any voltage spikes that could damage electronic components.
2. Do not allow the system voltage to go above 16 volts during full fielding.
3. Jumper the battery off another vehicle positive post to positive post and negative post to a ground away from the battery.
4. Disconnect the battery from the vehicle when using a battery charger.
5. Never remove the alternator or starter with the battery connected to the vehicle.

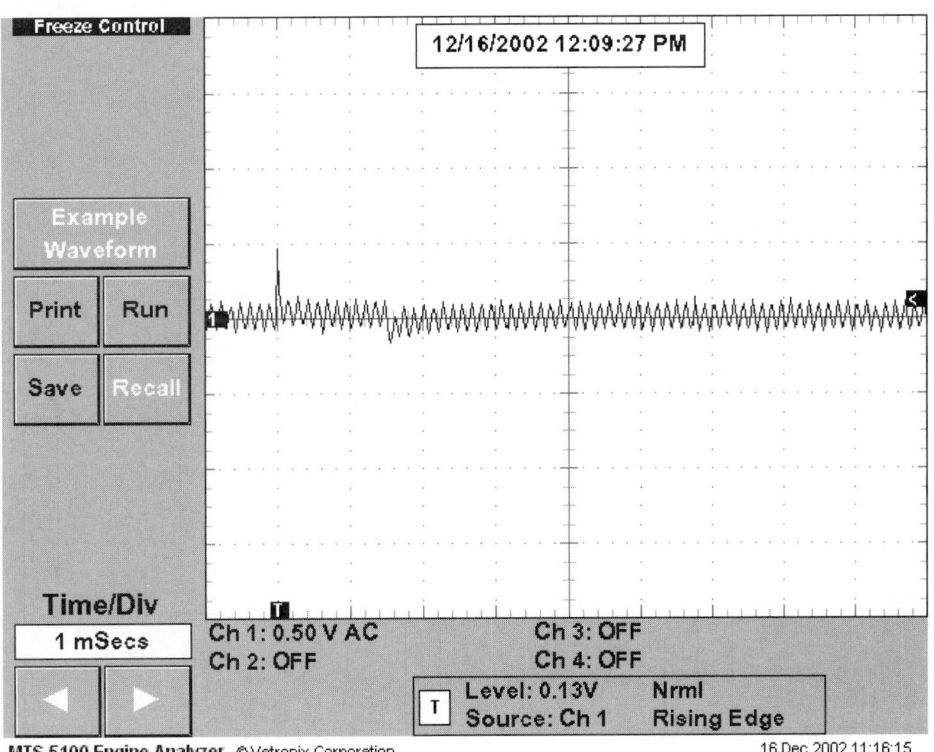

Figure 22. A DSO showing an acceptable AC level.

Summary

- Testing the battery is the first part of charging system diagnosis.
- The alternator should be capable of producing its rated output when a load is applied to the battery.
- The regulator should be capable of full fielding the alternator.
- The pulse width of the field current should increase as alternator output increases.
- When the alternator is full fielded, the PWM of the field should be 100 percent.
- Alternators with externally mounted regulators can be full fielded.
- An A circuit field is grounded for full-field testing.
- A B circuit field is powered (B+) for full-field testing.
- Never disconnect or reconnect electronic components with the ignition key on.
- Never operate an alternator with the battery disconnected.
- The voltage setting for the regulator is easily tested.
- Charging system voltage drops can cause either an overcharge or an undercharge condition.
- A maximum of 250 millivolts AC (0.250 volt) can come out of an alternator.

Review Questions

1. A battery that is overcharged can be a result of
   A. a loose drive belt
   B. a poor regulator ground
   C. a burned out diode
   D. resistance in the field circuit
2. An alternator should never be operated open circuit (battery disconnected) because this could
   A. overcharge the battery
   B. burn out electronic components
   C. ruin the alternator stator
   D. result in poor high-speed driving
3. A battery that is undercharged might be a result of
   A. a poor regulator ground
   B. an open-sensing wire to the regulator
   C. resistance between the output terminal and the battery
   D. high-speed driving
4. To full field an alternator, Technician A says to run a jumper wire from the F terminal of the regulator connector to ground on an A circuit. Technician B says to run a jumper wire from the F terminal of the regulator connector to B+ on a B circuit. Who is correct?
   A. Technician A only
   B. Technician B only
   C. Both Technician A and Technician B
   D. Neither Technician A nor Technician B
5. An alternator output test shows 45 amps at 13 volts. Full fielding produces 55 amps at 13 volts. This indicates
   A. nothing
   B. that the alternator is faulty
   C. regulator circuit problems
   D. that both the alternator and the regulator should be replaced
6. Battery voltage is 19.6 volts while the vehicle is running. Alternator voltage is 19.9 volts and regulator voltage is 14.3 volts. Technician A says that this indicates a voltage drop between the regulator and battery. Technician B says that this indicates resistance that is causing the output voltage to rise above specification. Who is correct?
   A. Technician A only
   B. Technician B only
   C. Both Technician A and Technician B
   D. Neither Technician A nor Technician B
7. An externally regulated alternator produces 55 amps all the time. Full fielding it produces 55 amps also. This indicates that
   A. the regulator is running the alternator full fielded all the time
   B. the alternator is faulty
   C. the output wire has excessive resistance
   D. not enough information is given
8. Computer-controlled charging systems are full fielded by
   A. special equipment
   B. jumping a special terminal to ground at the computer
   C. grounding or jumping to B+ the F terminal at the alternator with the computer disconnected
   D. none of the above
9. The battery cable should be disconnected before removing any electrical unit.
   A. True
   B. False
10. A faulty battery can cause excessive alternator output.
   A. True
   B. False

11. During full-fielding testing it is all right to allow the system voltage to go as high as it can.
    A. True
    B. False
12. Less than rated output during full fielding indicates need for a new regulator.
    A. True
    B. False
13. A voltage regulator whose voltage is too low can cause sulfated batteries.
    A. True
    B. False
14. Diode pattern checks should be done any time electronic components are being replaced.
    A. True
    B. False
15. An equal voltage on both sides of an indicator lamp will cause it to glow dimly.
    A. True
    B. False
16. Any sensing wire voltage drops will increase the system voltage by the amount of the drop.
    A. True
    B. False

Ignition Systems

SECTION OBJECTIVES

At the conclusion of this section, you should be able to:

- Understand the purpose of the ignition system.
- Recognize the difference between resistor and nonresistor spark plugs.
- Read a used spark plug.
- Set a spark plug gap.
- R & R a spark plug.
- Replace wires, caps, and rotors.
- Use a DSO to check secondary firing voltage.
- Use a DSO to check module functions.
- Use a DSO to check burn time.
- Find the cause of excessive firing voltage or reduced burn time.
- Understand how the different ignition systems function.
- Check the advancing system.
- Check the input sensor(s) with a DSO.
- Set base timing, if applicable.
- Use a DSO and obtain a pattern from an input sensor.
- Use a current probe to check primary current flow.
- Use a dual-trace DSO for comparing a current waveform with a voltage waveform.
- Test the PCM advancing system with a DSO.
- Use the appropriate test tool for different systems.

The optical ignition sensor generates a signal that can be as high as 6000 times a second. This makes testing by anything but a DSO very difficult.

Chapter 46 Secondary Ignition Systems

Introduction

The business end of the ignition system is inside the cylinder and is the spark plugs. We look at the secondary section of the ignition system beginning at the ignition coil and ending at the spark plugs. In Chapter 48, we look at how this system develops the spark that is designed to ignite the air:fuel mixture inside the cylinder. Recent years have seen many changes in the ignition system, especially in the development of the spark. We look at the systems in two categories: those with distributors and those without. As we look at the various systems on the road, always keep in mind the basic and fundamental goal of providing a spark inside the cylinder that will last long enough to ignite the air:fuel mixture. Without adequate ignition, power is reduced, mileage is reduced, and, most importantly, the vehicle emissions increase.

Before we begin to look at the individual components that make up the ignition system of today, a brief overview might prove helpful. The basic purpose of the system will be to supply a long enough high-voltage spark at the correct time to the compressed charge of fuel and air. It is important that you realize that these two purposes (long, high-voltage spark and correct time) are equally important. The spark must have a voltage that is high enough and lasts long enough to ignite the mixture with the least likelihood of a misfire. This ignition must occur at the correct time relative to the piston position or the amount of power might be reduced. In addition, the level of pollutants leaving the tailpipe is usually increased if the ignition timing (when the spark is delivered) is not correct.

OVERVIEW OF IGNITION SYSTEMS

The ignition system has seen tremendous changes in recent times. We have seen the system move from mechanical control to totally electronic and, in most cases today, computer control. In our discussion, we separate the system into two halves: primary and secondary. We then look at the different components that can be included. It is important to note that many different styles of ignition systems are on the market today. You might not find all of the components present in each system. In Chapters 47, 49, 51, and 53, we discuss how each component is tested and verified.

The job of the ignition system should be kept in mind as we begin to look at individual components. A high voltage sparks to ignite the mixture at the correct time. Let us look at an item that all ignition systems have—the spark plugs. Look at **Figure 1**. It is a drawing of a typical spark plug. The top terminal will be attached to the ignition system high voltage, and the steel shell will be threaded into the combustion chamber. The center electrode is insulated from the steel shell and the side or ground electrode is attached to the shell. This places the side electrode at ground potential and the center electrode at ignition potential. The air gap between the two electrodes is where the high voltage will arc to ground and ignite the mixture. This is the business end of the plug. Notice that an insulator, usually porcelain, insulates the center electrode and the shell from one another. This insulator must be capable of withstanding thousands of volts. Remember the object of the system. We must deliver the voltage to the air gap. Any insulation breakdown before the air gap will result in a misfire. A couple of design features make the plug the correct one for a particular vehicle.

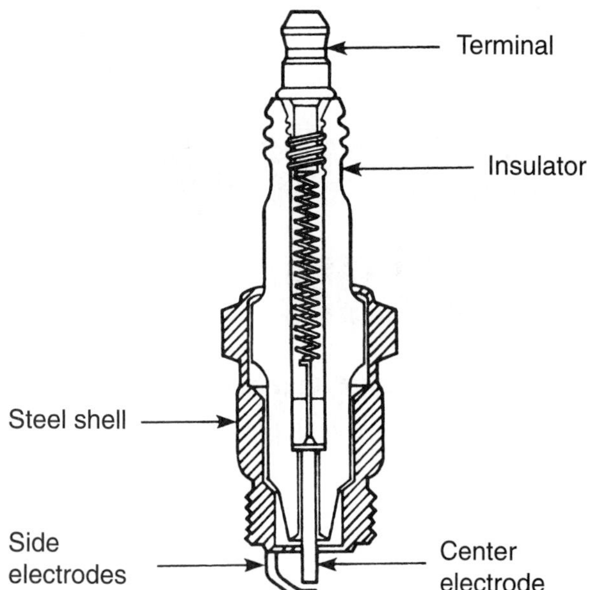

Figure 1. A typical spark plug.

> *Interesting Fact*
>
> *A look through a parts catalog will show you that there are thousands of different types of plugs with only one correct one for the vehicle you are repairing. Do not assume that the current plug in the engine is correct. Look it up and verify.*

Reach

The reach of the plug is the length of the threaded portion of the plug steel shell. The object of it is to ensure that the air gap is correctly positioned inside the combustion chamber. **Figure 2** illustrates the incorrect reach. In certain instances, the incorrect reach (too long) could result in extensive damage to the engine when the piston hits the plug. The reach must match the cylinder head design. Do not be surprised if the entire reach is not threaded.

Heat Range

Heat range is another design feature. Simply, it is the temperature that the plug will operate at under normal driving conditions. If a plug tip is operated too cool, it will tend to foul. Fouling is the process where mixture residue, gas, or oil sticks to the surface of the plug, eventually filling in the air gap and eliminating the arcing. This shuts down the cylinder, reducing the engine's power, and pumps raw fuel out the exhaust system into the air. If a plug tip is operated too hot, the electrode life will be greatly reduced. The electrodes melt or burn away. This increases the air gap and forces the ignition system to produce a higher voltage. The larger air gap can be the cause of a misfire, also. Ideally, the temperature of the tip will be between the fouling level and the overheating level. This is the heat range of the plug. **Figure 3** shows that the internal distance from the tip of the plug to cooling water is what will determine the heat range of the plug. The range that the plug manufacturer recommends for a particular engine will ensure that the tip temperature will stay within the safe zone during driving conditions. Tip temperature will rise as the vehicle travels at highway speeds and fall as the vehicle slows down. The number of sparks per second and the mixture temperature are what change the plug's temperature. Plug heat range is not something that technicians change. Typically, the correct numbered plug for the vehicle will be the correct heat range. Always rely on the vehicle manufacturer to determine which plug is the correct one.

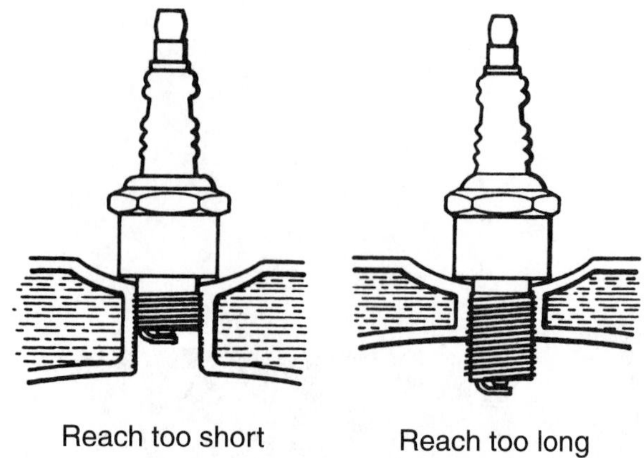

Figure 2. The plug's reach should allow the tip to just enter the combustion chamber.

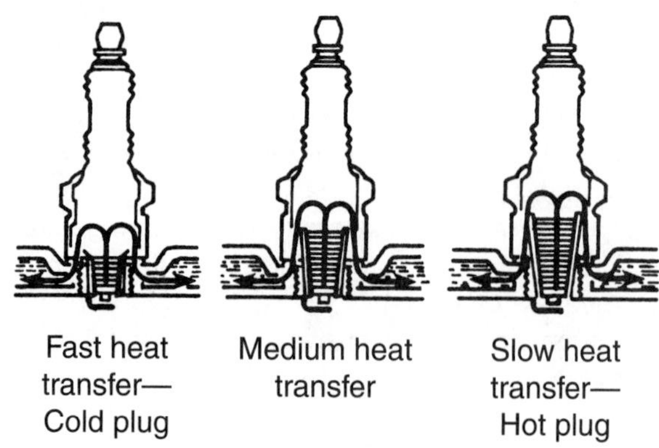

Figure 3. The heat range of the plug is determined by its heat-transferring ability.

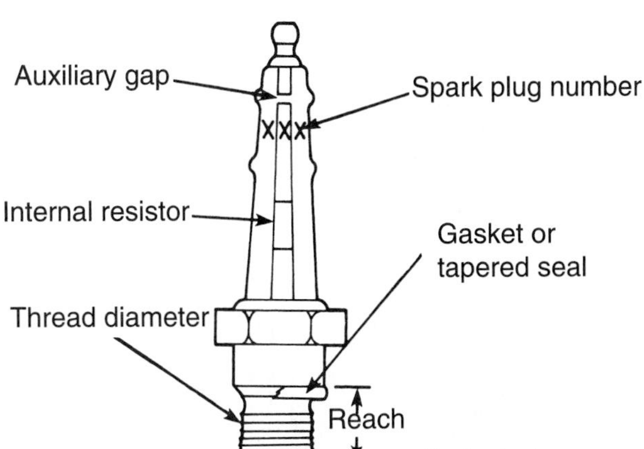

Figure 4. Most vehicles use resistor plugs.

RESISTOR AND NONRESISTOR PLUGS

The plug we have looked at so far has a straight conductor between the terminal and the center electrode. With the increased use of on-board electronics, these plugs have been replaced with resistor-style plugs shown in **Figure 4**. The additional resistor between the terminal and the center electrode will have two effects. First, it will raise up the firing and spark line voltage slightly, and second, it will suppress voltage spikes or AC while the plug is firing. This reduced AC produces less radio frequency interference. Always use a resistor plug if one is specified. Never replace a resistor plug with a nonresistor plug because many of the components currently in use today operate at radio frequency. The additional radio frequency produced by the plug could interfere with the operation of a computer sensor, radio, or clock.

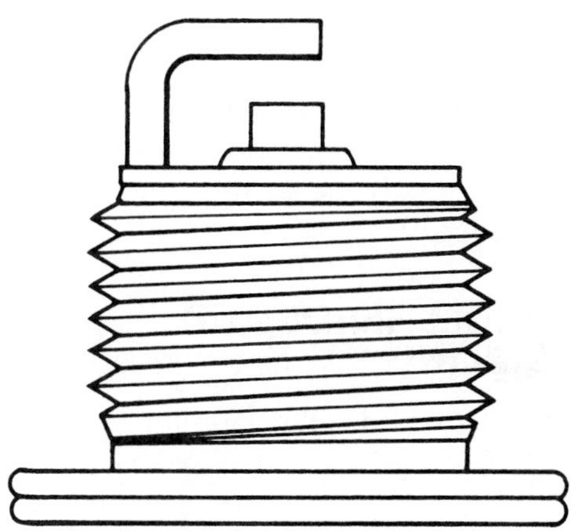

Figure 5. A new plug with a squared-off center and ground electrode.

GAP

The gap of a spark plug is the distance in thousandths of an inch (millimeters) between the ground or side electrode and the center electrode. The importance of this gap cannot be overstated. The gap will have a direct effect on the operation of the ignition system and on the ability of the plug to fire. When the plug is new, the electrode surfaces are flat, clean, and squared off **(Figure 5)**. Once the miles begin to build on the plug, the electrodes take on a rounded, rough texture **(Figure 6)**. The air gap will also get larger as the plug wears. Sometimes an old plug might have its gap changed by as much as 0.020 inch (0.5 millimeter). This increased gap is harder to fire efficiently, thereby forcing the rest of the ignition system to work harder. It is important that the air gap be set correctly when installing new plugs. Plugs are generally not found correctly gapped right out of the box. A plug feeler gauge **(Figure 7)**

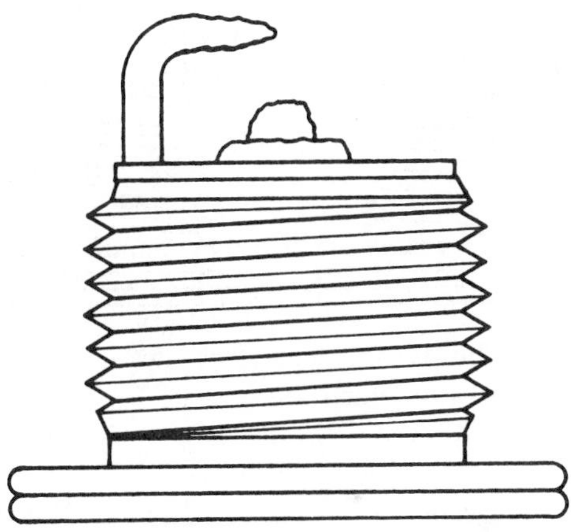

Figure 6. The plug gap will erode through use.

Figure 7. Spark plug feeler gauges.

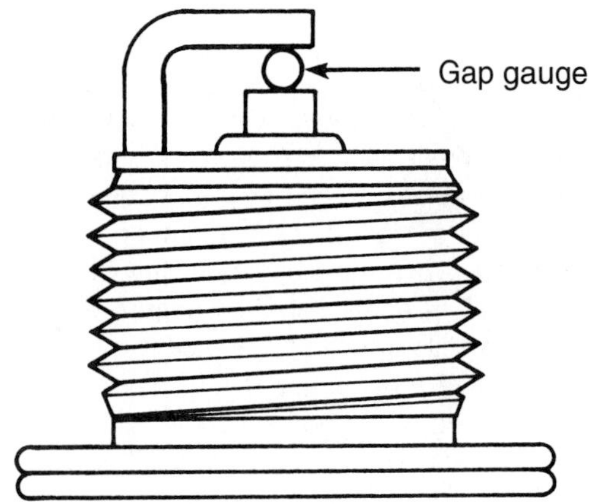

Figure 8. A correct gap results in a squared-off ground electrode.

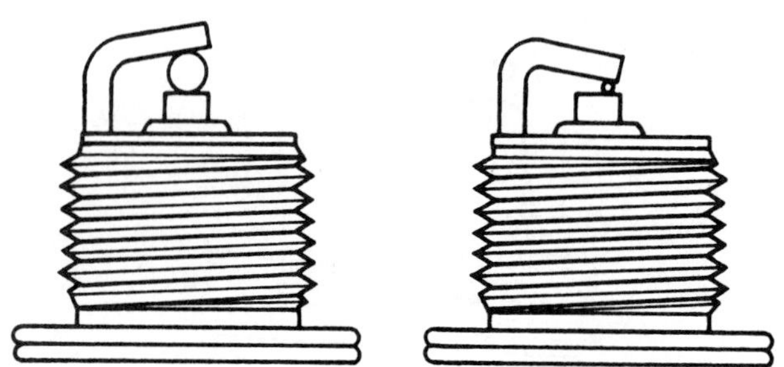

Figure 9. The wrong gap puts an angle to the ground electrode.

has rounded wires of the correct diameter. The ground electrode is bent down until the wire gauge passes through the gap, touching both the center and the side electrode. It is important for you to realize that the many different styles of plugs will require many different gap settings. Plugs should always be set to the correct gap as determined by the manufacturer. Once set, a correct plug gap will retain the squareness **(Figure 8)**. If, however, the wrong plug or wrong gap is used, the plug might look like those in **Figure 9**. In either case, the plug will not have the life expectancy required. Make sure that the plug side electrode is perpendicular to the center electrode after gapping. If it is not, recheck the part number of the plug and the manufacturer's gap. One or both of the two are wrong.

THREAD AND SEALING

Plugs of various thread sizes are available and in use today. The same principle as in heat range should guide you. Use what the vehicle manufacturer specifies. If the threads on the replacement plug are different from the plugs removed, recheck the part number to ensure that you are correct. In addition, you should realize that the plug

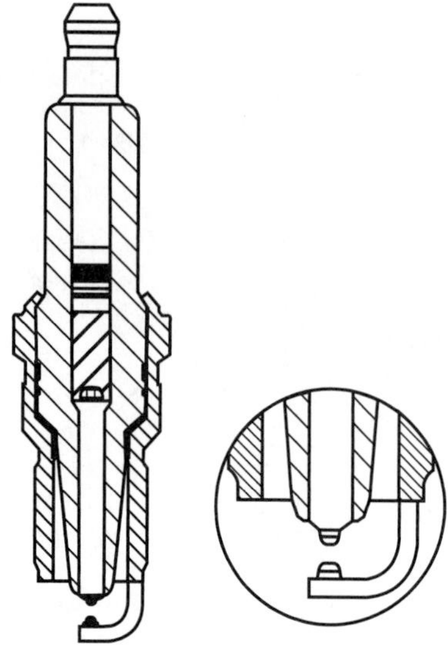

Figure 10. Platinum is sometimes used to extend the plug's life.

and cylinder head might seal with a gasket or with a tapered seat. The torque (twisting effort) that is required to seal either plug is very specific. Overtorque can result in damage to the plug or cylinder head, especially if it is aluminum. Overtightening will also reduce the heat range of the plug and tend to increase the probability of fouling. Undertorque has the opposite effect. The heat has a harder time leaving the plug, and the tip has a greater tendency to burn. The plug threads might also allow some blowby of hot exhaust gases on the power stroke. These gases will destroy the cylinder head threads.

The majority of plugs today use the tapered seat for sealing. You will, however, find a substantial number of plugs that use the gasket. This gasket is usually hollow and seals by being crushed down between the plug and the cylinder head. New gaskets are required each time a plug is installed in the head. The increased use of aluminum cylinder heads has greatly reduced the use of gaskets for sealing. In addition, most plugs in use by the vehicle manufacturers today are considered to have extended life. There are many different methods of extending the life of the plug. Some use copper electrodes, whereas others have precious metals such as the small platinum pads shown in **Figure 10**.

ALUMINUM HEADS AND HELI-COILING

With the increased use of aluminum cylinder heads in vehicles today, the technician must pay attention to detail when installing or removing spark plugs. The steel shell of the plug can destroy the aluminum threads of the cylinder head if it is overtightened or removed with the engine still hot. For this reason, never remove a spark plug from an aluminum cylinder head if the engine is still warm. Allowing it to cool down completely will ensure that heat expansion has not locked the plug into the head.

You Should Know *The torque of the plug is extremely important. It helps to determine the heat range of the plug and guarantees that cylinder head damage does not occur.*

The increased head damage seen frequently today can also be lessened with the use of powdered graphite on the plug threads. A small amount of the dry compound will not affect the heat range and will help to prevent thread damage.

If you encounter a damaged spark plug thread hole in the cylinder head, a process of installing an insert thread could possibly repair it. **Heli-coiling** is just such a process. The old threads of the head are drilled away, and the hole is rethreaded to a larger size. A special steel insert is then installed into the larger threads of the hole with a special tool called a **mandrel**. When the mandrel is removed, the steel insert remains in the head. It is not within the scope of this book to go into machine shop process such as heli-coiling. Refer to a cylinder head machining manual. It is important to note that the cylinder head usually is removed for this machining process. The repair is effective and does save the cylinder head. The need for the repair can frequently be traced to the removal of the spark plug while the engine was hot or to the overtightening of the plug. In addition, if the customer has not had the vehicle tuned and the plugs replaced at reasonable mileage intervals, you might experience the steel threads of the plug destroying the cylinder head threads when you remove the plugs. Spark plug mileage intervals must not be exceeded or a costly repair might be required.

READING A USED PLUG

Technicians must have the ability to look at an old set of spark plugs and determine the basic conditions that the plug has been operating under. Plug manufacturers publish wall charts that list the more common looking plugs. **Figure 11** shows what some of the more common engine conditions do to spark plugs. The key here is to look for differences between cylinders first, because this might indicate that the engine could not be tuned. For example, an oil-fouled plug on one cylinder indicates that oil is entering the combustion chamber past the rings or down past the valve guides. Major engine work might be necessary to restore the vehicle's drivability. A tune up and plug replacement might help but might not cure the entire problem. A careful explanation to your customer will help outline the options. Customers frequently are under the impression that a tune up will cure anything. Reading the plugs as they leave the cylinder helps us determine the course of action we will recommend.

Another example involving all cylinders might be a set of plugs that are carbon fouled. The customer might bring the vehicle to you again for a drivability or emission concern. Carbon forms on the plug if the amount of gas in the combustion chamber is greater than it should be. The system is running rich, and fuel system maintenance is required. Installing another set of plugs will probably improve the drivability and possibly reduce the emissions, but only until the new set gets carbon fouled. By reading the carbon condition on the removed plugs, you can alert the customer to the need for fuel system maintenance in addition to the need for plugs. We outline the removal, inspection, and replacement of spark plugs in Chapter 47.

WIRES

Now that we have discussed just how we are going to use the high voltage developed by the ignition system at the spark plugs, we will begin to look at the transmission of this high voltage. Ignition cables or plug wires will deliver this high voltage from the ignition coil to the plugs, on most systems. Some systems do not use plug wires and we will look at them later. Let us zero in on these wires and examine the most common types available, in addition to looking at just how the cable of today has evolved. Years ago, ignition wire was simply a conductor surrounded by lots of quality insulation. The insulation was necessary because the ignition system was developing as high as 20,000 volts. The size or diameter of the conductor was small because the current was extremely low. Remember that the amount of current flowing determines the size of the conductor, whereas the voltage applied determines the amount of insulation necessary. For this reason, our typical solid-core plug wire was composed mostly of insulation to ensure that the ignition current would reach the spark plugs. This wire was popular in vehicles until radios began to appear as an option.

One of the difficulties with the solid-core wire was that it produced strong magnetic fields around itself. These fields would find their way into the radio and produce a pop or click that could be heard each time that the plug fired. It was not uncommon, at highway speeds, to be unable to hear the radio because the popping would become a buzz, which would be louder than the music. The noise that was finding its way into the radio was called **radio frequency interference**, or **RFI** for short. The tendency of the plug voltage to rise and fall rapidly once the plug fired was the cause of the noise. These alternating currents would have to be suppressed or reduced if radio was

A normal spark plug.

A worn spark plug.

A cold- or carbon-fouled spark plug.

A wet- or oil-fouled spark plug.

A splash-fouled spark plug.

A plug with a bridged gap.

An overheated spark plug.

A plug with preignition damage.

Figure 11. Used spark plugs can indicate cylinder conditions. *(Courtesy of Champion Spark Plug Company)*

to become popular in vehicles. The earliest attempts at suppression were costly because they involved placing grounded metal shields around the entire ignition system. This was effective and can still be found on plastic or fiberglass vehicles such as the Chevrolet Corvette **(Figure 12)**. Setting up a grounded field around the high voltage prevents the RFI from getting into the radio circuits.

Eventually, the high cost of RFI shielding gave way to the most popular wire, **TVRS** wire. TVRS stands for **television and radio suppression**. TVRS wire is found in two varieties. One type is composed of a nylon strand with small pieces of conductive carbon impregnated into it. **Figure 13** shows a simple drawing of the wire. The carbon offers resistance to the spark and drops the voltage as it travels to the plug. This voltage drop along the wire is minimal but will have the effect of suppressing the RFI because the alternating currents usually present are prevented. In addition, the level of voltage along the entire length of the cable is slightly different and will have a lower field strength than with solid-core wire where the voltage is the same or stronger. For example, if a plug is firing at a level of 5000 volts with solid-core wire, the entire length of the wire will have an equal and, therefore, strong, force field. If, however, a plug is firing with TVRS wire at 5000 volts, the coil end of the wire could have as much as 7000 volts. The 2000 volt drop along the length of the wire (including the cap and rotor) puts each section at a different level, thus reducing the total field strength. Resistor

Figure 12. A Corvette suppression cover.

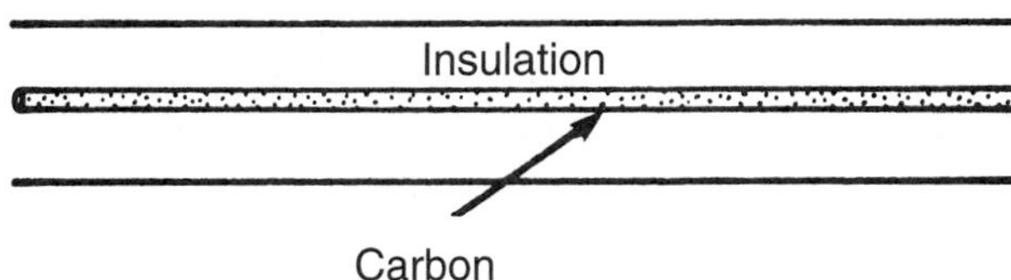

Figure 13. Carbon gives the resistance required in some TVRS plug wires.

wire, as TVRS is sometimes referred to, will have resistance per inch of a specified amount. The amount will differ but is usually approximately 4,000 ohms per foot. A 2-foot wire will, therefore, have 8000 ohms of resistance. A 3-foot wire equals 12,000 ohms, and so on. For the exact specification, consult a manual for the vehicle. TVRS cables should not be replaced with solid-core wire because the RFI might occur in some critical vehicle system or be objectionable to the customer.

MAGNETIC SUPPRESSION

Over the years, there have been many methods of trying both to suppress RFI and to make components less likely to be affected by the interference. One of the more recent has been the introduction of a solid-core wire that is wrapped in a conductive outer shell. **Figure 14** shows the principle behind this type of suppression. The wire

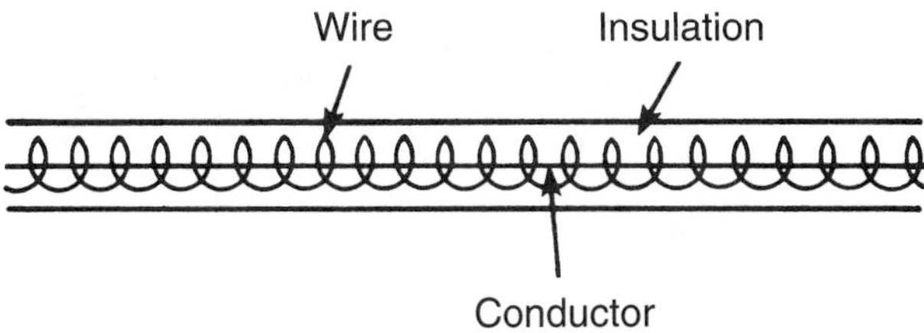

Figure 14. A magnetic suppression ignition cable.

wound around the spark-carrying conductor shields the magnetic force field that is produced. No effort is made to alter the spark, as is the case with resistive suppression. Instead, the wire itself prevents the RFI from leaving the surface of the wire. An ohmmeter placed at the two ends of this type of wire will typically read zero ohms of resistance. If the ohmmeter were to read anything other than zero, the wire is probably faulty.

This is a good time to point out the need for knowing and using manufacturers' specifications. Ignition cables might look exactly alike and yet they could either possess thousands of ohms per inch or have a specification of zero ohms for the entire length. Always refer to the manufacturer's specification if you are using an ohmmeter to diagnose a faulty cable. The exception to this is the use of an oscilloscope. A bad wire will usually result in a high firing line, as you will see when we learn to use a scope in ignition diagnosis.

The insulation surrounding the ignition wire is just as important as the wire itself. The ability of the wire to deliver the high voltage to the plug is dependent on the insulating ability of both the jacket and the boots or caps at the two ends of the cable. Some of the modern ignition systems have the capability of producing up to 100,000 volts. Electricity is always looking for the path of least resistance, which with poor insulation might be through the side of the wire over to the engine block. Delivering the spark to the engine block results in a dead cylinder, rough running, increased hydrocarbon emissions, and reduced mileage. In addition, with smaller engines the power is so greatly affected by the loss of one cylinder that the safety of the driver and passengers could be at risk. Delivery of the spark to the plug is the job of the insulation, and it obviously must be capable of insulating up to the limit of the ignition system. In addition, it should be capable of doing it under adverse conditions, such as –20°F, 110°F, or extreme dampness. When we look at the use of a scope, we will discuss testing of insulation in depth.

The connections at the ends of the ignition cables are usually made with insulated crimped-on metal terminals **(Figure 15)**. Removing a plug wire should be done only by grasping this metal terminal. Never pull on the wire or the

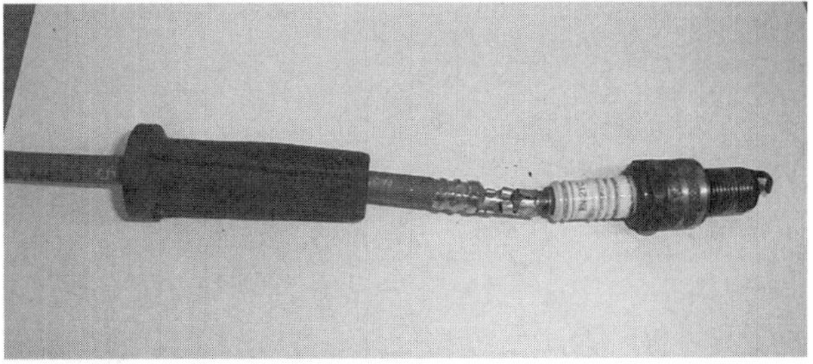

Figure 15. A plug wire boot pulled back to expose the connector.

end might break off. In addition, most manufacturers recommend the use of silicone dielectric compound in the boots to prevent their deterioration. Silicone dielectric compound does not conduct electricity and is used to lubricate, transfer heat, and suppress RFI. The compound also helps in preventing the boot from sticking and burning on the insulation of the plug. Plug wires are best removed by twisting slightly and pulling on the terminal end. Additional compound should be added to the boots to aid in their future removal. Your toolbox should contain a tube of dielectric compound because it is used frequently on modern vehicles.

CAP AND ROTOR

On many of the ignition systems in use today, a single ignition coil is used for a multiple number of spark plugs. The spark produced by this single coil is transferred to the correct cylinder through the use of the cap and rotor. **Figure 16** shows some of the different types of rotors used. For the most part, their principles are the same. A single coil wire brings the high voltage to the center of the distributor cap where it is transferred to the center of the rotor. The rotor is attached to the distributor shaft and turns at camshaft speed. Because it is turning, it will act like a rotating bridge and bring the high voltage to the outer distributor cap terminals, which are in turn connected to the individual plug wires. In this manner, a single ignition coil can supply all the spark plugs with high voltage in a sequence called the **firing order**. The manufacturer determines this specific order of spark firings. As each one of the cylinders is following its own intake, compression, power, and then exhaust sequence, the spark will be delivered at the correct time when the combustion chamber is full of compressed fuel and air. The placement of the ignition wires and the direction the distributor shaft turns set this firing order. For example, let us assume we are replacing the wires on a V-6 engine with a 165432 firing order. This means that the cylinder the manufacturer has numbered as number 1 will fire first. Then, 60 degrees of camshaft rotation later (a total of 120 degrees of crankshaft rotation), the next cylinder, number 6, will be ready for the spark plug to fire and ignite the mixture. Again, 60 degrees later, cylinder number 5 will be fired and so on until all cylinders have fired. The entire sequence of all plugs firing will take two complete crankshaft rotations with the individual power impulses spread out 120 degrees apart **(Figure 17)**. The wire positions on the outer edge of the cap will determine the spark plug firing order. The wires must always be correctly positioned or the spark might be delivered to the incorrect plug. By incorrect, we mean that the cylinder is not ready for the spark. Perhaps it is on the exhaust stroke or on the intake stroke. It is important to note that there are many different styles of ignition systems and many different firing orders. Always make sure that you are rewiring the cap in the correct order, and verify by looking up the specifications for the vehicle by year, make, and model.

The end of the rotor is a spot that does not touch the distributor cap terminal. There is a small air gap of approximately 0.002 inch (0.2 mm). This air gap is necessary because the rotor would destroy itself if it came into contact with the cap inserts. Generally, the arcing of high volt-

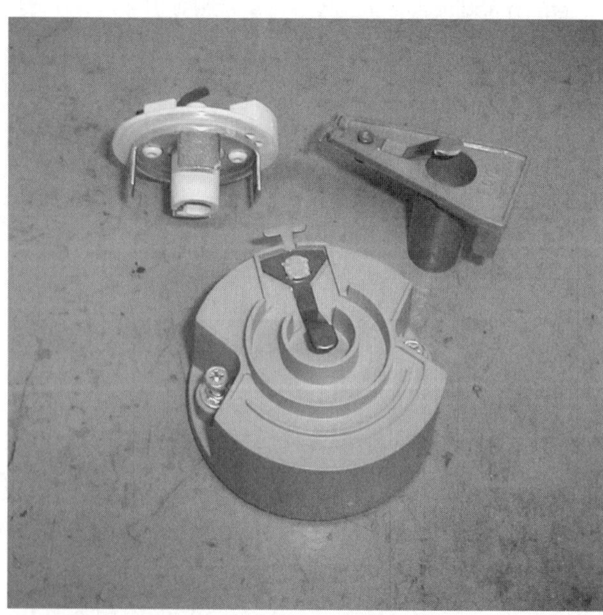

Figure 16. Different types of rotors.

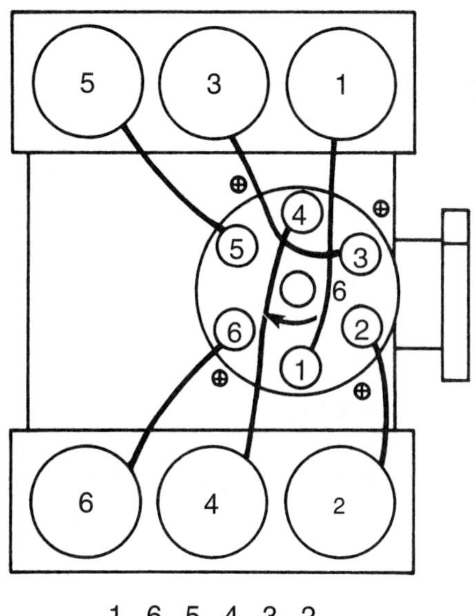

Figure 17. A firing order for a V-6.

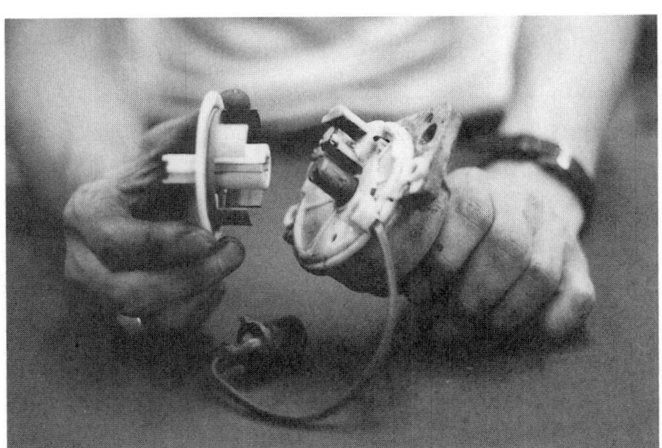

Figure 18. A rotor with Hall-effect shutters.

age that takes place at this air gap eventually causes sufficient corrosion that the cap and rotor require replacement. Scope testing allows us to pinpoint excessive corrosion caused by arcing. Some manufacturers recommend coating the end of the rotor with silicone dielectric compound to cut down on the arcing. This helps in the prevention of RFI and extends the life of the cap and rotor. In addition, some caps and rotors are utilized for other functions. **Figure 18** illustrates this with a rotor that has metal shutters attached. We study this in Chapter 48. Also, some rotors have multiple ends. **Figure 19** shows a whisker rotor and a dual-plane rotor.

Distributor caps on some older General Motors vehicles housed the ignition coil, as shown in **Figure 20**. This GM high-energy ignition coil is housed inside the top of the distributor cap. The bottom of the coil feeds the rotor directly, thus eliminating the need for a coil wire. A small carbon button is placed between the ignition coil output and the center of the rotor. This is not a current system for General Motors, but an adaptation of it is in use on some foreign vehicles.

The job of the cap and rotor is extremely important. It must deliver the high voltage to the correct plug without allowing the spark to find an alternate or short path to ground. High-quality caps and rotors have the ability to insulate at least 50,000 volts and last many thousands of miles. As mentioned previously, they are diagnosed usually with an oscilloscope and replaced as needed.

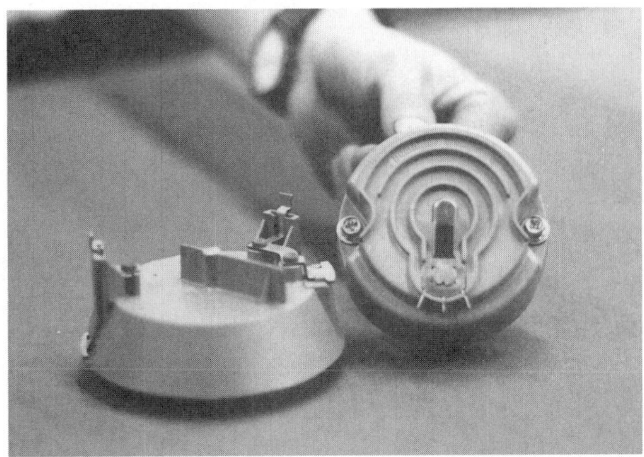

Figure 19. A dual-plane and whisker rotor.

Figure 20. Older General Motors vehicles housed the ignition coil in the distributor cap.

Summary

- Spark plugs may be sealed with a gasket or a tapered seat.
- Plug gaps, heat range, and reach are all determined by the vehicle manufacturer and should not be changed by the technician.
- If the plug is operated too cold, it may foul.
- If the plug is operated too hot, it may burn out prematurely.
- The torque of the plug helps to determine its heat range.
- Used spark plugs can be "read" to help diagnose engine conditions.
- Wires can have resistive or magnetic suppression characteristics.
- Resistive wires may have as much as 4000 ohms per foot.
- Magnetic suppression wires usually have no resistance.
- Caps and rotors are used to deliver the ignition current from a single coil to the individual spark plugs.

Review Questions

1. Which of the following components is not part of the secondary circuit?
   A. spark plug wires
   B. distributor cap
   C. rotor
   D. ignition module
2. Too high of a heat range spark plug will
   A. increase performance
   B. decrease plug life
   C. reduce ignition firing voltage
   D. increase the chances of fouling
3. Technician A states that TVRS wires may have resistance values of as high as 4000 ohms per foot. Technician B states that TVRS wires might use magnetic suppression techniques. Who is correct?
   A. Technician A only
   B. Technician B only
   C. Both Technician A and Technician B
   D. Neither Technician A nor Technician B
4. A spark plug is removed from a high-mileage vehicle. It has black sooty deposits completely covering the electrodes and the insulator. Technician A states that the vehicle is running too rich. Technician B states that the plug is the wrong one for the vehicle. Who is correct?
   A. Technician A only
   B. Technician B only
   C. Both Technician A and Technician B
   D. Neither Technician A nor Technician B
5. Spark plugs should never be removed from an aluminum cylinder head that is still hot.
   A. True
   B. False
6. A colder-than-specification plug will usually result in fouling.
   A. True
   B. False
7. The cap and rotor form a rotating bridge between the one-coil wire and the multiple plug wires.
   A. True
   B. False

Chapter 47

Servicing the Secondary Ignition System

Introduction

The secondary ignition system is one of the most frequently maintained systems on the modern vehicle. Plugs, wires, caps, and rotors account for quite a bit of required maintenance on a vehicle that has a distributor. In this chapter, we look at some of this required maintenance in addition to looking at trade-acceptable methods of diagnosing problems. Keep in mind that the function of the system is to deliver to the cylinder a spark that will ignite the mixture. If the secondary system fails to deliver the spark, the cylinder will misfire and allow unburned hydrocarbons (fuel) to leave the engine. Not only is this bad for the environment, but, if left unrepaired for an extended period of time, the unburned fuel will destroy the catalytic converter. If the failed component is the coil wire, cap, or rotor, it is possible that the vehicle will not even start. You can see that the system performance is critical to the efficient running of the engine and was never designed to last the life of the vehicle. Preventive maintenance is the key to its performance.

REMOVING AND REPLACING SPARK PLUGS

Removing and replacing plugs is a common procedure that is designed to not only replace a worn out plug, but to give access to a cylinder and allow for the reading of the old plug. Reading of plugs is covered in Chapter 46. Do not hesitate to go back to that chapter and review the information. Access to the cylinder might be necessary to run a mechanical compression test or use a scope to view inside the cylinder for diagnostic purposes. Most technicians use electronic compression testing, but there are times when a mechanical test is called for and can be run only with the plug removed.

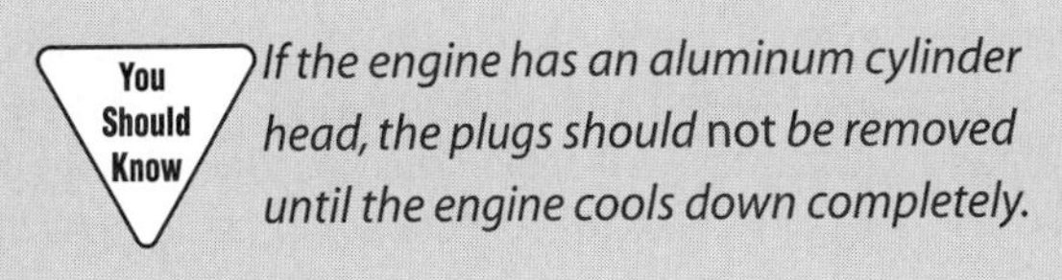

If the engine has an aluminum cylinder head, the plugs should not *be removed until the engine cools down completely.*

The spark plugs usually will be removed with a deepwell socket specially designed for plugs. What makes it special? It will be a 6-point 5/8 or 13/16 deepwell socket with a sponge rubber padding deep in the socket. **Figure 1** shows the pad removed from the socket. The top of a spark plug is porcelain and could crack. The padding protects the plug during removal and installation. In addition, the padding will hold the plug so that you can thread it in by hand before tightening it to the correct torque. Additionally, before removing plugs, inspect the cavity around the plug for dirt and grit. If the valve cover gasket has been leaking, the oil around the plug will have collected road

Figure 1. A pad protects the spark plug and holds it firmly.

dirt. If you do not clean this dirt out before removing the plug, it is possible that some of it will wind up in the cylinder or on the plug seat. Dirt in the cylinder could damage the piston or cylinder walls, and dirt on the plug seat will prevent correct sealing and heat transfer between the plug and cylinder head. Use some compressed air to clean out the plug cavity if it is dirty.

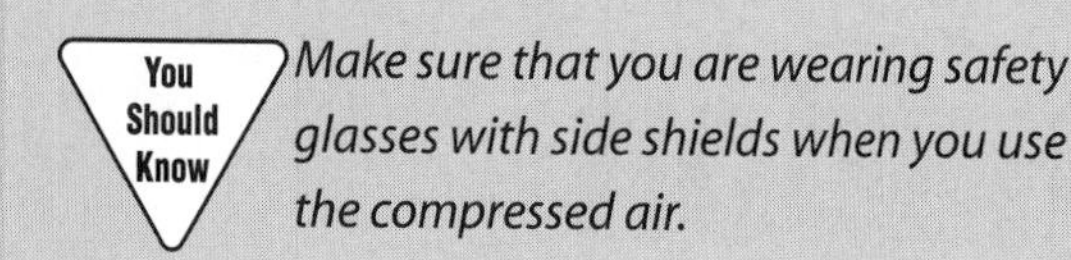

Make sure that you are wearing safety glasses with side shields when you use the compressed air.

In some cases, the cavity may be so full that you will have to put some solvent around the plug to get it clean. Keep in mind that the valve cover gasket also needs to be replaced if it is the source of oil.

If there is a possibility that you could mix up the plug wires, label them before their removal. They must be installed back on the same cylinder position or serious damage might occur. Remember that the firing order on a distributed vehicle is determined by the position of the wire, both in the cap and on the plug. Both must be correct or the firing order of the engine will be changed. If, for instance, you install the wires incorrectly and the spark is delivered to a cylinder that has the intake valve still open, a backfire into the intake manifold could occur. The backfire could damage sensors or fuel injectors and could possibly start a fire. Customers do not like to look out the service window and see their vehicles on fire. Label the wires so you know where they came from. If you are only going to replace the plugs, do so one at a time. Remove the wire, R & R the plug, and reinstall the wire. You will be assured that the wires are on the correct plug.

SETTING THE SPARK PLUG AIR GAP

After the plug has been removed, read it for signs of oil consumption, excessive fuel, misfire, etc. Note on the shop ticket your findings so that the customer can be informed of additional work that may be required on a cylinder-by-cylinder basis.

The next thing that must be done is to set the air gap on the new plugs. Even if spark plugs have been advertised as pregapped, they must be checked or regapped before installation using a feeler gauge.

One of the earliest ignition tools was the feeler gauge. Whenever we wish to measure the distance between two items, we use a feeler gauge. It comes in three basic designs: flat steel, flat brass, and round. The thickness of the two flat feeler gauges is normally printed on them. One is made of steel; the other one is made of brass. The brass one is used if the object that is being measured is magnetic. The nonmagnetic properties of the brass feeler gauge make it

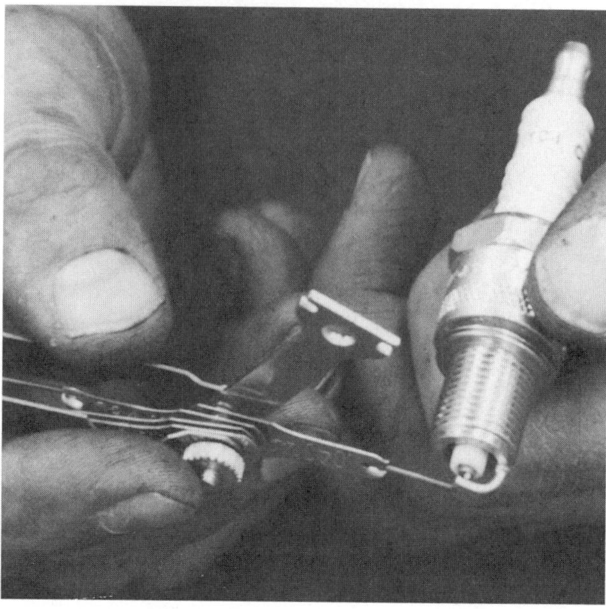

Figure 2. Use only a wire gauge to set plug air gaps.

the ideal choice when a magnetic field is present. The round feeler gauge is the type we use to set spark plugs. The diameter of the wire is printed on the holder and should match the specification of the air gap between the center electrode and the side electrode. **Figure 2** shows the air gap of a plug being measured with the wire gauge. The wire should pass through the gap with slight effort. This indicates that the plugs are gapped to the diameter of the gauge. Most wire gauges have a bend bar attached that is useful if the air gap is not correct. By bending the side electrode slightly, we should be able to get the air gap within specification. Remember from Chapter 46 that the finished air gap should put the center electrode and the side electrode at right angles to one another. **Figure 3** shows what a plug's air gap looks like if it is gapped either too small or too large. Usually, an incorrect looking plug is the result of either the wrong gap specification or the wrong plug for the vehicle. In either case, verify both the plug and the gap before proceeding. The wrong plug can cause some strange drivability problems. Do not assume that the correct plug is in the vehicle either, because this assumption can cause you lots of lost time in diagnosis. Always verify the component if there is any doubt in your mind. Also remember that new plugs should always be gapped, or at least checked if they are pregapped, before they are installed in an engine. Their gap will be close, right out of the box, but exactness is important on today's vehicles. The right plug correctly gapped will function as it was designed for the recommended tune-up interval.

After the plugs are gapped correctly, they can be installed in the cylinder head to the correct torque. Remember that torque is the twisting effort that will ensure the correct heat range and sealing ability of the plug. Using a

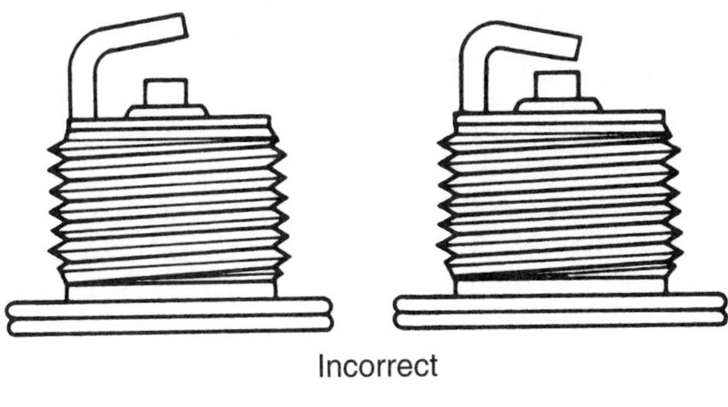

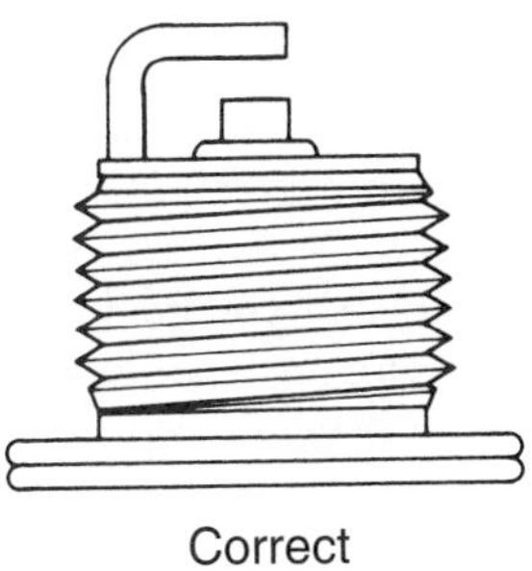

Figure 3. A correct gap will increase the ability to ignite the mixture.

torque wrench to tighten the plug will also assure you that the threads will not be damaged, especially in an aluminum cylinder head. The use of graphite or antiseize compound on plug threads is recommended for aluminum heads.

REPLACING IGNITION WIRES

Replacing ignition wires is not that difficult on most engines. The routing of the wires is the most important component of the replacement. The vehicle manufacturer routes wires so that they will not interfere with other electronic devices and are protected from pulleys or from exhaust manifolds that might burn the wires. There is a high level of RFI around a plug wire, even one that is of the TVRS type, so the position of the wire is important. The best method is to observe how the manufacturer has positioned the wires and duplicate this with the new ones.

The boots that will go over the plugs should have some silicone grease inside before their installation. Wires should be replaced one by one. Do not rip off all of the wires and then try to figure out the firing order. If you replace only one wire, it will be in the correct place. By removing one wire at a time, you will also be able to determine if the length of the new one is correct. Wire sets will have different length wires that match the overall length of the old ones. Nothing is worse than a wire that is too long or too short. Too short is an obvious problem of not getting to the plug, but too long can be a routing problem that results in interference because the wire is closer to something than the original. Take the time to make sure the new wires look correct and are routed away from potential problems **(Figure 4)**.

Figure 4. Wires should be routed correctly.

REPLACING IGNITION CAPS AND ROTORS

Generally, ignition caps and rotors are changed with any wire change; however, situations of excessive firing voltage or faulty insulation can result in more frequent replacements. We will look at how to diagnose faulty caps and rotors later. Replacement should be done in stages by moving the wires, one wire at a time, to the same position on the new cap that they were in on the old cap. Do not make the mistake of removing all of the wires from the old cap and then trying to figure out where they are supposed to go on the new one. Remember that the position of the wires in the distributor cap determines the firing order, and it must be correct. There are no exceptions. **Figure 5** shows

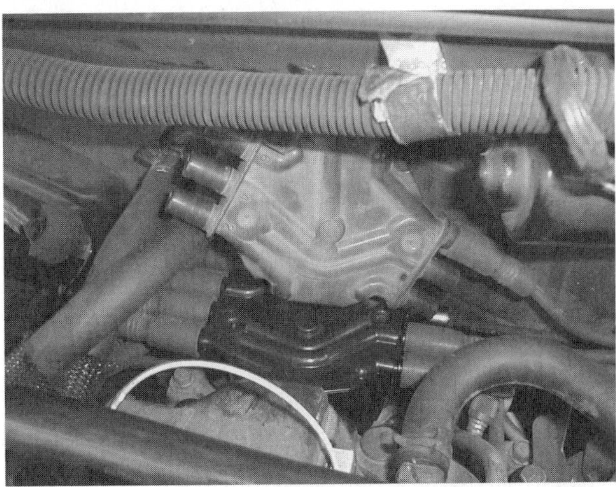

Figure 5. When replacing a cap, transfer the wires one by one.

a new cap alongside of an old one with the wires being transferred one at a time. The new cap is an exact replacement for the old and is being held in the same position next to the old. In this way, the technician will not get confused and change the firing order.

The camshaft and its relationship to the crankshaft mechanically determine the firing order of the engine. The distributor or the position of the ignition wires cannot change it.

Replacing the rotor is probably the easiest of ignition operations on most vehicles. Rotors are either pushed onto the distributor shaft or they are held down with screws. **Figure 6** shows the rotor from a General Motors V-6 before the replacement. Remove the screws, making sure that you note the direction that the rotor tip is facing, and remove the rotor. The new one should be an exact replacement and when you are finished should have the rotor tip pointing in the same direction as the old one. Make sure the rotor is screwed down solidly to the shaft, then put the cap back in place and you are finished. Many technicians will replace the rotor with the spark plugs and replace the wires as needed. However, with the long tune-up interval of many newer vehicles, by the time plugs are required, wires, cap, and rotor may also be needed.

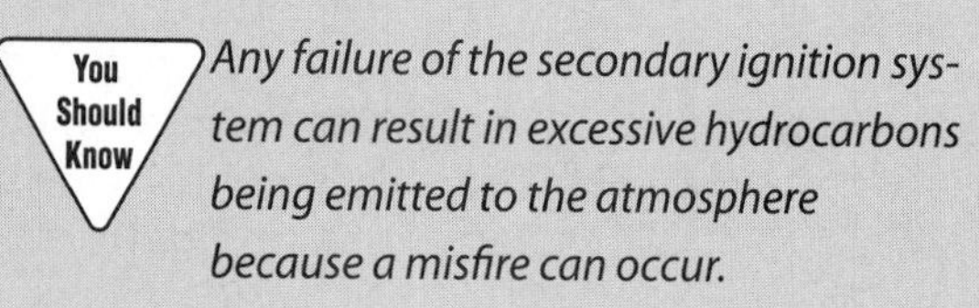

Any failure of the secondary ignition system can result in excessive hydrocarbons being emitted to the atmosphere because a misfire can occur.

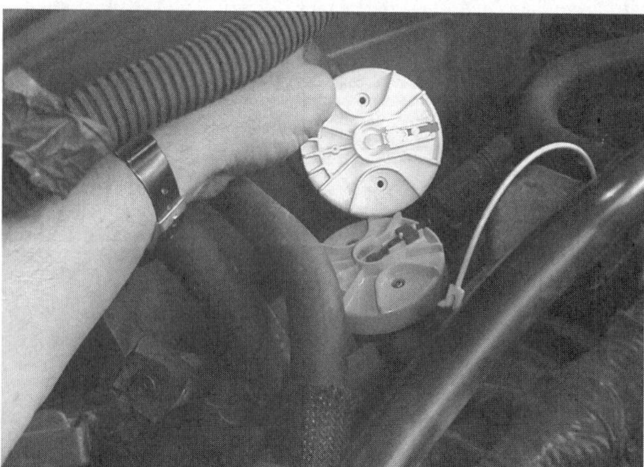

Figure 6. Make sure the new rotor matches the old.

USING A DSO TO CHECK SECONDARY FIRING VOLTAGE

No discussion of ignition system testing would be complete without a comprehensive discussion of an oscilloscope. For our purposes, we use a DSO because it has been introduced to you in previous chapters. Some analog scopes are still in some service centers, but their use is rapidly diminishing. The principles in their use, however, are the same as those of the DSO or digital engine analyzer. The DSO gives us the opportunity to look inside the ignition system, including the cylinder. The DSO is one of the most used testers in modern repair centers. The results are accepted by all of the major manufacturers worldwide for warranty claims. Do not be misled, however. The DSO will not make you a better technician. It will not normally give you information that you could not have found by other methods. The difference, and one of the main advantages to its use, comes from the speed and accuracy with which a diagnosis can be made. A good technician can, for instance, find a faulty spark plug wire with an ohmmeter or a substitute wire. The same technician will also find the faulty wire with a DSO but will probably find it much faster. Automotive technicians operate with speed and quality as the cornerstone of their business. The best technicians will starve if they are slow. The DSO can be the edge that the good technician needs to become a great technician. In addition, DSO testing is a form of quality control that allows the technicians to check their tune ups for accuracy. The scope should also be used for preventive maintenance. The same procedure with which we verify the accuracy of the tune up can be applied to a preventive maintenance checkup to determine exactly what the customer needs. The scope, or DSO, will not make you a better technician. You must have the basics and a good working knowledge of ignition systems or the patterns will probably not make a great deal of sense. The scope is also not a replacement for common sense. You should not need an expensive scope to tell you that the ignition wires are oil soaked and require replacement. Visual inspection and common sense are still two of the best diagnostic tools to use.

Before we look at ignition system test procedures, let us review the information presented in an earlier chapter about scopes. You will remember that an oscilloscope is nothing more than a voltmeter with a time frame built into it **(Figure 7)**. A voltmeter with a clock attached is how some technicians refer to a scope. The time frame, or clock function, is important because of the tremendous speed with which the ignition system produces and delivers sparks. This speed makes the traditional voltmeter almost worthless. For example, a four-cylinder vehicle running at 1500 rpm, or approximately 55 mph, is producing 83 sparks per second. Even if our meter could follow the voltage changes, our eyes and brain probably could not. This

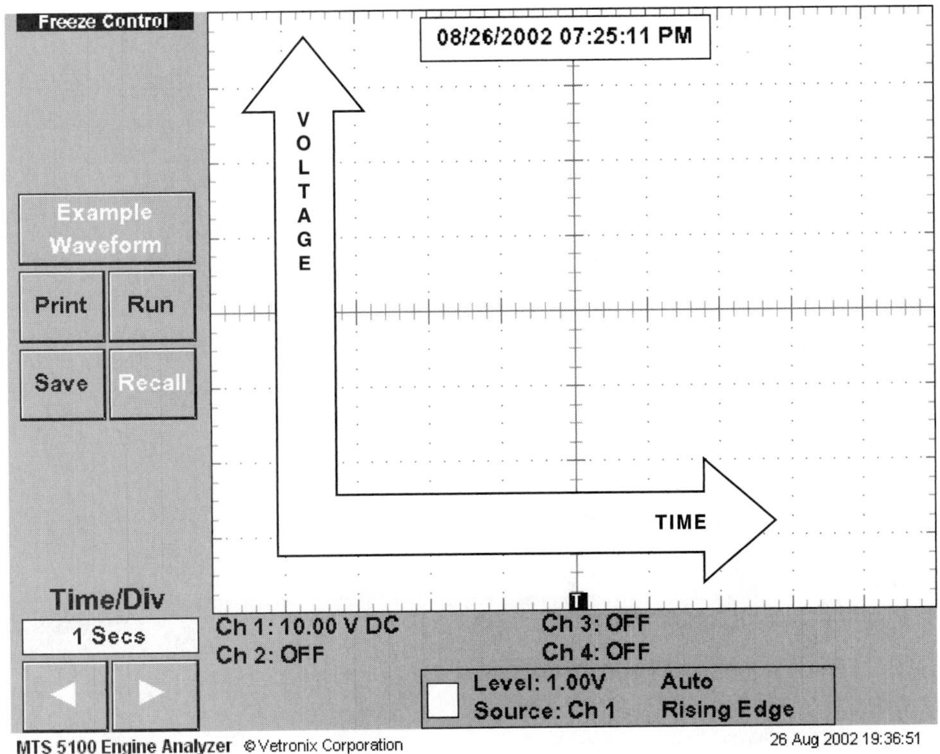

Figure 7. DSOs give a voltage reading over a period of time.

makes the DSO valuable because we can display the changing voltages on a screen and, if necessary, hold them for later diagnosis, somewhat like an electronic piece of graph paper with both positive and negative voltages displayed. **Figure 8** shows the DSO's screen. Notice the zero line toward the bottom of the screen. Our trace is showing zero volts right now. Voltage traces that are above the zero line are positive, whereas those that are below the zero line are negative. As is the case with any voltmeter, the scale, or range, of readable voltages must be determined and set. In most cases, the highest scale is used when dealing with ignition systems. Usually, this scale is 2,000 volts per division for a total of 16,000 volts worth of screen available for use. In addition, remember that the screen is read from left to right, with readings on the left of the screen appearing before those that are on the right side. The millisecond scale located in the lower left of the screen is used to give the pattern a time reference. Keep in mind that not all scopes will have the same capabilities or the same scales. Different manufacturers' scope screens might look different from that shown. Do not hesitate to review the literature supplied with the unit to familiarize yourself with the functions of it.

Most scopes currently in use in the automotive field have some available patterns that one can use to look in detail at different sections of the ignition system. Both primary and secondary patterns can usually be displayed in one of two different styles of patterns. In addition, many scopes have the ability to zero in on one cylinder for detail inspection. Let us look at each one separately, realizing that not every scope has the same patterns or calls them by the same name.

Display, or parade, is shown in **Figure 9**. All of the ignition firings are in a row in the firing order. Remember, firing order is the order in which each cylinder will have its spark plug fired. Our illustration shows an eight-cylinder engine with a firing order of 18436572. We have triggered the scope off cylinder 1, so the left-most pattern is from cylinder 1. This upward line represents the initial firing of the spark plug. Do not forget that in display the scope is read from left to right. The next pattern usually available is called raster or stacked. Notice **Figure 10**. The cylinders are again labeled with the same firing order, and the patterns are read from the bottom up with the first cylinder in the firing order on the bottom of the screen. You can see that this pattern will be useful in comparing all the cylinders against one another.

Figure 11 is from a six-cylinder distributed vehicle and will show us the firing voltage of the ignition system. The six long lines on the screen represent the firing voltage. In our example, they reach up to 12–15 kilovolts. This means that cylinder 1 is firing at 13,000 volts. Before we analyze whether this is good or bad, let us look at the cause of this upward turn of the DSO trace. An upward line, you will remember, signifies voltage. This upward line is indicative of the voltage that is necessary to start the spark. The ignition

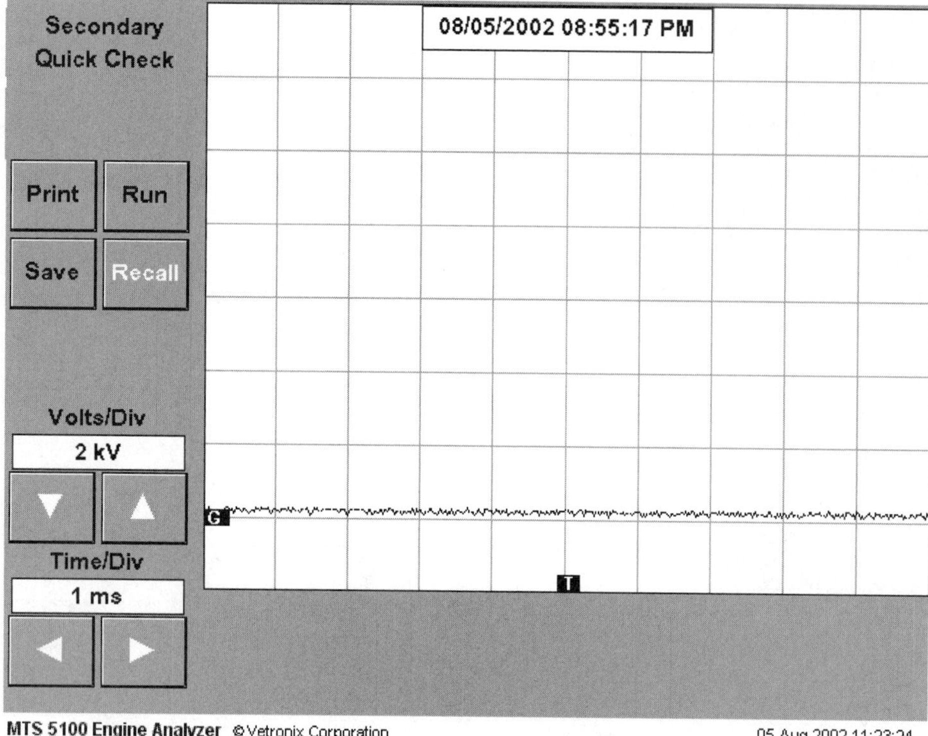

Figure 8. DSO setup for ignition secondary testing.

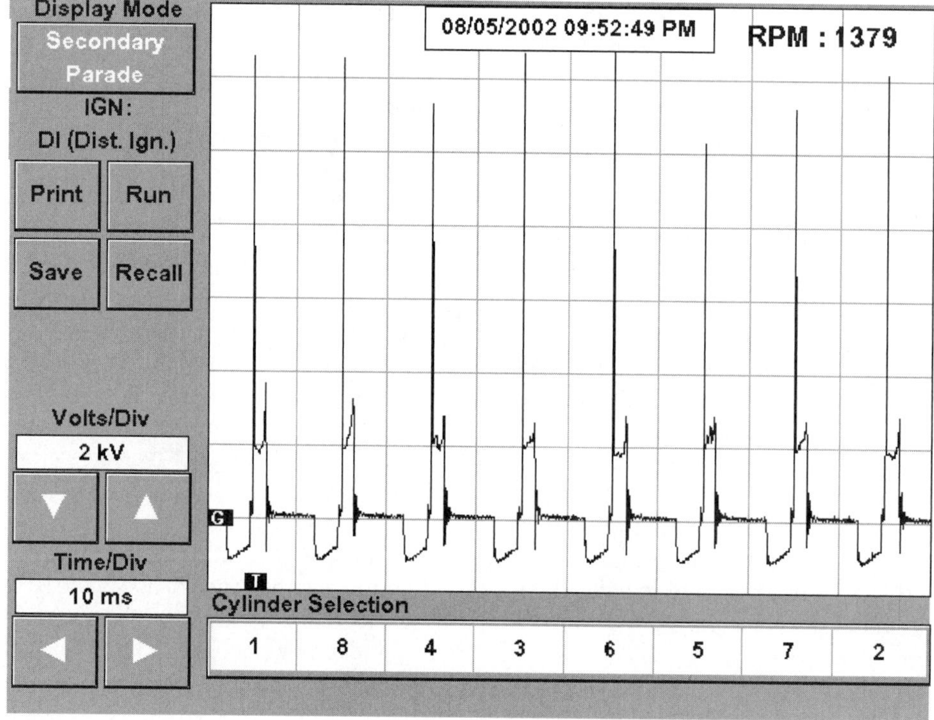

Figure 9. The display pattern for a V-8.

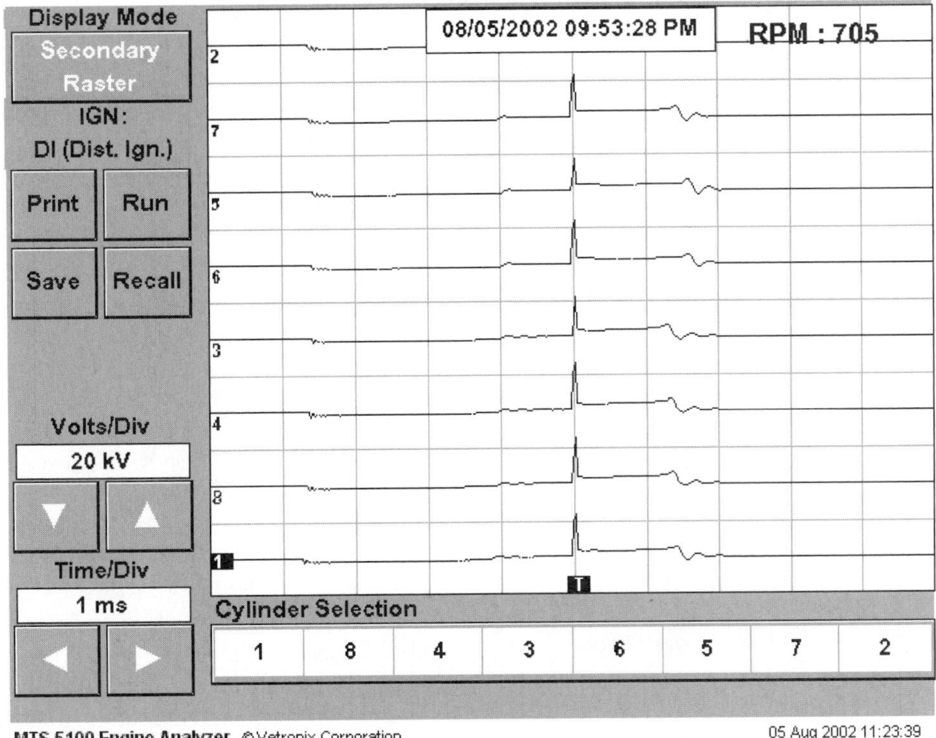

Figure 10. The secondary raster pattern for a V-8.

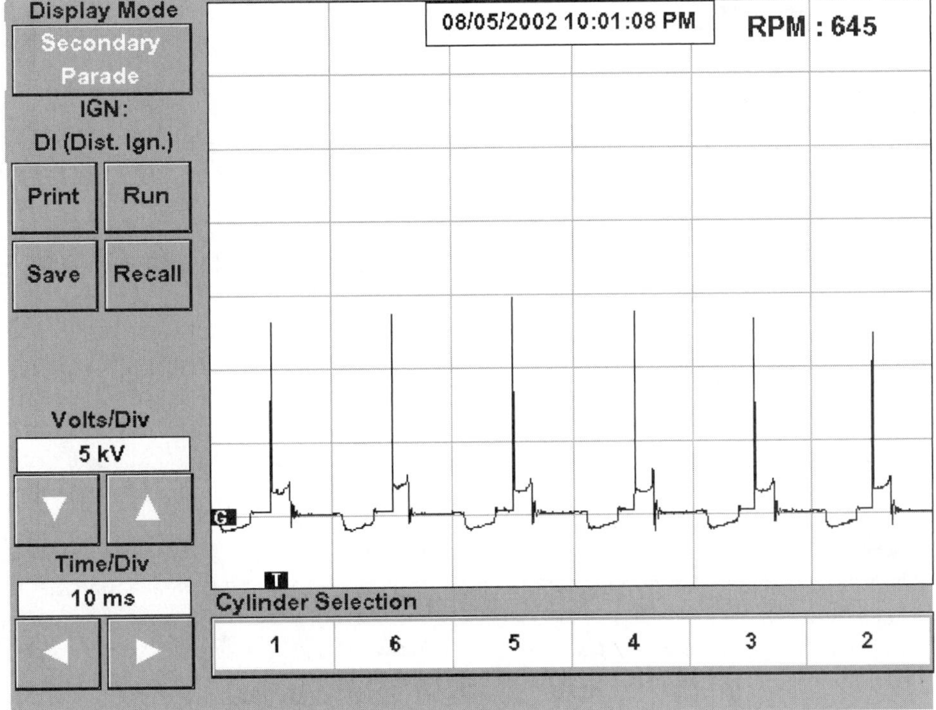

Figure 11. The secondary display for a V-6.

coil's pressure rises up to the point where it can overcome all of the resistance of the secondary circuit. This resistance includes the wires, cap, rotor, and plugs but also the conditions inside the cylinder. A rich mixture represents reduced resistance, whereas a lean mixture represents increased resistance. The greater this total resistance, the higher the voltage; the lower this resistance, the lower the voltage. You will see when you begin testing with the scope that typically it takes approximately 6–12 kilovolts (6,000–12,000 volts) to overcome this resistance and get our spark to begin. This line is called the firing line. Do not forget that it represents the voltage necessary to overcome all of the secondary resistance. In addition, it should show that all cylinders are firing at about the same level. It is trade standard that the firing voltages should be within 3 kilovolts of one another.

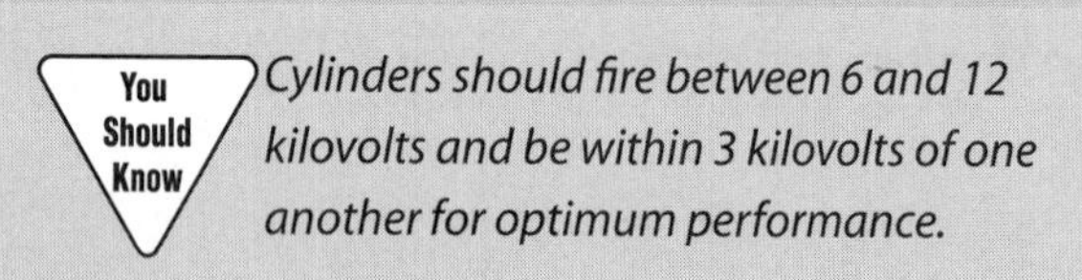

Cylinders should fire between 6 and 12 kilovolts and be within 3 kilovolts of one another for optimum performance.

What do we do if the firing voltages are not within specification?

USING THE DSO TO CHECK THE CAUSE OF EXCESSIVE SECONDARY VOLTAGE

If we have determined that the firing voltage is not within specification, we have to take note of whether all cylinders are affected or just one. If all cylinders exhibit high firing voltage, the cause will be something that is common to all cylinders. The coil wire, the rotor, or the conditions inside the cylinders are the most common causes. If, however, only one cylinder is high and the rest normal, a plug or plug wire is the most likely cause. **Figure 12** shows all cylinders firing high based on our 6–12-kilovolt specification. To diagnose this requires that we turn off the vehicle, remove a plug wire, and connect it directly to ground using a ground adapter. Restart the vehicle and look at the DSO. Did one firing line drop down? If yes, the conditions inside the cylinder are the cause. Let us apply this testing procedure to our high firing lines of Figure 12. **Figure 13** is the DSO with one cylinder to ground. There is virtually no difference. As a matter of fact, it is impossible to figure out which wire has been grounded. Apparently, the conditions within the cylinder are *not* the cause of all of the firing lines being excessively high. It is unlikely that all of the plug wires are faulty and have excessive resistance. This leads us to the cap/rotor or the coil wire. Determining which is easy. With the engine off, remove the coil wire from the cap and ground it. Capture a DSO pattern with the engine cranking. You will probably have to move the trigger to the coil wire. If the pattern is still high, the coil wire has excessive resistance. However, if the pattern is low or if the scope will not even trigger and display a trace, the coil wire is good and the problem lies within the cap and rotor. By putting the coil wire (distributor end) to ground, you have eliminated the resistance of the cap/rotor, wires, plugs, and cylinder. The only resistance still in the circuit is the coil wire. The coil wire should not have sufficient resistance to cause any firing voltage, so if virtually anything shows on the screen, you have a bad coil wire. **Figure 14** shows the results of cranking the engine with the coil wire to ground. We show only one pattern because the trigger is placed around the one coil wire. Notice that we still have voltage showing. This is an example of a coil wire with excessive resistance. The cure for the condition is easy: replace the coil wire and retest.

Let us look at another example from another six-cylinder vehicle. **Figure 15** shows the firing lines with one just about off the screen. Which cylinder is this? It is 5 in the firing order, or cylinder number 3. Remove the plug wire from cylinder 3 with the engine off and ground the wire. Restart the vehicle and look at the DSO. By now you should have figured out that if the spark plug or the cylinder were the cause of the excessive voltage, the firing line would drop, as shown in **Figure 16**. If the plug and cylinder are not the cause, the firing line will remain elevated, as shown in **Figure 17**. With the firing line still excessively high, we would turn the vehicle off and ground the distributor cap end of the plug wire. Restart and look at the DSO. If it drops, it is a wire with excessive resistance; if it stays high, it is a distributor cap with excessive resistance. What would your conclusion be in **Figure 18** with the ground placed in the cap? This looks like a wire with excessive resistance. Replace it and recheck with the DSO. It makes sense to replace all of the wires in some cases even though only one is showing excessive firing voltage. If the vehicle has high mileage or is more than 5 years old, it is likely that if you replace just one wire, in a short period of time another one might develop the same problem. Now may be a good time for a preventive maintenance secondary ignition tune up. Replacement of all of the wires, the cap, the rotor, and the plugs may be the most cost-effective approach for the customer.

Some DSOs have an additional pattern type that might be valuable under certain conditions: firing kilovolts in bar graph format. **Figure 19** illustrates this type of pattern with four lines of digital data in addition to the bar graph. Look at cylinder 1. The data beneath the bar shows that the minimum (Min) was 11 kilovolts, the maximum (Max) was 15 kilovolts, and the average (Avg) was 13.7 kilovolts. Some technicians find that this format gives them more information than the usual scope display. Some of the analysis has been done for them and printed in data format. The key is to put the data in the format that makes the most sense to you and allows for the easiest analysis.

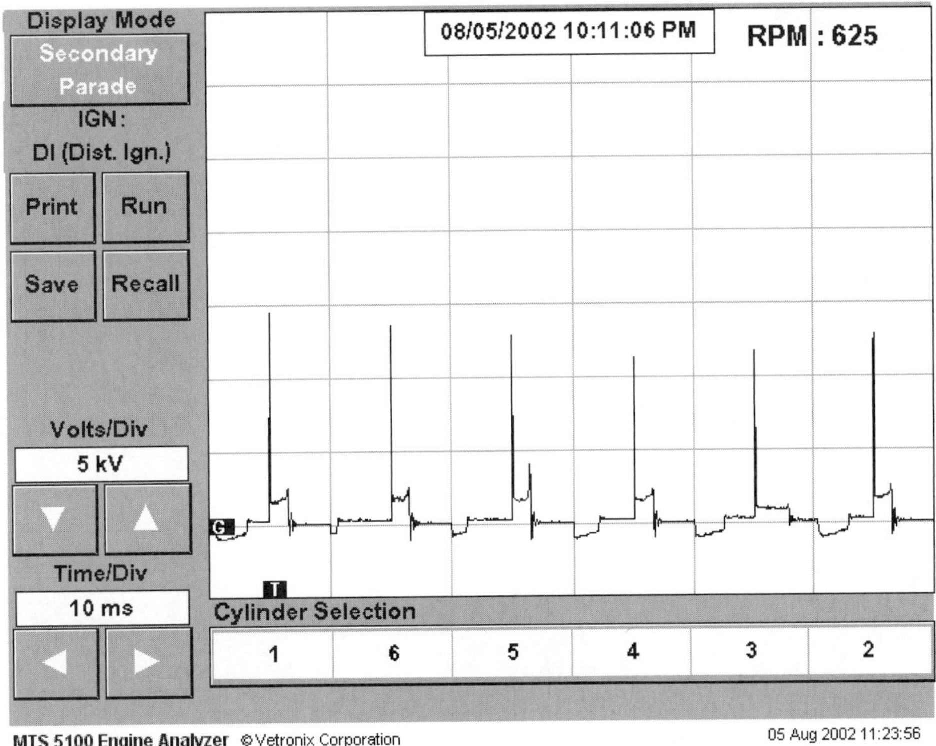

Figure 12. All cylinders firing slightly high.

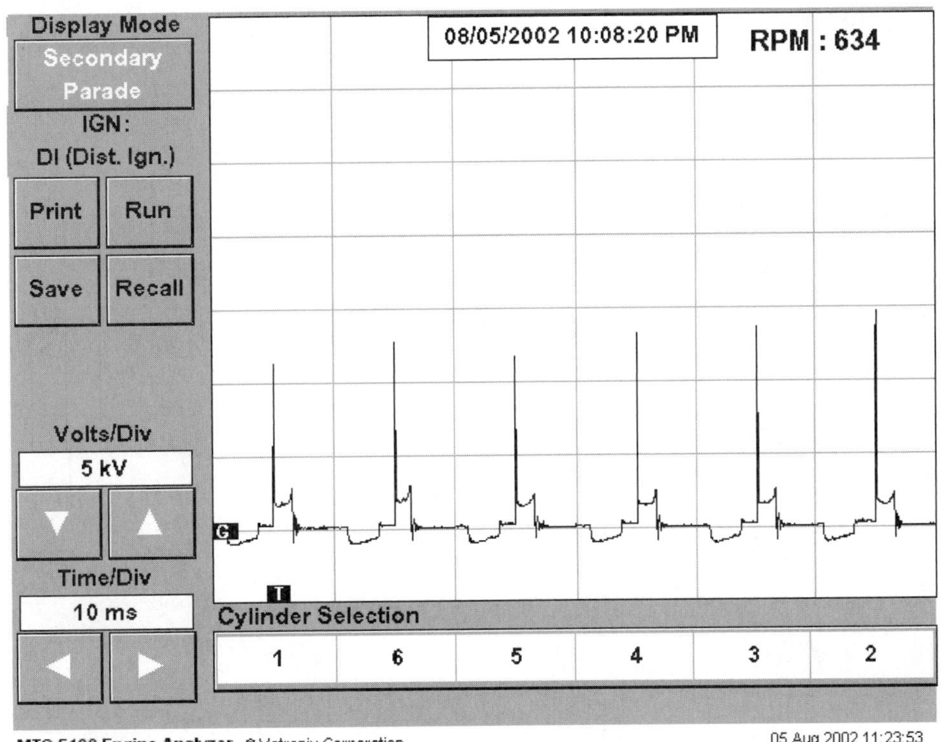

Figure 13. Excessive resistance still shows with one plug wire to ground.

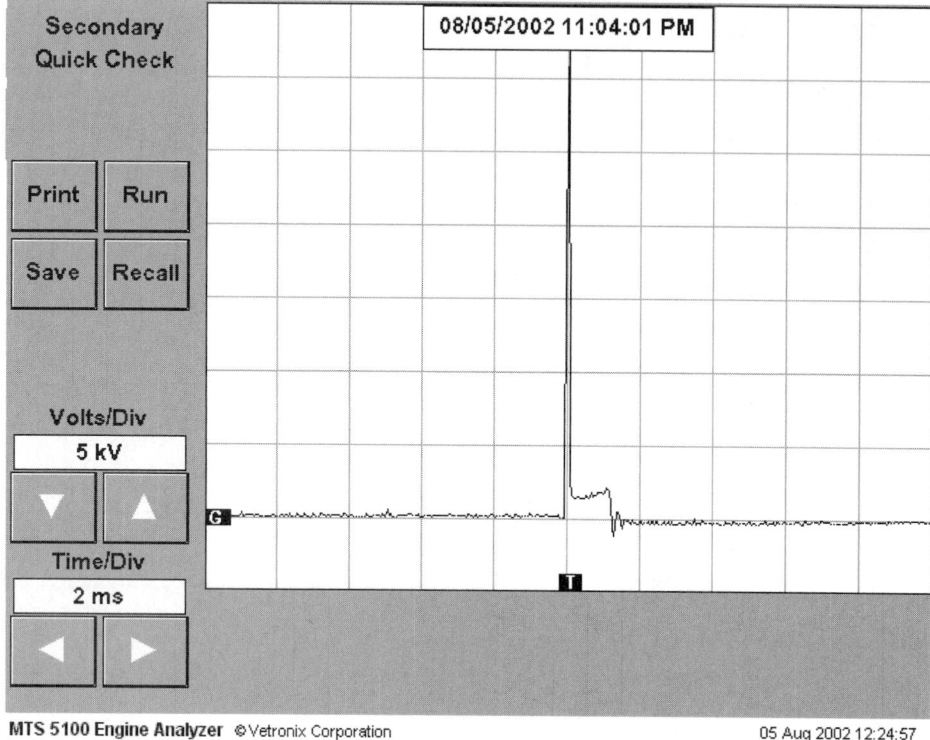

Figure 14. Cranking with a high resistance coil wire held on ground.

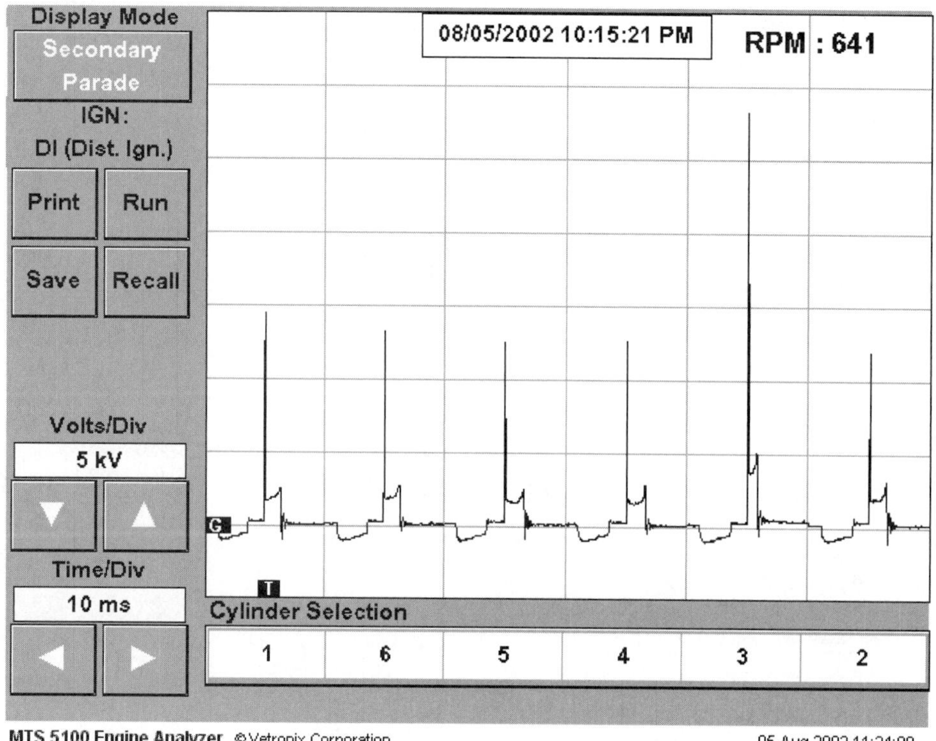

Figure 15. One cylinder showing excessive resistance.

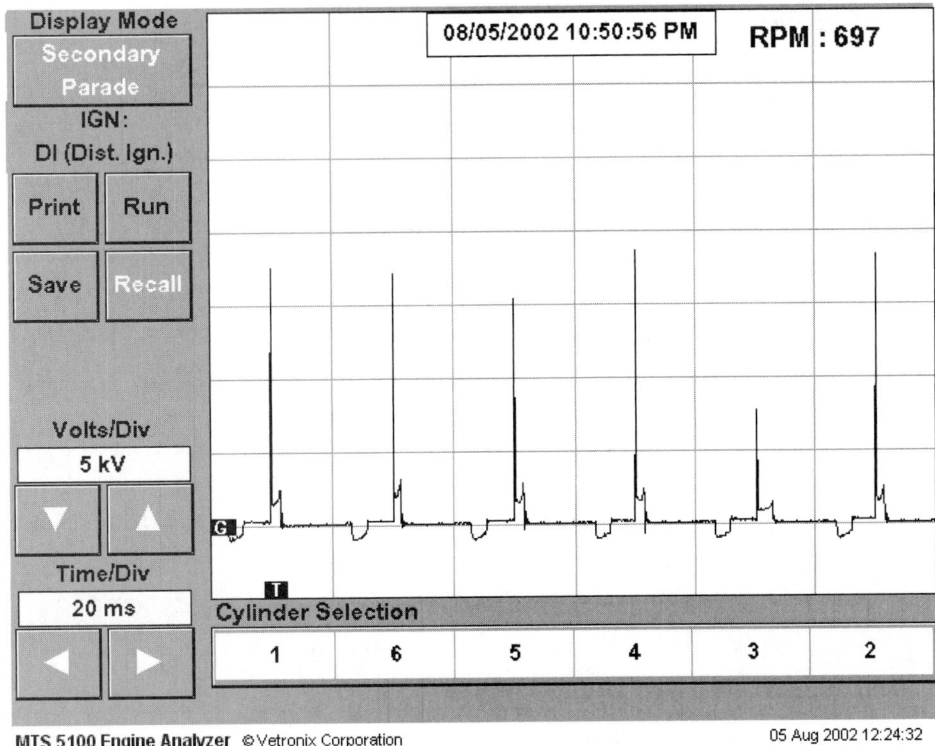

Figure 16. Cylinder 3 with plug wire held to ground.

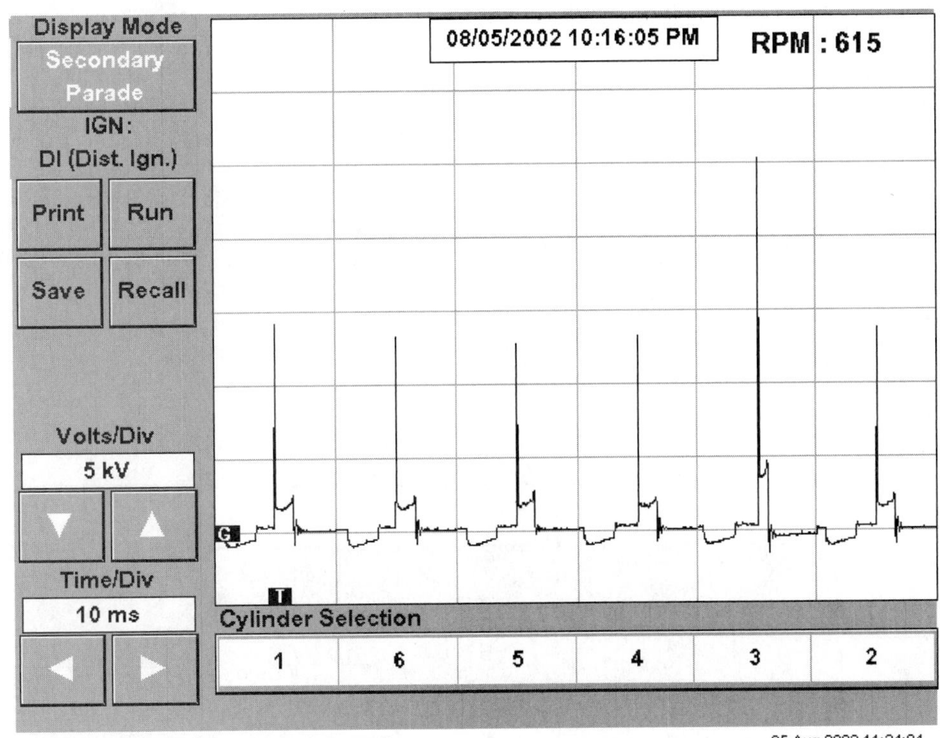

Figure 17. Excessive resistance in a plug wire will allow high voltage with the wire held on ground.

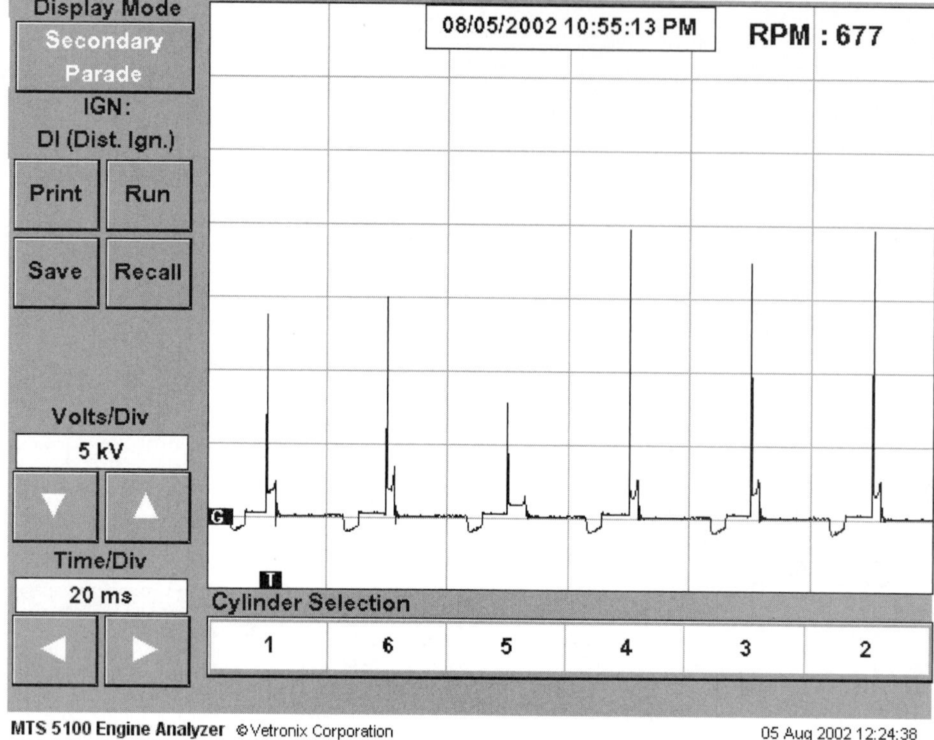

Figure 18. A pattern with the number 5 distributor cap terminal grounded.

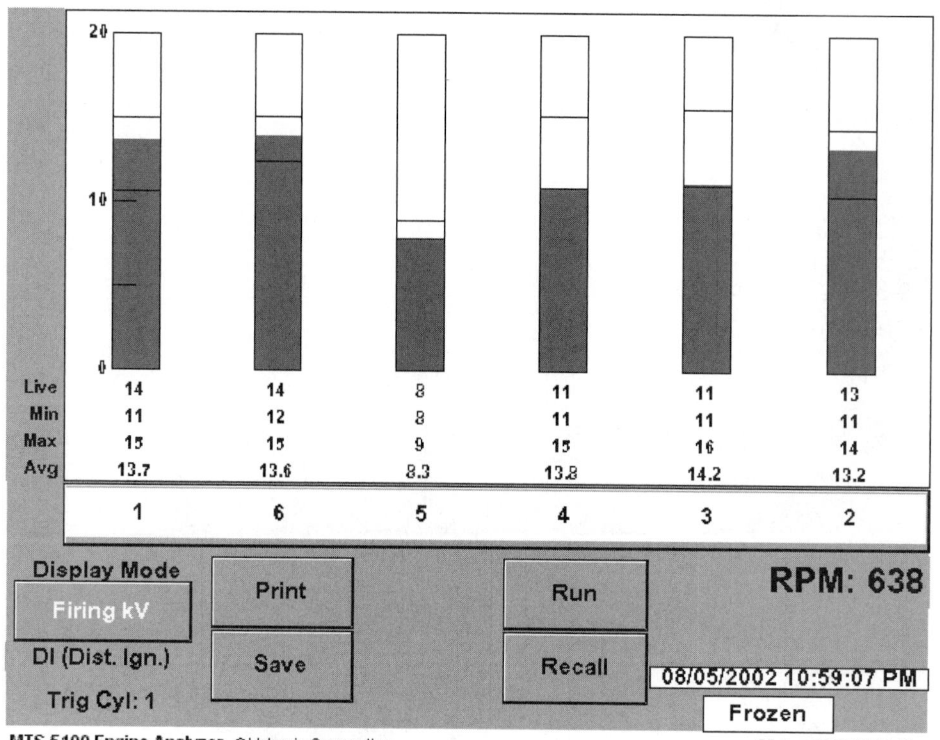

Figure 19. A bar graph of firing voltage information.

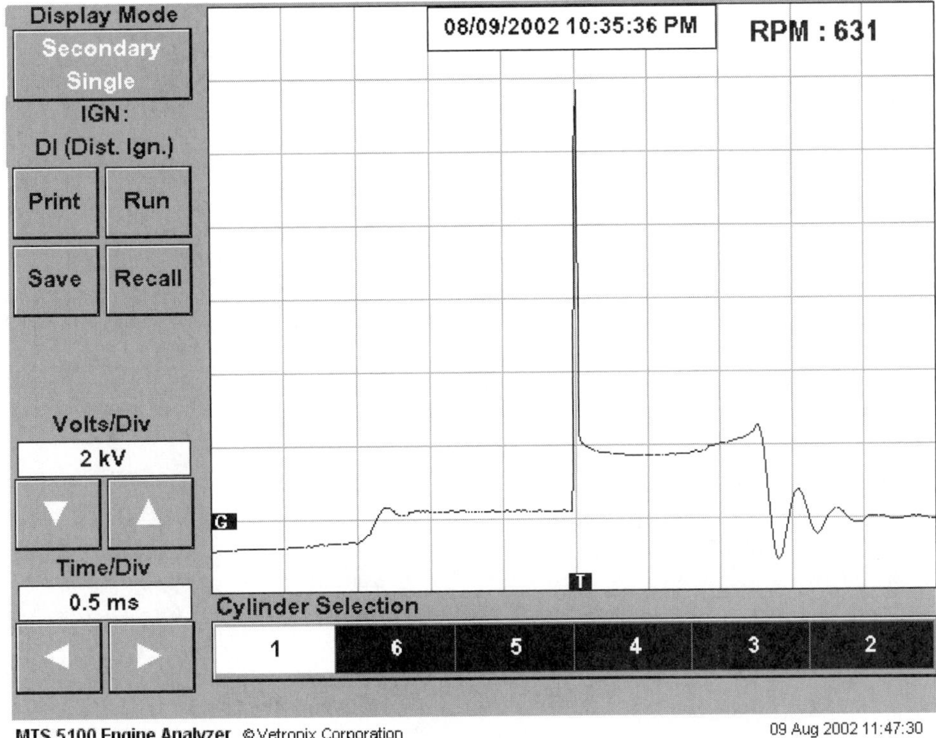

Figure 20. A display of cylinder 1 burn time, 1.25 milliseconds.

MEASURING HOW LONG THE SPARK LASTS

The level of voltage that produced the spark is an important test. Another important test is how long the spark lasts. Most manufacturers specify that the spark should last at least 0.8 millisecond. That is not a very long time, but it has been proven in emission testing that if the spark does not last at least 0.8 millisecond, the cylinder is more likely to misfire and hydrocarbons to increase. We will be able to measure this **burn time** by looking at the spark line on either a primary pattern or a secondary pattern. Immediately after the firing line will be a horizontal line that ends in some oscillations. It is the length of this line that we measure. How long did the spark last in **Figure 20**? Remember that it is the length of the horizontal line after the firing line that we are concerned with. Notice that the lower left of the screen shows the time per division as 0.5 millisecond. You can see that the spark lasted about 1.25 milliseconds. Some vehicle manufacturers also specify that the burn time should not exceed 2.0 milliseconds; however, this number is not nearly as important as the minimum of 0.8 millisecond.

USING THE DSO TO CHECK FOR INSULATION BREAKDOWN

If the firing lines are all correct, the last check of the secondary ignition system we should do is to look at the quality of the insulation. A visual inspection frequently reveals cracks in the cap, burns in the rotor, or hard, brittle wires that need to be replaced. One additional check involves providing a damp path for electricity to follow while you are watching the firing voltage on the DSO. Using a water bottle, spray a mist of water on the secondary components **(Figure 21)**. If a miss develops or if the pattern changes radically, the insulation is breaking down and should be replaced.

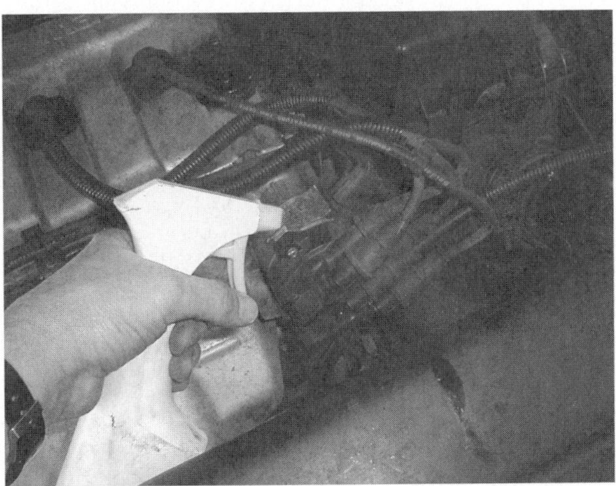

Figure 21. Use a water mist to indicate insulation problems.

Summary

- Spark plugs should not be removed from an aluminum cylinder head that is still hot.
- Spark plugs should be gapped with a wire gauge.
- Ignition wires should be replaced if their resistance is excessive or their insulation has failed.
- The firing order of the engine is determined by the placement of the wires in the ignition cap.
- Plug wires should be replaced one at a time to ensure that the firing order is not changed.
- Excessive secondary resistance will increase the firing voltage.
- Normal firing voltage will be 6–12 kilovolts with no more than a 3-kilovolt variance between the highest and lowest.
- Insulation breakdown can usually be observed on a DSO by misting down the wires with water.

Review Questions

1. Brass feeler gauges are generally used to set
   A. spark plug gap
   B. Hall effect sensor gap
   C. magnetic pickup air gap
   D. none of the above
2. A firing line that is excessively high on only one cylinder can be indicative of a(n)
   A. shorted rotor
   B. open coil wire
   C. lean air:fuel ratio
   D. open plug wire
3. A four-cylinder engine had the following firing voltages: 6, 12, 7, and 7k kilovolts. Technician A states that all plugs are firing within the normal kilovolt range. Technician B states that there is too much variance between the highest and lowest. Who is correct?
   A. Technician A only
   B. Technician B only
   C. Both Technician A and Technician B
   D. Neither Technician A nor Technician B
4. A DSO display will normally show cylinders firing from
   A. left to right in the firing order
   B. right to left in the firing order
   C. left to right in the engine numbering order
   D. right to left in the engine numbering order
5. If the resistance of a secondary ignition circuit increases, the firing line will
   A. decrease
   B. remain the same
   C. change position
   D. increase
6. Excessive secondary resistance will increase the firing voltage.
   A. True
   B. False
7. Reduced secondary resistance will decrease the firing voltage.
   A. True
   B. False
8. A wire feeler gauge should be used to set plug air gap.
   A. True
   B. False
9. Dirt around a spark plug will change the firing voltage.
   A. True
   B. False
10. Plug wires are designed to last the life of the vehicle.
    A. True
    B. False
11. Burn times should last in excess of 0.8 millisecond.
    A. True
    B. False
12. How is the test for excessive secondary resistance done using a DSO?
13. What would cause a single firing line to be excessively high?
14. What would cause all of the firing lines to be excessively high?

Chapter 48 Primary Ignition Systems

Introduction

In Chapter 46 and Chapter 47 we took a look at the secondary ignition system and how to diagnose common problems. It is important to note that the secondary ignition functions only if the primary system functions. It is within the primary system that the spark is developed and controlled. In addition, the timing of the spark, or when we deliver the spark to the secondary system, is controlled. It is not within the parameters of this book to look at breaker point ignition systems. They have not been on vehicles since the early 1970s. We begin our discussion with electronic ignition with a distributor. We also look at the input devices that the ignition module will need to function and look at the computer control that is so common today. The basic purpose of the system is to supply a long-lasting, high-voltage spark at the correct time to the compressed charge of fuel and air. It is important that you realize that these two purposes (long-lasting, high-voltage spark and correct time) are equally important. The spark must have a voltage that is high enough and lasts long enough to ignite the mixture with the least likelihood of a misfire. This ignition must occur at the correct time relative to the piston position or the amount of power might be reduced. In addition, the level of pollutants leaving the tailpipe is usually increased if the ignition timing (when the spark is delivered) is not correct.

IGNITION COILS

The ignition coil is the first component we discuss. It is found on all spark ignition systems. There will be at least one per vehicle and frequently as many as one per cylinder. Their function and operating principles are essentially the same on all systems, however. The ignition coil is necessary because the 12.6 volts of the battery will not be sufficient to jump the air gap of the spark plugs. The ignition coil will transform the low voltage, high amperage from the battery into high voltage, low amperage. Just how high the high voltage will be depends on many factors, such as plug gap, air:fuel ratio, plug wire resistance, timing, cylinder pressure, and so on. In addition, the type of ignition system in use will have a large bearing on the level of high voltage that is available. Keep in mind that we need the high voltage because the extreme resistance of the air gap will require quite a bit of pressure to push electrons across. The air gap, the mixture, and everything from the ignition coil to the plug ground will be viewed by the coil as resistance. As this resistance increases, the voltage will have to increase if a spark is to take place. You will remember that Ohm's Law stated that resistance will reduce voltage, and yet I have just said that additional resistance after the coil output will increase voltage. These two statements are not as opposite as you might think, because the ignition coil output in watts will remain essentially the same. The resistance will have the effect of driving up the voltage, while the amperage or primary on-time is reduced. This transformer effect will take place within the ignition coil. Let us build a simple transformer or coil and analyze how it will produce the high voltage required.

We will start by wrapping some heavy wire around a soft iron core **(Figure 1)**. This will be called the **primary winding** and it will have the job of giving the coil a strong magnetic field. You should remember that soft iron is used because it is a good conductor of magnetic lines of force. When we put B+ to one end of the primary winding and ground to the other end, we will have a strong electromagnet. The strength of the field will be dependent on two

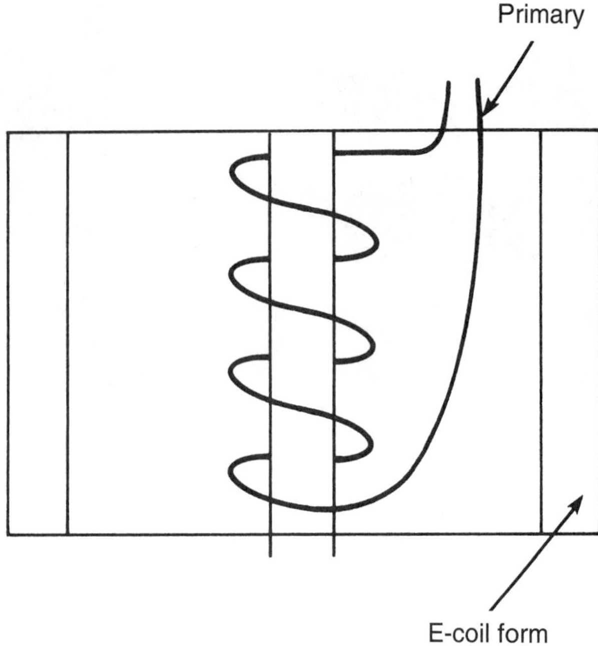

Figure 1. The primary has hundreds of turns of heavy wire.

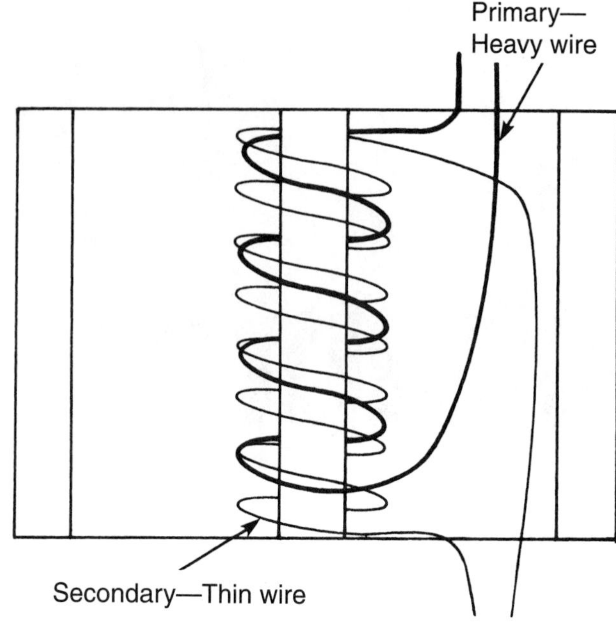

Figure 2. The secondary has thousands of turns of thin wire.

things: the amount and size of the conductor. The longer and thicker the primary wire is, the stronger the field will be. The overall resistance of the primary winding and the module or PCM will determine the amount of current that will flow through it. Keep this in mind later as we look at the different types of systems in use today. One of their major differences is in the amount of current they will draw. This is called **primary current**.

Now let us take some additional wire and wrap it around this same iron core, as shown in **Figure 2**. Notice that we have used many more turns of wire than we used in the primary winding. Also, this winding will ordinarily use smaller diameter wire than the primary. This winding is called the **secondary winding**. One end of the secondary winding will be connected to the spark plugs through the cap, rotor, and plug wires. The other end of the secondary will be connected to either ground or the primary winding. Typically, the ratio of primary to secondary windings is in the order of a few hundred to several thousand. The primary winding typically has approximately 1 ohm of resistance and, remember, is composed of a few hundred turns of relatively thick wire. The secondary winding usually has in excess of 5,000 ohms of resistance and is composed of thousands of turns of thin wire.

The action between the primary and secondary windings will result in the high-voltage spark we will need at the plugs. Let us follow the action through a typical ignition coil and see how we can increase our battery voltage. In sequence, the action looks like this:

1. A primary control device (ignition module) turns on primary current.
2. Current flowing builds the magnetic field around the soft iron core.
3. The coil's magnetic field becomes fully developed (saturated).
4. The primary control device opens the primary circuit.
5. The magnetic field collapses.
6. The moving magnetic lines of force cut across the primary and the secondary windings.
7. The moving magnetic lines of force induce high voltage into the secondary.
8. This high voltage pushes current to the plugs.

This sequence will become second nature to you, especially when you begin to use the DSO on primary circuits. Taking each one in sequence and analyzing it will increase your understanding of the coil's operation. A primary control device turns on primary current. This first action is important because we want the magnetic field to build. The electricity flowing through the primary winding will do just this. Our primary winding will become an electromagnet. The use of the term "primary control device" might be new to you, but simply we will want to turn the current on and off with a switch called an ignition module. We will look at different styles of primary control devices soon, but now let us get back to our sequence. As current flows through the primary winding, the coil's magnetic field builds. This is called **saturation time**. Once the magnetic field is fully saturated, as shown in **Figure 3**, the ignition module will open the circuit. This interruption of the primary current eliminates the electromagnetic field, and the lines of force go back to their point of origination—the soft iron core. To get to the core, the lines of force will have to cut across both the primary and secondary windings. These moving lines of force cutting across conductors should

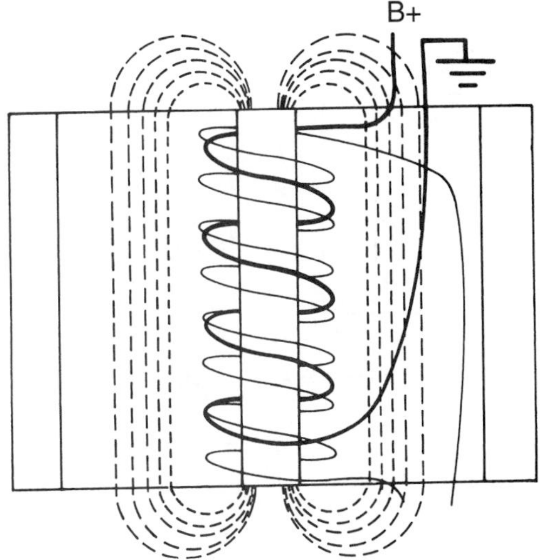

Figure 3. Current flowing through the primary causes a strong magnetic field to develop.

sound familiar to you because they are the same conditions that generated electricity in our study of the charging system. Here is where the additional turns of wire in the secondary will come into play. The amount of induced voltage in the primary and the secondary is relative to the number of turns of wire in each. Generally, a couple of hundred volts will be induced in the primary **(Figure 4)**. However, in the secondary, the thousands of turns of wire are affected to a greater degree by the moving lines of force. Here is where the transformation takes place. The motion between the fast moving lines of force and the thousands of turns of wire can boost the voltage up to as much as 50,000 volts or more. This high-voltage potential will pull the needed current from either the primary circuit or from ground, depending on where the secondary circuit is connected.

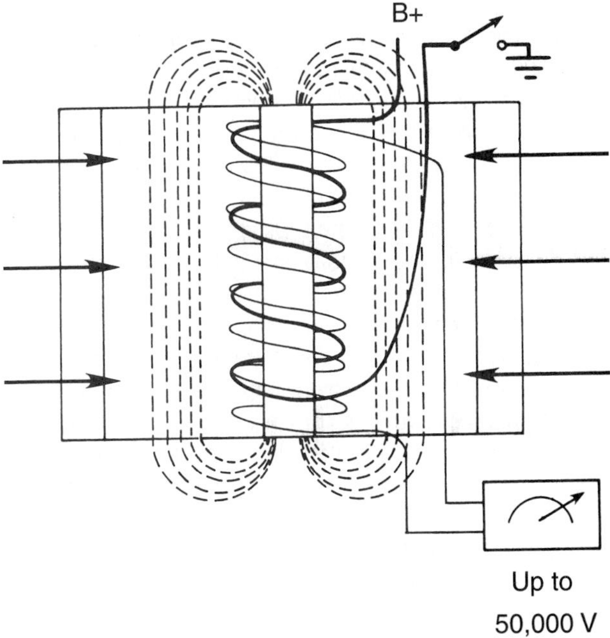

Figure 4. When the primary is opened, the collapsing magnetic field induces high voltage in the secondary.

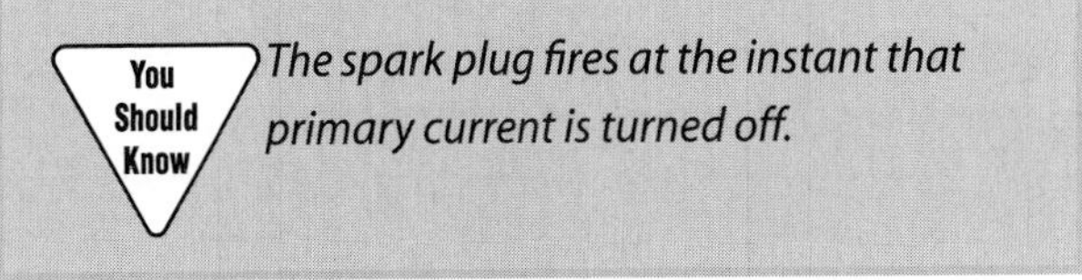

The spark plug fires at the instant that primary current is turned off.

A couple of points should be emphasized here. Saturation time, or the time that the magnetic field takes to completely build, is dependent on the amount of current flowing through the primary circuit. Demands on the ignition system over the years have seen the amount of primary current steadily increase from as low as 2 amps years ago to as high as 10 amps on some of the modern electronic systems currently in use. In addition, the term **dwell** should become part of your vocabulary. Dwell is the amount of time, in camshaft degrees, that primary current flows through the ignition coil. On some systems, it is variable. It is the time in degrees between primary current on and primary current off. It can be a fixed value such as 30 degrees or it can vary with speed, for example, 15 degrees at idle and 35 degrees at 4,500 rpm. We will cover this more later. Remember also that this sequence will repeat over and over for each spark plug.

PRIMARY RESISTANCE

The use of some type of primary resistance is found in many ignition systems in use today. It is important to note that this primary resistance can be external, as shown in **Figure 5**, or the most common style, a variable resistance electronically controlled within the ignition module. Let us look at the why of primary resistance before we look at the how.

The amount of primary current that flows through the ignition coil will determine how long the spark lasts and at what voltage potential. Many systems, especially those that are older, were designed around getting the vehicle started. How much spark was needed was calculated, and the primary circuit was designed around it. During cranking, the coil input voltage is naturally lower (10 volts) because of the starter motor's draw. When the vehicle is running, the charging system boosts the coil input voltage up to approximately 14 volts. Resistance added to the primary will drop the charging voltage down to about the level of starting (10 volts). Older vehicles did this with external resistance added in series. The resistance was in use during the running of the engine and bypassed during cranking. The use of external resistance has been reduced greatly. Few manufacturers use

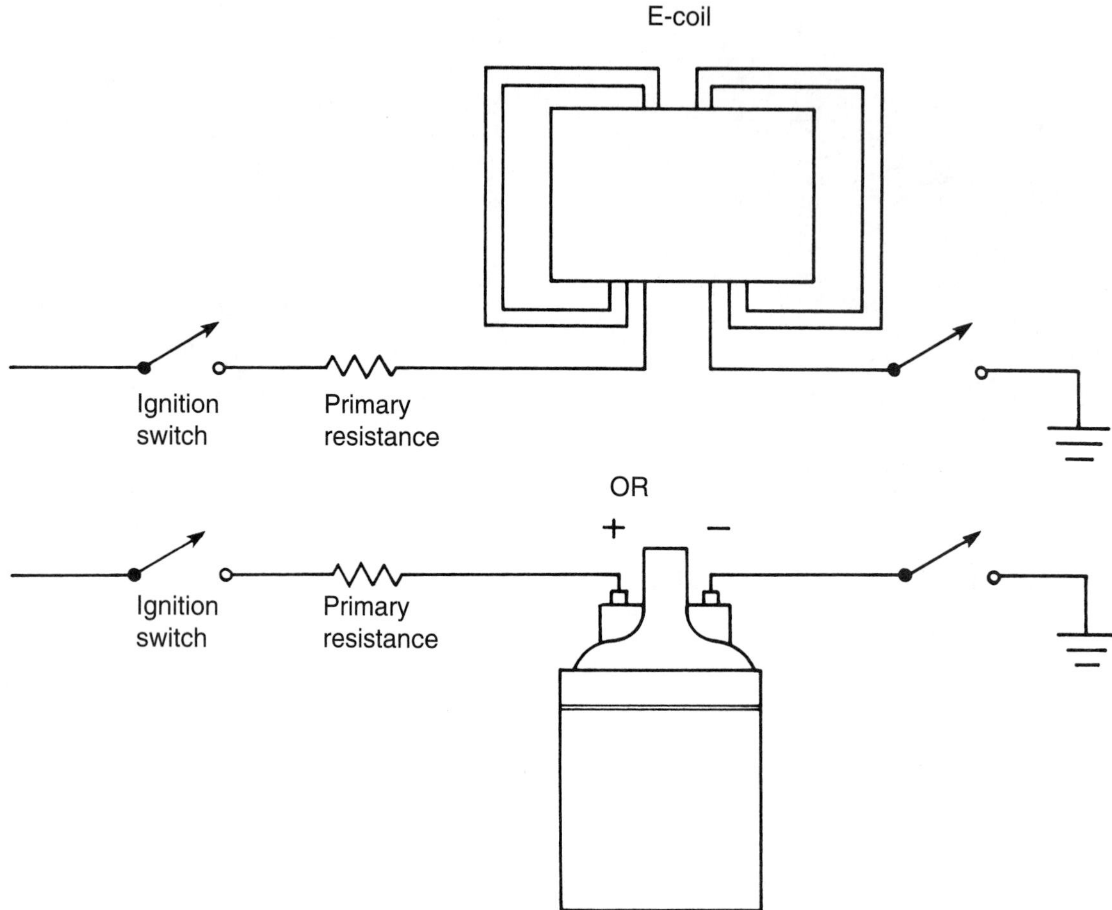

Figure 5. Older vehicles used external primary resistance.

it on their models today. Instead, a current-limiting type of electronically controlled resistance is used. With modern electronics, coil saturation is monitored. Once the coil is fully saturated, the electronic ignition module cuts back on the current with a process called **current limit**. This process greatly increases the life of the coil and yet keeps the available voltage high in case it is needed. In Chapter 49, we will see how this circuit is used and tested.

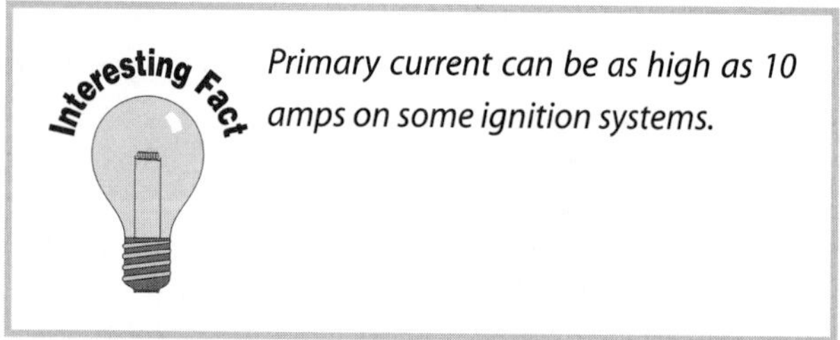

Primary current can be as high as 10 amps on some ignition systems.

PRIMARY CONTROL DEVICES

Up to this point, we have discussed the components that are common to most ignition systems. Ignition coils and secondary circuits are, for the most part, the same from manufacturer to manufacturer, including the import market. The difference in ignition systems is realized when one begins to look at the different types of ignition modules and the different methods of inputting the information that the module needs to function.

A block diagram of the typical transistor ignition circuit **(Figure 6)** shows that the primary current flows through the emitter/collector, while some type of a piston position sensor feeds into a control circuit that will turn the base of the switching transistor on and off. In principle, most systems operate in this manner. The only distinguishing feature is the piston position sensor. In common use are three different types: the magnetic AC generator, the Hall effect sensor, or the photodiode sensor. We also find systems on the road that have more than one of the sensors. They are usually limited to computer systems, which we will discuss later. For the most part, the majority of the electronic ignition systems with a distributor on the road use one sensor to indicate piston position. Let us look at each of the sensors separately and analyze what each signal looks like to the electronic

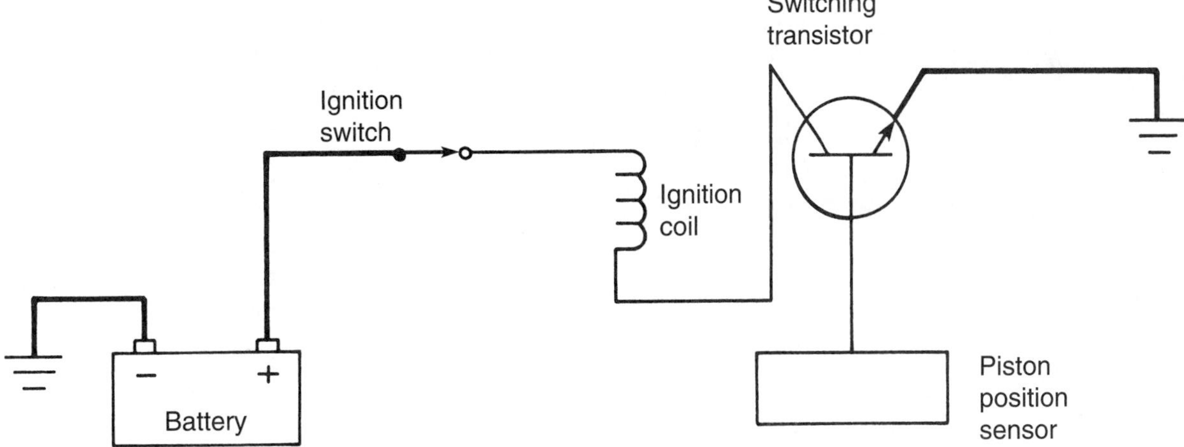

Figure 6. A typical electronic ignition (simplified).

ignition module. Keep in mind that the object of any of the sensors is to tell the ignition module the position of the pistons. This information will either directly turn the primary switching transistor on and off or feed its position information to the engine computer. You will see in this chapter how computer-controlled ignition systems use this information to determine just when to fire the plugs.

Let us take a look at a simple transistor circuit and walk through the process using information from the piston position sensor. Remember that on any system when the coil's magnetic field collapses, the plug fires. This is called timing. In our simple electronic ignition system **(Figure 7)**, the information from the piston position sensor is used to apply a ground to the base of the switching transistor. This ground turns on the emitter/collector circuit, and the primary current flows through the ignition coil. After the coil is saturated, the circuit removes the ground from the base of the transistor. Primary current stops flowing through the emitter/collector and the coil primary winding. The field collapses and induces the secondary winding. Timing occurs at the same time. The main advantage to this style of electronic ignition is the ability to have primary current greatly increased over that which can normally flow through a nonelectronic switch.

In recent years some of the manufacturers turned over the ignition module function to the vehicle PCM. The function and operation are basically the same as those of the external module and still require an input from a piston position sensor.

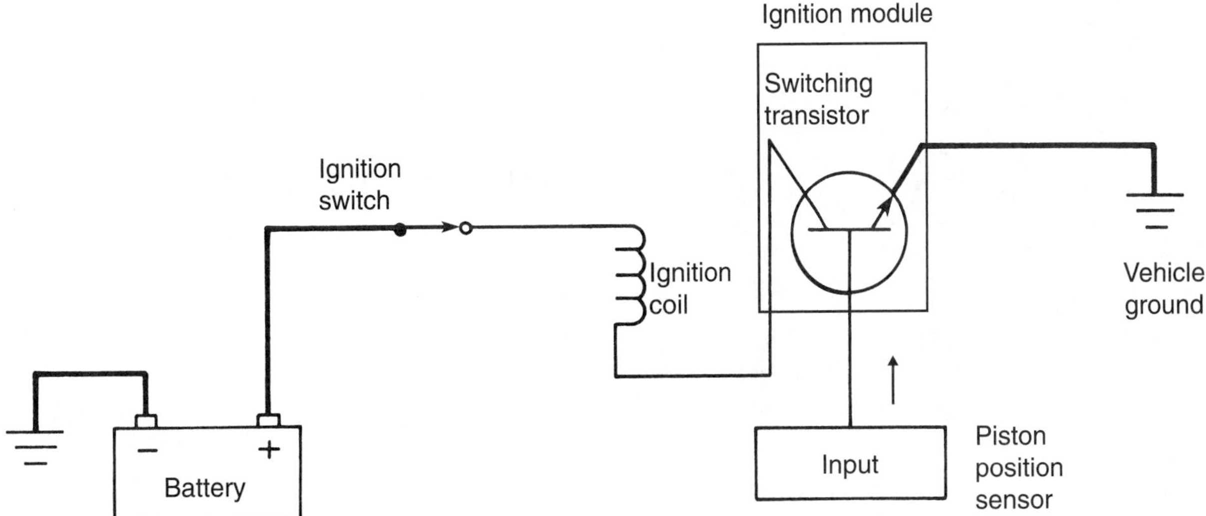

Figure 7. Primary current flow.

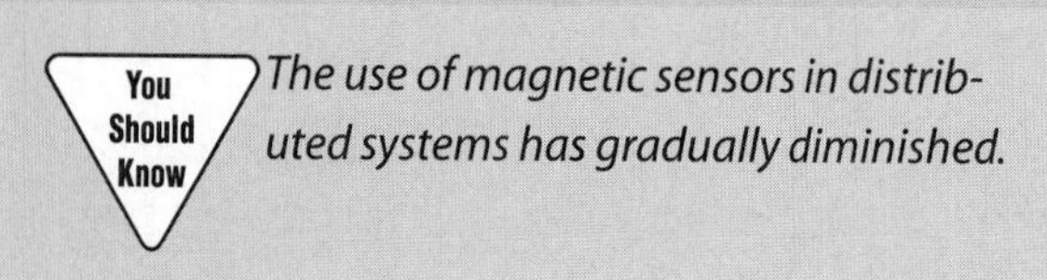

The use of magnetic sensors in distributed systems has gradually diminished.

INPUT SENSORS (PISTON POSITION SENSORS)

The magnetic AC generator sensor is a popular type of piston position sensor in both foreign and United States vehicles. In recent years, its use has diminished, but many of them are still on the road. The principle behind this style is similar to that of an alternator. Within our vehicle's alternator, a moving magnetic field generates an AC. In our typical ignition sensor, the generation of AC will usually be done inside the distributor. Because the distributor is driven off the engine, piston positions can be sensed and the timing determined. Look at **Figure 8**. It shows how the moving distributor will concentrate the magnetic line of force until the armature lines up with a stator tooth. As the armature passes the stator, the lines of force will unconcentrate until the next armature tooth approaches. Approaching armature teeth concentrate the lines of force, whereas passing armature teeth unconcentrate the lines of force. What we have here are moving lines of force, just like our alternator has. The assembly also consists of a coil of wire wound around the end of the stator tooth. The magnetic lines of force are moving across this coil of wire. Moving magnetic lines of force cutting across a coil of wire will generate a voltage, and it is this voltage that will be sent to the ignition module. The signal will be a type of sine wave with both a voltage change and a change in polarity evident when the armature is directly in line with the stator tooth, as shown in **Figure 9**. Most electronic ignition systems fire the plugs just as this polarity changes from positive to negative. The AC signal is amplified by the module and then applied to the base of the transistor switching circuit to turn primary

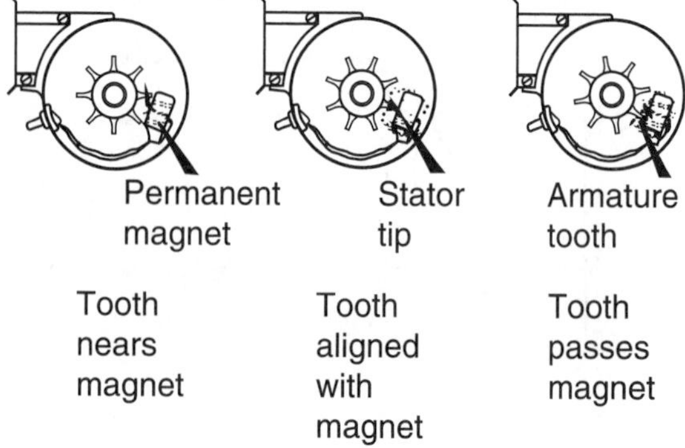

Figure 8. As the distributor turns, an AC signal will be generated.

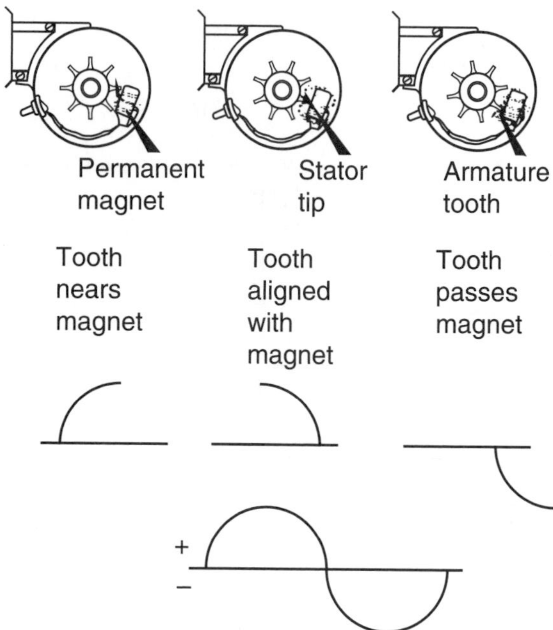

Figure 9. Most systems fire the plug when the voltage polarity shifts from positive to negative.

current and time the spark on and off. The strength of the AC signal is also dependent on the speed of the distributor. During cranking, the signal may be as low as 250 millivolts. With a signal this low, connections must be very good. The slightest resistance might reduce the signal lower than the trigger point of the module.

It is important to note that the operation of the stator and armature is found in virtually every manufacturer's design. The terminology "stator" and "armature" might not appear, however. For instance, General Motors calls the components that generate the AC signal the pickup coil and pole piece, whereas Chrysler calls them the reluctor and pickup coil. Their operation is essentially the same, as you will see in the diagnosis section of Chapter 49. Magnetic generators usually have two wires leading from the unit to the ignition module.

HALL EFFECT SENSORS

Another type of crankshaft or piston position sensor in use is the Hall effect sensor. It is different from those sensors previously studied because it does not generate an AC signal from a moving magnetic field. Instead, its signal is a rise in voltage followed by a drop in voltage. This signal that changes up, then down, then up, and so on, is generated by a sensor such as that shown in **Figure 10**. The shutters connected to the distributor shaft are used as shields between the two sides of the sensor. One side of the sensor has a permanent magnet. When the magnetic lines of force are

Figure 10. A distributor with a Hall effect sensor, shutters, and windows.

Figure 11. An optical sensor for a four-cylinder engine.

allowed to reach the other side of the sensor, a relatively low voltage is passed out (near zero volts on some models). As the shutter passes into the magnetic field, it blocks the sensor, and the voltage rises sharply coming out of the sensor element. If the element is not exposed to a magnetic field, it produces a voltage; if it is exposed to a magnetic field, it does not produce a voltage. The shutters, in this example, are used to block the magnetic field from the sensor. Hall effect sensors are used on many vehicles. They are different from magnetic AC generators in two ways, generally. First, they require an input voltage source, and second, their output is not dependent on engine speed. At any speed, even stopped, a Hall effect sensor will generate the high-voltage (input voltage) and then low-voltage (zero) change discussed. If an input voltage is applied to a Hall effect sensor, the output signal will depend on the shutter position, not on its movement as in the AC generator style.

OPTICAL CRANKSHAFT POSITION SENSORS

The final type of crankshaft position sensor that we look at is the optical style. It has had limited use but remains an option for the manufacturer. It consists of a light-emitting diode that will produce a small, sharp beam of light when B+ is applied to it. The beam of light is focused through holes onto a photoelectric cell that generates a voltage when light hits it. Let us follow the sequence of events using the distributor shown in **Figure 11**. The light-emitting diode is on top of the rotor plate, and the photo diode cell is below the rotor plate. The small slits in the rotor plate are placed at 1-degree spacing—360 slits around the edge of the plate. As the plate rotates with the engine's distributor, the photo diode cell will alternately see light and then not see light. The cell will generate a small voltage and then no voltage. This on-off pulsing voltage is used to control a 5-volt signal that is sent to the computer to determine both engine speed and piston position. Notice that **Figure 12** shows the signal that will be sent off the sensor to the computer. Also notice that our example has two photo diode cells, one for the 360-degree signal and one for the 60-degree signal. The 60-degree signal corresponds to the number of cylinders and camshaft degrees. The 360-degree signal is also in camshaft degrees. This unit would be for a six-cylinder engine with a cylinder firing every 120 degrees of crankshaft rotation.

The SAE has defined through its J-1930 that crankshaft position sensors are called CKP and camshaft position sensors are called CMP.

In addition, this type of sensor is sometimes used at the speedometer to determine vehicle speed for the computer, as shown in **Figure 13**. The output from a **light-emitting diode (LED)**-style sensor is a pulse voltage similar to that of a Hall effect sensor. Its level is also not dependent on speed. The number of pulses will change with speed but not the output level.

IGNITION MODULE FUNCTIONS

All electronic ignition modules have a variety of functions that they perform. Currently, modules have two, three, or four functions. All modules have at least two functions, that of turning the primary current for the coil on and then off. It is this turning on and then off that produces the spark.

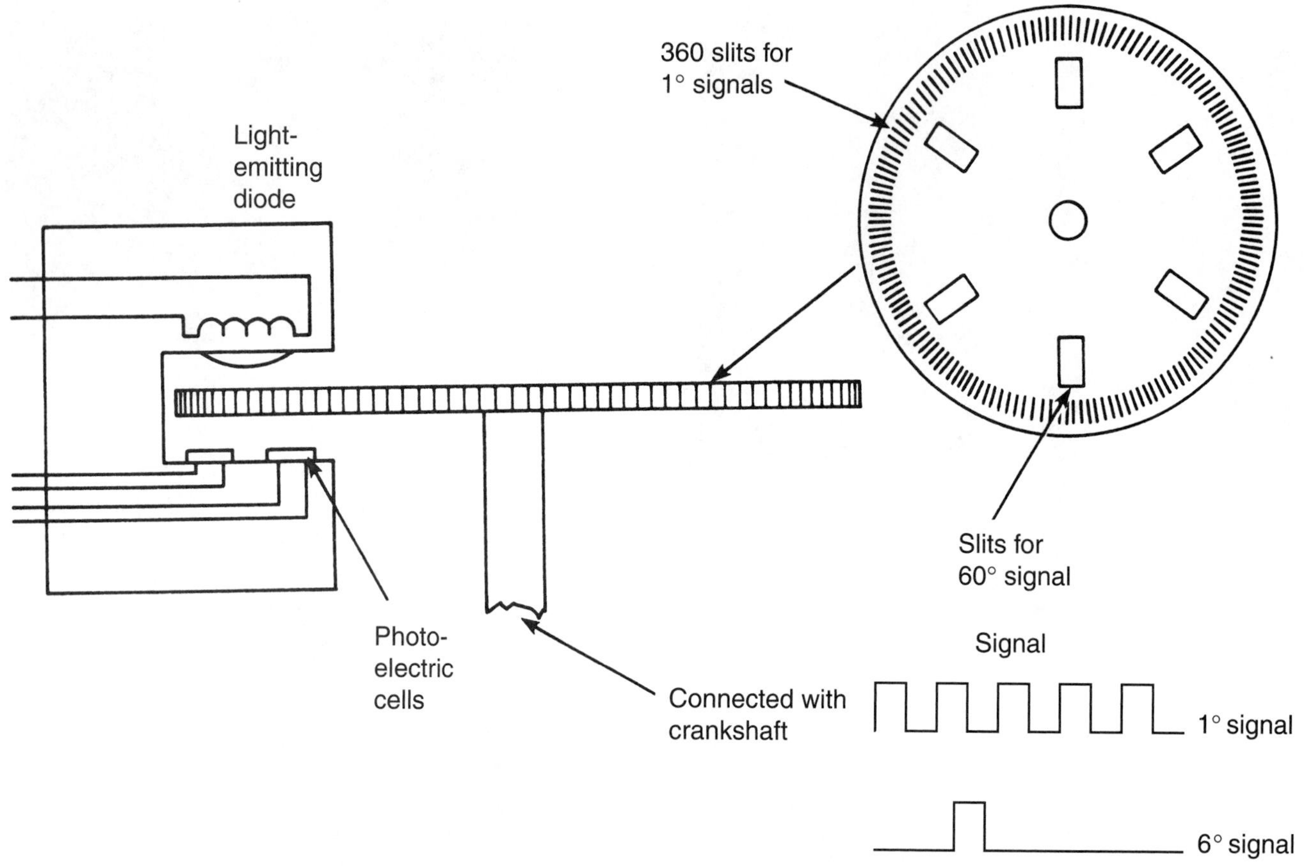

Figure 12. An optical sensor generating two outputs.

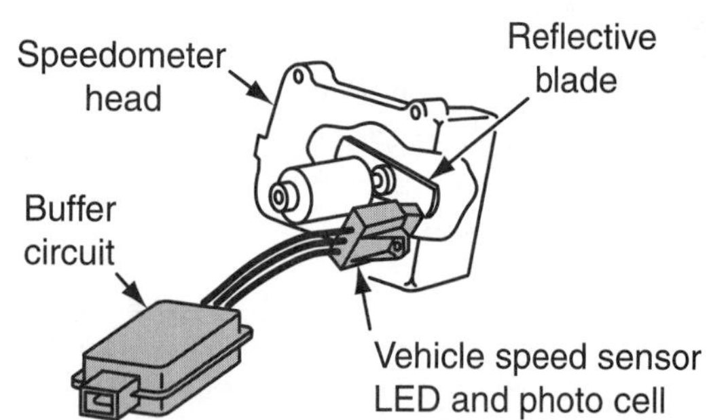

Figure 13. A vehicle speed sensor that uses an optical signal to indicate mph.

If, for some reason, the module cannot do the basic two functions the vehicle will not run. Whether or not the module has any of the other available functions depends on the manufacturer. All functions are readily seen on an ignition analyzer or a DSO.

The first optional function is that of current limit. The current limit function is found on ignition systems that do not have a primary resistor or some other method to control primary current. Chapter 49 shows what the pattern will look like with current limit, but let us analyze its function. When primary current is turned on, the magnetic field builds until the coil is fully saturated. With the current-limit function, once full saturation has taken place, the module will cut back on primary current. This function increases the life of the coil by keeping it cooler. You can recognize a system that has current limit by looking on the wiring diagram. If no primary resistor is seen, usually the system has the current-limit function or uses a ramp and fire system, in which the current is turned on for a specific period of time that is usually less than that required for full saturation.

Another function of many modern systems is variable dwell. On modules with this function, the dwell will be short at low engine speeds and longer as the engine speeds up. Dwell, you will remember, is the length of time in degrees of camshaft that primary current flows through the coil primary. A long enough dwell will result in sufficient coil saturation to attain full secondary output. Keeping the current turned on longer than is necessary accomplishes little at the plug end but can greatly decrease the life of the coil. As engine speed is increased, coil saturation time is automatically decreased if the dwell remains the same. For example, a dwell of 15 degrees allows for a certain amount of coil saturation at a speed of 1,000 rpm. At 2,000 rpm, if the dwell were to remain the same, the amount of available coil saturation time would have decreased by 50 percent. This level of saturation might not be sufficient for full coil output. If, however, the dwell were to increase to 30 degrees at 2,000 rpm, the amount of clock time would remain the same. For this reason, many of the manufacturers build their modules to have a variable dwell function.

To summarize, all ignition modules will have at a minimum two functions: primary current on and primary current off. In addition, they might have the current-limit function to replace the primary resistor and/or they might contain the variable dwell function. These function options, like the crankshaft position sensor's options, are found on a variety of different applications. They are easy to identify with an oscilloscope. General Motors and Ford have typically used all four functions in their modules for vehicles with a distributor. Once we switch over to distributorless, the situation changes.

ADVANCING SYSTEMS

Up to this point, we have mentioned timing only briefly and defined it as the moment that we delivered the spark to the cylinder. It is now time to discuss just how this is accomplished. We begin with a discussion of a mechanical system that has been around for a long time and is still found on some vehicles and end with the computer-controlled timing systems found on virtually all current production vehicles. Before we get into the specifics, let us discuss why we need any type of a timing system at all. If you remember, your four-stroke theory—intake, compression, power, and exhaust—assumed that between compression and power was a spark that would ignite the mixture and produce the power. In theory, the peak combustion power of the burning mixture was to be delivered to the piston just as it passed TDC at the beginning of the power stroke. This still remains the goal of most modern timing systems: to deliver peak combustive power to the piston just after TDC on the power stroke. The problem, however, is that the mixture will not develop peak combustive power at the same time as the spark occurs **(Figure 14)**. The mixture will have to be ignited and burn for a period of time before sufficient fuel will be delivering power. This time between ignition and peak combustive power is the reason for the sometimes complicated components in the mechanical advance mechanisms that are considered to be part of the ignition system. Let us define some common terms that will be part of our discussion. You already know that TDC is top dead center, the point where the piston is as high in the cylinder as it will go and stops to reverse its direction. During the four strokes, there are two TDCs: one between compression and power, and one between exhaust and intake. Anything that

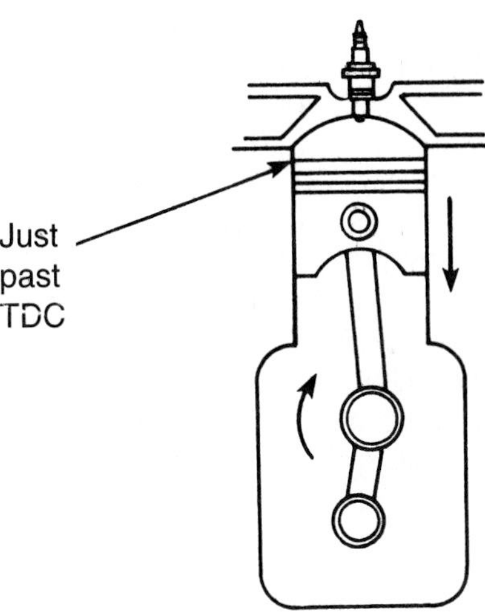

Figure 14. Most efficient power needs to be developed just past TDC.

occurs before TDC (for instance, during the compression stroke) is said to occur **before top dead center (BTDC)** and is usually expressed in crankshaft degrees **(Figure 15)**. For example, a spark that is delivered to the cylinder 10 degrees BTDC refers to the position of the piston during the last part of the compression stroke. Because each stroke is equal to 180 degrees of crankshaft rotation, you can tell that the compression stroke is just about over. The piston is on its way up and just about to begin the power stroke. Anything that occurs **after top dead center** is referred to as **ATDC** and again is usually expressed in degrees of crankshaft. Getting these two abbreviations mixed up while working on a vehicle can cause some terrible drivability complaints. Another term to define is **advance**. To advance the spark is to deliver it sooner. A change in timing from 10 degrees BTDC to 20

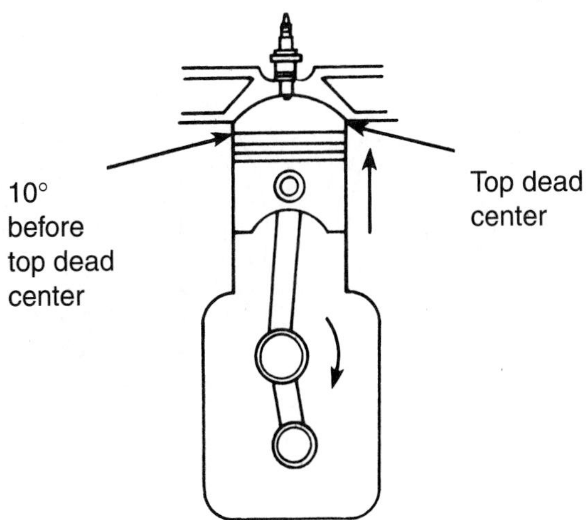

Figure 15. Ignition must occur before top dead center.

degrees BTDC is an advance of 10 degrees. The opposite of advance is **retard**, and it means that the spark is being delivered later. A change in timing from 20 degrees BTDC to 10 degrees BTDC is a retard of 10 degrees. Notice that the spark is still being delivered BTDC. As we look at different advancing and retarding systems, keep in mind the basic idea of timing. The spark will generally arrive before TDC so that peak combustive pressure will push the piston down on the power stroke. Also, keep in mind that anything that is done to change the amount of time it will take between ignition and peak combustive power will have to be compensated for in the advancing systems of the vehicle. The vehicle's ignition system will be constantly compensating for changes in the time between ignition and the power stroke.

The conditions that will need to be compensated for are generally engine speed and engine load. These two have the greatest effect on the burning time (time between ignition and peak combustive power). In addition, many modern systems compensate for altitude and engine temperature and also adjust the timing based on the air:fuel ratio. We will start with the mechanical systems, compensating for speed and load only, and move into the computerized systems.

MECHANICAL ADVANCE

The earliest form of speed-compensated advancing system was the mechanical advance. It was designed during the early years of the automobile and, although it is not currently being produced, is still found on many vehicles. It operates on the simple principle of centrifugal force. Centrifugal force is directly related to speed and so can be calibrated to give an increased advance with speed. **Figure 16** shows a mechanical advance unit in conjunction with an AC generator–style position sensor. As the distributor speed increases, the centrifugal weights fly out and put the stator ahead of the shaft. The number of degrees of advance is equal to how far the stator is ahead of the distributor shaft. The faster the distributor turns, the farther ahead the stator gets and the more advanced the spark is. As the engine slows down, the amount of centrifugal force decreases and the timing retards because the advance springs pull the weights back.

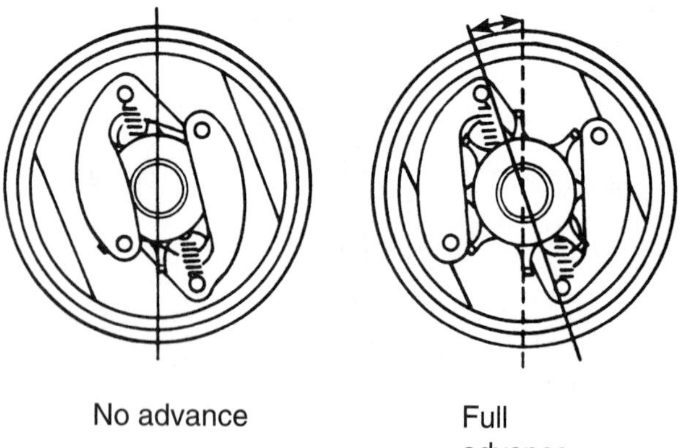

Figure 16. Older mechanical systems used weights and springs to advance the spark for increasing speed.

Typically, the mechanical advance system is relatively smooth operating, giving an equal advance with speed. The graph in **Figure 17** shows that at 500 distributor rpm the distributor should give an additional 2 degrees of distributor advance. Both of these numbers need to be converted to crankshaft degrees by multiplying them by 2. The distributor is traveling at half engine speed, so 1 degree of distributor is equal to 2 degrees of crankshaft. Based on this information, the timing will change (advance) by 4 degrees at 1,000 engine rpm. What will happen at 2,000 engine rpm? The chart shows that at 1,000 distributor rpm an additional 6 degrees of advance will take place. This will convert to 12 degrees crankshaft at 2,000 rpm. Notice that we have said additional degrees of advance, meaning additional degrees over something called **base timing**. Base timing is the minimum timing required by the engine at the lowest rpm that it will run at—idle. We cover more of this in Chapter 49 when we look at how timing and advance systems are checked.

VACUUM ADVANCE

Many people think of the timing change due to piston speed as the same thing one does when hunting. If a hunter wishes to shoot a duck that is flying, the hunter must lead the duck so that the shot reaches the spot where the bird is. The hunter aims ahead of the flying bird. The piston moving through the cylinder is using fundamentally the same idea. The spark must lead the piston, compensating for its speed. Double the speed, and you need to advance the timing twice as far. The principle is the same as our flying duck who suddenly starts flying faster. To have duck dinner, we will have to lead the bird even more to compensate for the increased speed. The increased advance supplied by the centrifugal weights will compensate for increased piston speed and will do it very well if the same volume of air and fuel is constantly in the cylinder. The time between ignition and peak combustive force is dependent on the amount of fuel in the cylinder. The greater the amount of gas and air, the faster the burning will be. The greatest volume of combustible mixture will be in the cylinder if the driver of the vehicle has the accelerator fully depressed. This opens up the throttle and allows the engine to breathe in the most air and fuel. For this reason, the mechanical advance is usually calibrated for the timing necessary with the throttle wide open. Anything less than WOT will reduce the volume of fuel and air and slow down the burning. Remember the goal of the advancing systems: to deliver peak combustive pressure to the piston

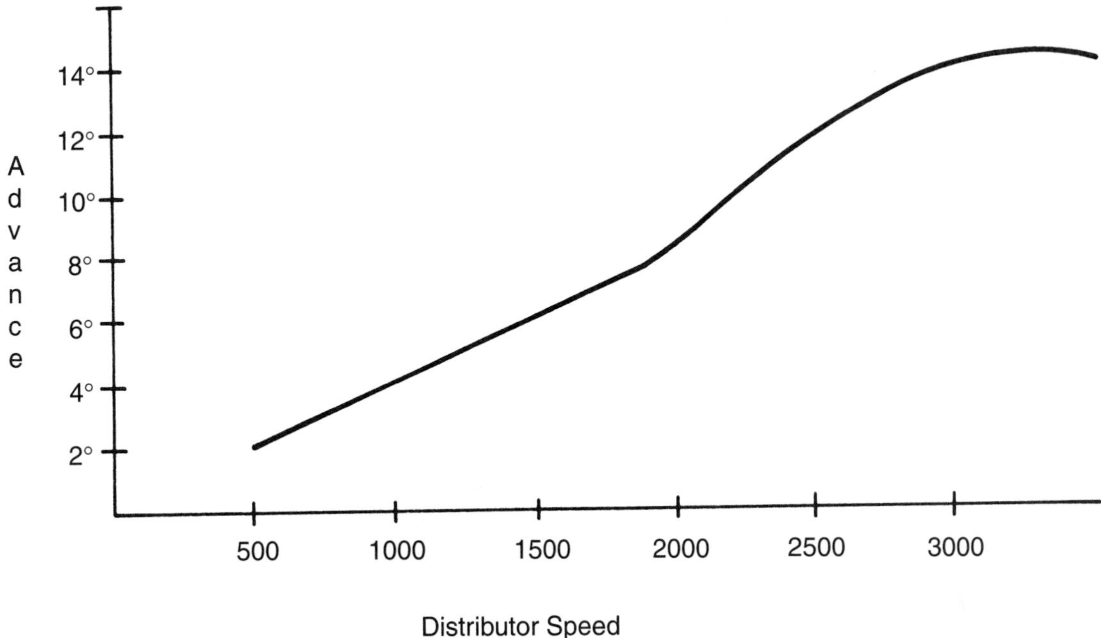

Figure 17. The spark is advanced based on distributor speed.

just as it passes TDC **(Figure 18)**. When do you think we will deliver peak combustive pressure if the throttle is partially open with only a mechanical advance? Let us analyze this question and see if we can figure it out together. If the mechanical advance is calibrated for WOT and the throttle is partially open, the burning will be slower, so the peak combustive pressure will be delivered later. The pressure will strike the piston after it has passed TDC and is on its way down. This will reduce both the power and the economy. Do not hesitate to review the last few sentences if the conclusion is not clear at this point.

So, we have found out that a mechanical advancing system will not be calibrated correctly except for WOT conditions. Obviously, most people cannot drive with their foot to the floor all the time, so the manufacturer generally supplies an additional advancing system that will compensate for less than WOT conditions. This system is called the **vacuum advance**, and it will use engine vacuum to pull on a diaphragm. **Figure 19** shows that the diaphragm is attached to the pickup coil plate. The move-

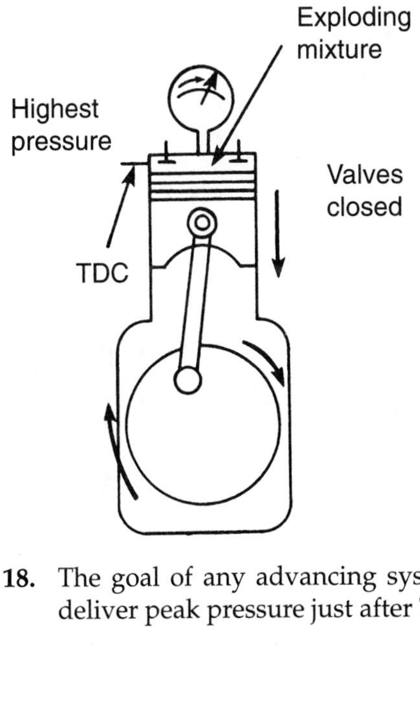

Exploding mixture

Highest pressure

Valves closed

TDC

Figure 18. The goal of any advancing system is to deliver peak pressure just after TDC.

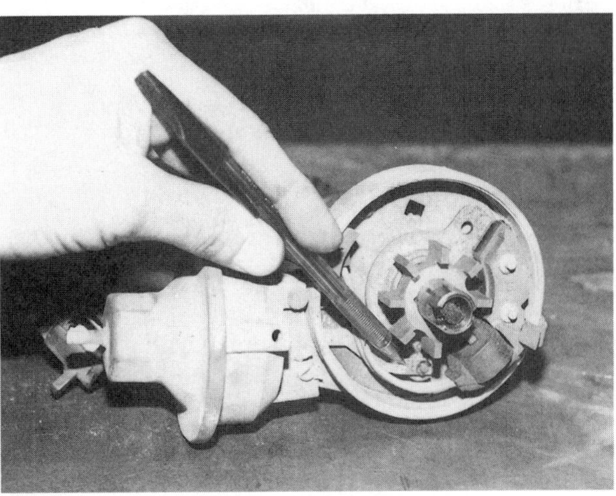

Figure 19. The vacuum diaphragm is connected to the pickup coil.

ment of the plate will pull the pickup coil in the opposite direction that the distributor shaft is rotating. This will advance the spark proportionately to engine vacuum. The greater the vacuum, the more the advance; the lower the vacuum, the less the advance. Vacuum and quantity of air:fuel mixture in the cylinder are the same. As the amount of air:fuel is increased (larger throttle opening), the vacuum is decreased, and as the amount of air:fuel is decreased (smaller throttle opening), the vacuum is increased **(Figure 20)**. The vacuum advance can be connected to a port that has vacuum any time the engine is on. This port is called a source of manifold vacuum. The other choice for advance vacuum is usually a port or hole above the throttle plates. It will be a source of vacuum when the engine is above idle **(Figure 21)**. This type of vacuum is called **ported vacuum** and has vacuum present only when the port is exposed by opening the throttle. In this manner, the vehicle systems will always give the correct timing for both the speed (mechanical advance) and the quantity of air:fuel (vacuum advance). Keep in mind that these two systems work in addition to the initial timing. You will see in Chapter 49 that these systems can be checked against specifications with an advance timing light.

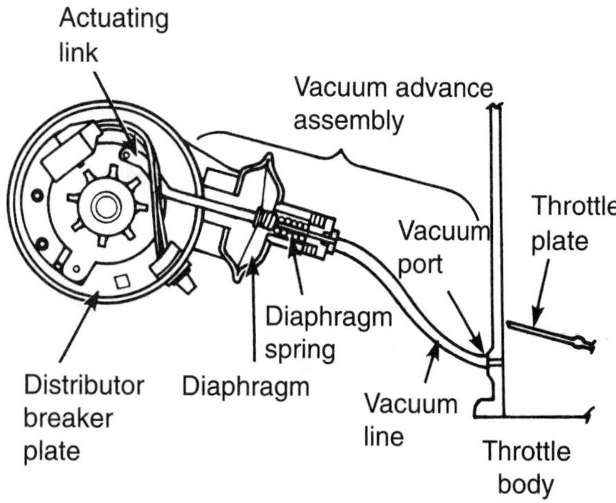

Figure 20. Manifold vacuum applied to the vacuum advance assembly.

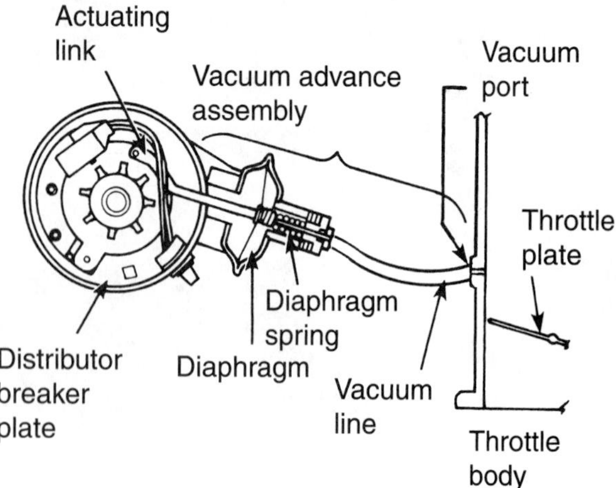

Figure 21. Ported vacuum applied to the vacuum advance assembly.

COMPUTERIZED ADVANCING SYSTEMS

Since the early 1980s, mechanical advancing systems have been replaced by electronic systems. The basic function remains the same, however: to deliver the spark at the optimum time that will allow peak combustive pressure to hit the piston just as it passes TDC on its way down during the power stroke. The advantage to an electronic system is increased accuracy over mechanical systems. The spark timing is still adjusted based on at least speed and load. In addition, the vehicle's PCM has all of the inputs that will be necessary for timing adjustments. These same inputs are being used to determine the other functions of the PCM. We look at two common systems, one from Ford and one from General Motors. We will not look at the Chrysler system because the ignition module is part of the PCM. All of the timing changes occur within the processor and are delivered to the module section. This eliminates our ability to even see these timing signals.

Ford TFI with PCM Timing

Ford has a simple system that uses a modified **thick film integrated (TFI)** module positioned on the distributor or near it. Follow along with **Figure 22.** The sensor inside the distributor is generally a Hall-effect sensor and goes by either the **profile ignition pickup (PIP)** sensor or the J-1930 term **camshaft position sensor (CMP)**. While the vehicle is cranking, the signal from the CMP goes directly to the ignition

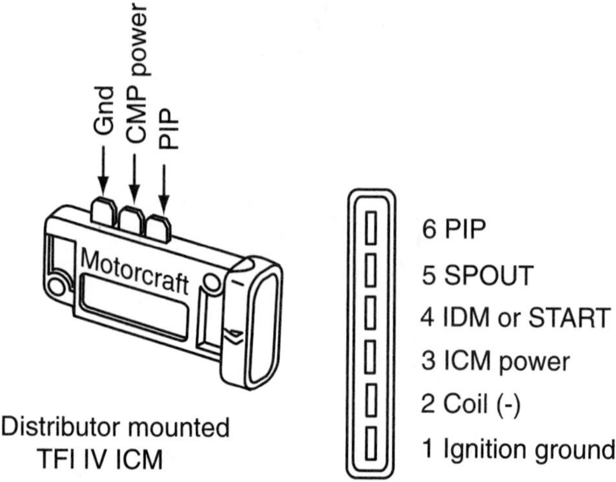

Figure 22. TFI ignition with computer advance.

module and is used to fire the ignition coil at base timing. The signal also is directed to the PCM along the PIP wire, which is terminal 6 on the figure. The PIP signal is used by the PCM to know engine speed (rpm) and piston position. Other systems, such as fuel injection, need to know rpm and position so that the engine will start. Once the engine is running, the PCM will send the modified timing signal back to the ignition module along the SPOUT wire (pin 5). When a SPOUT signal is available to the ignition module, it will be used to control the on/off of primary current to the ignition coil. The CMP (PIP) is still sending information about rpm and position to the PCM, but the PCM's SPOUT signal has taken over the timing function. An interesting additional signal that is found on some of the TFI systems is the **ignition diagnostic monitor (IDM)** signal. Remember that when the ignition coil discharges a rather large spike is produced at the coil negative terminal. This spike is sent to the PCM on the IDM wire. The PCM sees this spike (up to 400 volts) as an indication that the ignition coil fired. In Chapter 49 we look at these signals with a DSO and see how the system is diagnosed.

General Motors' Electronic Timing Control

General Motors' system is similar to Ford's in that a signal from the PCM is sent to the ignition module and used for timing. Refer to **Figure 23**. Notice that there are eight terminals on the ignition module. Let us concentrate on the four that connect the ignition module to the PCM. They are the bypass circuit, the **electronic spark timing (EST)** circuit, the reference signal (rpm), and a common ground that puts the module at PCM ground potential. With the engine cranking, a reference, or tachometer, signal is available to the PCM for fuel control. Once the engine starts, the PCM calculates the amount of advance necessary and sends a timing signal to the module via the EST circuit. In addition, the PCM generates a 5-volt signal that is sent to the module along the bypass circuit. Once the module sees the 5-volt bypass, it will use the EST signal rather than the reference signal to control the ignition coil. By comparison, Ford indicates to the ignition module that the PCM will control timing and sends the advance timing information along the SPOUT circuit, a single wire with a dual purpose. General Motors uses two circuits, the bypass and the EST, to accomplish the same thing. Both vehicles control the timing with a signal that starts in the PCM and is delivered to the ignition module. The ignition module still controls the ignition coil ground but relies on the PCM for the timing signal.

Recent years have seen the increased use of knock sensors by the manufacturers of computer-controlled systems. The theory behind them is that they are generally

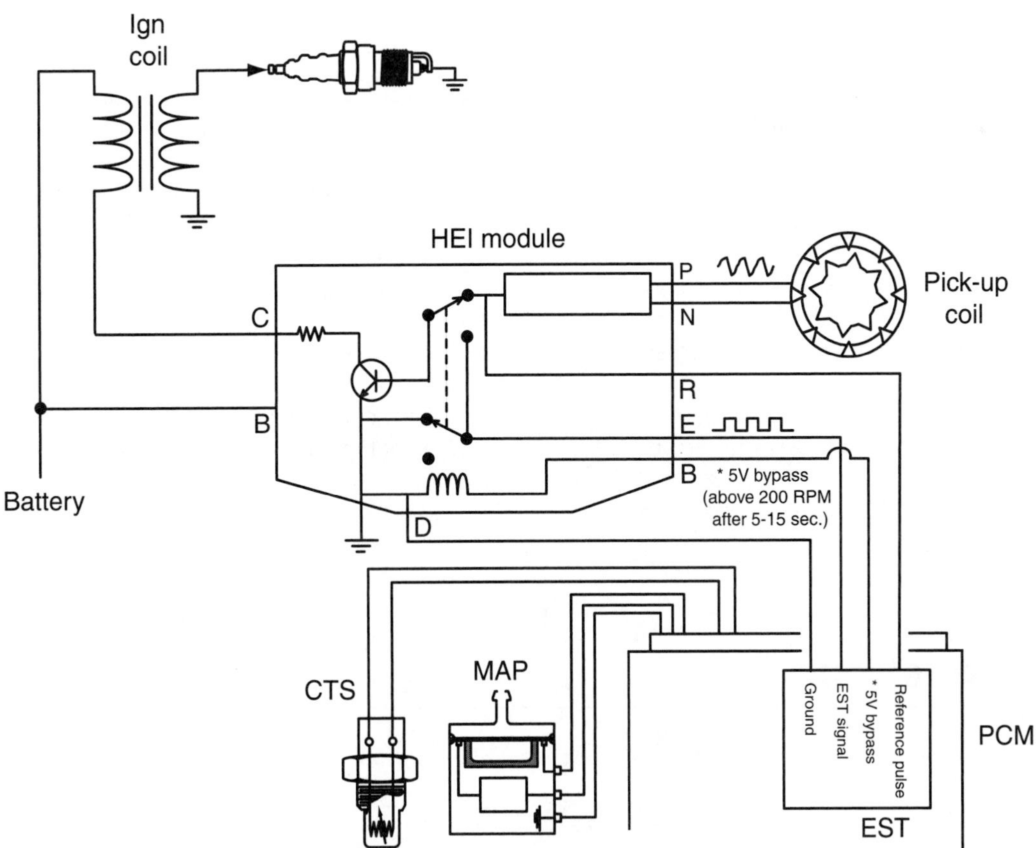

Figure 23. General Motors' 7 pin computer-controlled timing.

piezoelectric crystals tuned to recognize the frequency of an engine knocking. This pulse that is generated by the crystal is sent to the computer, which retards the timing until the knocking is eliminated. The circuit for a typical knock sensor is shown in **Figure 24**. This circuit allows the maximum advance just short of knocking. This is the timing that will give the greatest economy and the best drivability.

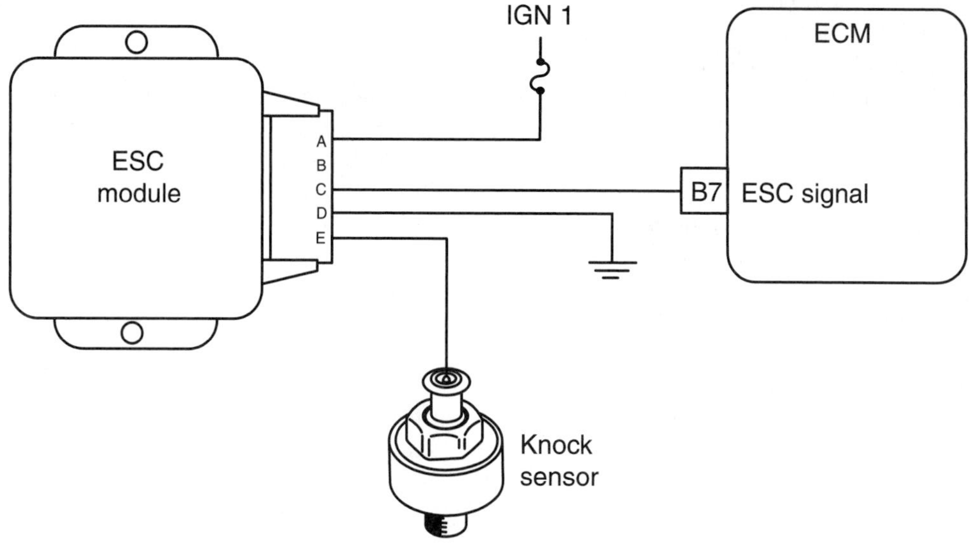

Figure 24. Knock sensors generate a signal that is used to retard the timing.

Summary

- The ignition coil is used to increase the charging/battery voltage to a level that will jump the air gap of the spark plugs.
- Primary current is used to develop a strong magnetic field.
- The spark plugs are fired when primary current is turned off.
- Dwell is the time measured in camshaft degrees that primary current is on.
- Primary coil resistance can be as low as 1 ohm.
- Secondary coil resistance is usually more than 5,000 ohms.
- The primary control device opens and closes the primary circuit.
- The ignition module is the most common primary control device.
- Primary current on, primary current off, variable dwell, and current limit are module functions.
- Magnetic, Hall effect, or optical sensors are used to indicate the position of the pistons.
- A magnetic sensor generates an AC sine wave.
- A Hall effect sensor generates a pulsing DC signal.
- An optical sensor generates a pulsing DC signal.
- Advancing systems are used to deliver the spark at different times relative to piston position.
- Most vehicles on the road use computer advancing systems.

Review Questions

1. When primary current is turned off, the coil's magnetic field
   A. builds
   B. collapses
   C. impedes
   D. relucts
2. Which winding directly supplies the spark plug with high voltage?
   A. primary
   B. secondary
   C. neither primary nor secondary
   D. both the primary and the secondary

3. Which of the following is not part of the primary circuit?
   A. rotor
   B. control module
   C. condenser
   D. points
4. Vacuum advance compensates for varying
   A. engine speeds
   B. weather conditions
   C. engine size differences
   D. loads
5. Which of the following components is not part of the secondary circuit?
   A. spark plug wires
   B. distributor cap
   C. rotor
   D. condenser
6. Normally, mechanical advance will
   A. advance the spark with higher vacuum
   B. advance the spark with additional speed
   C. cause little change in timing
   D. compensate for load and weather conditions
7. Current limit is found on electronic ignition systems that are missing the
   A. coil secondary
   B. coil primary
   C. Hall effect sensor
   D. primary ignition resistor
8. The ignition module
   A. turns the magnetic pickup on and off
   B. turns primary current on and off
   C. changes the ignition timing
   D. raises and lowers the secondary voltage
9. Vacuum advance units advance the spark through the distributor weights.
   A. True
   B. False
10. Variable dwell will result in a shorter primary current on-time, at high speeds higher than low speeds.
    A. True
    B. False
11. A knock sensor will retard the ignition timing until knock is eliminated.
    A. True
    B. False
12. The spark is delivered to the plug just as primary current is turned on.
    A. True
    B. False
13. Computer-controlled ignition systems adjust ignition timing through the use of sensors rather than by a vacuum and mechanical advance.
    A. True
    B. False
14. Current limit is a function found frequently on modern electronic ignition systems.
    A. True
    B. False

Chapter 49 Diagnosing and Servicing Distributed Primary Ignition Systems

Introduction

Up to this point, we have looked at the servicing of only the secondary ignition system, and although it is very important, it is not the only consideration when servicing the vehicle. In Chapter 48 we looked at how the spark was generated and controlled within the primary circuit. It is now time to examine the servicing procedures of the primary circuit. We begin with noncomputerized vehicles and the setting of base timing. You will see how we use a DSO to check the primary system both from a voltage standpoint and from a current standpoint. We will make use of the current probe connected to our DSO to get information about the flow of current. We also look at the diagnosis of piston position sensors and end the chapter with a simple no-spark diagnosis.

SETTING BASE TIMING

The timing light is still one of the most important tools of the professional technician. Its use as a diagnostic tool cannot be overstated. Most vehicles produced today have no means to adjust or check timing because they lack a distributor. However, many vehicles still have distributors and adjustable timing. We concentrate, in this chapter, on engines that have distributors. Most professional technicians use an adaptation of the basic light called the advance timing light or meter. We begin our discussion with the simplest form, however. The standard timing light that is available and in common use today has three connections, two for the battery and one for the spark plug **(Figure 1)**. The plug connection is usually inductive and designed to be placed around the number 1 spark plug wire. When cylinder 1 fires, the timing light's strobe light is also fired. The strobe stops or freezes the action of the vibration dampener or flywheel where the timing marks are located. **Figure 2** shows what a set of timing marks looks like on typical engines. Notice that the vibration dampener might have a single line notched onto the surface. The vibration dampener will be connected to the crankshaft and will be spinning at engine rpm. The small plate with numbers is attached to the timing cover or the front of the engine. Notice the 0 on the plate.

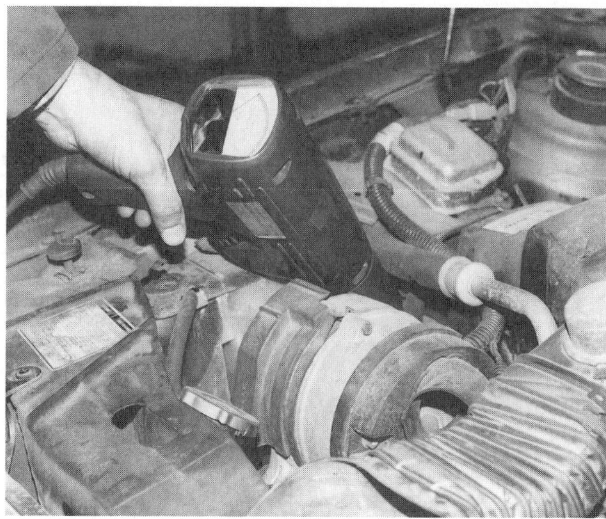

Figure 1. A timing light.

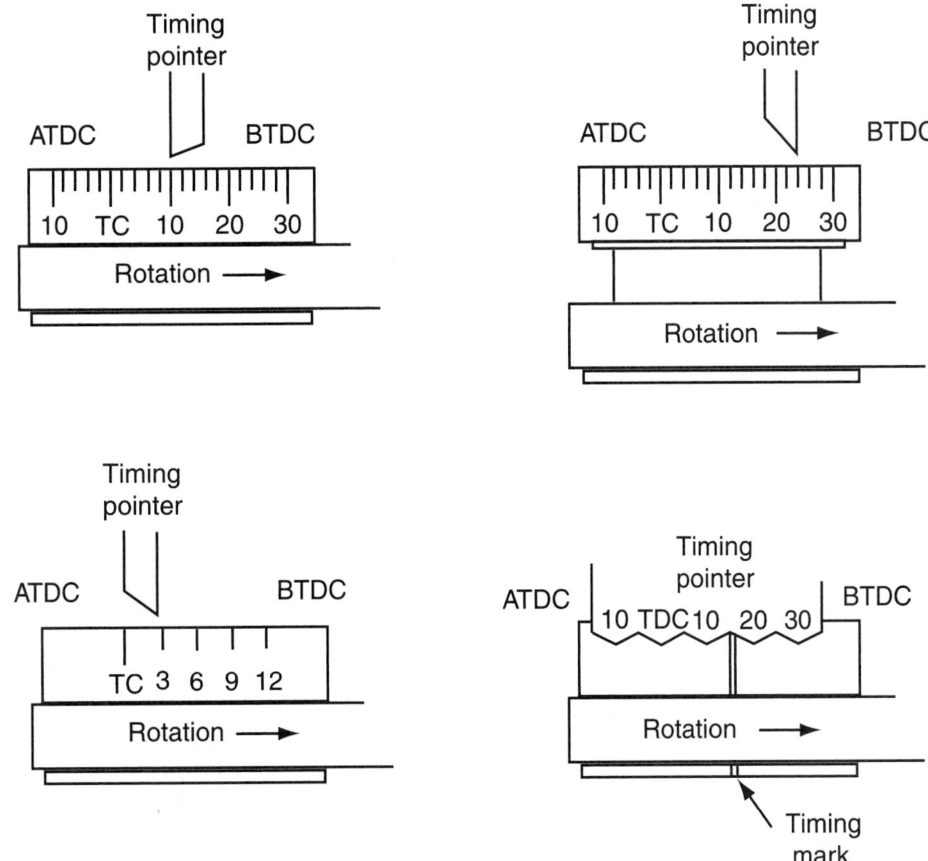

Figure 2. Different styles of timing marks.

When the notch on the vibration dampener lines up with the 0, cylinder 1 will be in TDC position. The strobe of the timing light will freeze the position of the vibration dampener and allow us to note what position cylinder 1 is in relative to TDC. This is called timing the engine, and it is basic to just about all drivability complaints. Timing must be correct. Let us use an example and see how and what we might do to time a noncomputerized vehicle. The specification for this engine read: 9 degrees BTDC at 650 rpm in drive. Vacuum advance disconnected and plugged.

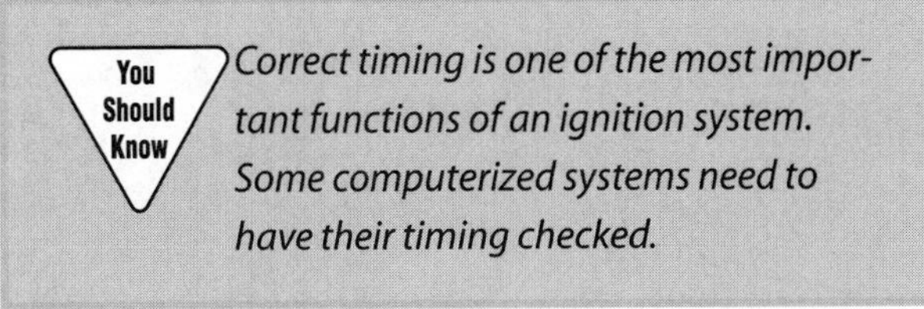

Correct timing is one of the most important functions of an ignition system. Some computerized systems need to have their timing checked.

Do you remember the relationship between speed and timing that we discussed in Chapter 48? As the engine speed increased, the timing was usually advanced. This is the reason for the timing check being done to the specific speed of 650 rpm. The manufacturer is specifying a speed below the point where the mechanical advance will begin to operate. The manufacturer is also saying that the point of ignition must be 9 degrees before the piston reaches TDC. This will allow the burning process to reach its peak combustible pressure by the time the piston reaches the end of the compression stroke and is on its way down on the power stroke. In addition, the speed of 650 rpm should be set with the automatic transmission in drive. The last piece of information that is important is that the vacuum advance is disconnected and plugged. Why do you think this is necessary? Remember that manufacturers have two choices of vacuum for the advance. One choice is above the throttle plates. This is called ported vacuum and will be present only with the engine off idle. At idle, no vacuum is present at a ported source. The other choice is to have manifold vacuum as the source for the vacuum advance. Manifold vacuum will be present any time the engine is running at something less than full throttle. From the specs, it appears that manifold vacuum is being used as the source for the vacuum advance and the manufacturer wants the initial timing to be done with no vacuum advance present. This is why manufacturers have specified that the advance be disconnected and plugged on most engines.

If we have met all of the conditions that the manufacturer has specified, we should expect the timing to be 9 degrees before TDC. Shining the timing light down on the marks and vibration dampener should show the marks as illustrated in **Figure 3**. If this were the case, we would say that the timing is correct. What do you think we would do if the timing marks lined up as in **Figure 4**? Notice that the vibration mark is lined up with the number 3 on the timing plate. We are off by 6 degrees. The timing is retarded by 6 degrees. Remember, retarded timing is timing that is behind what it should be. The opposite is advanced timing, which means timing that is ahead of where it should be. In this example, the specification is 9 BTDC and the actual is 3 BTDC. We will have to change the timing and advance it by 6 degrees to make it correct. This will be done by loosening the distributor hold-down bolt and turning the distributor housing until the number 9 is in line with the vibration dampener mark. Once the engine is timed, we can retighten the hold-down bolt and recheck. It is important to note that changing timing might have an effect on engine speed. This will mean that you will have to readjust the engine idle speed back to the specification. When you are completely finished, all conditions of the specifications should be met. The timing marks should line up at the 9-degree BTDC position with the engine speed at 650 rpm and the distributor vacuum line disconnected and plugged. If the conditions are met, the vacuum line can be reconnected. You have just set or checked timing. Remember that every engine has different conditions that must be met. Do not assume that all engines are timing exactly the same. In addition, realize that the timing is set ordinarily on cylinder 1 with the assumption that all other cylinders will have the same timing. Some vehicles use a different cylinder for the timing check. Do not assume that all engines are the same.

In recent years, the computer has taken on the responsibility of timing along with fuel control. This will allow the computer to compute the required timing and change it to match the conditions. This does not mean that initial timing is not set or checked. Many computerized vehicles with distributed ignition still have adjustable timing. The computer makes the assumption that initial timing is correct and computes running timing changes off initial timing. Timing a computer-controlled engine is just as easy as timing a noncomputer-controlled engine. The only difference is that the computer must be told to not control timing. **Figure 5** shows the SPOUT wire located at the distributor of a computer-controlled Ford vehicle. SPOUT stands for spark out. By disconnecting the connector, we will eliminate the computer from the timing circuit. The engine will now be running in a nonadvanced base or initial timing mode. We can now turn the distributor until the timing marks line up just as we did on a noncomputerized advance system. Once the initial timing is correct, the SPOUT is reconnected and the computer takes over the timing function. Notice that the procedure is easier because we do not have to worry about the speed the engine is running at. With no centrifugal advance system changing the timing with speed changes, the engine can be timed at almost any speed.

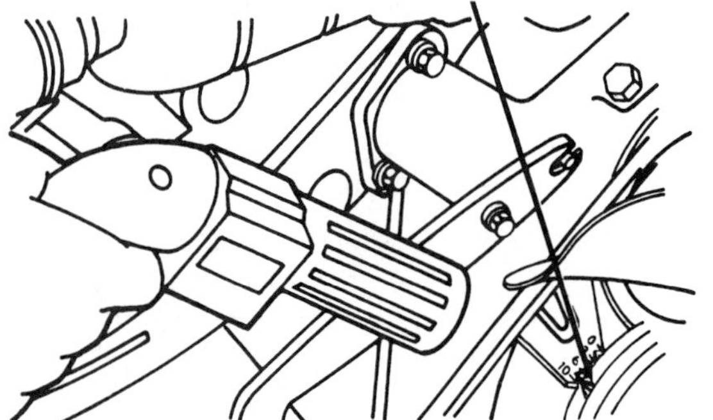

Should advance smoothly (Note
maximum advance for later reference.)

Figure 3. Timing marks at 9 degrees BTDC.

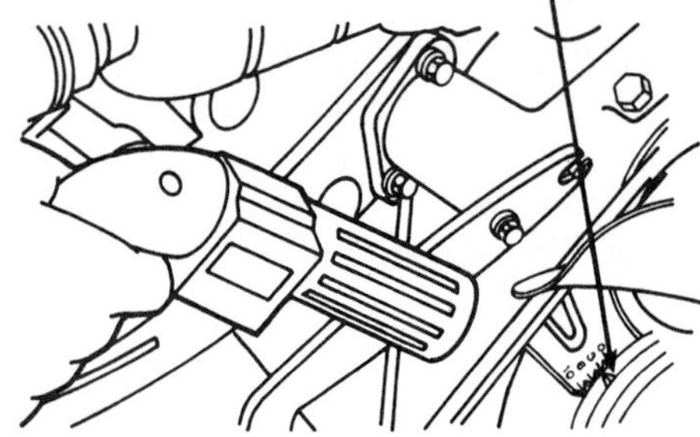

Should advance smoothly (Note
maximum advance for later reference.)

Figure 4. Timing marks at 3 degrees BTDC.

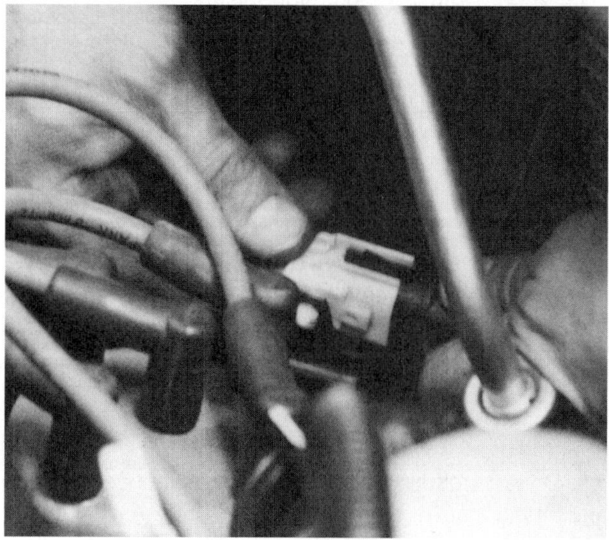

Figure 5. Opening the SPOUT connector puts the engine at base timing.

If base timing is not correct in a computerized system, the entire advance curve will be incorrect by the amount of the error.

You should begin to realize that each manufacturer will have different procedures that must be followed exactly. Always make sure that you are following the correct procedure for the year, make, and model of vehicle that you are working on. Most computerized systems will change timing from a base or initial timing position. If the initial timing is incorrect, virtually all calculations will also be incorrect and a drivability complaint might be present. Generally, overadvanced timing can cause reduction in fuel economy and a ping or knock, whereas retarded timing can cause hesitations and reduced fuel economy. Greatly overadvanced timing can also do engine damage because the ping or knock is indicative of uncontrolled combustion and greatly raised cylinder temperatures. Extreme temperatures can even melt pistons. Make sure that initial timing is correct. It is basic for a properly running engine.

CHECKING THE ADVANCING SYSTEM

The basic timing light is necessary and should be used frequently to either set initial timing or determine if it is correct. It does, however, have its limitations if the technician wishes to check something other than initial timing. If, for example, the customer complaint is poor economy and high-speed performance, the technician might decide that a check of the advancing systems might be helpful. A normal inductive pickup timing light would be limited to an idle-only timing check. This is when the advance timing light is useful. **Figure 6** shows a common advance timing light. Notice that the back of the light has a meter calibrated from 0 to 45 degrees. The knob that is part of the handle controls this meter and the amount of delay between the time the spark plug fires and the time that the light flashes. By turning the knob, the technician will be able to delay the flash of the light by the number of engine degrees shown on the back of the meter. The delay of the light allows the use of the engine's timing marks as a reference and enables the technician to note the timing change. Let us consider an example and try to explain just how the light allows us to read how many degrees advance the engine is running at.

The procedure is simple and starts with a check of the initial timing and a reset if necessary. Let us assume that the engine is timed to 9 degrees BTDC at 650 rpm and we wish to determine if the mechanical advance will advance the timing by an additional 10 degrees by the time the engine has reached 1500 rpm. With the vacuum advance disconnected and plugged, we would raise the engine speed up to 1500 rpm and turn the knob on the timing light handle until the timing marks again line up at the 9-degree spot. The meter on the back will now tell us how many degrees we have delayed the flash of light. This will then be the amount of mechanical advance at 1500 rpm. Note that the timing light just delays the flash; it does not change the actual ignition timing. With an advance timing light, we can advance curve the distributor against the manufacturer's specifications. Advance curving is just a fancy way of saying we have checked the mechanical advance at various speeds and the vacuum advance at various vacuum levels and compared both with the specifications supplied by the manufacturer. The comparison allows us to determine if the drivability complaint is advance related. Look at **Table 1** and try to figure out what the complaint might be.

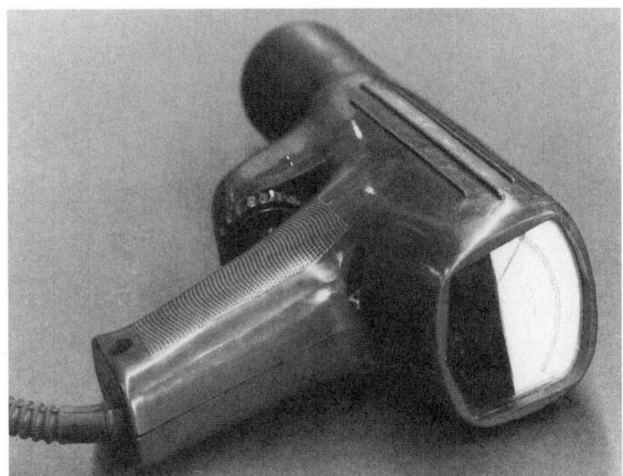

Figure 6. An advance timing light.

The graph in **Figure 7** is for the same vehicle. Notice that the dashed line is the specification, whereas the heavy line is the actual. What kind of driving condition might be experienced by the customer with this mechanical advance? Remember what we have repeatedly mentioned: retarded timing can result in hesitations and reduced economy, whereas advanced timing can result in pinging. In the midrange of 1,000–2,000 rpm, the actual timing is advanced

Speed	Actual	Specification
650	9 BTDC	9 BTDC
1000	16 BTDC	12 BTDC
1500	25 BTDC	15 BTDC
2000	30 BTDC	20 BTDC
2500	30 BTDC	28 BTDC

Table 1. Timing change versus specification.

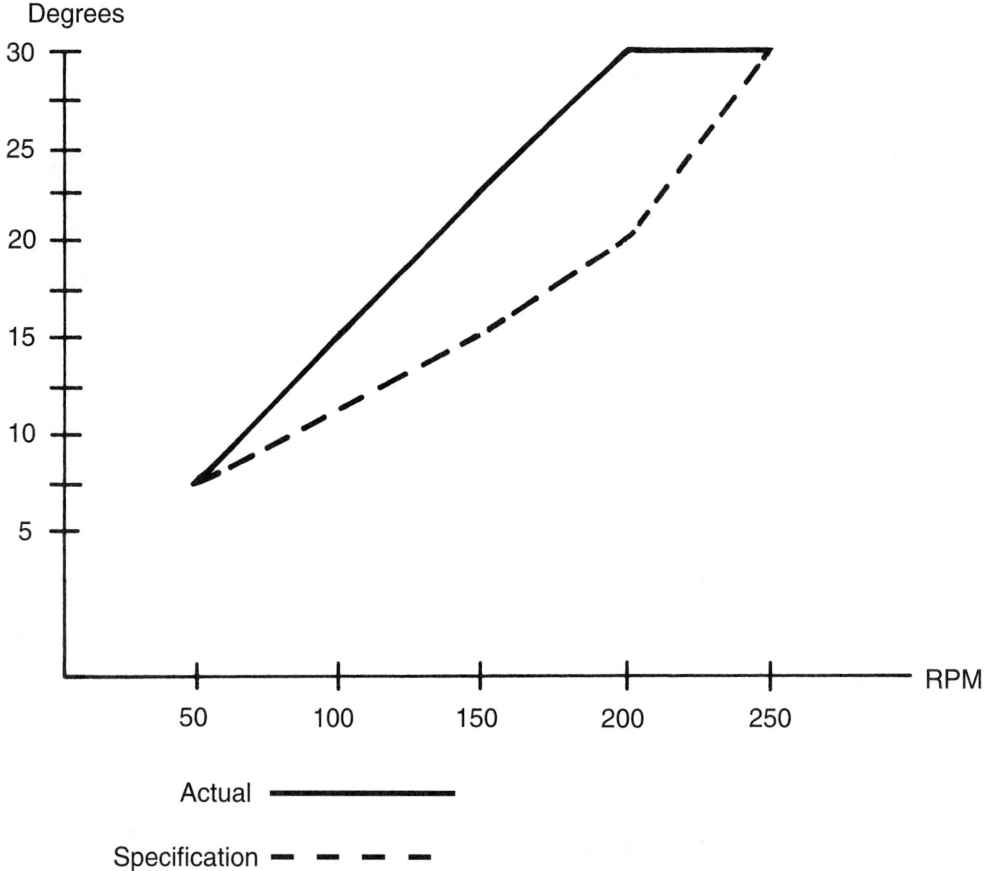

Figure 7. With the advance not to specification, midrange performance might suffer.

by as much as 10 degrees with the low end (650 rpm) correct and the upper end (1500 rpm) nearly correct (off by 2 degrees). We would expect this vehicle to exhibit a ping in the midrange, which gradually goes away as the customer drives faster and faster.

We can check the vacuum advance the same way. Use a vacuum hand pump to draw on the diaphragm **(Figure 8)** until the vacuum specification is achieved, and turn the timing light knob until the timing marks line up again. The amount of flash delay gives us the amount of vacuum advance. Let us look at **Table 2**, which is similar to the table we examined for mechanical advance.

Vacuum	Actual	Specification
0 in Hg	0 BTDC	0 BTDC
5 in Hg	2 BTDC	4 BTDC
10 in Hg	3 BTDC	9 BTDC
15 in Hg	3 BTDC	18 BTDC
20 in Hg	4 BTDC	28 BTDC

Table 2. Timing change versus specification for vacuum advance.

Figure 8. A vacuum pump can test the advance diaphragm.

What driving condition might the customer experience? Keep in mind that the only purpose of the vacuum advance is to improve economy under conditions of high vacuum. High vacuum is achieved during cruising or light loads and is measured in inches of mercury (in Hg). Our actual timing advance is quite a bit less than the specification. It is retarded from where it belongs and might result in hesitations or reduced economy.

The repair of the two vehicles just tested is simple once the diagnosis is complete. The first mechanical advance problem might require a new set of weights and springs, whereas the second vehicle will require a new vacuum advance unit and a check of the movement of the pickup. In most cases today, a replacement distributor would be installed in the vehicle because the cost of repairing it might exceed the cost of a rebuilt unit.

USING A DSO FOR PRIMARY CIRCUIT TESTING

The discussion in Chapter 47 centered on using the DSO with an ignition adapter to look at the secondary system. We emphasized looking at firing voltage and how long the spark lasted. Now it is time to begin looking at the primary system with the DSO. The purpose here is to highlight what we should look for when we connect our DSO up to the primary. Even though the secondary is the business end of the ignition system with the spark plugs connected, it is worthless without a fully functioning primary. It really does not matter whether you are trying to fix a distributed or a distributorless system a correctly functioning primary is mandatory.

The primary ignition is composed of the ignition coil, the battery, and the ignition module. We also need the signal to the ignition module of when to fire the plugs. Let us look at the ignition module briefly before we analyze a DSO pattern. Most ignition modules found on distributed vehicles today have at least two functions and many have four functions. They will:

1. Turn on primary ignition coil current.
2. Limit the amount of primary coil current.
3. Vary the dwell (the amount of time that current is on).
4. Turn off primary current to fire the plug.

It is possible, but not too probable, that an ignition module could lose one or more of these functions and still be able to fire the plugs. Realistically, a failure of any one of the four functions will usually result in a tow. You will see burned-out ignition modules in the field. You need to analyze a primary ignition DSO pattern looking for the future cause of failure. You also need to analyze the pattern of a vehicle that is not in fuel control looking for the cause of misfire. Remember that any misfire will result in increased hydrocarbons and usually a failed emission test. So what makes up a good functioning ignition primary DSO pattern, and what can it tell us?

Figure 9 shows a good primary ignition pattern. Let us analyze the pattern in sections and look at what each section can tell us. The DSO input lead has been placed on the negative terminal of the coil primary and its negative lead has been placed on a clean engine ground (or the battery). Follow your DSO manufacturer's recommenda-

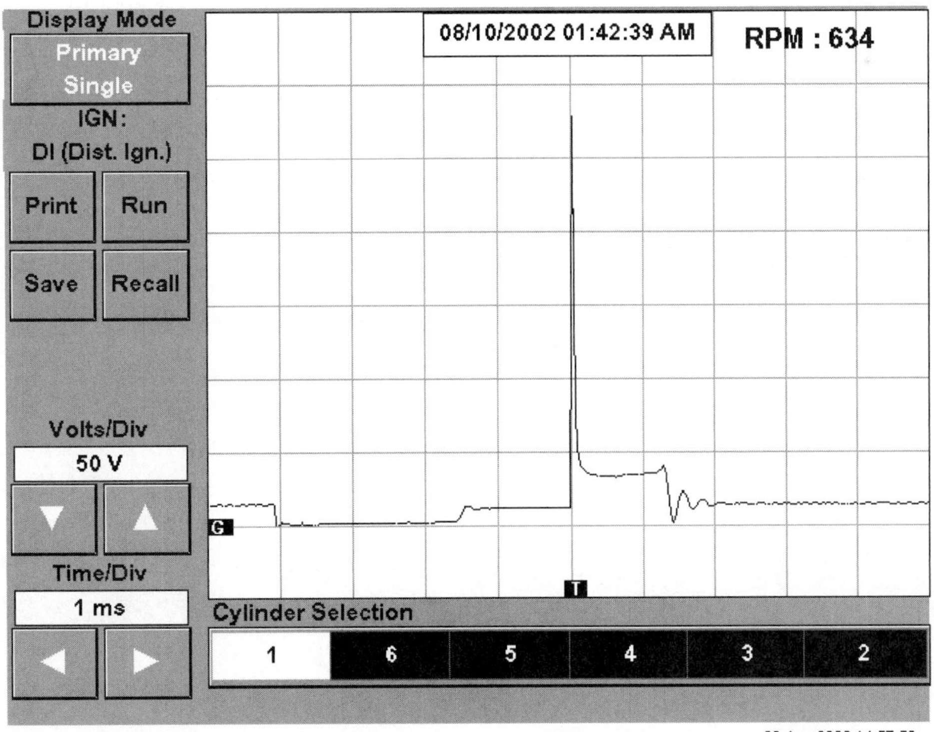

Figure 9. A cylinder 1 primary pattern.

tions regarding ideal settings. This MTS 5100 has been set to 1 millisecond and 50 volts per division, giving us a total time across the screen of 10 milliseconds and a total voltage of +350 volts, from the ground mark. Play around with the settings to get the clearest screen that will show the entire pattern with sufficient detail. Make sure that you do not chop off any voltage spikes and that you have an entire pattern on the screen. If your DSO is fully automatic, it will take care of these details by itself. As you analyze the pattern, keep in mind that there is a sequence of events shown in Figure 9, beginning on the left with primary current being turned on, then current limit kicking in, the coil firing, and primary current turning on again for the next cylinder in the firing order. Between the two ends is the information that we need to look at.

Figure 10 shows a firing section. This section will give us useful bits of information. It will tell us how much primary voltage is being induced when we fire the plug. This is usually referred to as the primary induced voltage. You will typically find primary induced voltages of 200–400 volts. Higher is all right, but lower is not. If the induced voltage is too low, it can indicate that the coil is not being fed with sufficient B+ or the coil has a problem and should be replaced. In our example, this primary circuit is generating more than 250 volts. Some DSOs give the ability to view the primary inductive kick in a bar graph format, as shown in **Figure 11**.

Now let us turn our efforts toward the length of time that the spark lasts. Virtually all electronic ignition systems run sufficient coil primary current to have the spark last between 0.8 and 2.0 milliseconds. Keep in mind that all plugs should fire for approximately the same time. Make sure that you look at each plug separately. If a plug is firing for less than 0.8 millisecond, it will usually misfire. It may not be a dead misfire, but it will contribute to increased HC, especially under load. The causes of decreased burn times are the same as those for decreased induced primary voltage: not enough B+ to the coil or a faulty coil. We will add to this list of causes a module that is not functioning correctly. Once primary current has been turned on, the DSO will give us a good idea of module function and how good it is. Burn time is the length of the spark line, which in our example begins at the "T" line. The T, or trigger point, is in the middle of the trace at the beginning of the 6 division. The trigger point is where we find the induced primary voltage and the beginning of the spark line. The length of the spark line tells us how long the spark lasts. In our example, **Figure 12**, you can see that we have a 275-volt induced primary voltage. Each division is 1 millisecond, so it looks like we have a burn time of 1.25 milliseconds. Both of these are within specifications.

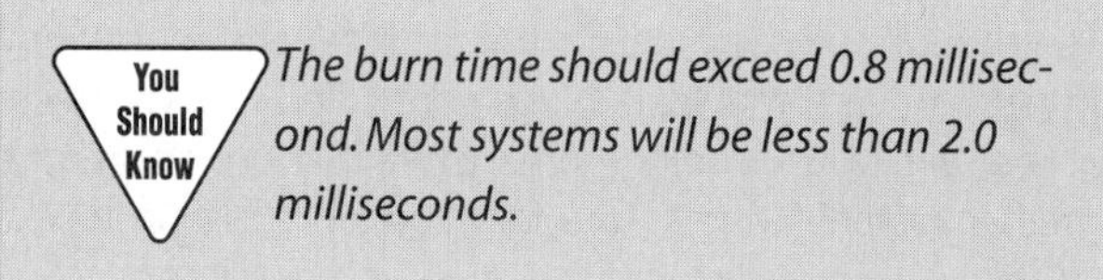

The burn time should exceed 0.8 millisecond. Most systems will be less than 2.0 milliseconds.

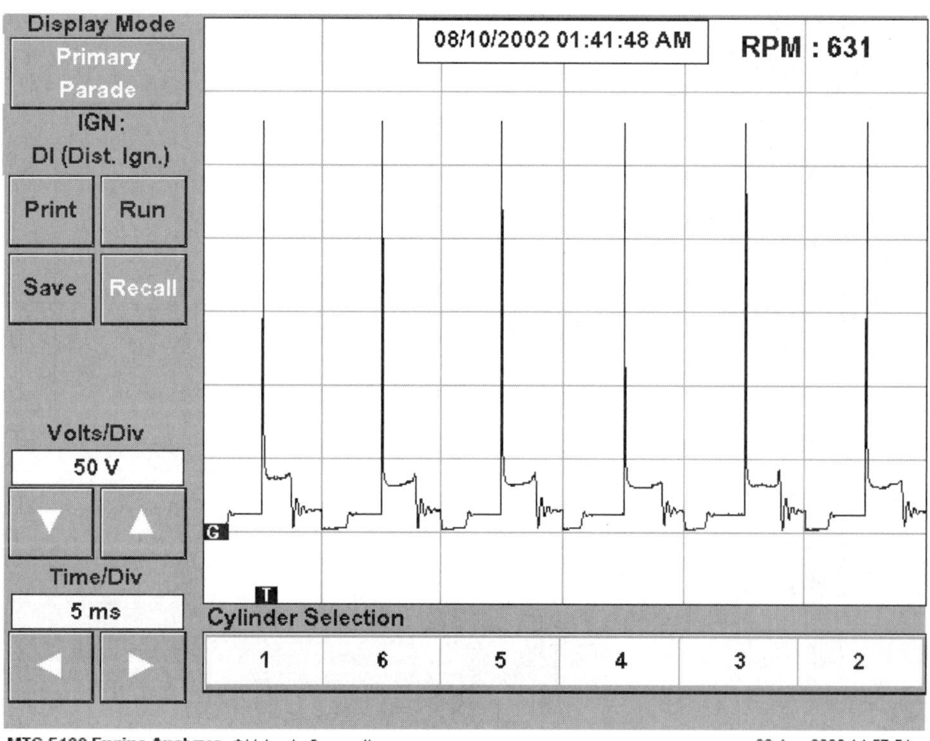

Figure 10. This pattern shows induced primary voltage.

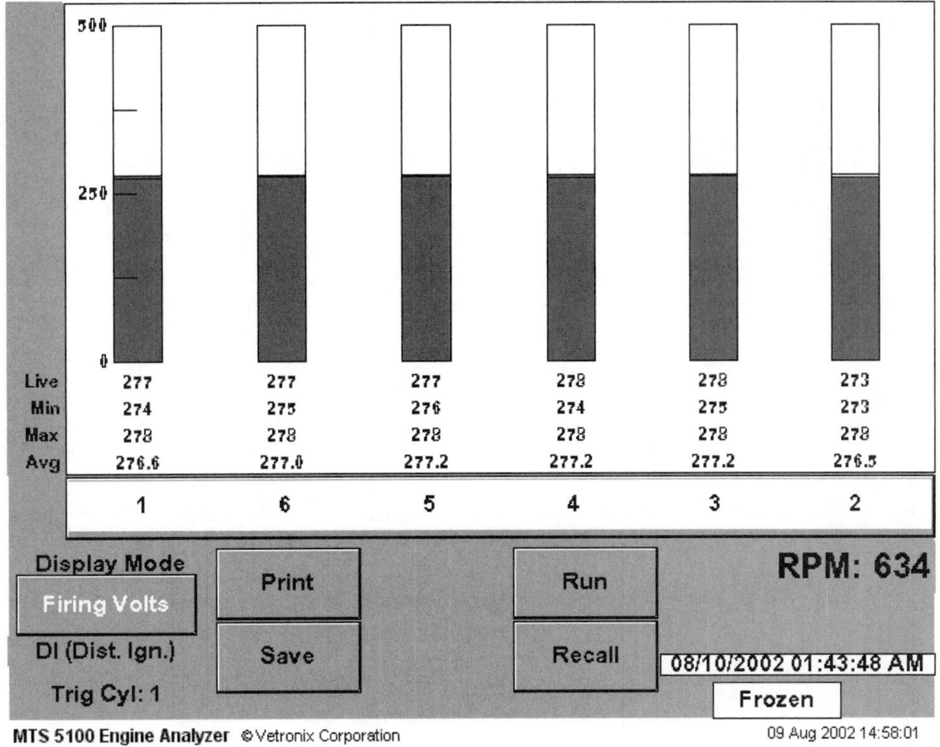

Figure 11. Induced primary voltage in bar graph format.

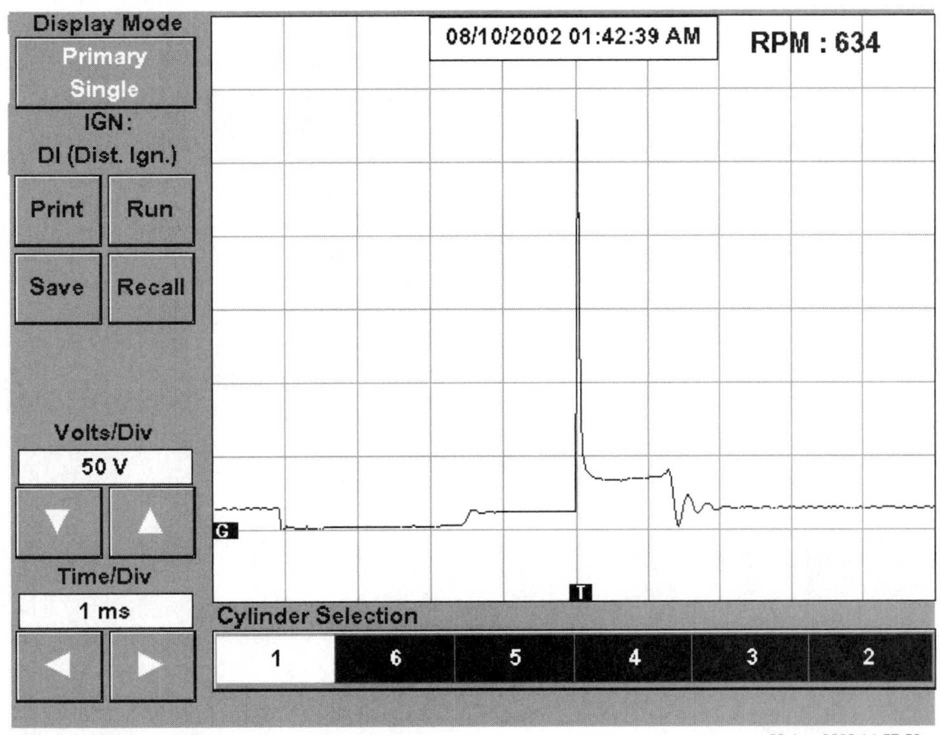

Figure 12. A DSO showing 1.25 milliseconds burn time.

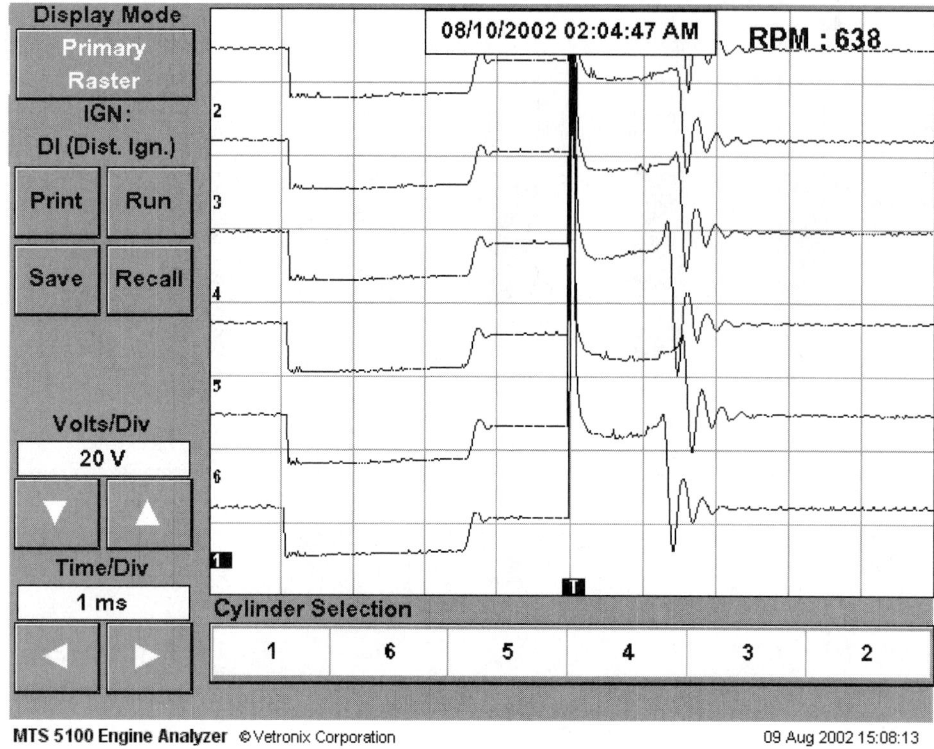

Figure 13. After the spark, the remaining energy disappears in 3–5 oscillations.

The slowly diminishing lines represent the next piece of valuable information after burn time. **Figure 13** is a raster pattern that again shows the 1.25 milliseconds of burn time. Ignore the lower voltage scale of only 20 volts on the left of the screen. Sometimes you may have to decrease the volts per division settings to get additional detail. At this point, we are not concerned with the firing of the coil because we have looked at it already. Instead, look at the oscillations that follow the burn time. These diminishing oscillations represent the remaining coil energy after the burn occurred that is being dissipated through the combined capacitance of the ignition coil and the ignition module. The odds are good that you have a shorted coil if these oscillations are missing. Generally, you should find 3–5 oscillations. We have about 4, so that is acceptable.

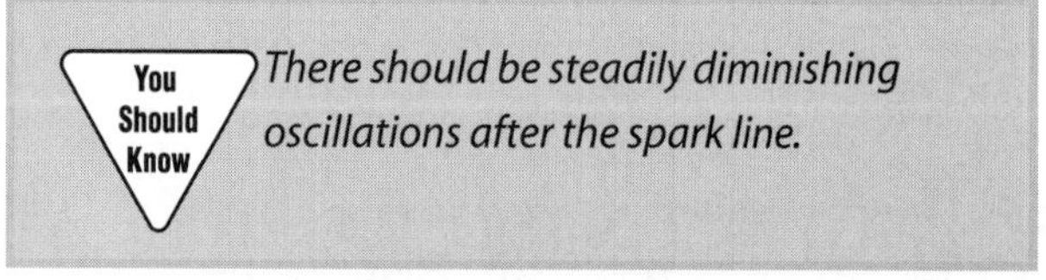

There should be steadily diminishing oscillations after the spark line.

We can refer to the same Figure 13 to see the next part of this primary pattern we need to look at, the dwell section. Dwell, or the amount of time that primary current flows, will vary with speed, so rev the engine up a little and make sure that the dwell lengthens as speed goes up and shortens as it slows down. This shows us that the all-important variable dwell function of the module is present. Once the ignition coil is fully charged or saturated, it is ready for the next firing. At this point, the module should throttle back on primary current. It takes a lot of current to saturate the coil's magnetic field but little to maintain it. The slight rise in the pattern that is about in the middle of the screen represents this throttling back or current limit occurring during the dwell period. Just after the first division, the trace takes a turn downward. This is the beginning of dwell. Once primary current has saturated the coil primary, the current will be throttled back and results in the rise or bump seen in the fourth division. The distance from primary current on until current limit has different specifications. On most vehicles, the distance or time will *exceed* 2.00–4.00 milliseconds, for example, mid-1990s General Motors products (with a distributor) that will be in excess of 2.5 milliseconds. Consider this example from a 1998 Blazer with a 4.3-liter engine (Figure 13). Remember that each division is equal to 1.0 millisecond and it looks like a bit less than four divisions. Primary current is on full for about 2.75 milliseconds. With a specification of 2.5 milliseconds as the minimum, our situation is all right. If the time is insufficient (less than 2.0–4.0 milliseconds, depending on the vehicle), either you have a module problem or the ignition coil is faulty. A lower-than-specification resistance

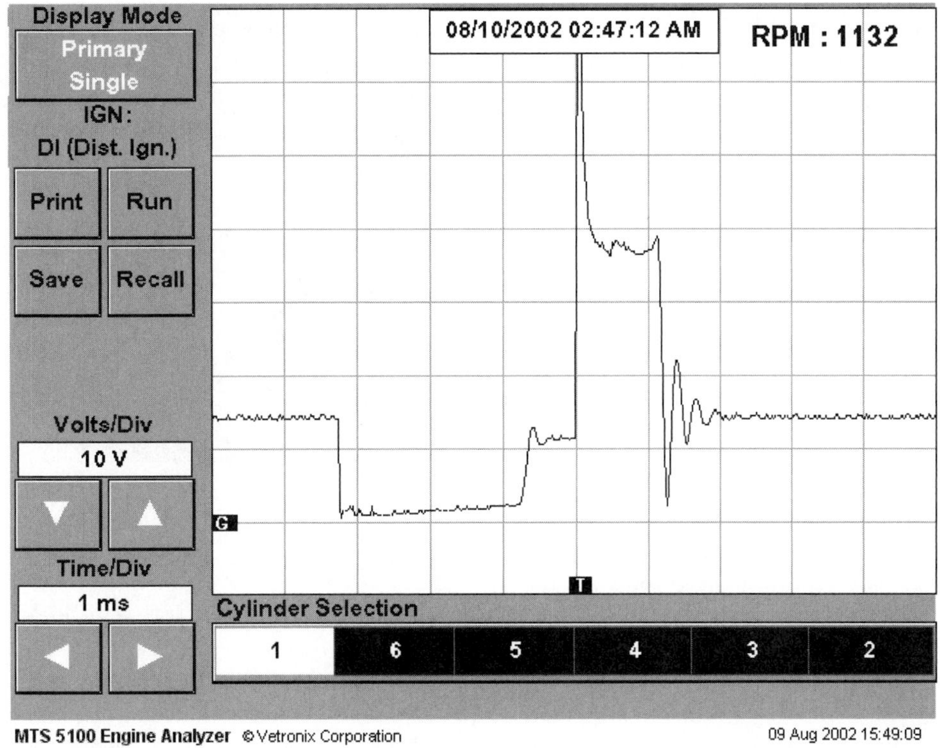

Figure 14. The second division shows power to coil; the third division shows ground applied to the coil.

coil primary will increase the current flow and result in a faster current limit function, and result in a lower-than-specification value. Usually, this will burn out the ignition module, and replacing just the module will not fix the problem. By checking the time between primary on until current limit, you will be able to pick out those lower than specification primary resistances. Always verify with your ohmmeter.

Figure 14 will give us the opportunity to check the power and grounds of the system. The module and coil are effective only if the coil has sufficient voltage applied to it. By looking at the voltage just before the point where primary current is turned on, we can measure the applied voltage to the ignition coil. With 10 volts per division and slightly less than one and a half divisions, it looks like we are in the 14-volt open circuit area. Once the firing of the coil takes place, the + terminal voltage will show on the DSO. If the voltage is low, the DSO input lead should be moved to the coil input and a remeasurement taken. The voltage of the line immediately preceding the primary current on signal is the applied voltage to the coil. The majority of ignition systems today run full B+ voltage to coils, so 14 volts is well within charging system voltage specifications.

The last bit of information we can obtain from a primary DSO pattern is the ability of the module to supply a good solid ground to the coil at the beginning of the dwell section. To get maximum current, we need to have a good ground applied by the module. The downward turn of the trace at the beginning of the dwell should get down to, or very close to, zero volts. In our example, the line is slightly off zero volts, which could indicate that there is some unwanted resistance on the ground side of the circuit.

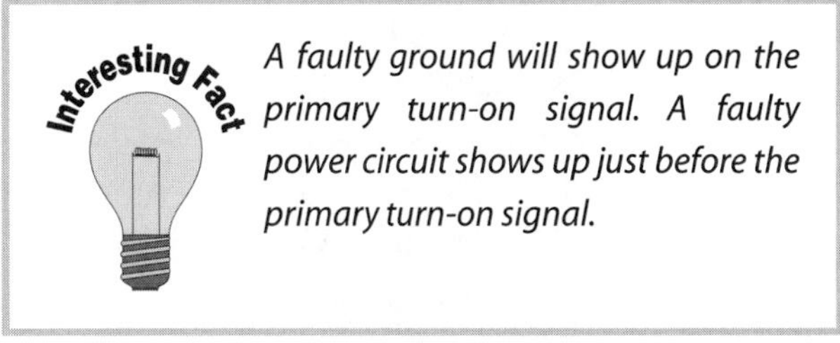

A faulty ground will show up on the primary turn-on signal. A faulty power circuit shows up just before the primary turn-on signal.

By analyzing the DSO primary coil pattern, we can look quickly at the coil, the power to it, the module functions, and general ignition function. It is easy with today's DSO. Just remember that burn time should be between 0.8 and 2.0 milliseconds, and the time from primary current until current limit should be between 2.0 and 4.0 milliseconds. Do not hesitate to spend some time with the primary. It can give you a tremendous amount of information quickly. Especially look at it for drivability or emission problems.

USING A CURRENT PROBE TO DIAGNOSE PRIMARY CIRCUITS

The use of current probes is steadily increasing because of the additional information that they can give to the technician. In addition, some systems on the road today do not allow us to get a primary voltage pattern, so diagnosing using current becomes a natural second choice. The connection is straightforward. The current probe is clamped around either of the primary coil wires. It is sometimes helpful to put the trigger lead around the other coil lead because it will allow the pattern to be triggered. Triggering the pattern locks it down on the screen and makes capturing it much simpler. **Figure 15** shows this setup. The DSO is set to capture a primary current trace once the conditions of the trigger have been met. **Figure 16** is the screen capture of primary current flow or ramp. Notice that the Ch 1 setting has changed from a voltage to a current value of 2.00 A DC per division. So each division will equal 2 amps of current. At the beginning of the trace on the left, no current is flowing. At the end of the second division, the DSO is showing a ramping up of current. The trace rises at about a 45-degree angle. This is the current increasing because the module has turned on the ground circuit. At the top of the ramp, the trace flattens out. This is when the current limit function cuts back on primary current. Count how many divisions of current there are from the zero line to the flattop section. There are about 3.75 divisions. This means that this ignition coil primary is drawing 7.5 amps of current, which is typical for this ignition system. We can also get the on-time information from the current waveform. Measure the time (1 millisecond per division) from the beginning of current flow until the current limit flattop. It is about 2.6 milliseconds, which is just about the same as our voltage waveform. One other thing that you may have noticed from the current waveform is the little oscillations that are present when primary current just turns on. When the current just turns on, there is a CEMF generated, such as our starter motor that prevents the current flow from

Figure 15. A current probe and trigger on the coil primary wire.

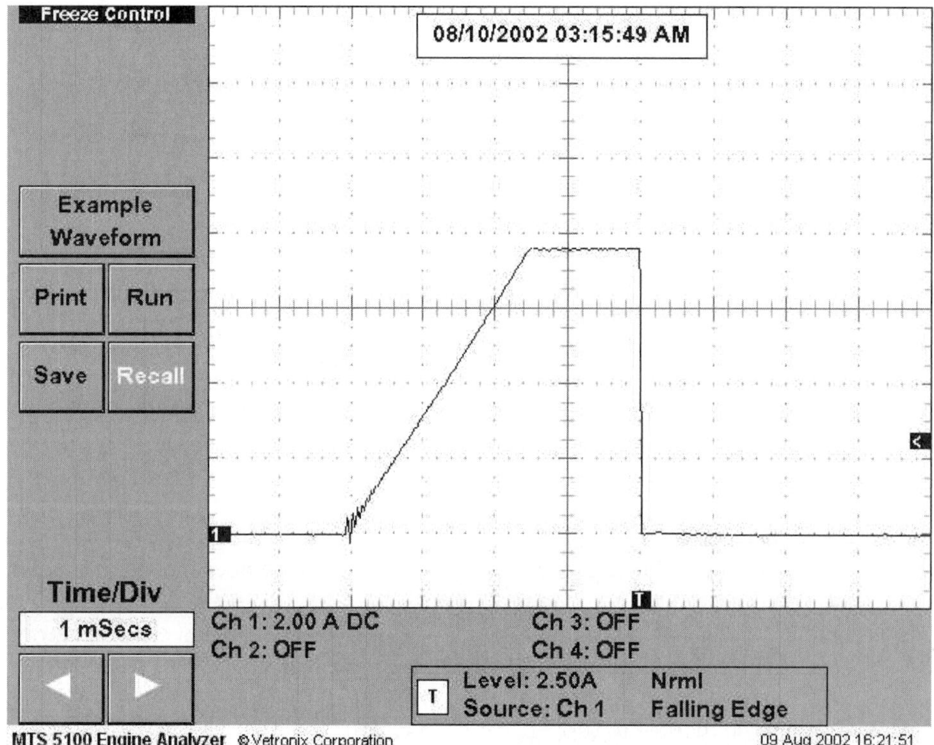

Figure 16. A primary trace showing current flow during the dwell period.

ramping up smoothly. Once the initial current surge is over, the flow flattens out. This is normal and is seen on most ignition coils.

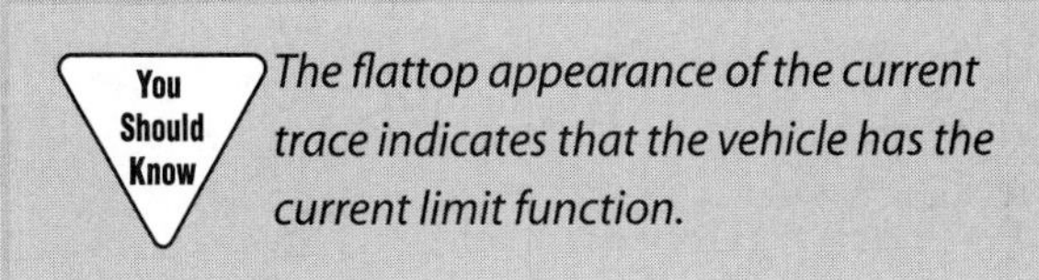

The flattop appearance of the current trace indicates that the vehicle has the current limit function.

The combination of looking at a current waveform and a voltage waveform from the same ignition system gives technicians all of the available information they should need to find and correct any problems. **Figure 17** shows both patterns: voltage on the bottom (channel 2) and current on the top (channel 1). Some technicians find that seeing both at the same time helps them analyze the relationship between voltage and current.

CHECKING INPUT SENSORS WITH A DSO

Remember that there are three basic input sensors that you need to be able to test. The first thing to realize is which type is installed on the vehicle. If the vehicle uses an AC generator, connect the leads of the DSO to the pickup coil leads. The scope needs to be on a low scale, typically approximately 100 millivolts per division. When the distributor shaft is spun, the coil will have an AC voltage generated in it, which will show on the screen. **Figure 18** shows the AC sine wave from a pickup coil with the distributor turning. You can see that the wave off the typical pickup unit will not be exactly a sine wave, but it will consist of a positive and a negative pulse and be apparent even at cranking speeds. Most units will generate at least 250 millivolts while cranking. If the pickup coil cannot generate sufficient voltage, the module will not be able to trigger and the vehicle will not run. Always adjust the volts per division to be able to capture the entire waveform. Notice that we have set the DSO up for 5 volts per division to be able to capture the waveform. Another method of measuring this AC signal is with a simple AC voltmeter. Again the pickup unit is connected to the voltmeter leads, and the meter is placed on the lowest range. The meter will register the AC voltage during cranking. Measure this voltage as close to the ignition module as possible, because it is the module that must see this voltage if it is to turn off primary current and fire the plugs.

To accurately test a Hall effect sensor, we can again look at the voltage that is changing from a high level to a lower level as the magnetic field is blocked and then allowed to pass into the sensor. This is not a sine wave signal because the voltage never changes polarity but is pulled down lower than the source. A square wave like that shown in **Figure 19** will be produced by a properly

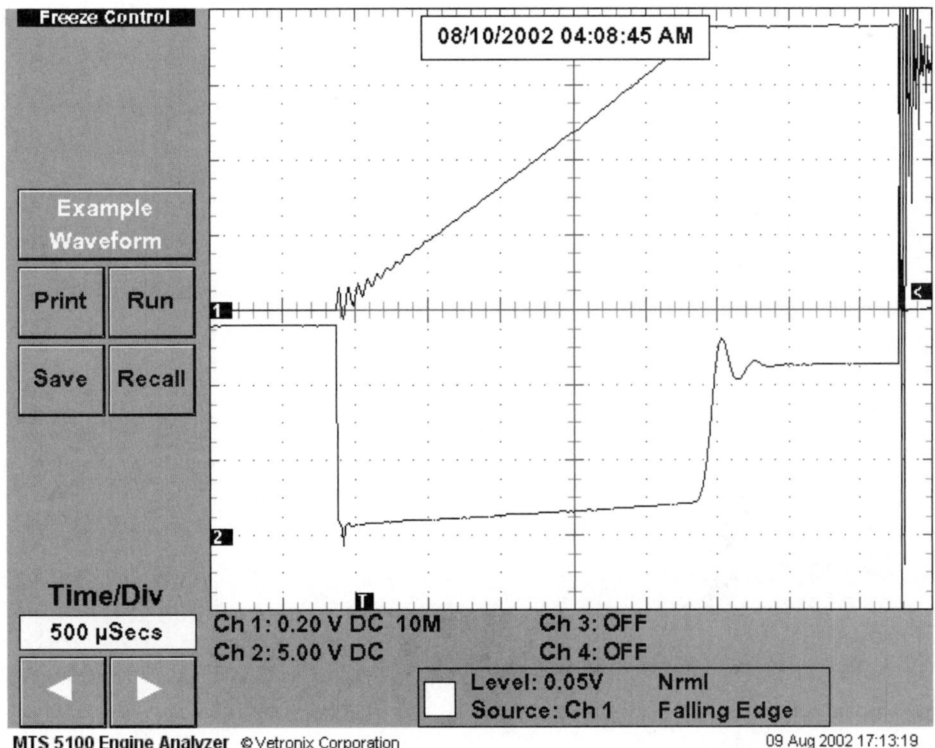

Figure 17. Dual trace, voltage on the bottom and current on top.

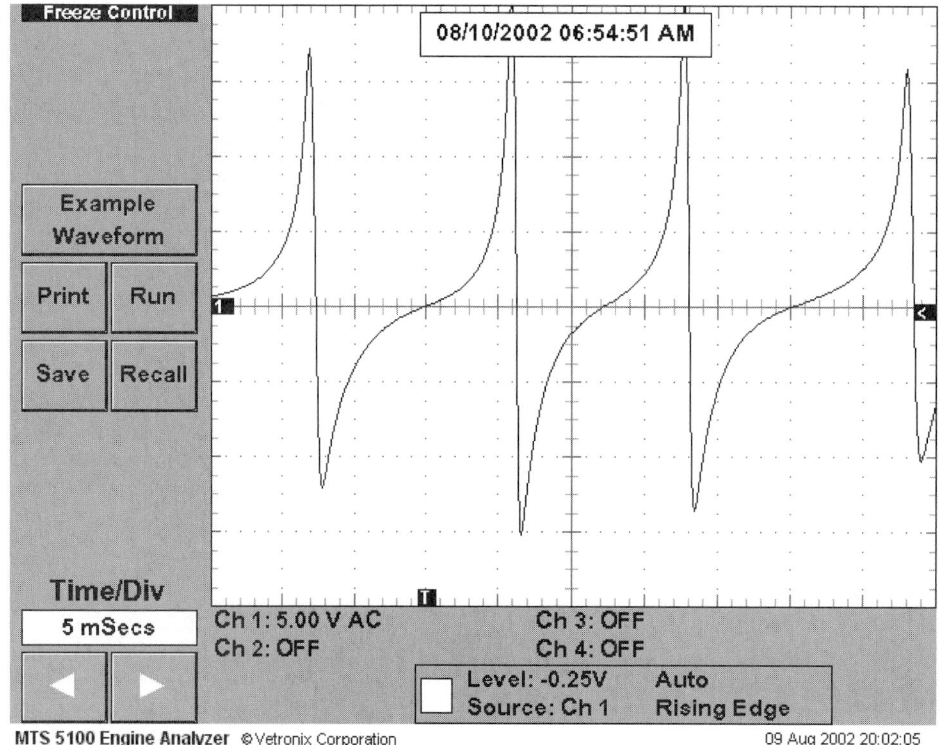

Figure 18. An AC sine wave from a magnetic sensor.

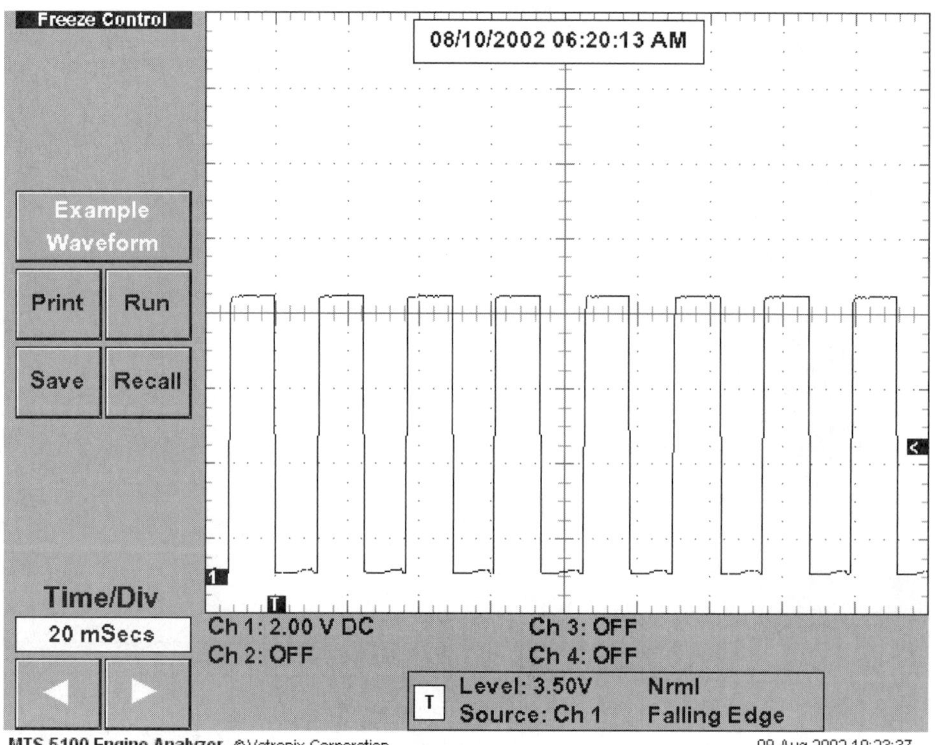

Figure 19. A Hall effect sensor pattern.

functioning Hall effect sensor. Remember that an input voltage, usually the same as the electrical system, must be present at the sensor before it can produce the square wave that will give piston position to the system. It is best to test the sensor by backprobing the connector at the module and cranking the vehicle over. Independently testing the sensor is difficult because you will have to supply input voltage and ground and then turn the distributor or the engine (for crankshaft sensors). The output can be measured on the DSO.

The procedure for testing an optical sensor is the same as for a Hall effect. The output should be an alternate high voltage (usually approximately 5 volts) and zero volts with the output being a square wave.

NO-SPARK TESTING

A good diagnosis opportunity will frequently appear in shops today. The engine cranks well and has fuel but lacks spark. The technician is faced with the problem of picking out the failed component, replacing it, and trying to determine why it failed so that a repeat repair will not be necessary in a short period of time. Electronic ignition is extremely reliable and has some expensive components. The high cost of each item makes a hit-and-miss approach, where we replace the most likely cause of the no spark, unacceptable, especially to most customers. In addition, if the module is faulty and replaced, most customers will want some assurance that the module will not require replacement at some future time. The inconvenience of having to face the day without the use of the vehicle makes customers seek out a repair that will last. Fix it right the first time.

There are many different methods of diagnosing a no-spark condition, in addition to many different types of equipment that claim to make the technician's job easier and faster. Generally speaking, the most readily available test tools work out the best. They might not necessarily be the fastest, but they frequently are the most reliable. The test procedure outlined here makes use of two pieces of equipment—a 12-volt test light and a DSO—and visual inspection. It also ignores the direct testing of the module and instead tests all of the other components in the system. If all other components are found to be within specification, the untested module is replaced. This procedure allows you to test the items that can be tested with the test light and the DSO. Most modules require expensive testers that usually work on only selected systems. By ignoring the module until we have tested the rest of the system, we will be indirectly testing it. If you do the procedure exactly the same each time, you will get very good and fast at it. Your accuracy will improve naturally. Do not change the sequence of the procedure, and keep in mind that we do not want to connect or disconnect the module with the ignition key on. The spike that might occur could destroy an otherwise good module, leaving you with the original cause of the no-spark in addition to a burned module.

Procedure for No-Start, No-Spark

1. Locate a wiring diagram that shows the ignition system in detail. An example of a current system is shown in **Figure 20**. Notice that the diagram shows the module, the ignition coil, the optical sensors, and all the connectors. The color of the wires is also noted for location assistance.
2. Use your 12-volt test light to determine if the system has power during both run and crank positions. Use the wiring diagram and find the correct wires. In our example, the run terminal of the ignition switch runs to the ignition coil and is pink. The same pink wire is paralleled to the ignition coil module. With the ignition key off, disconnect the connector at the module. The connector has four wires. Once the connector is disconnected, turn the key on and probe the pink cavity, testing for power. The test light should light brightly, indicating power. Do the same with the Bat terminal of the coil. It should also light the test light with the key in the run position. Once we have determined that we have B+ available with the key in the run position, we need to determine that B+ is also available with the key in the start position. The same pink wire brings power during cranking, so it is easy to turn the key to start and check for power. If you disconnect the S terminal of the solenoid, the engine will not crank. Power should also be present at the Bat terminal of the coil. If B+ is not present at either test point, the ignition system cannot function. A repair must be completed after the exact spot of the open circuit is located. B+ must be present before you go on.
3. Next we go to the ignition coil and test for a pulsing test light with the light on the negative side of the coil by backprobing the DK GRN wire. If the system is functioning correctly, the ignition module will pull the negative of the coil to ground, turning the light off. As the module opens up the circuit, the light will go back on. If the light pulses, the primary is functional and our problem lies in the secondary.
4. The white wire in the module connector comes from the powertrain module and will be a pulsing signal similar to a Hall effect sensor. This signal is easily tested with your DSO on 1 volt per division.
5. The next component to test is the piston position sensor, which in this case is an optical sensor. Use the DSO and check the signal on the high-resolution and the low-resolution signal (RED/BLK and PPL/WHT). Notice that the RED wire is power to the sensor and the PNK/BLK wire is the ground.

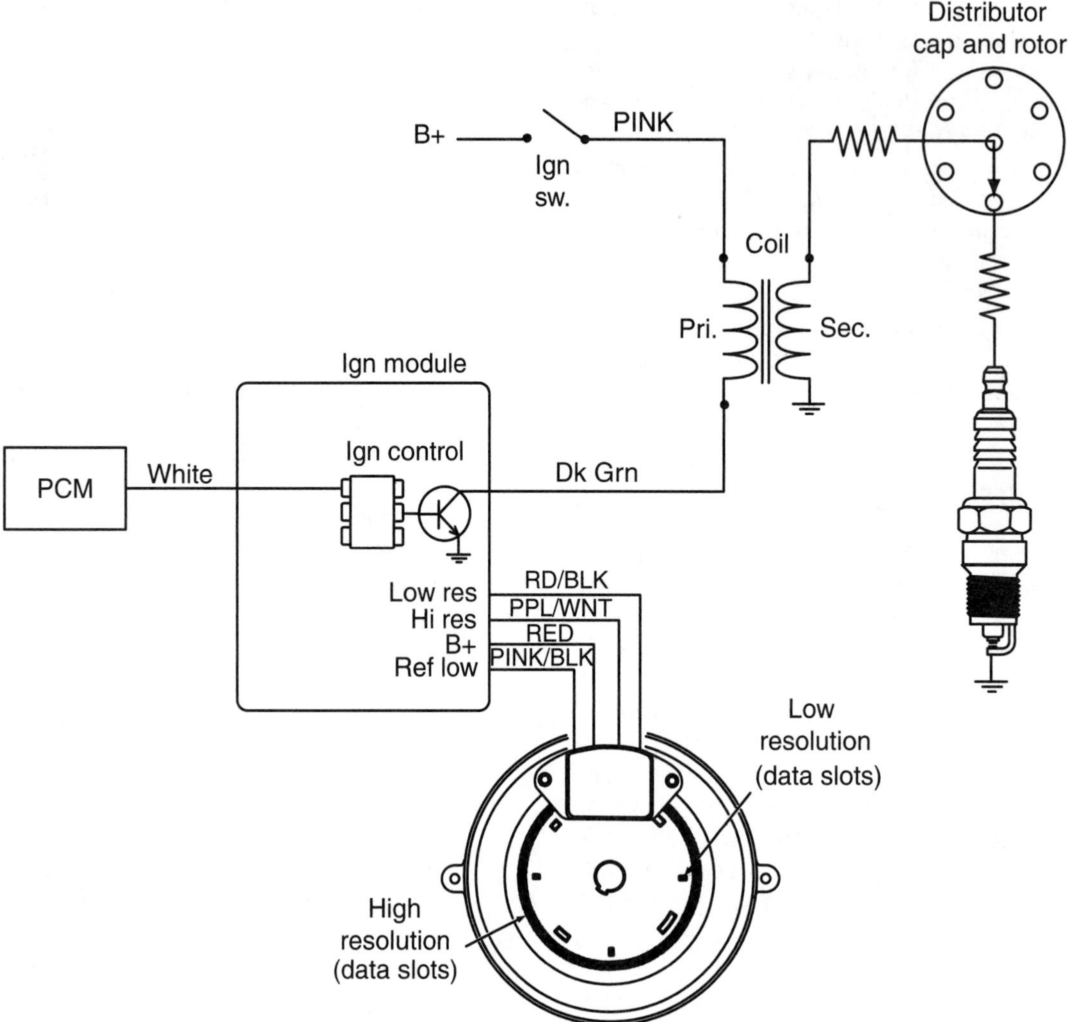

Figure 20. An optical sensor ignition system.

6. Use an ohmmeter on the ignition coil primary and secondary to see if it is to specification. Many ignition coils are such a low resistance that they are best tested by substitution of a known good coil.

With those six tests, you will be able to determine the validity of the major components with the exception of the module. If all of the tests were good, we would know that the module is not doing its job and that it should be replaced. By systematically testing around the module, we verify that all inputs are functional and that the coil is also functional. Use any manual that will give both a wiring diagram and the necessary specifications.

Summary

- Base timing is easily checked with a timing light.
- The advance function can be checked with the advanced timing light.
- The timing is changed by loosening the distributor hold-down and turning the distributor.
- A computerized advancing system needs to be put into base timing mode to check timing.
- Primary inductive voltage should be between 200 and 400 volts.
- Spark burn time should be between 0.8 and 2.0 milliseconds.
- Diminishing oscillations should follow the spark line.
- Most ignition systems will turn on primary current for 2.5 or 3.0 milliseconds minimum.

- When primary current turns on, the voltage trace should drop to zero volts.
- Current traces can show the maximum amount of current.
- The DSO can be used to check input sensors.
- A magnetic sensor will generate an AC sine wave.
- A Hall effect sensor will generate a pulsing DC signal.
- An optical sensor will generate a pulsing DC signal.

Review Questions

1. Initial timing on a noncomputerized vehicle is generally set at a specified speed with the
   A. transmission in reverse
   B. vacuum advance connected
   C. engine at 1,000 rpm
   D. vacuum advance disconnected and plugged
2. Mechanical advance is checked with a(n)
   A. advance timing light and a tachometer
   B. oscilloscope
   C. inductive timing light and a dwellmeter
   D. feeler gauge and a spring tension tester
3. A four-function ignition system is being tested on a DSO. Technician A says that variable dwell will be observed between the primary current on and off signals. Technician B says that the quality of the ground can be seen just before primary current on. Who is correct?
   A. Technician A only
   B. Technician B only
   C. Both Technician A and Technician B
   D. Neither Technician A nor Technician B
4. The current limit blip will appear
   A. once the coil is fully saturated
   B. before the coil is fully saturated
   C. after the cylinder plug fires
   D. just after the primary current-on signal
5. A primary ignition winding is being tested. Its resistance is three times the specification. This will cause
   A. longer spark lines
   B. longer dwell
   C. reduced secondary output
   D. nothing
6. Less than 0.5 millisecond of the spark is observed. Technician A says that this will increase the chances of misfire. Technician B says that this might be caused by the firing voltages being excessive. Who is correct?
   A. Technician A only
   B. Technician B only
   C. Both Technician A and Technician B
   D. Neither Technician A nor Technician B
7. Hall effect voltages will increase with speed.
   A. True
   B. False
8. Magnetic pickup voltages will increase with speed.
   A. True
   B. False
9. Initial timing must be correct on computer-controlled vehicles or the entire curve will be off.
   A. True
   B. False
10. To initially time a computer-controlled vehicle, the engine must be at a specific speed and the vacuum advance must be disconnected.
    A. True
    B. False
11. Current limit blips are found only on systems without primary resistors.
    A. True
    B. False
12. Excessive coil primary resistance will decrease the available voltage.
    A. True
    B. False

Chapter 50 Distributorless Ignition Secondary Circuits

Introduction

Up to this point, we have looked at a steadily evolving ignition system. If we start with breaker points, moving into electronic ignition, E-coils, and computer control of timing advance curves, we can see electronics have taken over mechanical devices. In addition, we can see service intervals have increased, from 10,000 miles for the old fashioned tune up to 100,000 miles or more between spark plug changes. Obviously, the majority of the consumers are pleased with increased service intervals, even if they are not aware of the evolution of the ignition system. However, even with the many changes studied so far, the major automotive manufacturers were still faced with minor problems centered around the distributor. As a mechanical device, it is subject to wear and can be responsible for unwanted timing change. In addition, consumers or uninformed technicians could easily set timing different from the specifications and increase tailpipe emissions. Longer emission warranties and more concern for "total" control of timing made distributorless ignition (EI) a natural evolution for the ignition engineers. Distributorless ignition was abbreviated as DIS until the SAE J-1930 term went into effect. DIS became EI. In addition, EI allows for greatly increased available dwell. A six-cylinder vehicle with EI can have as much as 720 degrees of available dwell (measured at the crankshaft), whereas the same vehicle with a distributor would have a maximum of 120 degrees. With the development of the EI, there is one less mechanical part to wear out, one less place where high ignition secondary voltage might find a path to ground, and one less component to have an effect on emissions.

In this chapter, we look at three of the common EI systems on the road. You will see that their introduction does not mean that you will have to start over and learn volumes of new information. Instead, you will find that the technology used is basically the same as in computer-controlled ignition systems, which have been around since the early 1980s. The manufacturers took existing technology, gave it a new twist, and adapted it to the EI systems of today. We will look at the most popular systems on the road and divide them up by the way they connect the coil(s) to the plugs. This approach should greatly simplify learning the systems. As usual, do not go on in your reading until the section just finished makes sense. The systems are not complicated if taken in small, bite-size pieces, which is exactly how we are going to approach them.

WASTE SPARK SYSTEMS

In the ignition systems that we have studied and repaired in past chapters, one end of the ignition coil's secondary winding was connected to either ground or the primary winding. The other end was connected to the spark plug through a cap, rotor, and wires combination. **Figure 1** and **Figure 2** illustrate these different methods of wiring an ignition coil's secondary winding. The early years of ignition systems saw the majority of manufacturers using secondary connected to primary wiring, as diagrammed in Figure 1. In the mid 1970s, some of the manufacturers began using a grounded secondary, as illustrated in Figure 2. Both of

Waste spark systems fire both companion plugs at the same time.

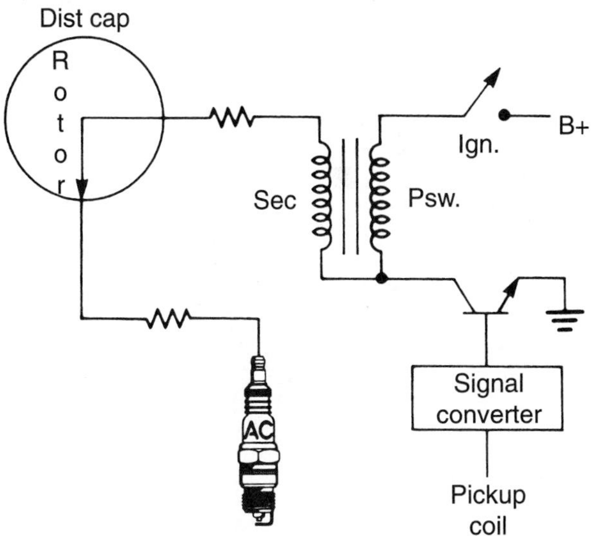

Figure 1. A coil secondary winding connected to the primary.

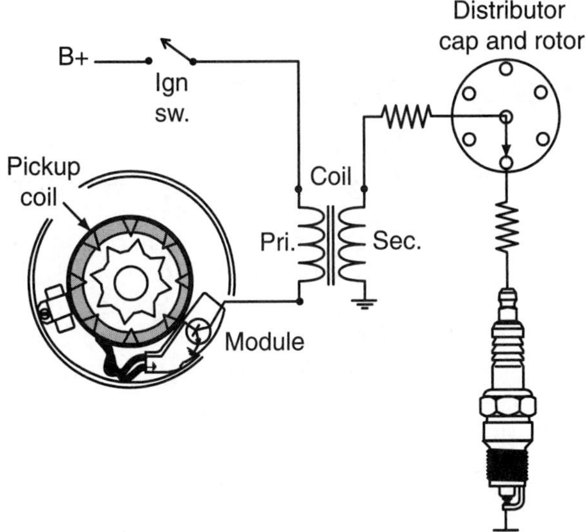

Figure 2. A coil secondary connected to ground.

these methods result in the same high-voltage spark being delivered to the plug; however, grounding the secondary appeared to have less RFI on some applications.

Look at **Figure 3**. This is the typical EI system, in which both ends of the secondary winding are attached to spark plugs. The two plugs are in companion cylinders (cylinders whose pistons are in the same position, moving in the same direction, but on different strokes). Cylinders on intake and power or compression and exhaust are examples of companion cylinders. Notice that the secondary winding is not connected electrically to the primary as in early ignition systems. This is a series circuit where both plugs will fire at the same time when the coil has reached sufficient voltage to

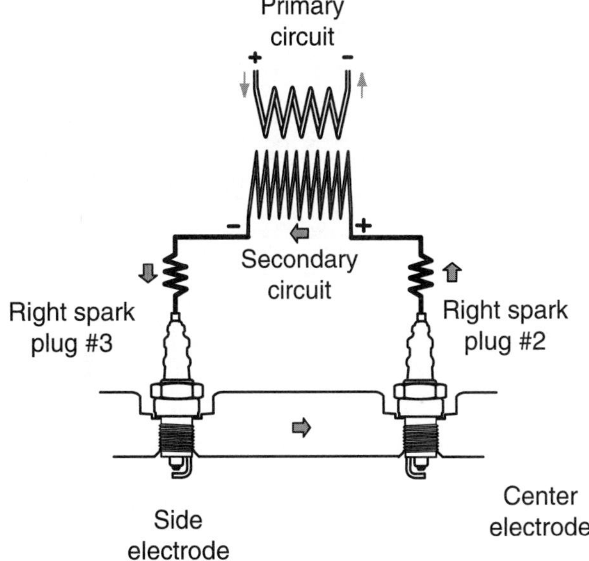

Figure 3. Each end of the secondary winding is connected to a spark plug.

overcome all the resistance of the circuit, including the two plugs. In this case, the cylinder head is the wire or conductor between the two plugs. We are firing both plugs at the same time in companion cylinders. The high compression and combustible mixture in the cylinder, which is on the compression stroke, will require a high voltage. We are able to see this on an ignition scope when we accelerate. The added pressure in the cylinder of the additional mixture during acceleration increases the firing voltage. The opposite is also true. If less pressure is placed on the plug, it will require less voltage to fire because its effective resistance has been reduced. This means that the plug firing on exhaust will fire at a lower voltage.

You should have figured out that one plug will have current flow from the ground electrode to the center electrode, whereas the other plug will have current flow from the center electrode to the ground electrode. In ignition systems of the past, coil polarity was extremely important. Plugs fired with opposite polarity required greater voltage, which was not available. For years, technicians checked to make sure that the firing line went straight up when using a scope. Firing lines going down indicated reverse coil polarity and a problem. With only 20 kilovolts, available misfires occurred with reverse coil polarity. However, with EI, available voltages are in excess

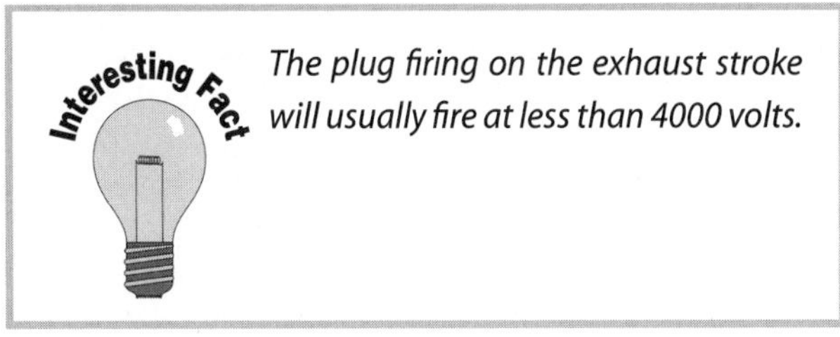

The plug firing on the exhaust stroke will usually fire at less than 4000 volts.

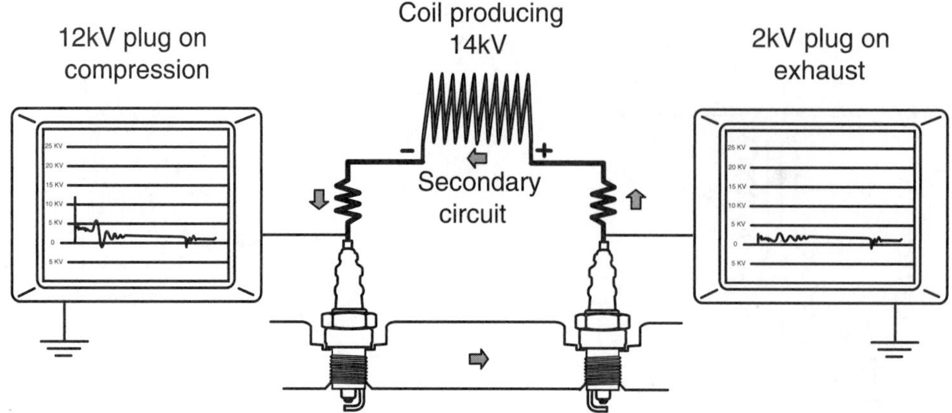

Figure 4. Coil voltage will divide across the two plugs.

of 50 kilovolts, which is enough coil power to fire just about anything, including a plug with reverse polarity.

With this style of EI, a single coil is required for each two cylinders, for example, two coils for a four-cylinder engine, three coils for a six-cylinder engine, and so on.

Keep in mind that the function and control of the primary will be exactly the same as it was in past ignition systems. Rapidly build and collapse a strong magnetic field across the secondary winding to produce the high voltage necessary to fire the plug, or in this case both plugs. In addition, realize that the total voltage required of the coil secondary will be the total of both spark plugs. If the igniting plug requires 12,000 volts and the waste plug requires 2000 volts, then the ignition coil secondary will produce 14,000 volts **(Figure 4)**:

12,000 volts + 2000 volts = 14,000 volts

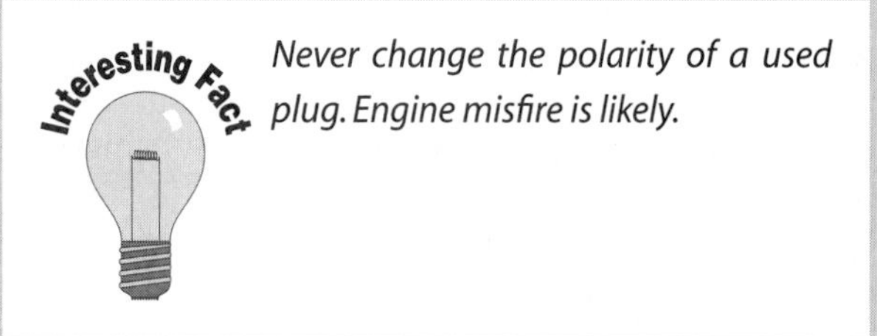

Never change the polarity of a used plug. Engine misfire is likely.

Some scopes on the market allow you to read the individual voltages at each plug on compression and on exhaust. The DSO that we will be using does not display exhaust firings because they are "waste" and generally serve no purpose. If the scope does display waste voltages, you should recognize that additional resistance might cause the waste voltage to rise. The compression voltage is important because it ignites the mixture. It is also important that you realize that an individual spark plug will fire alternately between igniting the mixture and waste spark. The plug polarity does not change. If it is a positive plug, it will fire in a positive direction whether it is waste or igniting the mixture. The polarity of the plug is determined when it is installed. Its polarity should not be changed by reinstalling it in a different cylinder. A positive plug will wear the center electrode more than the ground. A negative plug will wear the ground more than the center.

COIL-NEAR-PLUG SYSTEMS

Another type of system needs to be discussed and analyzed because it is very popular: the coil-near-plug variety. **Figure 5** illustrates this system. Notice that each cylinder has its own ignition coil and a short little ignition secondary wire that connects the coil to the spark plug. With a single coil for each plug, there is no waste spark to worry about. Recognize, however, that the primary circuit for a coil-near-plug or a coil-on-plug system is more complicated because it involves twice as many triggering devices (modules) as a waste spark ignition system. One of the benefits to the system is the ability to completely control each cylinder separately. In theory, each cylinder can have its own timing control. The ignition modules can be

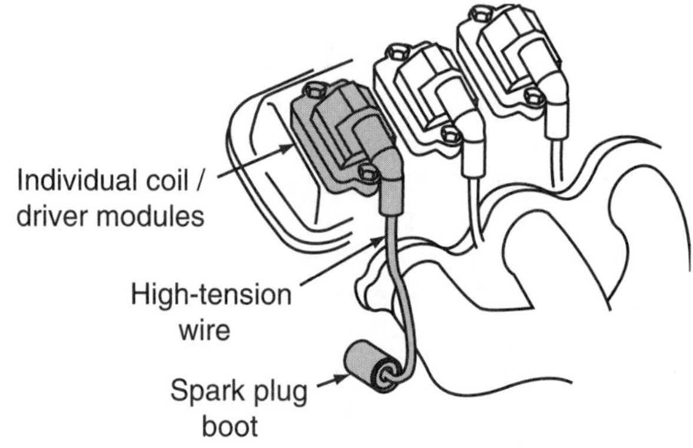

Figure 5. The ignition coil and module are one assembly in the General Motors coil-near-plug system.

Figure 6. The heat sink on top of the coil cools the ignition module.

remotely mounted, be part of the PCM, or in some cases be a part of the ignition coil. If the module is part of the coil, as seen in **Figure 6**, you will have to rely on secondary voltage and primary current for diagnosis. The ability to get a primary voltage pattern is eliminated. The PCM will still receive signals from various sensors to indicate piston position and speed. In addition, all of these systems have PCM timing control and eliminate the need to set base timing. The position of the piston position sensor or a signal from the PCM will determine base timing.

COIL-ON-PLUG SYSTEMS

Figure 7 shows a Ford coil-on-plug system. Notice that there are no ignition modules as in the General Motors coil-near-plug system that we just looked at. The ignition module function has been taken over by the PCM. This example

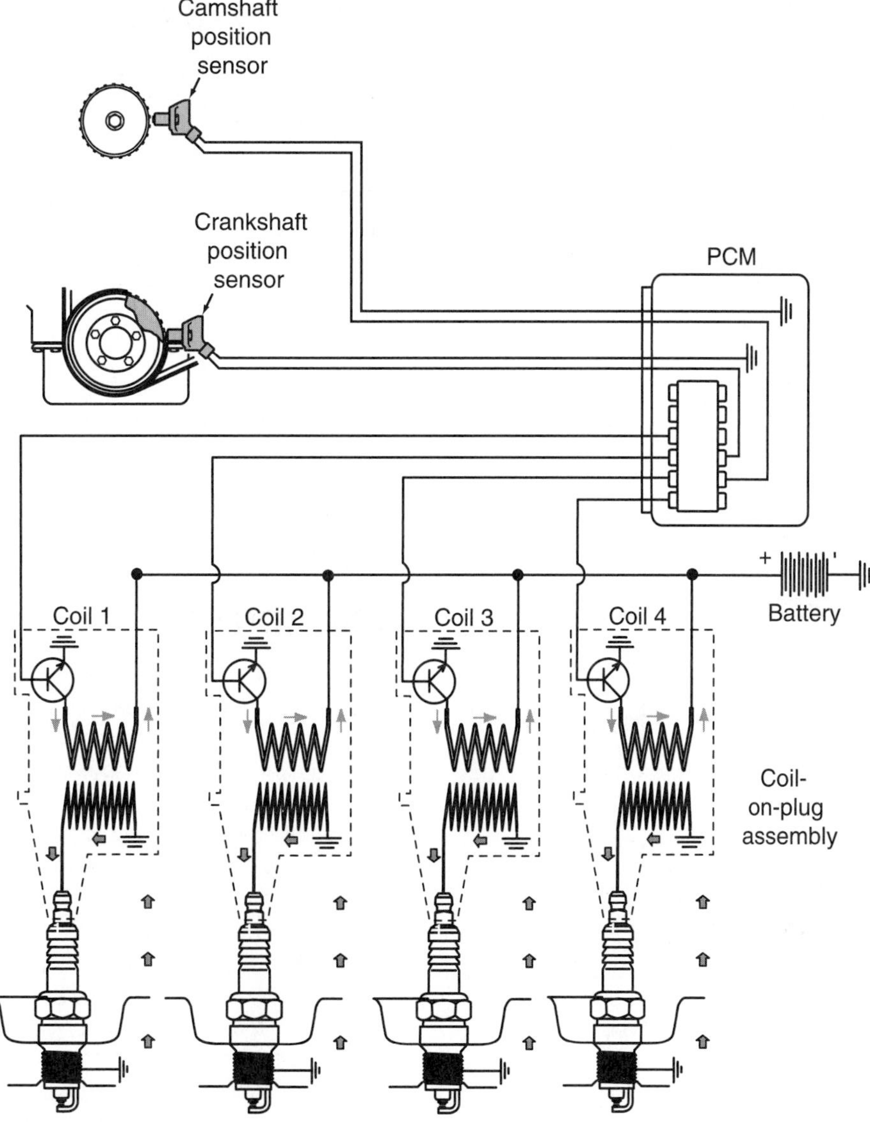

Figure 7. Coil-on-plug systems eliminate plug wires.

is for a four-cylinder system, so there are four coils, and this means that the PCM has four separate primary current drivers. Each coil requires its own driver.

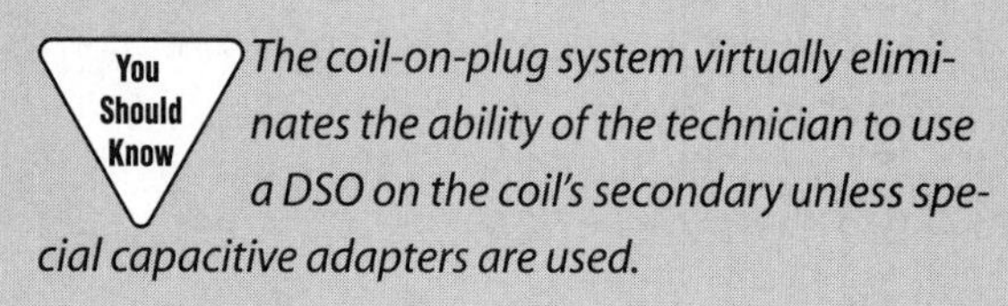

The coil-on-plug system virtually eliminates the ability of the technician to use a DSO on the coil's secondary unless special capacitive adapters are used.

This system has virtually no secondary to worry about. There are no wires because the coil is directly connected to the spark plugs. It is very difficult, on some engines, to obtain a secondary ignition pattern for this system. Traditional wire connectors will not work. Special adapters are required to allow for secondary DSO pattern analysis. To diagnose this system, you will rely on primary voltage patterns and primary current waveforms. More than enough information is available through primary analysis to diagnose and repair the system. We will look at using the primary for diagnosis in Chapter 53.

SPARK PLUG USE

El systems typically use the highest quality extended-life spark plugs. With some manufacturers recommending tune-up intervals in the 100,000-mile range, it makes sense that the plugs will be of the coated variety. We discussed platinum-tip plugs in Chapter 46, and **Figure 8** shows AC plugs that are considered to be an extended-life variety. The small tip and very low resistance platinum allow it to have an extended life. There is much debate among technicians regarding the 100,000-mile tune up, and realistically few vehicles will be able to make it the full interval. It is important that you recognize that more frequent plug replacement is normal because of the driving conditions of the vehicle. A vehicle that spends part or all of its time in low-speed operation or in temperature extremes will never make 100,000 miles before it needs plugs. The loading of the vehicle also comes into play. Consider the vehicle that tows a camper or boat part of the time, or the pickup truck that is hauling heavy equipment. Most vehicles will not make the tune-up interval and will need your skills of diagnosis and repair more frequently. Do not forget that spark plug gap, heat range, and reach are just as important with an El system as with a distributed system. The correct plug must always be installed. Make sure you look it up.

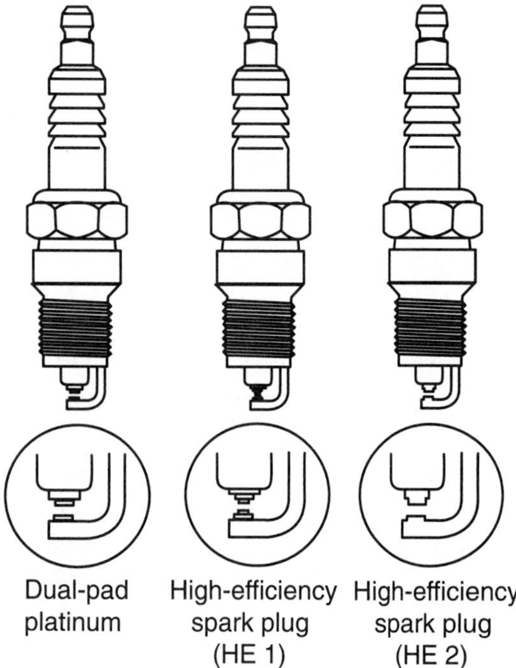

Figure 8. Extended-life spark plugs.

IGNITION WIRES

Ignition wires are basically the same as for distributed vehicles in all but the coil-on-plug variety. They are mostly magnetic suppression with the center conductor wrapped with an additional conductor to help eliminate RFI. We can measure the resistance with an ohmmeter or use the information from the DSO to diagnose them. It is important to note that when El began to show up on vehicles, in the early 1990s, the quality of ignition cable and connector boots improved. The longer cable required for a waste spark system allows the most opportunity for high-voltage leakage to nearby metal. The very short (couple of inches) cable used with a coil-near-plug system eliminates most of this problem. Ignition cables will usually last the life of the plugs. If customers try to go the full 100,000 miles between plug changes, it is likely that the secondary ignition cables will need to be replaced also.

IGNITION COILS

Ignition coils used in El systems are different from those used in a distributed system (DI). You have probably figured out that a waste spark system and a coil-on-plug system cannot use the same coil, so there must be differences. Let us look at waste spark systems first.

Within the category of El that uses waste spark, two different types of ignition coils are in use: type 1 and type 2. The difference is in how the coil is manufactured. If the three coils for a six-cylinder vehicle are formed together, as shown in **Figure 9**, the coil is considered to be a type 1. However, if the three are separate, the system is a type 2, as seen in **Figure 10**. Remember that the secondary of each coil is connected to a plug wire and a spark plug. There is no real positive connection to the coil because both plugs are grounded, and there is no electrical connection to the primary.

Figure 9. A type 1 coil pack for a six-cylinder engine.

The coil-near-plug ignition coil is probably the closest thing to a DI coil because one end of the secondary is connected to the plug wire. The other end of the secondary can be connected to the engine ground or to the primary of the coil **(Figure 11)**. From a functional standpoint, all three ignition types use the same coil principles: turn on primary current for a sufficient period of time to saturate the coil winding, turn off the primary current and allow the magnetic field to collapse across the secondary and generate the high voltage, and use the high voltage to fire one plug (coil near or on plug) or two plugs (waste spark). The major difference comes in the ability to use a DSO on the secondary. As we have mentioned before, the coil-on-plug system does not easily allow for secondary pattern analysis. In Chapter 53, you will see that we will rely on primary signals for the needed information.

Figure 10. A type 2 coil pack with individual coils.

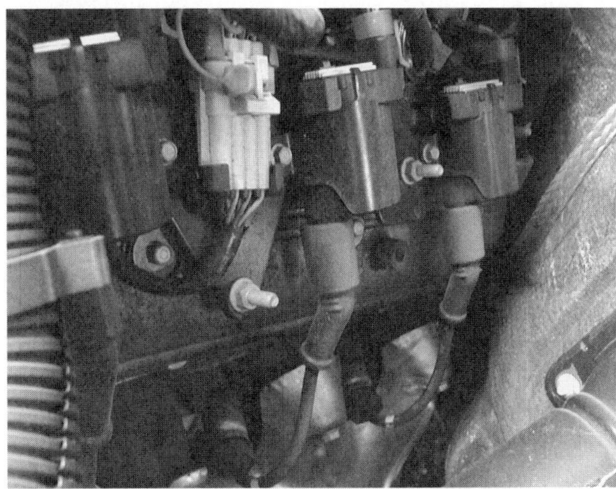

Figure 11. Short plug wires connect the coil's secondary to the spark plugs.

Summary

- The EI allows for more available dwell time (time that primary current flows).
- In a waste spark system:
  1. The secondary path relies on spark plugs for continuity.
  2. The two ends of the secondary coil are connected to spark plugs.
  3. One plug fires positively and one negatively.
  4. The plugs that are connected to one coil are in companion cylinders.
  5. The plug on compression will fire at a higher voltage than the one on exhaust.
- In a coil-near-plug system:
  1. Each spark plug has its own ignition coil.
  2. Each coil may be controlled by a remote or internal module or be PCM controlled.
  3. A short ignition cable connects the coil to the spark plug.
- In a coil-on-plug system:
  1. There are no ignition secondary cables.
  2. There is one coil per plug.
  3. Getting a DSO secondary pattern is very difficult.
  4. The PCM generally has the ignition module function.

Review Questions

1. Technician A states that most EI systems have no method of setting ignition timing. Technician B states that advance curves are determined by the PCM. Who is correct?
   A. Technician A only
   B. Technician B only
   C. Both Technician A and Technician B
   D. Neither Technician A nor Technician B
2. Waste spark refers to the
   A. plug that is firing while the cylinder is on the compression stroke
   B. timing change that the computer will produce
   C. reference pulse that the computer will use to determine engine speed
   D. plug that will fire while the cylinder is on the exhaust stroke
3. Companion cylinders are cylinders that are
   A. in the same position
   B. going in the same direction
   C. on opposite strokes, that is, intake/power or compression/exhaust
   D. all of the above
4. Firing voltage for the plug on the compression stroke is
   A. lower than for the plug on the exhaust stroke
   B. higher than for the plug on the exhaust stroke
   C. the same for both plugs
   D. always 6 kilovolts
5. Technician A states that on some EI systems it is difficult or impossible to get a secondary pattern. Technician B states that on waste spark systems the use of plug wires allows for secondary testing. Who is correct?
   A. Technician A only
   B. Technician B only
   C. Both Technician A and Technician B
   D. Neither Technician A nor Technician B
6. Companion cylinders are always on the same stroke.
   A. True
   B. False
7. Waste spark is the spark for the cylinder that is on the exhaust stroke.
   A. True
   B. False
8. EI can use a single coil for each pair of cylinders.
   A. True
   B. False
9. EI usually has no method of setting timing.
   A. True
   B. False
10. Coil near plug uses one coil for every two cylinders.
   A. True
   B. False
11. The ignition module function might be accomplished by the PCM.
   A. True
   B. False

Chapter 51

Diagnosing and Servicing the Secondary Ignition System on a Distributorless Vehicle

Introduction

The secondary ignition system for an EI vehicle is not that much different from the secondary of a DI vehicle. We still have to provide a path for the current to reach the spark plugs and at the same time have good enough insulation to make sure that the relatively high voltage does not find an alternative path to ground. Again, keep in mind that misfires cannot be allowed on any vehicle today because of the hydrocarbons that are emitted to the atmosphere. With this very high voltage available, it is always a good idea to look for additional paths, which might develop over the years. Moisture can become a problem and can give the high voltage a shorter path than the plugs offer. In Chapter 47, we discussed using a spray bottle of water to simulate high moisture levels, which might become a path to ground. This technique works equally well on EI vehicles. Do not forget that some EI systems use a spark plug on both sides of the secondary winding. Any problems on either side of the winding will usually have an effect on both plugs and cylinders.

USING AN IGNITION SCOPE FOR SECONDARY TESTING

Whether you are using an automotive oscilloscope or a DSO with ignition capability, the testing will be the same. The scope or DSO is still a valuable tool when preventive maintenance–checking an EI system, when looking for intermittent problems, or when checking up on the tune up you have just performed. Even some older scopes can be adapted for use on EI vehicles. **Figure 1** shows a six-cylinder type 2 ignition system with a pickup connected to each spark plug wire at the coils. If your scope does not have EI capability, use the pattern pickup around each plug

Figure 1. A six-cylinder type 2 with pickups on each plug wire.

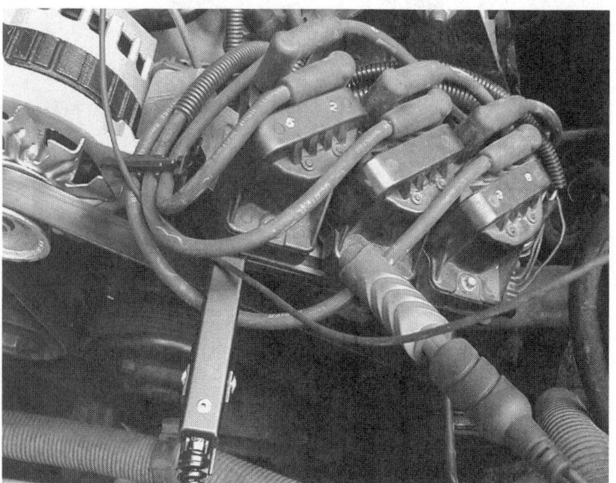

Figure 2. A single pattern pickup can be moved from plug wire to plug wire.

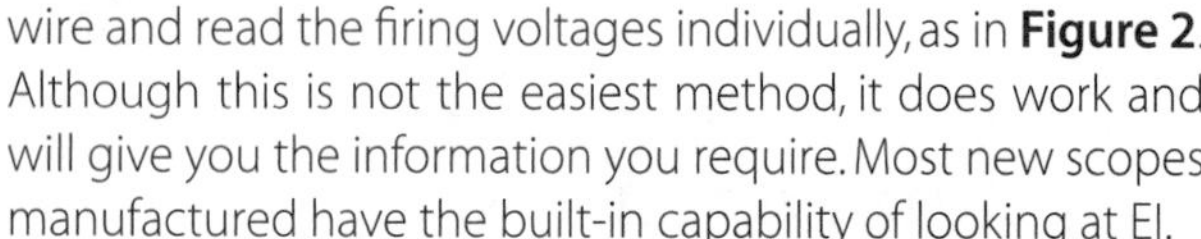

Figure 3. A spark tester can stress the system or find a weak coil.

wire and read the firing voltages individually, as in **Figure 2**. Although this is not the easiest method, it does work and will give you the information you require. Most new scopes manufactured have the built-in capability of looking at EI.

Do not let the obvious escape you. Look over the coils, wires, plug boots, and other connections that might be loose, corroded, or disconnected. Pull off the plug wires at the coil and at the plugs. Look for signs of failure. If the wire falls apart in your hands, you have found a problem. It may not be the one that brought the vehicle into your shop, but it is a problem nonetheless. You should not need some fancy piece of diagnostic equipment to detect simple problems. Just make sure you open your eyes and look for the problem. Do not forget to check the connection at the ignition module and at any sensors used.

For any testing, we must make sure that we understand that most modern EI systems can produce more voltage than their own insulation can stand. Do not, under any circumstances, open circuit one of these plug wires. Some of the systems are capable of producing close to 100,000 volts. You will destroy something with voltages this high. Instead, rely on a spark plug tester, as we have done in the past. If it has an adjustable air gap, set it for approximately 40,000 volts. If the system is capable of 40 kilovolts, it is good. There is no need to stress it any more than this. **Figure 3** shows the spark tester installed in the plug wire of a type 2 General Motors waste spark system. It has been set to 40,000 volts and the vehicle cranked. This should stress the coil and give us the information that we need about this coil. Remember that if multiple coils are present, they must all be tested. In addition, both towers should be separately tested. They are individual coils, and it is possible to have failure on one, while the other one(s) are fine. Do not crank the engine over for excessive periods of time during this test. Do it just long enough to get a scope reading and see the quality of the spark at the tester. While the engine is cranking, the fuel system will be pumping fuel into the manifold. If extensive cranking is done, you will flood the engine, and then you will have two problems—the one the vehicle came in with and a flooded engine. If you are dealing with a fuel injection system that is easily disconnected, disconnect the injectors to prevent flooding during this cranking test. Do not forget to test each coil pack. If you find a weak coil, place your test plug in the other plug wire from the suspected coil and retest. This will isolate the coil from a possibly bad plug wire. This procedure is most frequently done on a no-start situation or a misfire.

The manner in which you connect the ignition system to the DSO depends on the DSO manufacturer's recommendation and the plug polarity. If you are connecting to all plug wires at the same time, you must observe the plug's polarity or your scope will read the voltage incorrectly. **Figure 4** shows a generic hookup guide for the six-cylinder vehicle we are testing. Notice that cylinders 1, 2, and 3 have positive polarity and must have red leads connected to them. Cylinders 4, 5, and 6 are negative and need to have the black leads connected to the plugs. Connect the trigger or sync to cylinder 1 and the DSO's ground and you are ready to begin testing. Crank the engine over or start it. The results will direct you to the correct coil. Normally, with a 40,000-kilovolt plug gap in the system, the scope will show you the actual voltage. Adjust the volts per division until you can see the top of the firing line. In **Figure 5**, the coil is producing approximately 34 kilovolts. This figure was produced with a set of EI adapters installed, so each plug has its own pickup.

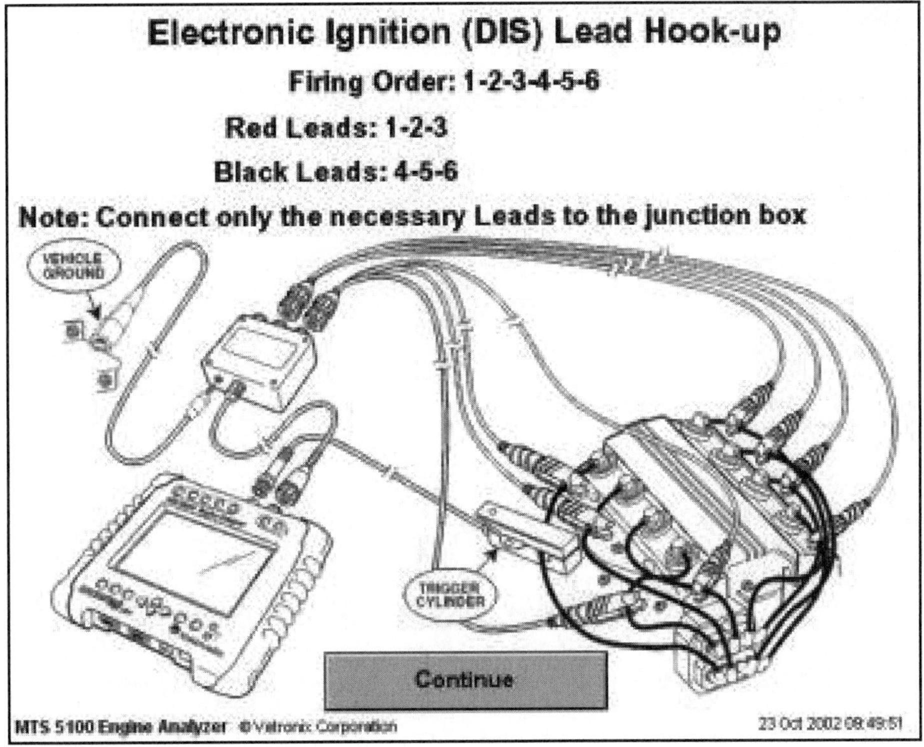

Figure 4. A lead connection guide showing spark plug polarity.

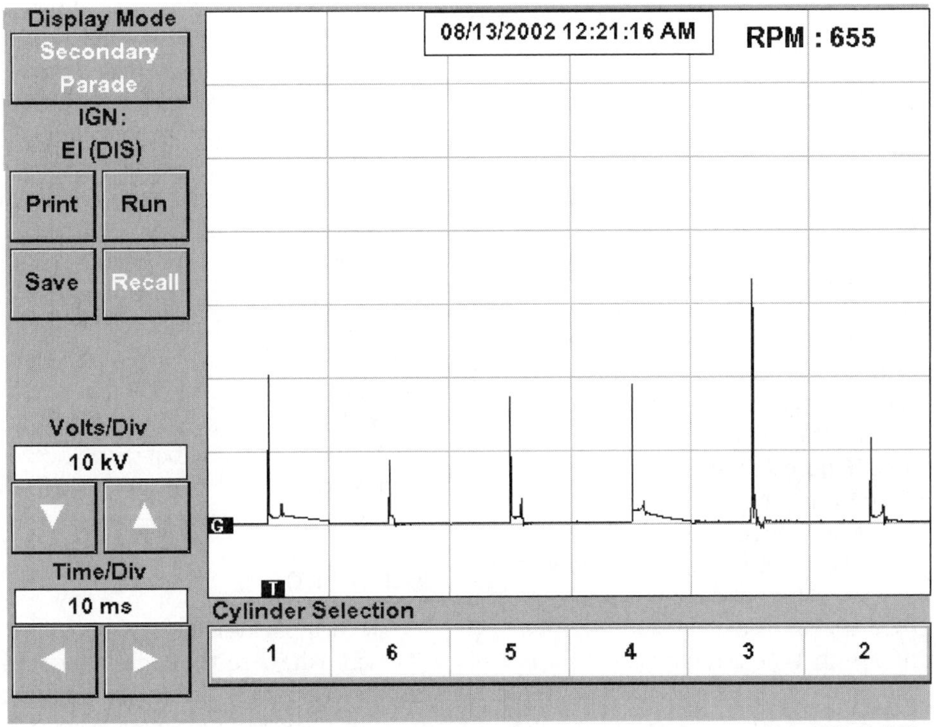

Figure 5. Cylinder 3 firing a spark tester at 34 kilovolts.

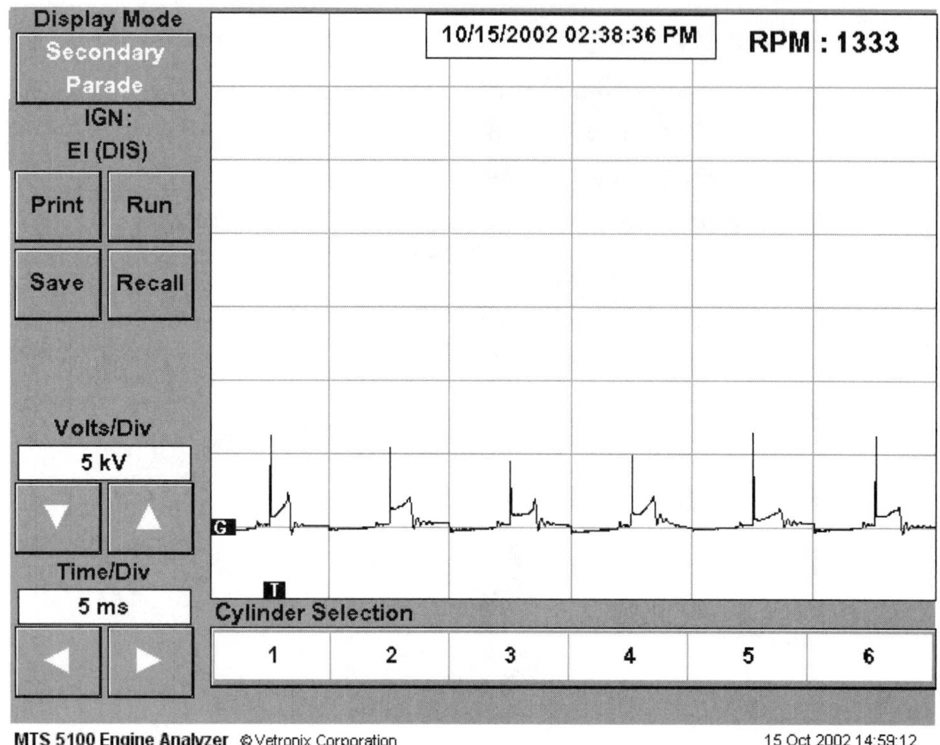

Figure 6. A six-cylinder display with cylinder firing voltages.

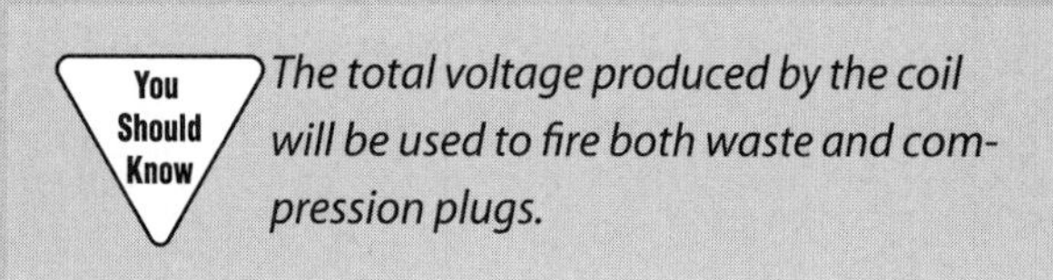

The total voltage produced by the coil will be used to fire both waste and compression plugs.

Firing Voltage

Your scope will probably show plug firing on the compression stroke identified as **Figure 6** shows. The individual cylinder number is printed on the bottom of the trace. Cylinder 1 is firing at the highest level of approximately 7.0 kilovolts, while cylinder 2 is at approximately 4 kilovolts. Many DSOs will allow firing kilovolts to be displayed in another format. **Figure 7** shows the same engine firing voltage in bar graph format. Notice that the cylinders appear to be about the same. Cylinder 1 is the highest with an average of 7.1 kilovolts, and cylinder 2 is the lowest with an average of 5.0 kilovolts. The minimum, the maximum, and the average appear below the bar graph for each cylinder. Notice also that the bar graphs appear in the firing order, which on this engine is the same as the cylinder numbering. If your DSO or scope will give both compression and exhaust firings, switch over and look at the data.

Figure 8 shows the power (compression) and the waste (exhaust) voltages. Notice that the power voltages are substantially higher than the waste voltages. This is normal. The compression within the cylinder makes it harder for the current to flow and forces the coil to produce higher voltages. When the plug is on the exhaust stroke it takes substantially less voltage to push the current across the plug gap. It is the combined voltage that the coil must produce. It is normal for the waste spark to be approximately 2 kilovolts or less. The lowest firing voltage displayed is cylinder 2, which is the second one in the firing order, whereas the first cylinder fires at the highest voltage. Observing the waste spark can be useful because it can indicate wire or plug problems. The range in our firing voltage (power stroke) is from a low of 5.0 kilovolts to a high of 7.0 kilovolts. The same specification applies here as that we saw on distributed vehicles: less than a 3-kilovolt range. Additionally, all cylinders should be firing at less than 12 kilovolts. This vehicle is in specification on both counts.

Burn Time

Remember the importance of correct burn time that we stressed in the DI section? The importance of burn time is no less for an EI vehicle. The amount of burn time, or spark time, must be correct or the vehicle will misfire. Most EI systems use the same 0.8-millisecond burn time as their minimum. For this reading, the raster type of display is the easiest to see.

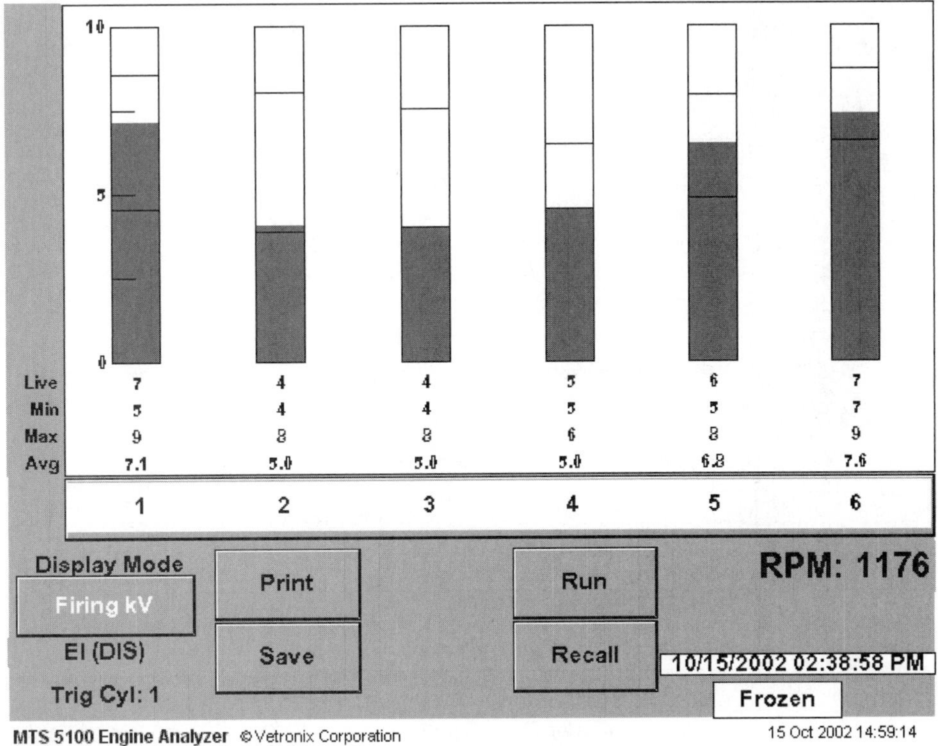

Figure 7. Six-cylinder firing voltage in bar graph format.

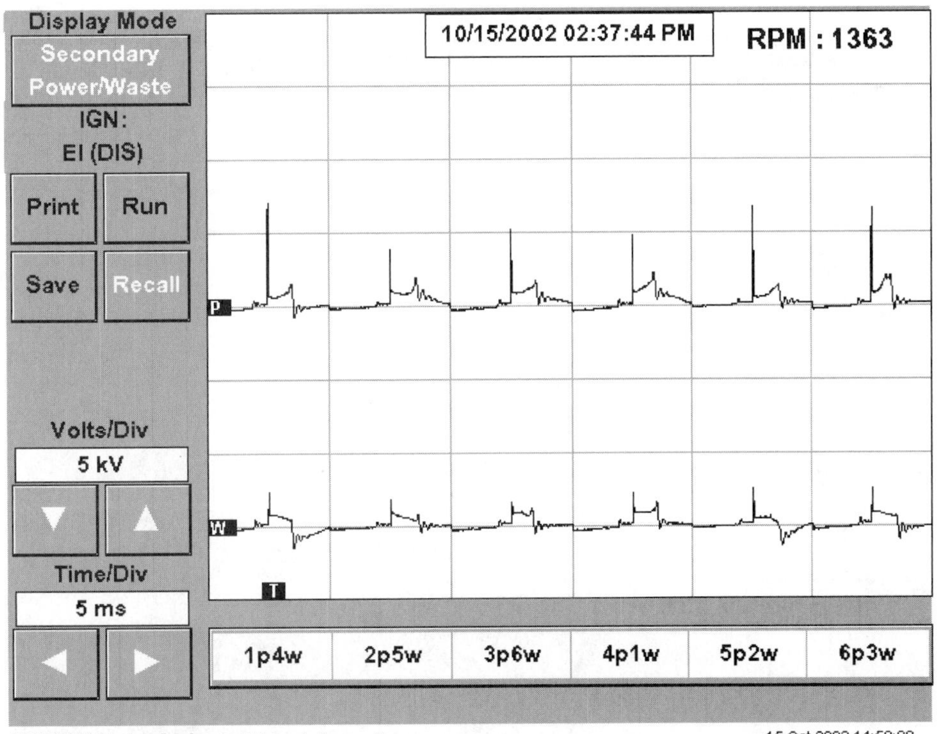

Figure 8. Both power and waste firings can be displayed.

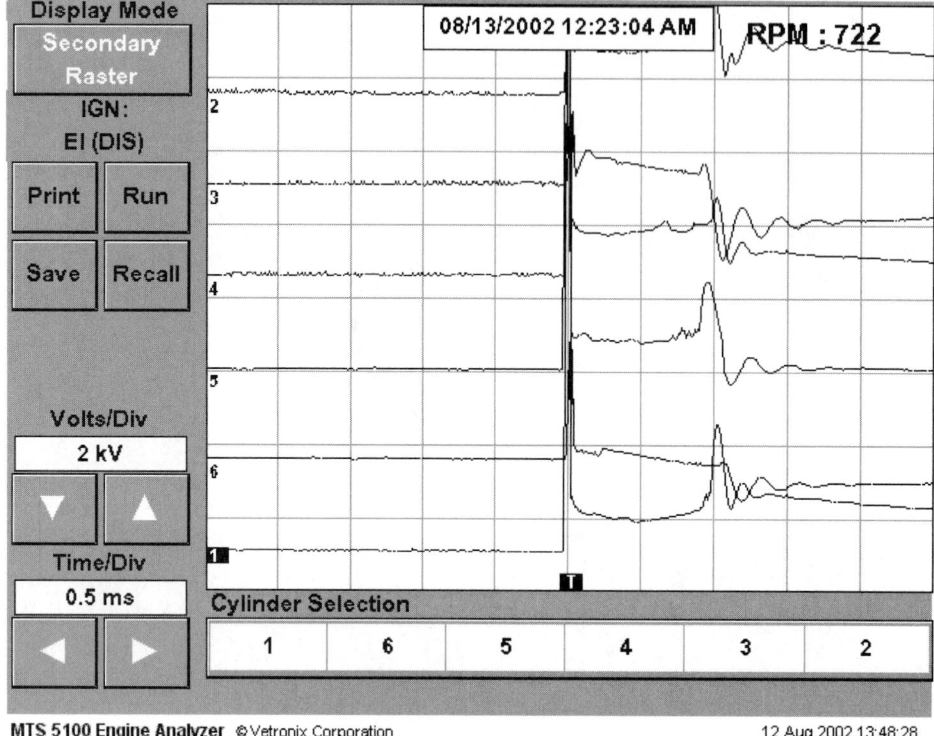

Figure 9. A raster pattern shows approximately 1-millisecond burn time.

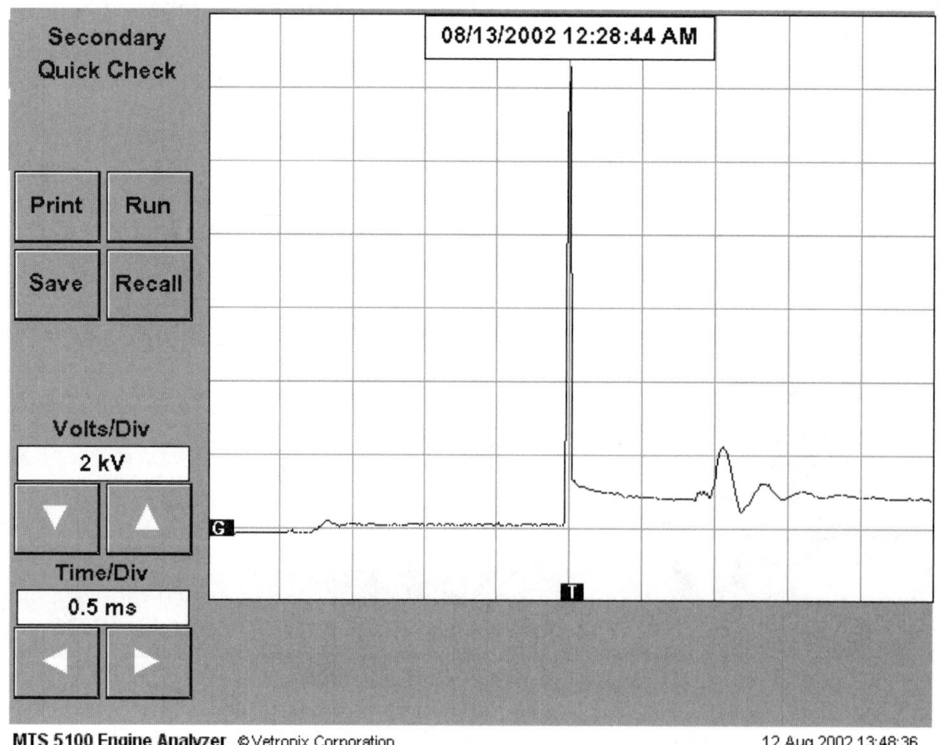

Figure 10. A single-cylinder pattern firing at 13 kilovolts.

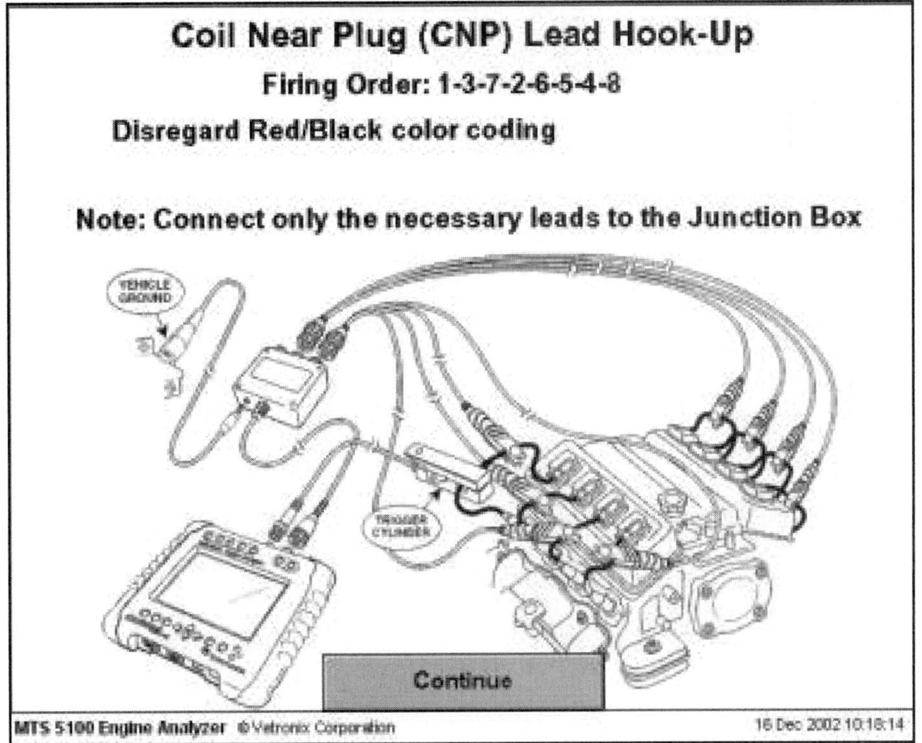

Figure 11. A coil-near-plug lead connection guide.

Figure 9 shows all cylinders firing at the T point (trigger), and the spark lines that follow all seem to be about two divisions long. The time/Div has been set for 0.5 millisecond, so two divisions equal about 1 millisecond. Although this is adequate, it is just barely inside the specification window.

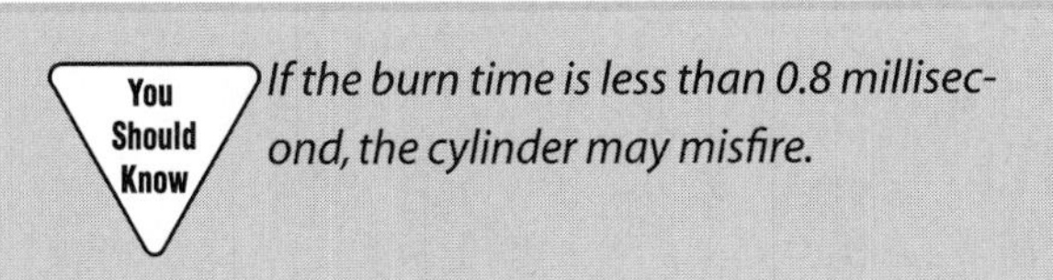

If the burn time is less than 0.8 millisecond, the cylinder may misfire.

This vehicle can be analyzed with a single scope connector, and the same information can be seen. Figure 2 showed the single pickup and the trigger lead around one plug wire. This will result in the one-cylinder secondary pattern that you see in **Figure 10**. This plug is firing at approximately 13 kilovolts and has a burn time of 1 millisecond based on the 2-kilovolt Volts/Div and the 0.5-millisecond Time/Div settings of the DSO. The disadvantage to this type of pattern is that the single cylinder does not allow any comparison. To get all of the data, you have to move the pickup and the trigger to the next cylinder, keeping track of the data along the way. The advantage to the single-cylinder setup is the speed with which the pattern can be displayed. There is no need to know the polarity of firing order until you find something wrong.

COIL-NEAR-PLUG SYSTEMS

The setup for the coil-near-plug system is the same as for the waste spark system. If the DSO allows for all coil connections, the setup will look like that shown in **Figure 11**. Notice that the top of the screen indicates that polarity is no longer important. All plugs in a coil-near-plug or a coil-on-plug system fire at the same polarity. **Figure 12** shows how three cylinders of a V-8 look with the individual pickups connected.

Figure 12. Individual pickups for a coil-near-plug system.

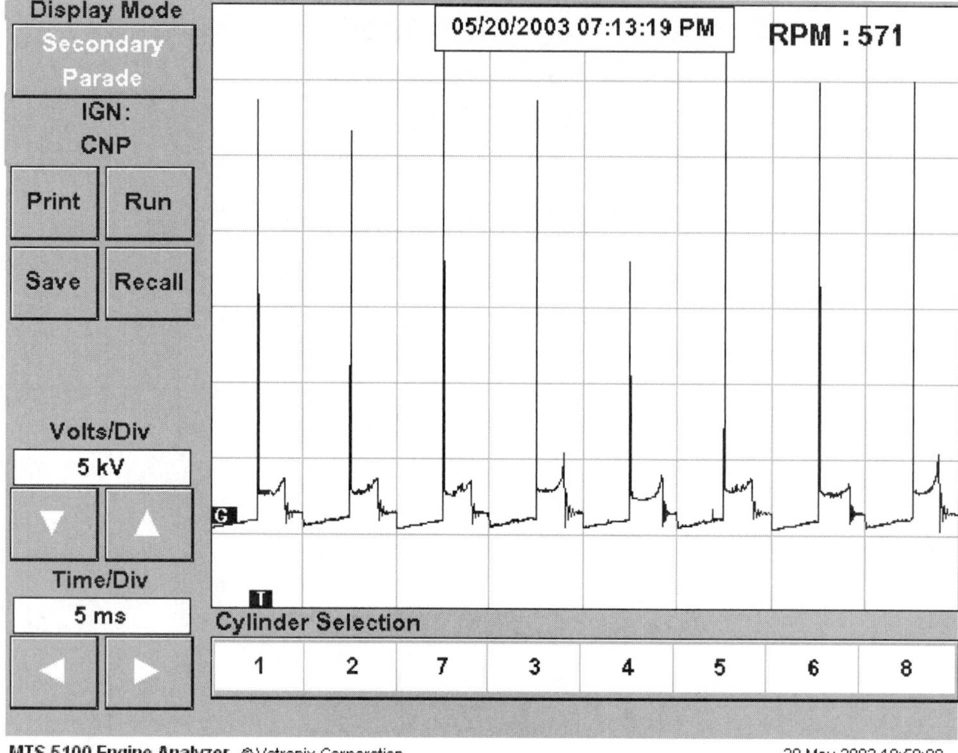

Figure 13. An eight-cylinder display for a coil-near-plug system.

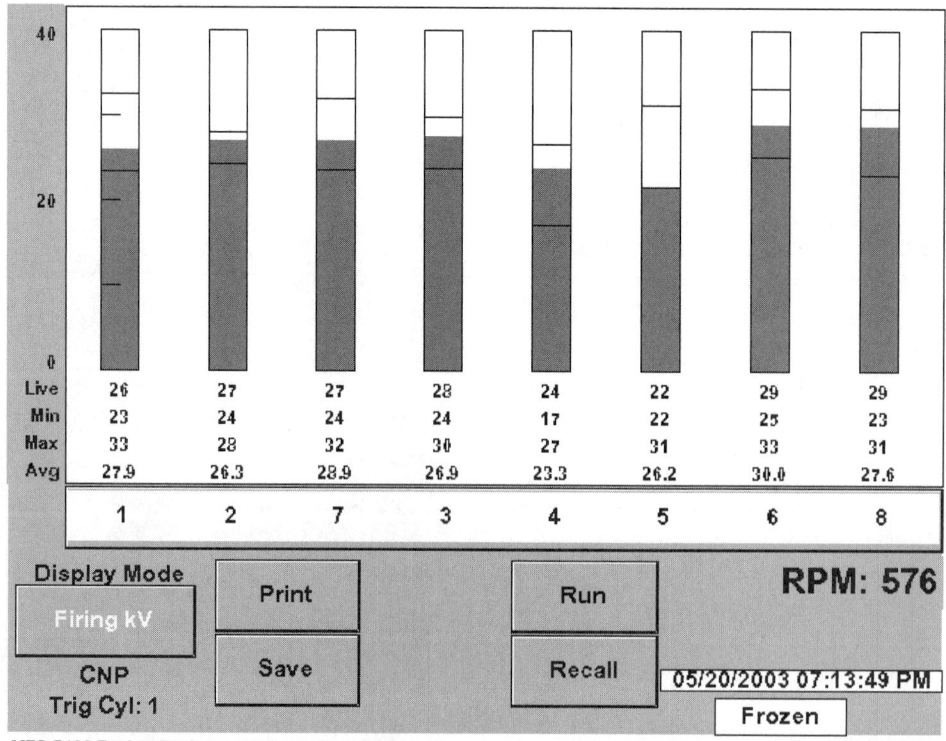

Figure 14. Firing voltage in bar graph format for a coil-near-plug system.

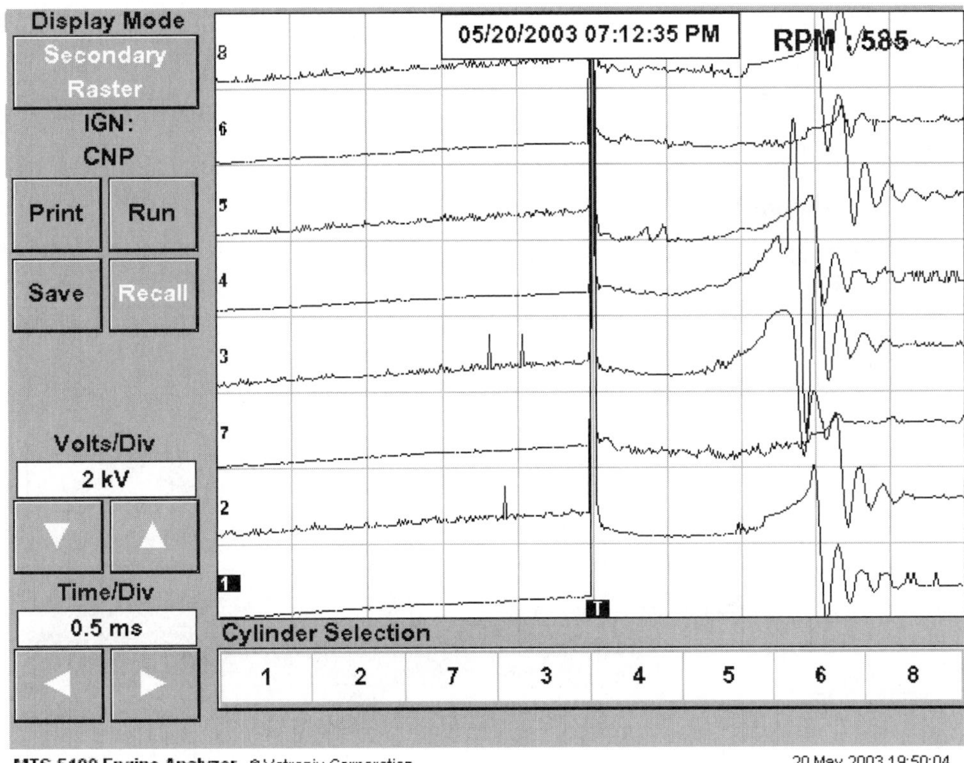

Figure 15. A V-8 coil-near-plug raster pattern.

The results of using the DSO with all cylinders connected will again show up in a parade format, as in **Figure 13**. Notice that the firing lines all appear to be quite high with the lowest firing at just under 30 kilovolts and the highest at approximately 34 kilovolts. A 4-kilovolt variance is out of specification, as is the 30-kilovolt average. This vehicle is in need of some secondary resistance diagnosis. The procedure is the same as that described for a DI vehicle. Putting the DSO into bar graph mode shows the excessive firing voltages graphically **(Figure 14)**. With the firing voltages as high as they are, you might be tempted to think that the burn times will be very small. **Figure 15** shows that burn times are approximately 1.5 milliseconds for all cylinders, well above the minimum specification of 0.8 millisecond. This ignition system is capable of producing plenty of spark time even with the excessive secondary voltages. You will see why in Chapter 52, when we will look at the primary circuit.

Another option here is to look at each cylinder individually with a single pickup, as shown in **Figure 16**, and get a pattern, which shows a firing voltage of 26 kilovolts and a burn time of 1.5 milliseconds—good burn time, but too much voltage **(Figure 17)**.

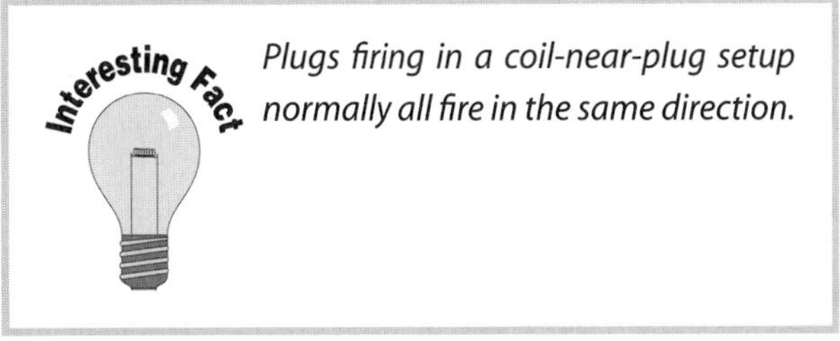

Plugs firing in a coil-near-plug setup normally all fire in the same direction.

COIL-ON-PLUG SYSTEMS

Most coil-on-plug systems do not allow for easy secondary diagnosis. As we mentioned earlier, we will rely on primary pattern analysis, which we will go over in Chapter 53.

Figure 16. A single-pickup coil-near-plug system.

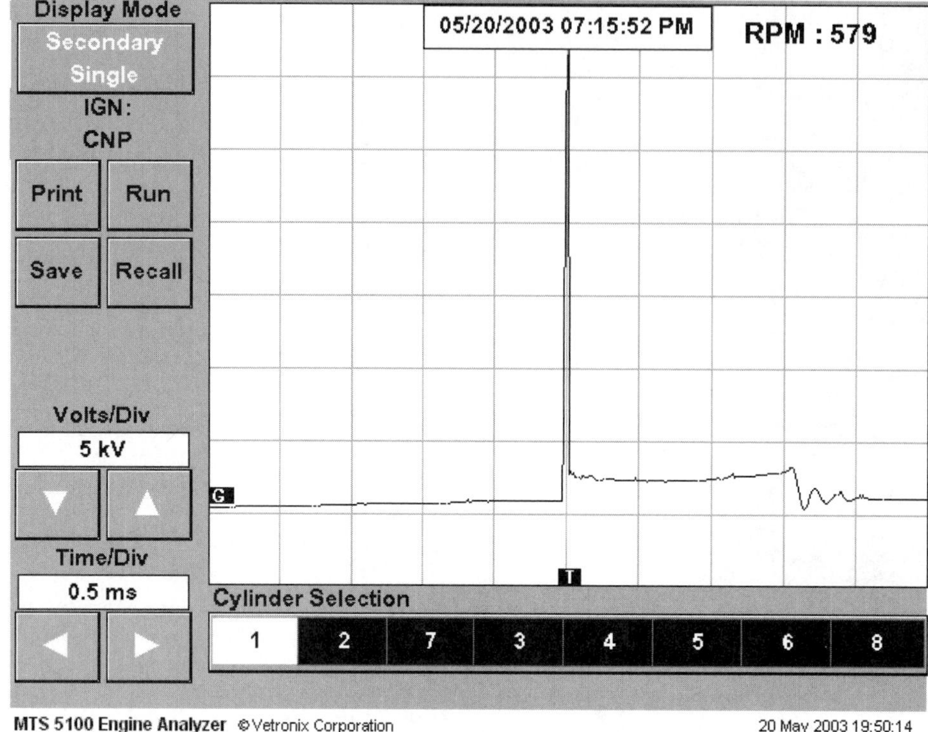

Figure 17. A single-cylinder display for a coil-near-plug system.

Summary

- EI vehicles require the use of special adapters to view all cylinders on a DSO or scope.
- Modern EI systems may have up to 100,000-volt capability.
- Secondary ignition testing should be done for firing voltage and burn time.
- The firing lines should all be within 3 kilovolts of one another and be between 6 and 12 kilovolts.
- Coil-on-plug systems usually have no easy method to test the secondary side of the ignition.
- A spark tester should be set to 40,000 volts and put in a plug wire to test for coil output.
- The plugs that fire in a positive direction should have the red pickups placed around the plug wires.
- The plugs that fire in a negative direction should have the black pickups placed around the plug wires.
- Some DSO screens allow both the power and waste plug firings to be shown together.
- Burn time should exceed 0.8 millisecond and will usually be less than 2.0 milliseconds.

Review Questions

1. Technician A states that the firing voltage for an EI vehicle should be between 6 and 12 kilovolts. Technician B states that the burn time should exceed 2.0 milliseconds. Who is correct?
   A. Technician A only
   B. Technician B only
   C. Both Technician A and Technician B
   D. Neither Technician A nor Technician B
2. The total firing voltage of a waste spark EI is the
   A. power voltage minus the waste voltage
   B. waste voltage minus the power voltage
   C. power voltage plus waste voltage
   D. none of the above
3. Secondary resistance will
   A. raise up the firing voltage
   B. lower the firing voltage
   C. increase the burn time
   D. have no impact on burn time or firing voltage

4. The order that the DSO will display the firing voltages in is
   A. cylinder number sequence
   B. all the positive plugs first, then the negative
   C. all the negative plugs first, then the positive
   D. in the firing order of the engine
5. Excessive resistance will lower the firing voltage.
   A. True
   B. False
6. The burn time for an EI system should exceed 2.0 milliseconds.
   A. True
   B. False
7. A higher-than-normal firing voltage will reduce the burn time.
   A. True
   B. False
8. The total voltage of the P and W plugs is the coil output.
   A. True
   B. False

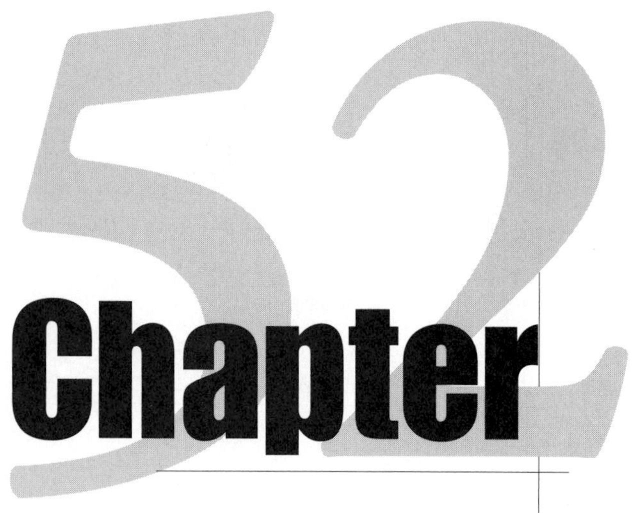

Chapter 52 Distributorless Ignition Primary Circuits

Introduction

At just about the same time that EI ignition came on the market, the manufacturers began to put significant changes into the primary circuit. Do not forget that EI is the preferred SAE J-1930 term for an ignition system that does not have a distributor. DI refers to systems with a distributor. In this chapter, we take a close look at how the manufacturer controls primary current through the ignition coil and the inputs necessary for this control. In addition, we look at the use of the PCM for ignition primary current control, a function that is usually accomplished by the module. As was the case in Chapter 51, you will not have to learn volumes of new information but rather will review the function of sensors and determine which ones are used on a particular system and how primary current is controlled. There really is very little new in terms of primary current control and sensor inputs. There are, however, significant differences in how the systems function. Throughout Chapter 52 and Chapter 53, it is assumed that the secondary circuit is functional and has the ability to generate the spark.

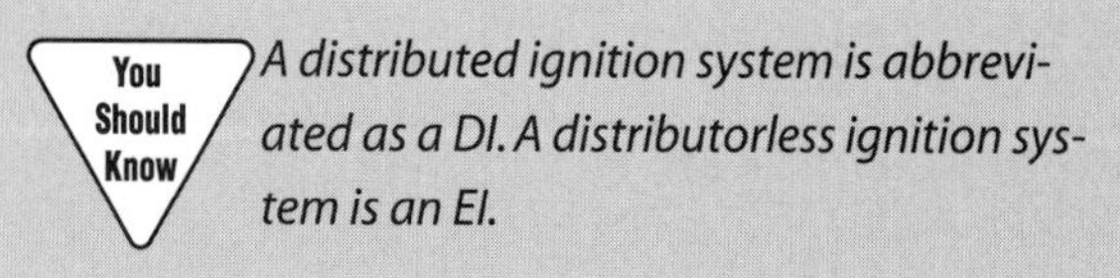

A distributed ignition system is abbreviated as a DI. A distributorless ignition system is an EI.

PRIMARY CURRENT FLOW

Primary current for most EI systems is substantially higher than it was for a DI system. For this reason, do not try to use your ohmmeter on the ignition coil primary. Its resistance is usually so close to zero that measuring inaccuracies are common. You will see in Chapter 53 that the current probe on the primary is the easiest and most accurate method to use.

Current flow will take one of three different forms. Let us look at each one and use various manufacturers as examples. The three forms are: a coil primary with a remote ignition module, a coil primary with PCM control, and integrated modules that are part of the ignition coil.

A Coil Primary with a Remote Ignition Module

Figure 1 shows a common General Motors four-cylinder ignition system. Notice that there is a single sensor that will send inputs to the PCM. We will look at these inputs later. This is a waste spark EI setup for a four-cylinder vehicle. Cylinders 1 and 4 are fired off one coil whereas cylinders 2 and 3 are fired off the other. The power for the primary comes through terminal A on the module, which is ignition switched and has a 10-amp fuse. Both coils are wired in parallel but have separate ground circuits. The coil drivers inside the module control the bottom terminal (negative) of each coil. Notice that the input sensor (a CKP) feeds its signal directly into the PCM. This means that the PCM is totally in control of ignition timing, including base timing. The wiring diagram is the most important piece of information we have. It tells us that the module turns the coils on and then off based on two separate signals (ignition control A and ignition control B) that come from the PCM. It tells us that a failure of either one of the coil drivers will shut down two cylinders. It also tells us that the vehicle needs to have the sig-

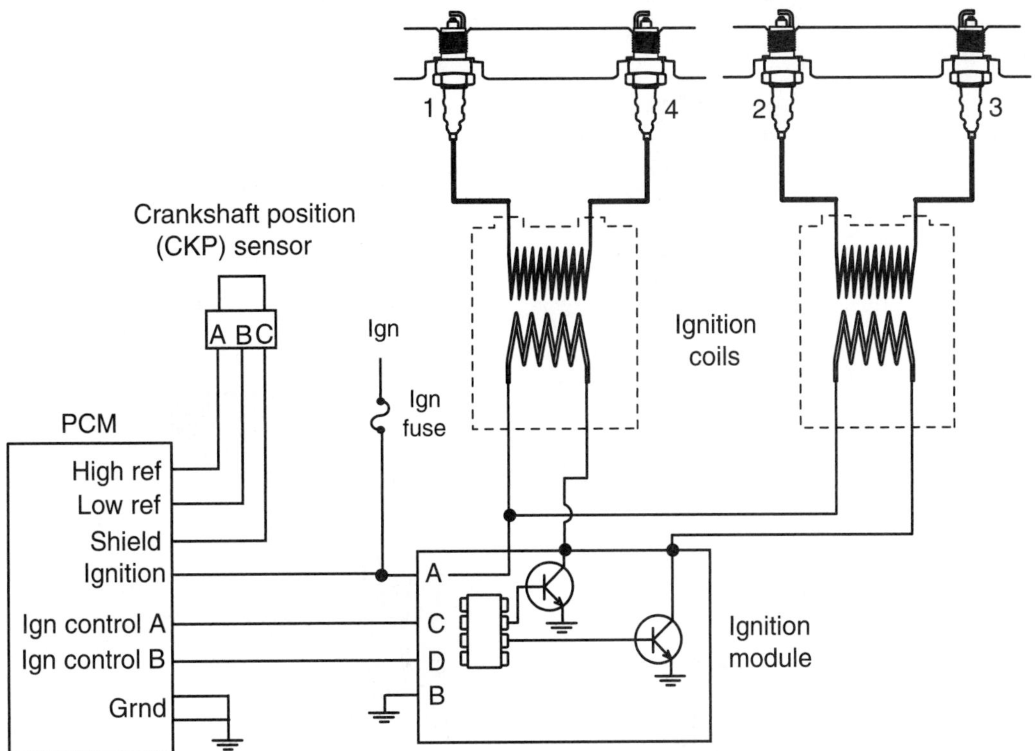

Figure 1. A single CKP sends information directly to the PCM.

nal from the PCM to even start. A DSO pattern of the primary will show that some of these systems are current limit types, while others are not. We will come back to this system when we look at input signals. Most of these systems have the coils bolted down to the module, which makes getting a primary voltage pattern almost impossible. We will rely on secondary voltage and primary current off terminal A for any diagnostics.

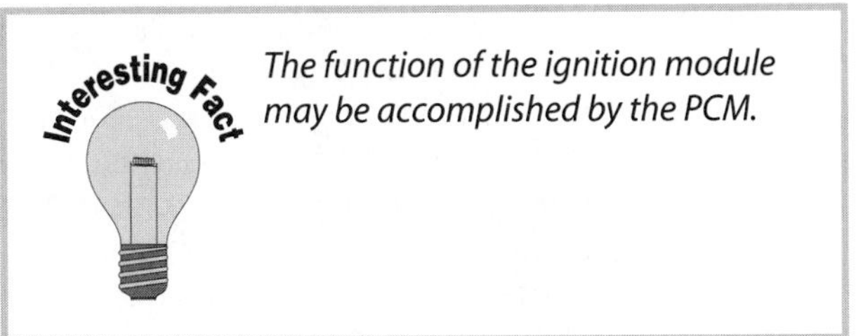

The function of the ignition module may be accomplished by the PCM.

A Coil Primary with PCM Control

The second system we look at is from Ford Motor Company and is called the integrated electronic distributorless ignition system **(Figure 2)**. What is the main difference here? You can see that we are missing the ignition module. There are three wires on the coil pack primary and four for secondary plug wires. The primary has power (VBATT) and two primary negatives that go off to the PCM (Coil 1 and Coil 2). The coil drivers are located within the PCM and the control of coil primary current will be done by the PCM based on the single input of the CKP. This is a waste spark system with a type 1 coil pack. Remember that a type 1 coil has both coils contained in one single molded pack. If one fails, both will be replaced. This system will allow us to get both a primary voltage pattern and a current trace because there is a primary wire running between the coils and the PCM, which has the module function.

Integrated Modules That Are Part of the Ignition Coil

Coil-on-plug or coil-near-plug systems lose the waste spark. The energy that was wasted can now be used at the plug. **Figure 3** illustrates a General Motors system that uses a coil/module for each plug. There is a spark plug wire, so we will be able to get a secondary pattern, like we did in Chapter 51. However, there is no way that we can capture a primary voltage pattern because the ignition module is part of the ignition coil. Each coil/module has five wires connected to it. They are the spark plug wires, a common ground, a B+ feed, an ignition control signal, and a reference low. Reference low connects the coil/module to the PCM and is a type of ground. Notice that all of the B+ feed circuits, ground circuits, and the reference low circuits are paralleled together. A failure of these circuits might result in a complete no-spark. The main control of

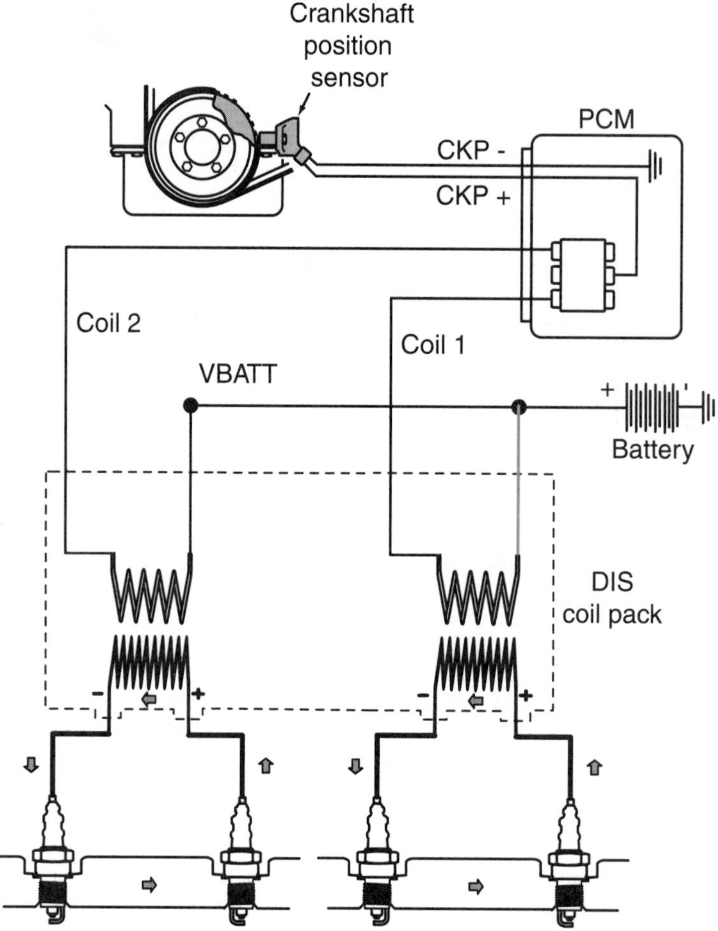

Figure 2. The PCM has the module functions.

the circuit arrives at the coil/modules along the ignition control circuit. This digital signal comes from the PCM. Each coil/module has its own signal. The PCM is totally in control of timing, including base timing. This system also requires two input signals not shown on this diagram: the CKP and the CMP.

There is one power feed for the entire ignition system that comes from the underhood electrical center. It is a 15-amp fuse that is live when the ignition switch is in the run or start position.

POSITION SENSORS: CRANKSHAFT

Crankshaft position sensors are either magnetic pick-up style or Hall effect. Currently, photo diode–style sensors are not in use in EI systems.

Magnetic CKP

A magnetic sensor will generate a small AC signal that will increase with speed. **Figure 4** shows what the signal looks like from a single notch. In this example, General Motors calls the sensor a reluctor type. It is the reluctance of the sensor that will change and allow it to develop an AC signal, which is easily tested with a DSO. Let us use **Figure 5**, which is from a General Motors 2.8-liter V-6 engine, to look at the use of the signal. Notice that there are 7 notches in the reluctor. The reluctor is part of the crankshaft. As the reluctor spins, it generates 60-degree signals and one 10-degree signal that is labeled "sync." The sync pulse identifies the position of the crankshaft. The ignition module has been programmed to ignore signals 1, 3, 5, and 7. It fires the coil for cylinders 2 and 5 on pulse 2, coils 3 and 6 on pulse 4, and coils 1 and 4 on pulse 6. This is a waste spark system that has the sen-

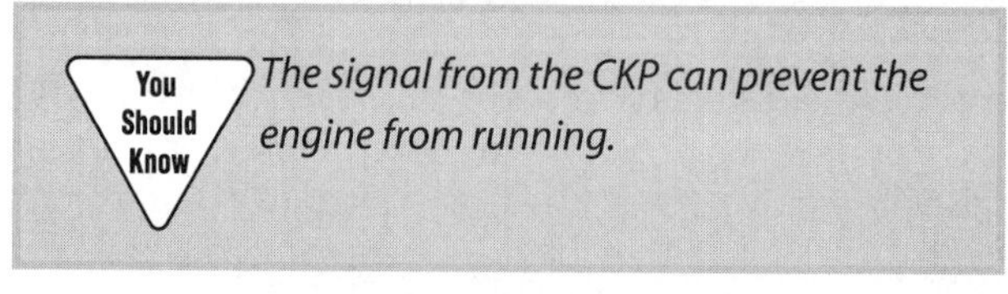

The signal from the CKP can prevent the engine from running.

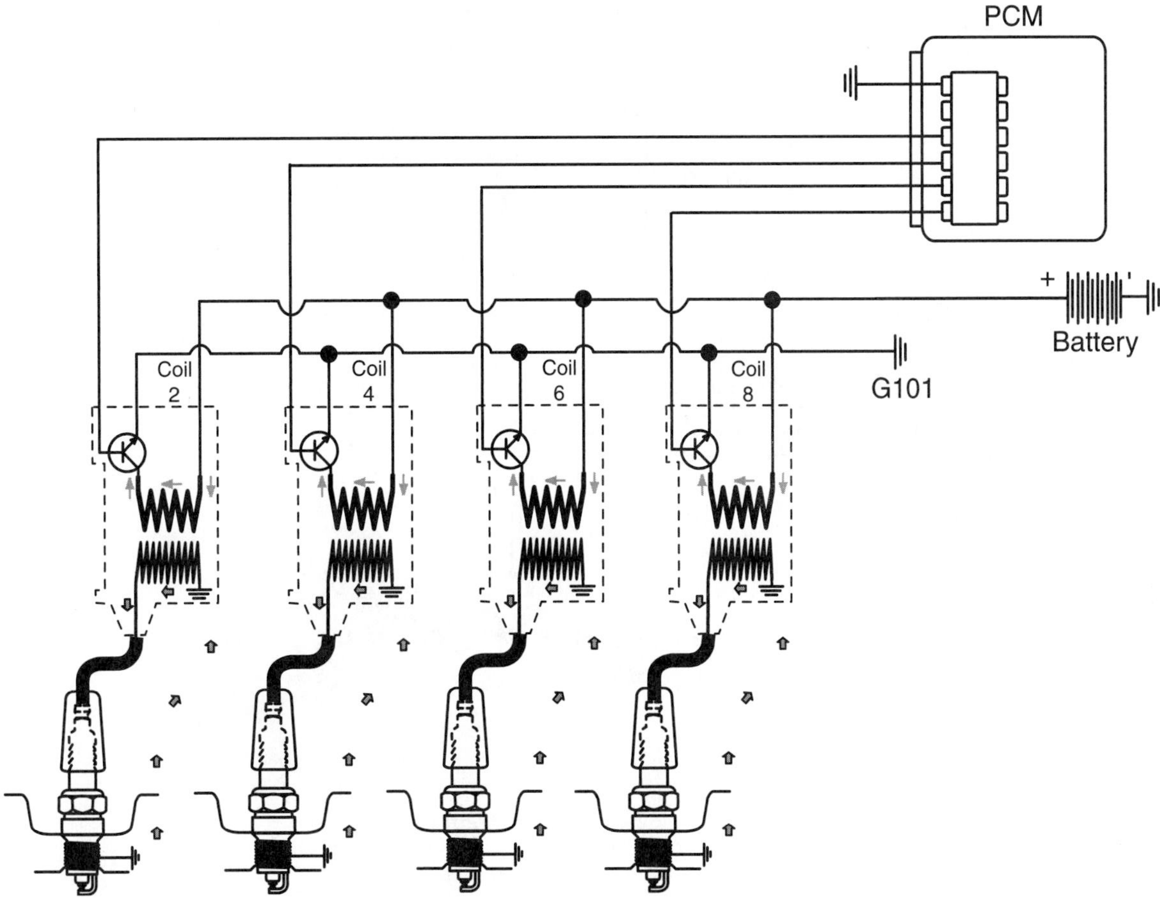

Figure 3. A General Motors coil-near-plug system with the coil and module in one unit.

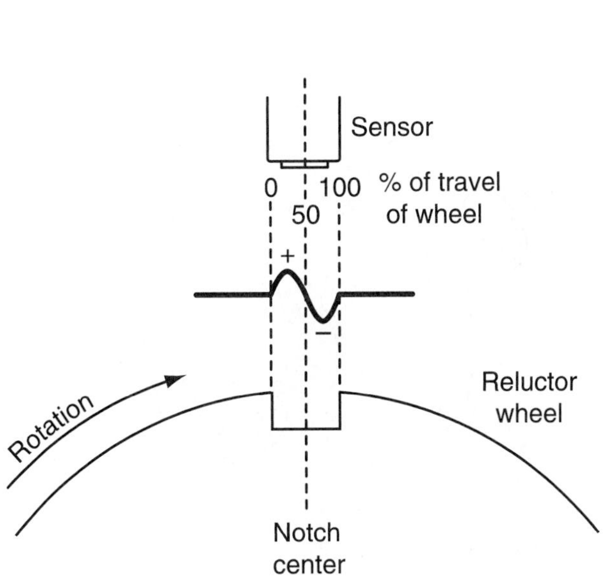

Figure 4. A magnetic sensor generates an AC signal.

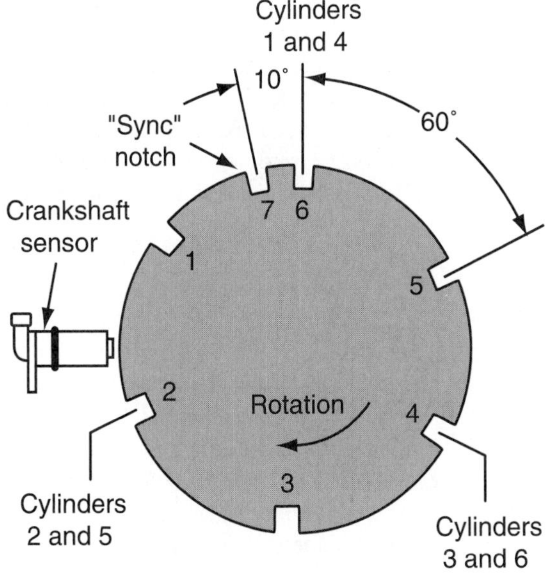

Figure 5. The spinning reluctor will generate 7 AC pulses per rotation.

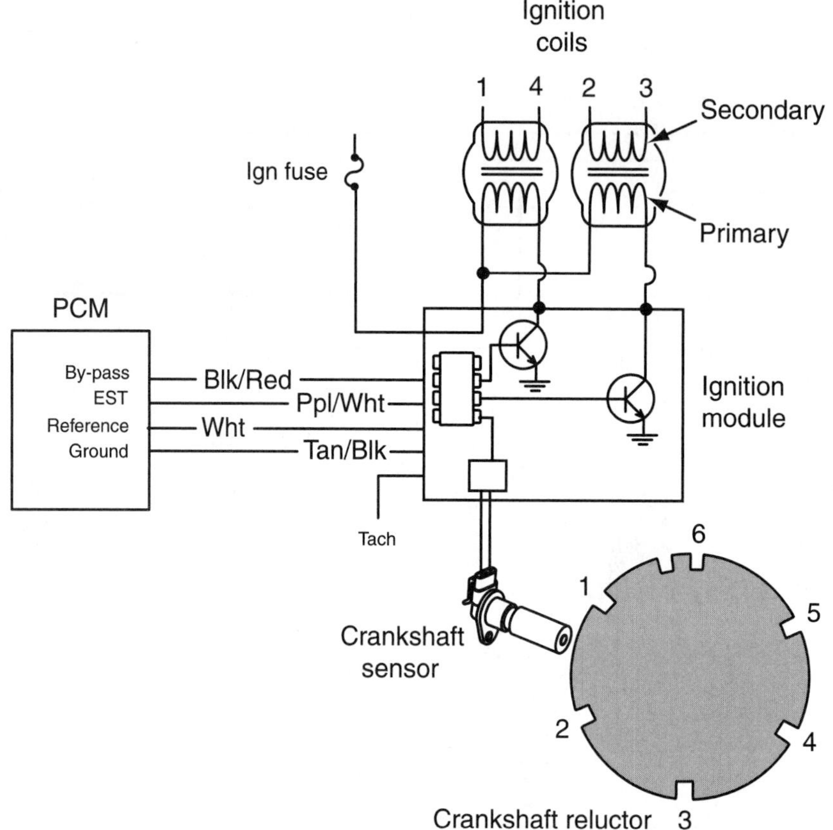

Figure 6. A waste spark system with the CKP connected to the module.

sor directly connected to the ignition module, as diagrammed in **Figure 6**.

Another example of a CKP that is magnetic is the variable reluctance sensor found on some Ford products **(Figure 7)**. It will generate a pulse every 10 degrees of crankshaft motion. There are 35 teeth on the vibration dampener spaced 10 degrees apart. One tooth is left off, which is why this sensor is sometimes referred to as the "missing tooth." The signal from it is shown in **Figure 8**.

Hall Effect CKP

The use of a Hall effect sensor as a CKP is an option for the manufacturer, and it comes in a variety of styles. The simplest style is shown in **Figure 9**. As the harmonic balancer spins, it generates a square wave signal every 120 degrees. Remember that a Hall effect sensor has three wires: power, ground, and a signal. A single Hall effect sensor has limited use and requires a CMP to function in most cases. Most of the manufacturers use a common magnet, common power, and ground and get two distinct signals

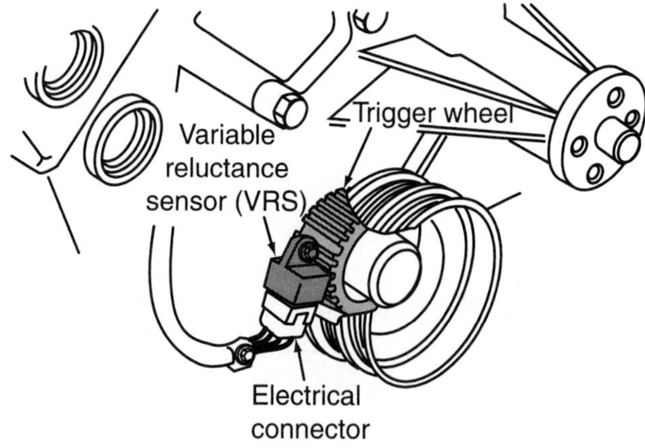

Figure 7. A missing-tooth-style CKP.

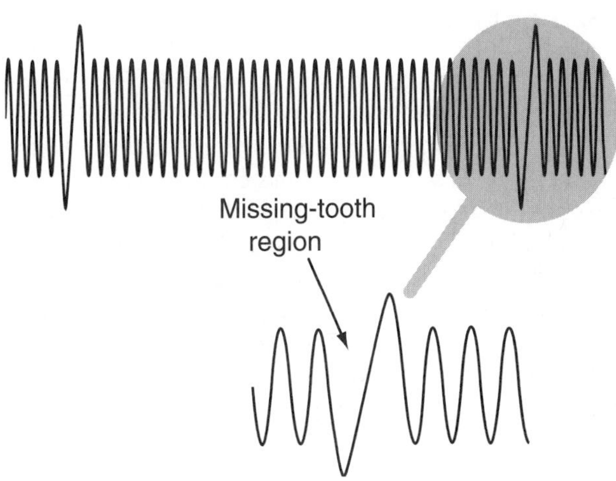

Figure 8. A generated AC signal that identifies piston position.

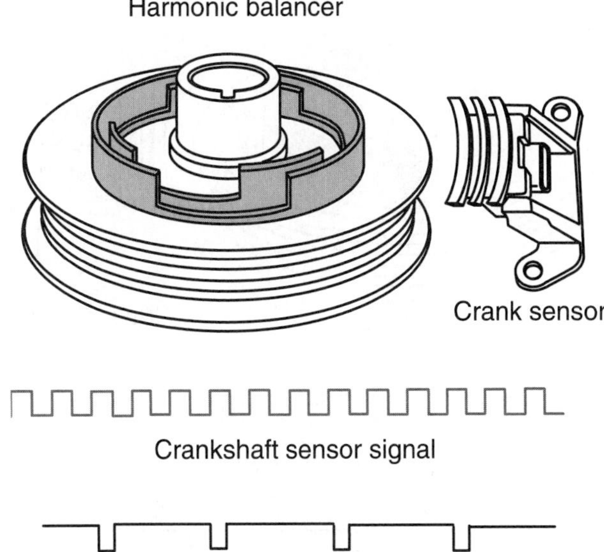

Figure 10. A two-output Hall effect sensor.

off the sensor. **Figure 10** shows a separate crank signal with a sync pulse coming off the CKP sensor. In this way, the signal both identifies speed (crank signal) and identifies position of the crankshaft (sync). A further adaptation of the system is illustrated in **Figure 11**. Notice that the inside vanes are different sizes. They represent a 10-degree movement, a 20-degree movement, and a 30-degree movement. The outside vane has 18 equally spaced windows. This is the popular 3X 18X sensor. The module will look at the relationship between the two sets of signals and determine position and speed quickly. This system is called "rapid start" and allows the engine to start in one-third of a rotation.

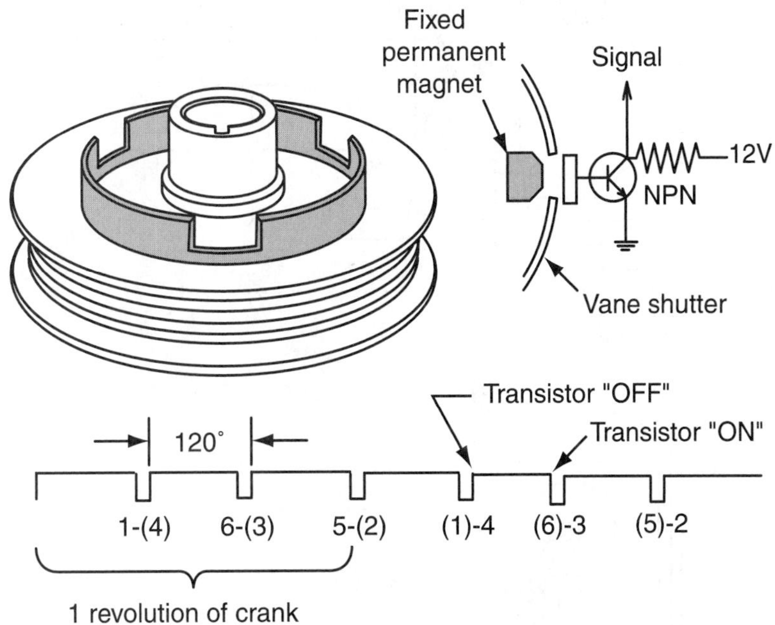

Figure 9. A Hall effect CKP, one output.

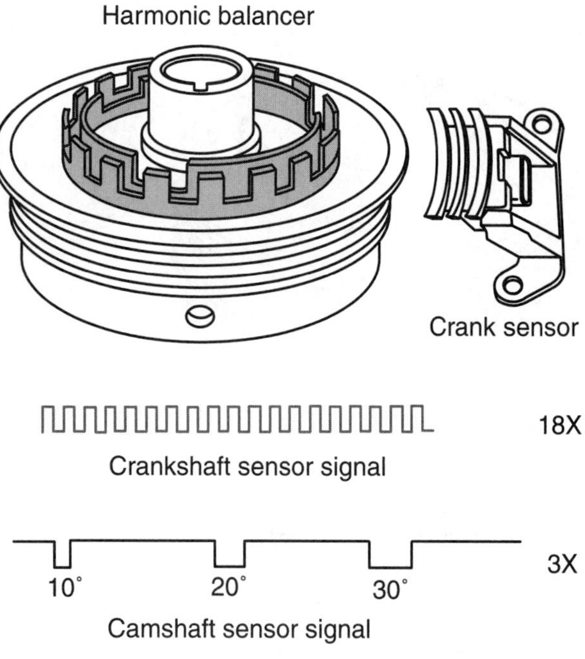

Figure 11. A common 3× 18× sensor.

CMP SENSORS

The first thing to consider is why a CMP might be needed. The answer goes to the requirement of the PCM and/or the ignition module to recognize which stroke the engine is on. For a waste spark system, the ignition system does not need to know the position of the camshaft, unless the fuel system is sequential. Sequential fuel injection will fire the injectors one at a time. Information regarding the cam will be used to determine which cylinder is on the intake stroke and in need of fuel. The CMP inputs are required in coil-near-plug and coil-on-plug systems because the spark will occur only during the last part of the compression stroke.

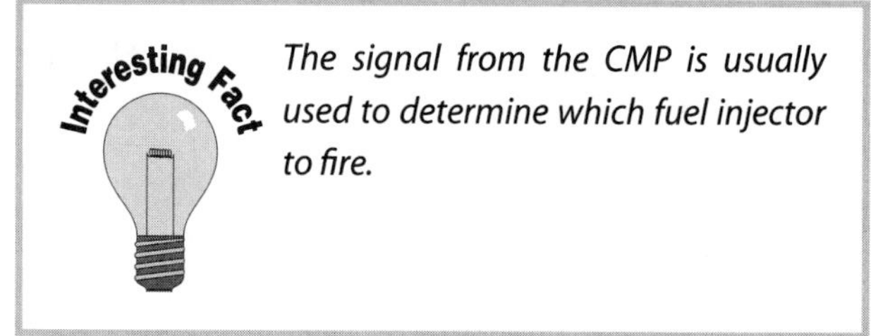

The signal from the CMP is usually used to determine which fuel injector to fire.

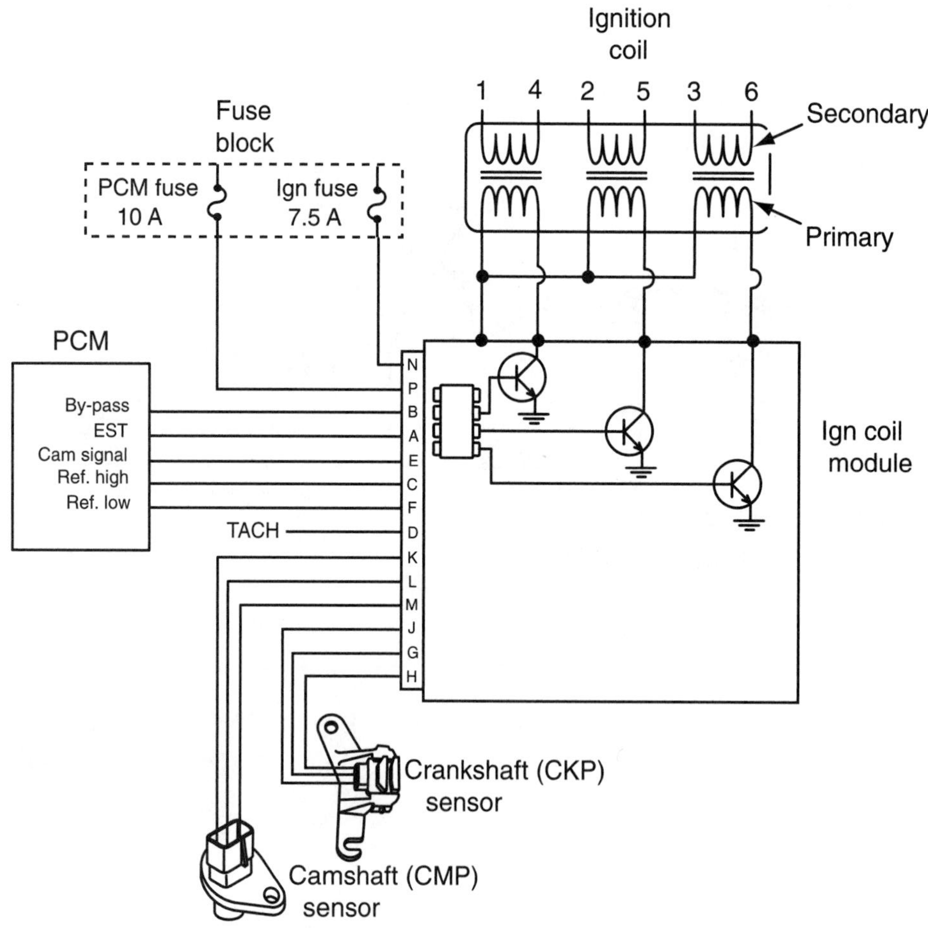

Figure 12. The CKP and CMP send position and speed information to the module.

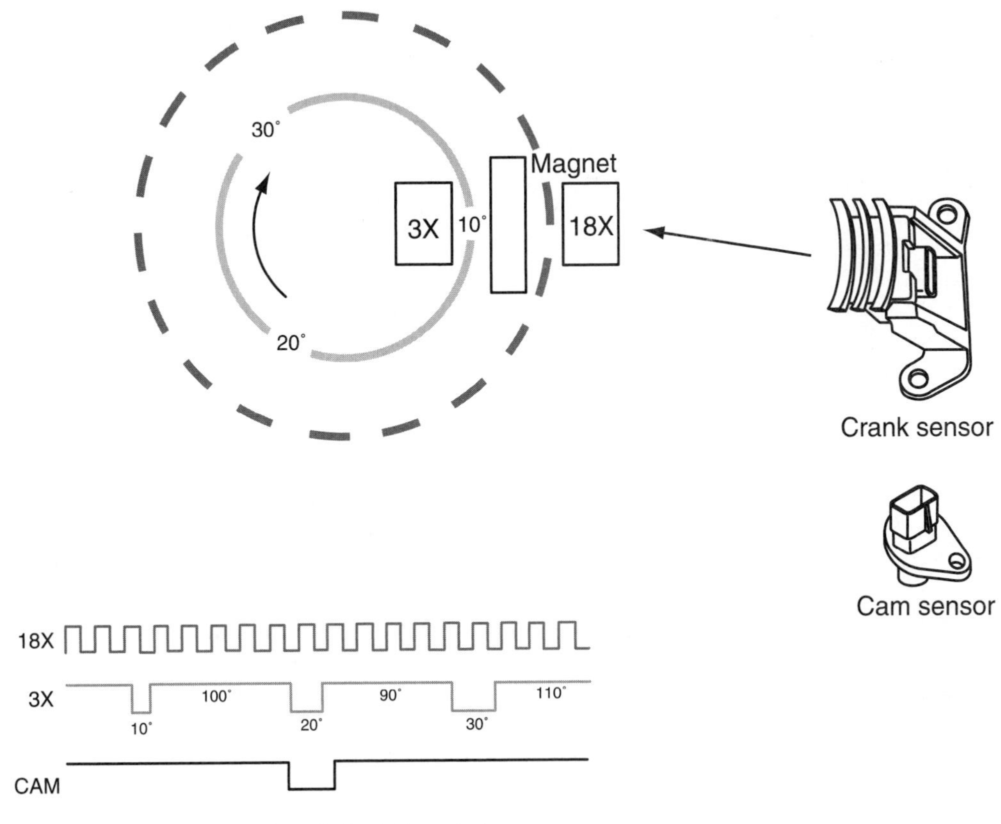

Figure 13. Piston position, rpm, and stroke information are sent to the module.

The CMP will supply information to the ignition module or to the PCM so that the correct coil can be fired. Again, CMPs come in two varieties, magnetic and Hall effect.

Hall Effect CMP

Figure 12 shows a CMP feeding its square wave digital signal to the General Motors ignition control module. This is the cam signal that goes with the 3X 18X sensor that we have looked at. Notice that the power and ground are shared with the CKP but that terminal L from the sensor. The "CMP sensor signal" goes directly into the module. From the module, the signal will go to the PCM. If all three signals were placed on the same figure, it would look like **Figure 13**.

Magnetic CMP

The use of magnetic CMPs is limited. An early General Motors EI system removed the distributor and replaced it with a cam sensor in a housing. **Figure 14** shows the sensor in its distributor-like housing. The gear at the bottom meshed with the camshaft and turned at one-half engine speed. Under the cap was a pickup coil with a timer core and pole piece. As the timer core spins, it generates the characteristic AC voltage signal. There are very limited applications that currently use a magnetic AC generator.

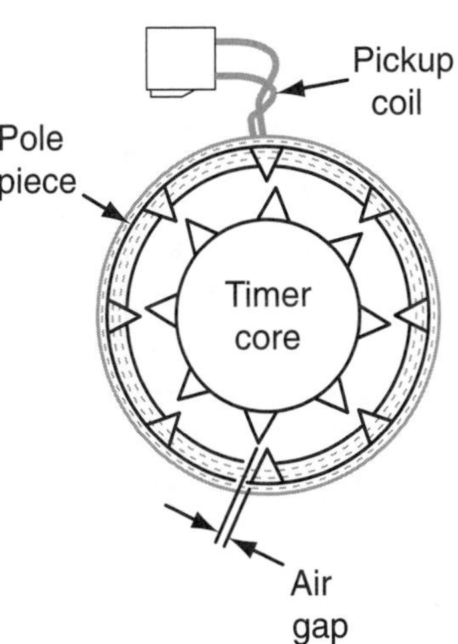

Figure 14. Some AC sensors are in distributor-like housings.

COMPUTERIZED TIMING

All EI systems are computer controlled. You have seen that some systems will use the PCM to take over module function. In all cases, however, the PCM will be responsible for timing changes. The timing will be controlled based on signals from the CKP, CMP, throttle position sensor, coolant temperature, and the mass air flow or manifold pressure sensor. As the PCM makes the determination of how many degrees of advance are needed, the signal will be sent to the module (if there is one) so that the module can adjust coil on/off time and change timing. Let us look at a common system and analyze how the timing signal is used. It is beyond the scope of this book to look at how the amount of timing is determined by looking at all of the input sensors. We will use the EI system for a four-cylinder vehicle with a magnetic sensor. Refer to **Figure 15** and follow the sequence of events. Initially the signal from the CKP arrives at the ignition module and is used to fire the coils directly. In addition, the signal flows through the module to the PCM on the PPL/WHT wire and is called reference (ref high). Does this sound familiar? This is basically the same system as the DI system we looked at in Chapter 48. Once the engine starts and the PCM has calculated the timing change required, the signal is sent to the module along the WHT wire (ignition control). The module is switched over to using the signal from the PCM for timing rather than the signal from the CKP by the 5-volt signal going to the module from the PCM along the TAN/BLK wire (bypass). As long as 5 volts is on the bypass, the module will use the ignition control signal for timing and ignore the CKP signal.

COMPUTERIZED TIMING RETARD

The last point of this chapter is the use of the **knock sensor (KS)**. The advance curve is programmed into the PCM, but under some conditions might need to be changed. Fuel octane, extreme temperatures, or unusual loads are some of the reasons why the timing might need to be "adjusted." Most systems on the road today look to the signal from the KS to determine if the timing is too far advanced. Timing that is advanced too far will cause a ping or engine knock. Engine damage might occur if the ping or knock continues for an extended period of time. The timing must be retarded until the ping is eliminated. The KS circuit shown in **Figure 16** will generate a signal specific to the bank of the engine that will be used by the PCM to retard the timing. The usual type of KS is a piezoelectric crystal that will generate a signal at a specific frequency of vibration. It is "tuned" to the frequency of the ping. If there is no ping, the KS does nothing, but if the engine pings and vibrates at the predetermined frequency, the KS will generate an AC signal. It is this signal that the PCM will use.

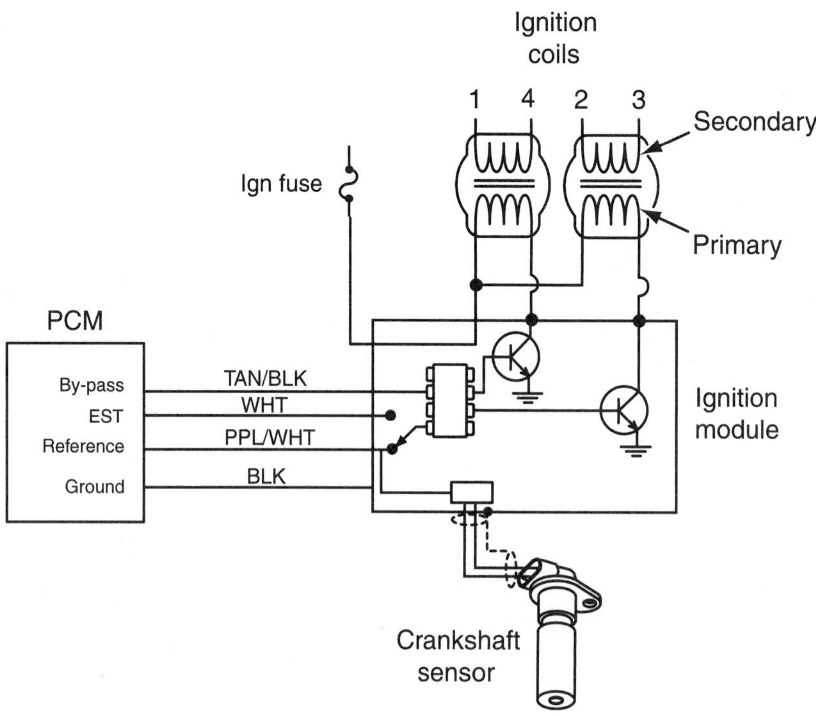

Figure 15. Timing advance is supplied to the module from the PCM.

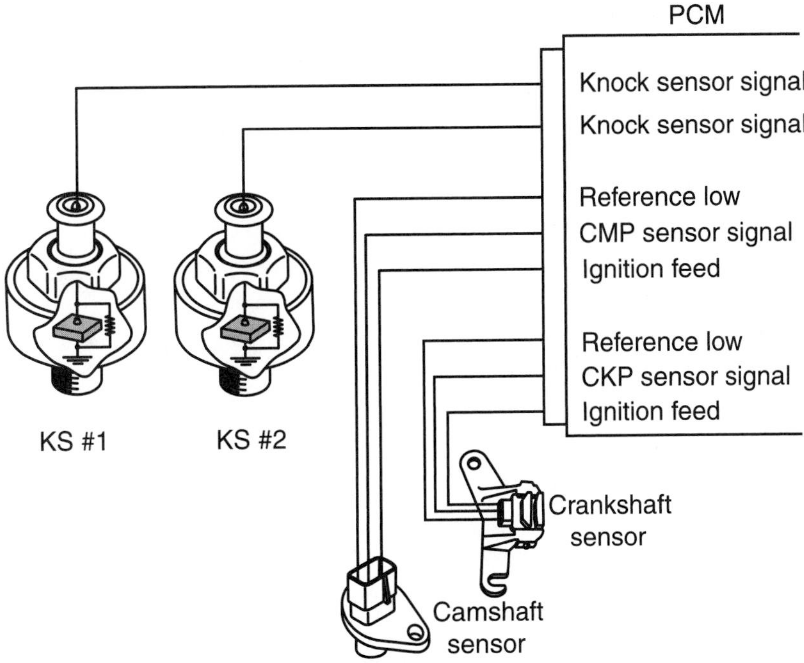

Figure 16. Knock sensor inputs are fed directly to the PCM.

Summary

- Primary current has steadily increased with EI systems.
- There are three forms of primary current control:
  1. Coil primary with an ignition module
  2. Coil primary with PCM control
  3. Integrated modules that are part of the coil
- The PCM might be totally in charge of timing, including base timing.
- The module or PCM will have one driver for two cylinders if it is a waste spark system.
- Coil-on-plug or coil-near-plug systems will have a driver for each coil.
- It is impossible to capture primary voltage patterns on some coil-near- or coil-on-plug systems.
- CKP sensors can be either Hall effect or magnetic.
- CMP sensors can be either Hall effect or magnetic.
- Magnetic sensors generate an AC waveform.
- Hall effect sensors generate a digital on/off signal.
- Cam sensors are found on sequential fuel injection systems.
- A 3X 18X sensor generates a full set of patterns for each rotation of the crank.
- The signal from the KS sensor is used by the PCM to retard the timing under conditions of ping or knock.

Review Questions

1. Technician A states that a waste spark system will have one coil driver for every two cylinders. Technician B states that the ignition module is in charge of calculating timing change. Who is correct?
   A. Technician A only
   B. Technician B only
   C. Both Technician A and Technician B
   D. Neither Technician A nor Technician B
2. The signal from a magnetic CMP is
   A. digital on/off DC voltage
   B. available once each crankshaft rotation
   C. an AC signal
   D. available twice for every camshaft rotation
3. The signal from the CMP is used to
   A. determine when to charge the coil primary
   B. fire the ignition coil
   C. tell the PCM engine speed
   D. determine valve position for sequential fuel injection

4. Primary voltage is available for analysis on all ignition systems.
   A. True
   B. False
5. A vehicle that is pinging needs to have the timing retarded.
   A. True
   B. False
6. The signal from the KS is usually digital.
   A. True
   B. False
7. Primary current flow is controlled by the PCM on all ignition systems.
   A. True
   B. False
8. Discuss the relationship between the CMP and the CKP.
9. Why is the ignition module eliminated on some ignition systems?
10. How does the ignition module know that the PCM wants to take over timing?

Diagnosing and Servicing the Primary Circuit on a Distributorless Ignition System

Introduction

Now that we have taken a look at how the spark is developed in a variety of EI systems, it is time to diagnose and service the primary circuit. Keep in mind that our assortment of tools available includes the DSO and the current probe. Many of the systems on the road today do not allow for primary voltage testing, so the current probe will become our tool of choice. For example, General Motors does not make an EI system that allows easy primary voltage testing. As in Chapter 52, we divide the systems into three types: waste spark, coil near plug, and coil on plug.

WASTE SPARK SYSTEMS

The primary circuit on a waste spark system can be tested using a variety of methods. The most frequent choice of technicians is to test using the DSO and observing a voltage pattern.

Voltage Testing

Many waste spark systems have a primary voltage waveform available. Let us use a Chrysler system and capture a voltage waveform. **Figure 1** shows the coil pack for a common four-cylinder vehicle with two coils remotely mounted. The ignition module function for this vehicle is inside the PCM. The three wires on the coil pack are power and each coil ground. Our primary voltage pattern will be available if we connect our positive DSO lead to either of the coil negative terminals. Each of these terminals goes to the PCM and is alternately switched to ground and off ground to fire the coil. **Figure 2** shows the pattern from coil 1 with the voltage setting high enough to see the inductive kick of about 115 volts. Our specification is between 200 and 400 volts, which is the same as for any other igni-

Figure 1. A four-cylinder type 1 ignition coil.

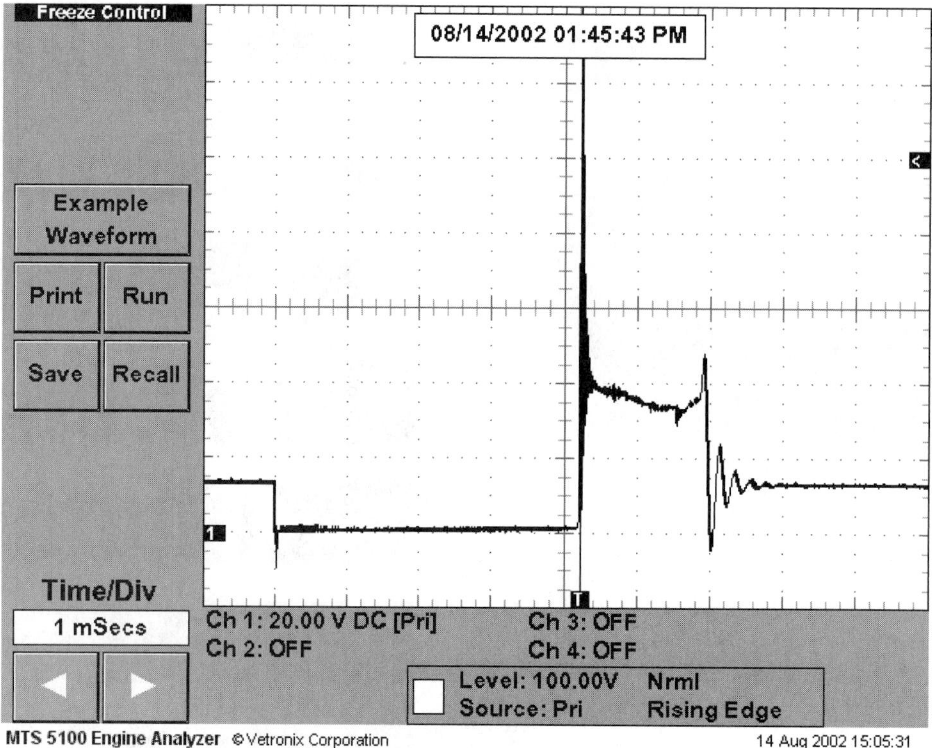

Figure 2. A voltage waveform from a type 1 coil at 20 volts per division.

tion coil. This inductive kick appears to be on the low side. Once we see that we have the correct inductive kick, we can change around the volts per division setting so that detail will be better displayed **(Figure 3)**. Ten volts per division usually shows the pattern well enough to display the details that we are looking for. The individual items that we need to check are:

1. Power to the coil.
2. System ground.
3. Primary on-time.
4. Burn time.
5. Coil capacitance oscillations.

There is no difference in checking the primary of a DI system compared with that of an EI system. We read the power to the coil just before the primary current turn-on signal or at the input of the coil and expect to find normal vehicle voltage of 13.5–14.5 volts. Our example shows approximately 14 volts, which is well within the specification. We read the ground just as the primary turn-on and should see zero volts. The main part of the downward turn does reach down to zero volts. The extra line that goes below zero does not indicate anything. The primary on-time should be in excess of 2.5 milliseconds (the time from primary on until either current limit or primary turnoff). This ignition system does not have current limit, so the time from primary current on until primary current off is 4.2 milliseconds and is acceptable. Burn time is approximately 1.6 milliseconds with a specification of 0.8–2.0 milliseconds. This also is well within specification. The last bit of information is the coil capacitance oscillations, and 3–5 should be visible. It looks like there are 4 after the burn line. Again, that is well within specification.

Current Probe Testing

The Chrysler system that we just finished testing with a voltage waveform can also be tested using a current waveform. As we have done before, place the current probe around the power lead to the coil pack. Play with the time and amps per division until both coils are represented on the screen as shown in **Figure 4**. Each division represents 1.00 amp of current. This coil is drawing more than 6 amps, which is typical for this system. Notice that the other coil draws approximately the same amount of current. Remember that this is a waste spark system with a type 1 coil pack. If one of the coils did not draw the same as the other, both coils would have to be replaced. By expanding the pattern, we can see more detail on either one of the coils. **Figure 5** shows that we have changed the time frame around to 500 microseconds, which is one-half of a millisecond. We can now see that the ramp up of current is steady and smooth. We can also see that the time from primary current on until current off is slightly more than eight divisions. If every

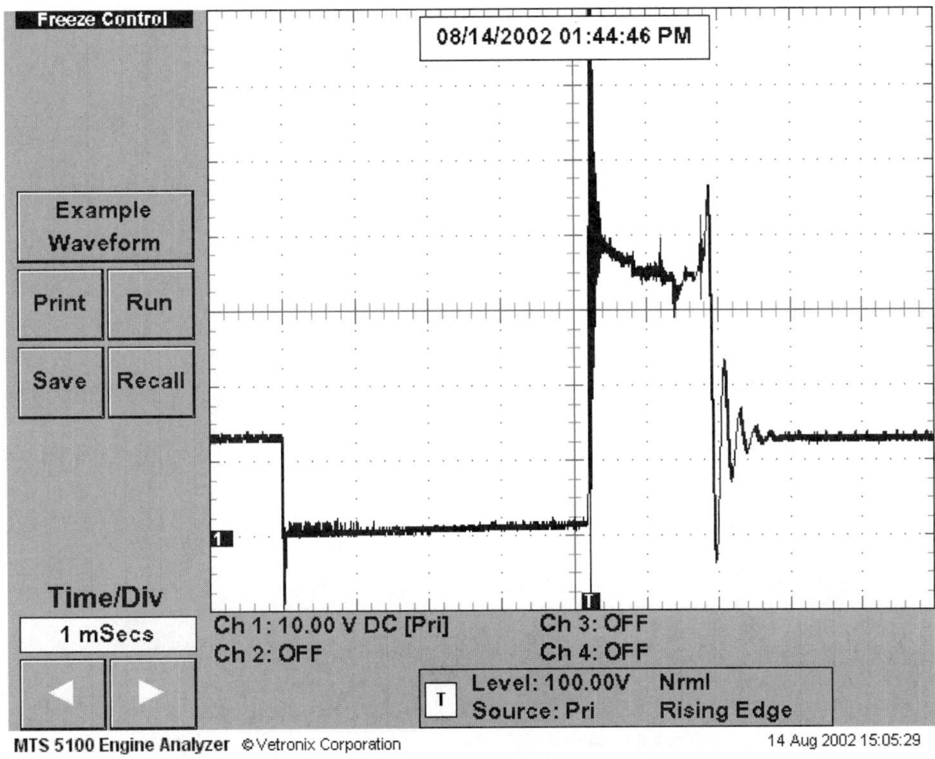

Figure 3. A primary voltage waveform from a type 1 coil at 10 volts per division.

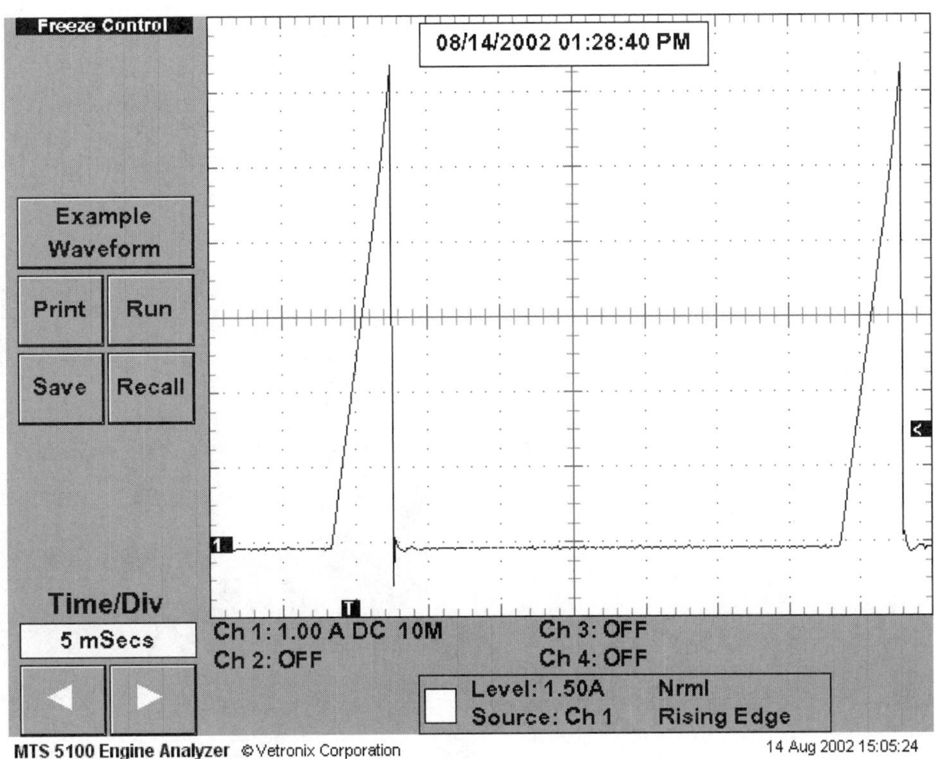

Figure 4. A current trace from a four-cylinder type 1 coil.

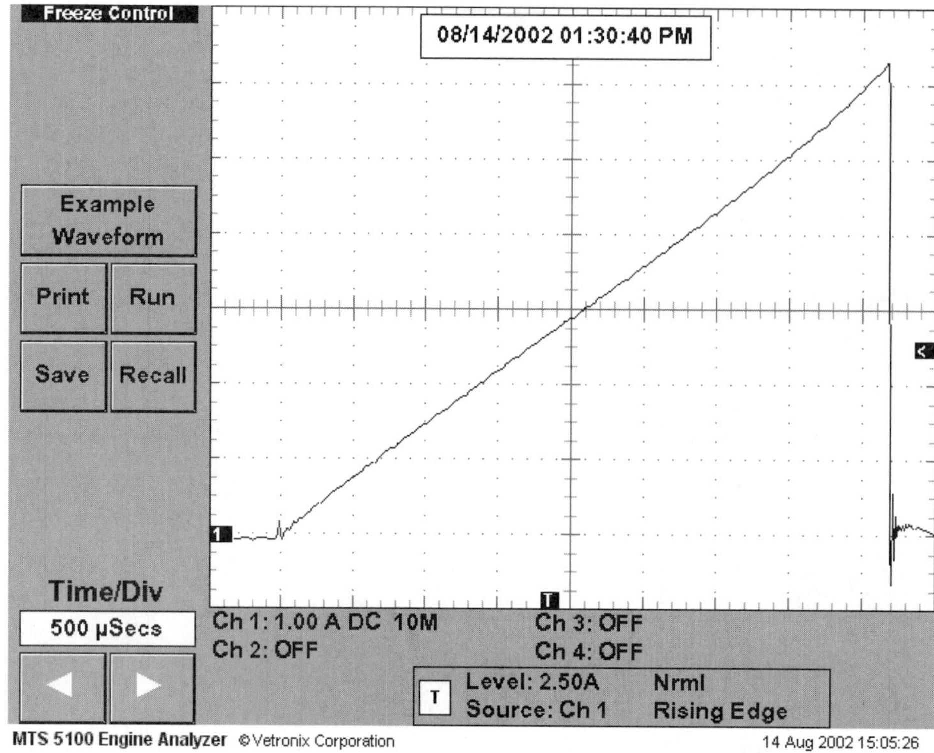

Figure 5. An expanded primary current trace.

division is 0.5 millisecond, then current remains flowing through the primary for a little more than 4 milliseconds, which is the same thing that our voltage waveform showed us. Do not forget that the specification is in excess of 2.5 milliseconds, so what we have found so far is acceptable. Also note that this vehicle does not have current limit. If it did have current limit, the current ramp would be flat-topped. The specification is for the time from primary current on until either current limit or primary current off. If we combine the information from secondary firing voltage with the primary voltage and current waveforms, we have an excellent representation of what is going on inside this ignition system.

Let us use an additional example. With a General Motors type 2 ignition system, it is virtually impossible to get a voltage waveform off the primary, so we will have to settle for a current trace. **Figure 6** shows the current probe around the power lead of the ignition module. The module is on the bottom of the coil pack with the primary connections hidden under the coils. This is why we have to settle for a current waveform. **Figure 7** shows the primary current flow through the three coils. Notice that each division is equal to 2.0 amps. These coils are drawing approximately 9 amps, which is a substantial amount of current. The flattops signify that this vehicle has current limit. The three patterns look the same in terms of ramping and the flattop, so we can expand the screen by decreasing our time base until we have good detail on one of the patterns. Our time base is 1 millisecond in **Figure 8**. By spreading out the waveform, we will be able to measure easily how long primary current was on. It looks like approximately 2.5 milliseconds from primary on until current limit. With a specification of 2.0–4.0 milliseconds, this vehicle just makes the specification. If the time was lower than the 2.0-millisecond specifi-

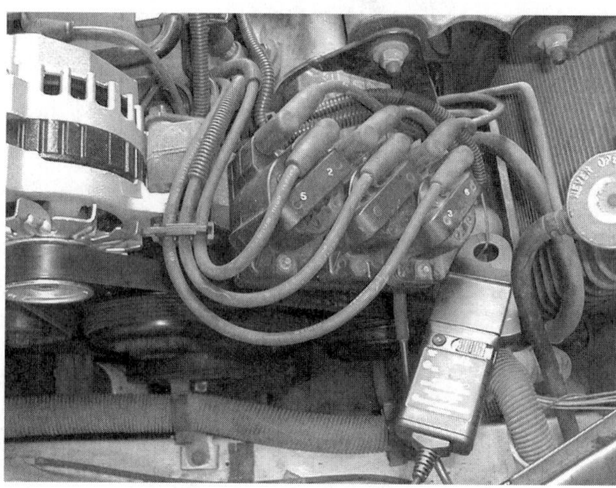

Figure 6. A current probe around coil power lead.

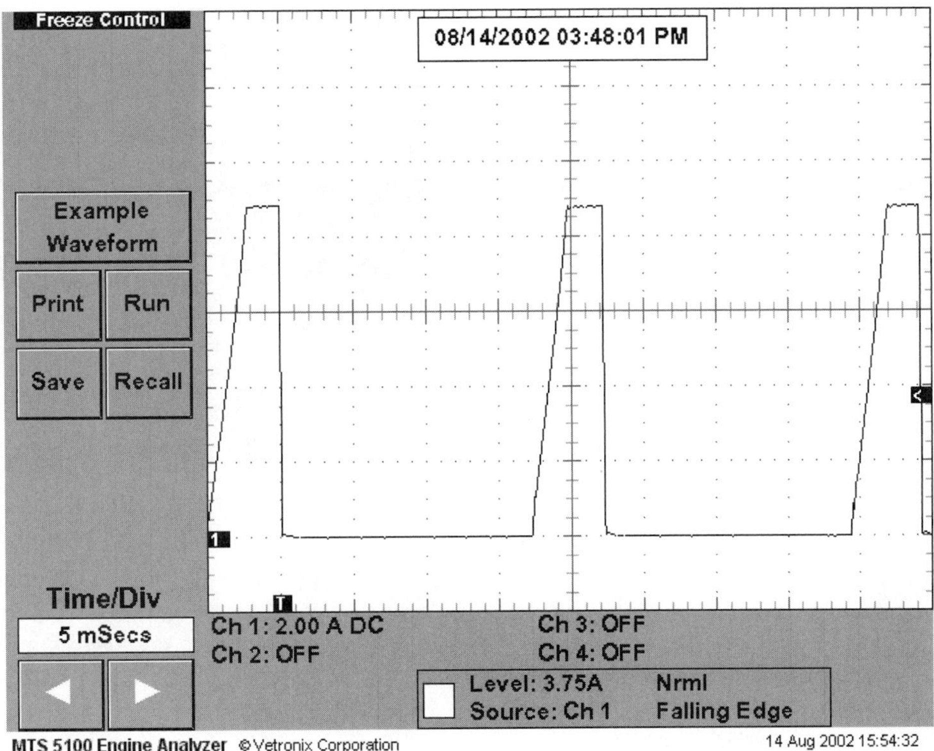

Figure 7. A V-6 current trace.

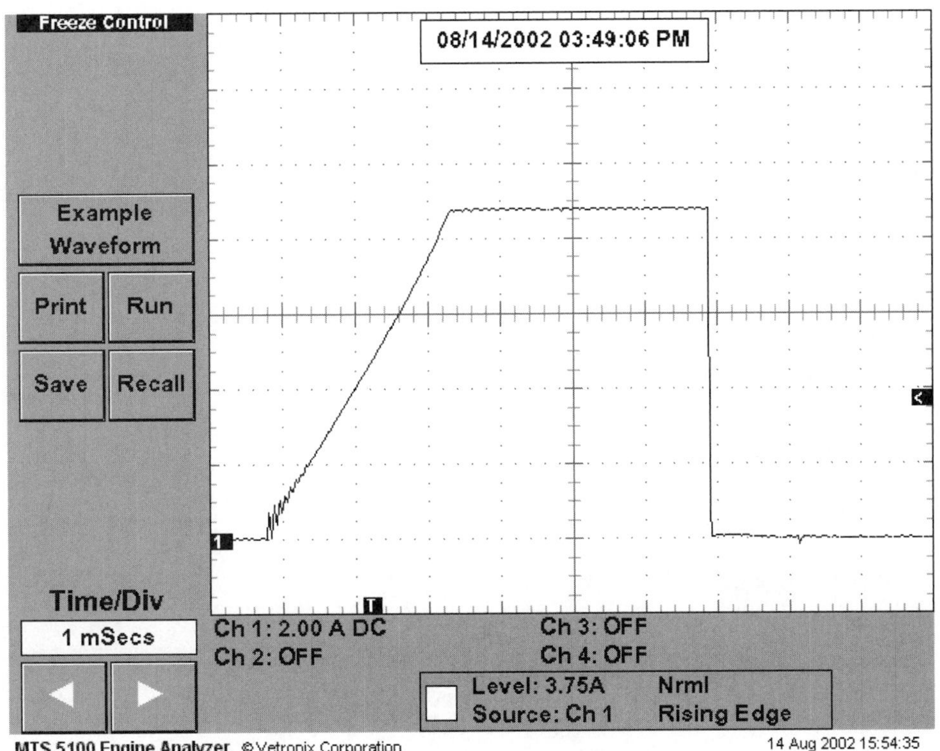

Figure 8. An expanded current trace.

cation, we would have to investigate further. Normally this is caused by the resistance of the coils being too low, forcing current limit to cut back too fast. The rest of the pattern looks good.

COIL-ON-PLUG SYSTEMS

The coil-on-plug systems that we looked at in Chapter 52 did not have a plug wire, so a secondary pattern is not available. Diagnosis will have to rely on a voltage trace and a current trace. The Ford system shown in **Figure 9** has the PCM doing the function of the ignition module. The two wires to each coil carry current through the primary. This is also a unique system because it will fire each spark plug three times at idle to ensure that the fuel ignites. Once the engine is running above an idle, the system will switch over to a normal one spark per cylinder.

Figure 9. Ford coil-on-plug coils.

Voltage Testing

Figure 10 shows the primary voltage pattern at idle for one coil. Notice that there are three firings in a row. Follow along with the figure as we analyze it along the same lines as we have done before. We are at 10 volts per division. The first line represents the power applied to the coil, and it looks like it is about 13 volts. When primary current turns on, the trace turns downward and just about hits the zero line. This indicates that the PCM ground is good. Primary current flows for about 1.5 milliseconds, and then we discharge the coil primary. After the inductive kick, we have about 0.5 millisecond of burn time before we turn on primary current again and keep it on for another 1 millisecond. Another firing and burn time of about 0.5 millisecond follows. Primary current is turned on again for about 0.8 millisecond, and then the final firing takes place with a burn time of about 1.0 millisecond. So the total burn time is approximately 2.0 milliseconds:

$$0.5 + 0.5 + 1.0 = 2.0 \text{ milliseconds}$$

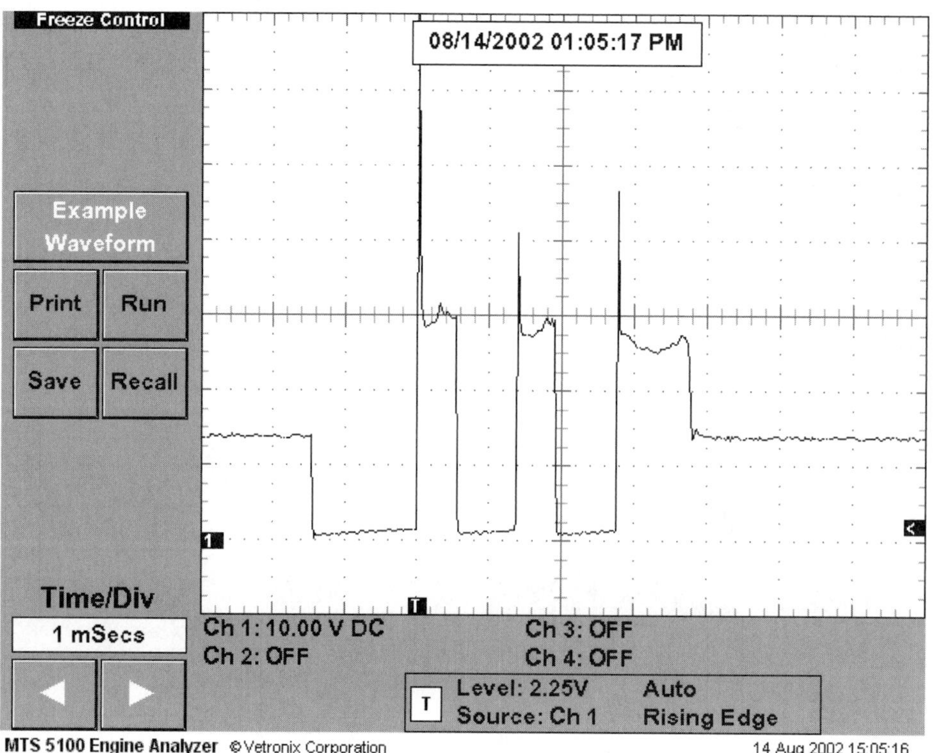

Figure 10. A primary voltage trace showing three firings at idle.

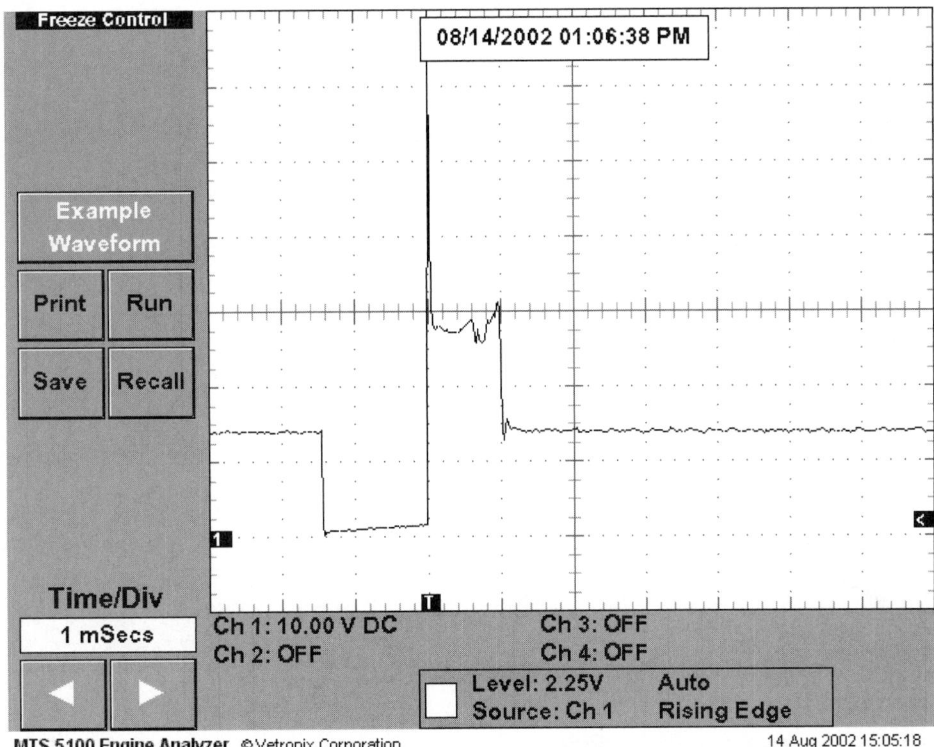

Figure 11. A primary voltage trace at 1200 rpm.

and the total on-time is approximately 3.3 milliseconds:

$$1.5 + 0.5 + 0.8 = 3.3 \text{ milliseconds}$$

Both are well within specification.

Figure 11 shows the same voltage waveform off idle. The three firings have become one long one. Primary current flows for a shorter period of time, approximately 1.5 milliseconds; the burn time is approximately 1.0 millisecond. The power and ground both look good. The coil capacitance oscillations are difficult to count, but it does show steadily diminishing oscillations, so everything looks good.

Current Probe Testing

Placing the current probe around either wire of an individual coil and running the engine at idle should show the three distinct turn-ons of the primary current. **Figure 12** shows the current trace. Notice that the system is drawing approximately 7 amps each time the primary is turned on. No flattop is seen here, so this is not a current limit system. The trace shows good ramping for each pulse that the PCM controls through the coil primary. Above idle, we should see that the three patterns become one. **Figure 13** does show this. Our on-time is approximately 1.5 milliseconds, and maximum current flow is approximately 7 amps at slightly above idle speed (approximately 1200 rpm).

The ability to capture both a voltage and a current waveform has allowed us to see the function of the ignition system. The only missing piece is the firing voltage of the secondary. Secondary firing voltage is difficult to capture here, so the only recourse is to pull the plug and visually inspect for carbon tracking or worn electrodes.

COIL-NEAR-PLUG SYSTEMS

Although there are a few different systems on the road, we will look at the General Motors coil-near-plug system that we examined in Chapter 52. Each individual coil has its own ignition module formed as one unit. Because the ignition module is part of the coil, it is impossible to get a primary voltage pattern. However, we can get a secondary voltage and a primary current pattern.

Current Probe Testing

Figure 14 shows the current probe test for a coil-near-plug system. We have placed the probe around the power lead to one of the coils. Remember that each coil must be individually tested before any conclusion is formed. Notice that this is the first General Motors system we have looked at that does not have the flattop indicative of a current limit function. Varying the dwell controls coil current. Figure out

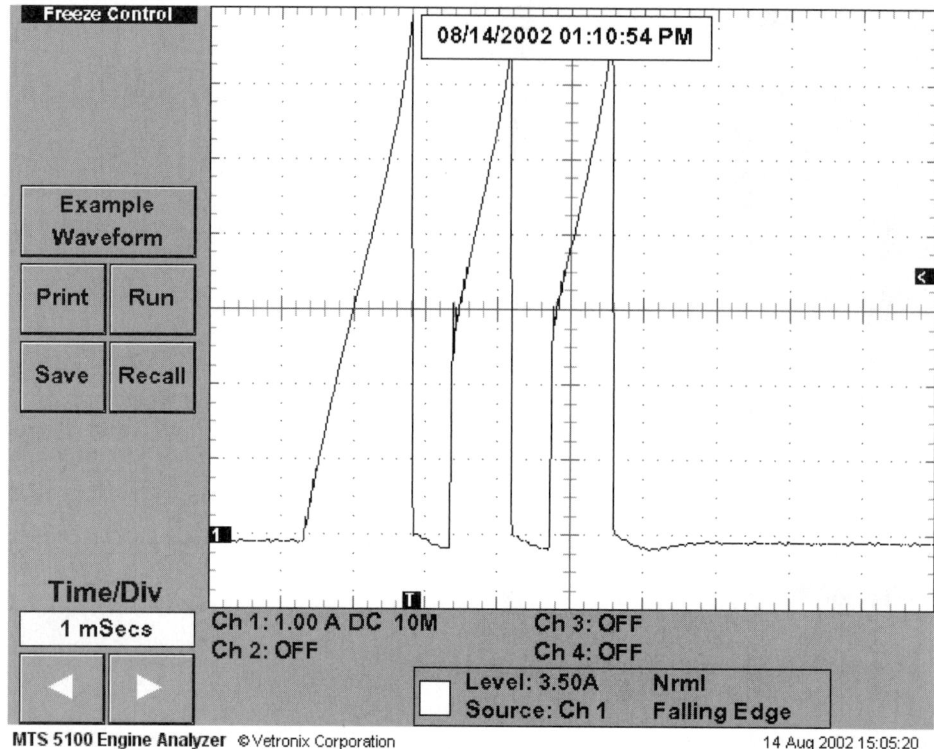

Figure 12. A primary current trace showing 3 pulses at idle.

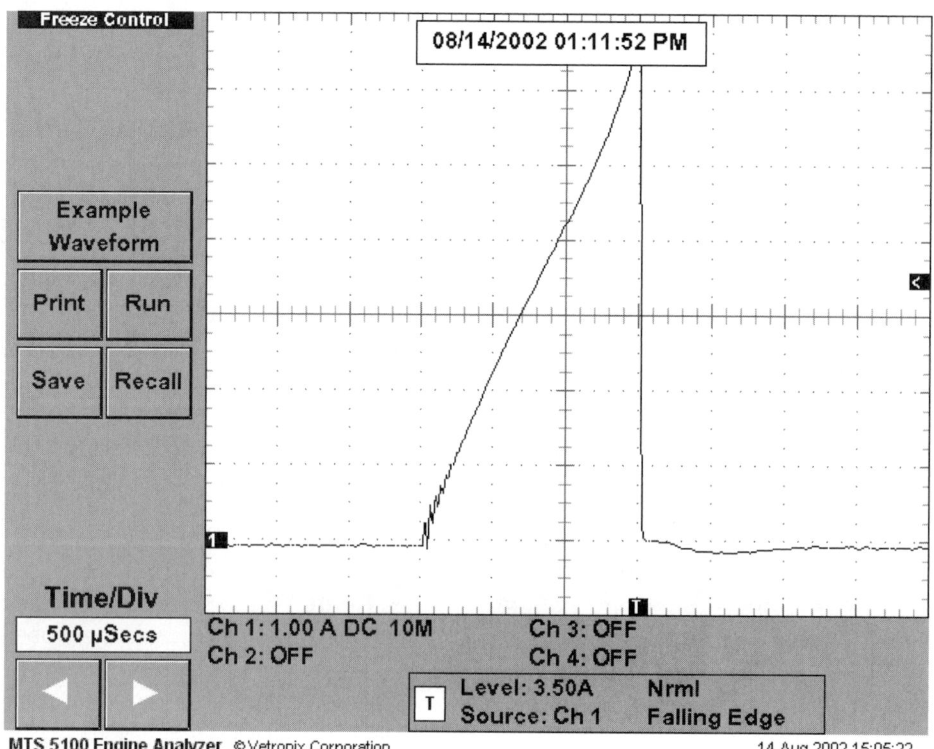

Figure 13. A primary current trace at 1200 rpm.

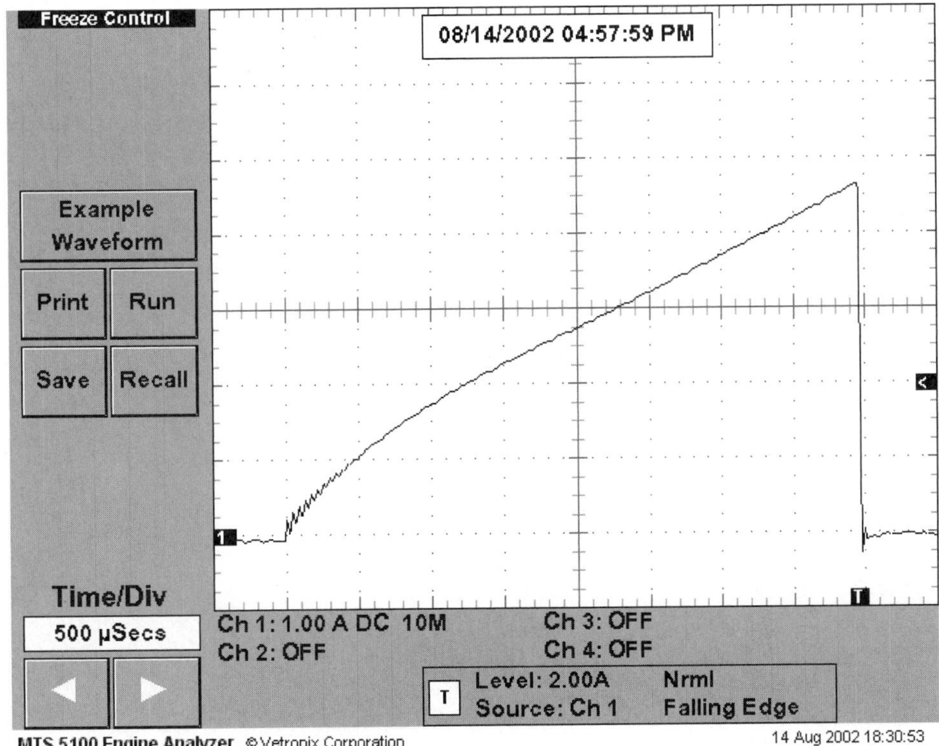

Figure 14. A coil-near-plug primary current trace.

the amount of current and the time that primary current was on. We are set for 1.00 amp per division and a time of 500 microseconds (0.5 milliseconds) per division. There are 4.75 divisions or 4.75 amps. Primary current was on for eight divisions of time, or 4.0 milliseconds:

0.5 milliseconds × 8 divisions = 4.0 milliseconds

This is plenty of current, a nice straight ramp, and an on-time that is adequate (more than 2.5 milliseconds). This is a good pattern.

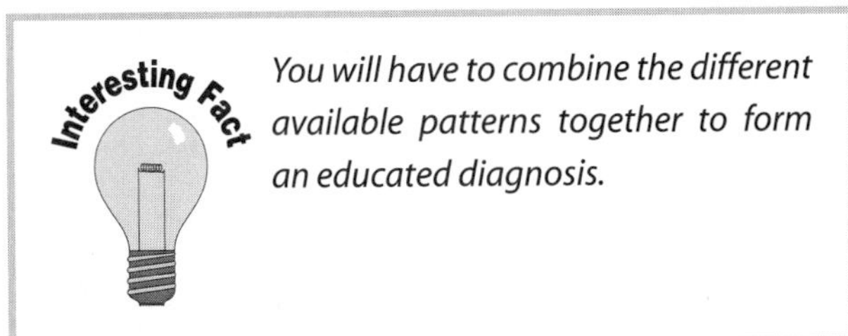

You will have to combine the different available patterns together to form an educated diagnosis.

Remember that on this system the secondary pattern must be used for firing voltage and burn time, whereas the current pattern is used for primary current on-time and amount of current. The combination of the two patterns gives us most of the information that we need to make an educated diagnosis.

After you initially look at a wiring diagram and have figured out what is available to you, use every bit of information, including current patterns.

INPUT SENSORS

The wiring diagram should give you basic information about the type of sensor that is in use, and you have only two choices: a magnetic sensor or a Hall effect sensor. With only two choices, it should be easy. Let us look at each one, so that you will see that diagnosis is not different from that of the distributed sensors.

Magnetic Sensors

Even though the use of magnetic sensors is limited, testing them is not hard. Only two wires are required. If the sensor is doing its job, it will generate at least a 250-millivolt AC signal during cranking. The signal will rise as engine speed increases. **Figure 15** shows a 4-pulse magnetic sensor output, with the DSO leads across the sensor. The sensor has been disconnected and the engine cranked. This is a very good sensor because it is generating more than 15 volts. If the sensor was connected and you made connection with one of the leads, the voltage would be lower, but the pattern would be the same. The triggering signal is the almost instantaneous drop from maximum positive voltage to maximum negative voltage. Remember that there are other applications of this type of sensor. Antilock brake wheel speed sensors, speedometers, and transmission output shafts are just some of the applications. This sensor is sometimes referred to as a **PM (permanent magnet) generator**.

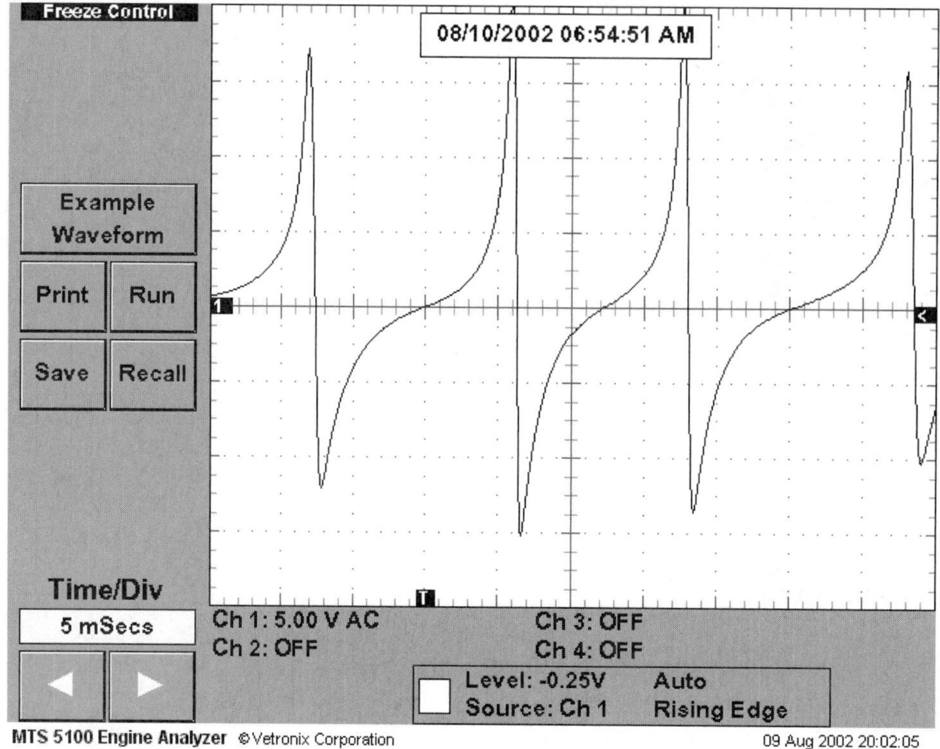

Figure 15. An AC voltage trace from a magnetic sensor.

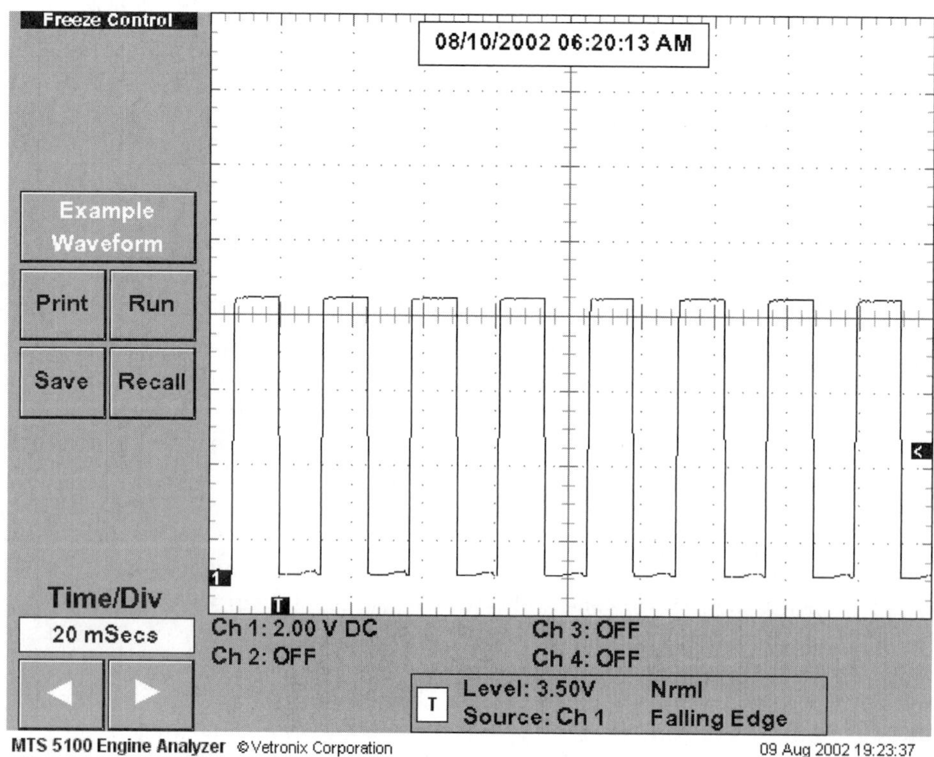

Figure 16. A functioning Hall effect sensor will generate a DC signal.

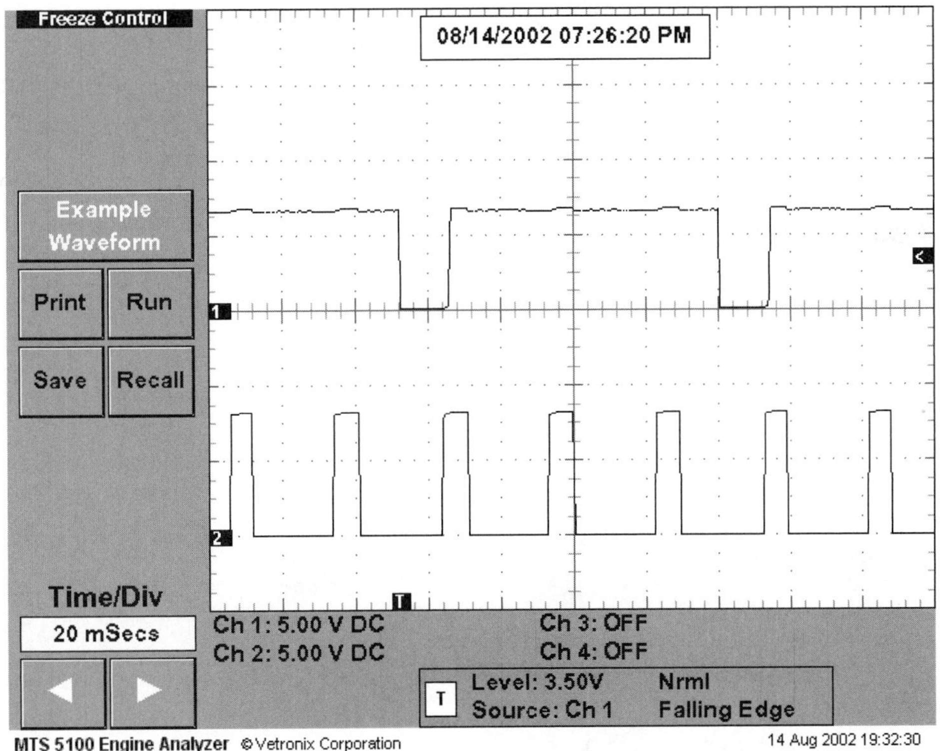

Figure 17. Outputs from 3X and 1X sensors.

Hall Effect Sensors

Hall effect sensors generate a square wave signal like that shown in **Figure 16**. The top of the trace shows a voltage of approximately 8 volts; this is the feed voltage. When the sensor turns on, the voltage drops to zero. The top of the trace is the same distance as the bottom, meaning that the windows and vanes where the signal is generated are the same width. The signal will not change amplitude with speed as a magnetic sensor does. It will remain the same throughout the entire speed range. The other advantage is that it generates a digital signal that is easily interpreted by the PCM or ignition module.

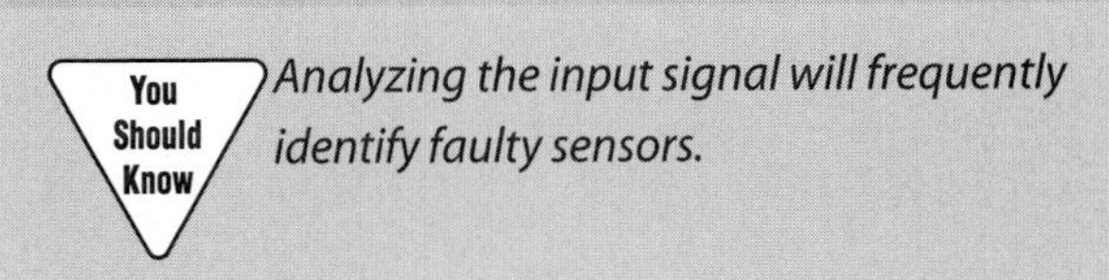

Analyzing the input signal will frequently identify faulty sensors.

Many vehicles use a multiple output such as that represented in **Figure 17**. This is a 3X 1X sensor that generates three signals out of one sensor and one signal out of the other during one rotation of the crankshaft. The relationship between the two is very specific. They are both fed off the same power source, of approximately 7 volts, and both are at zero volts when the sensor turns on, so this is a good pattern. The use of multiple sensors is popular, and their testing is not complicated, especially if your DSO has multiple channels. Putting the multiple signals up on the screen together helps with the diagnosis, especially in no-start conditions.

Summary

- Primary current flow can usually be captured on most ignition systems.
- Secondary voltage patterns may not be available on most coil-on-plug systems without special adapters.
- A primary pattern can show:
  1. Input coil voltage
  2. Coil ground
  3. Primary on-time
  4. Burn time
  5. Coil capacitance oscillations
- Primary current flow can show:
  1. On-time of the primary circuit
  2. Amount of current flowing through the ignition coil
  3. Resistance of coil (indirectly)

- Multiple coils can be compared by looking at the current flow from each.
- Magnetic sensors generate an AC signal.
- Hall effect sensors generate a DC signal.
- Multiple Hall effect sensors generate two different signals that are usually used to indicate piston position and speed.

Review Questions

1. Technician A states that a primary current pattern can show on-time. Technician B states that a primary voltage pattern can show burn time. Who is correct?
   A. Technician A only
   B. Technician B only
   C. Both Technician A and Technician B
   D. Neither Technician A nor Technician B
2. The primary coil on time is usually
   A. more than 0.8 millisecond
   B. Less than 2.0 milliseconds
   C. Greater than 3.0 milliseconds
   D. Less than 3.0 milliseconds
3. A type 2 coil for a V-6 shows 4 amps for the first coil, 4 amps for the second coil, and 1 amp for the third coil. Technician A states that the ignition module ground needs to be checked. Technician B states that the power feed to the coils needs to be checked. Who is correct?
   A. Technician A only
   B. Technician B only
   C. Both Technician A and Technician B
   D. Neither Technician A nor Technician B
4. From which of the following systems is it difficult to obtain a secondary voltage pattern?
   A. waste spark system
   B. distributed system
   C. coil-near-plug system
   D. coil-on-plug system
5. Burn time should exceed 0.8 millisecond.
   A. True
   B. False
6. The current level for multiple coils should be the same.
   A. True
   B. False
7. Coil capacitance oscillations should be more than 5.
   A. True
   B. False
8. The current probe is placed around either the power or ground connection to the ignition coil.
   A. True
   B. False
9. What systems benefit from diagnosis made with a current probe?
10. What types of patterns can be captured on a coil-on-plug system?

Accessories

SECTION OBJECTIVES

At the conclusion of this section, you should be able to:

- Trace various lighting circuits on wiring diagrams.
- Understand how the lighting circuits are diagnosed using a 12-volt test light.
- Test a light circuit.
- Test a relay.
- Test for poor ground conditions.
- Understand how the majority of rear defoggers function.
- Test rear defogger operation using a 12-volt test light and/or a DMM.
- Test horn circuit operation using a 12-volt test light.
- Test windshield wiper operation using a 12-volt test light and/or a DMM.
- Understand how reversible motor-driven accessories function.
- Diagnose a power problem.
- Diagnose a ground problem.
- Diagnose a switch problem.
- Diagnose a motor problem.
- Diagnose a multispeed cooling circuit.
- Diagnose a module-driven accessory.

On some vehicles, the "extra" accessories account for 50 percent of the consumer cost of the vehicle.

Chapter 54 Lighting Circuits

Introduction

Every vehicle has different methods of turning lights on and off, and in this chapter we indicate how this is done. Keep in mind the basic circuit construction that we have used up to this point and this chapter will not be difficult for you.

FOG LIGHTS

Let us start with fog lights. **Figure 1** shows the headlamps and fog-light circuit for a General Motors product and is representative of what is available. The two fog-lamp circuits are in the center left section of the diagram. The left lamp receives its power from S240, which comes down from the 15-amp fuse in the middle of the fuse block. The right-hand fog lamp receives its power from a different 15-amp fuse in the right side of the block. Notice that the bottoms of the fog lamps join together at S122 (PPL wires). The single PPL wire continues down to the load side of the fog lamp relay. If the contacts are closed, the circuit continues through the normally closed contacts of the fog-lamp cutoff relay. Why would we put two relays in series? Think about the function of each relay. One is trying to turn the fog lamps on while the other is trying to turn them off. The switch at the bottom of the diagram is the key to understanding the circuit. If the headlamp switch is placed in the high-beam position, a ground will be applied to the fog-lamp cutoff relay control coil, turning the fog lamps off. In other words, the fog lamps cannot be on if the high-beam headlamps are on. The choice is either one, but not both. The use of two relays in series is common when certain conditions need to be present before something goes on or off. Without the diagram, following this circuit would have been impossible.

TAILLIGHTS

Taillights are part of the headlight or exterior lights, as **Figure 2** shows. Do not let the maze of wires intimidate you. The exterior lamp circuit still comes down to a parallel circuit. The easiest way to follow most wiring diagrams is to start at the load and work your way back toward the battery. This vehicle has taillights that are combination run and stop. The turn-signal function powers another bulb, which we will look at later. Let us first turn on the parking lights or taillights. The bottom of the diagram shows the back of the vehicle. Notice the seven common points above the lamps. This is a good example of a parallel circuit. Power one bulb and all the rest are powered. We will use the lower right tail/stop light 1. Always look at the grounds first. The black wire common points at S403 and eventually makes its way to G304, which is under the left rear seat. Notice that all of the rear lights are carried through the same ground, G304. This would be an obvious place to start diagnosis if none of the rear lights were functional. G304 is also the ground for the turn signals. Back at the bulb, we can see that there are two wires coming in to a double-filament bulb. Let us follow the BLK/YEL wire. It runs up and joins two other BLK/YEL wires at S401(near the right taillights). Continue following up from S401 and you will come to S309. Notice that the other tail lamps are also joining at this same splice. From S309 a single wire heads up. We are now in the series part of this circuit. Seven lights have been common pointed from one BLK/YEL wire. If you got lost, start over. You should be able to follow from any of the tail lamps and eventually come up to S309. Power for this splice is coming

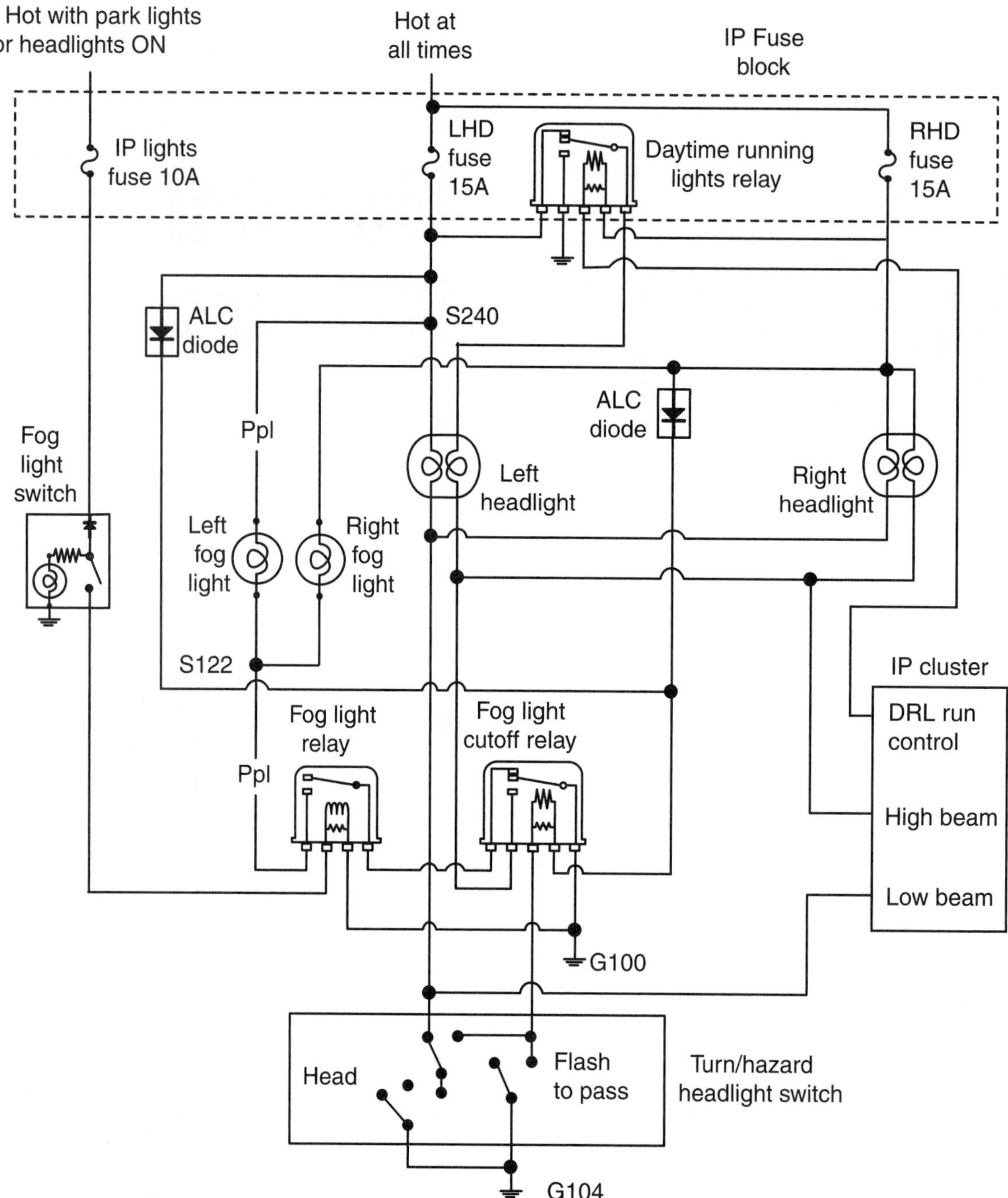

Figure 1. A fog-light circuit that uses two relays.

down from the junction block that has the park lamp relay. When the load side of the relay closes power from fuse 7, a 20-amp fuse will power all of the tail lamps. Getting the park lamp relay to close will take power and ground at the relay control coil. It looks like power will be applied from the same number 7 fuse. Consider the control coil ground. This is your introduction to body control modules. Many vehicles use a module to control lighting functions. It will take a voltage signal from the headlamp switch to indicate that the parking lamps need to be on. Notice the four different resistors within the switch. Each resistance is different and will result in a different voltage signal to the module. The module will interpret the voltage as switch position, and if there is a request for parking lamps, the relay will receive a ground from the module.

While we are on the subject of tail lamps, let us hit the brake pedal to turn on the stop lamps. Again, start at the same bulbs in the rear of the vehicle and follow the other color wire, WHT/TAN. Notice that all of the double-filament sockets have the WHT/TAN wire in them. All of

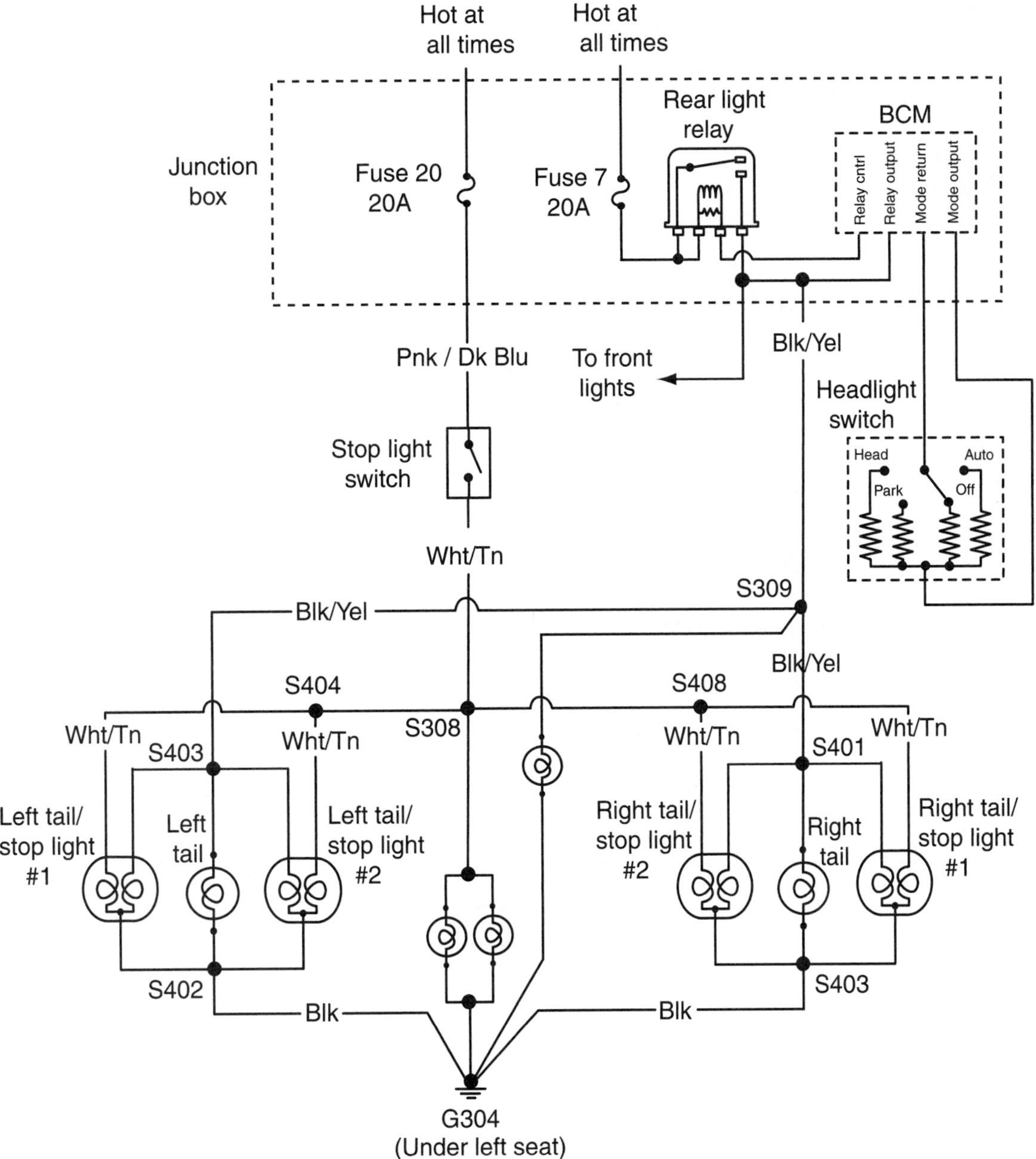

Figure 2. An exterior lamp circuit.

these wires join together at S308. We have an additional set of wires at this common point that come from the center high mounted stop lamp. This is the light you can see that is in the rear window and comes on only when the brake pedal is depressed. From S309, power comes from the stop-lamp switch, which is located at the top of the brake pedal.

> **You Should Know** *The adjustment of the stop-light switch should ensure that the brake lights are on only when the brake pedal is depressed.*

This is the main control for the brake lights. It will close when the driver applies the brakes. Even though the circuit looks complicated, it really is not. Power for the brake lights comes through the 20-amp fuse in the junction block, through the stop-light switch and down to the bulbs. With separate brake and turn signals, this circuit is one of the simpler ones on the road. We will come back to it when we look at combination brake and turn signals in the turn signal section of this chapter.

TURN SIGNALS

There are two different styles of turn signals. One style uses a combination brake and turn-signal lamp in the rear, and the other uses a dedicated turn lamp that is separate from the brake lamps. Let us look at the dedicated style and use a Ford circuit as an example. Again, do not let the diagram **(Figure 3)** overwhelm you. We will take it step by step. If you get lost, stop and back up to the point where you are not lost, and then continue.

Let us make a left turn, starting at the front turn-signal lamps on the left side of the vehicle. Find the left front park/turn-signal lamp on the left side of the diagram at the bottom. The brown wire will be for the parking lights, which do not concern us now. We need to follow the LT GRN/WHT wire because this is for the turn signal. It goes up the dia-

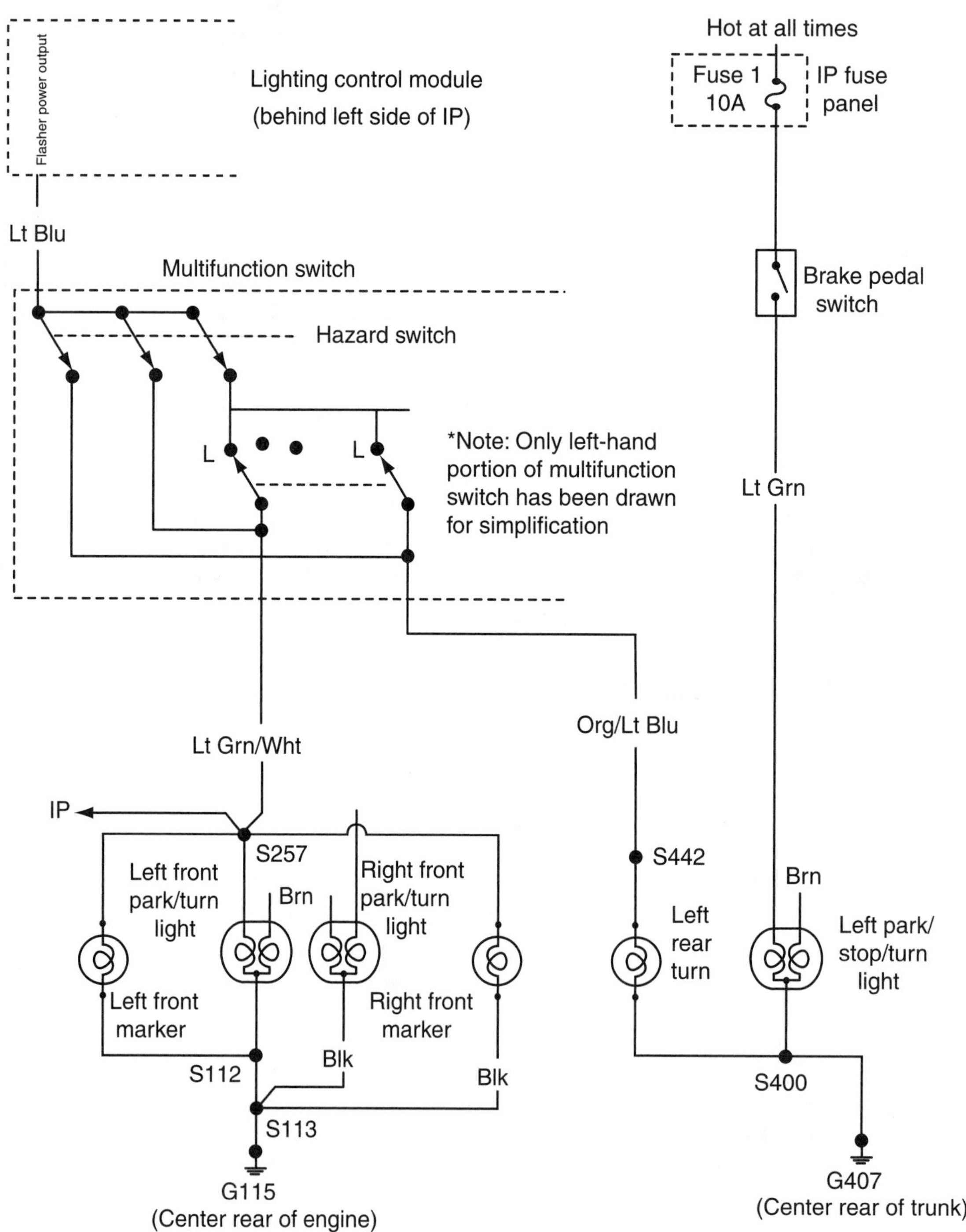

Figure 3. A dedicated brake/turn-signal circuit.

gram to common point S257. From here, the left side of the splice powers the left turn indicator in the instrument cluster. From the right of the splice, the same color wire goes up to the multifunction switch or turn-signal switch. The switch looks more complicated than it is. When the driver moves the signal lever to the left turn position, the two arrows directly above the LT GRN/WHT and ORG/LT BLU wires will move to the left. They both will come in contact with a different contact point. Follow these contact points through the other closed contacts and out the switch with an ORG/LT BLU wire. This wire comes from the lighting control module, which is behind the left side of the dash. This module controls the turn-signal function and the front left light blinks.

You Should Know *Vehicles are produced with either a dedicated brake circuit or a combination brake and turn-signal lamp circuit. Dedicated brake circuits require the addition of a special trailer module to be able to be connected to a tow-behind trailer.*

Consider the rear lights. Remember the LT GRN/ORG wire at the bottom of the multifunction switch? It will have the same pulsing current for turn signals on it that the front bulbs had. If you follow the wire down, it will common point at S442 where the rear turn-signal lamps are located. The ground for the rear is at the center rear of the trunk, whereas the ground for the front is located at the center rear of the engine.

If the vehicle has combination lamps in the rear the wiring is slightly different, as **Figure 4** shows. The difference is in the turn-signal switch and at the bulbs. Let us start in the rear of the vehicle on the right side of the diagram. Notice that the lamps have the notation of left tail/stop/turn light. When we speak of a combination lamp, we mean turn or stop. The same lamp filament will be used for either a stop- or turn-signal lamp. Determining which will be the function of the turn-signal switch. In the position drawn, the bulb filaments will be for stop. The brake switch near the top center of the diagram supplies the power through the closed turn-signal switch contacts. Both the YEL and the DK GRN wires are being powered from the brake switch. When the turn-signal switch moves, it will disconnect one wire from the brake switch and connect it to the turn-signal flasher PPL wire.

DAYTIME RUNNING LIGHTS

Figure 5 is a good example of **daytime running lights (DRL)**. In the left side of the junction block is a DRL module.

Many vehicles manufactured today come with DRL. Only the front of the vehicle has the DRL option. The rear lights function normally.

It has inputs of ignition, so it knows the vehicle is running, and park brake sense, so it knows if the parking brake is applied. If the vehicle is running and the lights are off, as they would be in daytime, the DRL module will turn on the low-beam headlights. Power for the lights and the module comes from fuse 15, which is 10 amps and located also in the junction block. The parking-brake switch turns off the DRL by applying a ground to the park-brake sense input. The module relies on the same ground as high-beam headlamp relay. Normally, the DRL module will not apply full power to the lights, but enough so that they are visible to oncoming traffic. Once the headlights are turned on in either the low or high position, relays supply full power to the lamps so that they will be on full brightness.

AUTOMATIC HEADLIGHTS

The last subject of this chapter is the automatic headlights. You will see that they are not complicated but involve another electronic module. The General Motors system in **Figure 6** uses a combination DRL and automatic-on headlights module under the dash. The only difference is in the ambient light sensor coming out of pins T and S from the module. It is located on the right top of the dash according to the diagram. The sensor is a photoelectric cell that will generate a signal when there is sufficient light. If the sensor sees light, the module turns off the headlights and turns on the DRL. If the light level is too low, sensor output drops, the DRL goes off, and the headlights come on. The headlights on reduced power are the DRL. Notice the normal headlight switch on the right side near the top of the diagram. The output from the switch goes into the module as an input and can come directly out to the headlights if the driver wishes to take over the control manually.

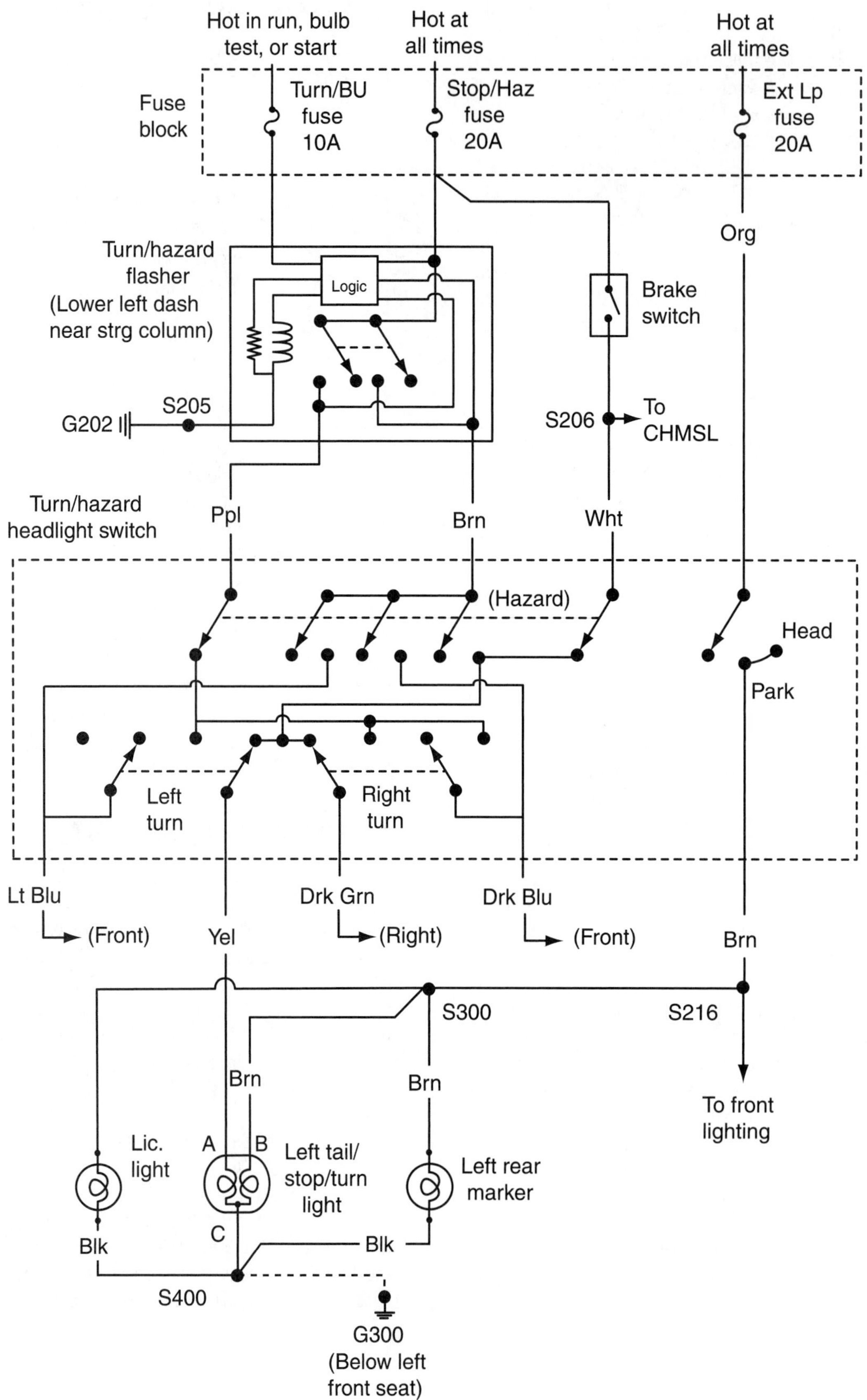

Figure 4. A combination brake/turn-signal circuit.

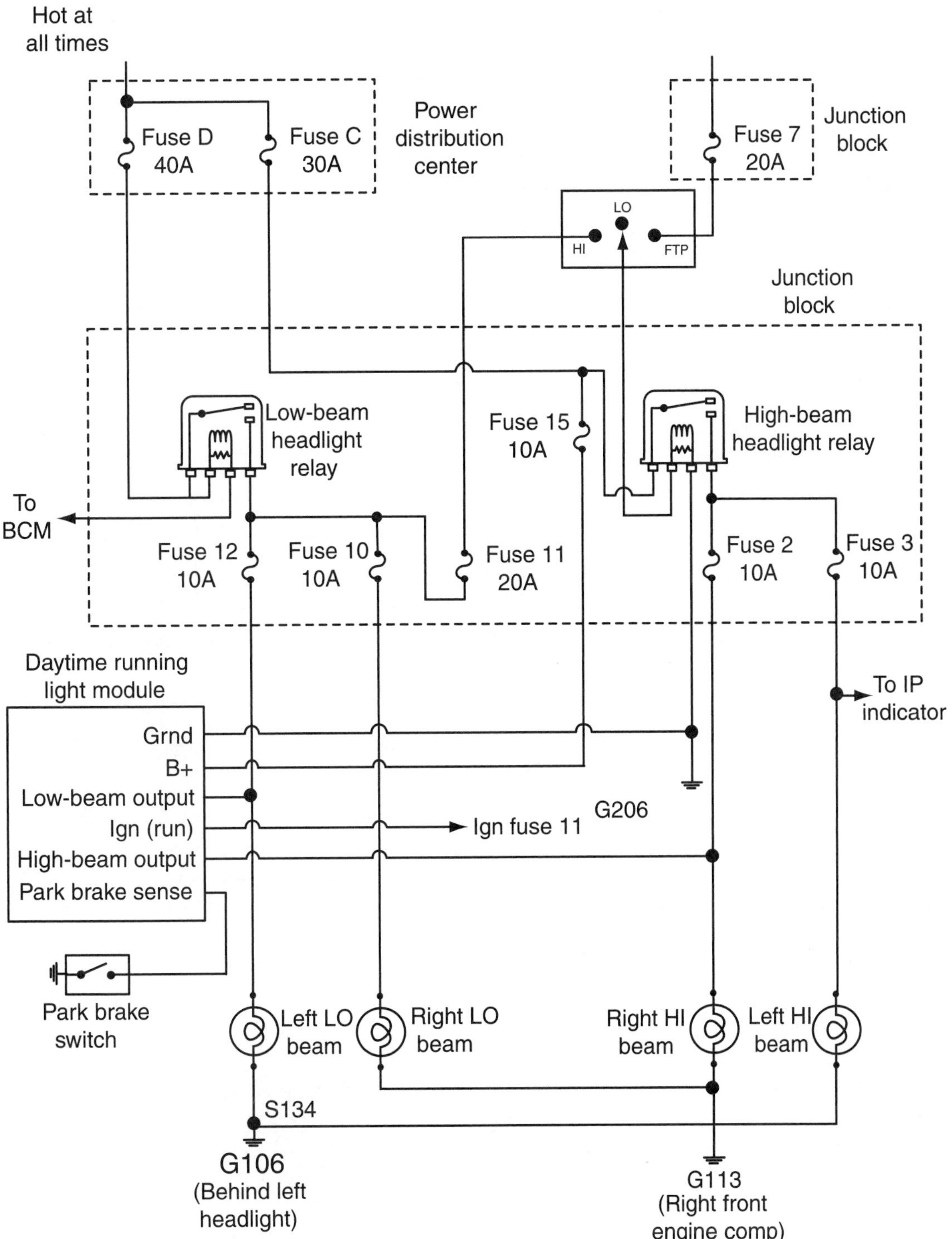

Figure 5. A DRL circuit.

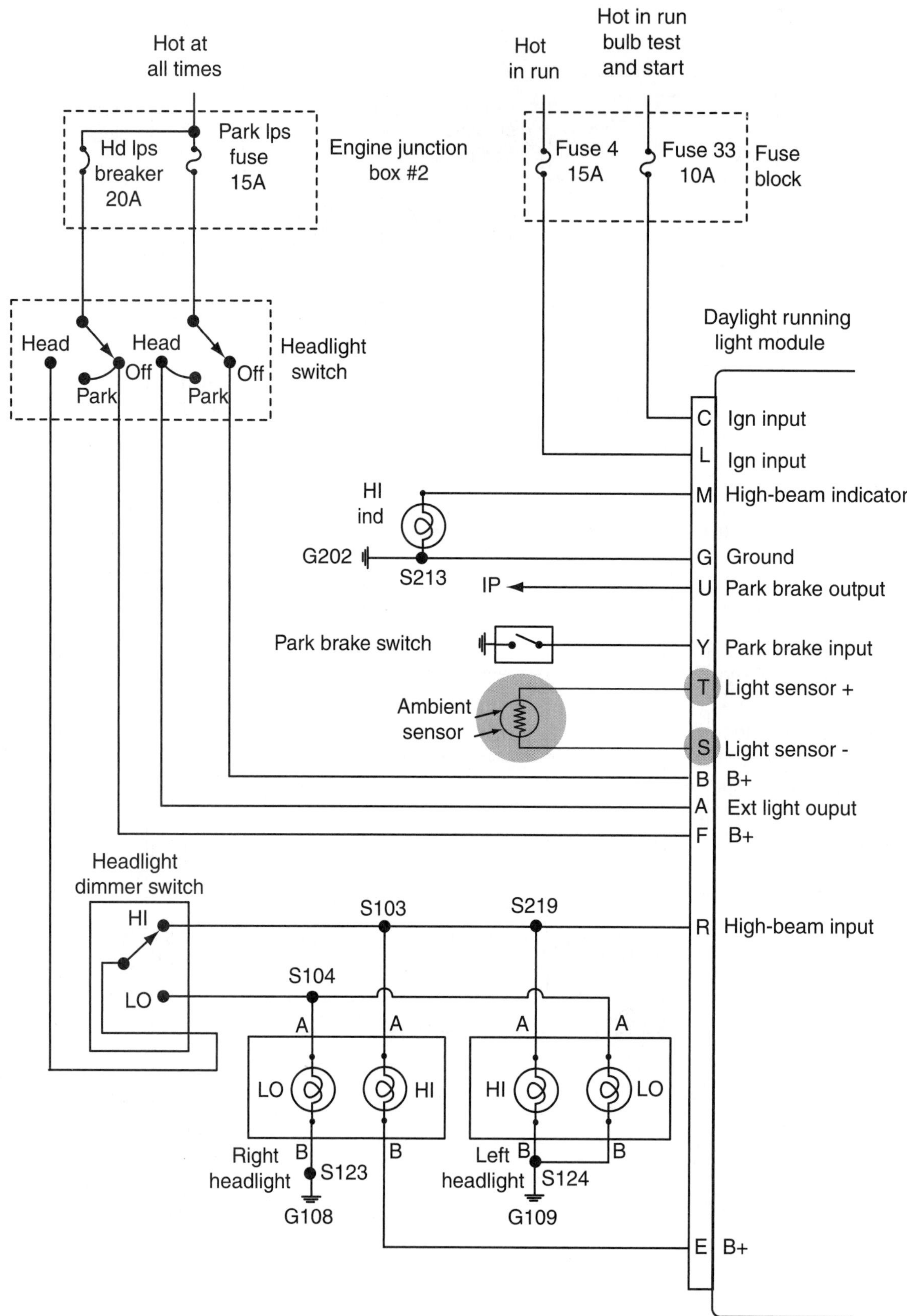

Figure 6. An automatic headlight circuit.

Summary

- Fog lights are usually independently controlled.
- All taillights are parallel wired together.
- The turn-signal brake lights can be independently wired together or be of the combination variety.
- Independent brake lights have a separate circuit from the stop-light switch.
- Most taillight assemblies use double-filament bulbs to accomplish two functions.
- The DRL are usually only on the front of the vehicle.
- The DRL are not in use if the normal headlights are on.
- Some lighting circuits will use a relay.

Review Questions

1. Technician A states that the fog lamps and the DRL are two separate circuits. Technician B states that relays might be in use for fog or DRL circuits. Who is correct?
   A. Technician A only
   B. Technician B only
   C. Both Technician A and Technician B
   D. Neither Technician A nor Technician B
2. When the turn signal is turned to the right and the brake is depressed, the right light blinks and the left is on solid. This system is a ____ system.
   A. DRL
   B. automatic
   C. combination
   D. relay style
3. The fog lights in the front of the vehicle are wired in
   A. series
   B. parallel
   C. either series or parallel depending on the manufacturer
   D. with the headlights
4. Automatic headlights depend on the signal from the ____ to turn on.
   A. PCM
   B. BCM
   C. driver
   D. photoelectric cell
5. A double-filament bulb is used only in a combination system.
   A. True
   B. False
6. The use of two relays in series allows for conditional control of the system.
   A. True
   B. False
7. What is the difference between a dedicated brake light circuit and a combination circuit?
8. Under what conditions would we want the fog lights to *not* be on?

Diagnosing Lighting Circuits

Introduction

We approach diagnosis differently in the three chapters on accessory diagnosis. Rather than going through a complete diagnostic on each system, we will place hypothetical problems into a section of the circuit and zero in on that section. By the time you have finished all three chapters, you will be able to go through a procedure on just about any simple accessory. Keep in mind that even the manufacturer's procedures do not usually test a module. Instead, they test around, looking at inputs and outputs. If all of the inputs are present and the outputs are functional, the obvious conclusion is that the module is not functional. This is the same procedure that we did in the ignition section of the text.

FOG LIGHTS

A vehicle arrives at your shop with one fog light functional and one not functional. We verify the problem by turning on the fog lamps, and the left one does not work. A quick walk through the wiring diagram, as in **Figure 1**, should reveal the following. The two lights are powered off two different fuses. The left light is paralleled with the left headlamp. We turn the headlights on and the left one works. The fuse is good, and we know it without crawling under the dash and testing it.

Make sure that you verify the customer complaint and remember that some of the modern circuits are very complicated. Your customer might not understand how to get the circuit to function correctly.

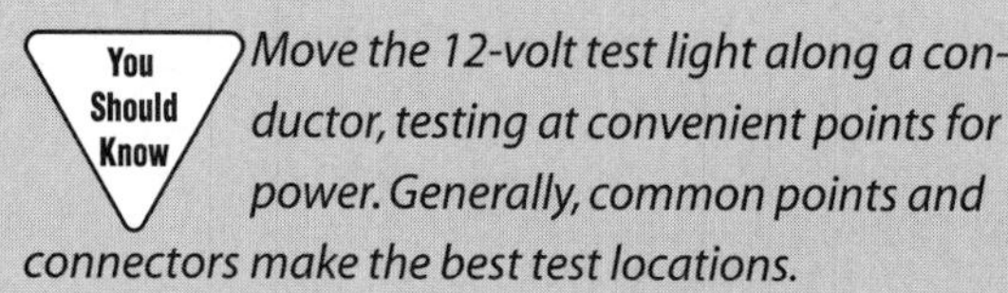

Move the 12-volt test light along a conductor, testing at convenient points for power. Generally, common points and connectors make the best test locations.

If we visually inspect the bulb or swap it out with the right one and it proves to be a good bulb, we know the problem has to be between S240 and S122. Why? If one of the lights works, the fuse, switch, relays, and series section of the wiring must be doing its job. In addition, G100 must be functional because both lights rely on the same ground. A large wiring diagram with many components has just been reduced down to three common points and some wire. **Figure 2** shows an expanded diagram of the suspect area. Get the 12-volt test light out and test for power at S240 on the ORG wire. If there is power, we move to the light and test for power at the ORG wire in the connector. If there is no power, a broken wire between S240 and the connector is the cause.

If there is power, then the problem appears to be on the ground side. Do you remember how we test the

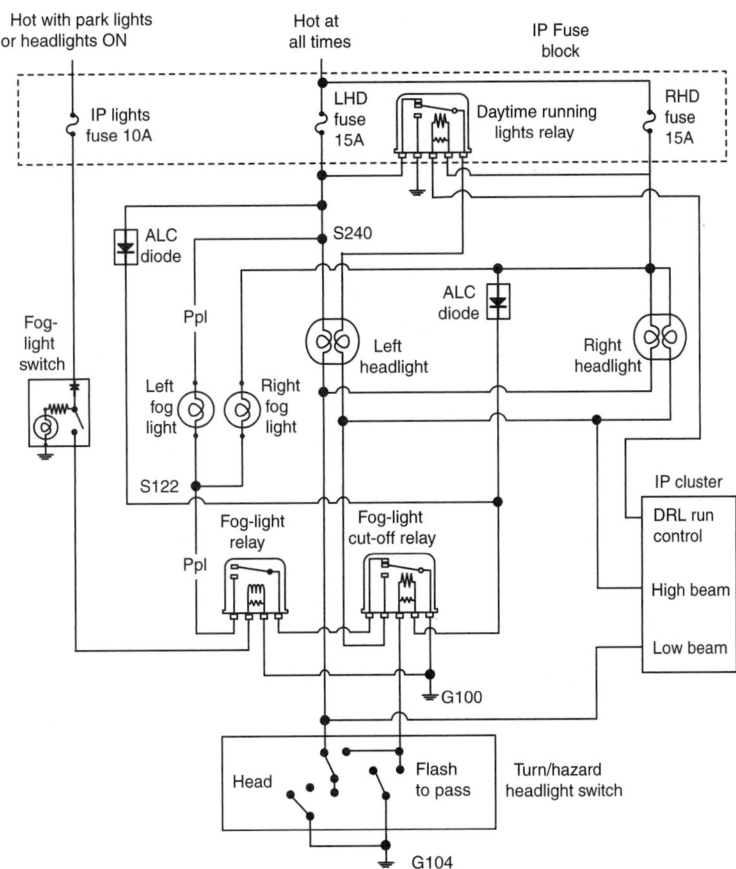

Figure 1. A fog-light circuit.

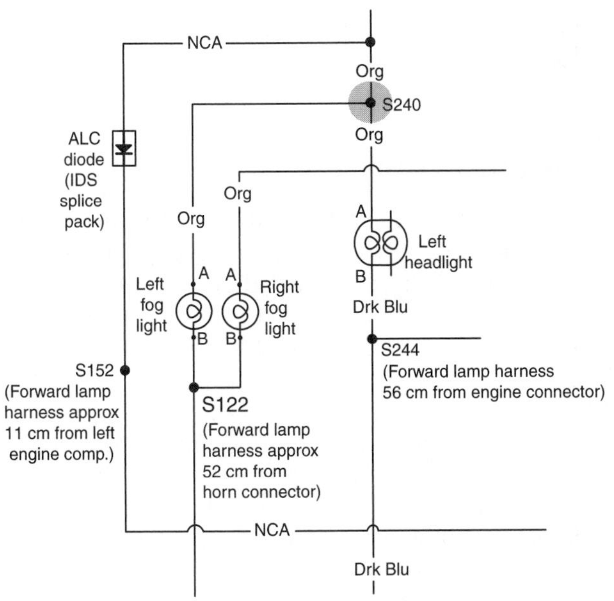

Figure 2. An expanded diagram of the two fog lamps.

ground? With the test light connected to power (perhaps at the battery), connect into the PPL side of the connector. The light will light if the PPL wire is ground. Follow along the PPL wire until we get to S122 looking for a ground. The broken wire will show up by using the diagram and the 12-volt test light and your own common sense.

TAILLIGHTS

A customer arrives at the shop with the complaint that the brake and taillights do not work. Again, the first thing we always do is verify the problem. Turn on the headlights. They work, but not the taillights. Turn on the turn signals. Nothing happens. One more check of the brake lights reveals that they do not function either. The problem has been verified and the customer is right. Logic tells us that the problem cannot be the bulbs, because there are too many for them all to be burned out. The wiring diagram as shown in **Figure 3** indicates that the brake-light circuit is separate from the run lights. The two circuits share only one

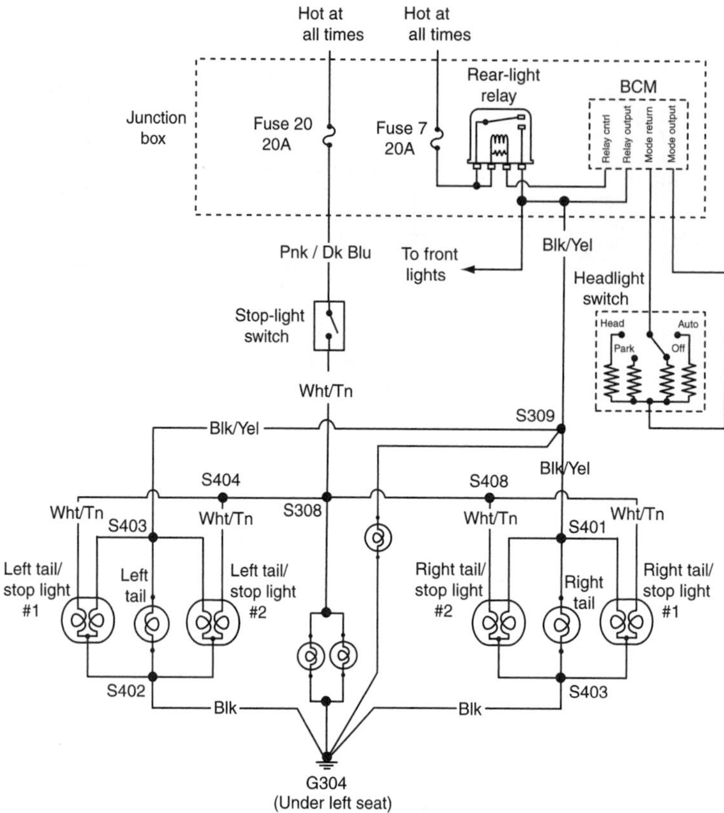

Figure 3. A taillight circuit.

thing—the ground. Expanding the diagram as in **Figure 4** indicates that there is only one ground, G304, which is located under the left rear seat.

Notice the dotted lines that connect the ground splices, S402, S405, and S403. The lines indicate that all of these splices are tied together. It should be relatively easy to follow the black wires in the trunk to the spot where they all join together.

Get out the 12-volt test light again and check for an open ground. Find power in the trunk and connect the clip lead of the light to it. Usually one of the bulb inputs can be the source with the running lights turned on. Now start at the splice and work your way to the ground under the seat. The light will come on when it "sees" a ground and be off with no ground. Repair will be to run a new ground wire, solder a broken wire, or replace a connector. Retest and you are finished.

You Should Know *Ground problems account for almost 25 percent of customer complaints. Do not ignore the ground circuit as you work your way through a circuit.*

Again, the hero in this example is the diagram, which allowed you to mentally test the circuit before even opening the trunk. You must systematically use the diagram to help you diagnose the problem. Do not forget to retest the circuits when you are finished.

TURN SIGNALS

The customer complaint is that the left-turn signals do not function. The wiring diagram shown in **Figure 5** indicates that the vehicle has a dedicated brake circuit and separate turn signals. Turn the KOEO, put the turn-signal lever to the right position, and check both the front and rear lights. Do they all function correctly? If yes, move the turn-signal lever to the left position. For our example, let us say that nothing works, not even the indicator lamp on the dash. A closer look at the multifunction switch **(Figure 6)** shows that the same flasher powers both sides and that there is only one source of power for the turn signals. This is beginning to look like a bad multifunction switch. One last check is to turn on the hazard lights. They work. Using the hazards proves that the connectors, wires, and bulbs on the left side of the vehicle function. The left-turn position of the multifunction switch is not being connected to

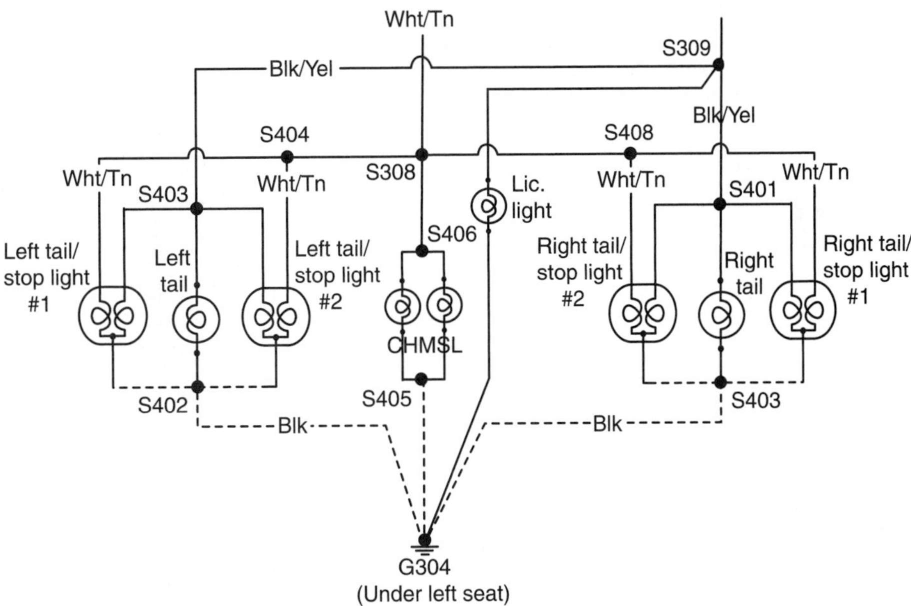

Figure 4. An expanded taillight circuit.

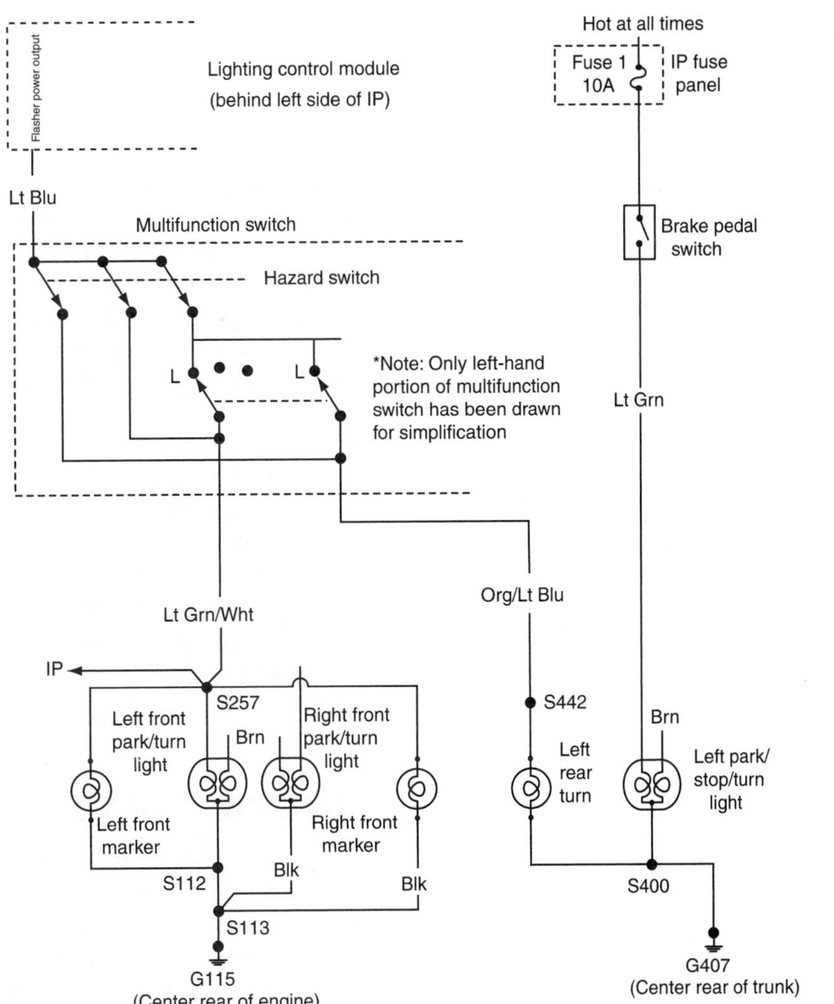

Figure 5. A turn-signal circuit.

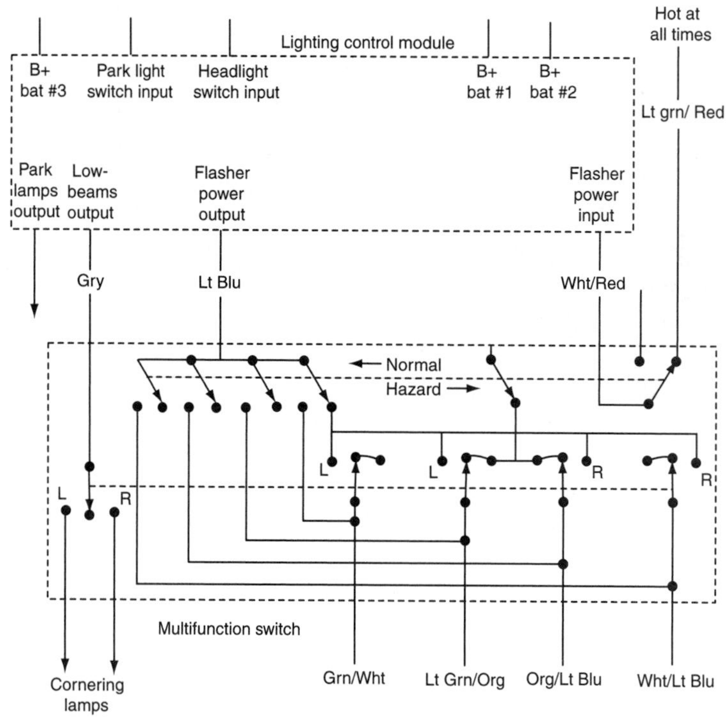

Figure 6. A multifunction switch.

the bulbs. This is a faulty multifunction switch and we proved it without using any piece of test equipment except the wiring diagram. Yes, the wiring diagram should be considered a piece of test equipment. There probably are not many tools that can save you as much time as the wiring diagram.

BRAKE LIGHTS

The customer complaint is that the brake lights do not function. After bringing the vehicle into the shop, the technician pulls the wiring diagram. **Figure 7** shows that this is a combination brake- and turn-signal system. The brake switch is closed by the brake pedal action and runs current through the turn-signal/hazard switch assembly. The most logical thing to do is turn on the turn signals and hazards to test the circuit wires and bulbs. Make sure that you look at the rear of the vehicle. They work but the brake lights do not. Go back to the wiring diagram for additional information. Power for the brake and the hazard lights comes from the same 20-amp fuse in the fuse block, and because the hazards function, it must be good. It looks like the problem has to be from the fuse, through the brake switch and into the turn-signal switch. **Figure 8** expands this part of the circuit for you. An ORG wire connects the fuse to the input of the brake switch, which is located on the brake-pedal support bracket. When the switch closes, power is brought down to the turn/hazard-switch assembly with a white wire. There appears to be one splice, S206, on the white wire. The 12-volt test light appears to be the tool of choice here, and we have the choice of starting at either end or the middle. Always choose the easiest-to-get-at place to start, which is probably the brake switch. According to the diagram, the ORG wire should be hot all of the time, and it is. Our test light lights when clipped to ground and touched to the ORG wire. Power is available to the brake switch. Backprobe the white wire and push on the brake pedal. If the switch works, the light should light. In our example it does not. We have a bad brake switch, which can be verified by using an ohmmeter or continuity indicator across the disconnected switch terminals. After replacing the switch, make sure you retest the entire system to make sure that it functions.

AUTOMATIC HEADLIGHTS

The customer complaint is that the automatic feature of the headlights does not work. **Figure 9** shows the wiring diagram for this circuit. This should be easy and come down to one of two components: the ambient light sensor or the module. First we need to verify the customer complaint and get some additional information. Manually turn on the headlights and make sure that they function. If they do,

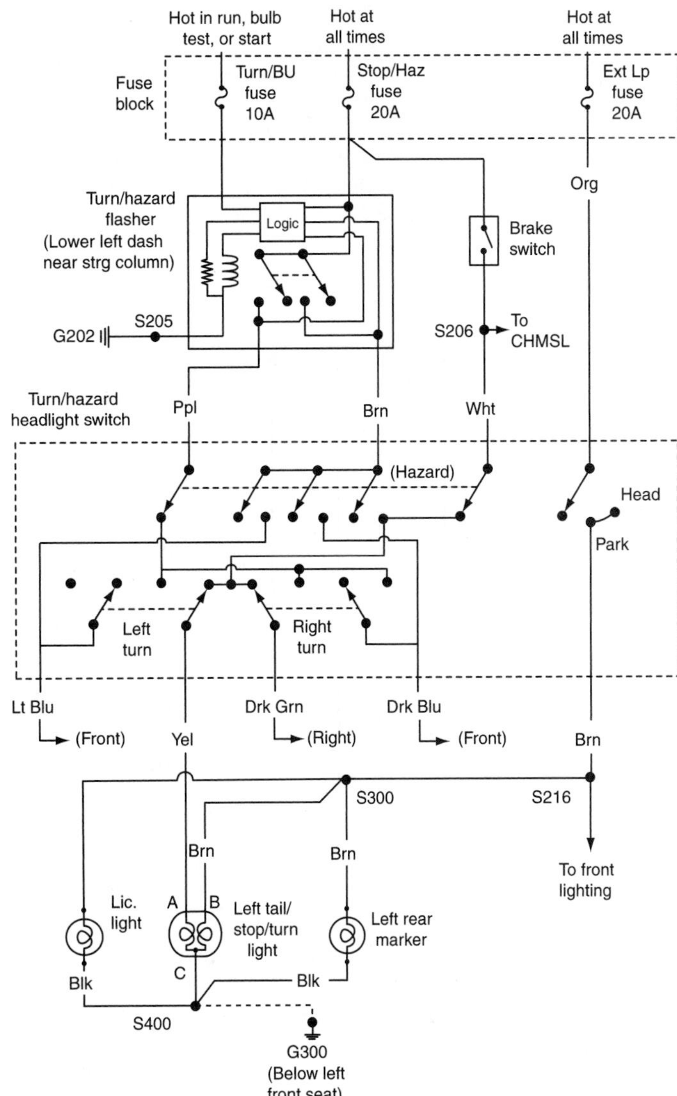

Figure 7. A combination brake-light circuit.

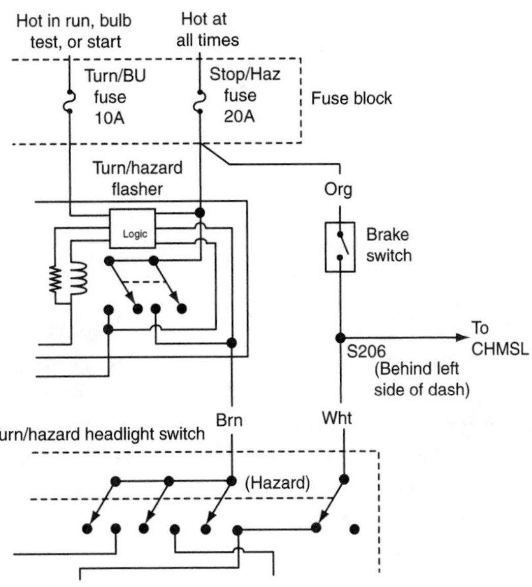

Figure 8. The brake switch connects to the turn/hazard switch.

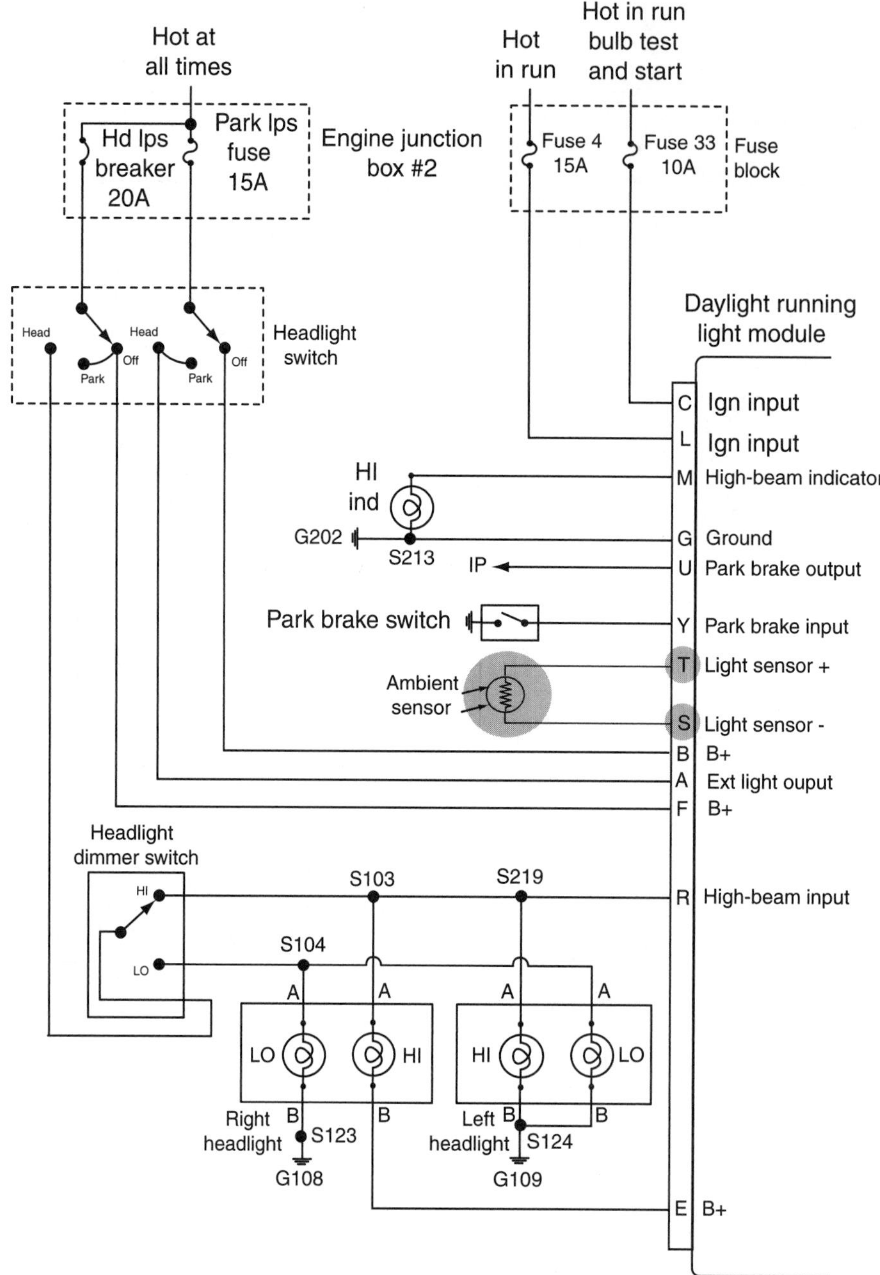

Figure 9. An automatic headlight circuit.

place a piece of cardboard or cloth over the sensor with the KOER. In a short period of time the lights should come on. If they do not, we need to focus on the sensor. The best place to do our testing is probably the module, which the diagram indicates is behind the left side of the dash on the electronic brake module bracket. If the module can easily be unbolted and dropped down to the floor, it will make testing easier **(Figure 10)**. Leave the 12-volt test light in the toolbox and get out the DMM or DSO. If we connect the DMM to terminals T and S, the sensor output in millivolts should be displayed. With the sensor covered up, the output should drop to zero. If the sensor behaves like it should, the module is not functioning and should be replaced. Remember that this vehicle has functioning manual headlights, which proves that the rest of the circuit is in good shape.

Again, the key to the circuit and how it functions is the wiring diagram. It not only gave us the information about how the circuit worked; it also gave us location codes, which are invaluable on today's vehicles.

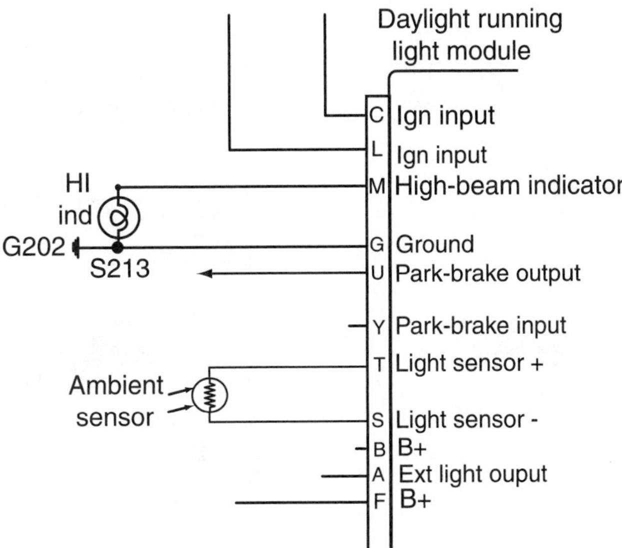

Figure 10. The ambient light sensor is wired to the automatic headlight module.

Summary

- Before testing, a circuit wiring diagram should be printed for the vehicle.
- The customer complaint should always be verified.
- All functions of the interrelated circuits should be tested.
- Look for common points or splices that are convenient test points.
- Use the 12-volt test light if no modules or electronics are involved.
- A ground can shut down a complete circuit.
- Use the DMM to test if modules are involved.
- Common-point diagnosis should be practiced.

Review Questions

1. One headlight works, but the other will not. The two lights share a common ground. Which of the following cannot be the problem?
   A. a bad ground
   B. a burned-out headlight
   C. a broken power wire to the off light
   D. a faulty headlight connector
2. A turn signal in the rear does not function on either side. Technician A states that the brake switch could be at fault. Technician B states that circuit ground might be open. Who is correct?
   A. Technician A only
   B. Technician B only
   C. Both Technician A and Technician B
   D. Neither Technician A nor Technician B
3. The ambient light sensor generates zero volts when a light is shined onto it. Technician A states that the light sensor might be at fault. Technician B states that the sensor might have a broken wire. Who is correct?
   A. Technician A only
   B. Technician B only
   C. Both Technician A and Technician B
   D. Neither Technician A nor Technician B
4. A connection has excessive resistance. This might cause a
   A. module to overheat
   B. load to burn out
   C. circuit to not function
   D. fuse to burn out
5. Ground connections can shut down a complete circuit.
   A. True
   B. False
6. The brake switch might run through the automatic headlight module.
   A. True
   B. False
7. Why is it important to verify the customer's complaint?
8. Why do we not usually test modules off the vehicle?

Defogger, Horn, and Windshield Wiper Circuits

Introduction

Although they are called accessories, most vehicles have a rear defogger and everyone has horn and wiper circuits. They are good examples of typical accessory wiring, though, and that is why we are going to cover them. As you work your way through these circuits, keep in mind that the manufacturers may not all wire every circuit the same. You will find differences from one vehicle line to another.

HORN CIRCUITS

We will start on an accessory that every vehicle has, the horn. This circuit is very similar across the various models and years. **Figure 1** illustrates a typical General Motors horn circuit using a relay, a horn switch in the steering column, and horns out in front of the grill. Even though the diagram shows only one horn, many vehicles have two. Notice that the horn is remote grounded to the lower left front of the engine compartment. Frequently, the horns are case grounded, but in recent years more plastic horns are showing up on vehicles. A plastic horn will require a remote ground. Notice also the splice, S115. Probably another component will be sharing the ground. This point may become important in diagnosing the circuit. Power for the horn comes through terminal B that connects the horn via a DK GRN wire to the horn relay, which is in the convenience center.

Interesting Fact

The current for the horn does not run through the steering assembly. A relay is most frequently used.

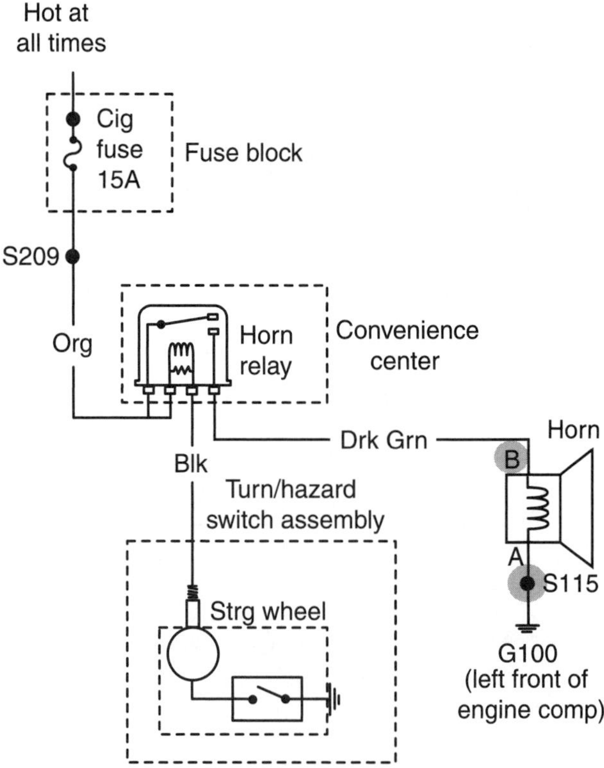

Figure 1. Relays are used in most horn circuits.

When the load contacts close, they must be applying power to the horn. As you follow the upper side of the contact to the left, power is applied to the relay through an ORG wire that comes from the 15-amp Cig fuse in the fuse block. The fuse is hot all of the time, so the horn should be live even with the key off. Leaving the load side, now let us look at the control side. Remember that a relay uses low control current to control the high current of, in this case, the horns. Notice that the same terminal of the relay and the same fuse are supplying power to the relay coil. The black wire leading down from the relay coil goes to the turn/hazard-headlight switch assembly, where slip rings and brushes allow the ground of the horn switch to be applied to the relay coil. That was not difficult. Most horn circuits are very similar to that shown. A wiring diagram will show you any differences.

WINDSHIELD WIPERS AND WASHERS

There are many differences in how the windshield wiper circuits function, and you will frequently find a simplified diagnosis chart in many of the repair manuals. We will give you an idea of how they work so that diagnosis can be accomplished without too much difficulty. **Figure 2** diagrams a common circuit involving a delay circuit. The delay circuit allows the wiper to wipe and then stop for a period of time that the driver determines. It is sometimes referred to as the mist circuit, because it works well when

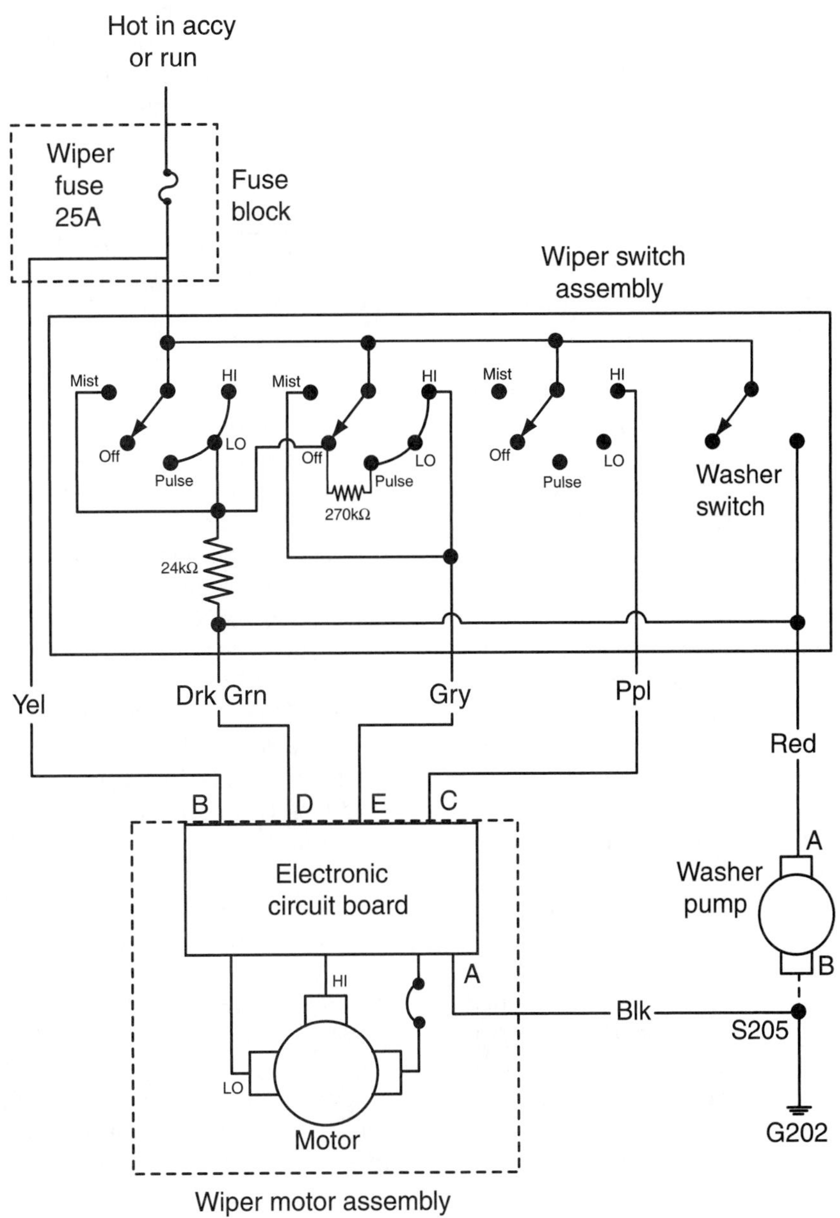

Figure 2. A common windshield wiper circuit.

it is not raining hard. This diagram also shows you how sometimes the wires are not shown inside the assembly. If the inputs are doing their job, the entire assembly is replaced.

Most windshield wiper circuits contain a circuit breaker to open the motor circuit if the blade is jammed or iced to the window.

Look at each wire coming from the assembly first. Terminal A is the system ground using a BLK wire that will common point to the washer pump and then to G202, which is on the left side of the dash. The washer is an easy circuit, so let us look at it now. When the washer switch is closed manually, power will be applied from the switch to the pump on the RED wire. Power for the switch came from the 25-amp wiper fuse in the fuse block. Let us get back to the motor assembly. A was the ground. B is power. A YEL wire comes directly from the same 25-amp fuse that powered the washer pump. Terminal C will be a signal to the assembly that the third wiper is in the HI position. Terminal D comes from the first wiper, and terminal E comes from the middle wiper. Notice that the terminal identifiers change at the wiper switch assembly. This can sometimes be confusing even with a diagram. Notice also the dotted line that connects all of the wipers inside the switch assembly. Remember that this means that all wipers move together as one unit. Let us start with the off position that the diagram is drawn in. There is no direct path through the switch to the motor assembly. The middle wiper does have continuity through the off position, but it is a 24,000-ohm resistor, so there is not much current flowing. When the electronic circuit board sees this very low voltage, it knows that the switch is in the off position. Each different position will send different voltages to the electronic circuit board so that it can interpret switch position. Consider the pulse position. The first wiper has continuity to the motor through the 24,000-ohm resistor. The second one has continuity through the 270-ohm resistor to the GRY wire to terminal E of the motor assembly. So now there are two inputs to the motor, one through 24,000 ohms and one through 270 ohms on two separate wires. Each position must have a different set of conditions on the three input wires from the switch so that the circuit board knows the switch position. Low changes around the resistance from the second wiper and brings full voltage to terminal E, and high brings voltage on terminal C. Notice that each position brings a different set of voltages to the motor assembly. It is the job of the electronic circuit board to interpret the signal. The switch varies the voltage to the board based on a series of different resistances. One last item to make a note of is that most windshield wiper circuits will have a circuit breaker within the motor assembly. If the wipers become jammed or iced to the window, the breaker will open and protect the motor.

DEFOGGER CIRCUITS

Rear window defogger circuits are basically the same across most manufacturers. When those circuits are energized, they will close a relay for a specific period of time. We will use a Ford product as our example in **Figure 3**. The rear window grid is in the lower right corner of the diagram. Notice that it is impossible to tell which side is positive and which is negative from the diagram. On a real vehicle, you would only have to use a 12-volt test light to tell which side is positive.

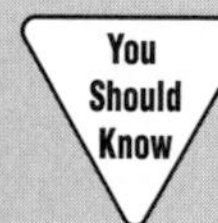

The window grid is a resistive heating element that will generate sufficient heat to remove fog or light ice from the window.

The defroster grid will receive its power and ground from the hidden antenna module. The antenna has nothing to do with the grid; it is just a convenient place for the manufacturer to make connections. Power for the grid will come from the rear window defrost relay in the engine compartment fuse box. The relay is hot all of the time because fuse 14, a 40-amp fuse, is hot. Notice that when the power comes out of the bottom of the relay, splice S252 will power an on-indicator LED, which is part of the defrost switch. The relay coil needs a ground to function because it has power from the same 40-amp fuse. Another module is in use here. The heated backlight relay output will time the on-period of the grid. A backlight is another name for the rear window. When the defrost on/off switch is turned to the on position, a momentary ground is applied to the relay control module, and the module applies the ground to the rear window defrost relay. The relay closes, turning on the grid and the indicator lamp on the dash. Once the preset time is over, the module will de-energize the relay, and the grid will shut down.

Let us look at another example and one that will use a body control module for the timing function. The Dodge circuit diagrammed in **Figure 4** is basically the same as the previous example, with one difference. On the left side of the diagram, the DK BLU/WHT comes out of the relay coil seeking a ground. With automatic air conditioning, the ground will be supplied by the BCM. The preset amount of

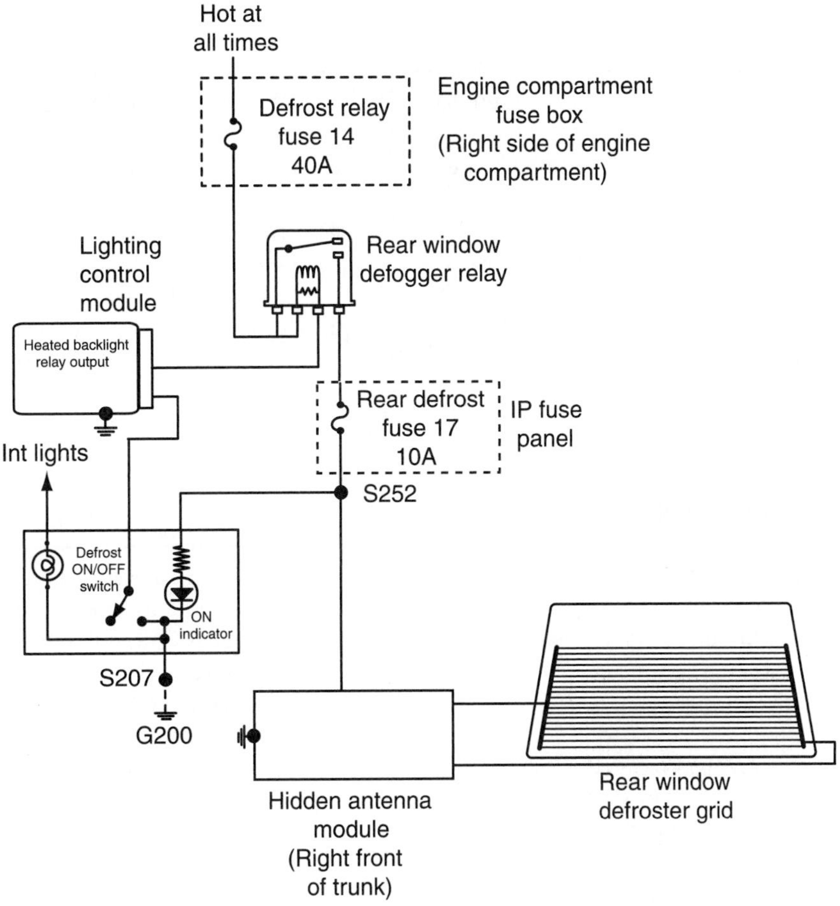

Figure 3. A rear defogger circuit that uses a module for timing control.

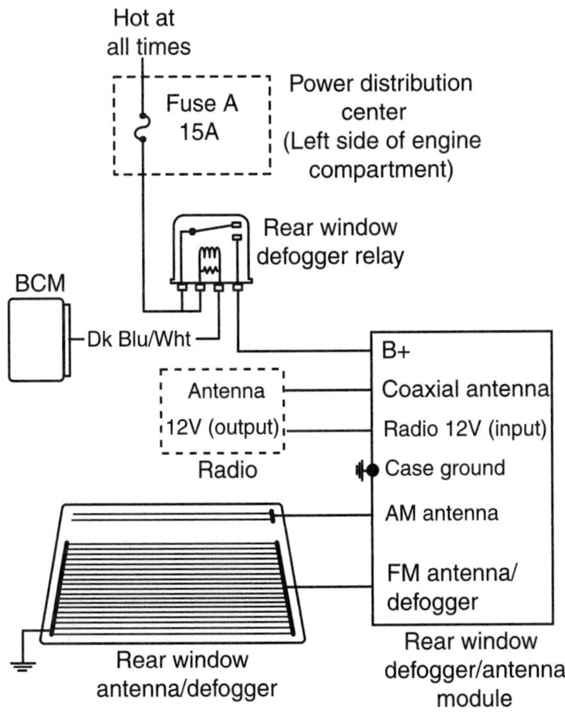

Figure 4. A rear defogger that uses a combination module.

time that the defogger is left on will be the responsibility of the BCM and will become a variable time based on outside temperature and other conditions that are inputs to the BCM. You may encounter a defogger module, or more recently the BCM that will leave the defogger on for a set amount of time. The BCM in this example also slightly raises up the idle and turns on the AC compressor so that additional dehumidification can take place. It makes sense in this age of multiple modules to interrelate items so they help one another get the job done.

Another noteworthy difference in this vehicle is that the FM radio antenna and defogger are one component. The grid on the rear window is being used for FM radio reception. This requires another module to allow both to function without interference between them. Located on the right pillar behind a trim panel is the rear window defogger/antenna module. This is a good example of two circuits having something in common that you might never have considered. The wiring diagram again has become your eyes into the circuit. Without it, you might never have realized the interconnection between the radio and the rear defogger.

Summary

- Horn circuits generally use a relay to supply current to the horn.
- Normally, the relay coil is powered off the same fuse as the horn.
- The ground for the horn relay will be supplied by the horn switch, usually in the steering wheel assembly.
- Windshield wipers usually use an electronic circuit board that will see different voltages from the switch. The different voltage is the signal for the board to change the wipers.
- Normally, a circuit breaker is inside the motor assembly in case the wiper blades get iced to the window.
- Rear defoggers will generally use a relay to turn on the large amount of current that the unit will draw.
- The defogger will be turned on for a preset period of time by a timer module, which may be built into the relay, or by the BCM.
- Some examples of the defogger and the FM antenna use some common components.

Review Questions

1. Technician A states that horn buttons in circuits with relays are B+ switched. Technician B states that horn buttons in circuits without relays are switches to ground. Who is correct?
   A. Technician A only
   B. Technician B only
   C. Both Technician A and Technician B
   D. Neither Technician A nor Technician B
2. Rear defoggers generally have a relay with a timer. This allows
   A. the defogger to operate for only a specific amount of time
   B. the defogger to function just until the rear window is clear
   C. the defogger to be independent of the ignition switch
   D. none of the above
3. The circuit breaker in most windshield wiper motor assemblies is designed to
   A. protect the motor from burning out
   B. open if the wiper is jammed
   C. turn the motor off if ice has frozen the blade to the windows
   D. all of the above
4. Technician A states that a separate module may supply the timing function for the rear-window defogger. Technician B states that the BCM might supply the timing function. Who is correct?
   A. Technician A only
   B. Technician B only
   C. Both Technician A and Technician B
   D. Neither Technician A nor Technician B
5. The horn button on a relay-equipped horn system will apply a ground to the relay coil to blow the horns.
   A. True
   B. False
6. Interval or intermittent wipers use a variable resistor to directly control current flow to the monitor.
   A. True
   B. False
7. Most accessory circuits contain relays.
   A. True
   B. False

Diagnosing Defogger, Horn, and Windshield Wiper Circuits

Introduction

We will follow the same format in this chapter that we did in Chapter 55. We will present problems with the circuits and use common sense and a process of elimination to diagnose with the help of the wiring diagram. We will rely on our 12-volt test light as the primary tool unless the circuit contains a module. Remember that the wiring diagram is the first and frequently the most important tool that we have available. It is sometimes useful to print a copy and color it with green for ground, red for positive, and blue for control. The coloring of the diagram does more than just make it look better. It helps in understanding how the circuit functions and what our test light or DMM should be reading at a particular point. The circuits that we will look at are the same ones that we introduced to you in Chapter 56.

HORN CIRCUIT DIAGNOSIS

The customer complaint is an intermittent horn. Sometimes it works and sometimes it does not. As we pull the vehicle into the bay, we hit the horn button a few times, and it does not always function. The next thing that is necessary is to pull the wiring diagram **(Figure 1)**. The circuit includes a relay; one load, the horn; and one control, the switch in the steering wheel.

The most logical approach will be to test the horn independent of the vehicle, and if it is all right, to look for excessive resistance. Resistance higher than specification can give us an intermittent condition. A broken wire or connector can also give the same condition. By removing the DK GRN wire from the horn and bringing over a jumper from the positive terminal of the battery, we hear the horn functioning normally. Each time we connect to it, it honks. We have also just tested the G100 connection. If it had excessive resistance, it could cause the problem, but it seems to be all right. We will save the horn switch,

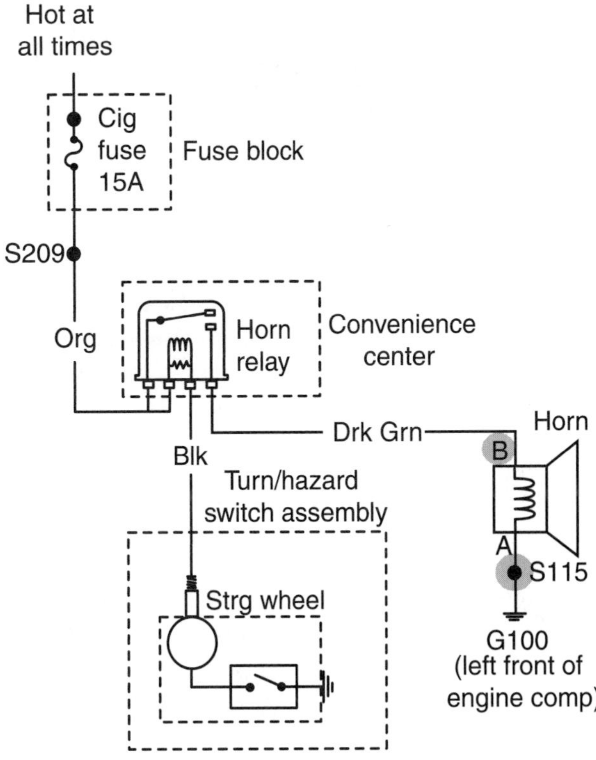

Figure 1. A common horn circuit.

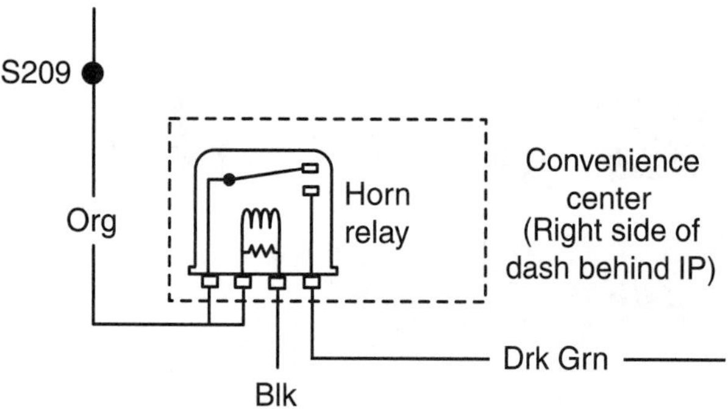

Figure 2. The horn relay controls the horn.

because it may be the most difficult component to get to. The relay is located in the convenience center and should be easy to remove and independently test. **Figure 2** is an expanded diagram of the relay. Let us review its function and operation.

A relay uses low current to control high current. In this example, the relay coil between the ORG and the BLK wires will have high resistance. The low current flowing through it is designed to generate a magnetic field that will close the contact points on the right side of the relay. The closed points will bring continuity between the ORG and the DK GRN wire, and the horn should honk—consistently.

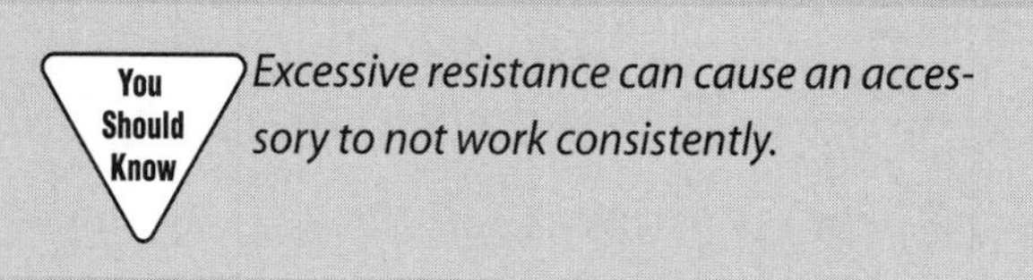

Excessive resistance can cause an accessory to not work consistently.

The manufacturer's specification for the relay coil is to have 50–70 ohms of resistance and zero ohms (continuity) across the contacts when they close. Either a coil problem or contact-point problem could give the intermittent condition that we have. Measuring the coil's resistance, we find 124 ohms, almost double the maximum. Let us power up the coil and put a continuity indicator across the contacts. Sometimes when power and ground are applied to the coil the contacts close, and sometimes they do not. Here is our

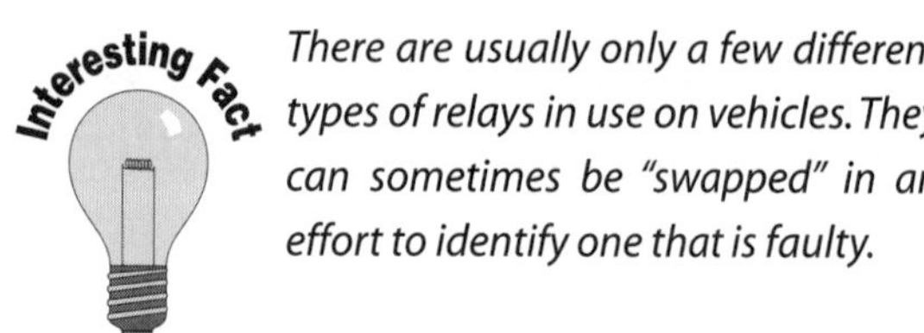

There are usually only a few different types of relays in use on vehicles. They can sometimes be "swapped" in an effort to identify one that is faulty.

intermittent horn function. The vehicle needs a new horn relay. It is also possible for a relay to "click" closed and yet have substantial resistance across the closed contacts. A voltage drop reading across the contacts will reveal excessive resistance.

After installing the new relay, retest the circuit before giving it back to the customer. Most of the time, if a circuit uses a relay, it can be independently tested.

Another tip is that most vehicles have many of the same relays. This allows the technician to move the suspected relay and substitute one from a known good circuit.

WINDSHIELD WIPER AND WASHER CIRCUIT DIAGNOSIS

The repair order for the vehicle states, "the WW will not function." If the circuit is completely dead, it actually simplifies the process. As we pull the vehicle into the stall, we turn on the windshield wipers and they are dead. Each position of the control has the same results. Let us get the wiring diagram out and prediagnose the circuit. **Figure 3** shows this General Motors vehicle's circuit. Notice that there is only one power lead from the 25-amp fuse in the fuse block and one ground, G202, located at the left side of the dash. Both of these are obvious places for an open to take place. No other accessory is on the fuse or the ground, so we are going to have to test it using the 12-volt test light. Notice, though, the power lead at the wiper motor assembly. If we remove the connector and probe the B terminal, we should see power, if the fuse is good. We have power, so the fuse is good. Next, check the ground. If we connect the test light into terminal B and pick up power, then the light should light if the probe is touched to terminal A. It lights, so we have determined that the two most likely causes of a completely dead windshield wiper assembly are both all right. It is now time to test the switch, and we will use a combination of the DMM on terminal D and the test light on terminals E and C. We will test the washer circuit later.

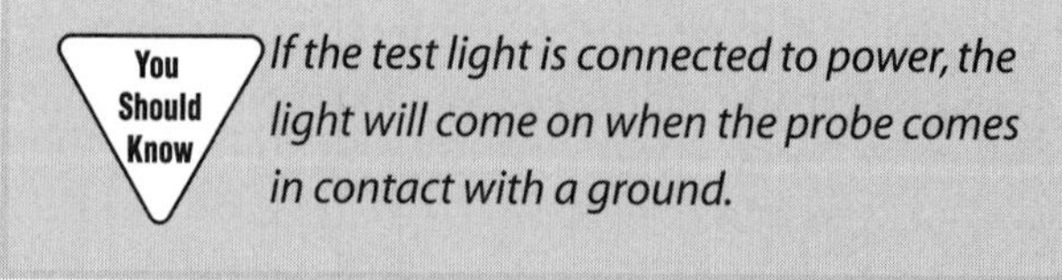

If the test light is connected to power, the light will come on when the probe comes in contact with a ground.

If you look closely at the switch assembly, notice that terminal C on the wiper motor receives full power in the high position and nothing in any of the other positions. The technician probes terminal C and with the switch on high and the KOEO. The light lights. What does this tell us? It gives us two bits of information. First, there is power from the fuse via the YEL wire to the switch, and second, the high position on the third wiper is functional.

Notice that on low and high the center wiper should have full power to terminal E. When it is probed, the light is on brightly in the low and high positions. This is beginning to look like a bad wiper motor assembly. Terminal D is more of a challenge to test because of the 24,000-ohm resistance that is wired in series. If you wire 24,000 ohms in series with the 12-volt test light, the light will not go on, proving nothing. Get out the DMM and wire it to terminal D. Turn the switch to any position, and the meter will show voltage. This is a functional switch. The wiper motor assembly needs to be replaced. One last check is to close the washer switch and make sure that the washer pump runs.

In this diagnosis, we have used our knowledge of Ohm's Law, the wiring diagram, a 12-volt test light, and a DMM to diagnose a faulty wiper motor assembly. It is now a simple task to replace the assembly with a new one or a remanufactured unit. Retest to make sure all functions are present and deliver the vehicle.

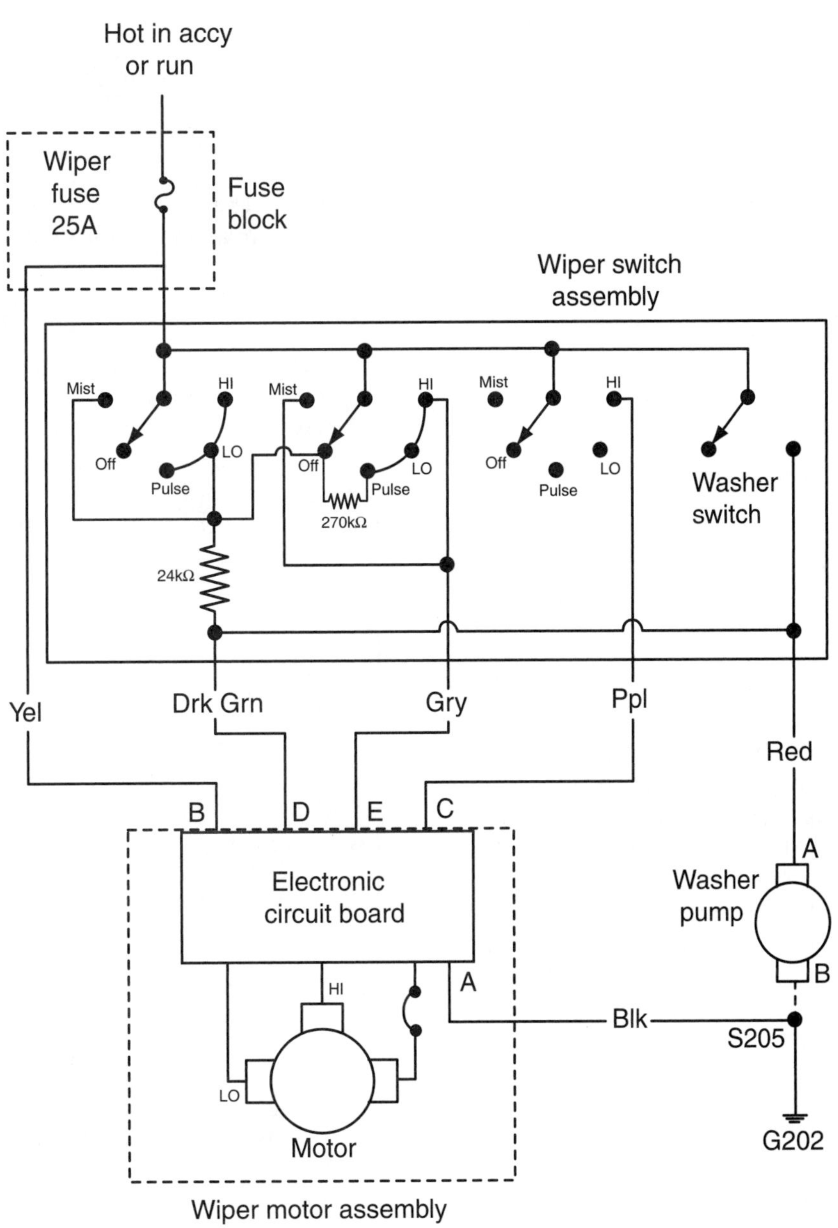

Figure 3. A common windshield wiper circuit.

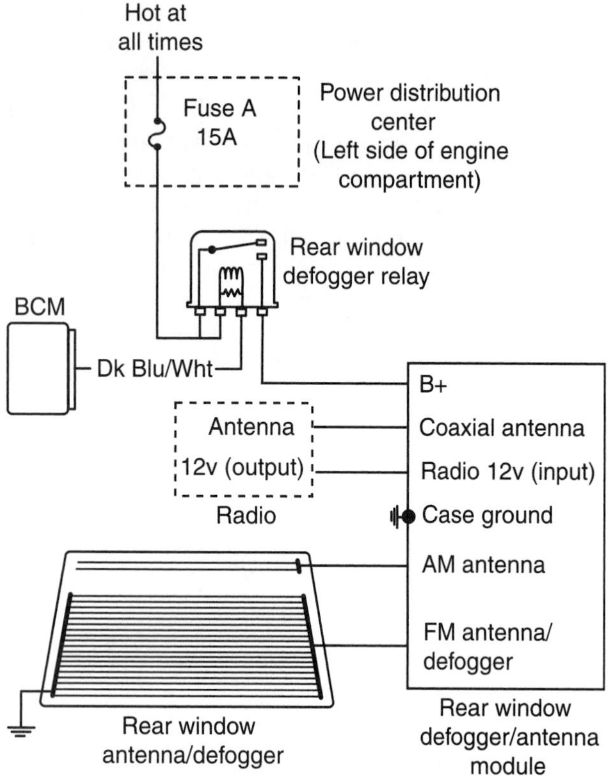

Figure 4. A rear defogger circuit.

DEFOGGER CIRCUIT DIAGNOSIS

The customer informs the service writer that, with the snow last week, she did not think her defogger was functional, although the dash light was on. With the vehicle in the service bay, we turn the KOEO and energize the rear defogger. The dash light turns on, but it is difficult to tell if the rear window is getting hot enough to melt light snow. Based on the diagram in **Figure 4**, the relay must be closed because the on indicator light is powered off the closed contact of the relay. Indirectly, we also know that the relay coil must be close to the resistance specification or it would not have worked. Here is where we will get out the DMM and measure the applied voltage to the defogger.

The rear window defogger is nothing but a set of resistances wired in parallel. They will draw current if voltage is available to push the current through the resistance. Remember that when current flows through resistance, heat may be produced. With the unit turned on and the KOEO, we should have full voltage applied to the grid. However, **Figure 5** shows that 9.28 volts is available across the grid. This is not nearly enough. The wattage developed will be insufficient to melt snow. It appears that the customer was correct. The voltage across the grid is low and could be the result of a bad ground or low applied voltage.

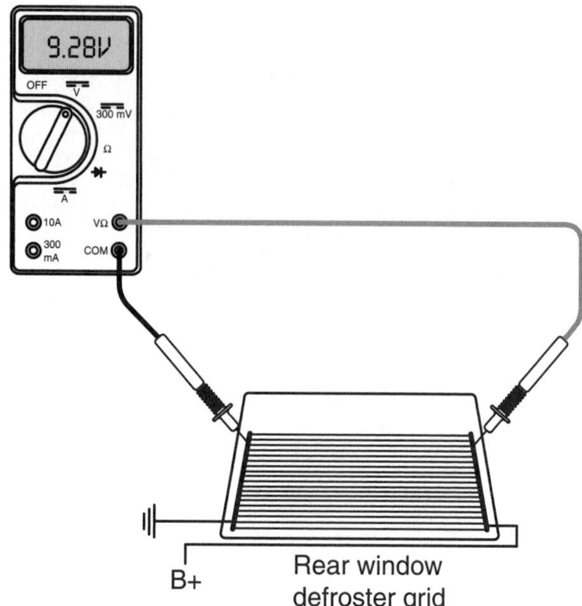

Figure 5. The voltmeter shows reduced voltage available to the defogger.

Figure 6 shows the setup for testing the grid ground. Our voltmeter shows 0.1 volt on the display. A good ground will have minimal voltage drop like this one. Let us move the voltmeter, as in **Figure 7**, so that we can look at the applied voltage. Notice that our meter reads 9.3 volts, which is too low. We need to zero in on the power from the relay

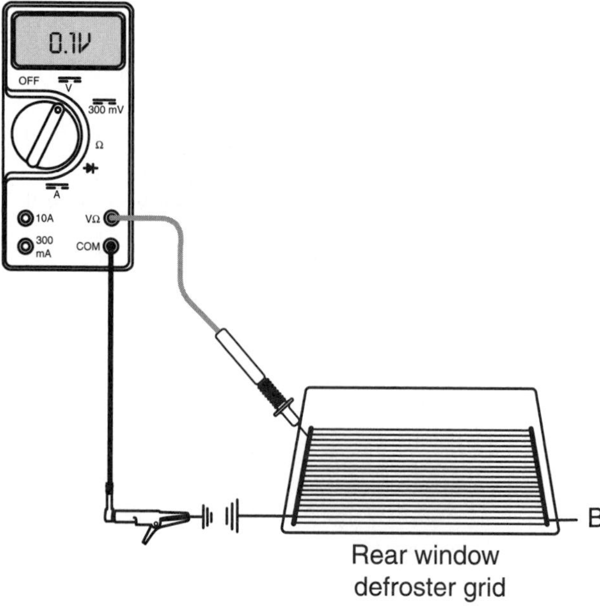

Figure 6. Using a voltmeter to test the ground circuit.

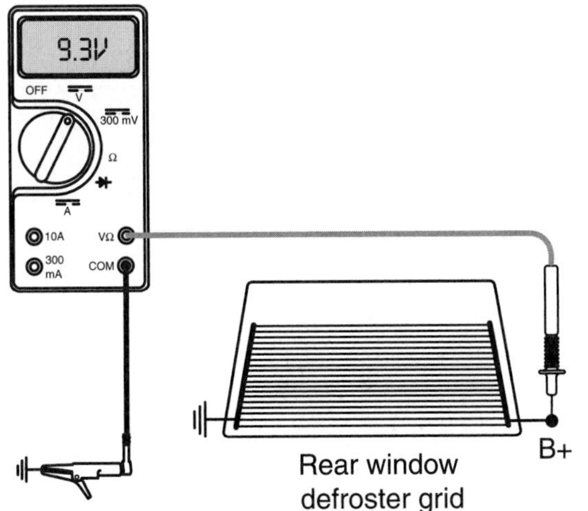

Figure 7. Using a voltmeter to test the input voltage.

to the grid. **Figure 8** is an expanded diagram of this B+ path, and, as always, we want to start from the easiest-to-get-at sections. We have one continuous wire from S252 near the multifunction switch breakout to the hidden antenna module. A breakout is where a group of wires leaves another harness. Connect the DMM to a good ground, and backprobe the BRN/LT BLU wire at the antenna module. The key should be on and the defogger energized. Make sure it stays on by checking the indicator light periodically. Our voltage at this point is still 9.3 volts. The resistance is still toward the front of the circuit. We need to find the BRN/LT BLU wire that is spliced to another BRN/LT BLU wire at S252. When we find it and connect the positive DMM lead to it, we read 12.4 volts. If 12.4 volts is available coming out of the relay and only 9.3 volts is available at the grid, we are losing 3.1 volts from the S252 to the grid. The repair might not be easy, but the diagnosis of excessive resistance was easy. We will need to find the resistance or replace the wire. A recheck with the DMM after the repair should show full B+ available to the grid.

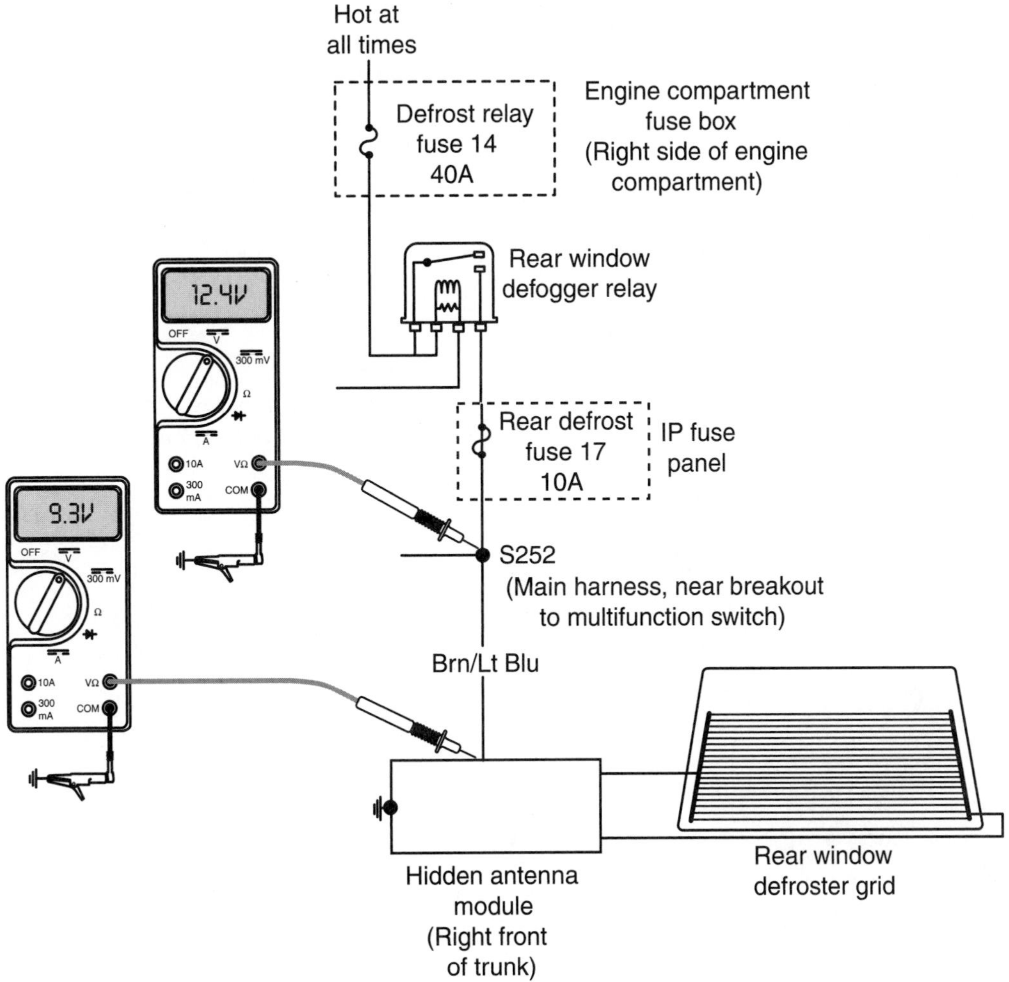

Figure 8. The two voltmeters indicate series resistance.

Summary

- Relays should be removed from the circuit and independently tested if possible.
- Circuits should be prediagnosed by using the customer information and the wiring diagram.
- Always try to duplicate the customer complaint.
- Always try all functions of the accessory.
- A windshield wiper switch can be diagnosed using a DMM and a 12-volt test light.
- Relays can be diagnosed using a continuity indicator and an ohmmeter.
- Electronic modules are not usually independently tested. Instead, they are tested around.
- Excessive resistance will cause a voltage drop, which will lower the efficiency of the device.
- A voltage drop occurs when current flows through resistance.
- A voltage drop can be on the power or ground side of the accessory circuit.
- A single accessory may have more than one source of power or ground.

Review Questions

1. Technician A states that most horn circuits use relay contacts to deliver current to the horn. Technician B states that most horn circuits have the horn switch apply a ground to the control coil of the horn relay. Who is correct?
   A. Technician A only
   B. Technician B only
   C. Both Technician A and Technician B
   D. Neither Technician A nor Technician B
2. A breakout is
   A. the section of the dashboard that has a hole
   B. a circuit breaker that is open
   C. the housing for all of the circuit breakers
   D. where a wire comes out of a harness
3. A DMM is measuring the voltage drop of the ground circuit. Technician A states that the maximum allowable drop should be battery voltage. Technician B states that the meter should be across the ground (in parallel). Who is correct?
   A. Technician A only
   B. Technician B only
   C. Both Technician A and Technician B
   D. Neither Technician A nor Technician B
4. The timer for most rear defoggers is
   A. part of the PCM
   B. an electronic timer
   C. part of the ground circuit for the load
   D. powered by the ignition circuit
5. Most circuit problems are on the ground side.
   A. True
   B. False
6. The windshield wiper circuit usually involves an electronic circuit board.
   A. True
   B. False
7. When the relay coil is not energized, the rear defogger is on.
   A. True
   B. False
8. Why do we use relays in many accessory circuits?
9. What impact does a voltage drop have on an accessory circuit?

Motor-Driven Accessories

Introduction

If you look at the modern vehicle, certain power accessories have just about become factory-installed normal accessories. Power windows fall into this category. The majority of the vehicles that you will encounter in the daily repair business have power windows. Power seats and mirrors are also beginning to become standard equipment on many vehicles. What do these and many other accessories have in common? It is that they all use DC motors. In most cases, the motors are reversible, so that the direction of the window or seat can be changed. We will start out with a look at the basics of the circuitry required to control a motor-driven accessory and then look at some specific examples using aftermarket wiring diagrams.

POWER WINDOWS

Power windows, whether they are for the side windows or the rear of a station wagon, are generally wired and controlled in the same manner. **Figure 1** shows a typical circuit. An ignition switched B+ is run through a circuit breaker to the master switch. A lock switch is paralleled off the breaker and becomes the source of B+ for the window switch. Conductors are run to the insulated reversible DC motor. Let us examine each component separately. Circuit breakers are generally used on power windows because they will open if an overload occurs. Winter weather might freeze the window closed. Without a circuit breaker to open, the motor might be damaged trying to move the window against the ice. As the ice is removed, the breaker will cool, close, and allow future window operation. The lock switch will prevent operation of the window switch if it is open, because it is the source of B+ for the switch. Generally, the lock switch is placed on the driver's door or on the dash to allow the driver control over the doors. This is especially important with small children in the back seat. The motor is usually an insulated directional style that will run up or down, depending on the direction of the applied B+ and ground. The master switch has its own B+ source and is the ground for the entire system. Notice that the style of switch used at the window and the master control is basically the same. The switch

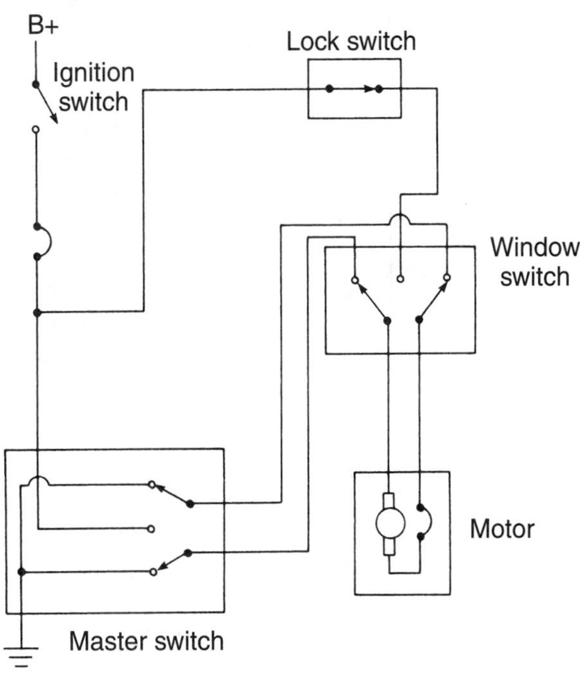

Figure 1. A simplified power window circuit.

will allow B+ to be directed through the motor from the center terminal and then flow back through the other side of the switch. The window switch and master switch are actually wired in series. When they are in a neutral position, as they are drawn, they are series connections. A problem with either switch will affect motor operation. A dead motor can be diagnosed with a 12-volt test light or voltmeter by first ensuring that B+ is available to both switches. Moving the switches should switch the B+ from one side to the other.

Let us take the principles of the first diagram and apply them to an actual wiring diagram for a Dodge. **Figure 2** shows a four-door vehicle with the express down feature. Express down is a popular feature that will lower the driver's window completely with the touch of the button. It is not necessary to hold the button.

> **You Should Know** *The express down feature is available only on the driver's window and requires the use of a special module.*

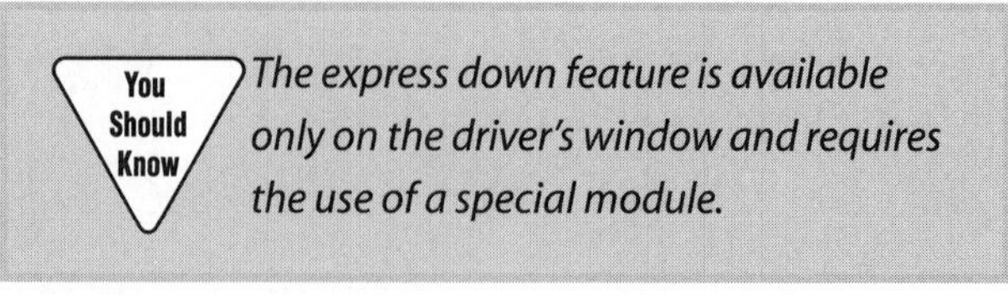

This is available only on the driver's door and requires the use of an express down module. Notice that the module has only four wires connected to it. It is replaced as a unit should it fail. Let us use the right rear window as our example. Notice that the driver's master door switch has put all of

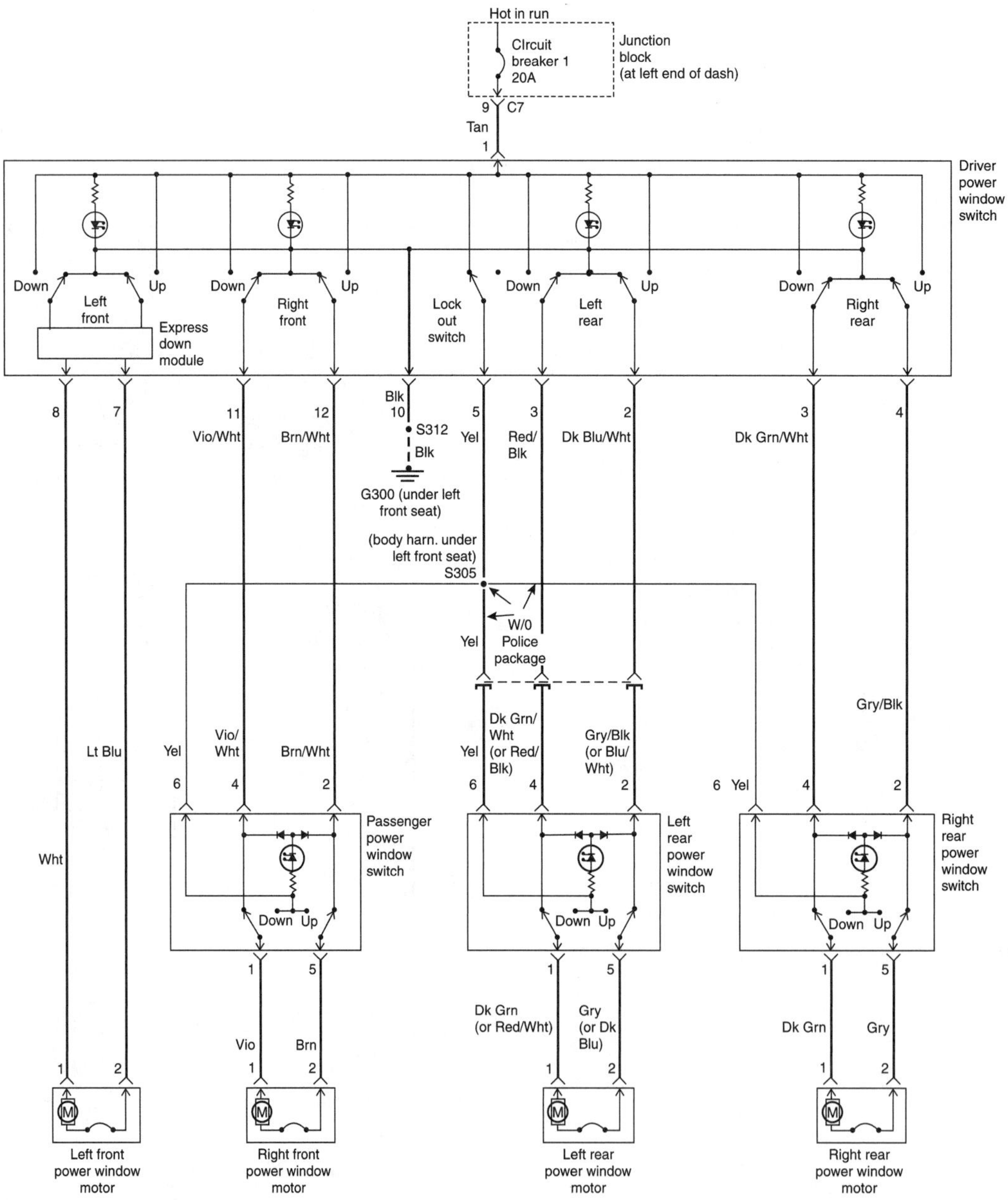

Figure 2. A common power window circuit.

the terminals at ground potential. The ground is G300, which is under the left front seat. If you follow the wires out of the master switch to the door switch, you will notice that they are all connected together and rest at ground potential. If the switch for the right rear window is moved to either the up or down position, the one terminal that moved will now be at B+ potential. If the one terminal remained in its neutral position, then it will be at ground potential from the master switch. One motor terminal at ground and the other at B+ should move the motor and the window. Power for the master switch comes in at terminal 1 (tan wire), while ground comes in from G300. At the window switch there is no need for a direct ground, as the master switch will supply it. However, power must be available to the door switches and come from the S305 splice where the three doors are common pointed together. Note that it takes both switches wired in series for the window motor to function. Again, the wiring diagram becomes our window into the circuit.

The LEDs that are at every switch are for illumination and are on any time the ignition switch is in the run position (KOER).

POWER SEATS

Power seat circuitry is straightforward and involves a DC reversible motor. The polarity is changed to reverse the direction of the motor. Let us use a Ford power seat with lumbar control as our example **(Figure 3)**. Notice that

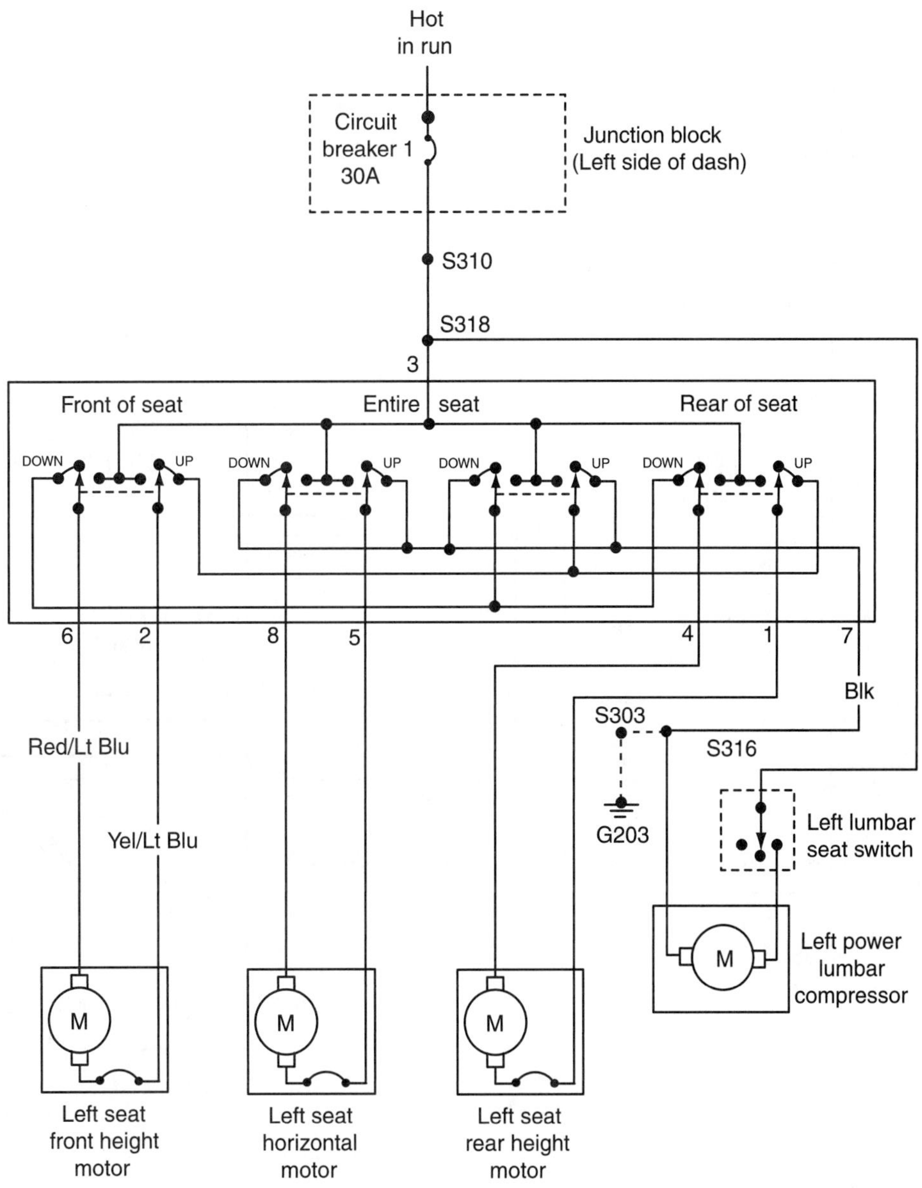

Figure 3. A power seat circuit.

ground are available to the switch but not to ..rs directly. This is the sign to us that the motors are ..ble. The switch will need to control the direction of ..ent and the direction of the seat. There are three ..otors for the seat and one for the lumbar control. The ..ront and rear of the seat can move independently, and then the whole seat can move forward and backward. Let us move the front of the seat up using the left-hand switch. Before we move the switch, note that all contacts are connected to one another. If you follow the path from any switch, you will eventually arrive at pin 7, a black wire that goes to G203, a ground behind the right kick panel. All of the motor wires are sitting at ground potential with the switches in the position they are drawn in. When the customer moves the front of the seat up, the switch on the left will move up and come in contact with the bar in the middle. Follow the bar connections up and across to pin 3, which is hot all of the time and uses a 30-amp fuse in the fuse/relay center. If the RED/LT BLU wire from the switch is now B+ and the YEL/LT BLU wire stays on ground, the motor will turn, and the front of the seat will go up. Reversing the switch just reverses the polarity of the motor feed wires and the seat front goes down. The lumbar motor is similar, but notice that it has a connection to a remote ground all of the time at S316. When power is applied from the inflate switch, the motor will run and inflate the lumbar bladder. Deflation is accomplished by opening a nonelectrical pressure valve and allowing the air to vent.

POWER MIRRORS

Let us use the same Ford vehicle and look at the power mirrors in **Figure 4**. The mirror motors are the same as any other DC reversible motor, so we will not cover them. The switch is the key to understanding the circuit. It is a four-switch joy stick style with eight contacts. In addi-

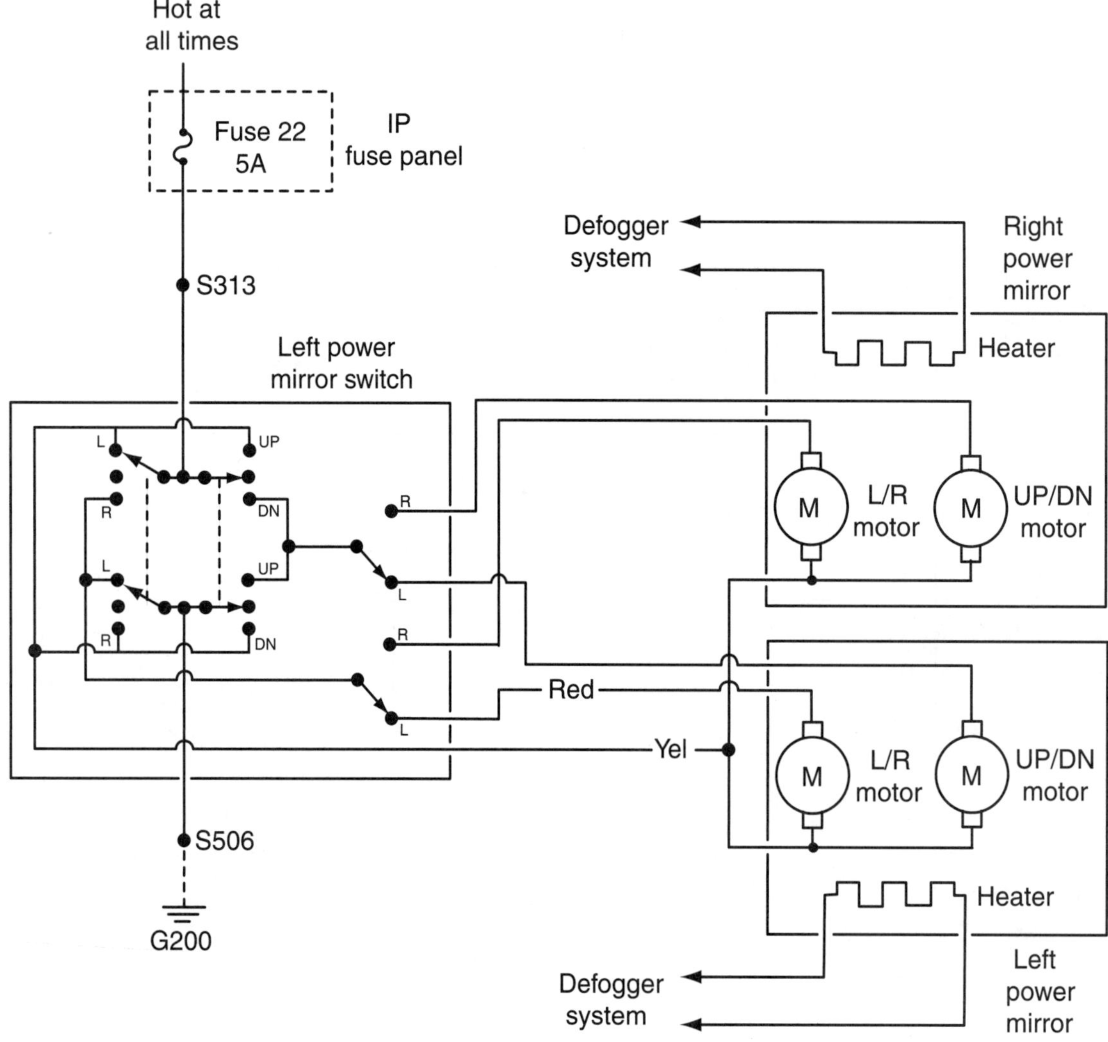

Figure 4. A power mirror circuit.

tion, there is a left/right switch that will choose which mirror the driver moves. Notice the dotted lines that connect the L/R side and the UP/DN side. Remember that a dotted line signifies that the wipers move together. If you move the R/L contact to the left, the upper and lower contact arms or wipers will move together. Let us move the upper set to the L side. The bottom wiper will also move to the L side. Look at the top switch first. If the arrow moves to the letter L, then L will become power, because the center of the arrow is powered off fuse 22, a 5-amp fuse in the instrument fuse panel. Consider the bottom wiper. If you follow the arrow side, it goes down to the ground G200 on the left kick panel. So the top L is now power and the bottom L is now ground. Follow the bottom L over to the L/R motor on the left side of the vehicle (RED wire). The top L goes over to the other side of the motor using the YEL wire. Once you analyze the diagram, it simplifies the diagnosis. This vehicle also has heaters in each mirror that are powered off the defogger system.

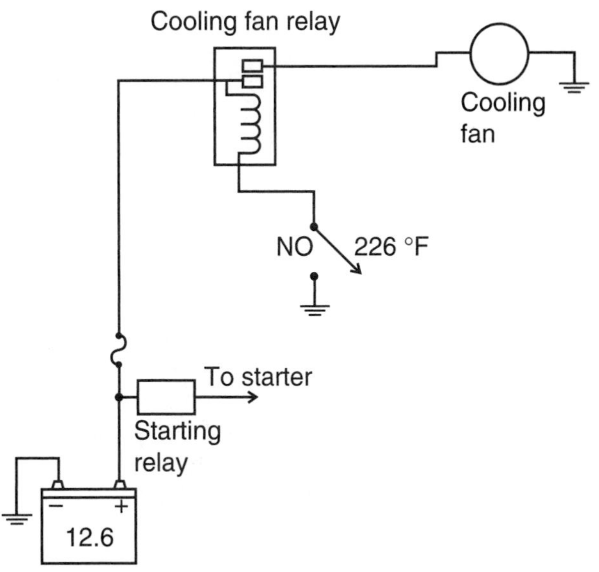

Figure 5. When the "NO/226°F" switch closes, the cooling fan will run.

ENGINE COOLING FANS

You cannot call the cooling fan an accessory, but it is an example of a motor-driven device. Two varieties are found in the field, the single speed and the multispeed. Some vehicles use more than one cooling fan, but the wiring is basically the same.

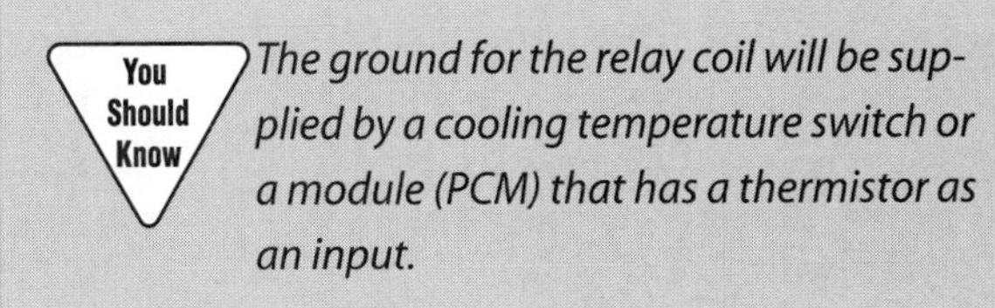

The ground for the relay coil will be supplied by a cooling temperature switch or a module (PCM) that has a thermistor as an input.

Figure 5 shows a simplified circuit with a normally open cooling switch that will close at 226° F. It will close the ground circuit for the control current coil of the relay. The power for the relay will come directly from the B+ terminal of the starter relay, which is hot all of the time. This design will allow the cooling fan to turn even if the ignition key is off. Let us take these principles and use them on a diagram for a General Motors single-speed cooling fan, as shown in **Figure 6**. Again, the circuit uses a cooling fan relay that gets its power from the 30-amp fuse in the fuse relay center. Notice, however, that the ground for the relay control current coil will be supplied by the PCM. This makes sense if you think about it, because the PCM has to know engine temperature to run the vehicle. In this way, a single input sensor, the ECT in the lower left of the diagram, supplies temperature information from its thermis-

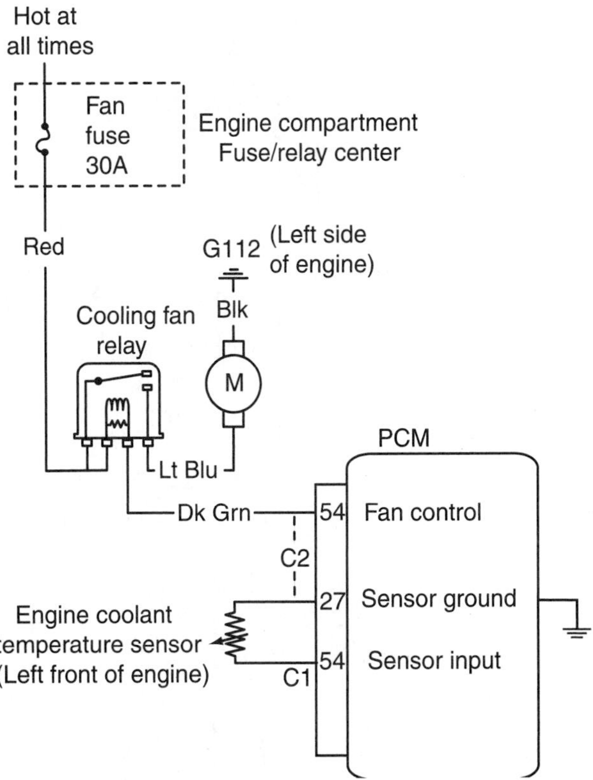

Figure 6. A single-speed cooling fan circuit.

tor directly into the PCM, and the PCM controls the ground side of the relay. If the ground is applied, the relay closes and the cooling fan runs. The temperature that causes the fan to turn on is programmed into the PCM, and this allows the fan to stay on until the temperature drops.

Figure 7 shows an adaptation of this circuit with a multispeed motor. Notice that the motor has three wires rather than the two we normally find. The DK BLU conductor is for low speed, and the ORG/LT BLU is for full speed. As each relay closes, it will bring B+ directly to the motor.

Notice that the low-speed relay gets its B+ from the 30-amp circuit breaker 17, whereas high speed uses fuse 13, which is a 50-amp fuse. Both protection devices are in the engine compartment fuse box. It is the control of the circuit that is unique. The positive side of each relay coil receives its power from the main PCM relay. This relay is closed any time the ignition key is on. Two different terminals of the PCM supply the individual grounds. When the temperature reaches the programmed level, the low-speed relay is grounded (the RED/ORG wire). If the temperature continues to rise or the AC compressor turns on, the second ground is applied on the LT GRN/PPL wire. This energizes the high-speed relay and the fan spins at high speed. When the AC system is on, more air is required through the condenser that is mounted in front of the radiator. This is the reason for the high-speed fan coming on. Even though the circuit is a bit more difficult, diagnosis will be easy, as you will see in Chapter 59.

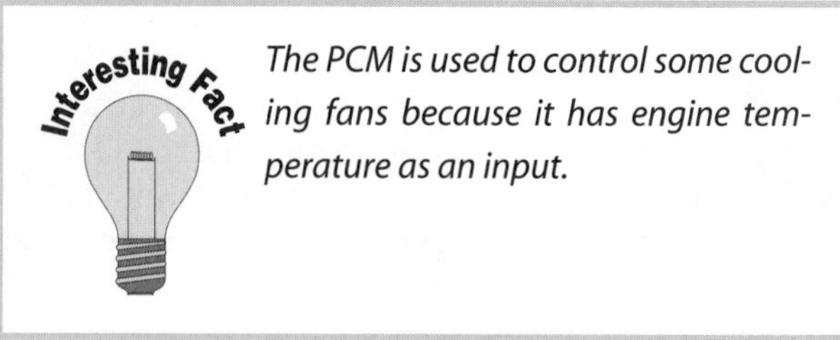

The PCM is used to control some cooling fans because it has engine temperature as an input.

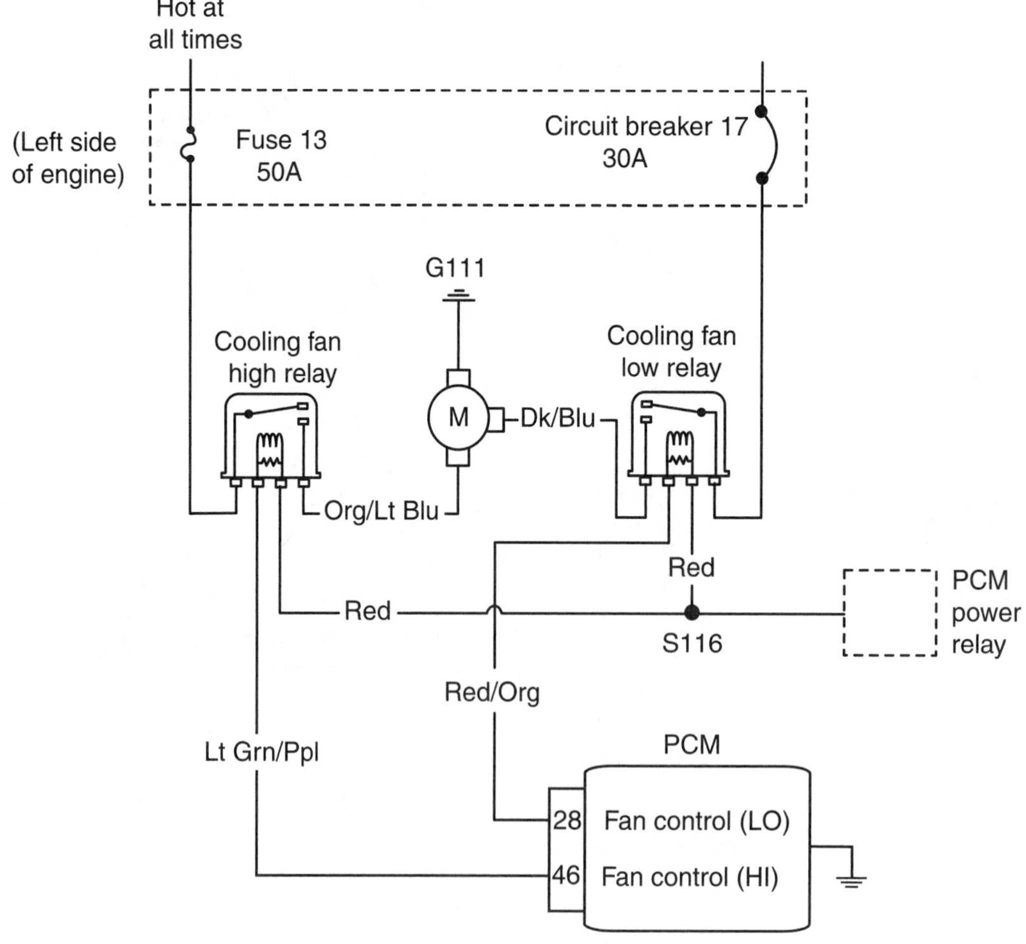

Figure 7. A two-speed cooling fan circuit.

Summary

- Many accessories use a DC reversible motor.
- If multiple switches are used, such as in a power window circuit, different switches usually supply power and ground.
- A motor will reverse its direction if the power and ground leads are reversed.
- Most motor circuits use a circuit breaker to protect the circuit.
- Many motor-driven accessories use a relay.
- Multispeed cooling fans are in use in some vehicles.

Review Questions

1. Technician A states that a power window circuit will include circuit breakers for the motor. Technician B states that the power and ground are reversed at the motor to change the window direction. Who is correct?
   A. Technician A only
   B. Technician B only
   C. Both Technician A and Technician B
   D. Neither Technician A nor Technician B
2. The high-speed function of the fan is turned on when what condition is present?
   A. The engine is overheating.
   B. The AC compressor is on.
   C. The high-speed relay is energized.
   D. All of the above
3. A PCM-controlled cooling fan will have the PCM responsible for the
   A. ground of the motor
   B. power to the motor
   C. ground of the cooling fan relay coil
   D. power of the cooling fan relay coil
4. The lock-out switch for a power window circuit normally opens the
   A. power (B+) to the master switch
   B. ground to the master switch
   C. ground to the door switches (other than the master)
   D. power (B+) to the door switches (other than the master)
5. If power is applied to both sides of a window motor, the window will
   A. go up
   B. go down
   C. be stopped
   D. none of the above
6. Circuit breakers are used in many motor-driven accessories.
   A. True
   B. False
7. Cooling fans use switches in all applications.
   A. True
   B. False
8. On power window systems, both switches are not necessary for the window to function.
   A. True
   B. False
9. What advantages are there to using the PCM to control cooling fans?
10. Why are circuit breakers in use on many motor-driven accessories?

Diagnosing Motor-Driven Accessories

Introduction

We continue here the diagnostic procedure that we introduced in the previous chapters. Remember that motor-driven accessories are generally DC reversible motors that will have the polarity changed by series switches. We will not look at every example, because they basically operate in a similar fashion. Instead, we will diagnose an open power, an open ground, a faulty switch, and a motor. In addition, we will examine the testing of an input device for a computer-controlled cooling fan.

DIAGNOSING AN OPEN POWER SOURCE

The repair ticket lists a power window problem, and the customer notes that none of the four windows works but all other accessories seem to be functional. As the vehicle is driven into the service bay, the technician tries to open the windows from the driver's seat. None of the windows function. Leaving the KOEO, the technician tries the passenger window switch, with no operation. Next, the wiring diagram is printed, as shown in **Figure 1**. The most likely cause of the windows not working is a power or ground problem because any other difficulty would shut down only one of the power windows. All motors share the same source of power and the same ground. The technician notes that all of the LED illumination lights are also off, verifying that the problem probably is power or ground. The tool of choice will be the 12-volt test light.

> **You Should Know** *A problem with either the power or ground circuit will shut down the accessory circuit.*

The diagram shows that power for the circuit comes from the ignition switch because of the "hot in run" notation above the circuit breaker. The location of the source of power is the left end of the dash in a junction block. The protection device is a circuit breaker. Remember from our previous discussions that circuit breakers are generally used when a motor is in use. The breaker will open up if the window is frozen and cannot move. Rather than burning out the motor, the circuit protection will open up. Once the motor cools, the breaker should shut. **Figure 2** shows an expanded view of the power circuit. We can test for power at the junction block or we can go to the TAN switch input wire. Let us start at the junction block because it is the easiest spot to get to. Both sides of the circuit breaker are live. Our test light lights on both sides with the KOEO. The circuit protector appears to be functional.

Most motor-driven power accessories use a circuit breaker for protection.

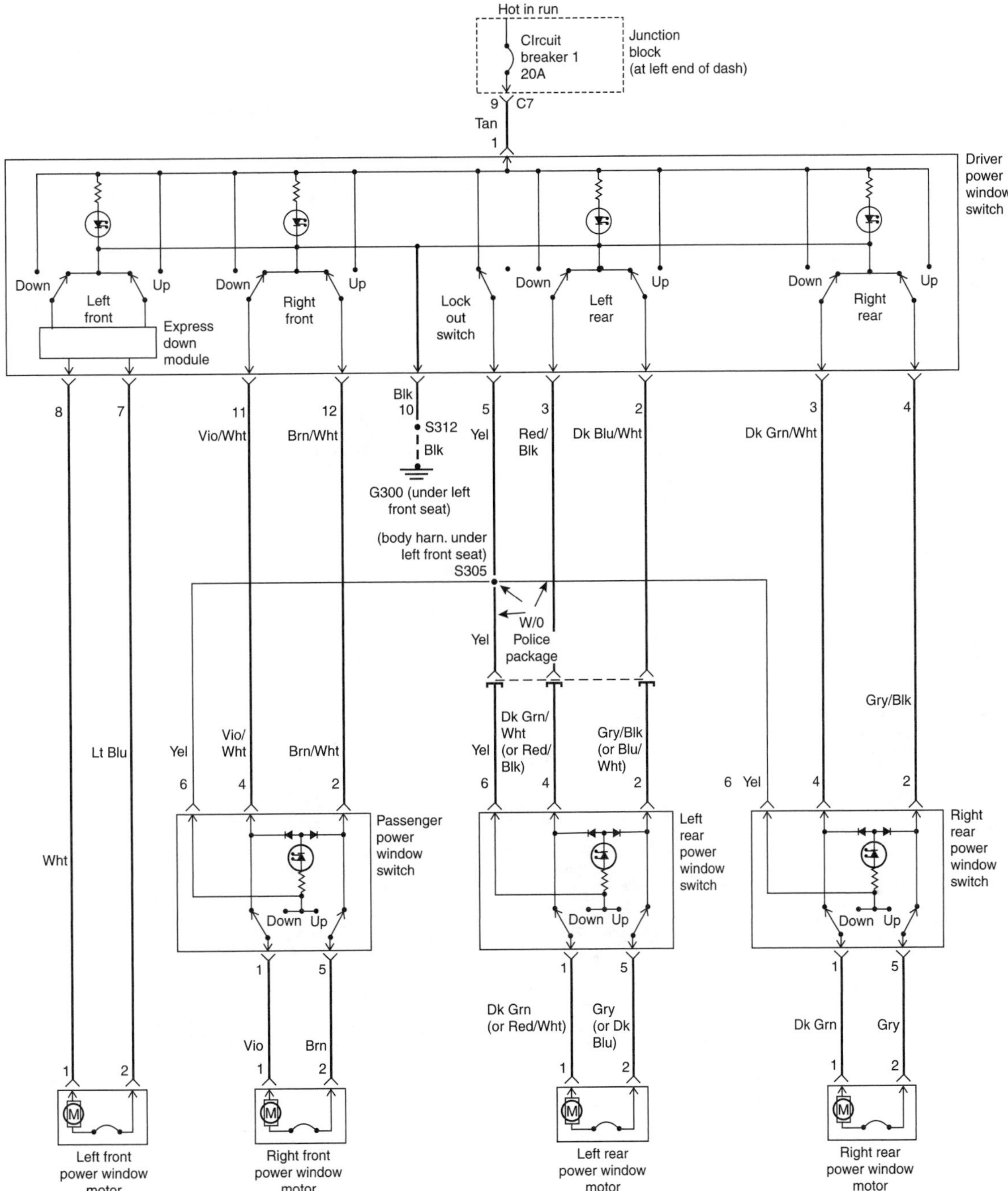

Figure 1. A power window circuit.

Where do we go next? The most logical place is the window switch that is located in the driver's door. With the switch removed, there should be power at the TAN wire terminal 1. Our 12-volt test light is off.

No power is available to the switch, and therefore the entire window circuit is dead. With power available from the circuit breaker and no power available at the window switch, there must be an open wire or connection in between. This is a frequent problem on older vehicles. The wires that go into the door break from the movement back and forth as the door is opened and closed. Installing a jumper across the open section of the power wire will repair the vehicle. The circuit is tested, and the vehicle released to the customer.

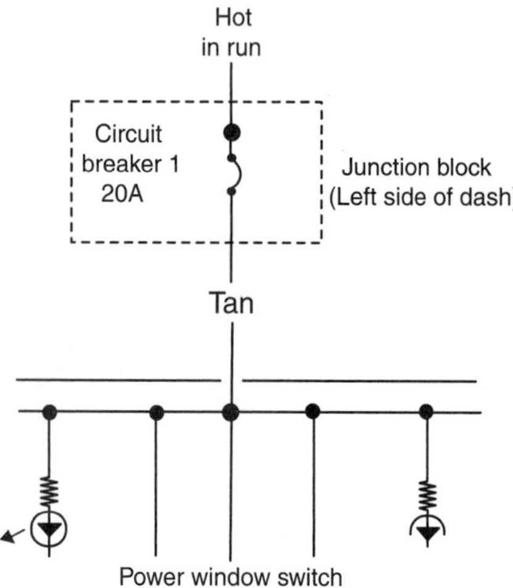

Figure 2. Power is supplied by a circuit breaker.

DIAGNOSING AN OPEN GROUND

A customer arrives at a repair shop and complains that the power seat on the vehicle will not function. The technician notes that only the driver's seat is power; the passenger seat is manually adjusted. It appears that none of the functions work, including the lumbar adjustment. This again is probably a power or ground problem. The wiring diagram is printed and shown in **Figure 3**. Note that the power and ground are shared between the seat motors and the lumbar compressor. Only a problem in the power or ground could cause the entire system to shut down. The engine compartment fuse/relay box contains the 30-amp circuit breaker that powers the circuit. In addition, each seat motor has an internal circuit breaker. The 12-volt test light is connected to circuit breaker ground and touched to both sides of the circuit breaker and lights. The fuse is good. Ter-

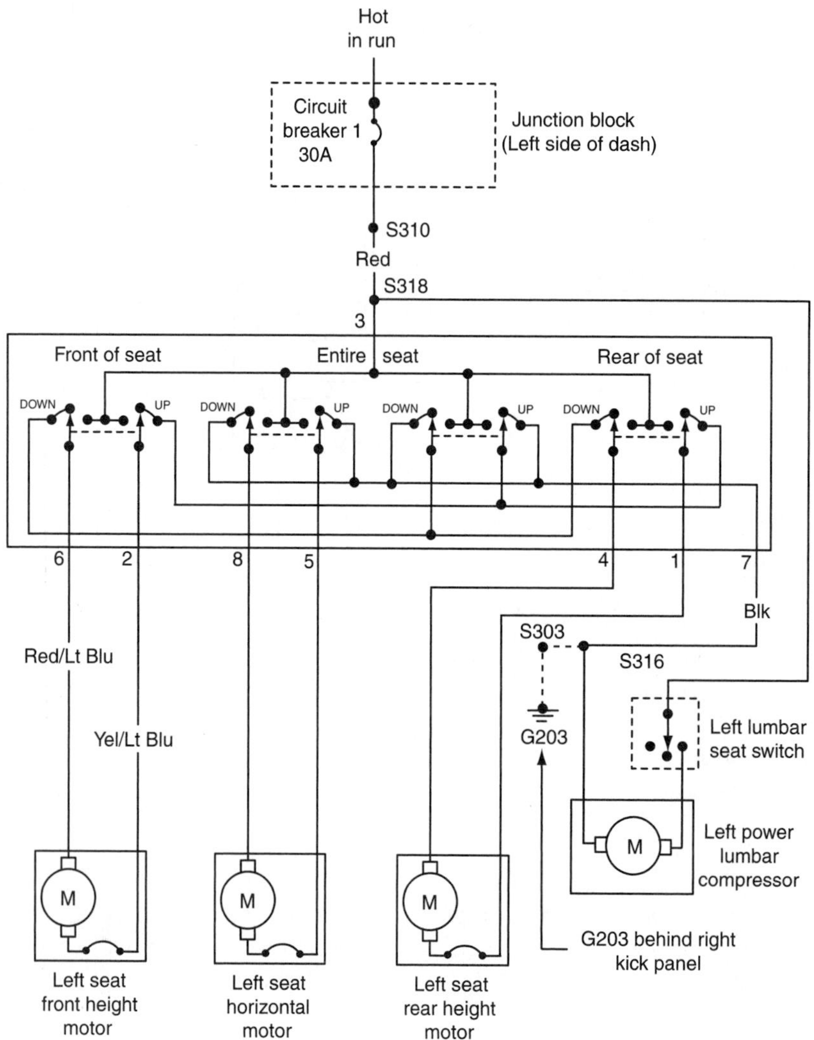

Figure 3. A power seat circuit.

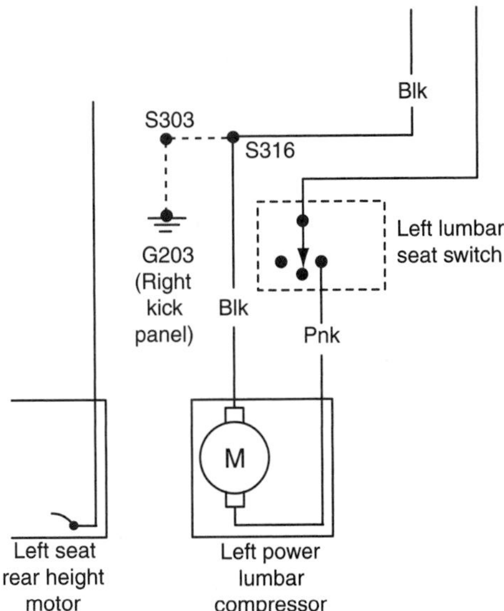

Figure 4. The power system uses a common ground.

minal 3 RED of the switch assembly is next tested for power, and the light again lights. Plenty of power is available to the circuit. We need to zero in on the ground side of the circuit, so the test light clip is placed on power. Now the light will light if it "sees" ground. Terminal 7, a BLK wire on the switch, should be ground. The light stays off when touched to the terminal. No ground is here. Go back to the diagram to determine the location of the ground. G203 is on the opposite side of the vehicle in the right kick panel. After the kick panel is removed, the ground is tested and it is all right. (The light is on.) A break is somewhere between the kick panel ground and the switch **(Figure 4)**. These two spots are a long way from one another, but the diagram shows us that there are two splices along the wire. S316 is probably near the power seat switch, and S303 is probably near the ground connection G203. If we follow up the ground wire from G203 until we come to a splice, we can test for ground. One side of the splice is ground, and the other is not. The splice is open. Splices are a frequent source of open circuits and are easily repaired. The splice is cut out and a crimp terminal is put in its place. Electrical tape is placed around the new splice, and the wire is pushed back into the conduit. The circuit is retested, including the lumbar, and the kick panel replaced. Another check of the completed repair reveals that this circuit works.

DIAGNOSING A FAULTY SWITCH

The customer repair order indicates that the passenger rear power window will not function and is stuck in the half-open position. As the vehicle is driven into the bay, the driver's window is lowered. The other back window and the passenger window also appear to function. An additional check of the power window operation using the rear door switch reveals that neither switch will move the window up or down **(Figure 5)**. Remember that on this vehicle both switches must be functional for the motor to move because the driver's switch supplies the ground path while the passenger rear window switch supplies the power. An open motor or circuit breaker at the motor could also cause that problem. There are four choices here that the wiring diagram indicates: the motor, the door switch, the master switch, or the wiring between. The motor and circuit breaker are one assembly. With two of the four possibilities in the rear passenger door, it will make sense to begin the diagnosis there. With the door panel removed, the switch terminals can be accessed. **Figure 6** shows an expanded view of the door switch and motor assembly. Terminal 6, a YEL wire, is a power source and is easily tested with the 12-volt test light. If the light is moved over to power, possibly from the YEL wire, the probe can be touched to either the DK GRN/WHT or the GRY/BLK wire. Both terminals turn the light on, indicating that the master switch is supplying the ground as it should. Terminal 1 is a motor terminal and should have ground with the switch in the neutral position and then switch over to a power (from the master switch) with the switch in the down position. When we backprobe the terminal, the test light is not on with the switch in the neutral position. Neither is terminal 5, the other motor terminal. With power on the YEL 6 terminal and no power on either of the DK GRN 1 or GRY 5 terminals, this switch cannot supply current to the motor and needs to be replaced. A jumper with power and another with ground connected to the motor moves the window. Our diagnosis is complete. A new switch will fix the vehicle.

DIAGNOSING A FAULTY MOTOR

A customer complains that the power window on the driver's door will not function. As the vehicle is driven into the bay, the technician tries all of the windows. They all work, with the exception of the driver's window. The wiring diagram for the vehicle is printed and shown in **Figure 7**. The driver's window switch is different from the other windows because it will supply both the power and the ground for the motor. The center terminal is a ground connection and each outside terminal is connected to power by the movable wiper arm that looks like an arrow. The polarity to the motor is determined by which wiper arm moves. We again take out the 12-volt test light and connect it across the disconnected motor terminals as shown in **Figure 8**.

When the switch is moved to either position, the light should come on if the switch is good and power and ground are available. The light does work in either switch

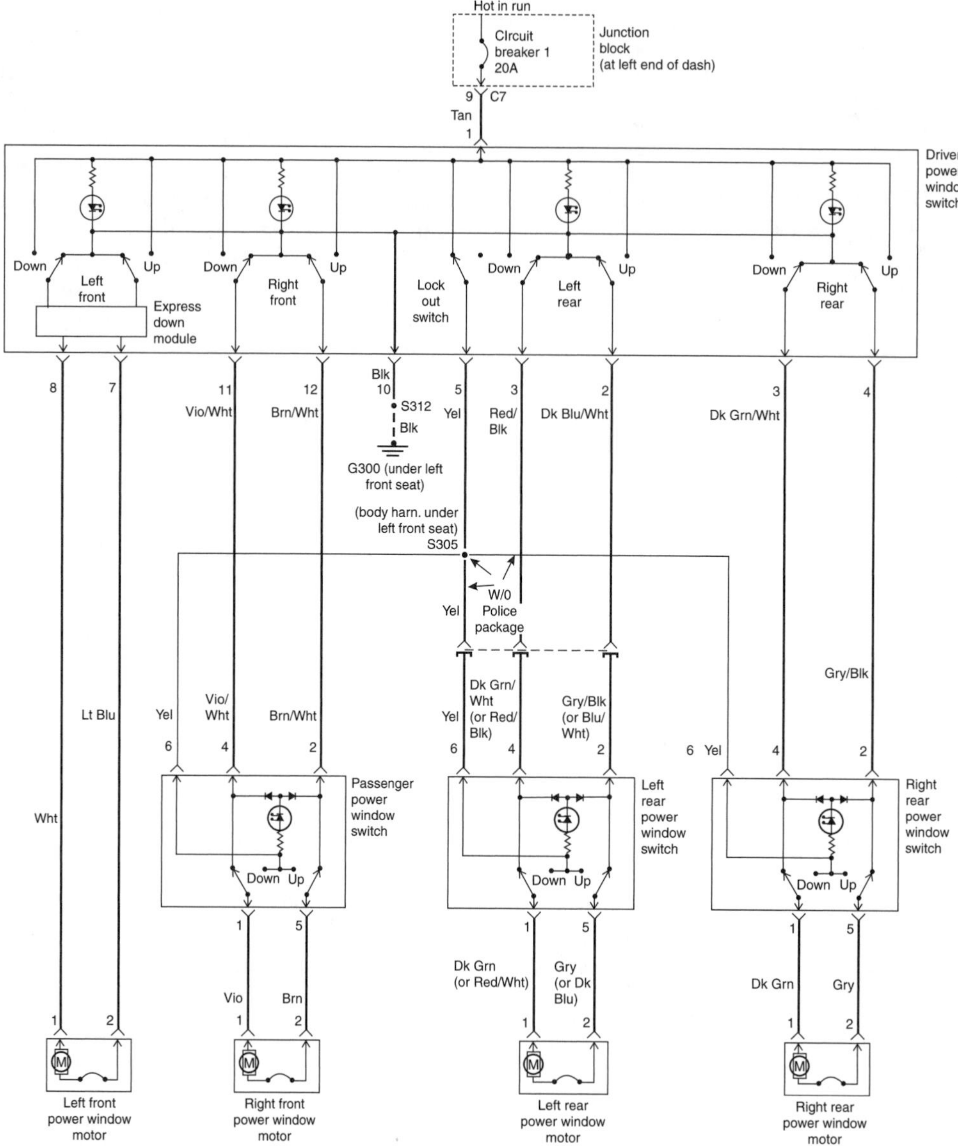

Figure 5. A power window circuit.

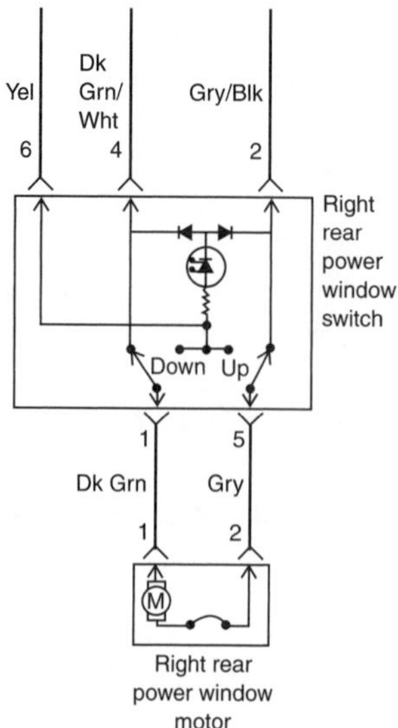

Figure 6. The control circuit for the right rear power window.

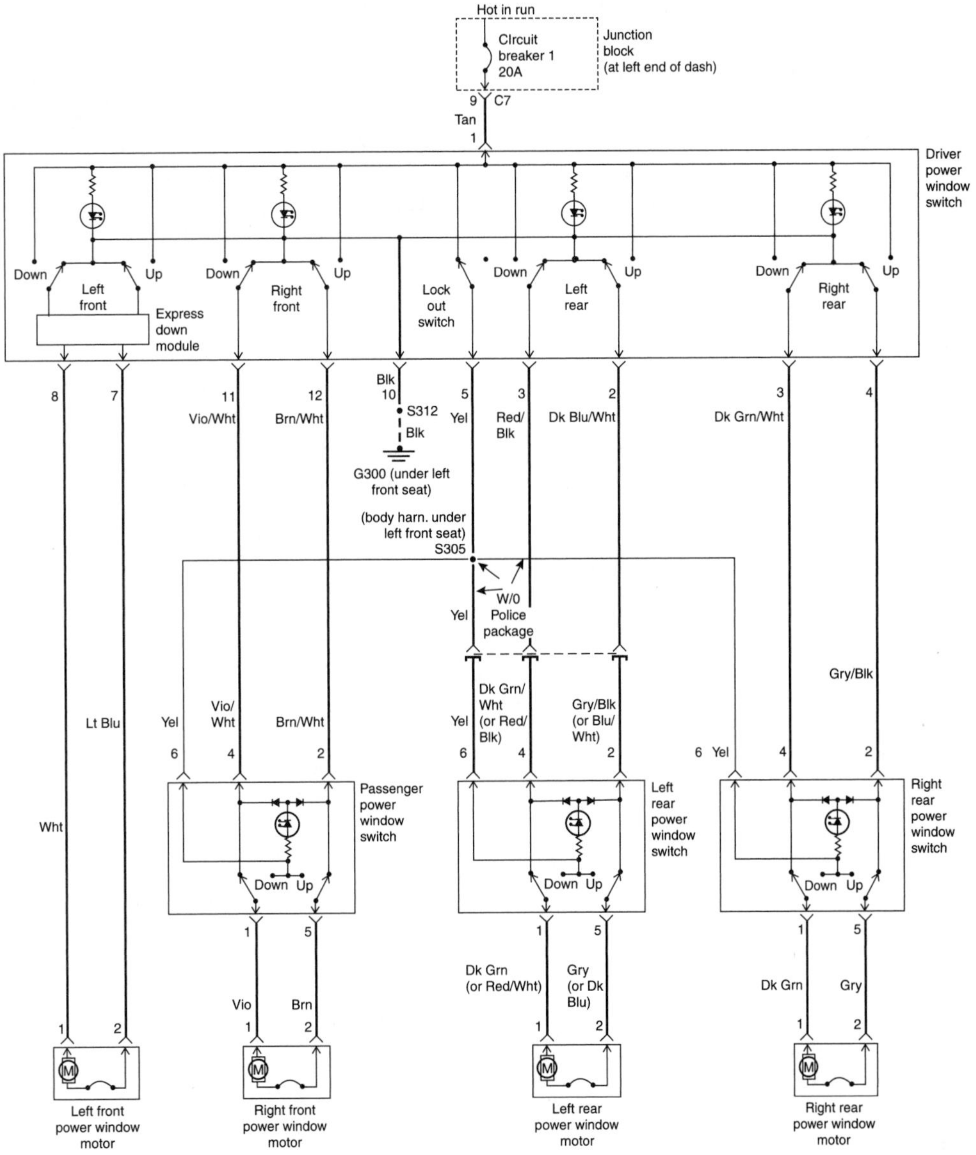

Figure 7. A power window circuit.

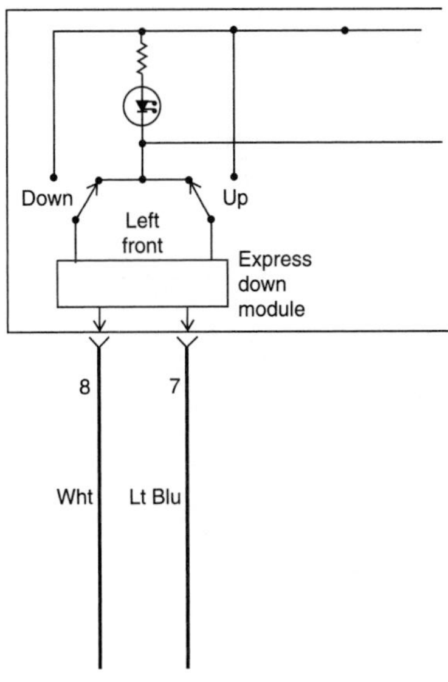

Figure 8. The left front window control

position. The motor assembly needs to be replaced. Most circuit breakers are part of the motor, and if they are the cause of the open circuit, the entire assembly is replaced, even though the motor might be all right. After the installation of the new motor assembly, the circuit is tested to make sure all power window circuits are functional, and the vehicle is returned to the customer.

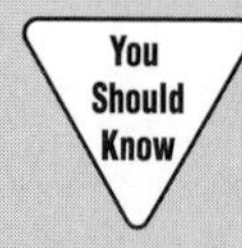

The 12-volt test light can take the place of the power motor. It will not draw as much current as the motor, but it will be a load to the circuit.

Summary

- A wiring diagram should always be printed before diagnosis.
- Always try to duplicate the customer complaint.
- Use the 12-volt test light connected to ground to test for power.
- Use the 12-volt test light connected to power to test for ground.
- The 12-volt test light can take the place of a load.
- A voltmeter should not be used in place of the load because it will not draw sufficient current to test the circuit adequately.

Review Questions

1. The right front window will not function on a power window vehicle. Technician A states that the window motor might be insulated from ground, causing the problem. Technician B states that either switch could be the problem. Who is correct?
   A. Technician A only
   B. Technician B only
   C. Both Technician A and Technician B
   D. Neither Technician A nor Technician B
2. A 12-volt test light is placed into the connector from a removed power window motor. The light is on with the switch in the up position but off with the switch in the down position. This indicates that the
   A. ground is faulty
   B. motor is open
   C. power is shorted
   D. switch is not functioning correctly
3. The power seat will not move in any direction. This is most likely an open
   A. power problem
   B. switch
   C. motor
   D. motor circuit breaker
4. To test for ground, connect the 12-volt test light to
   A. ground
   B. a motor
   C. power
   D. a switch
5. Use a DMM on volts to test for accessory power.
   A. True
   B. False
6. A 12-volt test light will load down the circuit.
   A. True
   B. False
7. An open ground could shut down a single power window.
   A. True
   B. False
8. Why is a 12-volt test light preferred over a DMM in certain applications?

Appendix

ASE PRACTICE EXAM FOR ELECTRICAL SYSTEMS

1. Technician A states that more work is accomplished with parallel circuitry. Technician B states that series resistances add up. Who is correct?
   A. Technician A only
   B. Technician B only
   C. Both Technician A and Technician B
   D. Neither Technician A nor Technician B
2. Adding additional resistance to a series circuit will
   A. have no effect on the current flow
   B. reduce the current flow
   C. increase the wattage
   D. increase the current flow
3. A taillight bulb is glowing dimly. A voltmeter across the bulb reads 7.4 volts. Technician A states that there must be a voltage drop somewhere else in the circuit. Technician B states that resistance in parallel must be present. Who is correct?
   A. Technician A only
   B. Technician B only
   C. Both Technician A and Technician B
   D. Neither Technician A nor Technician B
4. Technician A states that control current for a relay will be small in comparison to the contact current. Technician B states that the relay coil will have high resistance and the contacts will have no resistance. Who is correct?
   A. Technician A only
   B. Technician B only
   C. Both Technician A and Technician B
   D. Neither Technician A nor Technician B
5. Technician A says that a 12.6-volt OCV reading indicates that a battery needs recharging. Technician B says that a 1.265 hydrometer reading indicates that a battery needs recharging. Who is correct?
   A. Technician A only
   B. Technician B only
   C. Both Technician A and Technician B
   D. Neither Technician A nor Technician B
6. A 450-CCA battery is being load tested. The correct load will be
   A. 1,350 amps
   B. 450 amps
   C. 225 amps
   D. 990 amps
7. Technician A says that during a load test the battery voltage must not fall below 9.6 volts. Technician B says that the load applied should be three times the A/H rating. Who is correct?
   A. Technician A only
   B. Technician B only
   C. Both Technician A and Technician B
   D. Neither Technician A nor Technician B
8. Technician A states that slowing down the starter armature will result in increased amperage draw. Technician B states that speeding up the armature will result in decreased amperage draw. Who is correct?
   A. Technician A only
   B. Technician B only
   C. Both Technician A and Technician B
   D. Neither Technician A nor Technician B

9. An alternator should never be operated open-circuit (battery disconnected) because this could
   A. overcharge the battery
   B. burn out electronic components
   C. ruin the alternator stator
   D. result in poor high-speed driving
10. To full field an alternator, Technician A says to run a jumper wire from the F terminal of the regulator connector to ground on an A circuit. Technician B says to run a jumper wire from the F terminal of the regulator connector to B+ on a B circuit. Who is correct?
   A. Technician A only
   B. Technician B only
   C. Both Technician A and Technician B
   D. Neither Technician A nor Technician B
11. A primary ignition winding is being tested. Its resistance is three times the specification. This will cause
   A. longer spark lines
   B. longer dwell
   C. reduced secondary output
   D. nothing
12. Less than 0.5 millisecond of the spark is observed. Technician A says that this will increase the chances of misfire. Technician B says that this might be caused by the firing voltages being excessive. Who is correct?
   A. Technician A only
   B. Technician B only
   C. Both Technician A and Technician B
   D. Neither Technician A nor Technician B
13. An ohmmeter is connected to a conductor and reads infinity. This indicates
   A. nothing
   B. that the wire is open
   C. that the wire is shorted
   D. none of the above
14. An ohmmeter is placed on a load and reads OL. The meter is in the autorange position. Technician A states that this means infinity. Technician B states that no current will be able to flow through the load. Who is correct?
   A. Technician A only
   B. Technician B only
   C. Both Technician A and Technician B
   D. Neither Technician A nor Technician B
15. At one-half open throttle, the TP return voltage on a 5-volt computer system is 0.5 volt. Technician A states that this is normal. Technician B states that this is not normal. Who is correct?
   A. Technician A only
   B. Technician B only
   C. Both Technician A and Technician B
   D. Neither Technician A nor Technician B
16. A negative coefficient thermistor will have its total resistance
   A. go up as its temperature increases
   B. remain the same as its temperature increases
   C. go down as the temperature increases
   D. go down as the temperature decreases
17. The test light is placed on the M terminal and ground. When the key is turned to the start position, the light lights. Technician A states that this indicates that the circuit from the battery to the S terminal must be open. Technician B states that this indicates that the starter motor needs to be replaced. Who is correct?
   A. Technician A only
   B. Technician B only
   C. Both Technician A and Technician B
   D. Neither Technician A nor Technician B
18. Power is available to the load side and the control side of the relay. A ground is available to the control coil. The load contact will not close. What is wrong with the circuit?
   A. There is a bad ground connection.
   B. The ignition switch may be faulty.
   C. The starter solenoid is faulty.
   D. The relay is faulty.
19. Technician A states that most horn circuits use relay contacts to deliver current to the horn. Technician B states that most horn circuits have the horn switch apply a ground to the control coil of the horn relay. Who is correct?
   A. Technician A only
   B. Technician B only
   C. Both Technician A and Technician B
   D. Neither Technician A nor Technician B
20. Technician A states that a power window circuit will include circuit breakers for the motor. Technician B states that the power and ground are reversed at the motor to change the window direction. Who is correct?
   A. Technician A only
   B. Technician B only
   C. Both Technician A and Technician B
   D. Neither Technician A nor Technician B
21. If power is applied to both sides of a window motor, the window will
   A. go up
   B. go down
   C. be stopped
   D. none of the above
22. One headlight works, but the other will not. The two lights share a common ground. Which of the following cannot be the problem?
   A. a bad ground
   B. a burned-out headlight
   C. a broken power wire to the off light
   D. a faulty headlight connector

23. Which of the following is not a component common to all circuits?
    A. a fuse
    B. a control
    C. a load
    D. a module
24. A pintle bump is observed during the last part of the current ramp for a fuel injector. This will allow the cylinder to run
    A. richer than it should
    B. with a misfire
    C. leaner than it should
    D. without any fuel
25. Primary current for a distributed system is observed to be on for 1.3 milliseconds. This will probably
    A. increase the burn time
    B. cause the coil to not be fully saturated
    C. decrease the spark advance
    D. enrich the air: fuel ratio
26. The spark burn time on a four-cylinder vehicle is observed to be 0.6 millisecond. This might cause
    A. a richer-than-normal air: fuel ratio
    B. a leaner-than-normal air: fuel ratio
    C. decreased spark advance
    D. an increased likelihood of misfire
27. TVRS wires normally have a per foot resistance of
    A. 0 ohms
    B. 3 megaohms
    C. 4 kohms
    D. 4 ohms
28. A voltmeter is placed across the battery terminals with the engine cranking. If the system is functioning correctly, the voltmeter should read
    A. greater than 9.6 volts
    B. less than 9.6 volts
    C. 2,000 watts
    D. 12.6 volts
29. The three-minute charge test will test the ability of a battery to
    A. supply starting current
    B. maintain at least 9.6 volts during starting
    C. accept a charge
    D. supply current during a load test
30. Which of the following is not part of the charging system?
    A. a battery
    B. an alternator
    C. a regulator
    D. a module
31. A voltmeter across a battery with the engine running reads 12.4 volts. Technician A states that this could be caused by a faulty battery. Technician B states that this could be caused by a faulty alternator. Who is correct?
    A. Technician A only
    B. Technician B only
    C. Both Technician A and Technician B
    D. Neither Technician A nor Technician B
32. Two technicians are discussing ignition module functions. Technician A states that the ignition module controls the flow of current through the ignition coil. Technician B states that the module determines ignition advance. Who is correct?
    A. Technician A only
    B. Technician B only
    C. Both Technician A and Technician B
    D. Neither Technician A nor Technician B
33. Base timing is usually determined by
    A. the position of the distributor
    B. the position of the CKP
    C. the PCM
    D. any of the above
34. The timing signal for a vehicle with no distributor is generated by the
    A. CKP
    B. CMP
    C. ignition module
    D. PCM
35. A regulator circuit has resistance on the ground circuit. This will usually cause
    A. reduced alternator output
    B. reduced charging voltage
    C. increased charging voltage
    D. no charging current
36. Propane is added to the intake. The O_2S voltage should
    A. drop to between zero volts and 275 millivolts
    B. cause the frequency to increase
    C. cause the frequency to decrease
    D. increase the voltage between 800 millivolts and 1.0 volts

Bilingual Glossary

AC Alternating current.
AC *Corriente alterna.*

Advance Delivering the spark earlier to the cylinder.
Avance *Producir la chispa tempranamente hacia el cilindro.*

Advanced timing A spark that is delivered before the piston reaches TDC.
Avance de la regulación de tiempo *Chispa que se produce cuando el pistón alcanza el punto muerto superior.*

Alternator An AC (alternating current) generator that produces electrical current and forces it into the battery to recharge it.
Alternador *Generador de corriente alterna (AC) que produce corriente eléctrica y la fuerza en la batería para recargarla.*

Ammeter The instrument used to measure electrical current flow in a circuit.
Amperímetro *Instrumento que se usa para medir el flujo de corriente eléctrica en un circuito.*

Ampere The unit for measuring electrical current; usually called an amp.
Amperio *Unidad para medir la corriente eléctrica. Generalmente se conoce como amp.*

Armature The assembly that becomes the rotating magnetic field inside a motor.
Armadura *El ensamble que se convierte en el campo magnético giratorio dentro de un motor.*

ASD Automatic shutdown relay.
ASD *Relé de desconexión automática.*

ATDC After top dead center.
ATDC *Después del punto muerto superior.*

Auto-ranging A DMM that will automatically range to give the most accurate reading.
Automedición *DMM que mide de forma automática para proporcionar la lectura más precisa.*

Base The setting for ignition timing when the engine is at normal engine speed, load, and temperature.
Base *El ajuste para sincronización del encendido en el cual el motor está a velocidad, carga y temperatura normales.*

Base timing The timing that usually has the lowest advance that the vehicle requires. Rotating the distributor usually sets it.
Sincronización base *Sincronización que generalmente tiene el avance más bajo que requiere el vehículo. Hacer girar el distribuidor generalmente lo ajusta.*

BCM Body control module. The module that is used to control various body functions.
BCM *Módulo de control de la carrocería. Módulo que se usa para controlar funciones diversas de la carrocería.*

BTDC Before top dead center.
BTDC *Antes del punto muerto superior.*

Bullet connector A single wire connector.
Conector de bala *Conector de un solo alambre.*

Burn time How long the arc across the spark plug gap lasts; normally measured in milliseconds.
Tiempo de quemado *El tiempo de duración del arco al cruzar de lado a lado la brecha de la bujía, medido normalmente en milisegundos.*

Case ground A ground that is part of the metal component.
Tierra a chasis *Tierra que es parte del componente de metal.*

Cell The individual part of a battery.
Celda *Pieza individual de una batería.*

CEMF Counterelectromotive force; counter voltage.
CEMF *Fuerza contraelectromotriz; contravoltaje.*

Chip A group of electronic devices; an integrated circuit.
Chip *Grupo de dispositivos electrónicos; circuito integrado.*

Circuit An electrical path that will develop light, heat, and magnetic field, consisting of insulated conductors, control, load and circuit protection, and power source.
Circuito *Ruta eléctrica que desarrolla luz, calor, y campo magnético y que consta de conductores aislados, protección de control, carga y circuito, y fuente de alimentación eléctrica.*

Circuit breaker A circuit protection that will open the circuit when too much current flows.
Disyuntor *Protección de circuito que abre el circuito cuando fluye demasiada corriente.*

Circuit control The switch, relay, or module that turns on/off a circuit.
Control de circuito *El interruptor, relé o módulo que enciende o apaga un circuito.*

CKP Crankshaft position sensor.
CKP *Sensor de posición del cigüeñal.*

CMP Camshaft position sensor.
CMP *Sensor de posición del árbol de levas.*

Collector The part of the transistor that will have the load.
Colector *La parte del transistor que tiene la carga.*

Common point A point where more than one circuit is connected.
Punto común *Punto en el que se conecta más de un circuito.*

Commutator Part of the armature that allows current to flow from the brushes through the windings of the armature.
Conmutador *Pieza de la armadura que permite que la corriente fluya de las escobillas a través del devanado de la armadura.*

Conductor Material with a low resistance to the flow of current.
Conductor *Material con una baja resistencia al flujo de corriente.*

Connector A terminal where two wires are connected together.
Conector *Terminal donde se conectan dos alambres entre sí.*

Continuity An indication that a circuit is complete.
Continuidad *Indicación de que un circuito está completo.*

Control current The current that is used to control another circuit.
Corriente de control *Corriente que se usa para controlar otro circuito.*

CP Clutch position switch.
CP *Interruptor de posición del embrague.*

Crimp terminals Terminals that are connected to the end of a wire.
Terminales a presión *Terminales que se conectan al extremo de un alambre.*

CS Clutch switch.
CS *Interruptor del embrague.*

Current The number of electrons flowing past a given point in a given amount of time.
Corriente *Número de electrones que pasan por un punto determinado en una cantidad de tiempo determinada.*

Current limit A module function of many ignition systems. Once the ignition coil is fully saturated the module will reduce current flow through the current limit circuit.
Límite de corriente *Función de módulo de muchos sistemas de encendido. Una vez que la bobina de encendido queda completamente saturada, el módulo reduce el flujo de corriente a través del circuito de límite de corriente.*

Current probe A device that clamps around a conductor and senses current flow.
Sonda de corriente *Dispositivo que se sujeta alrededor de un conductor y mide el flujo de corriente.*

DC Direct current.
DC *Corriente directa.*

Delta winding An arrangement of alternator (AC generator) windings that is connected in parallel and looks like a triangle.
Devanado en triángulo *Disposición de los devanados del alternador (generador de corriente alterna) que se conecta en paralelo y tiene apariencia de triángulo.*

DI An ignition system that uses a distributor.
DI *Sistema de encendido que usa un distribuidor.*

Diode A simple semiconductor device that permits flow of electricity in one direction but not in the opposite direction.
Diodo *Dispositivo semiconductor sencillo que permite el flujo de electricidad en un sentido pero no en el sentido opuesto.*

Direct reading An ammeter that has all of the current in the circuit flowing through the meter.
Lectura directa *Amperímetro que hace que toda la corriente en el circuito fluya a través del medidor.*

Display The screen of a multimeter or scope.
Pantalla *Pantalla de un multímetro o dispositivo de medición.*

DMM Digital multimeter.
DMM *Multímetro digital.*

DPDT Double-pole, double-throw switch.
DPDT *Interruptor bipolar y bivanal.*

Drive pinion The end of the starter that will mesh with the flywheel.
Piñón actuado *El extremo del arrancador que se engrana con el volante.*

Driver A control circuit.
Controlador *Circuito de control.*

DRL Daytime running lights.
DRL *Luces de marcha diurna.*

DSO Digital storage oscilloscope.
DSO *Osciloscopio de almacenamiento digital.*

Dwell The amount of time that current flows.
Reposo *Cantidad de tiempo durante el cual fluye la corriente.*

ECM Electronic control module.
ECM *Módulo de control electrónico.*

ECT Engine coolant temperature sensor.
ECT *Sensor de temperatura del enfriador del motor.*

EI An ignition system that does not have a distributor.
EI *Sistema de encendido que no tiene un distribuidor.*

Electrolyte The acid/water in a battery.
Electrolito *El ácido o agua en una batería.*

Emitter The part of the transistor that is connected to power or ground.
Emisor *Parte del transistor que está conectado a la alimentación eléctrica o a tierra.*

EST Electronic spark timing.
EST *Sincronización electrónica de la chispa.*

Expanded splice A drawing that separates a large splice.
Empalme expandido *Trefilado que separa un empalme largo.*

False loads Resistance that is not designed into the circuit. Loose ground wires and corroded connections are examples of unwanted resistance that will change the flow of current and the voltage division of the circuit.
Cargas falsas *Resistencia que no está diseñada en el circuito. Los cables sueltos a tierra y las conexiones oxidadas son ejemplos de resistencia no deseada que cambian el flujo de corriente y la división de voltaje del circuito.*

Field current The current that flows through the rotor of an alternator (AC generator).
Corriente de campo *La corriente que fluye a través del rotor de un alternador (generador de corriente alterna).*

Firing order The order in which individual cylinders' spark plugs are fired.
Orden de encendido *Orden en el cual se disparan las bujías individuales.*

Flux density The concentration of the lines of force.
Densidad de flujo *Concentración de las líneas de fuerza.*

Flywheel The device used for starting that is connected to the rear of the crankshaft which has teeth around its circumference.
Volante *Dispositivo que se usa para arrancar conectado a la parte trasera del cigüeñal y que tiene dientes alrededor de la circunferencia.*

Forward current rating The amount of current that can pass easily through a diode.
Valores nominales de corriente de avance *Cantidad de corriente que puede pasar fácilmente a través de un diodo.*

Full field Having the regulator increase the PWM of the field to 100 percent, or full-field current.
Campo lleno *Cuando el regulador aumenta el PWM del campo a 100 por ciento o corriente de campo lleno.*

Fuse An electrical device used to protect a circuit against accidental overload or unit malfunction.
Fusible *Dispositivo eléctrico que se usa para proteger un circuito contra la sobrecarga accidental o mal funcionamiento de la unidad.*

Fusible link A type of fuse made of a special wire that melts to open a circuit when current draw is excessive.
Eslabón fusible *Tipo de fusible hecho de un alambre especial que se derrite para abrir un circuito cuando la llamada de corriente es excesiva.*

Ground The negatively charged side of a circuit. A ground can be a wire, the negative side of the battery, or even the vehicle chassis.
Tierra *El lado con carga negativa de un circuito. La tierra puede ser un cable, el lado negativo de la batería, o incluso el chasis del vehículo.*

Half-wave rectification When only the positive or negative pulse of an AC signal is changed into DC.
Rectificación de semionda *Cuando solo el pulso positivo o negativo de una señal AC cambia a DC.*

Hall effect sensor A device that produces a voltage pulse dependent on the presence of a magnetic field. Hall effect voltage varies as magnetic reluctance varies around a current-carrying semiconductor.
Sensor de efecto Hall *Dispositivo que produce un pulso de voltaje que depende de la presencia de un campo magnético. El voltaje de efecto Hall varía de acuerdo con la reluctancia magnética alrededor de un semiconductor que transporta corriente.*

HC Raw gas that was never ignited or burned.
HC *Gas crudo que nunca se ha encendido o quemado.*

Heli-coiling The process of rethreading a stripped out fastener.
Enroscado helicoidal *Proceso de volver a estriar un sujetador barrido.*

Hold-in windings The solenoid or relay winding that will hold the plunger into its starting position.
Devanados de sujeción *El devanado de solenoide o relé que sujeta el émbolo en la posición de arranque.*

Hydrometer A device used to check the state of charge of a cell in a battery by measuring the specific gravity of the electrolyte.
Hidrómetro *Dispositivo que se usa para verificar el estado de carga de una celda en una batería midiendo la densidad del electrolito.*

IAT Intake air temperature sensor.
IAT *Sensor de temperatura del aire de admisión.*

IDM Ignition diagnostic monitor.
IDM *Monitor de diagnóstico de encendido.*

Ignition bypass terminal The terminal which will bypass the ignition resistor on older ignition systems.
Terminal de desvío de encendido *Terminal que desvía el resistor de encendido en sistemas de encendido antiguos.*

Ignition module The device that controls primary current.
Módulo de encendido *Dispositivo que controla la corriente principal.*

Impedance The operating resistance of a component or piece of equipment. The higher the impedance, the lower the operating amperage.
Impedancia *La resistencia operativa de un componente o pieza de equipo. Cuanto mayor sea la impedancia, menor será el amperaje de operación.*

Induced voltage A voltage that is generated.
Voltaje inducido *Voltaje que se genera.*

Inductive pickup A probe that goes around a conductor and senses current flow.
Lector inductivo *Sonda que va alrededor de un conductor y que mide el flujo de corriente.*

Inductive probe An external electrical probe that will surround a conductor and generate a voltage signal proportional to the amount of current flowing.
Sonda inductiva *Sonda eléctrica externa que rodea a un conductor y que genera una señal de voltaje proporcional a la cantidad de corriente que fluye.*

Infinity More resistance than the meter can measure; usually an open circuit.
Infinidad *Más resistencia de la que puede registrar un aparato medidor. Normalmente un circuito abierto.*

Insulator Materials that have more than four electrons in their outer ring. With these materials, the force holding them in orbit is strong, and high voltages are needed to move them.
Aislante *Materiales que tienen más de cuatro electrones en el anillo exterior. Con estos materiales la fuerza que los mantiene en órbita es fuerte, y son necesarios voltajes altos para moverlos.*

Integrated circuit (IC) A large number of diodes, transistors, and other electronic components, all mounted on a single piece of semiconductor material and able to perform numerous functions; sometimes called a chip.
Circuito integrado (IC) *Gran número de diodos, transistores, y otros componentes electrónicos montados en una sola pieza de material semiconductor y capaz de realizar numerosas funciones; también conocido como chip.*

Kilo One thousand.
Kilo *Mil.*

KOEO Key on, engine off.
KOEO *Llave en encendido, motor apagado.*

KOER Key on, engine running.
KOER *Llave en encendido, motor en marcha.*

KS Knock sensor.
KS *Sensor de cascabeleo.*

LED Light-emitting diode.
LED *Diodo emisor de luz.*

Load In a circuit, the component whose resistance will produce light, heat, or magnetic field when current is pushed through it.
Carga *En un circuito, el componente cuya resistencia produce luz, calor o campo magnético al pasarle corriente.*

Mandrel A part of the tool used for rethreading.
Mandril *Parte de una herramienta que se usa para volver a enroscar.*

MAP sensor Manifold absolute pressure sensor.
MAP sensor *Sensor de presión absoluta del múltiple.*

Mega One million.
Mega *Un millón.*

Milli One-thousandth.
Mili *Una milésima.*

Millisecond One-thousandth of a second (0.001 second).
Milisegundo *Una milésima de segundo (0.001 segundo).*

MPMT Multiple-pole, multiple-throw switch.
MPMT *Interruptor multipolar y multivanal.*

MSDS Material safety data sheet.
MSDS *Hoja de Datos de Seguridad de Materiales.*

NC Normally closed; a set of relay or switch contacts that is closed with KOEO.
NC *Normalmente cerrado; conjunto de relés o contactos de interruptor que se cierra con la llave en encendido y el motor apagado.*

NO Normally open; a set of relay or switch contacts that is opened with KOEO.
NO *Normalmente abierto; conjunto de relés o contactos de interruptor que se abre con la llave en encendido y el motor apagado.*

O_2S Oxygen sensor.
O_2S *Sensor de oxígeno.*

OBD II On-Board Diagnostics, second generation.
OBD II *Diagnóstico a bordo, segunda generación.*

OCV Open-circuit voltage.
OCV *Voltaje de circuito abierto.*

Ohm A unit of measured electrical resistance.
Ohmio *Unidad de resistencia eléctrica medida.*

Ohm's Law The interrelationship of voltage, current, and resistance.
Ley de Ohm *Las interrelaciones de voltaje, corriente y resistencia.*

Ohmmeter The meter used to measure electrical resistance.
Óhmetro *Medidor que se usa para medir la resistencia eléctrica.*

Open circuit An electrical circuit that has a break in the wire.
Circuito abierto *Circuito eléctrico que tiene una rotura en el alambre.*

Optical refractometer A device to measure specific gravity.
Refractómetro óptico *Dispositivo para medir el peso específico.*

Oscilloscope A voltmeter that displays voltage over a period of time.
Osciloscopio *Voltímetro que muestra el voltaje a lo largo de un periodo.*

Overrunning clutch The part of the starter armature that will spin the flywheel.
Acoplamiento de rueda libre *Parte de la armadura del arrancador que hace girar el volante.*

Parallel circuit Type of circuit in which there is more than one path for the current to follow.
Circuito paralelo *Tipo de circuido en el cual existe más de una ruta para que fluya la corriente.*

PCM Powertrain control module.
PCM *Módulo de control de tren de potencia.*

Peak inverse voltage The voltage that the diode can withstand safely in its blocking mode.
Voltaje pico inverso *El voltaje que el diodo puede tolerar con seguridad en modo de bloqueo.*

Pickup coil A device that will generate an AC signal.
Bobina captadora *Dispositivo que genera una señal de corriente alterna.*

Pintle bump As the fuel injector pintle opens, it will generate a bump or hump on the current trace.
Tope de aguja *Al abrirse la aguja del inyector de combustible, se genera un tope o saliente en la trayectoria de la corriente.*

PID Parameter identification.
PID *Identificación de parámetro.*

PIP Profile ignition pickup.
PIP *Captación de encendido de perfil.*

Planetary gear The gearset that is used in many permanent magnet starters.
Engranaje planetario *El engranaje en muchos arrancadores de imán permanentes.*

PM generator Permanent magnet generator.
Generador PM *Generador de imán permanente.*

Polarity The particular state, either positive or negative, with reference to the two poles or to electrification.
Polaridad *Estado particular, ya sea positivo o negativo, con referencia a los dos polos o a la electrificación.*

Ported vacuum Vacuum port above the throttle plate.
Vacío por puerto *Puerto de vacío sobre el plato del acelerador.*

Potentiometer An electrical device that will vary the voltage based on its position.
Potenciómetro *Dispositivo eléctrico que hace variar el voltaje con base en su posición.*

Primary cell A nonrechargeable cell.
Celda primaria *Celda no recargable.*

Primary current The current that flows through the primary circuit of the coil that will generate the strong magnetic field needed for the generation of the ignition spark.
Corriente principal *Corriente que fluye a través del circuito principal de la bobina que generará el campo magnético fuerte necesario para la generación de la chispa de encendido.*

Primary winding Low-voltage winding in an ignition coil that draws current and produces a magnetic field.
Devanado primario *Devanado de bajo voltaje en una bobina de encendido que atrae corriente y produce un campo magnético.*

Pull-in windings The windings that will be responsible for pulling in the solenoid plunger.
Devanado de tracción *Devanados que es responsable de aplicar tracción al émbolo de solenoide.*

PWM Pulse width modulation.
PWM *Modulación de ancho de pulso.*

R&R Removing and replacing.
R&R *Extracción y sustitución.*

Range The maximum measurement that the meter can measure.
Rango *La medida máxima que puede registrar un aparato medidor.*

Raster A pattern that displays individual cylinders across the screen from bottom to top in rows.
Trama *Patrón que muestra cilindros individuales de lado a lado de la pantalla desde la parte inferior a la parte superior, en filas.*

Rectifier A group of diodes that changes AC into DC.
Rectificador *Grupo de diodos que convierte corriente alterna en corriente directa.*

Reduction gear Allows the starter armature to spin faster than the starter drive.
Engranaje reductor *Permite que la armadura del arrancador gire más rápido que el lado impulsor del arrancador.*

Reference voltage The small signal sent out from a vehicle's computer.
Voltaje por referencia *La pequeña señal que envía la computadora del vehículo.*

Refractometer A test instrument that measures the deflection, or bending, of a beam of light.
Refractómetro *Instrumento de prueba que mide el desvío o curvatura de un haz de luz.*

Regulator The device that controls the charging system.
Regulador *Dispositivo que controla el sistema de carga.*

Relay A device that uses low current to control high current.
Relé *Dispositivo que usa corriente baja para controlar la corriente alta.*

Remote ground A ground that uses a wire to connect the component to the ground.
Tierra remota *Tierra que usa un alambre para conectar el componente a tierra.*

Resistance The force that opposes the flow of electrons.
Resistencia *Fuerza que se opone al flujo de electrones.*

Retard To deliver the spark later to the combustion chamber.
Retardo *Producir la chispa posteriormente hacia la cámara de combustión.*

RFI Radio frequency interference.
RFI *Interferencia de radiofrecuencia.*

Ring gear The teeth around the flywheel of the engine.
Engranaje de anillo *Dientes alrededor del volante del motor.*

RMS Root mean squared; a specific AC voltage measurement.
RMS *Media cuadrática Medida específica de voltaje AC.*

Rotor The spinning part of the alternator (AC generator) that carries the magnetic field.
Rotor *Pieza giratoria del alternador (generador de corriente alterna) que transporta el campo magnético.*

Saturation time The time that it takes for maximum current to flow through a coil and build the maximum magnetic field.
Tiempo de saturación *Tiempo que tarda la corriente máxima en fluir a través del devanado y producir el campo magnético máximo.*

Scanner A device that will interface with a PCM and supply a read-out of data.
Escáner *Dispositivo que interactúa con un PCM y que proporciona datos de lectura.*

SCFH Standard cubic feet per hour.
SCFH *Pies cúbicos estándar por hora.*

Secondary cell A rechargeable cell.
Celda secundaria *Celda recargable.*

Secondary winding The part of the ignition coil that develops the high voltage required to fire the spark plugs when the primary winding collapses.
Devanado secundario *Pieza de la bobina de encendido que desarrolla el alto voltaje necesario para disparar las bujías cuando se colapsa el devanado primario.*

Sensing voltage The input into the regulator that is used to determine required output.
Voltaje sensor *La entrada en el regulador que se usa para determinar la salida requerida.*

Series circuit An electrical circuit that has only one path for current flow.
Circuito serie *Circuito eléctrico que solo tiene una ruta para el flujo de corriente.*

Series-parallel circuit A circuit that is a combination of a series circuit and parallel circuits.
Circuito serie-paralelo *Circuito que es una combinación de circuitos serie y paralelos.*

Short circuit A circuit defect that causes a decrease in circuit resistance and an increase in current flow.
Cortocircuito *Defecto del circuito que causa una disminución de la resistencia del circuito y un aumento del flujo de corriente.*

Shorting switch A switch that applies a ground.
Interruptor de puenteo *Interruptor que aplica una tierra.*

Slip rings Two smooth rings in a rotor that are insulated from the rotor shaft.
Anillos deslizantes *Dos anillos lisos en un rotor que están aislados del eje de rotor.*

Slope Whether the voltage is rising or falling at the trigger point on a DSO.
Pendiente *Subida o caída de voltaje en el punto de disparo de un osciloscopio de almacenamiento digital.*

Solenoid An electromagnetic device similar in operation to a relay, but movement of the armature or iron core changes electrical energy into mechanical energy.
Solenoide *Dispositivo electromagnético similar en operación a un relé, pero el movimiento de la armadura o núcleo de hierro cambia la energía eléctrica en energía mecánica.*

Solenoid shift A type of starter that uses the action of the solenoid to move the starter drive into mesh with the teeth of the flywheel.
Cambio de solenoide *Tipo de arrancador que usa la acción del solenoide para mover el lado impulsor del arrancador hasta que engrane con los dientes del volante.*

Solid State A device that has no moving parts but is used to control current flow.
Estado sólido *Dispositivo que no tiene partes móviles pero que se usa para controlar el flujo de corriente.*

SPDT Single-pole, double-throw switch.
SPDT *Interruptor unipolar y bivanal.*

Specific gravity The weight of a volume of any liquid divided by the weight of an equal volume of water at equal temperature and pressure; the ratio of the weight of any liquid to the weight of water, which has a specific gravity of 1.000.
Peso específico *El peso de un volumen o de cualquier líquido dividido entre el peso de un volumen igual de agua a temperatura y presión iguales; el índice de proporción del peso de cualquier líquido respecto al peso del agua, que tiene un peso específico de 1.000.*

Splice To join. Electrical wires can be joined by soldering or by using crimped connectors.
Empalme *Unión. Unión de cables eléctricos mediante el uso de soldadura o conectores a presión.*

SPOUT Spark out.
SPOUT *Chispa hacia afuera.*

SPST Single-pole, single-throw switch.
SPST *Interruptor unipolar y univanal.*

SRS Supplemental restraint system.
SRS *Sistema suplementario de restricción.*

Starter enable control A relay used in anti-theft circuits.
Control de activación del arrancador *Relé que se usa en circuitos contra robo.*

Starter relay The relay that energizes the starter.
Relé de arrancador *El relé que energiza el arrancador.*

Stator The stationary part of the alternator (AC generator) that develops the AC voltage.
Estator *Pieza fija del alternador (generador de corriente alterna) que desarrolla el voltaje de corriente alterna.*

Sulfation A condition that prevents the easy transfer of current out of the battery.
Sulfatación *Condición que evita que se transfiera con facilidad la corriente fuera de la batería.*

Sun gear The center of a planetary gearset.
Engranaje solar *Centro de un conjunto de engranaje planetario.*

Sync probe Another name for an external trigger.
Sonda sincronizada *Otro nombre para el disparador externo.*

TDC Top dead center.
TDC *Punto muerto superior.*

TFI Thick film integrated.
TFI *Integrado de película gruesa.*

Thermistor A type of variable resistor. A thermistor is designed to change in value as its temperature changes. Thermistors are used to provide compensating voltage in components or to determine temperature.
Termistor *Tipo de resistencia variable. El termistor está diseñados para cambiar de valor al elevarse su temperatura y se usan para proporcionar voltaje de compensación en componentes o para determinar temperaturas.*

Three-phase voltage Three-pulse voltage.
Voltaje trifásico *Voltaje de tres pulsos.*

Timing When the spark is delivered inside the combustion chamber.
Regulación de tiempo *Cuando la chispa se produce dentro de la cámara de combustión.*

Torque Twisting effort.
Par *Esfuerzo de torsion.*

TPS Throttle position sensor.
TPS *Sensor de posición del acelerador.*

TPS sweep The test of a replaced TPS.
Barrido de TPS *Prueba de un TPS reemplazado.*

TR Transmission range sensor or switch.
TR *Sensor o interruptor de rango de transmisión.*

Transistor An electronic device produced by joining three sections of semiconductor materials. A transistor is useful as a switching device, functioning as either a conductor or an insulator.
Transistor *Dispositivo electrónico que se produce al unir tres secciones de materiales semiconductores. Un transistor es muy útil como dispositivo de conmutación y funciona como conductor o aislante.*

Trigger The voltage that will start the display of a DSO.
Voltaje de activación *El voltaje que inicia la visualización de un osciloscopio de almacenamiento digital.*

Trigger position The spot on the screen where the pattern conditions of trigger and slope are met.
Posición del voltaje de activación *El punto en la pantalla donde se unen las condiciones del voltaje de activación y la pendiente.*

TSB Technical service bulletin.
TSB *Boletín de servicio técnico.*

TVRS Television and radio suppression.
TVRS *Supresión de señales de radio y televisión.*

Vacuum advance Advance timing under conditions of increased vacuum.
Avance por vacío *Sincronización de avance bajo condiciones de vacío aumentado.*

Valence orbit The outer orbit of electrons.
Órbita de valencia *Órbita exterior de los electrones.*

VAT Volts-amps tester.
VAT *Probador de voltios y amperios.*

Volt A unit of measurement of electromotive force. One volt of electromotive force applied steadily to a conductor of 1-ohm resistance produces a current of 1 ampere.
Voltio *Unidad de medida de la fuerza electromotriz. Un voltio de fuerza electromotriz aplicado uniformemente a un conductor de un ohmio de resistencia produce una corriente de un amperio.*

Voltage drop Voltage lost by the passage of electrical current through resistance.
Caída de voltaje *Voltaje que se pierde por el paso de corriente eléctrica a través de la resistencia.*

Voltmeter A tool used to measure the voltage available at any point in an electrical system.
Voltímetro *Herramienta que se usa para medir el voltaje disponible en cualquier punto de un sistema eléctrico.*

Waste spark The spark that produces no work on a distributorless ignition system.
Chispa residual *La chispa que no produce trabajo en un sistema de encendido sin distribuidor.*

Wattage The electrical means of monitoring how much work is being done.
Potencia en vatios *Método eléctrico de monitorear la cantidad de trabajo que se realiza.*

WOT Wide-open throttle.
WOT *Acelerador completamente abierto.*

Wye winding An arrangement of alternator (AC generator) windings that is connected in series and is in the shape of a "Y."
Devanado en estrella *Disposición de los devanados de un alternador (generador de corriente alterna) que se conecta en serie y que tiene forma de "Y."*

Zener diode A diode that will conduct when a specific voltage is achieved.
Diodo Zener *Diodo que conduce cuando se alcanza un voltaje específico.*

Index

P

R